Environmental systems

Philosophy, analysis and control

Environmental systems

Philosophy, analysis and control

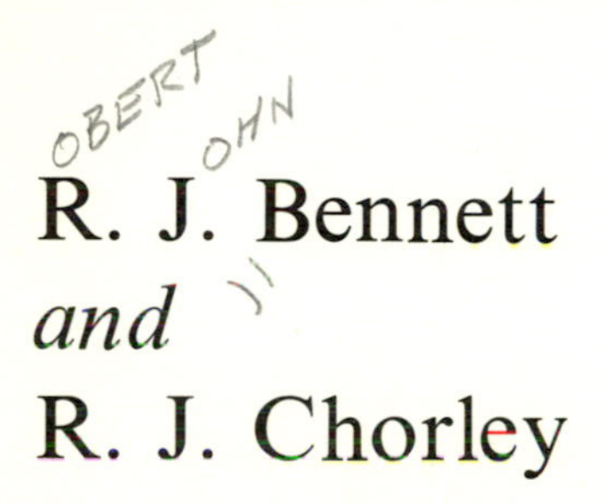

R. J. Bennett
and
R. J. Chorley

PRINCETON UNIVERSITY PRESS
PRINCETON, NEW JERSEY

Published by Princeton University Press, Princeton, New Jersey

LCC Number 78–55535
ISBN 0–691–08217–0

Printed in Great Britain

Contents

Acknowledgements viii
Preface xi

1 THE PHILOSOPHY OF ENVIRONMENTAL SYSTEMS 1
1 Systems and philosophy 1
2 Bases for theory 11
3 Man and environment 14

Part I Hard systems 23

2 SYSTEMS METHODS 25
1 The structure of systems 25
2 The mathematical representation of systems 47
3 Environmental systems analysis in practice 70
4 Conclusion 91

3 CONTROL SYSTEMS 94
1 The purpose of control systems 94
2 Types of control systems 98
3 Nested feedback control systems 111
4 Criteria for efficiency of control 115
5 Types of control strategies 122
6 Conclusion 150

4 SPACE-TIME SYSTEMS 152
1 Man in space and time 152
2 The structure of space-time systems 154
3 Purely spatial processes 182
4 Space-time control systems 188
5 Additional topics in space-time control systems 206
6 Conclusion 219

Part II Soft systems 221

5 COGNITIVE SYSTEMS 223
1 Mental functions and psychological models 223
2 Models of man and environment 234
3 Images 237
4 Environmental disturbances, shunting and memory 238
5 Belief and action 243
6 A general cognitive system 247

6 DECISION MAKING SYSTEMS 250
1 The decision making process 250
2 The decision making environment 268
3 The economics of decision making for control 274
4 Evaluation in decision making 281
5 Components of decision making systems 284
6 Spatial structure of decision making systems 310
7 Conclusion 313

Part III Complex systems 317

7 PHYSICO-ECOLOGICAL SYSTEMS 319
1 Mathematics and nature 319
2 Systems analysis 327
3 Systems synthesis 336
4 Spatial systems 363
5 Conclusion 397

8 SOCIO-ECONOMIC SYSTEMS 399
1 Introduction 399
2 Systems synthesis 406
3 Systems analysis 416
4 Control to objectives 436
5 Conclusion 459

Part IV Systems interfacing 465

9 SYSTEMS INTERFACING 467
1 Physico-ecological and socio-economic systems 467
2 Strategies of systems interaction 473
3 Environmental intervention 484
4 Environmental symbiosis 504
5 Problems of environmental symbiosis 527
6 Conclusion 538

10 CONCLUSION: FUTURE PROBLEMS 541
1 Technical problems 542
2 The demographic dilemma 543
3 Psychological difficulties 544
4 Epistemological transitions 546

APPENDICES 554
1 Simple matrix algebra 554
2 Derivation of optimum control equations 558
3 Notation (chapters 2, 3 and 4) 564

References 566
Index of persons 599
Subject index 607

Acknowledgements

The authors would like to thank the following editors, publishers, organizations and individuals for permission to reproduce figures, quotations and equations.

Editors

Ambio for 7.38, 9.23, 9.29, 9.30, 9.31 and 9.32; *Annals of the Association of American Geographers* for 2 quotations; *Annales Scientifiques de l'Ecole Normale Superièure* for 1 quotation; *Architectural Design* for 6.29; *British Journal of the Philosophy of Science* for 1 quotation; *Bulletin of Entomological Research* for 7.52; *Contemporary Psychology* for 1 quotation; *Dialectica* for 1 quotation; *Earth Surface Processes* for 7.7; *Economic Geography* for 3 quotations; *Economic Journal* for 8.12, 8.28 and 1 quotation; *Ecological Monographs* for 7.12; *Environment and Planning* for 6.30 and 8.26; *Environmental Science and Technology* for 7.39; *Geographical Review* for 9.10 and 1 quotation; *Human Relations* for 5.6; *Hydrosciences Bulletin* for 7.31; *IEEE* for 3.31, 9.1, 9.7 and equations; *Joint Automatic Control Conference* for 4.19; *Journal of Agricultural Economics* for 9.35; *Journal of Environmental Systems* for 6.34; *Journal of the Royal Statistical Society* for 1 quotation; *Meteorological Monographs* for 7.44; *New Scientist* for 4.26; *New Statesman* for 1 quotation; *Philosophy of Science* for 2 quotations; *Regional Studies* for 2.42, 2.43, 2.44 and 2.45; *Science* for 9.16; *Simulation* for 9.12; *Transactions of the American Geophysical Union* for 7.27 and 9.3; *Transactions of the Institute of British Geographers* for 9.4 and 9.9; *Water Resources Research* for 7.32, 7.34, 9.14, 9.17 and 9.20.

Publishers

Academic Press, London for 2.5, 2.6, 2.20, 2.27, 2.30, 7.10, 7.11, 7.12, 7.13, 7.14, 7.15, 7.17, 7.18, 7.45, 7.46, 7.47, 7.48, 8.27, 4 quotations, 2 tables and equations; Academic Press, New York for 2.35, 2.36, 8.24 and 8.29; Arnold, London for 8.19 and 4 quotations; Blackwell, Oxford for 2 quotations;

Cambridge University Press for 8.9 and 2 quotations; CBS International Publishing for 5.5; Chatto and Windus, London for 2 quotations; Collins, London for 1 quotation; Dover, New York for 2.35, 7.55A, 7.55B and 7.56; Dowden, Hutchinson and Ross, Stroudsburg for 7.19, 7.20, 9.1 and 1 quotation; Duckworth, London for 3 quotations; Earth Island, London for 9.26 and 3 quotations; Elsevier, New York for 4.14, 4.27, 7.3, 9.18 and 9.19; Faber and Faber, London for 6 quotations; Foreign Languages Publishing House, Moscow for 2 quotations; Freeman, Stanford for 1 quotation; George Allen and Unwin, London for 2 quotations; George Braziller, New York for 2 quotations; Harper and Row, London for 7.54; Harrap, London for 2.2 and 2.4; Harvard University Press, Cambridge, Mass. for 7.1, 7.49, 7,50, 8.14 and 2 quotations; Her Majesty's Stationery Office for 1 quotation; Holden-Day, San Francisco for 2.42 and 3.29; Holt, Rinehart and Winston, New York for 5.4; Hutchinson, London for 9.22, 9.24, 9.27, 9.34 and 5 quotations; Johns Hopkins, Baltimore for 9.5; Jossey-Bass, San Francisco for 8.1, 8.4 and 1 quotation; S. Karger, Basel for 3.3 and 6.33; Lexington Books for 6.1 and 6.2; Macmillan, London for 6.5, 6.11, 6.20, 6.21 and 2 quotations; Macmillan, New York for 7.2A, 8.7, 8.13, 8.14 and 1 quotation; McGraw-Hill, New York for 2.3, 2.15, 2.16, 2.19, 2.22, 2.28, 2.30, 2.33, 3.7, 3.9, 3.10, 4.16, 4.22A, 6.14, 6.15, 7.9A, 7.30, 7.33, 7.35, 2 quotations and equations; MIT Press for 1.4A, 1.4B, 3.27 and 4 quotations; Methuen, London for 6.2, 7.22, 7.42, 7.43 and 5 quotations; Minnesota University Press, Minneapolis for 2 quotations; Mouton, The Hague for 2 quotations; North Holland, Amsterdam for 3.2, 8.8 and 2 quotations; Oxford University Press for 2.9, 2.23, 7.5, 9.28 and 1 quotation; Penguin, Harmondsworth for 1.2 and 9 quotations; Pentech Press, London for 7.39 and 7.40; Pergamon Press, Oxford for 2.35, 2.26, 2.27, 2.32, 3.4, 3.17 and 2 quotations; Pion, London for 2.29, 2.38, 2.39, 4.1, 4.3, 4.4, 4.5, 4.6 and 4.7; Prentice-Hall, London for 1 quotation; Prentice-Hall, New Jersey for 2.12, 4.17, 6.1, 6.17, 7.8, 7.16, 8.3, 8.22, 8.23 and 1 quotation; Ronald Press, New York for 7.28; Routledge and Kegan Paul, London for 14 quotations; Saxon House, Farnborough for 7.41; Tavistock, London for 6.24 and 8.9; University of Michigan Press, Ann Arbor for 8.6 and 1 quotation; Wiley, London for 3.11, 6.23, 6.38, 7.26, 9.33 and 9 quotations; Wiley, New York for 2.1, 2.10, 3.8, 3.13, 5.1, 6.28, 6.31, 7.29, 8.25, 9.2 and 2 quotations; Wright-Allen Press, Cambridge, Mass. for 1 quotation; D. C. Heath and Co., Lexington, Mass. for 2 tables.

Organizations

American Geographical Society, New York for 1 quotation; Colston Research Society, Bristol University for 7.42; Department of Applied Economics, Cambridge University for 1 quotation; Department of Civil Engineering, Colorado State University for 7.21; Department of Civil Engineering, University of Newcastle-upon-Tyne for 7.36; Department of Civil Engineering, Stanford University for 7.22, 7.24 and 7.25; Department of

Geography, Northwestern University for 7.51; Department of Water Science and Engineering, University of California, Davis for 2.11; Division of Civil Engineering, University of Illinois for 7.4; Federal Pollution Control Administration, Washington D.C. for 7.37; Institute of Behavioral Science, University of Colorado for 6.32; Institute of Public Administration, New York for 6.18 and 6.22; Institute of Technology, University of Lund for 3.25 and 3.26; Polish Academy of Sciences, Warsaw for 9.8 and 9.15; Royal Town Planning Institute, London for 6.6; Smithsonian Institution, Washington D.C. for 7.54; Water Resources Center, University of California, Berkeley for 7.2B.

Individuals

Mr J. Altman, New York for 1 quotation; Dr M. Parkes, Environment Canada, Ottawa for 6.26; Dr Koppány György, Központi Elörejelzö Intézet, Budapest for 9.2; Mr J. M. de Unẽna, Department of Urban Design and Regional Planning, Edinburgh University for 6.25; Dr A. Warren, University College London for 9.11; Dr F. Wedgwood-Oppenheim, Institute of Local Government Studies, Birmingham for 6.27; Dr G. T. Wilson, Department of Control Engineering, University of Lancaster for 1 quotation.

Thanks are also due in respect of:

Editorial work – Dr R. J. M. More and Mrs E. A. Bennett;

Drawing – Mr R. W. Davidson, Mrs E. Greenwood, Mrs P. Lucas, Mr A. Shelley, Mr M. Young and Mr R. Versey;

Photography – Mr R. Coe and Mr C. Cromarty;

Advice on Mathematical Control Theory – Mr K. C. Tan;

Helpful general discussions – Dr J. G. U. Adams, Dr G. P. Chapman, Dr D. W. T. Crompton, Dr D. Evans, Mr D. Gregory, Dr J. Langton, Dr G. Manners, Dr D. Rand and Professor A. G. Wilson.

The authors accept responsibility for any errors remaining in the book.

Preface

There has been much recent debate concerning the nature of environmental systems. Much of this arises from recognition of the increased interaction between man and nature, and of the ascendancy of the former leading to the modification of the latter. This complex interfacing is presenting difficult problems for students of the environment, not the least of which has been the growing need to weld together the approaches of what we traditionally assume to be 'hard' disciplines with those we call 'soft'. It is over 400 years since Andreas Versalius wrote in the preface to *De humani corporis fabrica* (1543) of the parlous state of the study of the health of mankind that 'no slight inconvenience results from too great a separation between branches of study which serves for the perfection of one art . . . that those who have set before themselves the attainment of an art embrace one part of it to the neglect of the rest. . . . Such never achieve any notable result: they never attain their goal or succeed in basing their art upon a proper foundation.' We are now coming to recognize that the solutions to the future problems concerning the 'health of mankind' will require, in addition to aspects of medical sciences, study of the interfacing of man and nature for which no *one* art or science has yet proved adequate. There are philosophical and practical difficulties in sociobiological, biogeographical and genetic explanations which suggest that man, like other animals, is in the broadest sense powerless to change his destiny. Similarly suspect are the opposed views of the extreme sociologists, critical social theorists, economists, Marxists and neo-Marxists who suggest that man is completely free to determine his future evolution and social goals by merely specifying his price rules or system of social justice in order to allow infinite substitution between inexhaustible raw materials and energy sources.

This book has a number of aims, of which it would be appropriate to identify two at the outset. The first is to explore to what extent systems theory provides an interdisciplinary focus for those who are concerned with environmental matters, using the term in its widest context; and to what extent systems technology, as at present available or conceivable, provides a vehicle whereby these disciplines might converge on such a focus. The second aim, given the first, is to examine specifically the manner in which systems approaches aid in the development of an integrated theory relating social and economic theory, on the one hand, to physical and biological theory, on the other. The authors would be the first to admit that their work had fallen short of such ambitious aims, but they hope that the reader will agree that ample demonstration of the progress towards the attainment of their aims is that an otherwise bewildering body and range of scholarship can be viewed anew, hopefully in a much more coherent manner.

R. J. BENNETT
Department of Geography
University of Cambridge
Fellow of Fitzwilliam College

R. J. CHORLEY
Department of Geography
University of Cambridge
Fellow of Sidney Sussex College

1 The philosophy of environmental systems

Once upon a time there was a rajah in this region who called to a certain man and said: 'Gather together in one place all the men in Savatthi who were born blind ... and show them an elephant.'

'Very good, sire,' said the man, and did as he was told and said to them, 'O blind, such as this is an elephant' – and to one man he presented the head of the elephant, to another its ears, to another a tusk, to another the trunk, the foot, back, tail, and tuft of the tail, saying to each one that that was the elephant....

Thereupon, brethren, that rajah went up to the blind men and said to each: 'Tell me, what sort of thing is an elephant?' Thereupon those who had been presented with the head answered, 'Sire, an elephant is like a pot.' And those who had observed an ear only replied, 'An elephant is like a winnowing basket.' Those who had been presented with a tusk said it was a plowshare. Those who knew only the back, a mortar; the tail, a pestle; the tuft of the tail, just a broom.

Then they began to quarrel, shouting 'Yes, it is!' 'No, it is not!' 'An elephant is not that!' 'Yes, it's like that!' and so on, till they came to fisticuffs over the matter. Then, brethren, that rajah reflected deeply upon the scene.

Udana, IV, 6

Vicar: With God's help you have made this garden a thing of beauty.

Gardener: You should 'ave seen it when 'e 'ad it to 'isself!

1.1 Systems and philosophy

A system is a set of logical operations acting upon, and acted upon by, one or more *inputs*. These inputs lead to the production of *outputs* from the system and this process of *throughput* is capable of either sustaining the operational structure of the system, or of transforming it, perhaps catastrophically. Inputs, throughputs and outputs of systems can involve flows of mass, energy, information or ideas, depending on the manner in which the system is defined. The process by which this definition takes place is often difficult to identify, and philosophers have long distinguished between modes of thought which view systematic associations of phenomena as dictated by nature itself, or as identified by man for his convenience. The most distinguished precedents for the former approach are provided by Aristotle and Plato who believed that knowledge of the external world could be certain and eternal because it was concerned with the essential form and structure of unified parts of reality from which their determinate processes or throughputs could be deduced. Plato's vivid expression of the route to true understanding in the Simile of the Cave depicts Man as bound by perceptions of the 'shadows' of reality which can be dispelled only by pure thought and reason (fig. 1.1). In the Simile of the Divided Line this theme is further developed to a categorization of orders of understanding: from *knowledge* based on pure reason to *opinion* based on

Fig. 1.1 A pictorial representation of Plato's simile of the cave. Perception is limited to shadows which can be enlightened only by leaving the cave and seeing the real objects outside: a simile for the employment of reason.

shadowy understanding of the physical world (fig. 1.2) which should be unfettered to yield greater degrees of truth (Lee 1955). Aristotle believed that progress in the understanding of the world depends on one's ability to define the essential attributes of the forms with which we are presented and then, by examining the structures and interrelationships of these forms, to deduce the processes of which they are the tangible expressions (Grene 1963). He wrote, 'the principal object of natural philosophy is not the material elements, but their composition, and the totality of the form, independently of which they have no existence...' so that 'the structures found are always of determinate processes, functioning in determinate contexts'. A number of points of importance immediately emerge from this viewpoint: firstly, that reality should be studied in terms of operating entities; secondly, that these entities have objective and eternal existence, irrespective of the observer, and thus form the basis for the formulation of universal laws; and, thirdly, that the ultimate objects of study should not be the forms or *artefacts* produced by the operations of systems, but the operation itself. Aristotle's opinion that reality can be properly studied from the point of view of its self-justified parts has been termed one of 'immanent teleology'. Modern developments of the classical view combine either Francis Bacon's concern for empiricism and hypothesis testing or Popperian falsification with mathematical deduction as in Russell and Wittgenstein. This is based on universal doubt which is used to build up universal principles from initial tangibles (for example, one's own existence as in the Cartesian '*cogito ergo sum*'). From such axioms flow necessary conclusions. The aim is Aristotelian in seeking knowledge of the world independent of the observer, but uses the hypothetico-deductive method (or positivism) rather than pure reason.

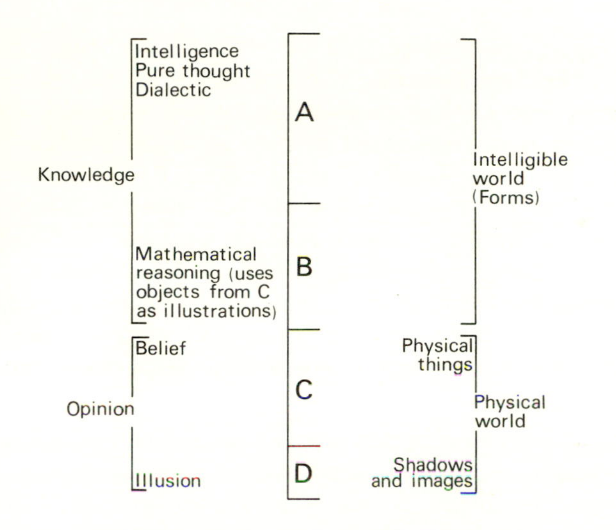

Fig. 1.2 The Platonic concept of the divided line illustrating the spectrum of knowledge from shadowy illusions to truth. (From Plato, *The Republic*, pt VII, after Lee 1955.)

Opposed to this view are those who either deny the objectivity of reality or point to its irrelevance when faced with the partiality of man's perception of it. This attitude, too, is dignified by antiquity, for the Epicureans believed that 'things are exactly as they appear to be to our senses, or rather as they would appear to be if our senses were slightly more accurate' (Latham 1951, p. 10). The attitude of Dewey is most expressive of the philosophy of the subjectivity of reality and he wrote, 'meaning . . . is not a psychic existence; it is primarily a property of behavior' (1958, p. 179). Although, with Aristotle, Dewey was concerned with the identification and examination of systematic entities, he believed in the essential subjectivity of the entities and wrote, 'inquiry is the controlled or directed transformation of an indeterminate situation into one that is so determinate in its constituent distinctions and relations as to convert the elements of the original situation into a unified whole' (quoted in Russell 1946, p. 851). The philosophy of Dewey has, indeed, been as influential among pure scientists as among social scientists, Whitehead (1957, p. 47) asserting that 'we must not slip into the fallacy of assuming that we are comparing a given world with given perceptions of it' and Bridgman (1954, p. 37) that 'it is in fact meaningless to try to separate observer and observed, or to speak of an object independent of an observer, or, for that matter of an observer in the absence of objects of observation.' Perhaps the most extreme statement of this subjective view of natural systems was given by Van Uexküll (1926, pp. ix–xv) as follows:

> Nature imparts no doctrines: she merely exhibits changes in her phenomena.... The sole authority for a doctrine is not Nature, but the investigator, who has himself answered his own questions ... the objects that surround us are constructed from the sense-qualities: and, indeed, one person uses some sense-qualities for the making of objects, and another uses others. So for him there are nothing but signs or indications for his subjective use, and they assert nothing whatever with regard to the phenomenon that is independent of him.... No attempt to discover the reality behind the world of appearance, i.e. by neglecting the subject, has ever come to anything, because the subject plays the decisive role in constructing the world of appearance, and on the far side of that world there is no world at all.

In the two foregoing examples one can see rather extreme statements of viewpoints which are currently active in environmental studies. The former, termed 'logical positivist', is clearly associated with the 'scientific' type investigations, whereas the latter, or 'subjective behaviourist', is identified with man-centred social or psychological studies. Indeed, it might be argued that there is a basic difference between the objects of these two types of investigation, in that the former uses the mind of man to illuminate an objective reality, whereas the latter employs man's subjective vision of reality to study the variety of imagination exhibited by the human mind. Perhaps this is a fundamental distinction between science and art (fig. 1.3). The paintings of Turner and Van Gogh tell us, after all, a great deal more about the character of the minds of these creative artists than about the landscapes of England and France. In geography the first wave of the 'New Geography' was dominantly logical positivist in character, whereas since the middle 1960s something of a behavioural backlash has occurred. For example, in a recent work on 'the affective bond between people and place' Tuan (1974) has drawn attention to the following key behavioural terms:

> *Perception*: 'Both the response of the senses to external stimuli and purposeful activity in which certain phenomena are clearly registered while others recede in the shade or are blacked out.'

Attitude: A cultural stance *vis-à-vis* the world.
World view: A belief system involving conceptualized experience, partly personal, largely social.

Tuan (1974, p. 248) concluded that man's environmental aspirations lie somewhere between the above approaches: 'Human beings have persistently searched for the ideal environment ... [drawing] on two antipodal images: the garden of innocence and the cosmos.... So we move from one to the other ... seeking a point of equilibrium that is not of this world.'

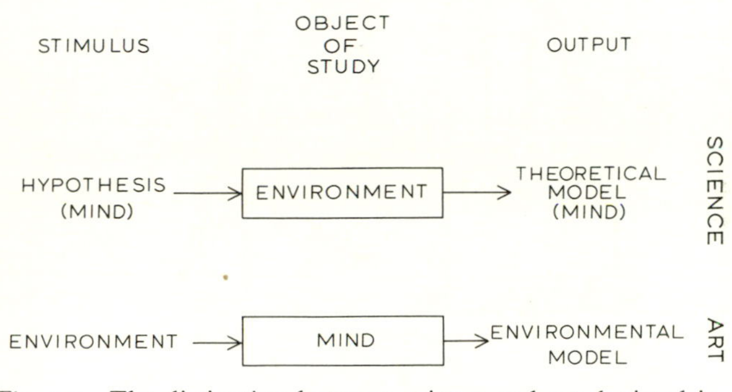

Fig. 1.3 The distinction between science and art depicted in terms of the object of study (system), the stimulus (input) and the resulting model (output).

It would be incorrect, however, to draw too broad a distinction between the structural and behavioural approaches, for both of them insist on the necessity of studying viable entities. From this point of view it is often of marginal significance whether these entities represent external natural truths or fortuitous erections of the mind of man. Again, both approaches stress the interdependence of the real world in its broadest sense, but recognize, with Aristotle, that the first necessary step in achieving knowledge is to separate the subject of study from the rest of existence (Grene 1963, p. 86). In systems terms this introduces us to the notion of *bounded* and *unbounded systems*. For our purposes a bounded system is one whose operation is entirely justified by its internally specified parts and relationships such that, if inputs occur, the resulting changes within the system and the outputs are capable of exact and unique prediction, given a perfect knowledge of the system. An input into a bounded system where output is inhibited is inevitably followed by a progressive equalization within the system and a breaking down of internal differentiation and of the structural hierarchies which are associated with the so-called increase in *entropy*. The level of entropy – i.e. of homogeneity, randomness or lack of free energy – thus uniquely describes the state of a bounded system. An unbounded system, on the other hand, is one whose internal operation is not solely controlled by input and output, but also by

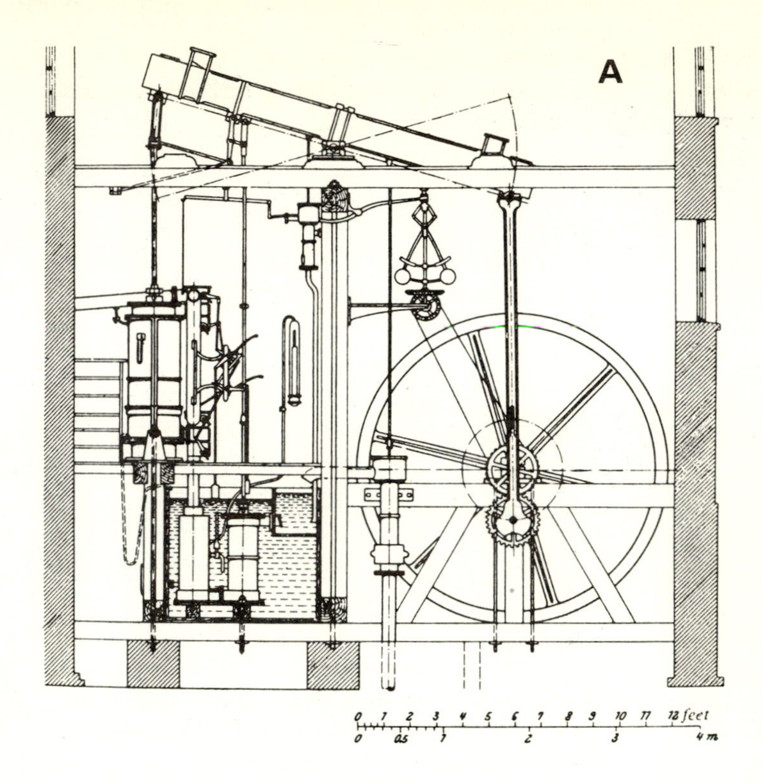
A
0 1 2 3 4 5 6 7 8 9 10 11 12 feet
0 0.5 1 2 3 4 m

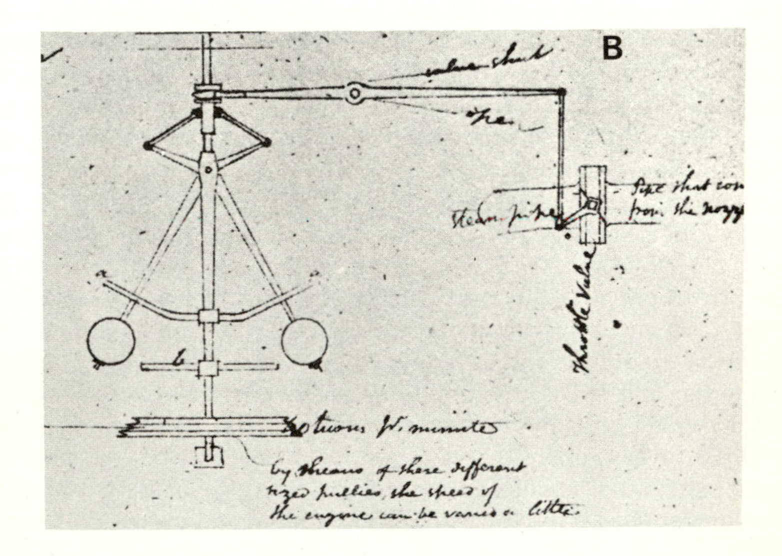
B
Steam pipe
Throttle Valve
by means of these different
sized pullies the speed of
the engine can be varied a little

other real world influences. An unbounded system cannot be considered apart from the other systems which intersect it, with which it shares components, and the operations of which make the effect of inputs into the open system to a degree indeterminate, both in respect of outputs and of internal organization of the system. Unbounded systems are thus sustained in a structured and hierarchically organized state by repeated inputs (i.e. applications of 'negative entropy'). It is clear from the foregoing that, with the possible exception of the universe itself, no truly bounded systems exist. However, from the problem-solving point of view it is convenient to assume that systems have been identified, either objectively or subjectively, and that their dominant behaviour can reasonably lead to the assumption of their bounding to all intents and purposes. Such assumptions form the basis of all observational and experimental scientific work. External influences are subsumed by the introduction of *error terms* which make up the deficiency between the actual and predicted behaviour of an unbounded system which has been investigated as if it were bounded.

A further aspect of systems analysis makes possible links with philosophical modes of thought and logical systems, and this involves the identification of the system's *executive* (Churchman 1971). The executive is that part of a system which determines: the functions of the other parts of the system, the circumstances under which each part is employed, and which is capable of judging system performance (Churchman 1971, p. 27). Judging or control systems have been an important inspiration to man's thinking at least since classical times (Mayr 1970) and the impetus has quickened with the nineteenth century development of the mathematical theory of control systems (Airy 1840, Maxwell 1867–8) to encompass abstract topological and morphogenetic features (Thom 1975). These mathematical developments discussed further in Chapters 3 and 4 derive from studies in biology (Lotka 1956), but more especially from the construction of automatic control devices such as wind machines (Lee 1747) and the Watt governor shown in fig. 1.4. This latter regulates the steam feed as input into a physical system (steam engine) designed to generate a given output (work). Mathematical developments also derive from the complex problems of inertial navigation of space vehicles, missile tracking and avoidance, and the complexities of control of large industrial plants. The executive in such systems determines the functions of subsystem components by manipulation of controllable physical inputs. The

---

Fig. 1.4 (*on facing page*)

A The Watt steam engine of 1789–1800 together with the centrifugal governor. (After Matschoss 1908, and Mayr 1970.)

B The centrifugal governor which maintains engine speed of rotation at a set level by automatic adjustments of the steam inlet valve. (From original drawing by Boulton and Watt in Dickinson and Jenkins 1927 and Mayr 1970.)

inspiration which mathematical control systems offer is as a means of determining attributes of controllability and observability, the categorization of the executive as the operator or steward over systems of nature, and the concentration on performance of systems defined by anthropocentric criteria. Hence they blur still further the distinction of Aristotelian abstract systems and perceptual subjectivism. The executive is the most important element in such systems.

A major function of the executive is to regulate and dispose inputs into the system and Churchman (1971, pp. 25–7) has restated Spinoza's (1677) four types of information processing as a 'taxonomy of inquiring systems', based on the role of the executive (fig. 1.5):

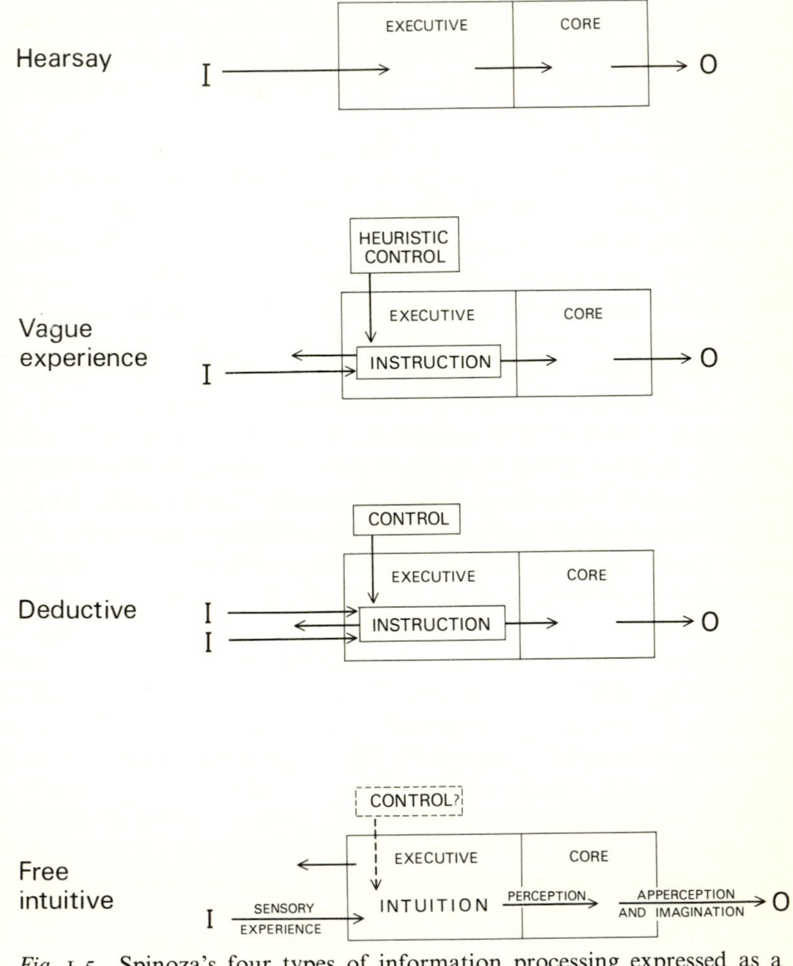

Fig. 1.5 Spinoza's four types of information processing expressed as a taxonomy of inquiring systems, stressing the role of the executive. (Suggested by Churchman 1971.)

(*a*) *Hearsay*: A system which will input, store and retrieve (i.e. output from core) anything which it is programmed to do. It has a weak executive routine operating on a simple classificatory basis.

(*b*) *Vague experience*: A system whose executive rules, however complex and flexible, are independent of the intellect and exhibit no intellectual curiosity. Pattern recognition machines are of this type.

(*c*) *Deductive*: A system whose executive is adaptable to the extent of deciding whether a method of proof is likely to succeed or not according to predetermined rules (e.g. game playing machines) but which cannot question the inputs or create new methods of proof.

(*d*) *Free intuitive*: A system based on intuition leading to knowledge which arises though the essence of the object perceived, such that the executive has a valid theory to explain why knowledge (i.e. inputs into the inquiring system) occurs. In such a system the executive has a key role in filtering and organizing the information inputs from the real world.

For Churchman (1971) it is possible to distinguish between major philosophical modes of thought, or 'inquiring systems', by employing a systems format based on the role of the executive (fig. 1.6). We might use the term *core* for the rest of the system which contains facilities for storage, manipulation and output of information processed and input by the executive.

The Aristotelian inquiring system is dominated by a search for the externally defined 'essential attributes' which form the structural keys to the definition of the total structure-process relationships which is the goal of scientific investigation. Locke's (1690) philosophy involves a very weak executive with outside influences (i.e. expert opinion) determining which information should be taken most seriously, and with the input being processed by devices having no built-in preconceptions of the world and no *a priori* information about nature. These devices, or logical operators, combine the information into categories from which are developed consistent expositions (Churchman 1971, pp. 33, 99, 177). The inquiring system of Leibniz (1714) can be viewed as the most conventionally 'scientific'. It has strong executive control in that it is assumed that the system possesses innate ideas (a similar notion to that of Plato) which act upon the input stream of information to filter it into *tautologies*, *self-contradictions* and *contingent truths*. These latter are admitted to the core where they are constructed into fact nets – i.e. into expanded sets of contingent truths, interlinked by appropriate relationships – by a process of testing, ranking, building and linking. The purpose of the Leibniz system is to develop ever-expanding fact nets by filtering contingent truths with the aid of the innate executive, without the control of which the uncurbed imagination could create fantastic and mythical fact nets of any kind or size. Leibniz believed that the object of philosophy is to produce ever larger and more improved fact nets which converge upon an optimum reality. In this he mirrors the contemporary scientist's belief that there is an accurate image of nature which is being converged upon by improved approximations of fact and theory. A very different inquiring system of Hegel and Kant held that an objective information input can only be obtained if the inquirer has chosen the correct 'image of the world' (*Weltanschauung*) or the way of viewing the entire system (Churchman 1971, pp. 35, 30–41, 95–7 and 170–7). The mind should be loaded with a wealth of information and experiences so as to generate a deep conviction as to the operation of the real world. This conviction is believed to be a necessary stage of inquiry into reality and naturally operates an exceptionally strong executive control on the information which is admitted

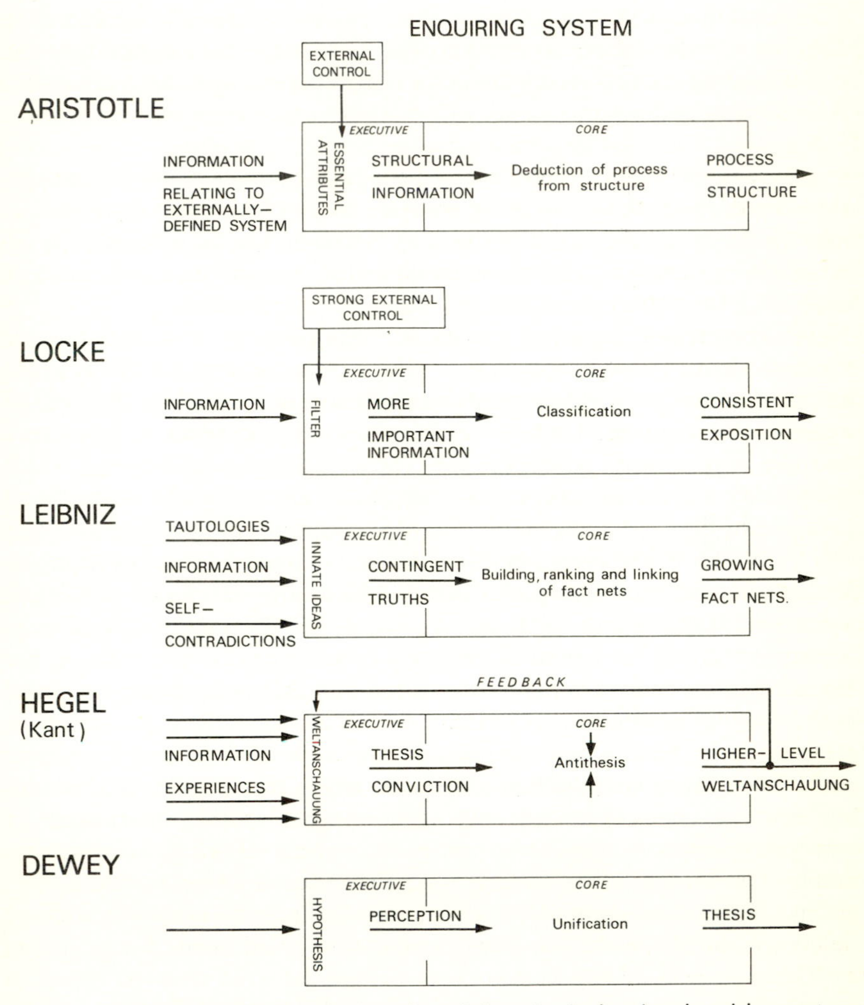

Fig. 1.6 The major philosophical modes of thought depicted as inquiring systems. (Suggested by Churchman 1971.)

to the core as being consistent with the appropriate *Weltanschauung*. Here it is interrogated as to why the inquirer holds his thesis with such conviction. The system then generates a diametrically opposed antithesis and conducts a confrontation, along the lines of Kantian dialectic, leading to a more refined and higher level of world image (or synthesis) which is then subjected to a process of *feedback* whereby the output of the inquiring system feeds back to influence subsequent inputs of information. Feedback may either reinforce the input in an explosive manner or dampen it. The philosophical system of Dewey has been compared with that of Hegel in that in an attempt to identify unified wholes, often by biological analogy, he erects initial hypotheses which act as strong executive controls over the input of information and its interpretation. As Santayana put it: 'In Dewey ... there is a ... tendency to dissolve the individual into his social functions, as well as everything substantial and actual into something relative and transitional' (quoted in Russell 1946, p. 855).

Fig. 1.6 is perhaps of added interest in that it is possible to use these inquiring systems to understand something of the development of attitudes to the environment by geographers in Britain and the United States. In this sense the Lockean system of the inter-war period was supplanted by the Aristotelian/Leibnizian system of the New Geography between 1950 and 1966, and this has been challenged by the Dewey behaviouralists and the Hegelian radicals during the past decade. In this respect it is interesting to speculate upon the similarities of the two last-named approaches with their strong executive control generated internally within the system. This, to some extent, rationalizes the parallel development of works so superficially different as those of Tuan (1974) and Harvey (1973).

1.2 Bases for theory

Intuitive attitudes to reality are difficult both to trace and analyse, yet they form the basis for data generation, hypothesis testing, and theory generation. These attitudes have been arranged in fig. 1.7 to form a rough spectrum from the eccentric, which concentrate upon the importance of bases external to man, to the egocentric, which are the most subjective. Although the teleological approach is most clearly associated with eighteenth-century attitudes to the environment and the conventionalist with the most modern subjective and radical attitudes, the order of this spectrum is not meant to imply any necessary sequence of time stages. Indeed, the cynic may derive some satisfaction from recognizing a significant identity between the teleological and conventionalist views! These intuitive bases may be defined as follows (Chorley 1976):

Teleological: Attitudes to reality aimed at relating observations to current views of the overall design of nature and concepts involving final causes. These are aimed at facilitating general views of the real world which satisfy both the dictates of the intellect as well as existing orthodoxy. Characteristic of this attitude is the philosophy of Bishop Berkeley which treated the whole of nature as a vast system of signs through which God teaches man how to behave, by informing, admonishing and directing (Passmore 1974, p. 15). Mainstream geological work in the eighteenth century and the geographical writings of Ritter are examples of this view.

Immanent: Attitudes which concentrate on the explanation of the characteristics of the real world in terms of their assumed indwelling or inherent nature. The geological interpretation of landforms and nineteenth-century geographical determinism present many features of immanence.

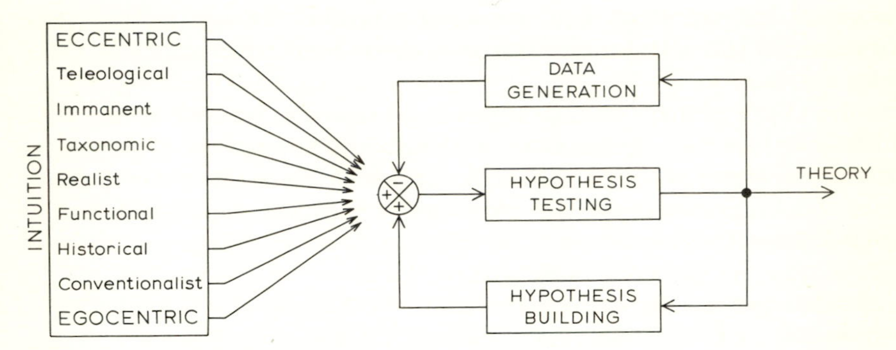

Fig. 1.7 The relations between intuitive attitudes, data generation, hypothesis building, hypothesis testing and theory construction.

Taxonomic: Attitudes which approach theory building by the grouping of phenomena into classes. Such an approach is invariably associated with an information explosion and its occurrence in time is to this extent scientifically fortuitous. When the taxonomy assumes a spatial flavour the most compelling environmental taxonomies emerge, as with the classic early twentieth-century French regional treatments. These and other spatially based taxonomies developed in association with a functionally based ecology and their functional flavour helps to explain some of the continuing popularity of regional intuitive attitudes.

Historical: Attitudes to reality based on a narrative reconstruction of a series of significant events assumed to have led up to the situation in question. This strong and intuitive basis for theory generation is often found to rely on allegorical, metaphysical or anthropomorphic underpinnings. Davisian polycyclic landform development theories and, to a significant extent, the theory of Darwin are examples of historically oriented theory construction.

Functional: Attitudes based on the belief that real world phenomena can be shown to be instances of repeatable and predictable regularities in which form and function can be assumed to be related. Theory of this type derives from the view that the discipline is empirically based, rational, objective, and aimed at providing explanation and prediction on the basis of observed regular relationships (Keat and Urry 1975, p. 4). The functional approach is most clearly associated with the logical positivism of mainstream scientific theory (Popper 1959, and Hempel 1965, 1967). In its extreme form it is based on the view that theory springs from objective observation, although this is less widely held now than hitherto (Popper 1959). However, the positivist approach is still associated with the view that there are contingent truths which cannot be known by *a priori* means and are not a matter of logical necessity. Scientific theories are thus concerned with the establishment of law-like relations linking observed phenomena in the 'constantly conjoined sequences of coexistence and succession' (Keat and Urry 1975, pp. 72–3, 74). In practice, the functional approach to environmental explanation has often been characterized by the use of surrogates for function, or process, which involve events of arbitrary magnitude and frequency, or even form itself. Functional explanation concentrates on form as the external manifestation or artefact of function, or process. Studies of process are thus viewed as means to a formal end. The first wave of the so-called 'new' geographical studies of the environment between about 1950 and 1965 were of this type.

Realist: At one extreme, this represents an extension of the functional or logical positivist approach. It is based on the view that real explanation involves something more than prediction based upon observed structural realities. This type of explanatory approach seeks to penetrate behind the external appearances of phenomena to the essences of mechanisms which necessitate them as the result of chains of causal connection (Keat and Urry 1975, pp. 5, 27–45). The realist thus considers himself to be concerned with the identification and investigation of the detailed, basic causal mechanisms and of the underlying structures of which the external gross forms are merely the often imperfectly representative artefacts. The development of systems applications in environmental studies may be viewed as part of this realistic shift, in that their concern has been increasingly for the flows of mass, energy and information which produce, sustain and transform the forms, rather than for the forms themselves. The search for underlying mechanisms has led human geographers away from their preoccupation with explanations based on the spatial application of classic economic and social theory. Of recent years there have been increasing numbers of geographers concerning themselves with disciplines such as hydrodynamics, thermodynamics, sociology, social anthropology, individual, social and cultural psychology. Among other tendencies, the revival of interest in historical geography and the decline in spatial stochastics can be understood in the context of realism. In an extreme form, realism is associated with the artistic, subjective, value-rich and qualitative tendencies in contemporary environmental attitudes.

Conventionalist: This attitude to reality is a natural extension of extreme realist views. It departs from logical positivism in the belief that no useful distinctions can be drawn between observation and theory and that the development of the latter does not inevitably follow the former. Although, as Wittgenstein points out, the resulting theories can be objectively tested and compared by observation and experiment. It is postulated that external reality cannot be usefully held to exist independently of one's theoretical beliefs, concepts and perception. Thus the adoption of theory is viewed as largely a matter of convention in which moral, aesthetic and utilitarian values play an essential part (Keat and Urry 1975, pp. 5, 46–70). The recent development of radical, committed, value-rich and utilitarian forms of environmental theories are in this sense conventionalist. It is interesting that Marxists view their explanatory base as realist, rather than conventionalist, in nature. Marxism holds that there is no distinction between facts and values, no separate existence for the products of human labour from the activity which produces them (a view which has interesting overtones of Vidal de la Blanche's view of regional geography and of William Morris's view of art), and no economic or social laws independent of historical structures. Kuhn (1970) has pointed to the revolutionary stages in science which bring the established paradigms into question and which 'require sociological and psychological categories for their explanation' and Kalakowski believed that 'the data of experience always leave scope for more than one explanatory hypothesis, and which one is to be chosen cannot be determined by experience' (Keat and Urry 1975, pp. 55, 60). As stated earlier, there is an interesting feedback from the conventionalist to the teleological basis for theory via the problem of 'teleological bias' (Sztompka 1974, p. 138).

1.3 Man and environment

It is more than two centuries since Rousseau in his *Discours sur les sciences et les arts* (1750) formulated a view of human development in terms of the paradox that mankind deteriorates as material civilization advances. Nevertheless this paradox increasingly perplexes those geographical methodologists who view the discipline in terms of man–environment relationships, and it is no accident that Glacken's (1967) important treatise on the relationships of man to the habitable earth terminates in the mid-nineteenth century. Glacken saw this hiatus largely in philosophical terms wherein the long-dominant concept of a 'designed earth', which forms the thesis of his work, was to be successively modified by the linked theories of evolution and ecology. Glacken (1967 pp. 549–50) showed how, by the nineteenth century, the older teleology of the 'web of God's design' had begun to decline, and has suggested how the work of Darwin was to replace it by considerations of environmental adaptations and interrelationships (the 'web of Life') which culminated in the twentieth-century concept of ecology. Many geographers have implied that man in society might be viewed as forming part of a complex ecological system, and, although George Perkins Marsh (1864) was impressed by man's ability to dominate the landscape, his was still essentially an ecological view of man's unity with his environment (Chorley 1973, pp. 157–9). The thesis of human ecology draws on the philosophy of Francis Bacon who believed that 'Nature can be conquered only by obeying her' (Chappell 1975, p. 161), and from the Epicurean view that 'though you drive Nature out with a pitchfork, she will still find her way back' (Horace, *Epistles*, I, *x*). It has been stated in a most extreme form by Hewitt and Hare (1973, p. 24) who advanced the view that geographers 'must seek nothing less than a generalization of ecosystem theory so that it includes the spatial activities of man'. It is easy to link this approach with the philosophy of Dewey via the concept of *transaction*—'observations of this general type sees man-in-action, not as something radically set over against an environing world nor yet as something acting "in" a world, but as action *of* and *in* the world in which man belongs as an integral constituent' (Dewey and Bentley 1949, p. 52).

One may take issue with Glacken's thesis from two main standpoints. Firstly, it is by no means clear that the ecological model to which Glacken's work seems to point is capable of accurately depicting man–environment relationships even in the nineteenth century, let alone now. The ecological model fails as a supposed key to the general understanding of the relations between modern society and nature, and therefore as a basis for contemporary environmental studies, because it casts social man in too subordinate and ineffectual a role. Although much interesting and vital research is at present being conducted into the ecological aspects of some more isolated societies, into less advanced agricultural communities and with regard to the origins of agriculture (see chapter 9), it is difficult to see how this analogy can profitably be extended to form a model which will encompass the activities of an important and rapidly expanding portion of social and economic life styles. Human geography is no simple extension of biogeography. Things have gone too far for that – if they were ever so. Even the Garden of Eden had its reptilian entrepreneur. Man's relation to nature is increasingly one of dominance and control, however lovers of nature may deplore it. Man is not passive, subject only to principles of physics and chemistry or Darwinian natural selection; his great success is not just genetic but is based also on orderly innovation, especially the 'noogenesis' of Teilhard de Chardin associated with the invention of language and knowledge which allows the permanence of sentient, responsible and creative man (Polanyi 1958, p. 387). Passmore (1974) has suggested that extreme ecologists are 'pessimists who hark back to the quasi-Arcadian life of the countryside', being possessed of a 'Romantic Passion for wild nature' which may be 'a selfish attitude adopted by a small group' (Matley 1975, pp. 533–4). Galbraith has cynically defined a conservationist as 'a man who concerns himself with the beauties of nature in roughly inverse proportion to the number of people who can enjoy them', and conservationists are sometimes condemned as middle class élitists who are 'attempting to impose sacrifices on a working class which is enjoying for the first time in human history a degree of affluence' (Passmore 1974, pp. 94, 105). It is by his very increase in numbers, resulting in the growth of material strength, proliferation of competitive demands and in the increasing complexity of organization which so impressed Rousseau, that social man is being more and more set apart from the physical and biological environment. Competition to exploit, control and consume all manner of terrestrial resources, including space, is impelling man continually to extend his environmental dominance. Many prognosticators have expressed the very real fear that the continued unbridled operation of positive socio-economic feedback loops, involving the exponential growth of population and of industrial capital, must lead to their ultimate arrest by the lagged operation of the negative feedback physico-environmental loops of pollution, resource exhaustion and famine (Forrester 1971, Meadows *et al.* 1971, pp. 156–7) (see Chapter 9). However, it is apparent that, far from encouraging man to relax his overall environmental grasp, such tendencies may result in proposals for increased planning and control. It seems clear, therefore, that geographers might temper their preoccupation with the ecological type of model in favour of the application of that of the control system; thus, adapting Glacken's terminology, 'God's design', having been replaced by 'Nature's design', may be in turn supplanted by 'Man's design'.

The second point with which one might take issue as to the implications of Glacken's work is that its concentration on the transition from a designed earth to one providing an ecological setting influencing, and influenced by, man ignores the other nineteenth century man–environment models which are also of contemporary significance. In fig. 1.8 we return to a simple designation of the *object* of investigation, the system, and the manner in which the transformation of throughputs is used to illuminate it. The Medieval and Renaissance teleological model, explored so extensively by Glacken, made God the object of study by means of demonstrating that an apparently chaotic Nature, when interpreted in terms of God's design, could be shown to be ordered. Man appeared in this system as partaking something of the image of God, rather than forming part of Nature, and, in the words of Milton, the object of natural philosophy was 'to justify the ways of God to man'. Passmore (1974) has termed this the attitude of 'man as a despot'. The Judaeo–Christian ethic, as for example embodied in Genesis 1:26, leads us to consider man as 'superior to nature, contemptuous of it, willing to use it for our slightest whim' (White 1967, p. 3767), and Calvin was assured that God created all things for man's sake. This view went back to that of the Stoics and of Aristotle, and to the belief of Descartes that nature is a great machine to be manipulated by man to suit his own ends; while Spencer's applications of Darwin's theory depicted man as struggling against nature thereby demonstrating his moral superiority and Freud (1930) defined the human ideal as 'combining with the rest of the human community, and taking up the attack on nature, thus forcing it to obey human will, under the guidance of science' (Passmore 1974, pp. 10–23).

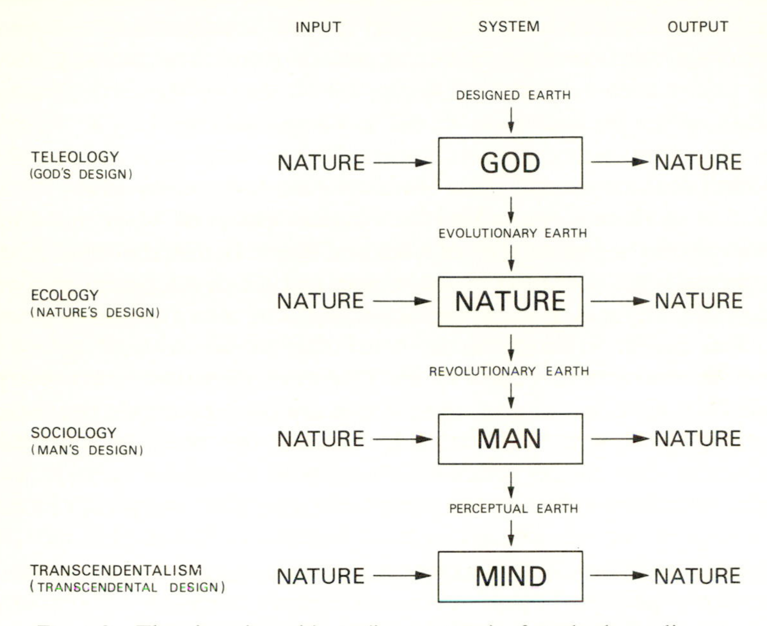

Fig. 1.8 The changing objects (i.e. systems) of study depending on views on the overall design of creation.

The ecological model aims at an understanding of the interrelationships of nature, which subsumes man, by taking observations of nature, structuring them into working systems in the manner envisaged by Dewey, and then testing these systems by further recourse to natural observations. This view of 'man as a steward' (Passmore 1974, pp. 28–53) was clearly stated by George Perkins Marsh (1864, p. 35) who believed that nature 'has left it within the power of man irreparably to derange the combinations of inorganic matter and of organic life, which through the night of aeons she had been proportioning and balancing, to prepare the earth for his habitation, when, in the fullness of time, his Creator should call him forth to enter into its possession.' However, as Passmore points out, the main support for the view of environmental stewardship does not come from the Christian ethic, despite the recent views of the German metaphysicians and Teilhard de Chardin that nature is in the process of perfection because of man's operations on it and that to cooperate with nature is to link hands with one's predecessors (Passmore 1974, pp. 34–5). Plato believed that, although man's soul was alien in a world of change and decay, it was the responsibility of the animate to care for the inanimate, and William James (1907) was concerned that man's efforts should deliver him from the 'nightmare of entropy'. It is thus clear that, although related, there is a distinction between the environmental models of man as a steward and man as a component of the ecosystem. This examination of nature's design, and man's ecological place within it, has much to commend it philosophically and emotionally, and it is no accident that, at a time of concern regarding shortages, pollution and environmental degradation, this model enjoys such popularity. Despite the decline in the anthropocentric Christian ethic, the ecological model is being challenged by at least two contemporary man–environment models, both of which have their origins in nineteenth century thought. (See Stoddart 1967 for counter view.)

The first of these is the radical sociological model of Marx and Engels, stemming from the philosophies of Kant and Hegel. In this, the object of study and concern is the condition of social man and the development of a world image which will lead to improvements in his state along certain doctrinaire lines. This model is based on the belief that theory without action is a misleading abstraction. Engels in his *Dialectics of Nature* (1873–83) propounded the view that natural phenomena are not immutable but form an ever-changing current. This attitude became the basis of the radical environment–man model wherein information about the natural world, suitably monitored by the system's executive, could be interpreted by social man to output an ever-revised view of nature which will play its role in advancing social vision and in promoting action leading to an improvement in the human condition. Engel's concern for man's despoiling of nature was solely on the grounds of the deprivation of later generations of men. Harvey's *Social Justice and the City* (1973) is the most distinguished of the current crop of radical works promoting Pope's dictum that 'the proper study of mankind is man'. For the Marxist (and even for many of his Russian predecessors), human occupation of the land has not been viewed as the ecological symbiosis of the Western geographer, but as a dialectic in which society confronts nature, and, by a mixture of force and coercion, achieves power over her: 'Under a planned socialist economy, the paths of natural processes progressively diverge from the natural and are transformed directionally' (quoted in Isachenko 1973, p. 196). The work by Gerasimov (1969, see 1976, p. 84) entitled 'There is a need for a general plan of the transformation of nature in our country' typifies much of the communist attitude to man–environment relationships. Anuchin (1972) distinguishes the external natural phenomena making up the 'landscape envelope' which, when locally acted upon by human society and 'saturated with the results of labour', becomes the 'geographical environment'. Although many Soviet environmentalists recognize the existence of geosystems, composed of the interrelations of 'landscape ecology' and 'human ecology' (Sochava 1971, pp. 279–82), they would regard them as essentially autonomous, although interrelated. In general they view the Western human ecological approach as 'idealistic' – in other words that, by attempting to treat natural and social regional processes in the same manner, one is forced towards the belief that reality is beyond the reach of the physical senses (i.e. the opposite of 'materialistic'). In this, of course, they are following the view of Engels that natural and social phenomena obey different laws (Chappell 1975). Despite Marx's (1971) rejection of the 'deification of nature' and his belief that 'nature becomes for the first time simply an object for mankind, purely a matter of utility', it is fair to say that many authors are now interpreting Marxist man–environment relations in a more complex manner (Schmidt 1971, Harvey 1974). For the modern Marxist theorist there is an essential difference between, on the one hand, the 'unalienated creation of nature' by a new social system composed of individuals who 'not only feel, but also know their unity with nature' (Engels 1873–83, p. 296) and who 'by thus acting on the external world and changing it . . . at the same time change . . . [their] own nature' (Marx 1967, p. 175), and, on the other hand, the alleged 'mindless exploitation and capitalism' which views nature as 'simply an object for mankind, purely a matter of utility' (Marx 1971, p. 94) (see Harvey 1974, pp. 266–7). Clearly the extent and proliferation of Marxist writings allows a variety of interpretations to be placed upon them.

The other important nineteenth-century model was based on the intuitive or 'transcendental' philosophy of nature mysticism, which had some impact on the work of Darwin. This can be interpreted in systems terms whereby the system under consideration is not the social condition of man but the attitude of the human mind. Perceptual images of nature are input, which both affect the mind of man and lead to an output of a view of nature which further reveals the mind to itself. As Emerson wrote in his diary in 1833, 'The purpose of life seems to be to acquaint man with himself. . . . The highest revelation is that God is in every man'. The humane study of *Topophilia* by Tuan (1974) is typical of a contemporary expression of this philosophy, which extends the ecology and territoriality of place to a general religious concept of 'geopiety' (Tuan 1976). This humanist conception reveals the depth of emotional bond between man and nature as man and place which Lowenthal (1975) has exampled as nostalgia, and which grows out of a 'reciprocation' of man in nature. The maturation of this view is unlikely to lie in the fields of geosophy and phenomenology. The lineage followed by Schumacher (1973) draws from Buddhism and corresponds in part to Plato's conception of justice involving the duality of civil rights and civil duties, whilst Tuan (1976) has pointed to ideas both in the Western ethic of the Roman *pietas* and the Eastern, Chinese ethic of *hsiao* central to which are family bonds, Confucian respect for natural order, and emphasis on a continuity of life through the community which necessitates both duty and obligation of man in nature.

In both the radical and humanist challenges to the nineteenth-century ecological model, man is the demiurge: we are led to a necessary consideration of the ethic by which man chooses to organize himself. There are at least two main approaches to the study of the ethics of man–environment interrelations. The first of these, the intuitionist ethic, derives from the twin lines of the Aristotelian specification of ideals and the Lockean concept of right and wrong formulated into logic and translated into social action either by way of social contracts as suggested by Rousseau, or as a sense of natural justice, as formulated by Kant, which appears in most Western constitutions (written or unwritten) and charters of rights. The difficulty in this approach is the emergence of the *a priori*: the mind is a black box which is not amenable to prediction or explanation, and hence neither are the goals, nor the organization of society, nor the ecological or genetic consequences of social conclusions. Explanation is prescriptive since the defined ideals of society become the norm (Wilson 1975).

The developmental-genetic ethic provides a second approach, evidenced, Wilson (1975) suggests, by the coming together of sociology and biology as sociobiology. This provides a new synthesis which is structuralist and in which stages of social evolution of environmental ethics (inferred by Piaget from the development of children) form a hierarchy from obedience, through conformity, duty, and legal constraint, to conscience: from 'should' to 'ought' statements (von Mises 1951, Piaget 1969). The mechanism is ethical behaviourism in which moral commitment in society to 'place' is entirely learned and conditioned within the current norms of society such that cognition itself becomes a function of socialization (Kohlberg 1969). Controversy surrounds the degree to which genetic, societal or environmental factors control cognition and behaviour, as discussed in chapter 5 and evidenced in the debate over Burt's (1953) studies of the inheritability of intelligence. It is clear, however, that the direction which sociobiology indicates is one in which man is seen as essentially no different from other organisms: *is* a part of nature. However Wilson recognizes that within modern man there is a polymorphism or moral ambivalence in which messages from emotive centres to the brain are in part the response of relict features (Pleistocene and pre-Pleistocene) of instincts of the hunter-gatherer and colonizer, and in part the nascent response to new situations, especially of urbanism, requiring learning, adaptation and creativity (Wilson 1975, pp. 563–5, 574–5). Thus sociobiology offers biogeographical explanation of only the lower levels of social emergence (Polanyi 1958, p. 399) – explanation of resourcefulness in achieving the foreseeable, and of ontogenetic maturation as in the Piagetian adaptive application of logical thought to new problems; old can-openers offer morsels of new understanding. Polanyi suggests the need for higher level explanation of true heuristic achievement giving the basis for anthropogenesis: a phylogenetic emergence requiring a high degree of originality and unprecedented operating principles giving rise to life of the mind guided by universal ethical standards.

The demand for creative assimilation of urban change has been identified by Barbara Ward (1976) as the most pressing problem to be overcome in maintaining *The Home of Man* in its spiritual and environmental sense, and one which is especially difficult to overcome in Third World countries where very rapid social changes have occurred. Desirable altruism in man–man and man–environment relations can be induced where a shift in values is called for, and this seems to be the ambition of Edward Wilson's sociobiology. If this revives arguments surrounding the work of Galton or Malthus or Burt, as presaged so vividly by Aldous Huxley, it is clear that whilst such aims are motivated by concepts such as that of the economist's social welfare and common good, or the Skinnerian psychologist's desire for the greatest happiness for all, the manner in which the happiness or overall welfare is defined is not at all clear (as, for example, in Arrow's impossibility theorem discussed in chapter 8) and deductions based on Polanyi's finalistic opportunity of affirmation of man's ultimate aims, on Benthamite 'felicific calculus', on utility maximization, and on welfare economics, are grossly inadequate.

The intuitionist and developmental ethics have the common characteristics of advocating levels of behavioural modification of man's position within the environment and of his use of its resources in both a qualitative and a quantitative sense, but it is the goals or ethic of this control rather than the manner in which it is achieved that creates the severest difficulties for defining theoretical bases for resolving Rousseau's (1750) paradox of the deterioration of mankind with material advances. If man now possesses an adequate, or at least a perfectable, technology of control which can be applied to ecological maintenance of the species, he will be capable of wielding it to good effect only after considerable reappraisal of the manner and scale at which he chooses to organize society, of our social ethic, and of our epistemology.

This book has been conceived as providing a unified multi-disciplinary approach to the interfacing of 'man' with 'nature', and, as intimated in the Preface, has a number of broad aims. Firstly, it is desired to explore the capacity of the systems approach to provide an inter-disciplinary focus on environmental structures and techniques. Secondly, we wish to examine the manner in which a systems approach aids in developing the interfacing of social and economic theory, on the one hand, with physical and biological theory, on the other. A third aim is to explore the implications of this interfacing in relation to the response of man to his current environmental dilemmas, and hence to expose the technological and social bases of the values which underlie man's use of natural resources. It is hoped to show that the systems approach provides a powerful vehicle for the statement of environmental situations of ever-growing temporal and spatial magnitude, and for reducing the areas of uncertainty in our increasingly complex decision-making arenas. The 'environment' is thus interpreted in the broadest sense to embrace physical, biological, man-made, social and economic reality.

The book is divided into four parts. The first of these develops the technical concepts by which systems can be conceived and described. Part II explores the cognitive foundations of decision making. In Part III a range of examples of physical, ecological, social and economic systems are approached from the point of view of their analysis, synthesis and spatial expression. The final part draws together the inter-disciplinary approaches by exploring the dilemmas which confront man's intervention in natural systems which underlie his 'living together' with nature.

Part I
Hard systems

2 Systems methods

Nature doesn't come as clean as you can think it.

(Alfred North Whitehead)

Much of chapter 1 was devoted to an exposition of the way in which systems analysis can be applied to achieve a better understanding of the different modes of thought which condition our attitude to the stream of information being received from the 'real world' of our environment. We have now to introduce a more formal 'systems' philosophy, based on the view that portions of the real world can be perceived as units which are able to maintain their identity in the face of inputs which generate both outputs and changes within the systems themselves (Van Gigch 1974, p. 51). In chapters 2, 3 and 4 we therefore present the principles of the treatment of so-called *hard systems* – those capable of specification, analysis and manipulation in a more or less rigorous and quantitative manner.

2.1 The structure of systems

The science of systems analysis has been developed to enable the structure and behaviour of systems to be explored. A system is composed of a number of *state variables* which are related to one another by *operators*, and are subjected to *inputs* (X) to produce *outputs* (Y). Hence a system may be characterized by the three elements of which it is composed: input, output, and translation operator linking the two.

The system input and output variables may be continuous or discrete. A system is said to be continuous if its input and output are measured as continuous variables which are defined everywhere on a given interval – for example, a time interval t ($-\infty < t < \infty$). Continuous variables are usually also considered differentiable on this interval and written for variable X as $X(t)$. A discrete system is one in which the input and output are defined only at discrete distinct points – for example, the integer range ($t = \ldots, -1, 0, 1, 2, \ldots$), usually restricted to a domain or range of points N. Such discrete variables may be the result of discrete events (single occurrences, for example, of purchases and sales) or, more commonly, may result from measurement only at a range of data points. These latter give rise to *sampled data systems*. Discrete observations of a variable X are defined by subscript notation as X_t, and the entire available sequence of observations by the bracketed term $\{X_t\}$. It is discrete rather than continuous systems which are of most frequent occurrence in environmental problems.

Operations with systems of discrete or continuous variables can be represented in a number of ways. If we define two discrete variable sequences $\{X_t\}$ and $\{Y_t\}$ (digital signals) as the respective system input and output, then the operation of an environmental system can be characterized by the transformation equation

$$Y_t = SX_t. \qquad (2.1.1)$$

The transformation *operator* S is termed the system *transfer function*. It is this element which determines the manner in which the system input $\{X_t\}$ is translated to become the system output $\{Y_t\}$ at any point in time t. Hence, the characterization of the system transfer function also serves to define the system itself. The transfer function *is* the system in the sense that it completely describes the process or changes induced by the operation of the system modulating system inputs to produce given system outputs. For the case S is unity, the system is trivial and has no discernible effect. In most cases of interest S is not unity and has a specific structure made up of a number of *parameters*, or constants, which determine the magnitude and shape of modulation induced by system operators. The parameters are the time constants and exponents which govern the behaviour of the system. They are frequently considered to be constant or *time invariant* over the period of analysis, but in complex environmental systems evolution and time variance are some of the most important modes of system behaviour (see sections 2.3 and 2.4). In these cases, the system parameters become variables which are themselves subject to constant or non-constant models and the division between system variables and system parameters becomes arbitrary (Kalman 1960). For the sake of simple exposition in the first two parts of this book, however, the transfer function and the parameters of which it is composed are usually treated as *constants* and constant exponents in each system equation relating input and output.

It is clear that systems can be defined in terms of equations wherein the output (Y) is a function of the input (X), which is subjected to the action of a number of operators, and where the relations between Y and X can be explained and predicted by estimating the values of the parameters associated with these operators. The latter characterize the system structure with sets of equations. For continuous systems the resulting system operators are described by differential equations, and for discrete systems by difference equations.

Equations are the axioms of the mathematical structure which provides the model. The form of the equations (e.g. regressions or power series) are based, firstly, upon mathematical and physical principles and, secondly, on the literature, personal knowledge and intelligent guesses made by the operator because of his familiarity with the characteristics of the system (Kowal 1971,

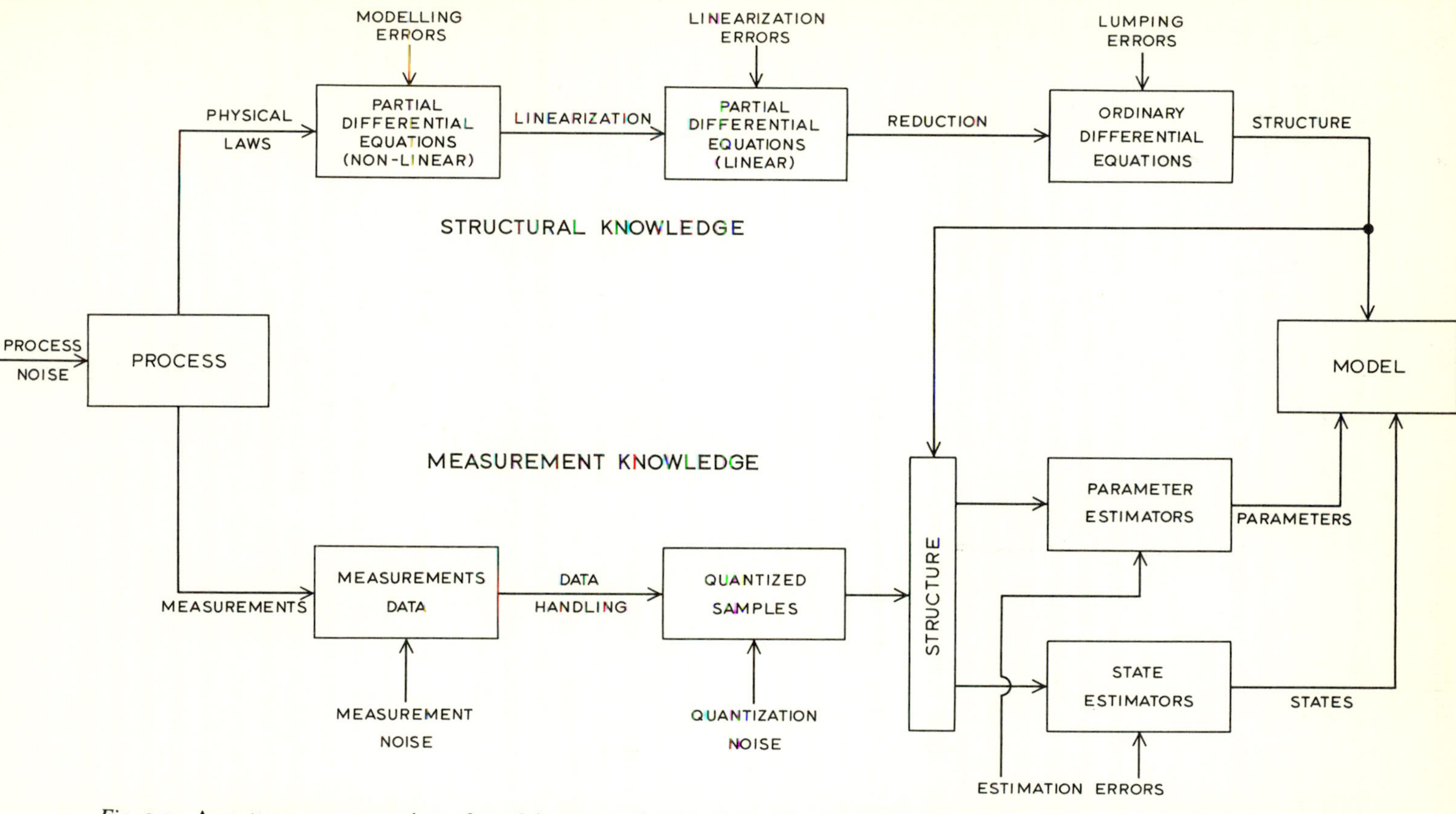

Fig. 2.1 A systems representation of model construction involving the development of structural knowledge and measurement knowledge. (From Eykhoff 1974.)

p. 137). Fig. 2.1 shows the relation between the simultaneous development of the structural knowledge about the system which comes from, on the one hand, an *a priori* postulate of the variables involved and the relations between them, and, on the other, *a posteriori* measurement knowledge from sampling values of the variables and their relationships in nature. This leads to the estimation of the parameters and the construction of a robust theoretical model of the system.

It is possible, by using conventional symbols for the operators, to represent systems by means of block diagrams. Block diagrams include systems components (i.e. state variables) and mathematical operators. They provide a topological, or structural, representation of the system which focuses attention on the parameters and subsystems (McFarland 1971), and which can be manipulated and analysed algebraically according to a set of rules (MacFarlane 1970, pp. 163–4). Of the many possible mathematical operators, the most commonly encountered are (fig. 2.2) (Talbot and Gessner 1973, p. 12, MacFarlane 1970, pp. 159–63):

1 Summation (i.e. algebraic summation).
2 Identity or starting point (i.e. channelling of a single variable into two or more paths of identical value).
3 Multiplier (i.e. algebraic multiplication).
4 Splitter (i.e. dividing an input into two or more parts).
5 Integrator (i.e. accumulating a total of the values which pass through it). The integrator forms the basis for systems which are expressed in terms of differential equations, and the number of integrators defines the *order* of a system. For systems which are being treated by difference equations, the role of the integrator is assumed by a time delay operator.
6 Time delay or shift.
7 Linear cascade. The system variables are commutative and associative such that the cascade of $X(t)$ into $Z(t)$ and $Z(t)$ into $Y(t)$ is equivalent to the cascade of $X(t)$ into $Y(t)$ direct.

The operation of simple systems consisting of a multiplier and an integrator, respectively, is shown in fig. 2.3.

Fig. 2.4 indicates the way in which block diagram manipulation using these operators can be developed. Complicated systems can be modelled and then collapsed into a single *transfer function*. For practical purposes it is often simpler to analyse block diagram models of systems by first converting them into *flow graphs*, which provide a more flexible representation, emphasizing the variables and depicting summation in a more natural manner (fig. 2.5). Flow graphs can be subjected to a series of algebraic manipulations (fig. 2.6) to produce a composite transfer function which expresses the overall behaviour of the system (fig. 2.3) (McFarland 1971, pp. 9–11). The manner in which block diagram, flow diagram, and algebraic operations can be applied to

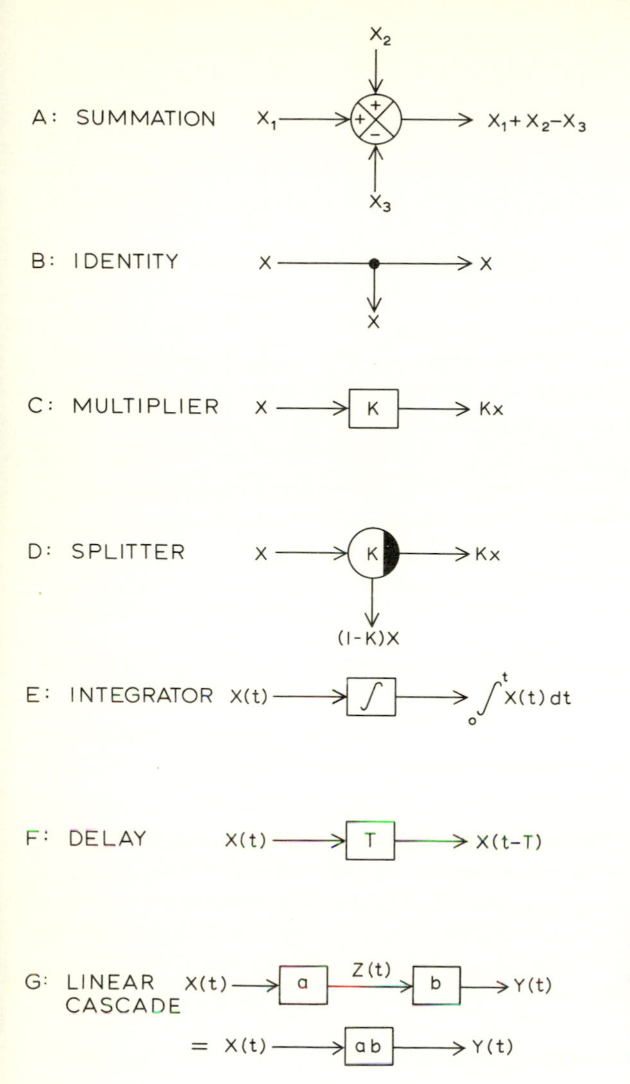

Fig. 2.2
Some common mathematical operators. (Partly after MacFarlane 1970, and Talbot and Gessner 1973.)

describing system input–output relationships via the transfer function is discussed below.

2.1.1 SYSTEM INPUT

The form of the input sequence X_t which enters the system will be as important as the system transfer function in determining the form of the system output observed. The behaviour of systems is largely conditioned by the inputs to

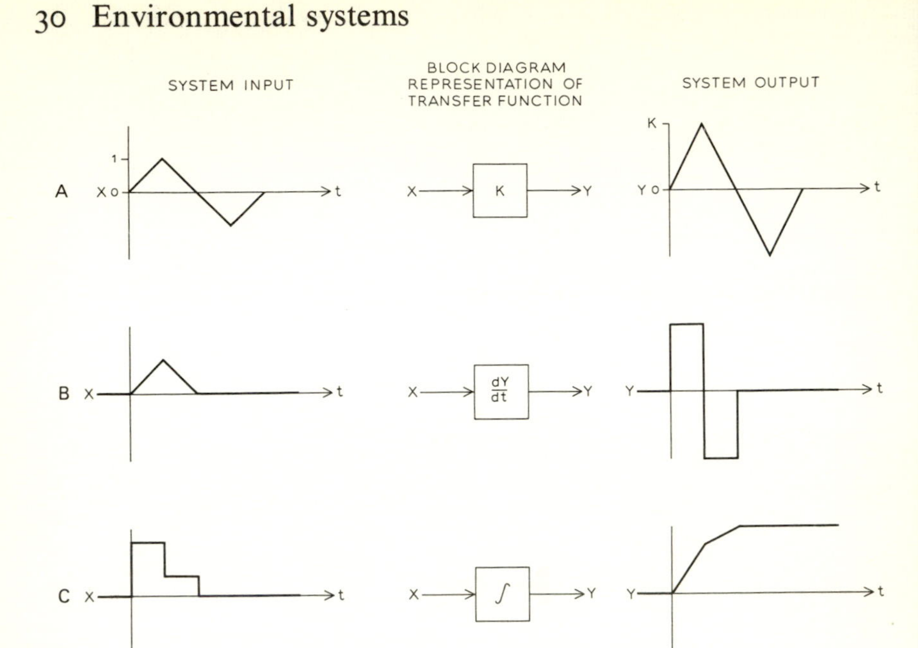

Fig. 2.3 The operation of three simple systems transfer functions: a multiplier, a differentiator and an integrator (or store). (After Blesser 1969.)

which they are subjected. These inputs may be classified broadly under five headings:

(1) *Transient impulse.* This represents a point stimulus into the system and is only of momentary, or of one sampling instant, duration. In continuous systems impulsive inputs can be represented by the delta function $\delta(t)$ defined by

$$\delta(t) = \begin{cases} 1, & t = \varepsilon \\ 0, & -\infty < t < \infty, t \neq \varepsilon \end{cases}$$

and as $\varepsilon \to 0$, $\delta(t) \to \infty$ at point ε, so as to satisfy

$$\int_{-\infty}^{\infty} \delta(t)\, dt = 1.$$

As examples of a reasonably isolated impulse we might consider the isolated rainstorm, or the unique technological innovation.

(2) *Unit step.* This represents an instantaneous change in input at time t from a zero level to a new constant (non-zero) level which is maintained for a length of time. This is represented by the unit function:

$$U(t - \tau) = \begin{cases} 0, & t < \tau, \\ 1, & t > \tau. \end{cases}$$

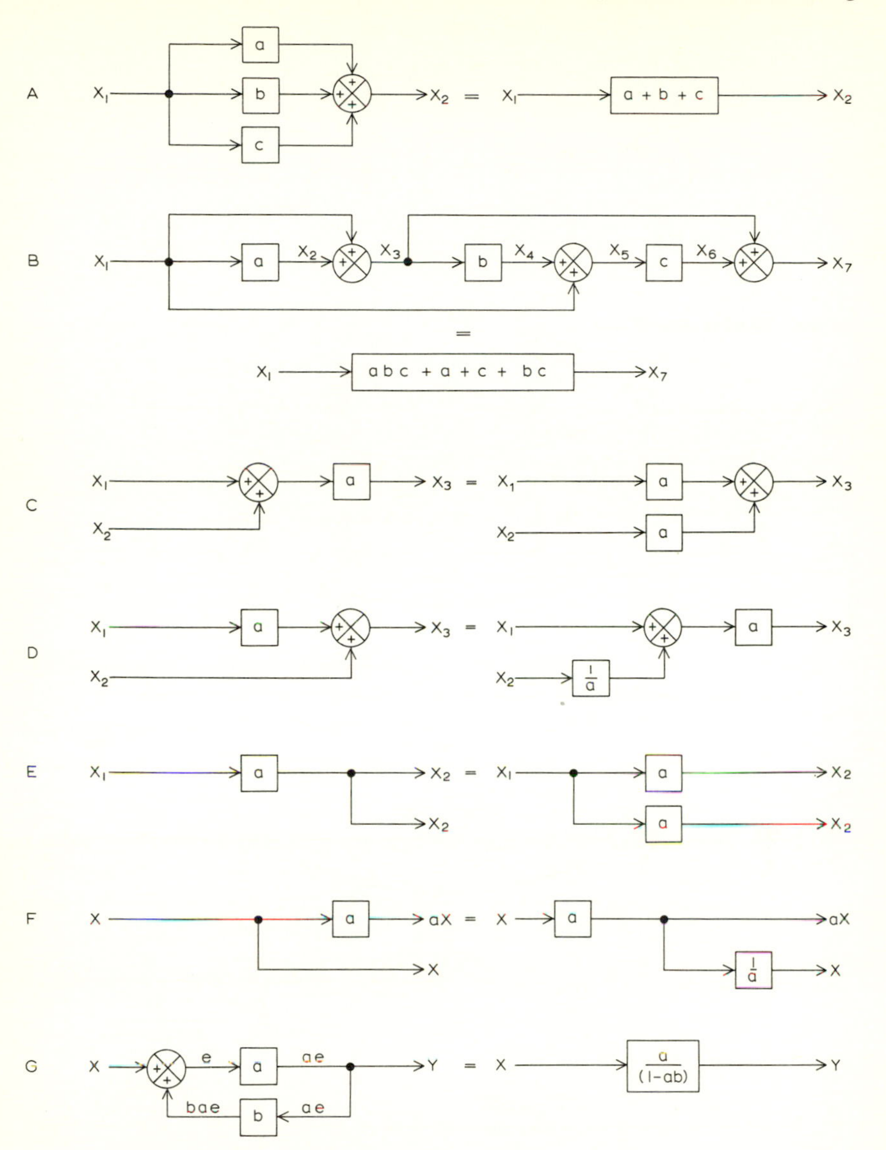

Fig. 2.4 Block diagram manipulations:
A Parallel combination.
B Cascade.
C Back-shift multiplier.
D Forward-shift multiplier.
E Forward-shift multiplier through an identity operator.
F Back-shift multiplier through an identity operator.
G Removal of a simple constantly oriented circuit.
(After MacFarlane 1970.)

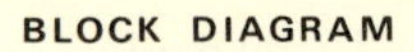

BLOCK DIAGRAM

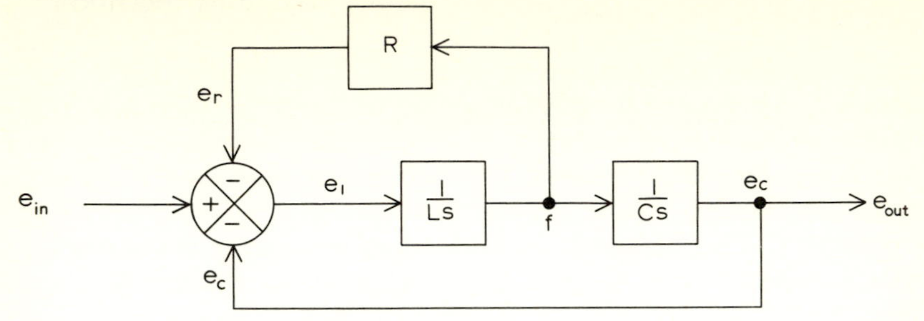

FLOW GRAPH DIAGRAM

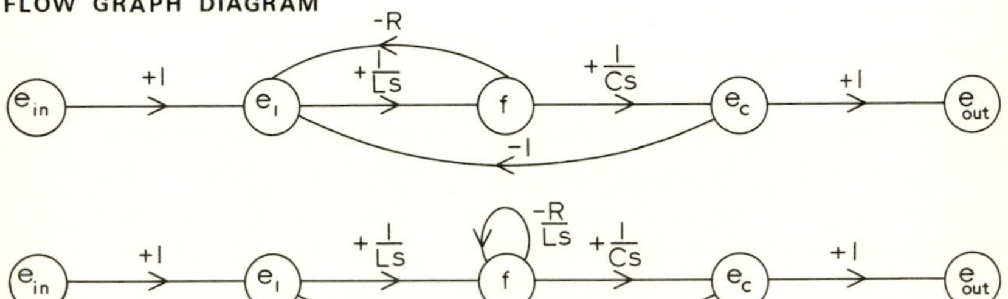

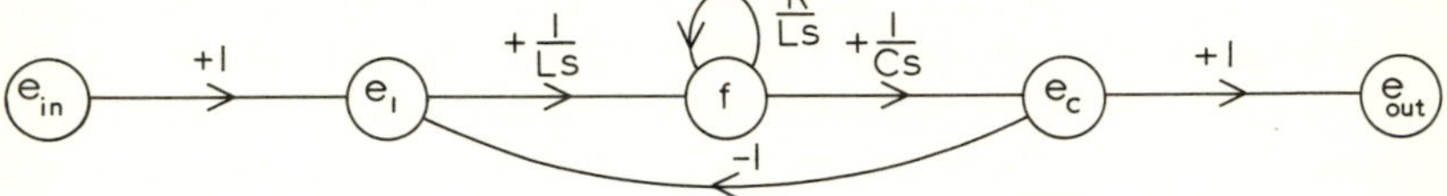

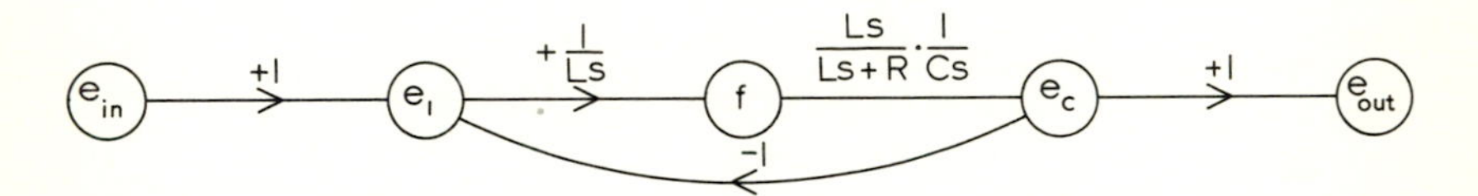

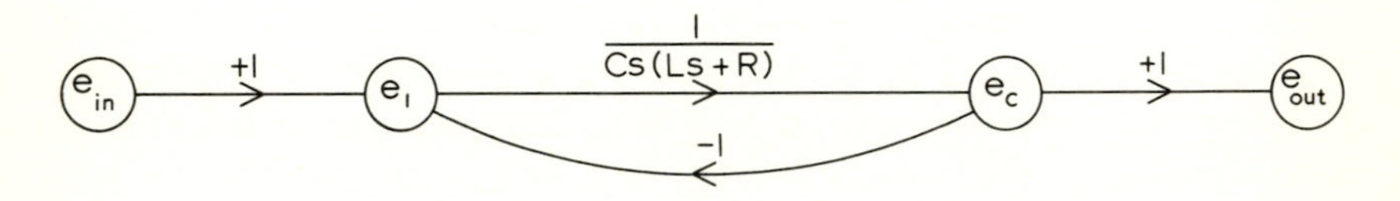

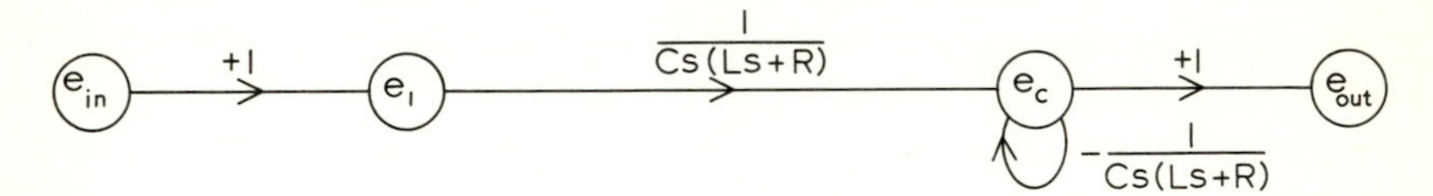

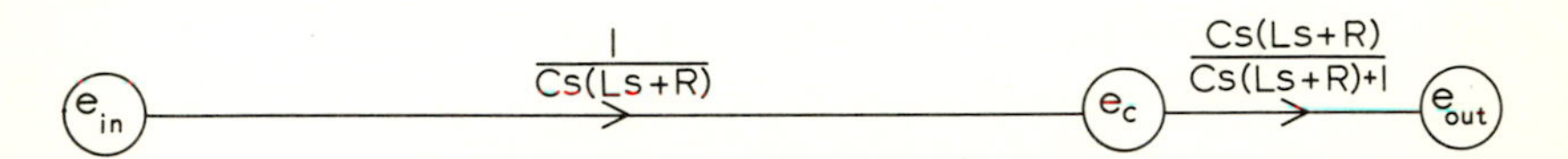

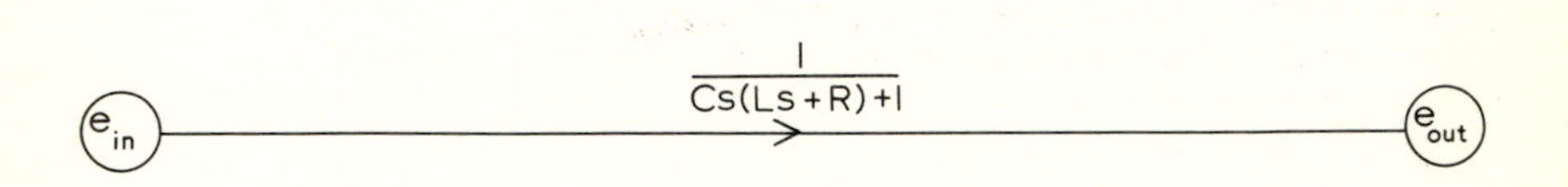

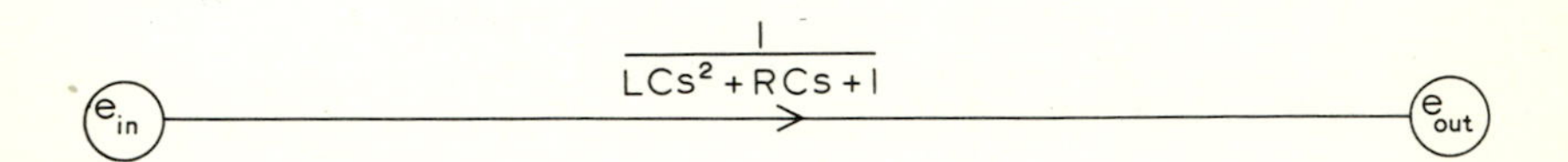

Such a response represents the 'turning on' of an input which is left on thereafter, an example in a socioeconomic system being the implementation of a policy control shift or change in an overall control variable.

(3) *Linear function (e.g. ramp)*. This is a constant increasing or decreasing trend as, for example, in linear growth. Many nonlinear growth functions may be converted to linear form by suitable transformation of coordinates, e.g. loglog, semilog, logistic, Gumbel plots, etc. (Elderton and Johnston 1969). These transformations apply strictly only under certain assumptions about the underlying input, e.g. the substitution of semilog plots for exponential (geometrical, multiplicative) inputs. Most environmental systems are subject to non-linear inputs and reduction to linear forms often induces a degree of distortion.

(4) *Periodic*. This is a regular oscillation, or a combination of such oscillations. In electrical engineering, for example, the response to a sinusoidal functional input is of great importance in the design of components. In environmental systems sinusoidal response should be examined in studies of the economic cycle and in biohydrological systems subject to seasonal, daily, and other periodic oscillation rhythms. The input in this case is represented by a sum of trigonometric sine and cosine components for frequency f as a fourier series:

$$X(t) = \int_0^{N/2} a(f) \cos(2\pi tf) + b(f) \sin(2\pi ft)\, df.$$

(5) *Stochastic*. This is a time stream of inputs which may be *stationary* or *nonstationary*, depending on whether the mean of the series is varying consistently, and which may be predominantly *random* in magnitude or exhibiting some consistent relations between members of the sequence (i.e. autocorrelation). Continuously varying stochastic inputs have been of most frequent concern to the time series analyst in statistics and econometrics. The input sequence is a sequence of impulses at each sampling instant which must be described in terms of its statistics of (*a*) central tendency, (*b*) dispersion, (*c*) intercorrelation or association properties between successive observations. These three statistics describe the mean, variance and autocorrelation of the input sequence, and are defined by the expectations:

$$E(X_t) = \overline{X} = \lim_{N \to \infty} \int_{-N}^{N} X(t)\, dt. \tag{2.1.2}$$

$$E(X_t - \overline{X})^2 = \sigma_x^2 = \lim_{N \to \infty} \int_{-N}^{N} (X(t) - \overline{X})^2\, dt. \tag{2.1.3}$$

$$E(X_t X_{t-\tau}) = R_{xx}(\tau) = \lim_{N \to \infty} \int_{-N}^{N} (X(t) - \overline{X})(X(t-\tau) - \overline{X})\, dt. \tag{2.1.4}$$

Fig. 2.5 (*on facing page*) Conversion of a block diagram to a flow graph and its simplification into a single transfer function. (After McFarland 1971.)

FLOW GRAPH ALGEBRAIC MANIPULATIONS

Multiplication

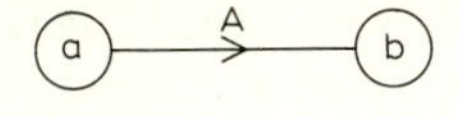

$b = Aa$

Summation

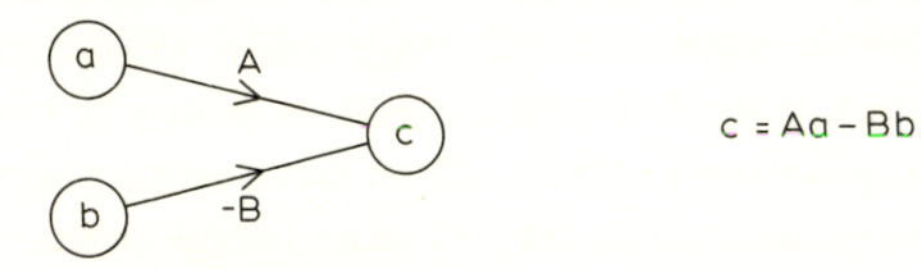

$c = Aa - Bb$

Elimination of cascade nodes

Insertion of dummy nodes

Unification of parallel branches

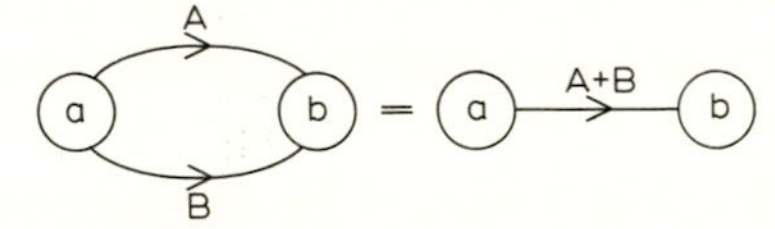

Branch splitting

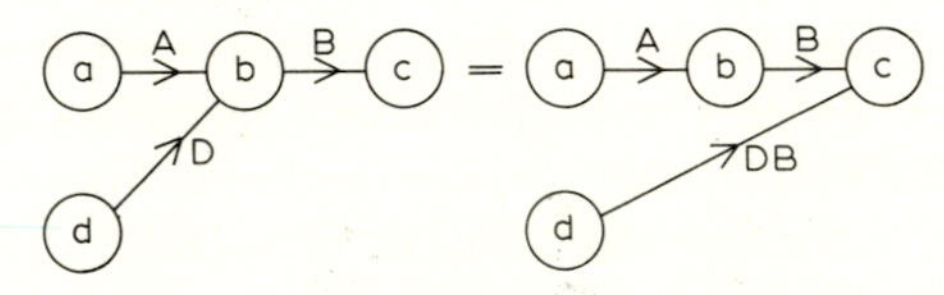

Closed loop replaced by self loop

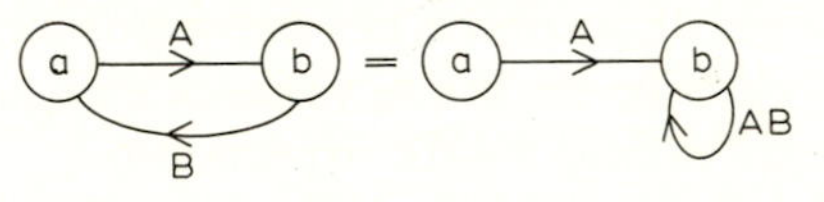

Removal of self loop

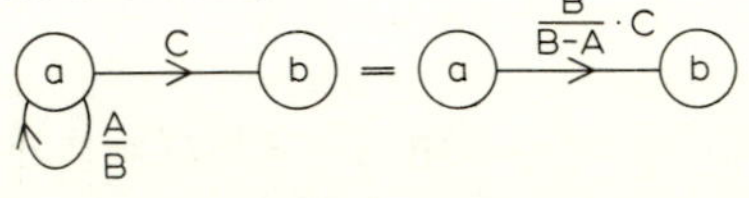

Fig. 2.6 Common algebraic manipulations of flow graphs. (After McFarland 1971.)

In many environmental systems the observed system input may be the result of all or any combination of these five forms. The classical time series model, for example, categorizes a given sample series into three elements: trend, seasonal, and stochastic component (Kendall 1973).

2.1.2 SYSTEM OUTPUT

The combination of the input and the transfer function characteristics govern the overall response of the system as observed in the output sequence $\{Y_t\}$. The characteristics of the system input (steps, impulses, etc.) cannot usually be observed in the actual system output since for a continuously exciting input into a system the responses to inputs at one time (t) have not decayed away before the input at the next time ($t + 1$) begins to produce an effect. Two specific system responses are of common interest: the impulse response and the step response.

The *impulse response* represents the response of the system process to unit input of unit duration. A typical form is shown in fig. 2.7. The *step response* of the system represents the response of the process to an input of unit magnitude of infinite duration. The form of this response for one example system is shown in fig. 2.7 together with the response to other function inputs (ramp and sinusoidal). The value of the impulse response function in systems analysis is that it gives a good description of process response to various magnitudes of inputs and provides a method of short-term prediction. Given the magnitude of inputs for each instant ($\ldots, t - 1, t, t + 1, t + 2, \ldots$) and the ordinates of the process impulse response, it is possible completely to specify the future behaviour of the output. This method is used particularly in hydrology where the unit hydrograph represents catchment impulse response and can be used to make flood predictions.

The step response of a system is used in many studies especially of control systems. It gives valuable information on process steady-state conditions, i.e. the form of process response resulting from continuous and unvarying input. Features of the step response which are particularly important in this control context are the steady-state error (a tolerance limit of $\pm\alpha\%$ of final value, usually $\pm 5\%$), the settling time (time to reach point before response is at tolerance of $\pm\alpha\%$ of the steady-state error) and the percentage overshoot (ratio of difference between maximum value of true response and the steady-state value to the steady-state value).

The models which are constructed to approximate the system outputs of real-world systems are of a great variety, but they are most usefully divided into two groups: linear and nonlinear.

A *linear system* is one in which none of its terms involves powers or products of the output, or dependent variable. The transfer function (S) remains constant for all magnitudes of input, i.e.

$$Y_t = SX_t, \quad S = \text{constant}\ (-\infty < X_t < \infty).$$

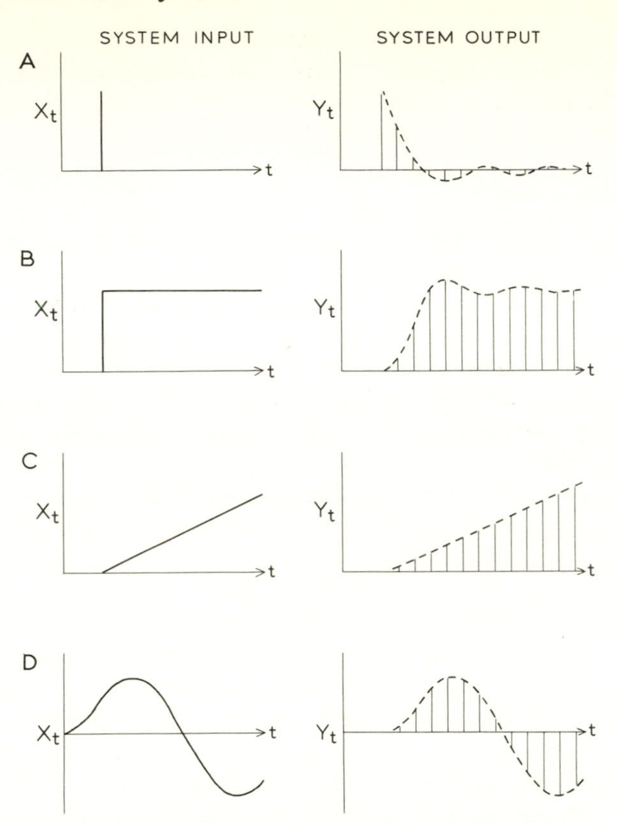

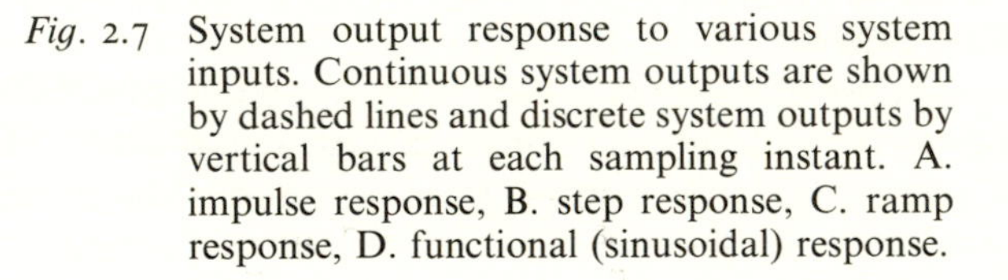

Fig. 2.7 System output response to various system inputs. Continuous system outputs are shown by dashed lines and discrete system outputs by vertical bars at each sampling instant. A. impulse response, B. step response, C. ramp response, D. functional (sinusoidal) response.

This is sometimes referred to as the additivity property or *law of superposition* which states that the response of $\{Y_t\}$ to $\{X_t\}$ remains the same for all values of the input stimulus. The superposition property of linear systems represents a simple and mathematically convenient device for approximating some environmental systems. For example, drainage basin runoff has been modelled for rainfall inputs in a linear manner, but it is now generally recognized that the runoff response is nonlinear (see chapter 7). Fig. 2.8 shows a set of impulsive rainfall inputs into a catchment. Each such input has an impulse response with the shape of the basin hydrograph, the magnitude of each hydrograph being an arithmetic scaling up or down dependent upon the magnitude of the rainfall. The runoff observed in the mouth of the basin will be a combination (termed a *convolution*) of the hydrograph responses to each

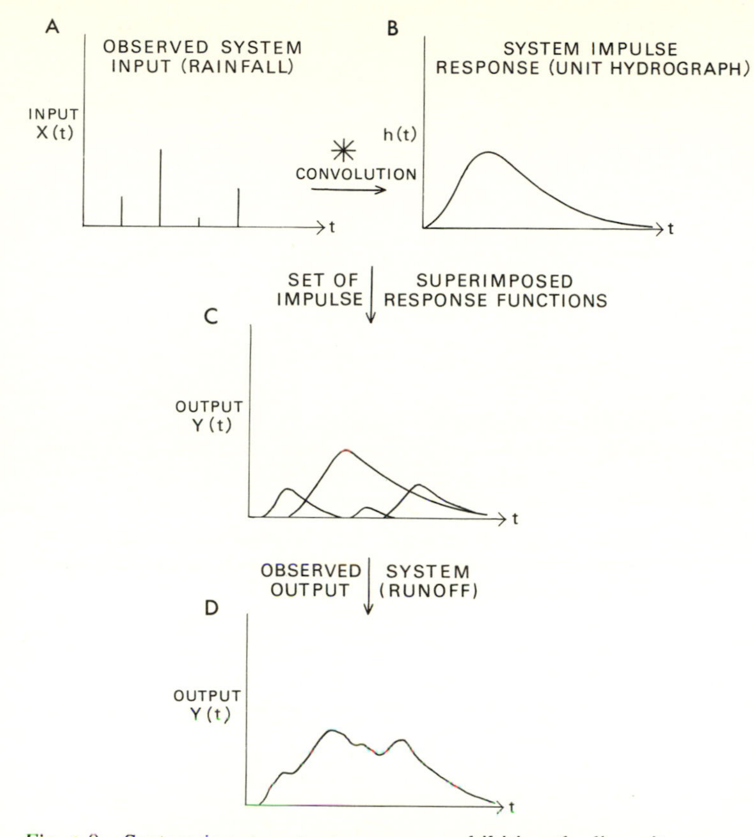

Fig. 2.8 System input–output response exhibiting the linearity property of the law of superposition in respect of the rainfall–runoff relation of a catchment with unit hydrograph ordinates given by $h(t)$. Convolution of system inputs A with impulse response B produces outputs C which, when summed, give the continuous output sequence D.

part of the precipitation input (Amorocho and Hart 1964). When the input or independent variable is decomposed into two or more components, the output will be the algebraic summation of the responses to each component (fig. 2.9). In other words, it responds to each of several inputs as if the others were not present (fig. 2.10).

A *nonlinear system* is one which produces an output which does not bear such a simple algebraic relation to the components of its inputs. The transfer function becomes a function of the magnitude of the input, i.e.

$$Y_t = S(X_t)X_t$$

and the value of $S(X_t)$ will vary depending upon the size of the input stimuli. Such processes illustrate the principle of *gestalt*, namely that the whole is

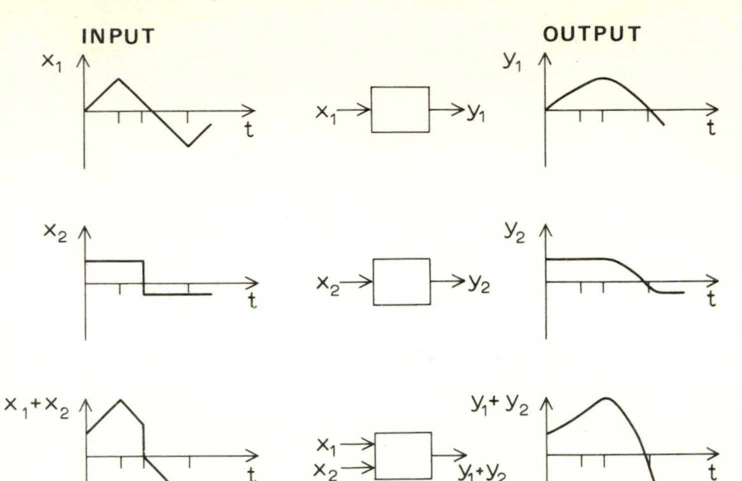

Fig. 2.9 Input–output relations for a simple linear system showing the output to be the algebraic summation of the responses to each input component. (After Jacobs 1974.)

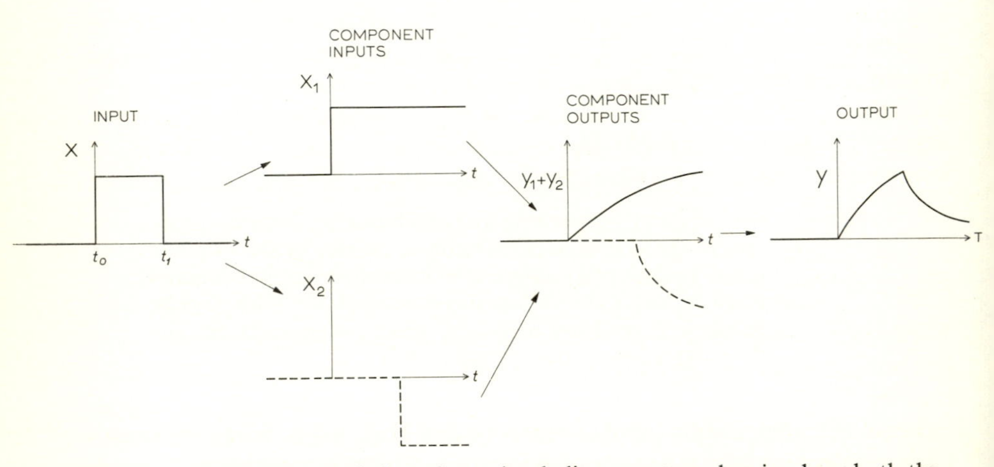

Fig. 2.10 Input–output relations for a simple linear system showing how both the input and the output can be viewed as superimposed components. (Partly after Deutsch 1967.)

operating in a manner which is not a simple reflexion of the sum of the constituents, but is some form of more complex function. Fig. 2.11 shows one example of a nonlinear system impulse response function where system output is a multiplicative function of system inputs into a river catchment. In algebraic terms, the effect of this nonlinear response is that changes in the level of output Y_t will differ for the same unit change in input X_t if the magnitude of the input differs. For example, a small magnitude change in X_t by one unit (e.g.

from 4 to 5) may result in a change in Y_t of 0·25 units, but a similar change in X_t by one unit at larger magnitudes of X_t (e.g. from 14 to 15) may result in a change in Y_t of 2·5 or ten times.

Unfortunately not only are nonlinear systems very difficult to analyse, but most environmental systems are of this type because almost all internal relationships are nonlinear.

Examples of this are drainage basin response to different magnitude storm (fig. 2.11; see chapter 7), economic elasticity of demand with income, etc. However, much of systems analysis is based on linear models despite the fact that the linear formulation is 'crude and often valid for only a limited range of model variables, including time' (Kowal 1971, p. 140). It is therefore common practice to try to approximate natural nonlinear systems by systems made up of linear components, the behaviour of which can be more effectively simulated and predicted. Such approximations induce a degree of distortion into resulting system simulations unless the system behaves in such a manner that analytical mathematical transformations yield exact linearized representations. For example, simple growth and cell division models of plant and animal communities correspond, at least in the early stages of development, to exponential growth as a function of time (chapter 9). This nonlinear multiplicative relationship can be reduced to a linear form in terms of the *logarithm* of the number of cells at any time. In such cases linear transformations induce no distortions. In other cases, linearized forms may be valuable *approximants* of environmental system behaviour even though a degree of distortion is induced. Hence the legitimacy of linear assumptions or linear transformations will depend upon the accuracy of the resulting model.

2.1.3 SYSTEM TRANSFER FUNCTION

The system transfer function S, which has been discussed in general terms up to this point, can take on an infinite variety of forms. In practice, however, most transfer functions that are observed in natural systems conform, or can be reduced to, a surprisingly small number of models. Each characteristic model form is defined by the magnitude, sign (positive or negative) and behaviour of the parameters (constants) which make up the transfer function. For both continuous and discrete systems most classes of system can be reduced to species of differential (continuous) or lag-difference (discrete) equations. Differential equations are composed of a set of system *differentiators* which relate the instantaneous rate of change of system output to the rate of change of system input. Difference equations are composed of a set of system *lag operators* which relate changes of the output to changes in system input over time. These two equation representations are mathematically transformable between each other such that any system can be represented in either form (provided that the alternative representation is meaningful).

LINEAR CONTRIBUTION

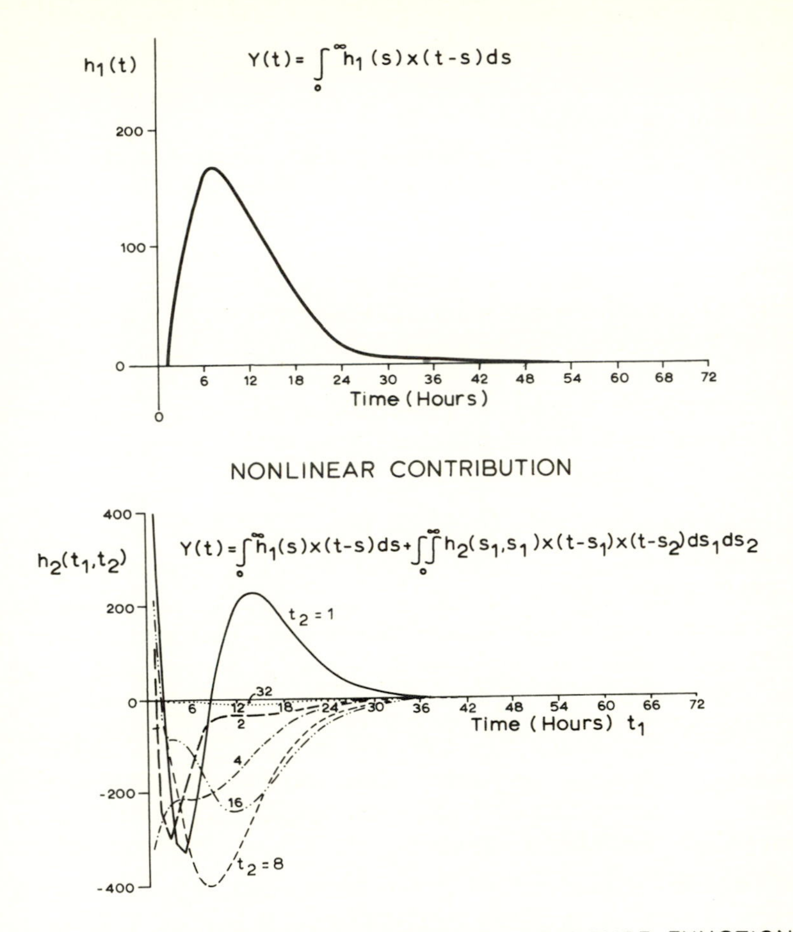

NONLINEAR (SECOND ORDER) IMPULSE RESPONSE FUNCTION

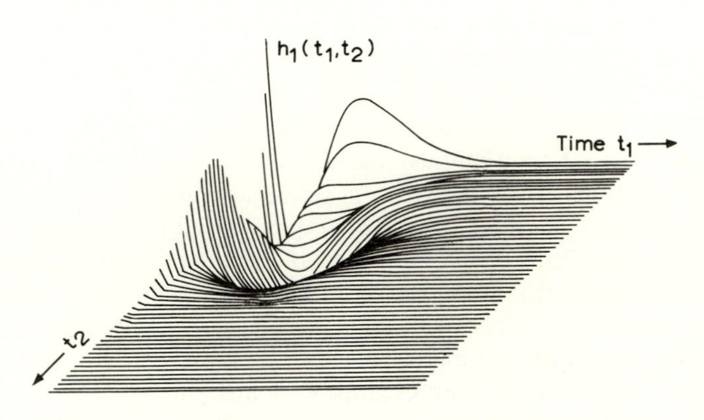

The number and form of the differential and lag operators define the structure of the system transfer function equation. Each class of equation has been given varying nomenclature (Box and Jenkins 1970, Kendall 1973, Bennett 1974, 1978), as summarized in table 2.1. Simple gain or multiplier processes are analogous to the univariate linear regression equation ($Y = a + bX$) and may or may not be associated with sets of delay or lag terms covering periods of dead time in which no response occurs. Distributed lag processes provide common models of environmental system behaviour in which a whole set of gain and delay terms governs the relation of current values of the dependent variable $\{Y_t\}$ or system output to the independent variable or system input $\{X_t\}$. System inputs produce responses which are delayed by various lag times and modulated by the gain parameters b_i. Distributed lag processes are analogous to the multiple regression equation ($Y = a + b_1X_1 + b_2X_2 + \cdots + b_nX_n$). The length of maximum lag q defines the *order* of the process, although some of the intervening terms b_i ($i = 0, 1, 2, \ldots, q$) may be zero.

A prominent feature of environmental systems is the existence of looped relationships which feed back or feed forward the influences of certain variables, bypassing others. The block diagram shown in fig. 2.13 depicts a system containing two such autoregressive or *feedback* loops, where outputs are split by identity operators and fed back to influence a subsequent input by means of a summation operator which acts either positively or negatively. The feedback loops are termed autoregressions (literally, self-regression) since they involve a lagged dependence of the system output on itself. The resulting regression is analogous to the multiple regression equation except that the independent variables are lagged dependent variables. This ambiguity of definition of dependent and independent variables creates a number of problems for the estimation of parameters (section 2.2). The effect of the autoregressive feedback, as to whether its influence produces stability or instability in the system output over time (fig. 2.12) depends, as we shall see, on whether the sign of the summation operator is negative (i.e. *negative feedback*) or positive (i.e. *positive feedback*) and on the values of any multiplier operators within the loop. A loop is said to be *stable* if its output in response to any input tends to a finite steady value after the input assumes a constant value. Fig. 2.13 shows the manipulation of simple negative and positive feedback loops to produce consolidated transfer functions. The negative feedback loop provides a high quality control mechanism, or servomechanism (see chapter 3). *Feedforward* or lead processes occur where a signal is fed forward, bypassing

Fig. 2.11 (*on facing page*) Nonlinear impulse response function expressed as a multiplicative function of input magnitude. This can be used to express different catchment hydrographs for different magnitude rainfall events. $h(\cdot)$ is discharge in cfs/inch/hour. (After Brandstetter and Amorocho 1970, for Petaluma Creek.)

part of the system, allowing another part of the system to anticipate changes of input. Although on the whole less common than feedback, feedforward is especially prominent in certain artificial control systems, in those modelling biological organisms and in human socio-economic systems.

Table 2.1

Continuous systems (differential equation)	Discrete systems (difference equation)	Type of model
$Y(t) = \beta X(t)$	$Y_t = bX_t$	Gain, multiplier and linear regression.
$Y(t) = \beta \frac{d^n X(t)}{dt^n}$	$Y_t = bX_{t-g}$	Shift, delay, lag or integrator.
$Y(t) = \beta_0 X(t) + \beta_1 \frac{dX(t)}{dt} + \beta_2 \frac{dX(t)}{dt^2} + \cdots + \beta_n \frac{d^n X(t)}{dt^n}$	$Y_t = b_0 X_t + b_1 X_{t-1} + b_2 X_{t-2} + \cdots + b_q X_{t-q}$	Distributed lags or integrators.
$Y(t) + \alpha_1 \frac{dY(t)}{dt} + \alpha_2 \frac{d^2 Y(t)}{dt^2} + \cdots + \alpha_n \frac{d^n Y(t)}{dt^n} = 0$	$Y_t + a_1 Y_{t-1} + a_2 Y_{t-2} + \cdots + a_p Y_{t-p} = 0$	Autoregression, feedback, memory or differentiators.
$Y(t) + \alpha_1 \frac{dY(t)}{dt} + \cdots + \alpha_n \frac{d^n Y(t)}{dt^n} = \beta_0 X(t) + \beta_1 \frac{dX(t)}{dt} + \cdots + \beta_n \frac{d^n X(t)}{dt^n}$	$Y_t + a_1 Y_{t-1} + \cdots + a_p Y_{t-p} = b_0 X_t + b_1 X_{t-1} + \cdots + b_q X_{t-q}$	General transfer function process or Padé series.
	$Y_t = b_0 X_t + b_1^+ X_{t+1} + \cdots + b_q^+ X_{t+q}$	Lead process.
	$Y_t = b_0 X_t + b_1 X_{t-1} + \cdots + b_q X_{t-q} + b_1^+ X_{t+1} + \cdots + b_q^+ X_{t+q}$	Lead-lag process.

The structure of system stability and feedback is governed in all most important characteristics by the nature of the autoregressive feedback elements that are present. These feedbacks of system output dependence on previous output values act as multipliers which have the ability to scale up or

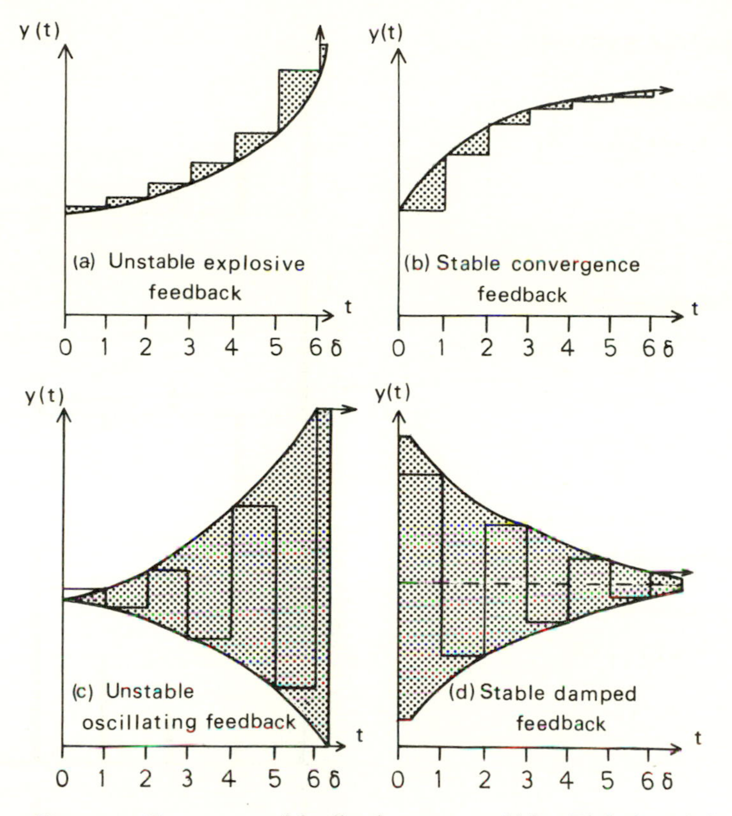

Fig. 2.12 Four types of feedback response. (After Blalock 1969.)

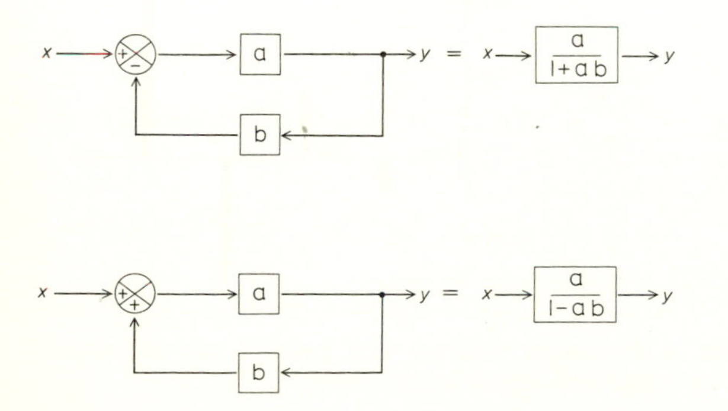

Fig. 2.13
The algebraic representation of simple negative and positive feedback loops as consolidated transfer functions.

down system response and, in certain cases, to generate continuous, self-perpetuating response. The simple response of various discrete autoregressive feedback systems is shown in fig. 2.14. It can be demonstrated that, in the general case, system stability is governed by the magnitude of the autoregressive parameters which scale the degree of dependence of the present

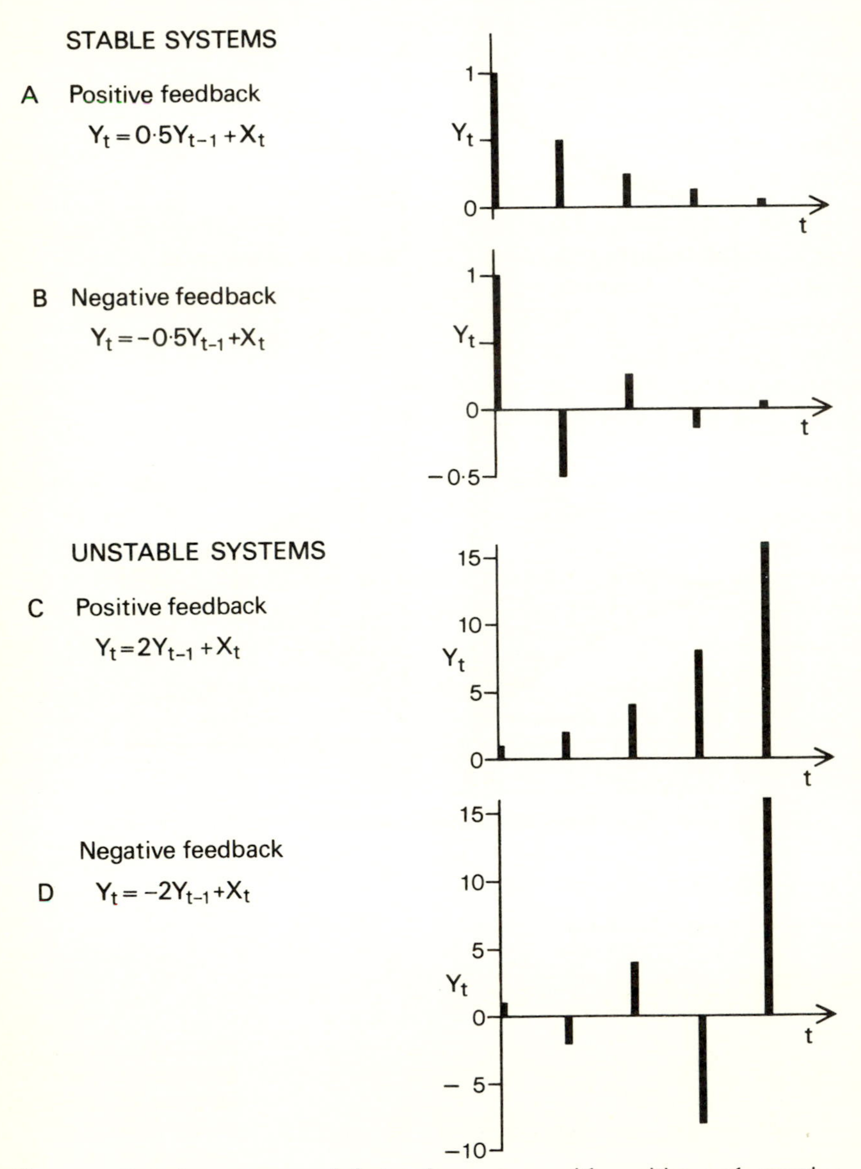

Fig. 2.14 Impulse response of first-order systems with positive and negative feedback, showing stable and unstable responses.

output or past output values. If these autoregressive multipliers are less than unity in absolute value, then the system is stable (i.e. the response declines to zero). If the autoregressive parameters are greater than unity, then the system is unstable (i.e. the response is explosive); but if any of the parameters is equal to one in absolute value, then the system is conditionally stable (i.e. the output is constant, neither declining nor growing). This can be more simply expressed in terms of the system equation characteristic polynomial discussed later (p. 59).

The structure and complexity of a system can be defined on the basis of the number of accumulators or *integrators* and the number of differentiators or lags which it possesses. For continuous inputs such systems can be defined by ordinary differential equations. The system *order* is now defined by the maximum degree of differential term.

(1) *Zero-order systems.* These contain no integrators or differentiators (and therefore no time lags) such that there is no time interval between the input and the output, which stream has been operated on by multipliers. Strictly speaking, such systems are of abstract interest only, since few systems can be regarded as producing simultaneous input–output response. However, some very rapidly operating electrical circuits, for example, may be approximated by zero order systems. Also purely spatial processes must be treated in this way (see chapter 4.4).

(2) *First-order systems.* These involve one integrator and one differentiator; fig. 2.15 shows two basic first-order systems with negative feedback loops, one having multiple inputs ($X_1(t)$, $X_2(t)$). T is the *time constant* of the system and is a measure of the total system parameter effects (defined as the ratio of the energy storage and dissipative parameters of the system – i.e. in this case $T = RC$). If Y is zero at time $t = 0$, the complete solution for the

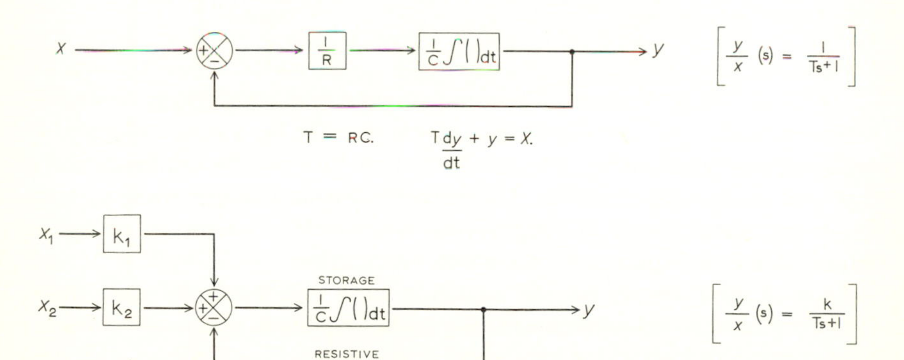

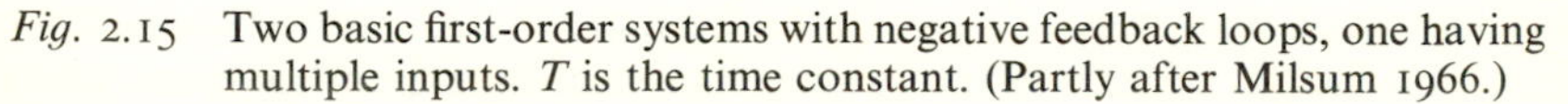

Fig. 2.15 Two basic first-order systems with negative feedback loops, one having multiple inputs. T is the time constant. (Partly after Milsum 1966.)

response at any time in the future is

$$Y(t) = X(1 - e^{-t/T}) \tag{2.1.5}$$

and the general solution for $X(t) = 0$ for all t is

$$Y(t) = e^{t/T}Y(0). \tag{2.1.6}$$

(3) *Second-order systems.* These contain two integrators and two differentiators; fig. 2.16 shows a basic second-order system with two nested negative feedback loops. As will be shown later, inputs into second order systems characteristically produce oscillatory outputs, the features of which are controlled by the *natural frequency* ($\omega = \sqrt{a/J}$) and the *damping ratio* ($\zeta = b/(2\sqrt{Ja})$). The time constant (T) in this instance equals $\sqrt{T_1 T_2}$, where $T_1 = b/a$ and $T_2 = J/b$.

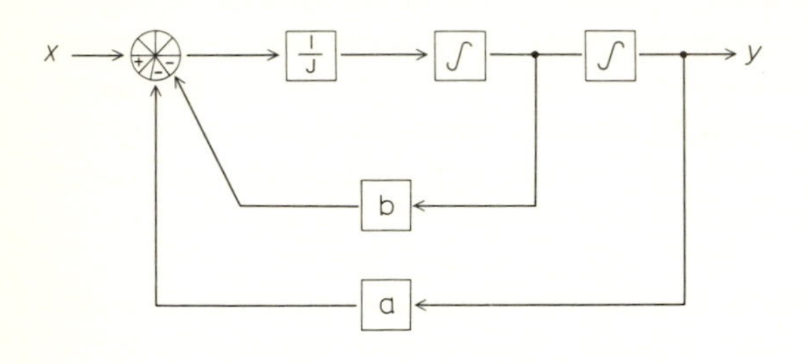

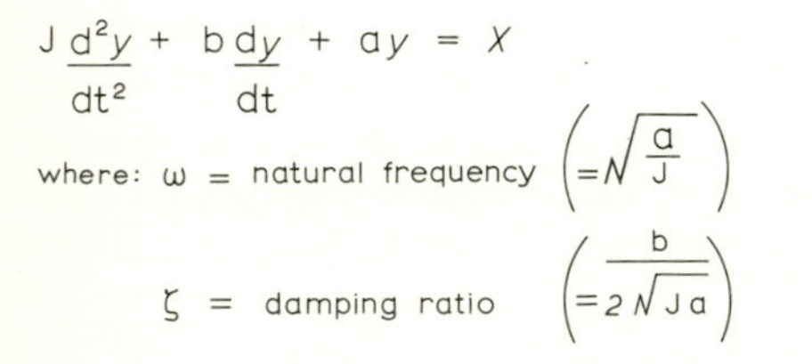

Fig. 2.16
A basic second-order system with two nested negative feedback loops, each containing a multiplier. (Partly after Milsum 1966.)

(4) *Higher-order systems.* These are obviously more difficult to analyse mathematically and are either treated empirically by means of digital or analogue computers or by approximating them by a suitable configuration of first- and second-order systems. The direct treatment of higher-order systems, involving nested loops of successively lower-order systems, is of special importance in complex control systems (see chapter 3).

Each of these system transfer functions links one input and one output. However, most environmental systems are not as simple as this and contain a large number of interacting inputs and outputs, as well as feedbacks between the two. Such *multivariate systems* of n multi-input and m multi-output variables have $(n \times m)$ transfer functions, as shown for some simple examples in fig. 2.17. They are special cases of higher-order systems and their mathematical treatment is more complex than that of univariate systems requiring the use of vectors to describe the set of inputs and outputs, and a

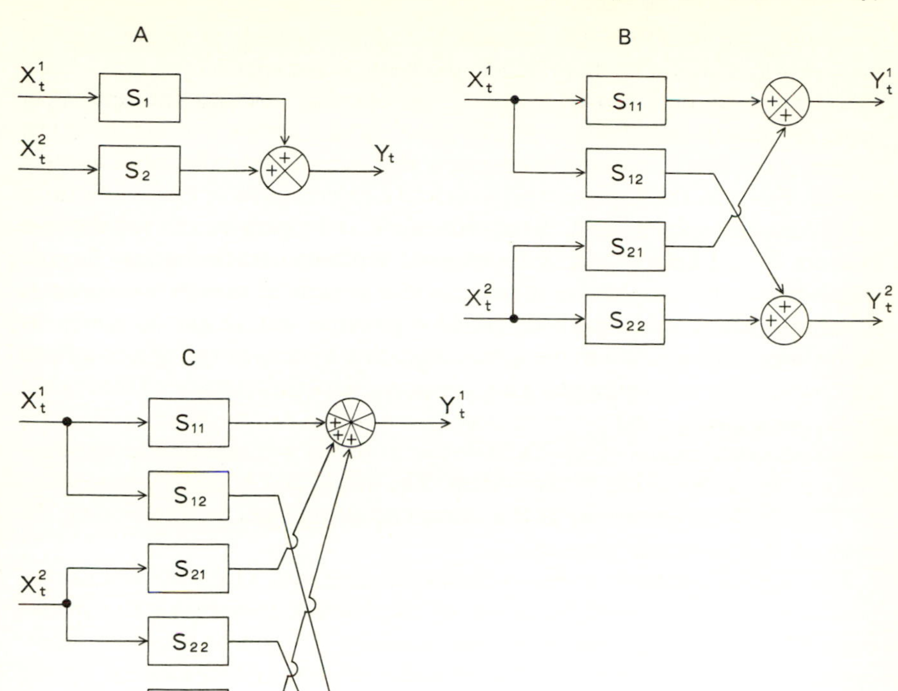

Fig. 2.17 Multivariate system transfer functions with n inputs and m outputs. Such systems have $(n \times m)$ transfer functions if each input affects each output (complete cross-coupling).

matrix to contain the $(n \times m)$ transfer function components. Simple algebraic operations with vectors and matrices are discussed in Appendix 1. The mathematical structure of such systems is discussed in the next section together with simple univariate first- and second-order systems, where it is demonstrated that even the most complex multivariate system can be reduced to sets of first-order systems which allow the separate component effects to be deciphered and analysed. The existence and nature of this reduction becomes especially important when it is required to understand and control the complex interlinked feedback loops that characterize environmental systems.

2.2 The mathematical representation of systems

Various forms of transfer function have been used in figs 2.4 to 2.6 and in the ensuing discussion with block diagrams and flow graphs. However, the definition of transfer function processes in terms of differential and difference equations, as in table 2.1, allows the development of more general mathematical solutions than block diagram manipulation permits. Such solution can be obtained by a number of classical calculus and analytical methods (Zadeh and Desoer 1963, Eykhoff 1974, Carslaw and Jaeger 1941), but the most useful in many instances is the operational calculus of control theory. This uses the *Laplace transform* and *Laplace operator* (s) for solutions to continuous systems and the *Z-transform* and operator (z) for discrete systems. The Laplace and Z-transforms represent mathematical devices which allow the simplification of dynamic system equations to algebraic equations. Many other methods are also possible, but all aim to allow the derivation of the *solution* to the system equation by transforming it to a form which is simpler to manipulate. This solution allows the response of the system to be calculated at any point in time given *initial and boundary conditions* which define the values of input and output at time $t = 0$, and on the spatial or other limits of definition of the system. The initial and boundary conditions are frequently as important as the system equation itself in determining the form of system output derived.

The Laplace operator (s) is a complex variable into which the function of time (f(t)) is transformed by means of the *Laplace transform* ($\mathscr{L}$) (Spiegel 1965, McFarland 1971, pp. 7–9). The function $f(t)$ will be a system input, output, or transfer function. Multiplication of this system element by the Laplace operator e^{-st} and integration yields:

$$\mathscr{L}\{f(t)\} = F(s) = \int_0^\infty f(t)\, e^{-st}\, dt. \tag{2.2.1}$$

This operation has convenient mathematical properties. Among the most important of these is that a convolution of system elements, which is an integral of a continuous series of multiplications (as shown in figs 2.8 and 2.11), can be simplified to multiplications of the Laplace transforms. After such a manipulation the inverse Laplace transform is then employed to obtain the algebraic solution in terms of the original system context (e.g. time t rather than s). Thus an integral operator can be substituted for the value $1/s$. Time-invariant systems can be transformed into Laplace systems which can be treated in an algebraic manner to solve the problem in terms of s. Then the inverse of the Laplace transform can be used to obtain the answer as a function of time.

Some common examples of Laplace transforms are given by the following:

1. For a unit impulse:

$$\begin{aligned}\mathscr{L}\{f(t)\} = F(s) &= \frac{1}{st}(1 - e^{-st}) \\ &= 1 - \frac{st}{2!} + \frac{(st)^2}{3!} - \cdots \\ &= 1.\end{aligned} \tag{2.2.2}$$

2. For a unit step (i.e., $f(t) = 1$, for all $t > 0$):

$$\mathscr{L}\{f(t)\} = F(s) = \int_0^\infty e^{-st}\,dt$$
$$= -\frac{1}{s}[e^{-st}]_0^\infty \qquad (2.2.3)$$
$$= \frac{1}{s}.$$

3. For a differentiator $(df(t)/dt)$:

$$\mathscr{L}\left\{\frac{df(t)}{dt}\right\} = F(s) = \int_0^\infty e^{-st}\frac{df(t)}{dt}$$
$$= [e^{-st}f(t)]_0^\infty + s\int_0^\infty e^{-st}f(t)\,dt$$
$$= s\int_0^\infty e^{-st}f(t)\,dt - f(0)$$
$$= sF(s) - F(0)$$

or,

$$F(s) = \frac{F(0)}{s-1} = \frac{1}{s-1}. \qquad (2.2.4)$$

Similarly, for example, for a ramp input $F(s) = 1/s^2$ and for a sine input $(\sin \omega t)$, $F(s) = \omega/(s^2 + \omega^2)$. A full table of Laplace transforms is given by Spiegel (1965, Appendix A and B). Fig. 2.18 gives an example of the Laplace transform of a simple negative-feedback system with one multiplier. The response to three system inputs are shown: impulse, step and ramp. In each case the system input $X(t)$ is Laplace transformed to give $X(s)$ as in equations (2.2.2) and (2.2.3). The system transfer function is then Laplace transformed to become $k/(s + k)$. Then the Laplace transform of the system output $Y(s)$ is given by the simple multiplication of the input and transfer function transforms (the integral equation has been removed). The inverse Laplace transform will then give the output in terms of the more familiar functions of time. Fig. 2.19 shows these output responses as paths over time.

The Z-operator (or unit shift transform) is an alternative method of solution for discrete systems defined (Jury 1964) by f_t and given by:

$$\mathscr{Z}\{f_t\} = F(z) = \sum_{t=0}^{\infty} f_t z^{-t} \qquad (2.2.5)$$

where

$$Y_t z = Y_{t-1}$$
$$Y_t z^2 = Y_{t-2}.$$

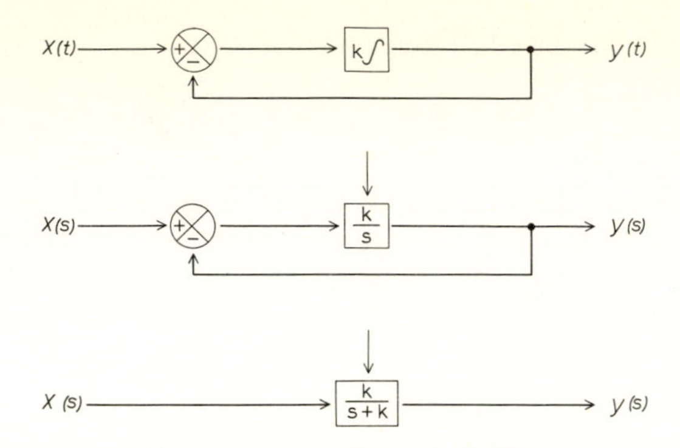

Fig. 2.18 The application of the Laplace transform to a simple negative feedback system to produce the following transfer function:

$$\frac{y(s)}{x(s)} = \frac{k}{s+k}$$

If:

Impulsive input: $X(s) = 1,\quad \therefore y(s) = \dfrac{k}{s+k},$

$$\therefore y(t) = \frac{k}{T}e^{-t/T}$$

Step input: $X(s) = \dfrac{1}{s},\quad \therefore y(s) = \dfrac{k}{s(s+k)},$

$$\therefore y(t) = k(1 - e^{-t/T})$$

Ramp input: $X(s) = \dfrac{1}{s^2},\quad \therefore y(s) = \dfrac{k}{s^2(s+k)},$

$$\therefore y(t) = k[t - T \times (1 - e^{-t/T})].$$

This has similar properties to the Laplace transform but yields the solution to the difference equations governing system response as a function of time. For example, the first-order discrete system governed by the difference equation:

$$Y_t = 0{\cdot}8Y_{t-1} + X_t$$

has a time response solution given by

$$\mathscr{Z}\{Y_t\} = \mathscr{Z}\{0{\cdot}8Y_{t-1}\} + \mathscr{Z}\{X_t\}$$

$$Y(z) = \frac{1}{(1 - 0{\cdot}8z)}X(z)$$

and inverse transform tables (Jury 1964) yield

$$Y_t = \sum_{t=-\infty}^{t} \sum_{k=0}^{\infty} 0{\cdot}8^k X_{t-k}$$

which is a geometrically declining (exponential) response function. Using the Z-operator it is now possible to express system stability conditions more precisely – namely, that the roots (poles or eigenvalues) of the characteristic polynomial formed from the system autoregressive elements should all be less than one in absolute value for system stability. Hence, the roots of

$$\begin{aligned}(1 - a_1 z - a_2 z^2 - a_3 z^3 - \cdots - a_p z^p) \\ = (z - \lambda_1)(z - \lambda_2)(z - \lambda_3) \ldots (z - \lambda_p)\end{aligned} \tag{2.2.6}$$

give the following system properties

$|\lambda_i| < 1, \quad (i = 1, 2, \ldots, p) \quad$ (stable system),

$|\lambda_i| > 1, \quad (i = 1, 2, \ldots, p) \quad$ (unstable system),

$|\lambda_i| = 1, \quad$ at least one i $\quad$ (conditionally stable system).

It is also possible to express the discrete system equations of table 2.1, e.g. $Y_t + a_1 Y_{t-1} + a_2 Y_{t+2} + \cdots + a_p Y_{t-p} = b_0 X_t + \cdots + b_q X_{t-q}$, in the more compact notation:

$$A(z) Y_t = B(z) X_t \tag{2.2.7A}$$

where

$$\begin{aligned}A(z) &= (1 + a_1 z + a_2 z^2 + \cdots + a_p z^p) \\ B(z) &= (b_0 + b_1 z + b_2 z^2 + \cdots + b_q z^q).\end{aligned} \tag{2.2.7B}$$

The relation between the Laplace and Z-operators is discussed by Jury (1964) and can be used to transform between the discrete and continuous domains (and *vice versa*). In all cases, however, such transformations are limited by the system characteristics themselves. For example, it is not meaningful to express systems governed by discrete events in continuous form (e.g. batches of orders or sales, unique events), but it is meaningful to re-express systems which have been observed in discrete form (e.g. by sampling) as continuous systems, although interpolation assumptions will usually be required (MacFarlane 1970).

2.2.1 FIRST-ORDER SYSTEMS

First-order systems are defined by the simple differential equation:

$$\frac{dY(t)}{dt} = aY(t). \tag{2.2.8}$$

The solution to this equation can be written using the Laplace transform definition (equation 2.2.4) directly to give the dynamic system equation in algebraic form:

$$sY(s) - Y(0) = aY(s)$$

and hence

$$Y(s) = \frac{Y(0)}{s-a} = e^{at}Y(0)$$

where the inverse Laplace transform $1/(s-a) = e^{at}$ has been used (Spiegel 1965). This solution to the first-order differential system is most important and governs the way in which the output $Y(t)$ evolves from initial conditions $Y(0)$ over time t (in this case without the effect of the input forcing $X(t)$ being considered).

It is instructive to consider the effect of adding to this system solution the effect of various input terms. Consider for the moment the effect of purely functional continuous inputs. Fig. 2.19 shows the relations between step,

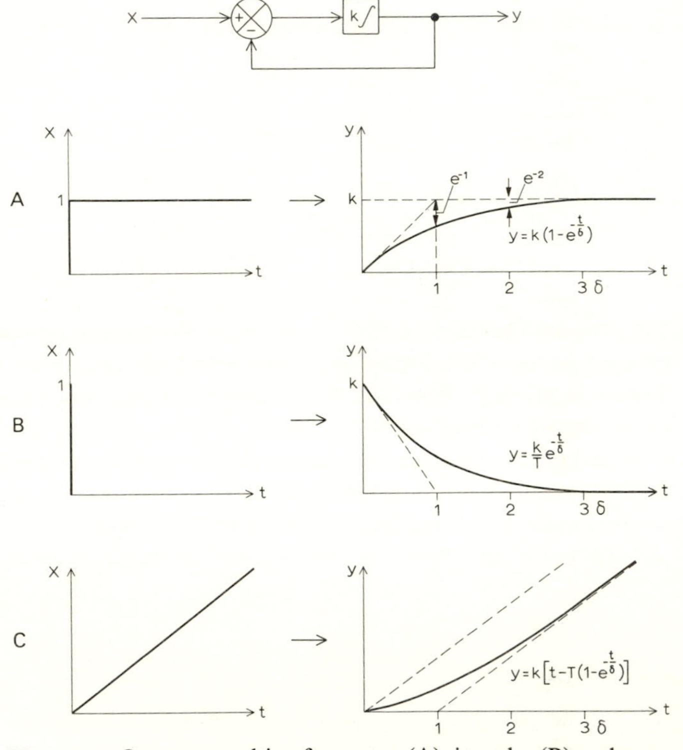

Fig. 2.19 Outputs resulting from step (A), impulse (B) and ramp (C) inputs into a simple first-order negative feedback linear system. (Partly after Milsum 1966.)

impulse and ramp inputs and the resulting outputs from a simple first-order linear system involving negative feedback, and develops from the inputs shown in fig. 2.7. For a periodic input the system output is the sine curve (fig. 2.20) having the form:

$$X(t) = A \sin \omega t$$

where:

A = amplitude,
ω = angular frequency (radians/sec) ($=2\pi f$),
f = frequency (cycles/sec),
T = time period of the cycle (in secs) ($1/T = \omega/2\pi = f$).

A sinusoidal input into a linear system will produce a sinusoidal output of the same frequency, but differing in amplitude and *phase* (fig. 2.21) (McFarland 1971, pp. 72–80). The change in amplitude is expressed by the *amplitude ratio* (*AR* – sometimes called the *gain*). The change in phase is termed the *phase shift*

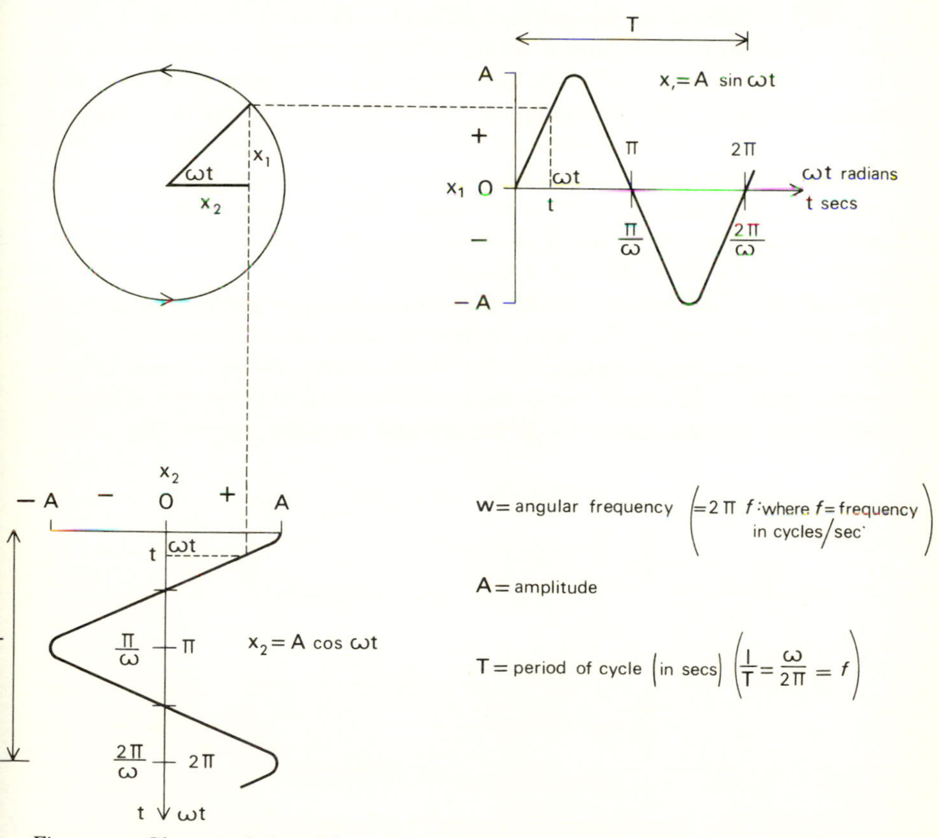

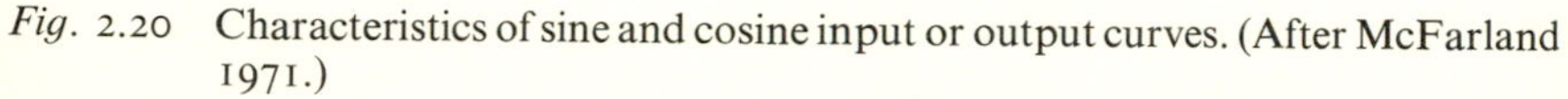
Fig. 2.20 Characteristics of sine and cosine input or output curves. (After McFarland 1971.)

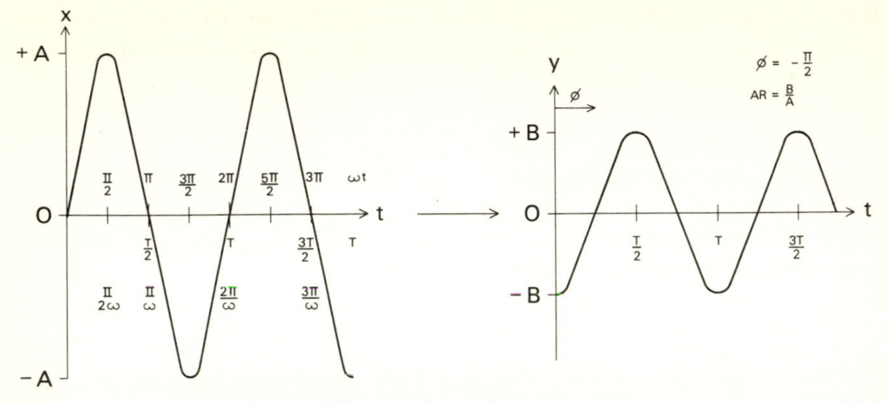

Fig. 2.21 A sinusoidal input (X) into a linear first-order system produces a sinusoidal output (Y) of the same frequency, but with differing amplitude and phase. AR = amplitude ratio or gain. ϕ = phase shift.

(ϕ), which is defined as the angular difference between the instantaneous values of the input and output waves. The phase shift is positive for an advance of the output (*phase lead*) and negative for the more usual case of a retardation (*phase lag*). The input of a sine wave into a first-order system characteristically involves a phase lag in the output of $\pi/2$ (or $\pi/2\omega$). It should be noted, however, that sinusoidal inputs into a linear system previously at rest need not produce an output of constant amplitude until a period of time has passed during which the effect of the input has been distributed evenly throughout the system (Milsum 1968, pp. 142–4).

The effects of an irregularly varying stochastic input stream into a first-order system are more difficult to analyse in terms of a continuous output, and it is here that the advantages of the linear assumption are especially well shown in that by the use of *fourier analysis* it is possible to approximate any irregular continuous input by superimposed periodic curves (fig. 2.22).

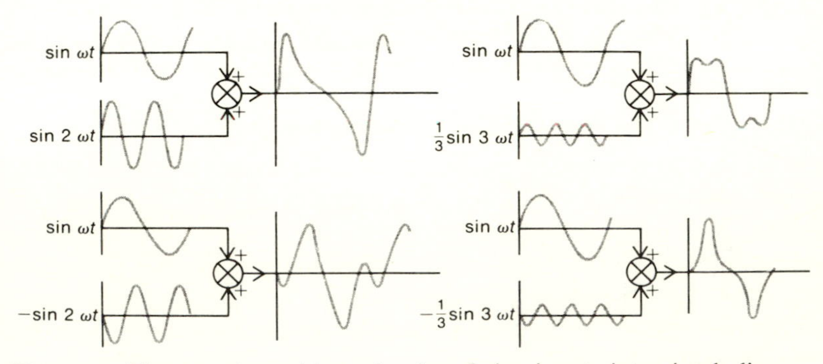

Fig. 2.22 The superimposition of pairs of sine inputs into simple linear systems to produce more complex output streams. (From Blesser 1969.)

However, as we have seen, such input streams commonly appear, or are treated, as discrete time values governed by difference equations, which for the first-order system are written:

$$a_1 Y_{t+1} + a_0 Y_t = X_t,$$

which expresses the first-order relation in terms of single lags, one data interval apart.

As was mentioned previously, the role of the integral accumulator in a system defined by a differential equation is assumed by the *time delay* or lag in a discrete time system. If $Y_0 = 0$, then the equation above has the solution:

$$Y_t = \frac{X_t}{a_1 + a_0}\left[1 - \left(-\frac{a_0}{a_1}\right)^t\right].$$

The *gain* of the system is a_0/a_1 and it is clear that this term determines the form of the output, such that a series of discrete unit pulses into a first-order system produces differing outputs depending on the sign and magnitude of the ratio. A negative value produces no *overshoot* (fig. 2.23) and the dotted envelope

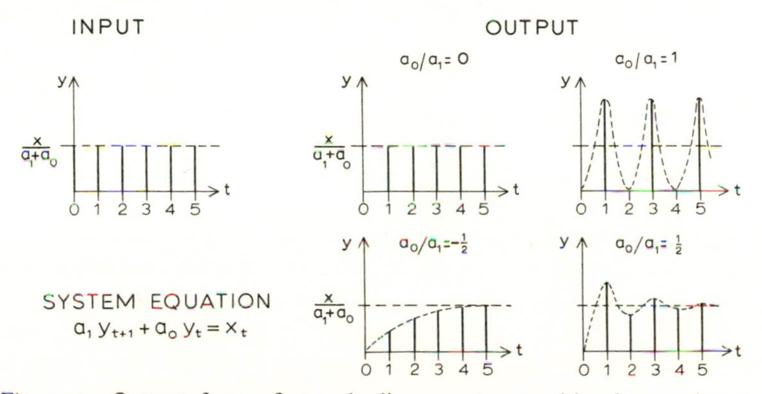

Fig. 2.23 Outputs from a first-order linear system resulting from an input stream of discrete time unit impulses. The form of the output is shown to depend upon the system gain a_0/a_1. (After Jacobs 1974.)

curve is that which would be produced if the step input entered a continuous-time first-order system (fig. 2.24). A positive value of a_0/a_1, which can be produced by a positive feedback, gives an overshoot oscillation – damped if the ratio is less than unity, stable if it is unity and unstable if it is greater than unity. This type of first-order system behaves in much the same way as a second-order system. If the gain is zero, the output follows the input exactly after a one-period time delay (Jacobs 1974, pp. 11–12). If a single unit pulse is fed into the above system and a_0 = unity, then the output will oscillate in a manner determined by the system gain $(1/a_1)$ (fig. 2.25). A positive feedback

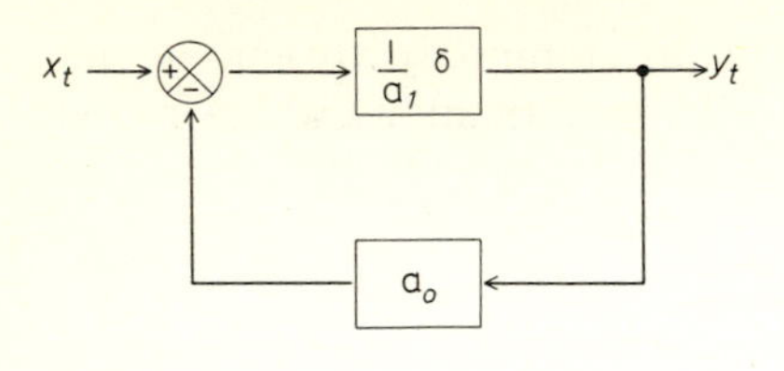

Fig. 2.24
A discrete-time first-order system, where δ is the time interval of the input data sampling.

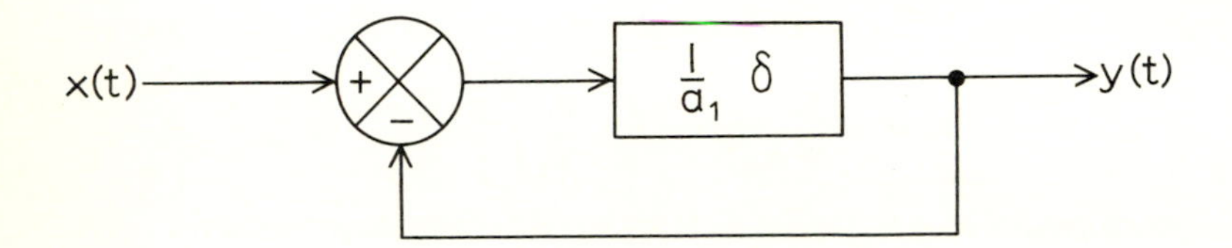

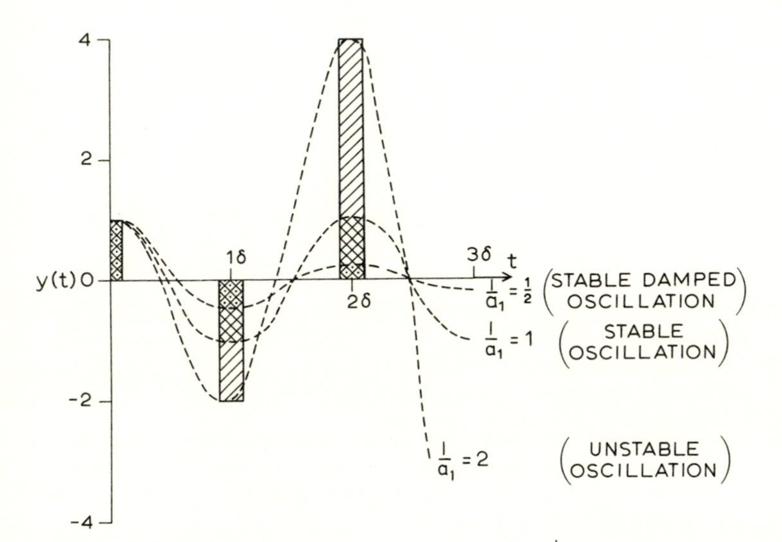

Fig. 2.25 Outputs (y) resulting from a unit impulse input (x) into a negative feedback first-order system of varying gain ($1/a_1$) with a time interval of the input data (δ). (Partly after Milsum 1968.)

loop produces outputs from a unit pulse which are stable for positive values of $1/a_1$ of less than unity and unstable for values greater than unity (fig. 2.26) (Milsum 1968, pp. 48–50). Where $1/a_1$ is also unity, the system is termed a *pure-time-delay system*. The input of a sine wave into a positive feedback pure-time-

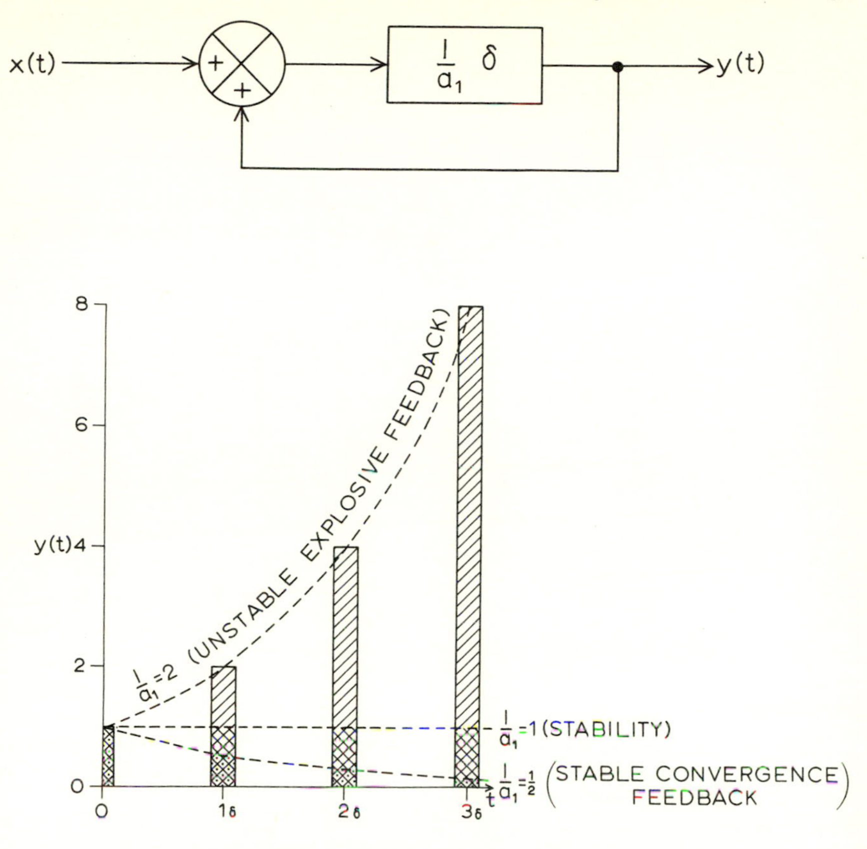

Fig. 2.26 Outputs (y) resulting from a unit impulse input (x) into a positive feedback first-order system of varying gain ($1/a_1$) with a time interval of the input data (δ). (Partly after Milsum 1968.)

delay system will lead to the output of an unstable oscillation of the same frequency after a one-period delay (fig. 2.27A). A similar input into a negative feedback system also produces an unstable oscillation after a one-period delay, but the oscillations are inverted and of half the frequency (fig. 2.27B) (Milsum 1968, pp. 51–2).

An important systems configuration exists when the output from one system forms the cascaded input for another system. Fig. 2.28 shows the progressive outputs obtained by cascading a unit impulse through a number of first-order linear systems. Such a configuration finds analogies with the flood routing treated in chapter 7 and, more generally, with the cascading principles implicit in space-time systems analysis (see chapter 4).

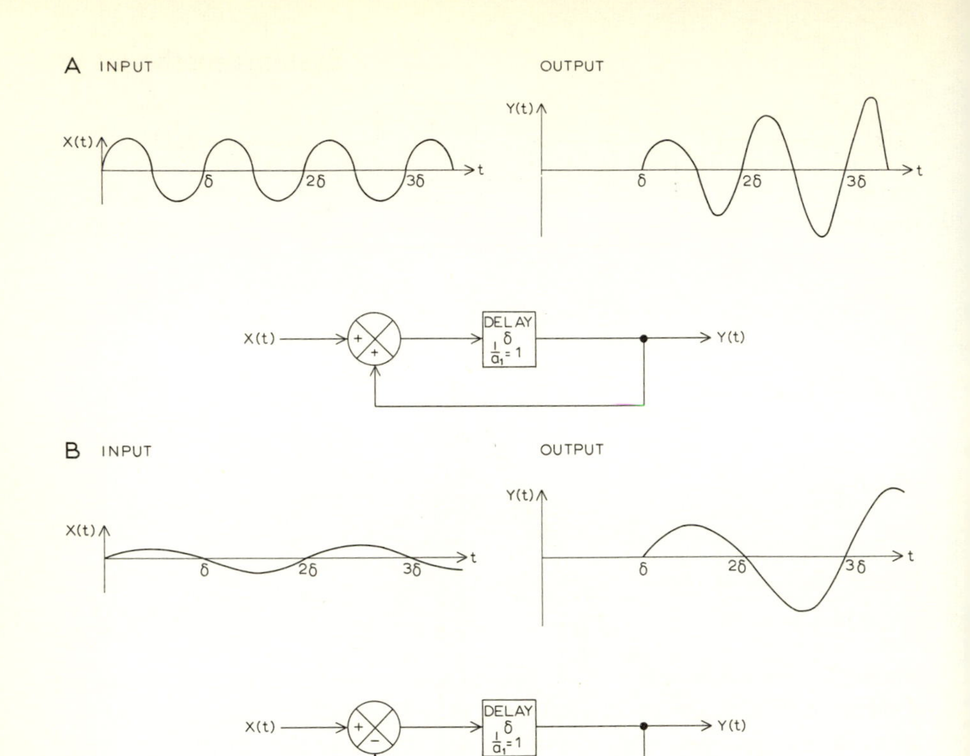

Fig. 2.27 The outputs resulting from the input of a sine wave into (A) a positive feedback and (B) a negative feedback pure-time-delay system. (Partly after Milsum 1968.)

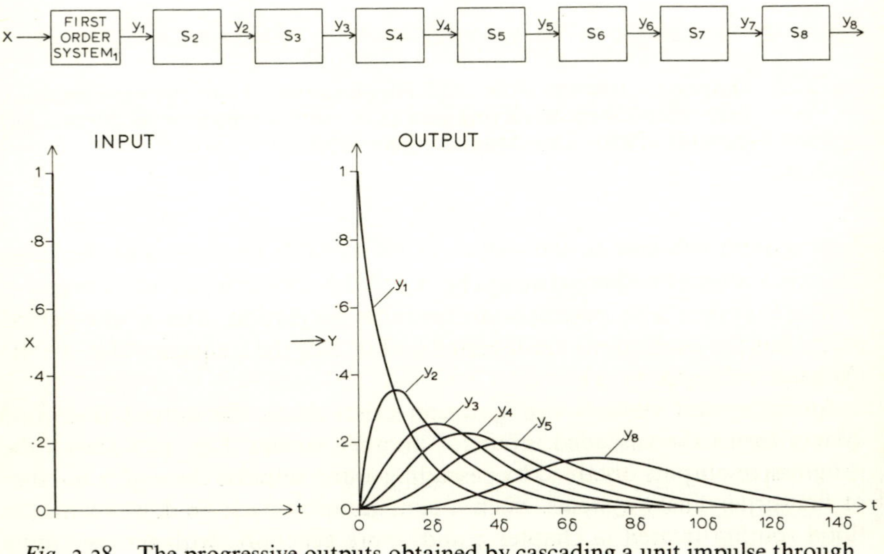

Fig. 2.28 The progressive outputs obtained by cascading a unit impulse through eight first-order linear systems. (After Milsum 1966.)

2.2.2 SECOND-ORDER SYSTEMS

Second-order differential equation systems can be solved by a number of methods (Carslaw and Jaeger 1941, MacFarlane 1970). One of the simplest is direct factorization. For example, the second-order system (with no effect of input forcing)

$$\frac{d^2 Y(t)}{dt^2} + 3\frac{dY(t)}{dt} + 2Y(t) = 0 \qquad (2.2.9)$$

is of the form

$$\lambda^2 + 3\lambda + 2 = 0$$

which has factors

$$(\lambda + 2)(\lambda + 1) = 0; \qquad \lambda = -2, -1.$$

Hence, the second-order system admits the solution:

$$Y(t) = (A\,e^{-2t} + B\,e^{-t})\,Y(0) \qquad (2.2.10)$$

where the constants A and B are determined from the initial and boundary conditions. The form of this solution, which expresses the system output impulse response $Y(t)$ at any time t, is analogous to that of the first-order system equations, but now represents the sum of two exponential functions. Special cases arise when the roots (λ) are equal or complex. In these conditions the solution becomes:

$$Y(t) = \begin{cases} (A + Bt)\,e^{-at} \\ (At + B)\,e^{-at} \end{cases} \quad \text{(equal roots)},$$

$$Y(t) = A\,e^{ait} + B\,e^{bit} \quad \text{(complex roots)}.$$

In the case of complex roots, (where $i = \sqrt{-1}$), the output response of the system becomes a set of sine and cosine functions. This arises from the equalities:

$$\cos t = \frac{e^{it} + e^{-it}}{2}$$

$$\sin t = \frac{e^{it} - e^{-it}}{2}.$$

Hence the form of system output response $Y(t)$ with time is governed by the magnitude and distribution of the roots of the system differential equation. A plot of the two roots (system poles) in the complex ($i\omega$, σ) plane shown in fig. 2.29 demonstrates this critical control of root location.

When the effect of input stimuli $X(t)$ into the system are considered, it is often useful to take a frequency domain viewpoint and to consider the response of the system to cyclic (usually sinusoidal) input stimuli. The

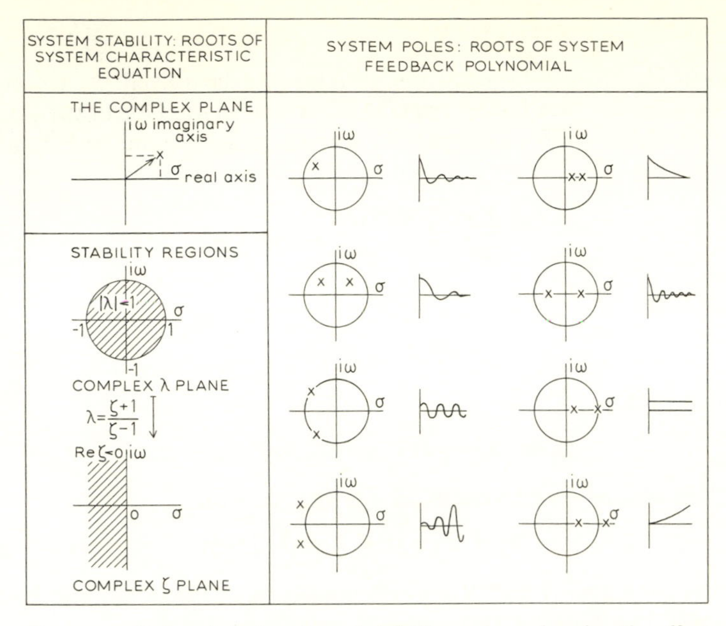

Fig. 2.29 Polar plot of roots and stability regions showing the effect on system response for a second-order transfer function. (After Bennett 1978.)

frequency components of systems are governed by the natural frequency (ω) and damping ratio (ζ) of its internal feedback loops. For a simple example, the outputs from a second-order system resulting from a unit pulse and a unit step are oscillatory, depending on the value of the damping ratio (ζ) as shown in fig. 2.30. Where ζ = zero the system is *undamped* and the output is a uniform oscillation. Where $\zeta <$ unity the system is *underdamped*; where $\zeta >$ unity it is *overdamped*; and where $\zeta =$ unity it is *critically damped*, reaching equilibrium after a time 2π (or $2\pi/\omega$) (fig. 2.30). A sinusoidal input into a second-order system produces a similar output with a phase lag of π (or π/ω). The standard second-order system can be expressed as a difference equation (fig. 2.31) and its behaviour can be predicted by using the curves in fig. 2.30 as envelope curves. The frequency response of systems becomes extremely important when oscillating signals dominate inputs. Such system inputs frequently occur for example in most electrical devices, and in environmental systems derive from annual and diurnal climatic rhythms (chapter 7) and more irregular socio-economic cycles (chapter 8). As with first-order systems, outputs resulting from discrete time inputs can be treated by the methods of time series analysis.

2.2.3 HIGHER-ORDER SYSTEMS

The solution to differential equations of higher order than the simple first- and

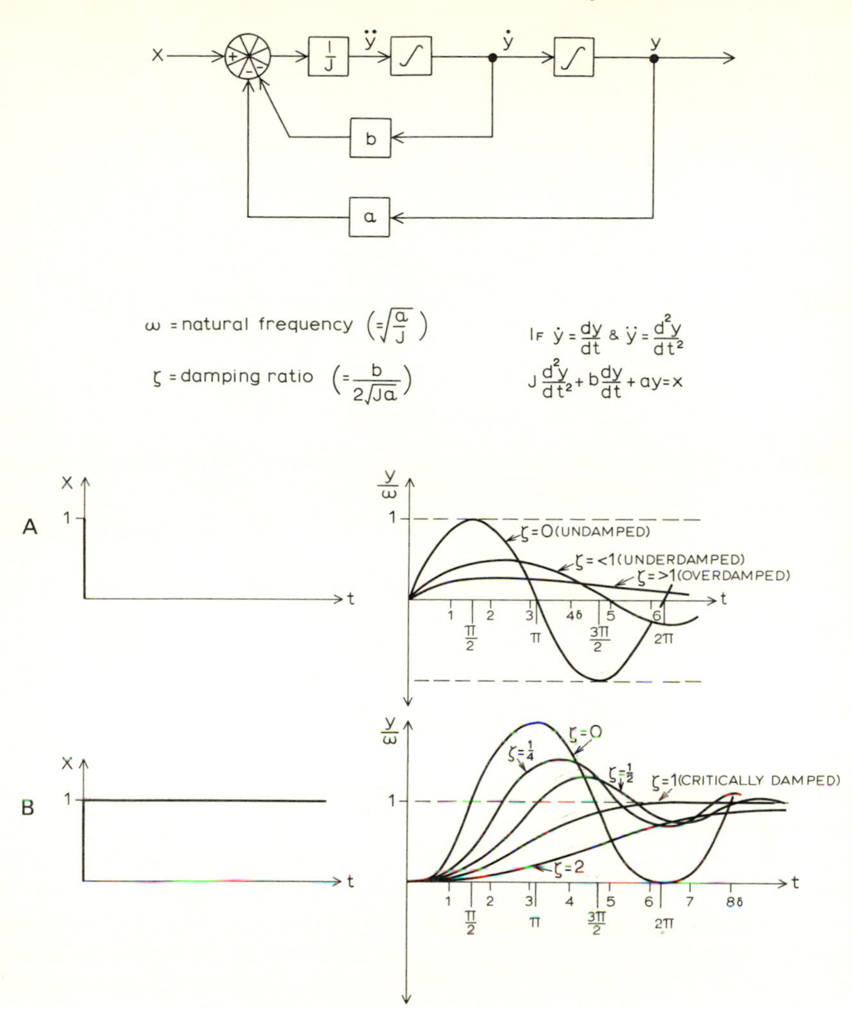

Fig. 2.30 Outputs (y/ω) from a second-order system resulting from a unit impulse and a unit step input (X), showing the effect of the damping ratio (ζ). (Partly after McFarland 1971 and Director and Rohrer 1972.)

second-order system discussed above will usually be very complex since a set of (p) unknowns (cf. table 2.1) and a set of ($p + 1$) initial conditions must be specified for the unforced (no input $X(t)$) response alone. It can be shown that analytical solutions will still have the form of a sum of exponentials and/or complex exponentials as for second-order systems, but the determination of the exact structure will be difficult. For this reason, the pth order system is frequently transformed to a set of p first-order systems each of which can be treated and solved as discussed in section 2.2.1. To accomplish this

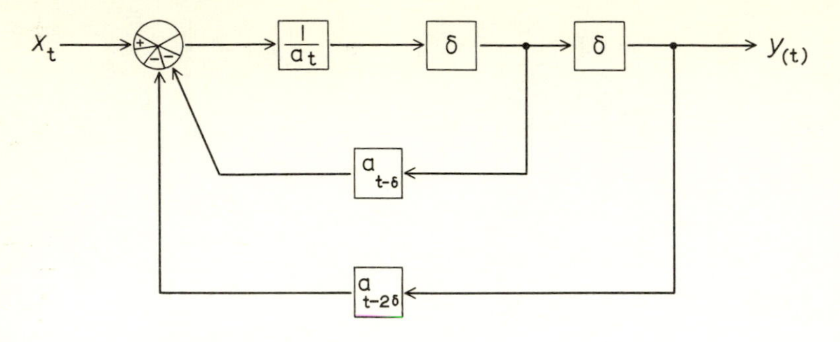

$$a_t y_{t+2\delta} + a_{t-\delta} y_{t+\delta} + a_{t-2\delta} y_t = X_t$$

Fig. 2.31 A discrete-time second-order system, where δ/t is the time interval between samples of the input data.

transformation a *state space* mathematical representation is adopted. For the pth order differential equation

$$\frac{d^p Y(t)}{dt^p} + a_p \frac{d^{p-1} Y(t)}{dt^{p-1}} + \cdots + a_2 \frac{dY(t)}{dt} + a_1 Y(t) = 0 \quad (2.2.11)$$

we may define a new set of *state variables* $Z(t)$. Each of these new variables is defined as follows:

$$\begin{aligned} Z_1(t) &= Y(t) \\ Z_2(t) &= dZ_1(t)/dt = dY(t)/dt \\ Z_3(t) &= dZ_2(t)/dt = d^2 Y(t)/dt^2 \\ &\vdots \\ Z_p(t) &= dZ_{p-1}(t)\, dt = d^{p-1} Y(t)/dt^{p-1}. \end{aligned} \quad (2.2.12)$$

Hence the pth-order differential equation (2.2.11) may be written as the much simpler first-order differential equation:

$$\frac{dZ_p(t)}{dt} + a_p Z_p(t) + a_{p-1} Z_{p-1}(t) + \cdots + a_2 Z_2(t) + a_1 Z_1(t) = 0$$

and

$$\dot{Z}_p(t) = -a_p Z_p(t) - a_{p-1} Z_{p-1}(t) - \cdots - a_2 Z_2(t) - a_1 Z_1(t).$$

It is convenient now to adopt a vector and matrix notation for the system equation. This allows the equation to be rewritten simply as a *state equation*:

$$\begin{bmatrix} \dot{Z}_1(t) \\ \dot{Z}_2(t) \\ \vdots \\ \dot{Z}_{p-1}(t) \\ \dot{Z}_p(t) \end{bmatrix} = \begin{bmatrix} 0 & 1 & 0 & \dots & 0 & 0 \\ 0 & 0 & 1 & \dots & 0 & 0 \\ \cdot & \cdot & \cdot & \cdot & \cdot & \cdot \\ 0 & 0 & 0 & \dots & 0 & 1 \\ -a_1 & -a_2 & -a_3 & \dots & -a_{p-1} & -a_p \end{bmatrix} \begin{bmatrix} Z_1(t) \\ Z_2(t) \\ \vdots \\ Z_{p-1}(t) \\ Z_p(t) \end{bmatrix}$$

or

$$\underset{p\times 1}{\dot{\mathbf{Z}}(t)} = \underset{p\times p}{\mathbf{A}}\ \underset{p\times 1}{\mathbf{Z}(t)} \qquad (2.2.13)$$

and a *measurement equation*:

$$Y(t) = \underset{1\times p}{\mathbf{E}'}\ \underset{p\times 1}{\mathbf{Z}(t)} \qquad (2.2.14)$$

where the measurement matrix $\mathbf{E}$ is defined by:

$$\mathbf{E}' = [1, 0, 0, \dots, 0].$$

The dot over the state variable denotes the derivative $(dZ_i(t)/dt)$ of that variable and the matrix $\mathbf{A}$ is termed the system *state transition matrix*. On substituting back from equation (2.2.12) it can be seen that this is identical to the original system differential equation but has a much simpler structure. In particular equation (2.2.14) has identical form to the first-order system (2.2.8), except that it is written in vector-matrix form. Since the solution to the first-order system equation $dY(t)/dt = aY(t)$ is given by $Y(t) = e^{at}Y(0)$, we might suspect that the solution to the matrix system (2.2.14) might be given by:

$$\mathbf{Z}(t) = e^{\mathbf{A}t}\mathbf{Z}(0). \qquad (2.2.15)$$

However, this solution will be valid only if it is possible to write and interpret the matrix exponential function $e^{\mathbf{A}t}$. Fortunately this is possible (Zadeh and Desoer 1963, Eykhoff 1974). As an example, consider the differential equation for a second-order system:

$$\frac{d^2Y(t)}{dt^2} + 2\frac{dY(t)}{dt} + 3Y(t) = 0.$$

This can be written in the state space form $\dot{\mathbf{Z}} = \mathbf{AZ}$ using

$$Z_1(t) = Y(t); \qquad Z_2(t) = dY/dt.$$

Then

$$\dot{Z}_2(t) + 2Z_2(t) + 3Z_1(t) = 0$$

and

$$\begin{bmatrix} \dot{Z}_1(t) \\ \dot{Z}_2(t) \end{bmatrix} = \begin{bmatrix} 0 & 1 \\ -2 & -3 \end{bmatrix} \begin{bmatrix} Z_1(t) \\ Z_2(t) \end{bmatrix}; \qquad Y(t) = [1 \quad 0] \begin{bmatrix} Z_1(t) \\ Z_2(t) \end{bmatrix}$$

In the first-order case the Laplace transform of the derivative gives the solution directly as equation (2.2.4):

$$sY(s) - Y(0) = AY(s).$$

We may write the analogous result for the matrix system and use the properties of matrices discussed in Appendix 1:

$$s\mathbf{Z}(s) - \mathbf{Z}(0) = \mathbf{AZ}(s)$$

or

$$[s\mathbf{I} - \mathbf{A}]\mathbf{Z}(s) = \mathbf{Z}(0)$$

then

$$\mathbf{Z}(s) = [s\mathbf{I} - \mathbf{A}]^{-1}\mathbf{Z}(0)$$

which, after taking the inverse Laplace transform, yields the desired expression for the solution of the output response of the system over time (Zadeh and Desoer 1963). Hence, for the second-order example equation (2.2.9) we have the following solution:

$$(s\mathbf{I} - \mathbf{A}) = \begin{bmatrix} s & 0 \\ 0 & s \end{bmatrix} - \begin{bmatrix} 0 & 1 \\ -2 & -3 \end{bmatrix} = \begin{bmatrix} s & -1 \\ 2 & s+3 \end{bmatrix}$$

$$(s\mathbf{I} \quad -\mathbf{A})^{-1} = \frac{1}{(s+1)(s+2)} \begin{bmatrix} s+3 & 1 \\ -2 & s \end{bmatrix} = \begin{bmatrix} \dfrac{s+3}{(s+1)(s+2)} & \dfrac{1}{(s+1)(s+2)} \\ \dfrac{-2}{(s+1)(s+2)} & \dfrac{s}{(s+1)(s+2)} \end{bmatrix}$$

$$\mathscr{L}^{-1}(s\mathbf{I} - \mathbf{A})^{-1} = \begin{bmatrix} 2e^{-t} - e^{-2t} & e^{-t} - e^{-2t} \\ 2e^{-2t} - e^{-t} & 2e^{-2t} - 2e^{-t} \end{bmatrix}$$

where use is made of the inverse Laplace transform (Spiegel 1965):

$$\frac{1}{s+a} = e^{at}$$

$$\frac{1}{(s-a)(s-b)} = \frac{e^{bt} - e^{at}}{b-a} \qquad a \neq b$$

$$\frac{s}{(s-a)(s-b)} = \frac{be^{bt} - ae^{at}}{b-a} \qquad a \neq b.$$

Hence the solution to the matrix of first-order differential equations is a matrix set of exponentials. From this we may write

$$Z_1(t) = (2\,e^{-t} - e^{-2t})Z_1(0) + (e^{-t} - e^{-2t})Z_2(0) \tag{2.2.17}$$

$$Z_2(t) = (2\,e^{-2t} - e^{-t})Z_1(0) + (2\,e^{-2t} - 2\,e^{-t})Z_2(0) \tag{2.2.18}$$

or

$$\mathbf{Z}(t) = e^{\mathbf{A}t}\mathbf{Z}(0)$$

which is the matrix exponential equivalent to the simple exponential solution for the first-order system (equation 2.2.8). For more complex systems than the example given here, the economies of the method become obvious.

When an input forcing $X(t)$ is considered to enter the system, the state space matrix differential equation for the system is written

$$\underset{p\times 1}{\dot{\mathbf{Z}}(t)} = \underset{p\times p}{\mathbf{A}}\;\underset{p\times 1}{\mathbf{Z}(t)} + \underset{q\times q}{\mathbf{B}}\;\underset{q\times 1}{\mathbf{X}(t)} \tag{2.2.19}$$

which has solution (Zadeh and Desoer 1963):

$$\mathbf{Z}(t) = e^{\mathbf{A}t}\mathbf{Z}(0) + \int_0^t e^{\mathbf{A}(t-\tau)}\mathbf{B}\mathbf{X}(\tau)\,d\tau \tag{2.2.20}$$

which requires the matrix convolution integral. The elementary derivation of these equations is discussed in more detail by Bennett (1978, chapter 2).

The strength of the state space representation is that with a multivariate system (m-inputs; n-outputs) this can be reduced to a set of $p \times (n \times m)$ first-order equations in an exactly analogous way. The resulting system matrix will be block diagonal if none of the $(n \times m)$ systems interact with each other, block triangular if a recursive structure is present (Blalock 1964, Duncan 1975), and full when each input and output interact. Some difficulties in using the state space representation may arise because of its non-uniqueness. The proper definition of the state equation (2.2.13) requires that each of the state variables is independent of each of the other state variables, i.e. they are orthogonal and A is a *canonical* form with full rank p (Bennett 1978 chapter 2). It will usually be the case that many different state space representations can give appropriate canonical forms. Each of these is mathematically *equivalent* to the others, but one form is usually to be preferred on grounds of its physical interpretability. In all the ensuing discussion, especially in chapters 7–9, it is assumed that the state variables are all physically interpretable.

Higher-order systems are most effectively treated by digital or analogue computing methods, especially when multivariate elements are present. The effect of *cascading* the outputs of systems to form the spatial inputs of others will be treated in chapter 4. For non-spatial systems it is instructive here to examine the behaviour of some simple systems configurations. A common type of system is the double-loop negative and/or positive feedback first-order system (fig. 2.32). The positive-feedback loop leads to an explosive growth of the initial input ($X(0)$), whereas the negative feedback loop causes this growth

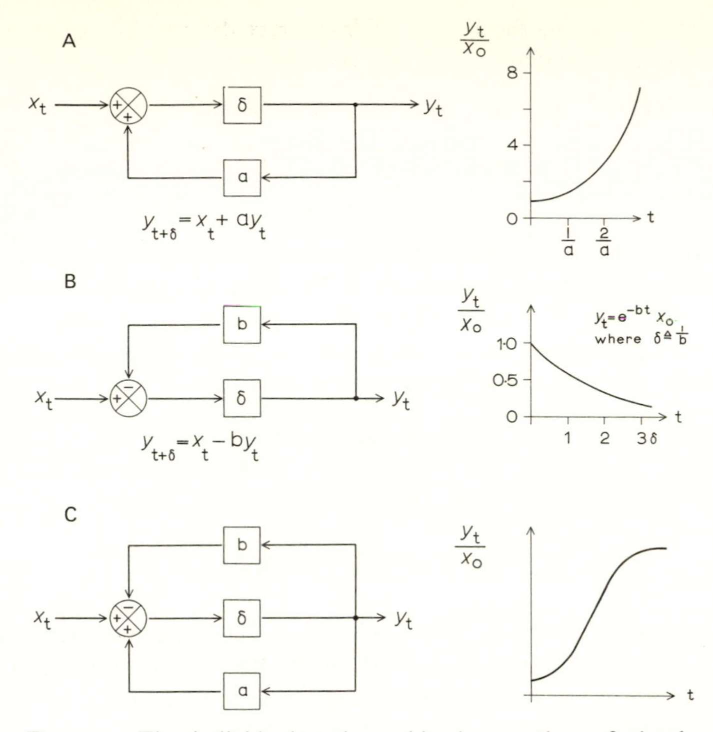

Fig. 2.32 The individual and combined operation of simple positive and negative feedback systems. (Partly after Milsum 1968.)

to decay. The effect of the combined loops is to produce a sigmoidal-type output from a single impulse input, the form of which depends on the magnitude of a and b (i.e. the gain). Another first-order configuration is that of the *lead-lag* (fig. 2.33) in which the operation of a negative feedback loop causes part of the output to lead (a) and part to lag (b).

The overall behaviour of dynamic systems can also be considered in the geometrical *state space* of dimension defined by the order of the state variable vector involved. This allows *global analysis* of system behaviour over all input and output feedback regimes (Thom 1975). We have already noted that the system response (i.e. differential equation solution) is dependent as much upon the initial conditions as upon the form of the system equation. Global analysis enables us to study the behaviour of any system equation over the entire range of system initial conditions. Taking the second-order univariate differential equation example discussed in section 2.22, the solution trajectory of the system over time can be sketched for each state variable as shown in fig. 2.34. Solutions deriving from other initial conditions should be calculated by the reader. The figure also shows the state space (in this case in two dimensions, as there are two state variables) trajectory of the system from

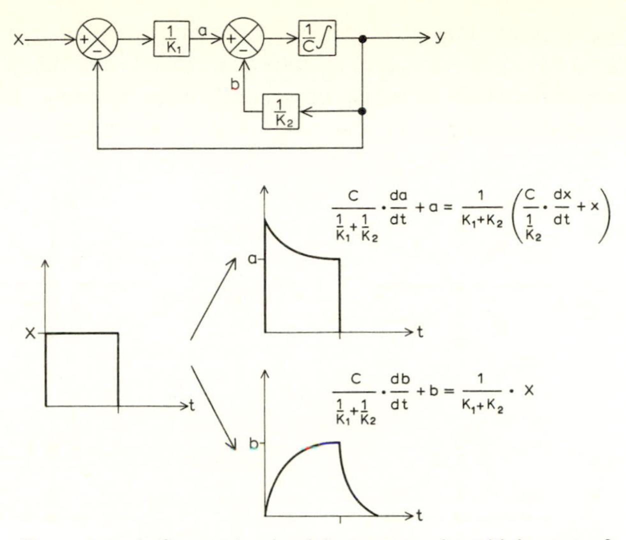

Fig. 2.33 A first-order lead-lag system in which part of the output leads (*a*) and part lags (*b*). (Partly after Milsum 1966.)

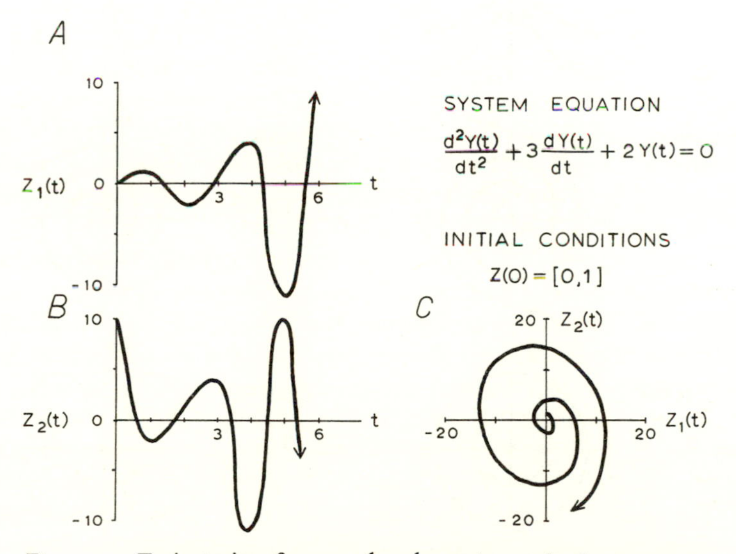

Fig. 2.34 Trajectories of a second-order system output response on each state variable Z_1 and Z_2 (A and B), together with (C) the trajectory in the state space of the two state variables combined.

given initial conditions. This geometrical representation of system behaviour can be extended to consideration of system trajectories over the global range of initial conditions. For a range of second-order systems these global trajectories are shown in fig. 2.35. The cases of sinks and nodes correspond to

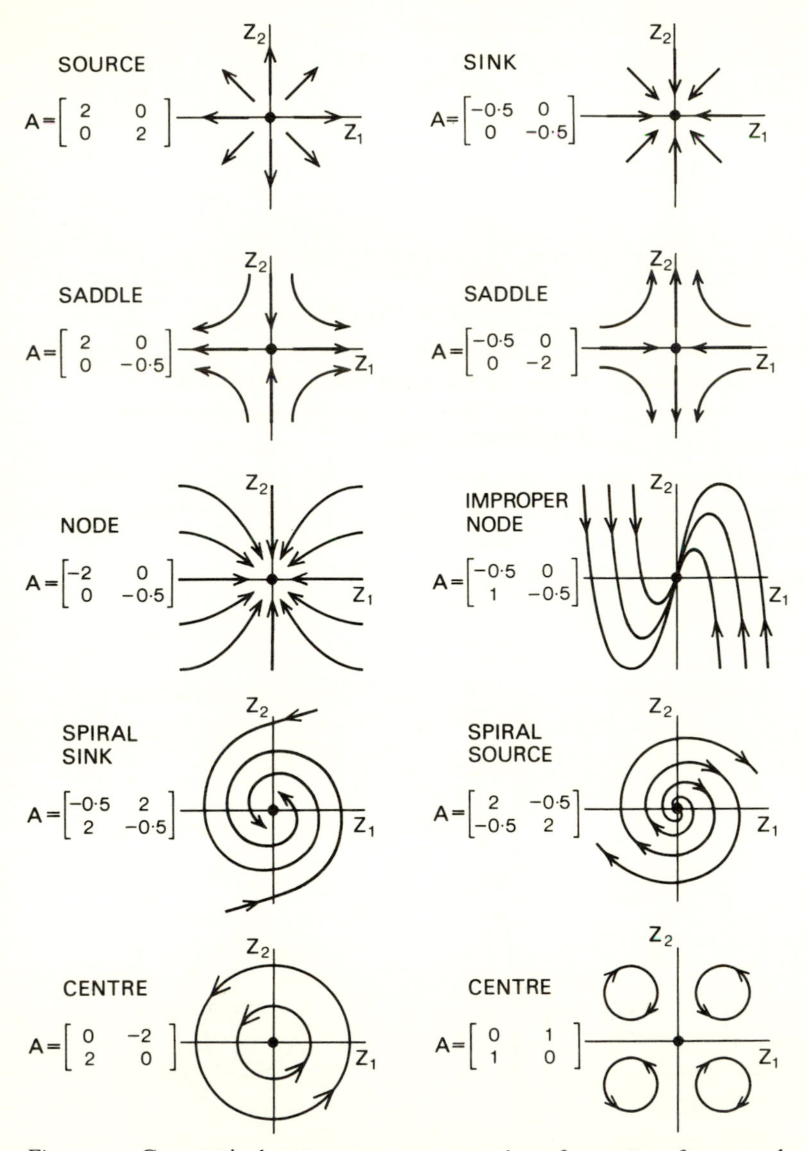

Fig. 2.35 Geometrical state-space representation of responses for second-order systems (with two state variables Z_i) having various parameter structures given by the system matrix A. (Partly after Lotka 1956, Hirsch and Smale 1974.)

stable systems, whilst the sources are unstable systems. Saddles are stable only under the special initial condition that one state variable equals zero, and centres are improper systems which are conditionally stable with neither increasing nor decreasing output response with time. In each case these results can be verified by calculation by the reader by defining an arbitrary vector of initial conditions $Z(0)$ for any system matrix **A**. From the figure it can be seen that the system output and state space trajectory are determined absolutely by the system matrix (i.e. state transition matrix) **A**; for diagonal **A** we have sources, sinks, saddles and nodes; for full **A** spirals, and so on. This behaviour can be best generalized (fig. 2.36) using the system matrix determinant, (Det (**A**)), and trace, (Tr (**A**)) for which

$$\mathrm{Tr}\,(\mathbf{A}) = \sum_{i=1}^{p} a_{ii} \quad \text{(see Appendix 1).}$$

Figs 2.34 and 2.35 should be interpreted as allowing the tracing of system behaviour over the global range of initial conditions to which the system may be subjected. Starting at any point in the arrowed trajectory in fig. 2.35, the path of system response in state space can be sketched. Since the state variables are transformed to become the system output, the trajectories shown in the figure give the output response with time: steadily declining (though perhaps with oscillations) in the case of a sink and node, and steadily increasing in the case of a source. The system shown in fig. 2.34 can be identified as a spiral source.

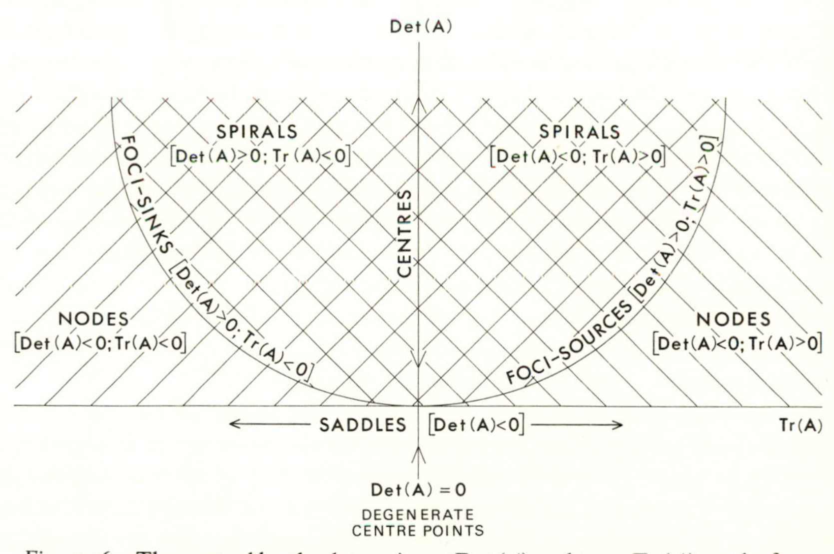

Fig. 2.36 The control by the determinant Det (A) and trace Tr (A) on the form of the state-space representation of system output responses. (After Hirsch and Smale 1974.) See Appendix 1 for definition of matrix determinant and trace.

2.3 Environmental systems analysis in practice

In most environmental systems the analyst has incomplete and partial knowledge of the system which he seeks to understand and control. The nature of his knowledge is often incremental in that greater insight is gained as analysis or control continues. But elements in the analysis of environmental systems will normally be necessarily indeterminate (or even counter-intuitive: Meadows *et al.* 1971), both because of the complexity of such systems and because of the intimacy of man-environment interrelations and constraints on understanding. As discussed in chapter 1, these limit the bounds of existential knowledge. In particular, man's knowledge of environmental systems is limited by:

(1) Methods of measurement and observation.
(2) The presence of stochastic variation largely resulting from (1) and the methods of system analysis adopted.
(3) The manner in which systems knowledge is obtained and accumulated; and the practical stages of systems analysis.

Each of these three restrictions is briefly discussed below. Further details may be sought in Bennett (1978).

2.3.1 OBSERVATIONAL RESTRICTIONS ON SYSTEMS ANALYSIS

It has been assumed up to this point that the structure of the system of interest is known to the analyst, together with its inputs and outputs. In most practical instances, however, one or more of the three system elements (input, output and transfer function) may be unknown, or only partially known. The various permutation possibilities of system information have been summarized by Bennett (1978 ch. 2) and are given in table 2.2. These range from perfect to partial knowledge. In many cases the presence of uncertainties or stochastic elements results in transitional states of partial knowledge between each of the eight cases. In other instances it may not be possible to determine exactly what form of knowledge is available: i.e. the degree of stochastic disturbance or inconsistency of the transfer function relationship is unknown, the degree of corruption of input and output by measurement errors is unknown, and the stability of all three over the simulation period, and certainly over any forecasting period of the future, is highly circumspect.

One especially important instance is that of case (6) which has been much analysed in the statistical analysis of time series (Kendall 1973). It is perhaps surprising in a case where information is available only on system output that it is possible to reach any systems analysis solution at all. The solution which is obtained is an *approximate* model. The approximation relies on the use of approximating polynomials (e.g. Padé approximants) and reproduces the form of system response as accurately as possible within statistical confidence limits and within the restrictions of the data available. The model does not

Table 2.2 The effects of system knowns and unknowns on the existence and nature of systems analysis (from Bennett 1978, ch. 2)

	Knowns	*Unknowns*	*Does solution exist?*	*Nature of solution*
1	Input, output, transfer function	—	Yes	*System description:* Any legitimate use of the model is possible.
2	Input, output	Transfer function	Yes	*System identification and estimation:* Input–output relationships (e.g. spectrum and correlations functions) are examined, models hypothesized, then parameters estimated and the model checked.
3	Input, transfer function	Output	Yes	*System convolution or forecasting:* Reproduction of system output within given probability bounds controlled by the level of system noises. Forecasting for reproduction over the future.
4	Output, transfer function	Input	Yes	*System deconvolution:* Inverse problem to (3), input reconstructed within probability bounds. Backforecasting is used outside sample period.
5	Input	Output, transfer function	No	Frequent in policy (control) applications and in man's environmental effects. System stimuli are known but the system and its outputs are indeterminate.
6	Output	Input, transfer function	Yes/No	*ARMA system identification and estimation:* Single series inference using autoregressive moving-average models which are autoprojective; it is not possible to differentiate system input and transfer function dynamics.
7	Transfer function	Input, output	Yes/No	*System simulation:* Simulation of the effect of system operation with hypothetical inputs is possible and may be useful in control policy assessment; but it is not possible to determine the inputs and outputs when they are not hypothetical.
8	—	Input, output, transfer function	No	No knowledge of any system element requires us to undertake measurement, construct theoretical models and formulate hypotheses.

necessarily, nor usually, correspond to the true form of system response (but the model and true system responses should be indistinguishable).

The output sequence $\{Y_t\}$ contains the following five elements:

(1) The response of $\{Y_t\}$ to $\{X_t\}$
(2) Correlation properties in $\{X_t\}$
(3) Errors in $\{X_t\}$
(4) Errors in $\{Y_t\}$
(5) Errors associated with the transfer function.

One solution to the problem of identifying the inputs and transfer function in case (6) (see p. 71), adopted initially by Wold (1938), Durbin (1960) and Quenouille (1947), has been to assume that X_t (the input) is a normally distributed independent sequence of observations. This is a minimum knowledge assumption. Independence requires the input to be a random variable. The normal distribution assumption corresponds to the occurrence properties of many natural events. However, in the case of extreme and infrequent natural events (storm inputs, earthquakes, surges, etc.) the normality assumption will not hold. Many environmental systems will also have deviations from normality induced by nonlinearity, giving skewed distributions (e.g. many populations of slope angles, species counts, etc.). Despite these drawbacks, which present very severe limitations in some environmental systems, in those environmental systems which are subject to inputs that are not too markedly affected by extreme events, the case (6) approximation works well in practice. If we represent the independent normal input sequence as a (new) random variable $\{e_t\}$, then the general transfer function model can be rewritten as the difference equation:

$$\begin{aligned} Y_t &= -c_1 Y_{t-1} - c_2 Y_{t-2} - \cdots - c_p Y_{t-p} + e_t + d_1 e_{t-1} + \cdots + d_r e_{t-r} \\ &= C^{-1}(z) D(z) e_t. \end{aligned} \tag{2.3.1}$$

This is an autoregressive moving-average (*ARMA*) system in which we have p autoregressive elements, as in the transfer function case, but r parameters which control lag dependence on a set of independent random variables $\{e_t\}$ termed the moving-average components. The physical interpretation of this model is that the $\{Y_t\}$ output sequence is dependent upon an 'ignored variable' or set of ignored variables which can be reproduced by the moving-average terms, plus internal memory (autoregressive) elements. All correlation properties of the system input are taken into the *ARMA* system parameters $\{c_i\}$ and $\{d_i\}$, i.e. c_i and d_i will be a combination (usually nonlinear) of the unknown system transfer function (S), and any structure (trend, cyclicity, etc.) existing in the input sequence.

In addition to the various instances of observational restrictions on systems analysis, theoretical restrictions also arise. These will affect the analyst's ability to observe and monitor input–output relationships whatever the extent of his knowledge. The structural properties of a system input, output and transfer function, which determine whether all elements of system response can be observed, are usually termed 'observability and controllability criteria' (Kalman 1960). Consider the system developed in fig. 2.37 with input X_t and output Y_t. In each case the operator S_i indicates the transfer function of a subsystem element, four elements in all. These four elements correspond to the four possible system configurations given by Gilbert (1963) so that:

S_1 is controllable and observable,
S_2 is controllable but unobservable,
S_3 is uncontrollable but observable,
S_4 is uncontrollable and unobservable.

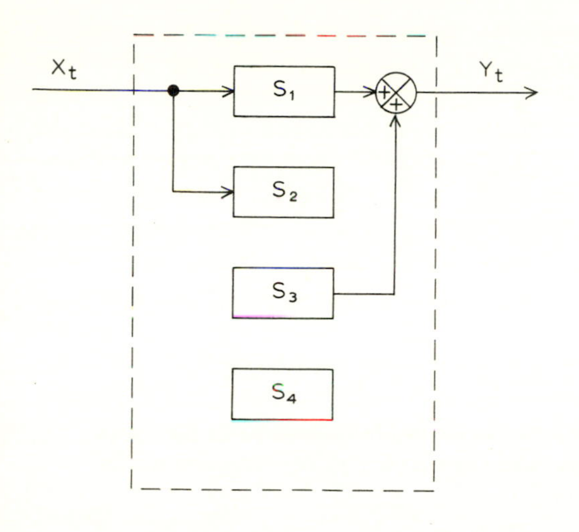

Fig. 2.37
Theoretical restrictions on the analysis, observability and controllability of systems.

Thus, controllability implies the ability of the control input to affect any state of the system and observability implies the ability of each state to influence the system output. The transfer function is an accurate representation of the system if, and only if, the system is controllable and observable, i.e.:

Controllability: A system is said to be state controllable if any initial state $Z(0)$ can be transformed to any final state $Z(t_f)$ in a finite time $t_f \geqslant 0$ for some input $X(t)$.
Observability: A system is said to be observable if every state $Z(0)$ can be exactly determined from measurements on the output $Y(t)$ over a finite interval of time $0 \leqslant t \leqslant t_f$.

Mathematical definitions of these terms are given in chapters 3 and 4 and are part of the minimal realization problem (Bennett 1978, ch. 2).

2.3.2 STOCHASTIC COMPONENTS IN SYSTEMS ANALYSIS

Stochastic elements or *noise* (Wiener 1949) can enter each of the three system elements: input, output and transfer function. Each of these sources of stochastic disturbances can be represented by the sequences $\{w_t\}$, $\{v_t\}$ and $\{n_t\}$, respectively, which can be described by difference equations, as shown in fig. 2.38 for transfer functions systems, and in fig. 2.39 for autoregressive moving-average systems. The physical interpretation of each source of noise is as follows (Bennett 1978):

(*i*) *Input noise*

This may be generated by either errors in measuring the input X_t, or by the presence of 'ignored variables'. Measurement errors are present in most

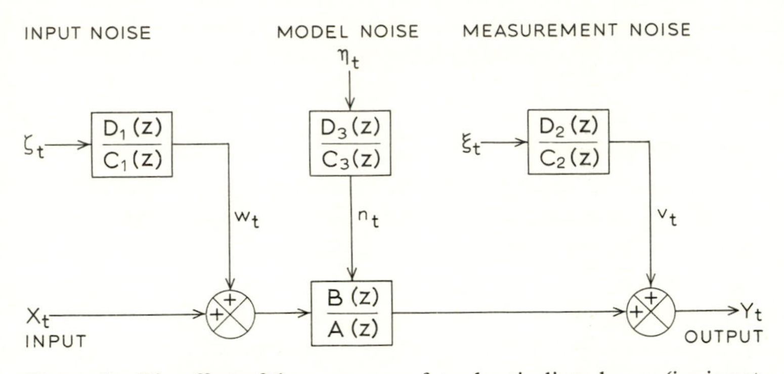

Fig. 2.38 The effect of three sources of stochastic disturbance (i.e. input, model and measurement noise) on a transfer function system. (After Bennett 1978.)

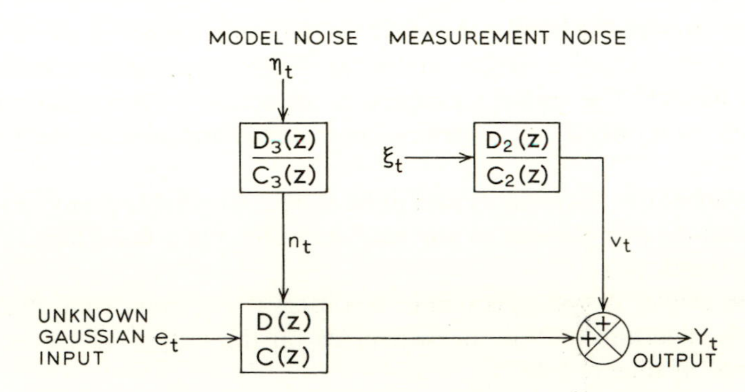

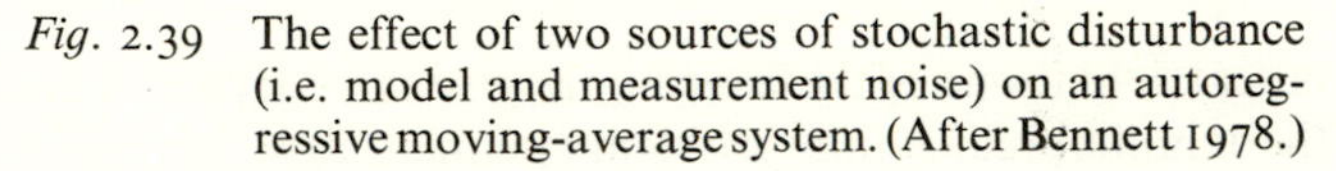

Fig. 2.39 The effect of two sources of stochastic disturbance (i.e. model and measurement noise) on an autoregressive moving-average system. (After Bennett 1978.)

physical apparatus (flow meters, rain gauges, etc.), in sources of data derived from census and from sampling (e.g. quadrat analysis), and in the use of the data when readings from instruments are employed. Thus it is rare that the input sequence of a system is known without some source of errors being present, however small. Ignored variables effects are particularly prevalent in the complex multivariate systems characterizing environmental problems in which the input sequence observed is only one of a number of stimuli which enter the system. This is so because most environmental models are only partial representations of one or two of the more important feedback loops which are singled out for attention (chapters 7 and 8). The measurement errors may well be unsystematic in character, but it will be possible to represent the ignored variables by an independent random sequence only in special cases. More generally the input noise process will be correlated. For example: systematic operator errors in measurement from, or situation of, a rain gauge or flow meter; perceptual bias in sampling or analysis; physical limits on accuracy of equipment, such as incorrect splitting of rain drops into rain gauges.

(*ii*) *Transfer function noise*

The system transfer function may be subject to noise input due to model misspecification, nonlinearity and nonstationarity. Specification errors arise from attempts to construct over-simple or over-general models – for example, the approximation of a nonlinear system by a linear model in hydrological catchment modelling, the ignoring of some multivariate elements in socio-economic models of the economy, and in world economy models (Forrester 1971). The effect of specification errors will in all cases lead to the entry of a systematic (correlated) noise sequence into the transfer function. Stochastic elements which enter the transfer function, due to real variation in the response of the system to different inputs at different times, are a special case of nonstationarity which may be either deterministic or random, and in some cases may represent the effects of specification errors.

(*iii*) *Output noise*

Errors which corrupt the measurement of the system output variable may, again, result from census and sampling variability, slight variability in the performance of physical apparatus, or in its use. There may also be specification and aggregation problems in that the $\{Y_t\}$ output sequence measures not only the system response but also includes responses to other transfer functions which have not been included in the analysis. The precise division between the effects of input, output and transfer function noise are usually difficult to determine in practice. In any complex environmental or socio-economic system it will be very difficult tc obtain continuous repetition of a result without error in the same way that it is possible to replicate an experiment in physics or engineering. Experimental controls may be tightened to reduce variability and error, but complete control will be impossible. Participants in an economic or social situation will always respond in a variable fashion. Variation in plant seeds or in animals in biological experiments can never be completely eliminated. The pattern of runoff response of a catchment will never be exactly the same for two storms. The effects of the presence of sources of stochastic variation are that, when the noise is correlated (i.e. coloured noise), it is necessary to model a set of noise models as well as systems models to predict or remove the stochastic effects. These are shown in figs 2.38 and 2.39. When the noise is uncorrelated (i.e. white noise), specific noise models are not required but the system analysis methods which can be adopted must be modified.

2.3.3 STAGES IN PRACTICAL SYSTEMS ANALYSIS

When the environmental system of interest is subject to observational restrictions arising from partial knowledge, partial hypotheses, theoretical controllability and observability limitations, or is corrupted by stochastic elements which introduce uncertainty into system input–output relations, it is necessary to undertake systems analysis procedures which differ from the mathematical solutions which have been discussed in section 2.2. It will not usually be possible to obtain analytical results or mathematical solutions, and models must be fitted or *calibrated* against real-world data derived from observing the operation of the environmental system. The stages of analysis required in many practical situations are summarized in fig. 2.40. The degree

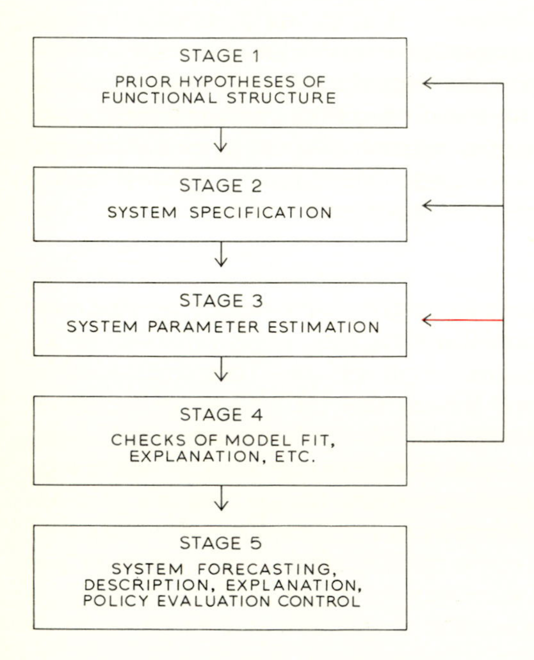

Fig. 2.40
Stages in practical system model generation, testing, use and control.

of prior knowledge that is available in system operation is used to prespecify the important variables and directions of linkage in the overall environmental model which it is desired to create. Stage two of systems analysis is concerned with specifying and identifying the nature of the transfer function governing the system input–output relations – i.e. how many $\{a_i\}$ and $\{b_i\}$ parameters exist? Stage three develops estimates (for example, by least squares regression in many cases) of the magnitude of the parameters making up the transfer function structure specified in stage two – i.e. what are the magnitudes of the $\{a_i\}$ and $\{b_i\}$ parameters? Stage four of the analysis is a 'checking' stage in which the efficiency and fit properties of the resulting parameter estimates of the transfer function model are tested in order to determine, firstly, if it is indeed a useful and successful explanatory device, and, secondly, if the resulting model accords well with accepted hypotheses of system structure. If it is found that the parameter estimates are inefficient or suboptimal in some fashion (for example, low significance levels, poor simulation properties, residual autocorrelation, etc.), then recourse must be made to the previous stages of analysis to determine if the initial hypotheses and the resulting model specifications and parameter estimates were correct. If the model is acceptable, then the final stage of analysis is to use the estimated model to generate forecasts, simulations, or explanations of the environmental system under study and to use these forecasts for environmental control, policy appraisal and decision analysis. This style of approach is similar to the Box and Jenkins (1970) formalized path of analysis for time series models, and is akin to the Blalock (1964) causal structure analysis procedure. Various approaches are summarized in Bennett (1978, chs 2 and 4). In chapter 1 we have already considered how man generates hypotheses about nature and develops his knowledge and understanding of it. In this section we develop specification (identification) and estimation procedures, and in chapters 3 and 6 show how the resulting environmental system model can be used in control and decision situations.

The *identification* or *specification* of environmental systems is the operation of choosing the most appropriate model and process order (length of maximum lag) from the class of models given in table 2.1. This can be undertaken in the time or frequency domains (Box and Jenkins 1970, Jenkins and Watts 1968) or by sequential estimation.

(i) Time domain specification

For stochastic systems the autocorrelation, partial autocorrelation and cross-correlation functions are used as diagnostics of the most appropriate orders for the model equations given in table 2.1. Estimates of the autocorrelation function (equation 2.1.4) can be derived from:

$$R_{yy}(k) = \frac{1}{N} \sum_{t=1}^{N-v} (Y_t - \overline{Y})(Y_{t+k} - \overline{Y})[\hat{\sigma}_y^2]^{-1} \qquad (2.3.2)$$

and the cross-correlation function is given by:

$$R_{yx}(k) = \frac{1}{N} \sum_{t=1}^{N-v} (X_t - \overline{X})(Y_{t+k} - \overline{Y})[\hat{\sigma}_y^2 \hat{\sigma}_x^2]^{-1}$$

$$\hat{\sigma}_y^2 = \frac{1}{N} \sum_{t=1}^{N} (Y_t - \overline{Y})^2 \qquad (2.3.3)$$

and σ_x^2 is similarly defined.

In each case the sample is over the range ($t = 1, 2, \ldots, N$), k is the lag involved and v is the maximum lag adopted. $\hat{\sigma}_y^2$ and σ_x^2 are the variance estimates, $\overline{Y}$ and $\overline{X}$ are the mean estimates of the respective output and input of the system, each averaged over the whole data set. The partial autocorrelation function $\{\phi_{kk}\}$ is given most simply by the recursive formula (Durbin 1960):

$$\phi_{k+1,k+1} = \frac{R_{yy}(k+1) - \sum_{j=1}^{k} \phi_{kj} R_{yy}(k+1-j)}{1 - \sum_{j=1}^{k} \phi_{kj} R_{yy}(j)}$$

where

$$\phi_{k+1,j} = \phi_{kj} - \phi_{k+1,k+1}\phi_{k,k-j+1}.$$

Further details are given in Kendall and Stuart (1966, chs 45 and 46), and Box and Jenkins (1970, ch. 3). The efficiency of alternative estimators is discussed in Jenkins and Watts (1968, ch. 5). Confidence levels can be placed on the estimates of each of these correlation functions (2.3.2) and (2.3.3) using formulae due to Bartlett (1946, 1966).

The autocorrelation gives a measure of persistence or dependence between values in a single series at a lag distance k apart. The partial autocorrelation function gives a measure of the significance in degree of explanation of each increment in the lag in the autocorrelation function. The cross-correlation function gives a measure of the persistence or dependence relation between the values of $\{Y_t\}$ and the values of the $\{X_t\}$ series over a range of lags ($k = 0, 1, 2, \ldots, v$). Fig. 2.41 demonstrates how the cross-correlation (lagged correlation) between $\{Y_t\}$ and $\{X_t\}$ series is calculated by lag shifts to give the system correlogram.

The importance of these various correlation functions is that the forms of the cross-correlation, autocorrelation, and partial autocorrelation functions are known for the range of mathematical models associated with the system equations given in table 2.1. The autocorrelation and partial autocorrelation functions are used for the identification of models of a single series of system output $\{Y_t\}$ when the $\{X_t\}$ input is not observed (i.e. autoregressive moving-average case), the case (6) problem discussed on pp. 70–1. The cross-correlation function is used for identification of the input–output transfer function between two series when both are observed. Diagrams giving the form of these functions for various low-order models are given in Box and Jenkins (1970) and summarized in fig. 2.42 for the autocorrelation and partial autocorrelation, and in fig. 2.43 for the cross-correlation function. A symmetry between the expected form of autocorrelation and cross-correlation functions given in these diagrams for any system model and the estimates of these functions from observation of the environmental system, allows the specification of the input–output transfer function of the process under study.

The correspondence between the graphs and fig. 2.42 and 2.43 must be judged by eye, although statistical test procedures are also available. These are based on confidence interval calculations (Bartlett 1946) and allow some objective measure to be imposed. But for low-order models (i.e. small number of lags) a set of generalizations exists. The autoregressive moving average models shown in fig. 2.42 have the following properties. For an autoregressive process of order p, the autocorrelation function tails off exponentially or with exponentially damping oscillations, whilst the partial autocorrelation is cut off after the lag p; a moving-average process of order q has autocorrelation cut off after lag q, and partial autocorrelation tailing off; for the mixed autoregressive moving average process, of order p and q respectively, both the autocorrelation and partial autocorrelation functions tail off with a mixture of exponentials and damped sinusoids after the first q–p lags for the autocorrelations, and after the first p–q lags for the partial autocorrelations. Similar generalizations hold for the cross-correlation functions between two series of input and output data. In fig. 2.43 the length of rise time in the cross-correlation function indicates the approximate lag order of the input $\{X_t\}$, whilst the persistence and degree of oscillation in the declining limb of the cross-correlations indicates the autoregressive order. This method of identification is subjective and open to a number of criticisms. The assumptions underlying the Bartlett (1946) and Quenouille (1947) confidence limits is of zero correlation beyond lag k, the assumed order. This assumption gives high standard error limits and makes it easier to accept that the process is uncorrelated beyond lag k. Hence, there is a high probability of making a type 2 error of inference which makes the method prone to ignoring long-period lags. As Hannan (1969) notes, it is more or less impossible to diagnose higher-order systems. Further criticisms are discussed in Box and Jenkins (1970), Kendall (1973) and Chatfield (1975). In most situations, however, there is little alternative to subjective methods of identification. Where alternatives do not exist, the method is rapid, simple, and extremely useful, especially in cases in which there is little prior knowledge for generating at least first hypotheses of time series model structure in a very rapid manner. Examples of the method are discussed in King *et al.* (1969), Box and Jenkins (1970), Jenkins and Watts (1968) and Bennett (1974).

(ii) Frequency domain identification

The frequency domain character of a system model can also be used for identification of the order of the process in the equations of table 2.1. For single

A Input X_t	B Lagged output Y_{t-1}	C Lagged output regression $0{\cdot}5Y_{t-1}$	D Observed output (A + C) Y_t	
0	0	0	0	
−0·22	0	0	−0·22	initial condition
0·58	−0·22	−0·11	0·47	
−0·51	0·47	0·23	−0·28	
0·38	−0·28	−0·14	0·24	
0·22	0·24	0·12	0·36	
−0·70	0·36	0·18	−0·52	
−0·22	−0·52	−0·26	−0·48	
−0·06	−0·48	−0·24	−0·30	
−0·40	−0·30	−0·15	−0·35	
0·82	−0·35	−0·17	0·65	
−0·71	0·65	0·32	−0·49	
mean $\overline{X} = -0{\cdot}06$			mean $\overline{Y} = -0{\cdot}07$	

SYSTEM $Y_t = 0{\cdot}5Y_{t-1} + X_t$

1 System input deviations $x_t = (X_t - \overline{X})$ (zero mean)	2 System output deviations $y_t = (Y_t - \overline{Y})$ (zero mean)	3 Lagged output y_{t+1}	4 Lagged output y_{t+2}	5 Cross-correlation (1 + 2) $x_t y_t$	6 Cross-correlation (1 + 3) $x_t y_{t+1}$	7 Cross-correlation (1 + 4) $x_t y_{t+2}$	8 Auto-correlation (2 + 3) $y_t y_{t+1}$	9 Auto-correlation (2 + 4) $y_t y_{t+2}$	
0	0	·	·	·	·	·	·	·	
0·64	0·54	·	·	·	·	·	·	·	
−0·45	−0·21	0·54	·	·	·	·	·	·	initial condition
0·44	0·31	−0·21	0·54	0·14	−0·09	0·24	−0·07	0·17	
0·28	0·43	0·31	−0·21	0·12	0·09	−0·06	0·13	−0·09	
−0·64	−0·45	0·43	0·31	0·29	−0·28	−0·20	−0·19	−0·14	
−0·16	−0·41	−0·45	0·43	0·07	−0·07	−0·07	0·18	−0·18	
0·00	−0·23	−0·41	−0·45	0·00	0·00	0·00	0·09	0·10	
−0·34	−0·28	−0·23	−0·41	0·10	0·08	0·14	0·06	0·11	
0·88	0·72	−0·28	−0·23	0·63	−0·25	−0·20	−0·20	−0·17	
−0·65	−0·42	0·72	−0·28	0·27	−0·47	0·18	−0·30	0·12	
$\sum x_t^2 =$ 2·019	$\sum y_t^2 =$ 1·478			$\sum x_t y_t =$ 1·612	$\sum x_t y_{t+1} =$ −0·845	$\sum x_t y_{t+2} =$ 0·031	$\sum y_t y_{t+1} =$ −0·286	$\sum y_t y_{t+2} =$ −0·069	

Variance:

cross-correlation:

autocorrelation:

$\sigma_x^2 = 0{\cdot}202$ $\quad R_{yx}(0) = \dfrac{0{\cdot}161}{\sqrt{0{\cdot}202 \times 0{\cdot}148}} = 0{\cdot}931 \quad R_{yy}(0) = 1{\cdot}0$

$\sigma_y^2 = 0{\cdot}148$ $\quad R_{yx}(1) = \dfrac{-0{\cdot}0845}{\sqrt{0{\cdot}202 \times 0{\cdot}148}} = -0{\cdot}489 \quad R_{yy}(1) = \dfrac{-0{\cdot}029}{0{\cdot}148} = -0{\cdot}193$

$R_{yx}(2) = \dfrac{0{\cdot}0031}{\sqrt{0{\cdot}202 \times 0{\cdot}148}} = 0{\cdot}018 \quad R_{yy}(2) = \dfrac{-0{\cdot}0069}{0{\cdot}148} = -0{\cdot}047$

Fig. 2.41 The cross-correlation (i.e. lagged correlation) between output (Y_t) and input (X_t) series calculated by lag shifts to produce the system correlogram.

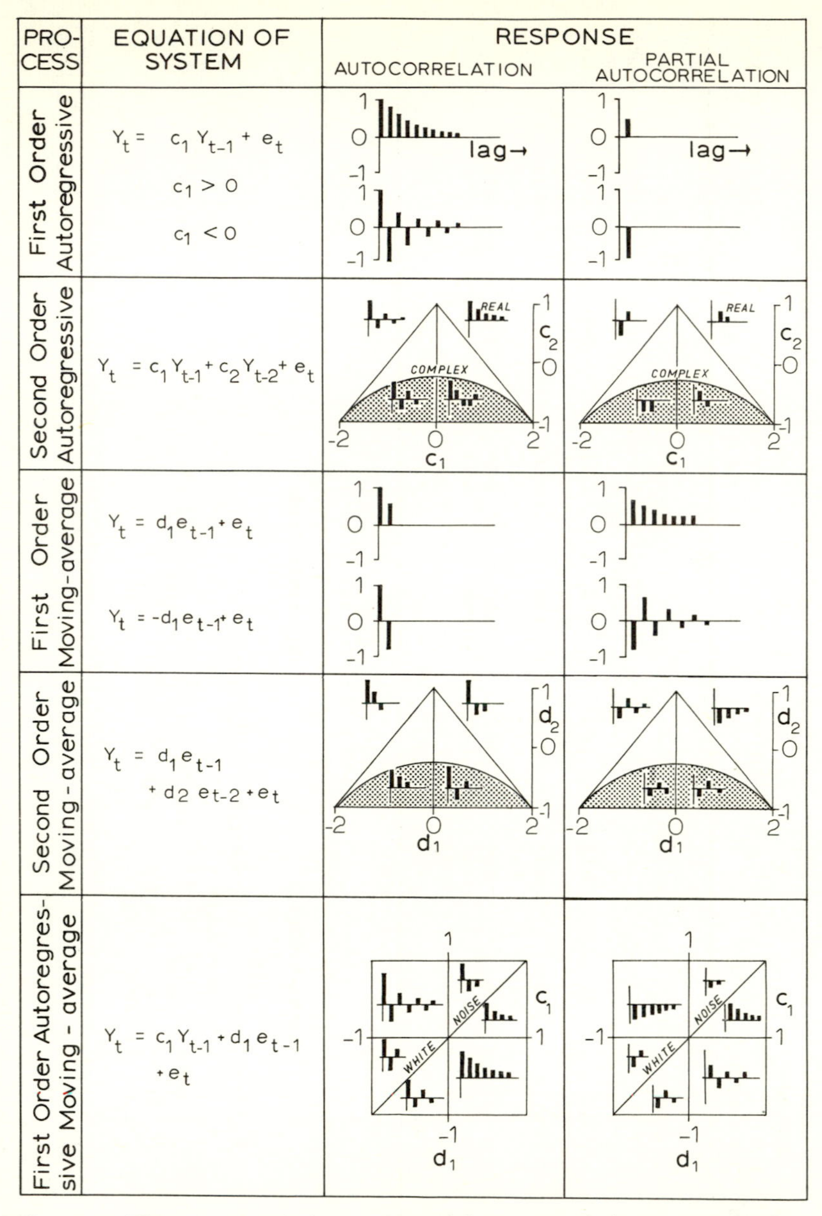

Fig. 2.42 The autocorrelation and partial autocorrelation responses for various low-order autoregressive moving-average systems. Comparison of these theoretical forms with those calculated for any system under study allows the recognition of most appropriate system model. (After Box and Jenkins 1970, Bennett 1974.)

PROCESS ORDER	EQUATION OF SYSTEM	IMPULSE RESPONSE
	SYSTEM INPUT	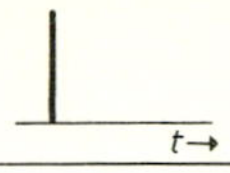 $t \rightarrow$
q=1 p=0	$y_t = b_o x_t$	
q=2 p=0	$y_t = b_o x_t + b_1 x_{t-1}$	
q=3 p=0	$y_t = b_o x_t + b_1 x_{t-1} + b_2 x_{t-2}$	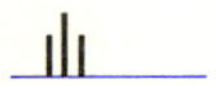
q=1 p=1	$y_t = b_o x_t + a_1 y_{t-1}$	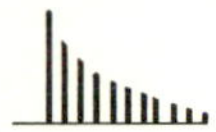
q=2 p=1	$y_t = b_o x_t + b_1 x_{t-1} + a_1 y_{t-1}$	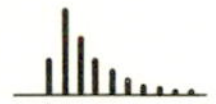
q=3 p=1	$y_t = b_o x_t + b_1 x_{t-1} + b_2 x_{t-2} + a\ y_t$	
q=1 p=2	$y_t = b_o x_t + a_1 y_{t-1} + a_2 y_{t-2}$	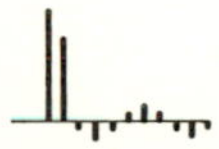
q=2 p=2	$y_t = b_o x_t + b_1 x_{t-1} + a_1 y_{t-1} + a_2 y_{t-2}$	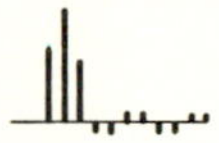
q=3 p=2	$y_t = b_o x_t + b_1 x_{t-1} + b_2 x_{t-2} + a_1 y_{t-1} + a_2 y_{t-2}$	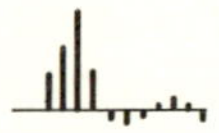
q=1 p=3	$y_t = b_o x_t + a_1 y_{t-1} + a_2 y_{t-2} + a_3 y_{t-3}$	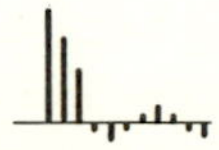

Fig. 2.43 The cross-correlation response (impulse response) for various low-order transfer function systems. Comparison of these theoretical forms with those observed in the system of interest allows initial hypotheses of model order to be formulated. (After Bennett 1974.)

series (i.e. autoregressive moving-average (*ARMA*) systems), the spectrum $S(f)$ for frequency f is adopted, and for input–output relations, the cross-spectrum. These are equivalent to the autocorrelation and cross-correlation functions in the time domain. Estimates of these functions are given in Granger and Hatanaka (1964), Jenkins and Watts (1968), and Rayner (1971), and have been used by Bassett and Tinline (1970), Haggett (1971) and Bennett (1974) for identification of system structures.

The spectrum is defined, for a sample size N, by:

$$\begin{aligned} S_y(f) &= \frac{1}{N}\int_{-N/2}^{N/2} Y(t)\, e^{-i2\pi fm/N}\, dm \\ &\approx \frac{2}{N}\sum_{m=0}^{N-1} Y_m \cos\frac{(2\pi fm)}{N} \end{aligned} \tag{2.3.4}$$

for frequency f, and sample points m, The typical spectra of various low-order *ARMA* models are discussed in Jenkins and Watts (1968) and are shown in fig. 2.44. There is a lack of uniqueness in the estimated form of the spectrum of the generating process and for most identification purposes the spectrum provides corroboration rather than diagnostic information. It is of most utility when periodic components are suspected, such as daily and seasonal rhythms in plant and climatological systems, or seasonal and longer-period economic cycles in the socioeconomic sphere.

Frequency domain identification of input–output relations is more useful and diagnostic. Here the gain, phase, and Nyquist plots are of greatest utility. The gain is defined by

$$G_{yx}(f) = \frac{R_{yx}(f)}{S_x(f)} \tag{2.3.5}$$

and the phase by

$$\phi_{yx}(f) = \frac{2N}{2\pi f}\arctan\left[\frac{A_q(f)}{A_c(f)}\right] \tag{2.3.6}$$

where $A_q(f)$ and $A_c(f)$ are the quadrature and cospectrum, respectively. The gain represents the mean proportionate increase in $\{Y_t\}$ at frequency f for a given change in $\{X_t\}$; and the phase represents the mean distance between maxima and minima in $\{Y_t\}$ and maxima and minima in $\{X_t\}$ at frequency f.

The Nyquist plot is defined in the complex plane by the coordinates (real and complex) given by:

$$\left[\frac{A_c(f)}{S_x(f)}, \frac{A_q(f)}{S_y(f)}\right].$$

The Nyquist plot is a composite of the information contained in the gain and phase. For any frequency f, the gain contributes the amplitude of curvature

PROCESS	EQUATION OF SYSTEM	SPECTRUM (NORMALIZED)
First Order Autoregressive	$Y_t = c_1 Y_{t-1} + e_t$	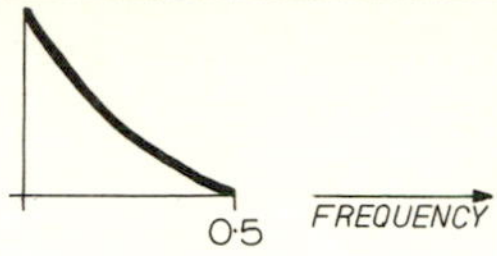 0·5 FREQUENCY
	$Y_t = -c_1 Y_{t-1} + e_t$	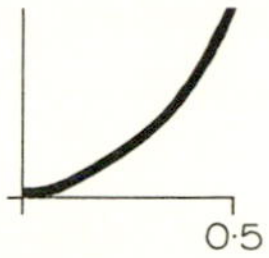 0·5
Second Order Autoregressive	$Y_t = c_1 Y_{t-1} + c_2 Y_{t-2} + e_t$	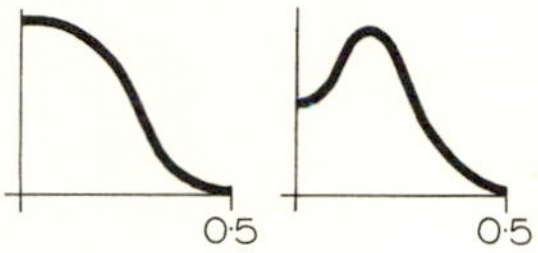 0·5 0·5
First Order Moving-average	$Y_t = d_1 e_{t-1} + e_t$	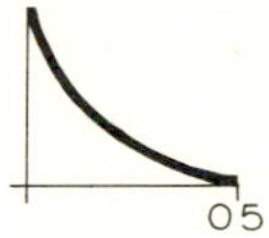 0·5
	$Y_t = -d_1 e_{t-1} + e_t$	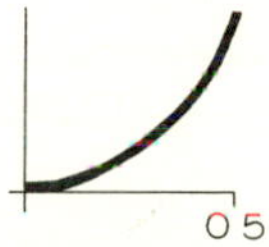 0·5
Second Order Moving-average	$Y_t = d_1 e_{t-1} + d_2 e_{t-2} + e_t$	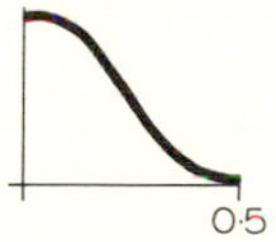 0·5
	$Y_t = -d_1 e_{t-1} + d_2 e_{t-2} + e_t$	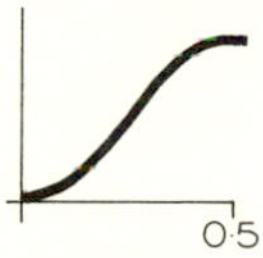 0·5
First Order Autoregressive Moving-average	$Y_t = c_1 Y_{t-1} + d_1 e_{t-1} + e_t$	 0·5

Fig. 2.44 The frequency-response spectra for various low-order autoregressive moving-average systems. Comparison with observed spectra allows the generation of hypotheses of model order. (After Bennett 1974.)

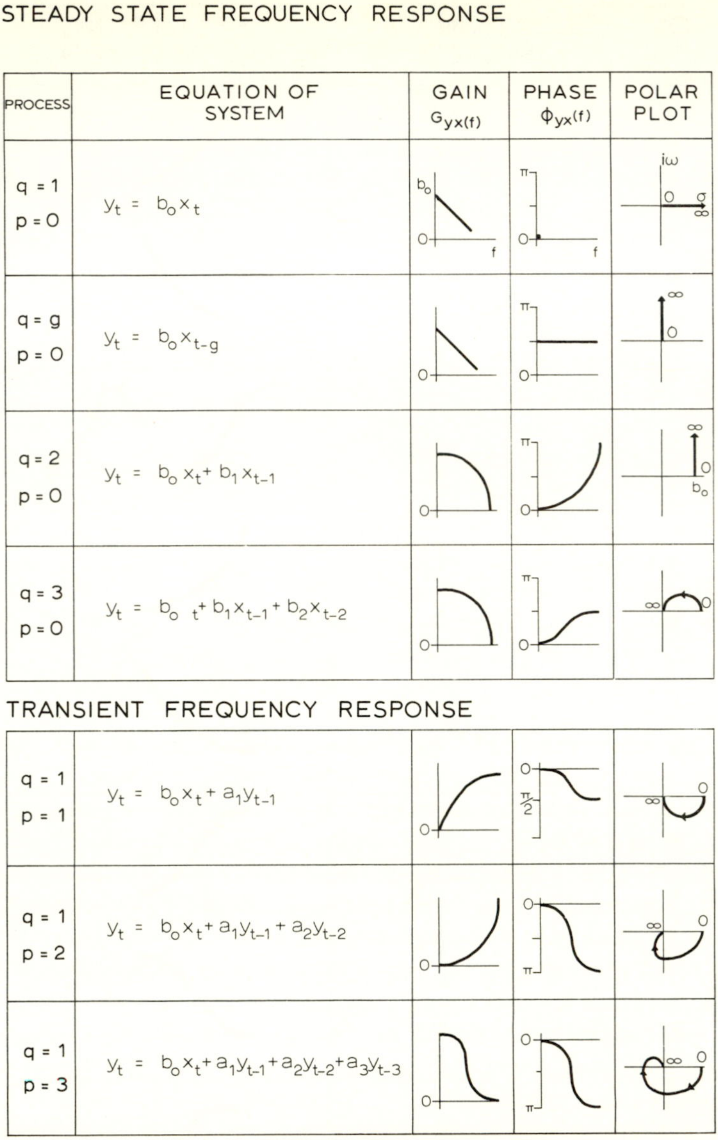

STEADY STATE FREQUENCY RESPONSE

PROCESS	EQUATION OF SYSTEM	GAIN $G_{yx(f)}$	PHASE $\Phi_{yx(f)}$	POLAR PLOT
q = 1 p = 0	$y_t = b_o x_t$	b_o, 0, f	π, 0, f	iω, 0, σ, ∞
q = g p = 0	$y_t = b_o x_{t-g}$	0	π, 0	∞, 0
q = 2 p = 0	$y_t = b_o x_t + b_1 x_{t-1}$	0	π, 0	∞, 0, b_o
q = 3 p = 0	$y_t = b_o \;{}_t + b_1 x_{t-1} + b_2 x_{t-2}$	0	π, 0	∞, 0

TRANSIENT FREQUENCY RESPONSE

PROCESS	EQUATION OF SYSTEM	GAIN	PHASE	POLAR PLOT
q = 1 p = 1	$y_t = b_o x_t + a_1 y_{t-1}$	0	0, $\frac{\pi}{2}$	∞, 0
q = 1 p = 2	$y_t = b_o x_t + a_1 y_{t-1} + a_2 y_{t-2}$	0	0, π	∞, 0
q = 1 p = 3	$y_t = b_o x_t + a_1 y_{t-1} + a_2 y_{t-2} + a_3 y_{t-3}$	0	0, π	∞, 0

Fig. 2.45 (*above and facing page*) The steady-state, transient and mixed frequency responses for various low-order transfer function processes shown in terms of Bode (gain and phase) and Nyquist (polar) plots. Comparison with estimated cross spectra generates hypotheses of model order. (After Bennett 1974.)

MIXED STEADY STATE AND TRANSIENT FREQUENCY RESPONSE

PROCESS	EQUATION OF SYSTEM	GAIN $G_{yx(f)}$	PHASE $\Phi_{yx(f)}$	POLAR PLOT
q = 2 p = 1	$y_t = b_0 x_t + b_1 x_{t-1} + a_1 y_{t-1}$	1, 0	$\pi/2$	∞, 0
q = 3 p = 1	$y_t = b_0 x_t + b_1 x_{t-1} + b_2 x_{t-2} + a_1 y_{t-1}$	1, 0	$\pi/2$	∞, 0
q = 2 p = 2	$y_t = b_0 x_t + b_1 x_{t-1} + a_1 y_{t-1} + a_2 y_{t-2}$	1, 0	0, $\pi/2$	∞, 0
q = 2 p = 3	$y_t = b_0 x_t + b_1 x_{t-1} + a_1 y_{t-1} + a_2 y_{t-2} + a_3 y_{t-3}$	0	0, $\pi/2$	∞, 0

and the phase the number of quadrants crossed in the complex plane. It, therefore, expresses the proportionate change and lag structure of the system transfer function. The form of the gain, phase, and Nyquist plots is given in Bennett (1974) and in fig. 2.45 for various low-order models. The effect of autoregressive output feedback lags gives increasing gain with frequency, and increasing curvature with lag, whilst phases are negative. The effect of input lags gives declining gain with frequency and positive phases. In the Nyquist plot the effect of increasing the autoregressive order is that each lag factor contributes a phase lag of 90°, or $\pi/2$, and moves the frequency response one quadrant in a negative direction in the complex plane. The effect of input lags is to give a phase lead, moving the frequency response one quadrant clockwise for every input lag. Spectral and cross-spectral analysis derive from the fourier transform of the system autocovariance sequence which decomposes any covariance stationary process into an infinite sum of *orthogonal* stochastic processes corresponding to each frequency component. Fig. 2.46 gives the input and output of a sampled data system corrupted by noise. The input autocorrelation and spectrum, and the input–output cross-correlation , gain, phase and Nyquist diagram are also shown in the figure. The system transfer function is composed of one gain and one autoregressive component. The autocorrelation and spectral plots of the input $\{X_t\}$ sequence show that this is uncorrelated random noise. The plots of the system input–output cross-

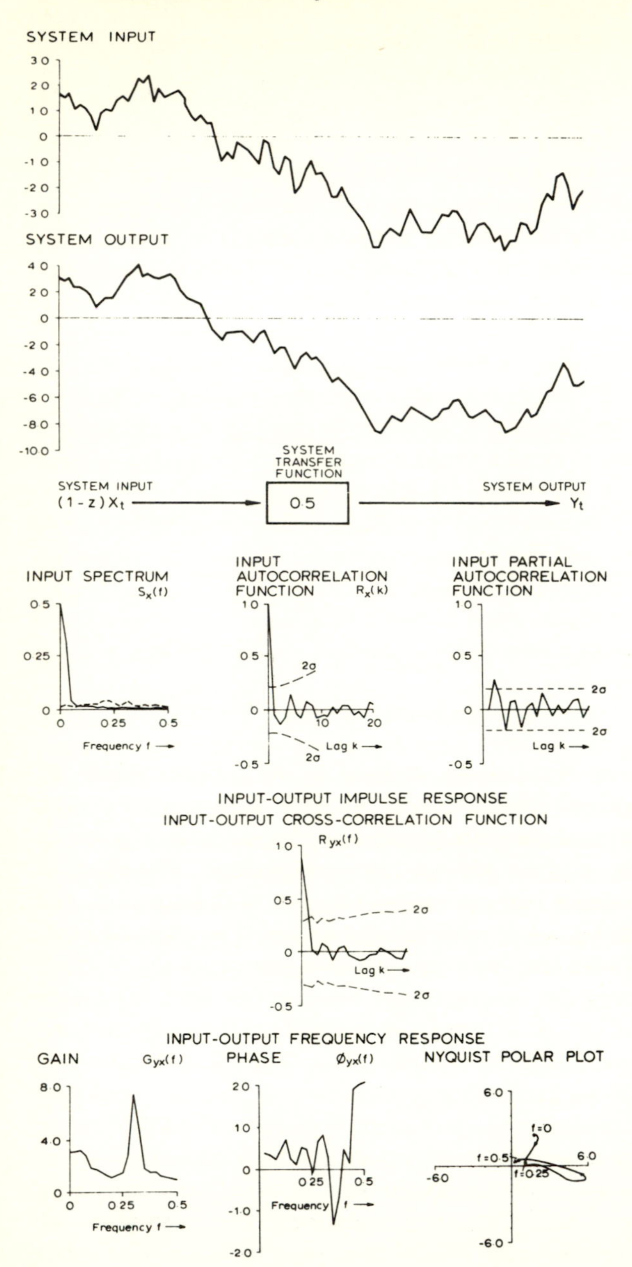

Fig. 2.46
The input and output streams of a sampled data system corrupted by noise. The empirical estimates of the input autocorrelation and spectrum show that the input is random (white noise). The cross-correlation and cross-spectral functions allow the generation of a hypothesis of a model with one distributed lag and two autoregressive lags.

correlation, gain, phase and Nyquist diagram should lead to a correct identification of the structure of the (known) transfer function. On comparing these plots with figs. 2.41 to 2.44 the correct transfer function is indeed indicated, although the effect of noise corruption and finite sample estimation leads to the appearance of some small (spurious) contributions within each diagram.

(*iii*) *Sequential estimation*

If estimation techniques are available an alternative method of system identification can be adopted based on sequential (iterative) estimation of a range of system models for various transfer function orders. The residuals and other properties of these models can then be compared in order to define an optimal model from the range considered (Whittle 1952, Åstrom 1965, Bennett 1974). Various tests of optimality have been proposed: residual variance, residual autocorrelation, total R^2, parameter standard errors, minimality and absence of collinearity. But it has been noted by Bennett (1974) and others that these diagnostics may be in conflict. For example, a good R^2 may be important in system forecasting applications but low system parameter standard errors will be more useful in inference and explanation. No simple resolution of these conflicts is available but a solution will lie in determining 'hill climbing' maxima and minima on the multicriteria objective function formed by each diagnostic of model optimality. How these solutions will fit with vaguer and more intuitive ideas of parsimony (Tukey 1961), in which an attempt is made to minimize the complexity of system structure, is extremely uncertain as yet.

Parameter estimation with environmental systems requires the pre-specification of the structure (variables, lags, delays, etc.) of the model which is to be used to describe the system. This specification can be derived either from the identification methods described above or (usually preferably) from the use of known model structures deriving from theory. Once the nature of the system equation has been specified, estimates of the magnitude of the system parameters yield a final model (subject to checking). It is first necessary to restate the system equation given in table 2.1 and equation (2.2.7) in a form amenable to statistical estimation. For the system (2.2.7) given by:

$$Y_t + a_1 Y_{t-1} + a_2 Y_{t-2} + \cdots + a_p Y_{t-p} = b_0 X_t + b_1 X_{t-1} + \cdots + b_q X_{t-q}$$

or

$$A(z) Y_t = B(z) X_t$$

we have

$$Y_t = \boldsymbol{\theta}' \mathbf{X}_t' \qquad (2.3.7)$$

using vector notation (Appendix 1) defined by

$$\begin{aligned} \boldsymbol{\theta}' &= [a_1 a_2 \ldots a_p b_0 b_1 \ldots b_q] \\ \mathbf{X}_t' &= [-Y_{t-1} - Y_{t-2} \cdots - Y_{t-p} X_t X_{t-1} \ldots X_{t-q}] \end{aligned} \qquad (2.3.8)$$

in which $\boldsymbol{\theta}$ is now a vector of system parameters and $\mathbf{X}_t$ a vector of system input and output feedback variables. In statistical estimation we wish to derive estimates of the parameters $\boldsymbol{\theta}$, given knowledge only of the system variables $\mathbf{X}_t$. In evaluating the properties and strengths of alternative estimation methods (estimators) which tackle the problem the following criteria are usually adopted (Johnston 1972):

(1) *Unbiasedness.* The property that *on average* the parameter estimates, $E(\hat{\boldsymbol{\theta}})$, are correct:

$$E(\hat{\boldsymbol{\theta}}) = \boldsymbol{\theta}$$

where $\boldsymbol{\theta}$ is the true parameter value. An estimator is *asymptotically unbiased* if it converges to the true parameter values in an infinite sample

$$\operatorname*{plim}_{N\to\infty} (E(\hat{\boldsymbol{\theta}}) - \boldsymbol{\theta}) = 0.$$

(2) *Consistency.* An estimator is consistent if the parameters are asymptotically unbiased and the parameter estimates converge on to the same distribution as the true parameters – i.e. to zero dispersion for deterministic parameters, or to the true population distribution for stochastic parameters. In addition, the errors must be normally distributed as $N \to \infty$. This is not a useful property in small samples where asymptotic convergence is irrelevant.

(3) *Efficiency.* The property that, in comparison with other estimators, a given estimate is unbiased and has minimum variance. Given two estimators, $\hat{\boldsymbol{\theta}}^*$ and $\hat{\boldsymbol{\theta}}$:

$$E(\hat{\boldsymbol{\theta}} - E(\hat{\boldsymbol{\theta}}))^2 < E(\hat{\boldsymbol{\theta}}^* - E(\hat{\boldsymbol{\theta}}^*))$$

or var $(\hat{\boldsymbol{\theta}}) <$ var $(\hat{\boldsymbol{\theta}}^*)$ (minimum variance property) and $E(\hat{\boldsymbol{\theta}}) = 0$ (unbiased property).

In many cases, least-squares regression (O*LS*) can be adopted. For the transfer function equation, for example, the O*LS* estimator, derived from information on only system input and output, is given by

$$\hat{\boldsymbol{\theta}} = [\mathbf{X}'\mathbf{X}]^{-1}\mathbf{XY} \tag{2.3.9}$$

where

$$\mathbf{X}' = [\mathbf{X}_1\mathbf{X}_2 \ldots \mathbf{X}_N]$$

$$\mathbf{X}_t' = [-Y_{t-1} - Y_{t-2} \ldots - Y_{t-p}X_tX_{t-1} \ldots X_{t-q}]$$

$$\mathbf{Y}' = [Y_1Y_2 \ldots Y_N]$$

$$\boldsymbol{\theta}' = [a_1a_2 \ldots a_pb_0b_1 \ldots b_q],$$

in which $\mathbf{X}$ and $\mathbf{Y}$ are vectors of the entire set of observations of the system input and output over the time period N for which information is available. For transfer function systems with no stochastic errors, the O*LS* estimator will be consistent and unbiased (Johnston 1972). It will probably be inefficient in small sample situations and white noise stochastic environments, and will be biased when measurement errors are present, whether these errors occur in $\{X_t\}$ or $\{Y_t\}$. It will also be biased when the residuals are correlated (i.e. with coloured noise) (see, for example, Johnston 1972). In these circumstances estimators as alternatives to *OLS* must be sought – for example, two-stage least-squares (2*SLS*), instrumental variables, maximum likelihood estimates (*MLE*), etc. The application of these alternative estimators is discussed in Bartlett (1966) and Johnston (1972).

For autoregressive moving-average systems, *OLS* estimates will be unbiased and consistent when autoregressive components alone are present (Mann and Wald 1943), but will be very inefficient in small sample situations. In moving-average and mixed autoregressive moving-average systems *OLS* cannot be adopted, and recourse must usually be made to maximum likelihood estimation or alternative methods (Johnston, 1972).

For the first-order system given in fig. 2.46, for example, the use of *OLS* estimation yields the following estimated systems equation

$$Y_t = (1 + Z)\underset{(0\cdot006)}{0\cdot49}\, X_t + e_t$$

where the term in brackets under the parameter is the standard error associated with that parameter. The sample size of this example ($N = 100$) is not large, but the resulting parameter estimate is very close to the known true system parameter (of 0·50) used to generate the system input and output. For systems which are affected by trends, nonstationary (adaptive) transfer functions, and nonlinear behaviour, special estimation and identification methods are required which are discussed more fully in Bennett (1978, ch. 5).

2.4 Conclusion

The exploration of environmental system response is a complex field which has been only briefly developed above, and some of the major mathematical and philosophical implications are explained more deeply in later chapters. Particularly important to note at this stage is the effect of practical constraints upon the knowledge we can obtain regarding the behaviour of environmental systems. One particularly important constraint is the effect of bias in estimation of system parameters which results from correlation between the system errors which usually represents the variables 'ignored' from the analysis. As a consequence of this and other constraints, the transfer function and autoregressive moving average structures adopted must usually be viewed as approximating polynomials which can reproduce system input–output relationships arbitrarily closely, but which *may* bear no relation to the true system generating parameters. This can be demonstrated, for example, by considering the manner in which the infinite series function $H(z)$ in the convolution term:

$$Y_t = \sum_{i=0}^{\infty} h_i z^i X_{t-i} = H(z)X_t$$

can be reproduced by the ratio of the finite transfer polynomials $B(z)/A(z)$, i.e.

$$Y_t = \frac{B(z)}{A(z)} X_t = H(z)X_t. \tag{2.4.1}$$

The distinction between the infinite series representation and the ratio of polynomials must be seen more as a function of the way we view the system than as a function of the system itself. As equation (2.4.1) stands, however, there is no unique definition for $A(z)$ and $B(z)$. For example, $gB(z)/gA(z)$ will satisfy the same equation for arbitrary scalar g.

The ratio of approximating polynomials allows what is in the real world an infinite system memory of all past history to be truncated to a finite memory (usually short) of time lags. Such approximating terms derive from Padé (1892) (see Dhrymes 1971), although alternative polynomial approximations based on Lagrange, Legendre, Chebychev, Jacobi, Laguerre and Hermite schemes can also be adopted. The higher the degree (order) of the approximating polynomials (and hence of the system transfer function), the better will be the correspondence to the parameters of the real generating process (see ch. 7). However, the rationale for using the polynomial approximation lies in its ability to give good approximations to system response with small values for p and q (the polynomial orders) – the principle of parsimony in parameterization (Tukey 1961, Box and Jenkins 1970). The reproduction of the lag dynamics of the generating process is a one-to-one correspondence with the real-world process only in the mathematical sense. In general, 'where the parameters of the approximating scheme are estimated we must bear in mind that if the general infinite lag structure is true, then by using the approximate structure we are committing a mis-specification error of generally unknown proportions' (Dhrymes 1971, p. 52). Hence, the symmetry between the transfer function model parameters and the real-world environmental system parameters can be established only by sound theoretical and empirical experience. The consequence of the way we view a system, and hence of the way we observe and measure it, is that our understanding of the system is limited by that view or measurement. As noted in chapter 1, the environment and man are inextricably interlocked and our environmental understanding necessarily indeterminate.

> Il est ordinairement inutile, sinon impossible, d'obtenir le résultat avec une complète exactitude. Le plus souvent, il suffit seulement d'en connaître une valeur approachée telle que l'erreur commise en adoptant cette valeur au lieu du résultat exact soit inferieure à une limite donnée *a priori*. . . . On concoit ainsi combien il est important de savoir obtenir des valeurs

approchées d'un nombre qui satisfassent à des conditions d'approximation impasées a l'avance. (Padé 1892, p. S.6.)

The limit set prior to an analysis derives from the stochastic errors present, which themselves derive from the way the system can be defined and specified. In the following chapters we show how systems analysis methods can be applied to environmental problems to push back this accuracy constraint. In particular, it will be shown how man's increasing degree of understanding of environmental systems can be used to increase the possibilities of control, and to reduce at each subsequent analysis the accuracy limit that must be accepted *a priori*.

3 Control systems

> So Man, who here seems principal alone,
> Perhaps acts second to some sphere unknown,
> Touches some wheel, or verges to some goal;
> 'Tis but a part we see, and not a whole'
>
> (A. Pope: *An Essay on Man*)

3.1 The purpose of control systems

In the control of systems two points must be clearly grasped. Firstly, that the understanding of a system does not necessarily imply that we can control it and, secondly, that control only has meaning in terms of some stated goal. Thus control systems can be interpreted in three main ways:

(1) As wholly or partly autonomous feedback mechanisms that control the environment. These systems are designed to operate on the input so as to yield the required output. This is most clearly exemplified by the early control systems which were designed to monitor a changing reference quantity and to keep the output equal to a fixed reference input in the presence of external disturbances (*disturbance inputs*) (as in a thermostat) (Grodins 1970, pp. 2–6). This commonly implies the use of tools (*control efficators*) and *energy sources* to control the environment by changing the future course of events towards a goal or selected future target state (*reference input*). Thus the conscious processes of control may consist of (Kelley 1968, pp. 10 and 41):

(*a*) *Perception*: the perception of many possible courses of events.

(*b*) *Conception*: the envisioning of two or more possible alternative future states, together with the events which would lead up to them.

(*c*) *Selection*: the selection of the required goal from a range of possibilities. This is a balance between the recognition of the existence of alternatives and the consequences of their selection. The question of goals is treated further in chapter 6, but it is important to recognize that they are of many types, including:

1. Externally imposed goals – e.g. the completion of plans imposed by higher authorities, etc.
2. Self-imposed goals – e.g. the fulfilment of one's own plans, what is viewed as 'destiny', etc.
3. Interim goals – e.g. those leading to long-term goals or, alternatively, those pursued in the absence of long-term goals.

4 Subconscious goals – e.g. preservation of the species, response to the collective unconscious (see the work of Jung in chapter 5), etc.
5 Conflicting goals – those which conflict with other goals of one's own or with those of others.

Using any one goal it is possible to construct an *objective function* describing the degree of achievement of the goal with any given level of control or action – for example, personal preference functions to a range of choices. Two or more goals can be simultaneously sought by constructing multicriteria objective functions. In general, however, there may be conflict between attaining several goals simultaneously – for example, travel cost minimization and maximization of environmental satisfaction in residential location. Conflict especially affects political control decisions. In such cases several alternative maxima may be available. In all cases it is important that goals are clearly defined so that it is clear for what aim or target the system is being controlled.

(*d*) *Programming*: the planning of control actions to institute a train of events leading to the selected goal.

(*e*) *Execution*: the monitoring and modification of the train of events until the selected goal is achieved.

(2) As an extension of the life process of man's conscious control over the environment by means of man-machine systems. Such systems combine in a complex way the capabilities of man and machine for optimal decision making. Indeed the man-machine is part of the general decision making structure discussed in detail in chapter 6. For the present it is sufficient to note that a man in the control system is able to employ his thinking and learning capabilities and this implies a multilevel approach to control. In this the machine carries out routine tasks that can be described and optimized mathematically and repeated automatically, the man intervening only in infrequent strategic and policy decisions as part of a 'mixed scanning' process as discussed in chapter 6. The human hand is an example of such a multilevel system. At a very low level there is automatic adjustment of the hand to the shape of an object to be grasped by feedback from the skin to the actuators, and the adjustment of the prehension force measuring weight (Tomovic 1969). At a higher (algorithmic) level coded coordination movements are employed, such as the grasping actions to make a fist, or sign. At the highest level of decision, the man is required only to determine the algorithm to be used – fist, sign, point, open hand, etc. – and the rest of the commands proceed automatically (Tomovic 1969).

Hence man-machines contain the first category of autonomous control systems discussed above as a special lower-level case.

(3) As the ability to change the system itself so that it yields the desired output from the existing input (Grodins 1970). Such changes are effected

either by a planned reorganization of the system or by manipulating the disturbance inputs so that the system is forced through an irrecoverable threshold and subjected to internal reorganization. This represents a restructuring of the decision levels, of the algorithms to be employed in achieving given system response, and of the nature of the goals towards which the system is being driven. Hence, within this control system man-machine and autonomous control systems are contained as lower-level cases.

The common characteristics of most control problems are shown in fig. 3.1 where the system which it is desired to control is made up of a transfer function process as discussed in chapter 2. But the system differs from that discussed in chapter 2 in that two system inputs are now considered: an *uncontrolled* input sequence $\{X_t\}$ which represents stimuli to system action (as used in chapter 2), and an additional *controlled* input sequence $\{U_t\}$. The controlled input sequence is the one which receives fullest attention in this chapter. This represents a variable (or set of variables in multivariate systems) that can be

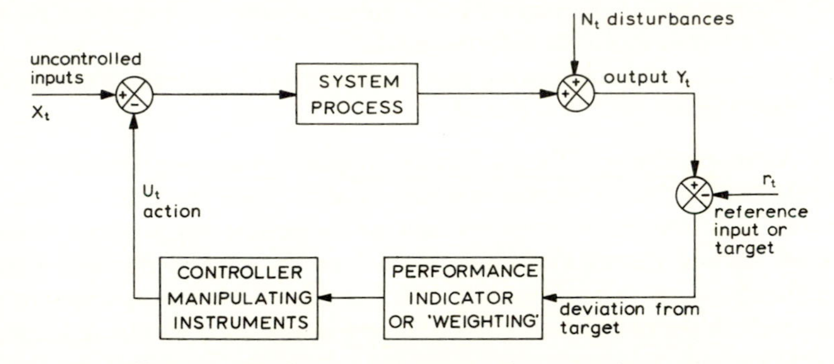

Fig. 3.1 The basic components of a control system.

manipulated by the controller and decision maker to influence the system under consideration and induce the output to approach more closely to 'desired' output settings. Variables amenable to such manipulation are termed *policy instruments*. In the overall control system any given output is compared with a *reference input* or *target* (r_t) which represents the desired response of the system. The difference between the real and desired output generates an 'error' sequence which, when weighted by the chosen performance index or objective function, is used by the controller to manipulate those variables which are amenable to control (or policy instruments $\{U_t\}$). The *performance index* is a 'figure of merit' by which deviations from the desired response are scaled.

This approach covers a wide variety of control systems that have been treated in economics by Tinbergen (1937), Dorfman (1969) and Peston (1973), in environmental control by Forrester (1971), and in control engineering by Kalman (1958, 1960). Within this broad approach it is possible to recognize two schools. On the one hand, the econometric school has usually asked the simulation question: what will be the value of a system endogenous variable given the value of a control instrument exogenous to the system? On the other hand, the control engineering approach has led to the question: at what value must the exogenous variable instruments be set in order to achieve a given value of the system endogenous variable? The latter approach gives the optimal solution to the problem of maximizing total system performance on any objective function. The optimal policy approach developed by Tinbergen (1937) is shown in fig. 3.2.

Much of the complexity of the resulting control will depend upon the structure of the system model:

$$\underset{m\times 1}{\mathbf{Z}_{t+1}} = \underset{m\times m}{\mathbf{A}}\ \underset{m\times 1}{\mathbf{Z}_t} + \underset{n\times n}{\mathbf{B}}\ \underset{n\times 1}{\mathbf{U}_t} + \underset{n\times r}{\mathbf{C}}\ \underset{r\times 1}{\mathbf{X}_t} \tag{3.1.1}$$

$$\underset{m\times 1}{\mathbf{Y}_t} = \underset{m\times 1}{\mathbf{E}}\ \underset{m\times 1}{\mathbf{Z}_t}$$

where $\mathbf{Z}_t$ is a vector of intermediate (state variables) (see chapter 2).

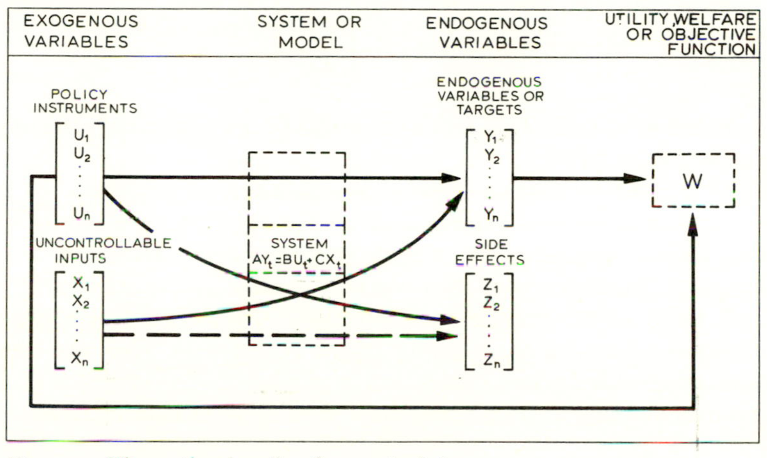

Fig. 3.2 The optimal policy for control, i.e.
Maximize

$$W = AY_t + BU_t.$$

Subject to the model

$$AY_t = BY_t + CX_t + \text{Boundary conditions.}$$

Then the optimal control is:

$$BU_t = AY_t - CX_t$$

or:

$$U_t = (B^{-1}A)Y_t - (B^{-1}C)X_t.$$

The exogenous variables are composed of two sets: those amenable to control and those which are not. The optimum control is a feedback strategy using a combination of the costs of control and system costs as an objective function, and using the system model as a linear constraint. (After Tinbergen 1937, Sengupta and Fox 1969.)

1. If both **A** and **B** are diagonal matrices then the achievement of each goal will be independent of the others, and each instrument has independent effects.
2. If **A** and **B** are triangular, then the model can be solved as a recursive causal chain (Blalock 1964), as a hierarchical policy design.
3. If **A** and **B** are full, the model must be solved by simultaneous equation methods.

In some cases it may not be possible to achieve control of this system in the way intended. Tinbergen (1937) has shown that, in general, the same number of policy instruments are required as the number of independent targets it is required to satisfy. This general condition for controllability can be stated as a condition on the matrix **Q** in

$$[\mathbf{B}, \mathbf{AB}, \ldots, \mathbf{A}^{n-1}\mathbf{B}] \begin{bmatrix} \mathbf{U}_t^{n-1} \\ \mathbf{U}_t^{n-2} \\ \vdots \\ \mathbf{U}_t^{0} \end{bmatrix} = \mathbf{QU}_t \qquad (3.1.2)$$

that it should be of rank n, i.e. that all controls U_t^i are capable of affecting the system (Wonham 1967).

In this chapter we discuss mainly the character of automatic or semi-automatic control systems. Such systems used in isolation are of limited utility in either physico-ecological or socio-economic systems. However, in combination with higher levels of man-machine control and restructuring, they perform the important routine and algorithmic tasks of control where only limited interaction with the decision maker is required. They are thus an extension of the *hard systems* discussed in chapter 2. The interface of these hard control systems with soft systems is developed in later chapters.

3.2 Types of control systems

The overall type of controlled system required in any environment is dependent not only on the control instruments available, but also upon the nature of the disturbances entering the system. These may vary from minor 'random' shifts from target levels to major structural shifts in the nature of the system it is sought to control. Schoeffler (1971) notes that the frequency of recurrence of disturbances is normally related to their magnitude, which has a parallel in terms of time series structures as shown in table 3.1. The general magnitude-frequency relation between environmental inputs into systems and their response is discussed further in chapters 7 and 9. Here we discuss the range of control systems which can be utilized to counteract a given disturbance structure. Fig. 3.3 shows a hierarchy of such control systems.

Table 3.1 Decomposition of disturbances according to frequency and magnitude of occurrence (after Schoeffler 1971, figs 8–13)

Increasing magnitude of disturbances (↑)	Increasing frequency of disturbances (↓)	Disturbances	Time series equivalent
		STRUCTURAL DISTURBANCES	Trend
		Economic changes	
		Schedule changes	Adaptive
		Component changes	
		Behaviour changes	
		PARAMETER DISTURBANCES	Seasonal
		Wearing of components	
		Aging	
		Parameter changes	
		System learning	
		OPERATING DISTURBANCES	Stochastic
		Load changes	
		Drift in operating point	
		Changes in set points	
		Shifts in class and sector forms	
		PROCESS DISTURBANCES	
		Pressure changes	
		Compensations	
		NOISE	Random noise
		Minor random and inexplicable shifts	

(A) *Uncontrolled systems.* These are extremely primitive systems which are susceptible to an uncontrolled stream of disturbance inputs which determine the system output. Disturbance inputs are analogous to the inputs of chapter 2 and may be viewed in continuous systems as a sampled spectrum of processes of differing magnitude which may have an internal correlation structure, or in discrete systems as a time stream of events which may exhibit correlation between successive lag or lead observations. Either case represents the input of stochastic noise disturbances which, as discussed in chapter 2, may affect the system input (ignored variable and measurement errors), the system output (measurement errors) or the system transfer function (misspecification and learning). Each will usually arise from the presence of 'ignored variables' left out of the analysis by decision as to system closure. The correlation structure (autocorrelation and spectrum) must be analysed by statistical methods or Markov type specifications.

(B) *Open-loop control systems.* In this instance some external control is exercised over the operation of the system to condition the output from the stream of disturbance inputs. The output of the system exercises no control

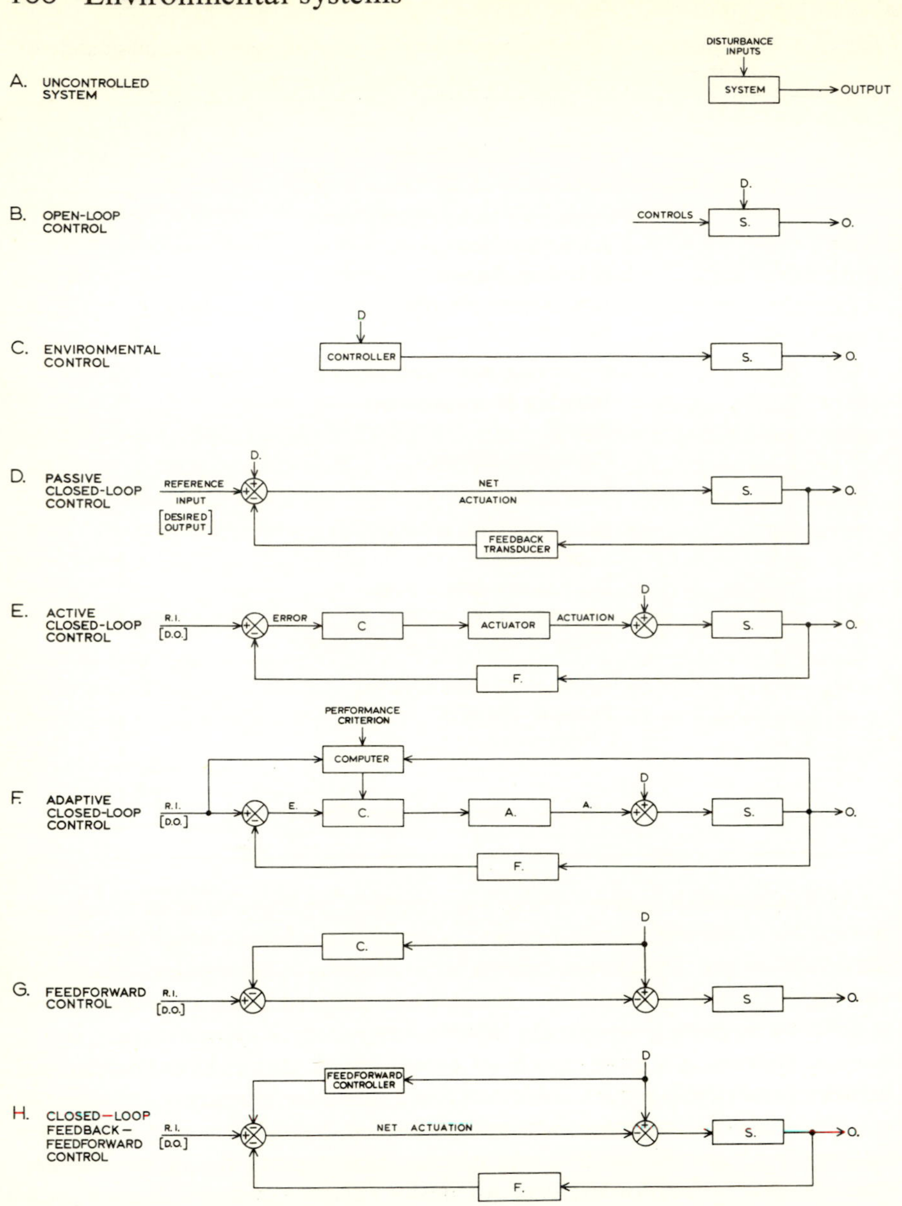

Fig. 3.3 Types of control systems. (Partly after Grodins 1970.)

over the inputs. However, the adjustment is not part of the normal operation of the controller. A set of traffic lights at a road intersection is an example of this type of system. The control system of colour changes, timing of relative length of red and green period, and repetition cycles are all *pre-set*, the calibration of each value being based upon experience of traffic density at the junction. The system can be adjusted only infrequently and is hence vulnerable to unexpected fluctuations and random disturbances. It is, however, very cheap and simple to operate.

(C) *Environmental control.* One simple method of improving open-loop control system performance is to control the disturbances in the system environment, by eliminating them or constraining each disturbance input to a design value. This method is cheap and simple if the disturbances are known and can be controlled, but in socio-economic and physico-ecological systems it is rarely possible to identify all possible disturbances and one is *never* able to control them all.

(D) *Passive closed-loop feedback control systems.* As has been shown in chapter 2, the negative feedback loop provides control in that it compensates against system disturbances (fig. 3.4). This control can be made more effective by, among other things, increasing the loop gain, but there is an upper limit above which the loop becomes unstable and breaks into sustained or even destructive oscillations (Milsum 1966, p. 45). Fig. 3.5 shows the passive open-loop control of a second-order system. Biological examples of such systems are: the pupillary mechanism ensuring a constant flux of light to the retina; the retinal system for light and dark adaptation of the retina to different light intensities; the thermoregulator of warm-blooded animals; enzyme formation in bacteria controlling metabolite flow against variations in carbon source; and regeneration and healing of wounds (Rosen 1975). These control systems thus introduce a number of new features for us:

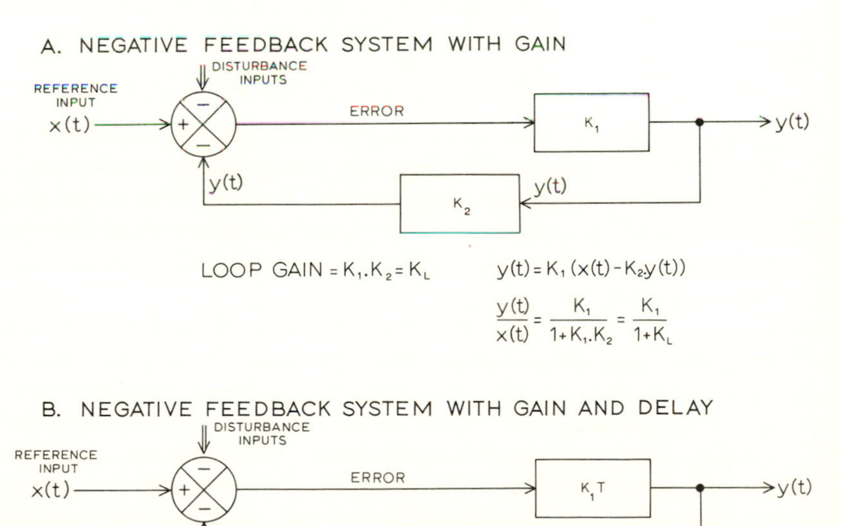

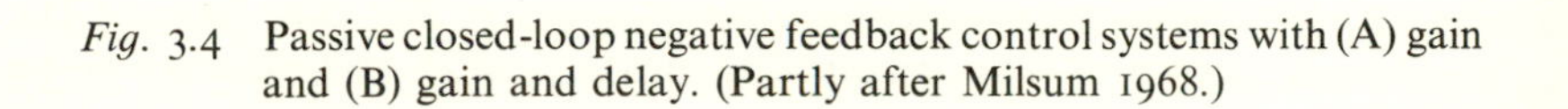

Fig. 3.4 Passive closed-loop negative feedback control systems with (A) gain and (B) gain and delay. (Partly after Milsum 1968.)

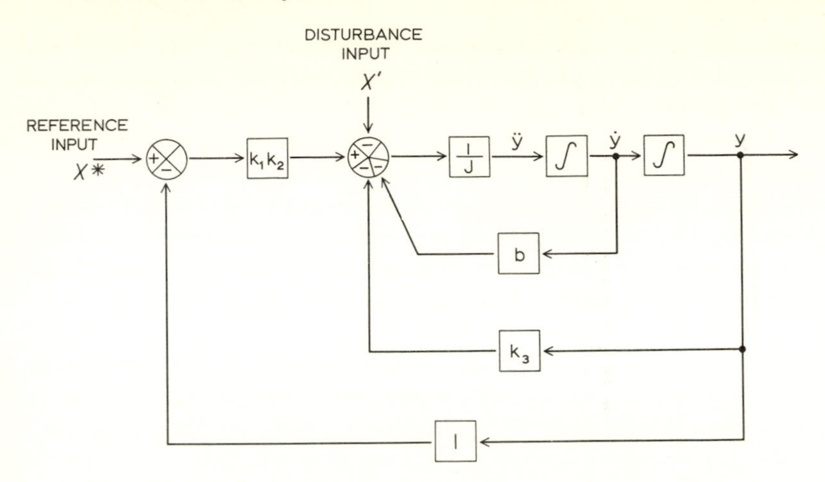

Fig. 3.5 Passive closed-loop control of a second-order system.

(*a*) A *reference input* or 'set point' (r_t) which, in practice, represents the target value setting at which it is desired to maintain the system. In the passive closed-loop case the goals are not forced on the system but arise from 'unpredictable' internal regulations such as homeostasis which determine the fitness of an organism in a particular ecological niche, for example. The reference input may have two forms. First, as a single or multiple constant figure setting the desired values for each controlled output:

$$r^* = r_t \quad (t = 1, 2, \ldots, N).$$

This gives the *regulator problem*. Alternatively, the reference input may be time-varying on a nominal trajectory giving a different value for $\{r_t\}$ for each time t over the planning time horizon. This is the *servomechanism* or *tracking* problem. Servomechanisms are a special form of model-reference system which allow continuous adjustment to a variable trajectory of goals.

(*b*) A negative *feedback loop* which allows the controlled output to be compared with the reference input by means of a *comparator* or summation operator. The loop contains a *feedback transducer* which has two functions: firstly, to sense the actual controlled output and, secondly, to translate it into a form compatible with the reference input.

(*c*) An *error* signal produced by the comparator which, under the action of continued disturbance inputs, leads to a *net actuation* of the system or change in system operation. The closed-loop control system is thus a very special form of feedback system which has the advantage over open-loop and environmental control systems that the value of the system output is continuously compared with the desired value (reference input). If we define the operation of the system S by the equation $Y(t) = SX(t)$, and the operation of the control regulator R as an adjustment of the input $X(t) = RY(t)$, then the total system representation of the closed-loop (system-plus-controller) is given by block diagram reduction as:

$$\begin{aligned} Y(t) &= S[X(t) + \Delta X(t)] = S[X(t) + RY(t)] \\ &= SX(t) + SRY(t). \end{aligned} \tag{3.2.1}$$

This can be written in terms of the input variable alone as:

$$Y(t) = \frac{S}{1 - SR} X(t) \tag{3.2.2}$$

where the operator $S/(1 - SR)$ is termed the closed-loop transfer function. The method pays no attention to the nature of the disturbances and will not be able to cope with extreme events which have not been anticipated. Also, since the system is a feedback one, no control action will be taken until *after* the deviation has occurred, and once the control action begins it will also be retarded by any lag and delay terms within the system itself. This leads to slow implementation of counteracting action. In many cases, the slowness of response may engender a disproportionate increase in control input causing the system output to *overshoot* the reference level which may give a set of subsequent oscillations back and forth – *hunting*. Alternatively, the control action may be too weak, resulting in a slow response time. In each case, the control system is limited since it is not possible to set a control loop *gain* until after the disturbance has occurred.

(E) *Active closed-loop feedback control systems.* These systems are similar to the passive systems except that human control is interjected between the comparator and the system. The reference input is, therefore, an operational target derived from the overall human goals and the objective function of the system. Whilst these goals may be formulated in intuitive as well as mathematical terms, the operational targets of the system are precise figures (with confidence bounds): for example, the acceptable biological oxygen demand level in a river or lake, the target level of gross national product in an economy (see parts II and III). Fig. 3.6 shows the structure and operation of a simple active closed-loop control system in which an attempt is made to control the height of water in a tank ($Y(t)$ – *controlled output*) at a desired constant value ($r(t)$ – *reference input*) in the face of *disturbance inputs* which consist of outflows from the tank at an uncontrolled and variable rate ($Q_0(t)$) (Milsum 1968). A float and connections (*feedback transducer*) relay the output to a visual scale (*comparator*) showing the deviation of the actual output from the desired input. The *controller* uses this error information to formulate a control strategy which is implemented by means of the manipulation of a tap

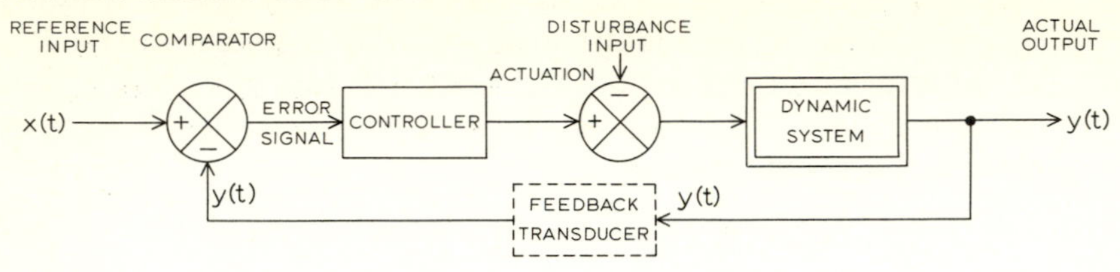

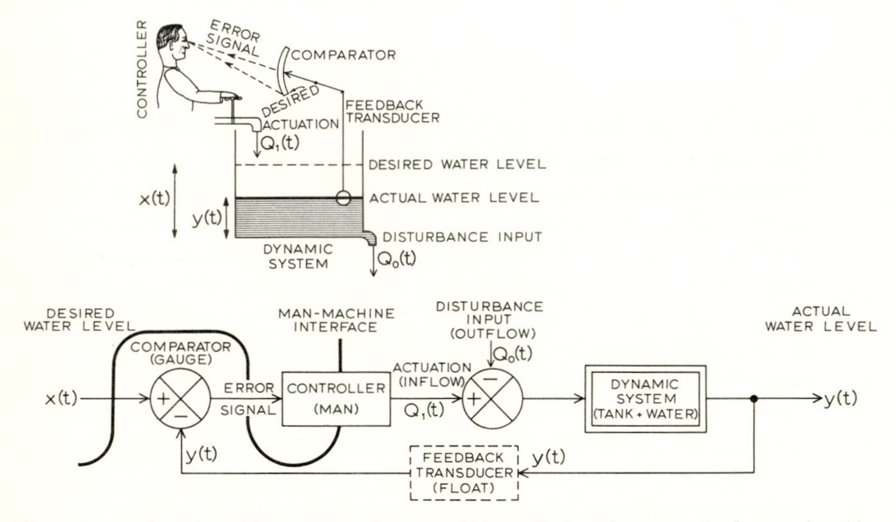

Fig. 3.6 Active closed-loop control system (A) applied to the control of water level in a tank (B).

(*actuator*). This takes the information from the controller, possibly amplifying it and translating it into another energy form, so leading to an *actuation* of the flow rate ($Q_1(t)$) into the tank. It should be noted that it is this actuation which is in competition with the disturbance inputs to yield the actual controlled output. Fig. 3.7 shows a similar system designed to keep a car at a constant speed in the face of the varying effects of gravity (disturbance inputs) as the gradient of the road changes. Figs 3.6 and 3.7 introduce two novel features to the concept of control: firstly, that of man-machine systems and, secondly, that of the harnessing of an external power source. As regards the external power source, it is sufficient to note here that, in terms of environmental control, 'power' may consist either of physical power or socio-economic power, the latter usually operating by means of the former. The availability of external power commonly conditions the extent to which the targets of the control system can be achieved in terms of actual output.

Man-machine systems have been defined as 'a combination of one or more

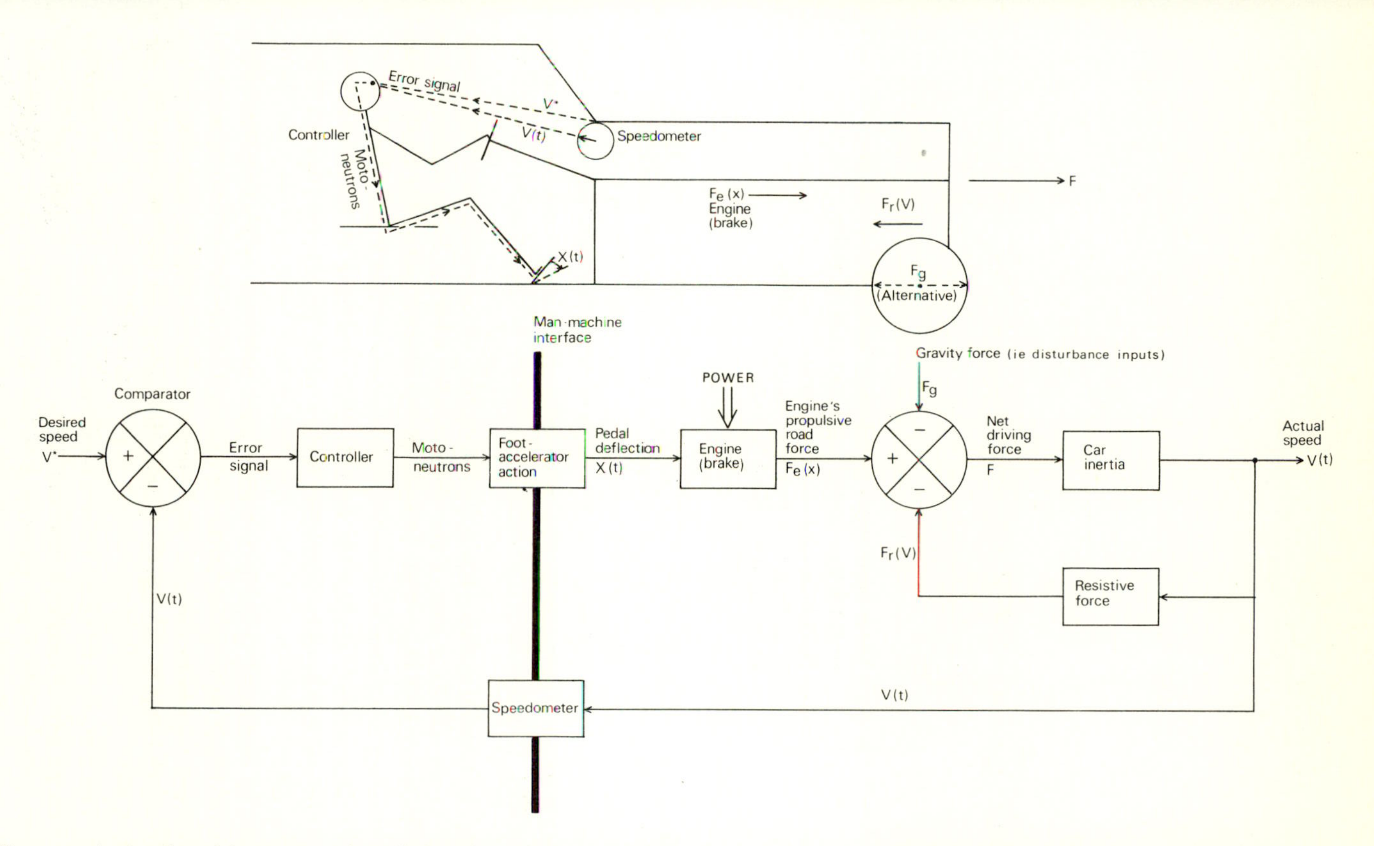

Fig. 3.7 Active closed-loop control applied to the maintenance of a car at a constant speed in the face of varying gravity disturbance inputs. (Partly after Milsum 1966.)

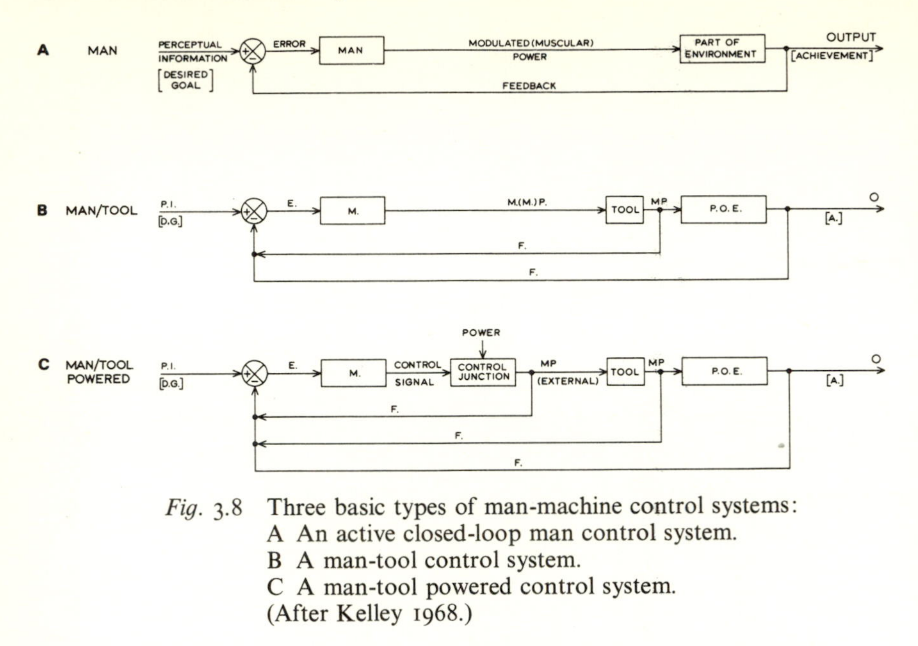

Fig. 3.8 Three basic types of man-machine control systems:
A An active closed-loop man control system.
B A man-tool control system.
C A man-tool powered control system.
(After Kelley 1968.)

human beings and one or more physical components interacting to bring about, from given inputs, some desired output within the constraints of a given environment' (McCormick 1970, p. 5). The study of such systems forms part of the study of *ergonomics* which deals with the customs, habits and laws of work as an approach to the problems of human work and control operations. It is possible to identify a hierarchy of man-machine systems:

(*a*) The manual system, in which man employs modulated muscular power to operate on a part of the environment which is being controlled to produce a change. This change is the output and is fed back to the human controller who employs his senses to check the actual output (achievement) against the desired goal, allowing an error-correcting control signal to be initiated (Kelley 1968, pp. 13–15) (figs 3.8A and 3.9A).

(*b*) The man-tool system, in which man operates as a *controller* using feedback information from the changed environment (outer loop) and from the operation of a tool (inner loop) to release muscular power which operates on the tool, and, through it, on the environment (fig. 3.8B).

(*c*) The man-tool powered or semi-automatic system. Here the controller uses a nest of feedback information to send a *control signal* to a *control junction* at which energy from a power source is released, regulated or modulated as a function of the information supplied from both the goal selection system (i.e. perceptual information and the desired goal) and

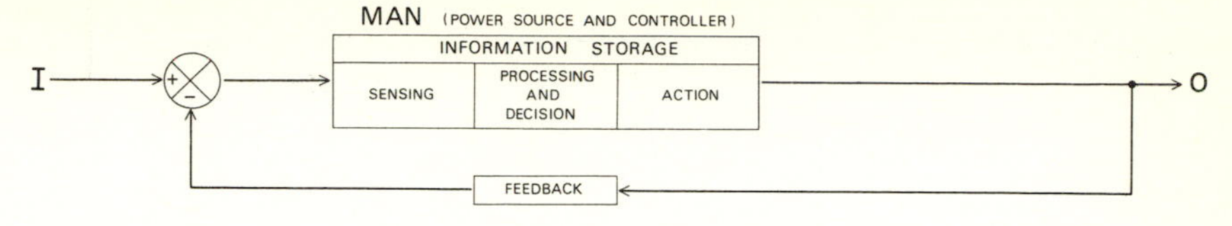

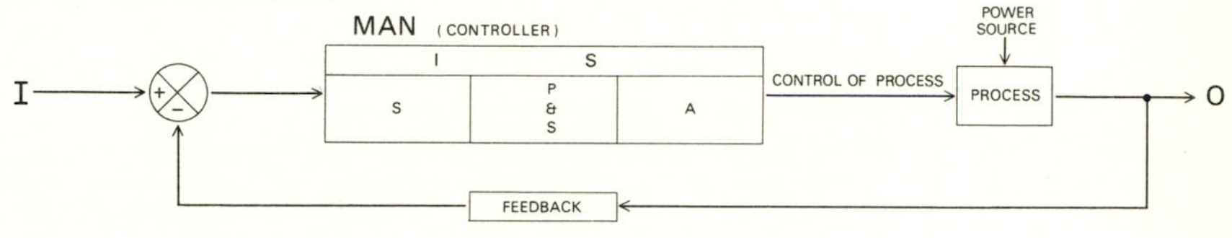

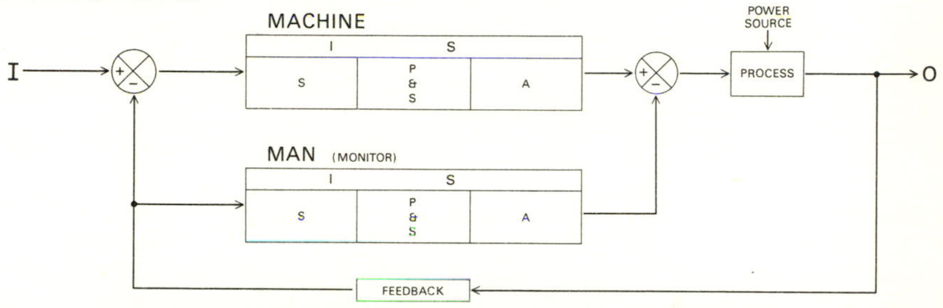

Fig. 3.9 A classification of man-machine control systems into (A) manual, (B) semi-automatic and (C) automatic, depending upon the relative roles of man and machine in the control operation. (After McCormick 1970.)

the feedback sensors. A usual feature of the control junction is that here large amounts of energy are controlled by the use of small amounts of energy (Kelley 1968, pp. 18–19) (figs 3.8C and 3.9B).

(*d*) The automated system. These are systems in which man and machines operate in parallel to improve efficiency by means of each of them contributing in a most efficient manner. Fig. 3.9C shows man monitoring operations controlled by the machine in a way which commonly involves the anticipation or forecasting of the rate of change of the output. Commonly a controller in a compensating system is involved in the triple operation (fig. 3.10) of:

1. Summing the effects of the system output over a period of time.
2. Estimating the rate of change of the output so as to anticipate the control necessary. This is termed *leading*.
3. Comparing the output with the reference input.

A compensating system involving two integrations and an operational delay (fig. 3.10B) is basically unstable and needs anticipation from the

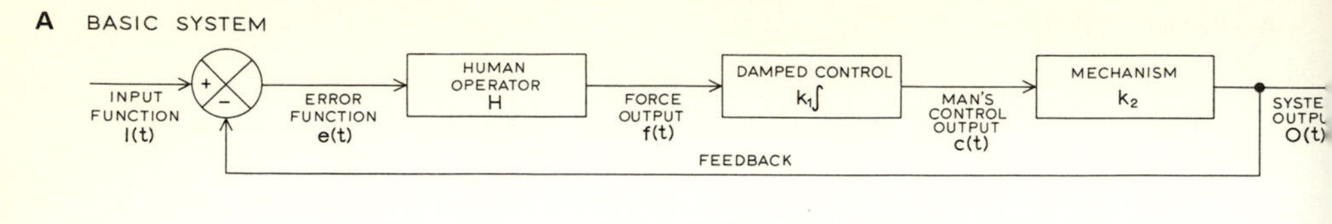

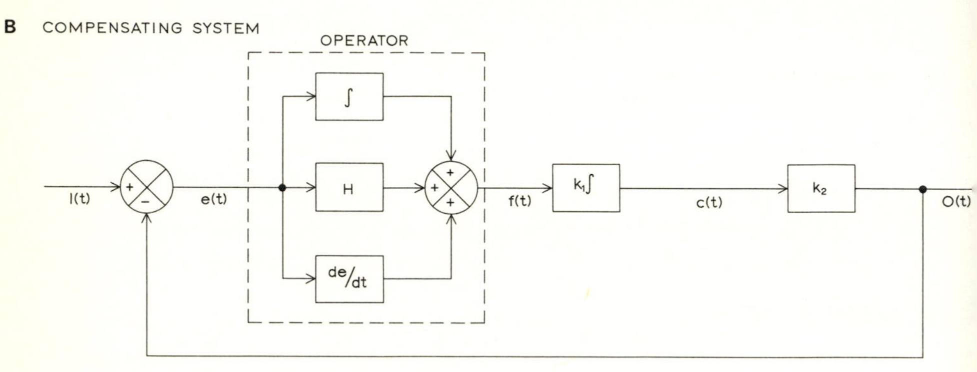

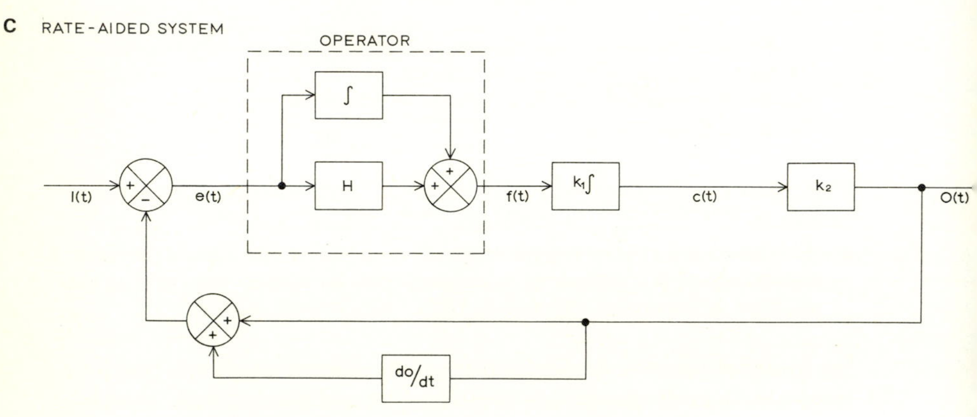

Fig. 3.10 The role of the human operator in control operations:
A The basic active closed-loop control system.
B The compensating system where the operator sums (integrates), leads (integrates) and compares.
C The rate-aided system where automated leading results in quickening.
(After Bekey 1970.)

controller. Automation commonly involves relieving the operator of either of these operations (termed *aiding*). The relief of the operator from the task of differentiation (i.e. estimating rates of change) or of the need to sense and utilize derivative information separately is termed *quickening* (fig. 3.10C) (McCormick 1970, p. 269), in which the controller is not presented with the actual system error but a display which includes some element proportional to the rate of change of the control variable. The logical extension of man-machine systems lies in

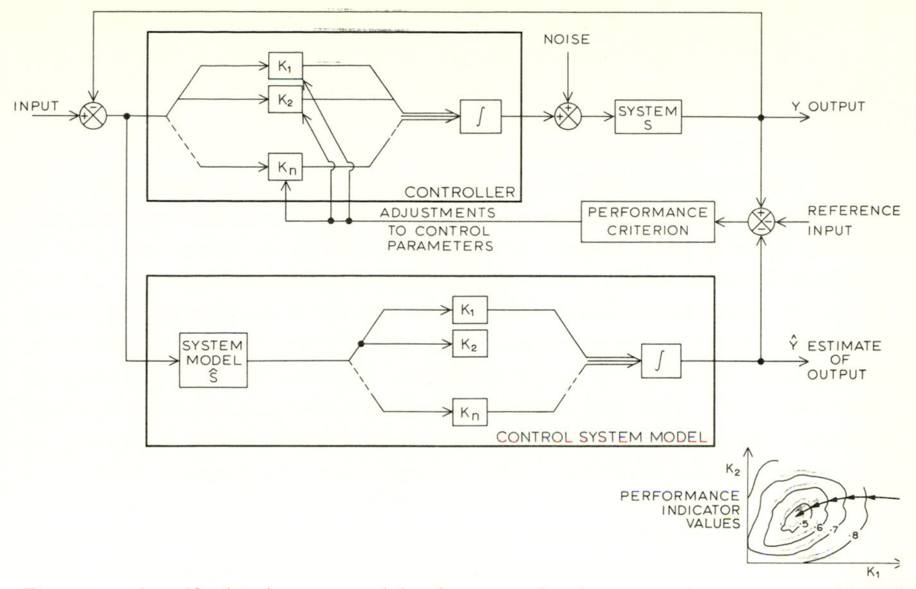

Fig. 3.11 A self-adaptive or model-reference adaptive control system capable of learning. Changes to system parameters are fed back using information derived from steepest descent on the performance indicator (shown in inset). (Partly after Davies 1970.)

the field of *bionics*, the study of life-like systems, which consists of the coalescence of biology and electronics. The major goals of bionics are to adapt man to the environment and *vice versa*, to extend man's physical and intellectual capabilities by prosthetic devices and to replace man by automata and intelligent machines (Van Gierke, Keidel and Oestreicher 1970, p. 23).

(F) *Adaptive closed-loop feedback control systems.* These are collaborative man-machine systems – for example, large computers with multiple access interaction, whereby the structure of the system itself is capable of modification so as to produce the required output from the given inputs. In terms of environmental control, the closest analogy is to be found in chapter 10 where revolutionary changes in decision making systems are treated. Since the structure is evolving in time, the structure of the controller should parallel and model this evolution. This has given rise to self-adaptive or model-reference adaptive control systems, as shown in fig. 3.11. Tracking of systems disturbances generates not only feedback control signals but also shifts in the structure of the control model, in that the control deviations are used to adjust the system parameters. Thus the overall control system performs the threefold function of identification of significance of disturbances, decision, and modification. Such systems can be automated only in simple cases using forms of model reference objective functions (performance indicators) as shown in fig. 3.11. This is an example of a learning system which moves successively down-gradient on the objective function (performance indicator) as shown. All such learning systems are gradient algorithms which adjust controller structure to minimize the deviations from reference input targets automatically – for example, the Van der Grinten technique (Davies 1970). More commonly in environmental systems the identification-decision-modification procedure must be controlled in terms of cognitive and decision making systems as discussed in chapters 5 and 6. A special category of feedback adaption occurs when the controller is able to modify its level of reference input (performance measure) as well as to modify control system response, which also can be fed back. The overall effect is to improve both the performance of the system and the definition of performance itself. This often takes the form of enlarging or modifying the class of environments in which control is maintained. A biological example is immunity when a 'memory' of past exposure is fed back to counteract the effects of re-exposure, the memory after one exposure being an adjustment of the performance set point. Environmental feedback adaptive control is frequent with immunity to antibiotic, insecticide and toxic materials (Rosen 1975) and is discussed further in chapters 7 and 9.

(G) *Feedforward control systems.* Each of the three feedback control systems – passive, active and adaptive, suffer from the same disadvantage that they correct for system deviations from the target values after a deviation has occurred. Whilst such control systems are very useful in industrial and military applications, they are far less appropriate to environmental systems since the fundamental human ability to anticipate is ignored. Hence, it is preferable in many cases to attempt to detect disturbances before they occur and before a deviation from the target value can begin. This is in contrast to the environment control system which attempts to *prevent* disturbances. The feedforward controller shown in fig. 3.3G detects a disturbance which will enter the system, and, before it has had time to affect system response, a modification in system input is made to counteract the system disturbance. Examples of this type of controller are common in many industrial processes where a fluctuation in the level of one raw material input into a product can be counteracted by change in another controllable variable, e.g. fluctuations in temperature into a chemical plant may be counteracted by changes in water feed as a control variable. Biological examples are the preparation of plants for winter on the basis of day length rather than ambient temperature; the movement of photophobic (light sensitive) mammals to low light areas to maintain high humidity; and reflex actions (Rosen 1975).

The main advantage of the feedforward controller is that it uses the human ability to anticipate the course of events. Thus, in detecting a disturbance before it occurs, it is theoretically possible to achieve perfect control. However, as in the case of environmental control, the method requires the identification and measurement of all input disturbances. This implies extensive knowledge of the system operation. The most serious disadvantage, however, is that there is no comparison of the system output with the reference target to determine how far the control system is effective. Feedforward control is therefore seldom used in isolation. It can be combined with each of the environment, passive and active feedback and adaptive closed-loop systems (discussed above; B to F).

(H) *Closed-loop feedback-feedforward control systems.* These represent the most important of the feedforward combination systems, and are most widely used in current practice. The feedback-feedforward schemes complement each other since serious deviations can be detected prior to their effect on the system, and feedback elements can counteract any imperfections in the feedforward scheme and the effects of other (unmeasured or unmeasurable) disturbances. The structure of this system is shown in fig. 3.3H. The net actuation of the control system is a combination of counteraction of anticipated disturbances and damping of past disturbances.

In practical systems control, it is usually necessary to build up a complex set of feedback, feedforward adaptive and environment control systems capable of representing the whole range of system disturbances and changes which may occur. This requires the development of hierarchical and nested control systems.

(J) *Adaptive feedback-feedforward control systems* (Universal adaptive systems). These control systems are capable of using past behaviour to modify future response and/or the set point performance measure. The internal model is capable of reprogramming itself and must employ additionally some feedback elements especially the use of the fed back error signal (Rosen 1975). Additionally the system has a set of 'effectors' (Kuhn 1974) with which it can manipulate the environment and create new sensory modalities for itself by fabricating instrumentation and sensors to gain new knowledge about environmental behaviour. Human behaviour and resource management of the man-environment interface are examples of extremely complex universal adaptive systems which continually modify goals, sensing and response characteristics (chapters 7, 8 and 9).

3.3 Nested feedback control systems

One of the major impediments to the development of a viable theory relating to environmental control systems has been the difficulties in modelling them in terms which allow the application of, firstly, the philosophy and, secondly, the techniques of control engineering. These difficulties have included:

1 The decomposition of environmental systems into subsystems simple enough to be approximated by linear models (see chapter 7).
2 The manner in which the simplified subsystems should be assembled and connected.

3 The specification of both the desired output and the reference input (see chapter 8).
4 The point of application of the power source.
5 The locations within the system at which the environmental disturbance inputs operate (see chapter 9).
6 The statistical spectrum exhibited by these disturbance inputs and the relationship between this spectrum and planning constraints.

A further problem facing the design of environmental control systems has been the emphasis on the economic and social decision making which precedes environmental planning. This complex and, to an extent, *ad hoc* process often leads to environmental intervention which is so arbitrary and heavy handed as to involve action which is not only environmentally unwarranted but economically profligate as well (see chapter 6). It is to be hoped that the application of control theory to environmental development will encourage planners to give increasing attention to the structural character of the systems which they wish to control so that human intervention can be deft as well as effective.

One possible link between environmental modelling and control engineering is suggested by nested closed-loop guidance systems, such as those which have been presented by Kelley (1968, pp. 27–34, 118–20). The loops are centred about the control junction, the location in the system at which external power is applied (fig. 3.12A), on the input side of which is a hierarchy of human controllers and on the output side a hierarchy of *control effectors* or *actuators*. The latter are subsystems of the system to be controlled and represent the elements which successively apply the energy provided by the power source to modify the environmental system. The effects of each effector interact with environmental disturbances and are fed back to the appropriate controller by means of sensors. Kelley thus views the control system as a nested hierarchy of successively higher-order systems in which the control of the smaller, more immediate events employing less energy in the inner lower-order systems brings about larger, more distant, events involving more energy in the outer higher-order system loops. Thus the output from the inner loops proceeds to the outer loops as the hierarchy of control is ascended. Although the processes of the inner loops are generally of lower magnitude and of higher frequency than those in the outer loops, control of the inner loops is more possible and, by influencing the rates of change of the higher-order loops, provides a wider range of choice in influencing the behaviour of the outer loops. The inner loops are susceptible to short-range planning in order to bring about longer-term changes in the outer loops, but it is also clear that, even without the introduced effects of environmental disturbance inputs, such planning usually involves complex and flexible inner loop restraints which are required to produce more simple longer-term changes in the outer loops. Fig. 3.12B illustrates this concept with reference to the conventional guidance

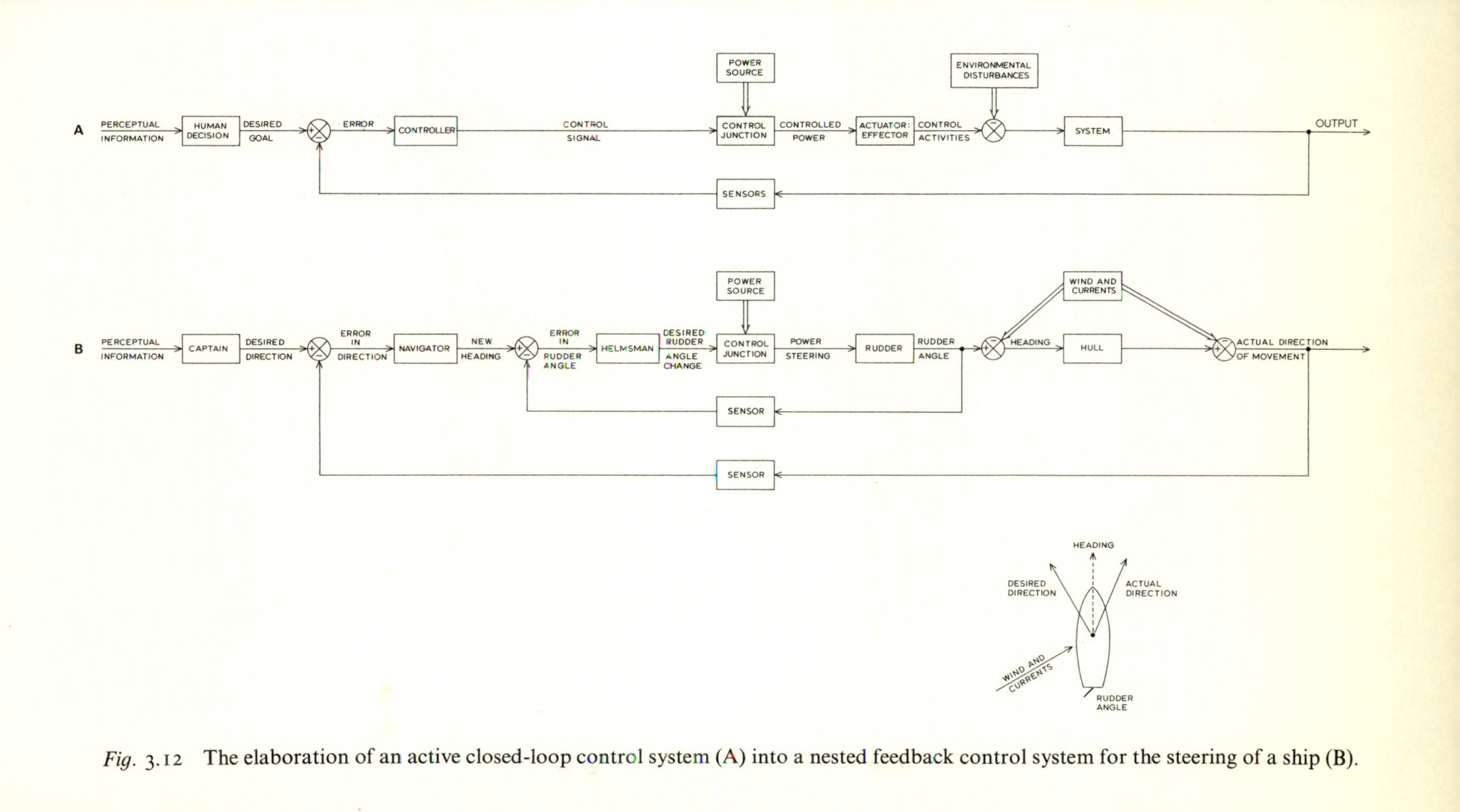

Fig. 3.12 The elaboration of an active closed-loop control system (A) into a nested feedback control system for the steering of a ship (B).

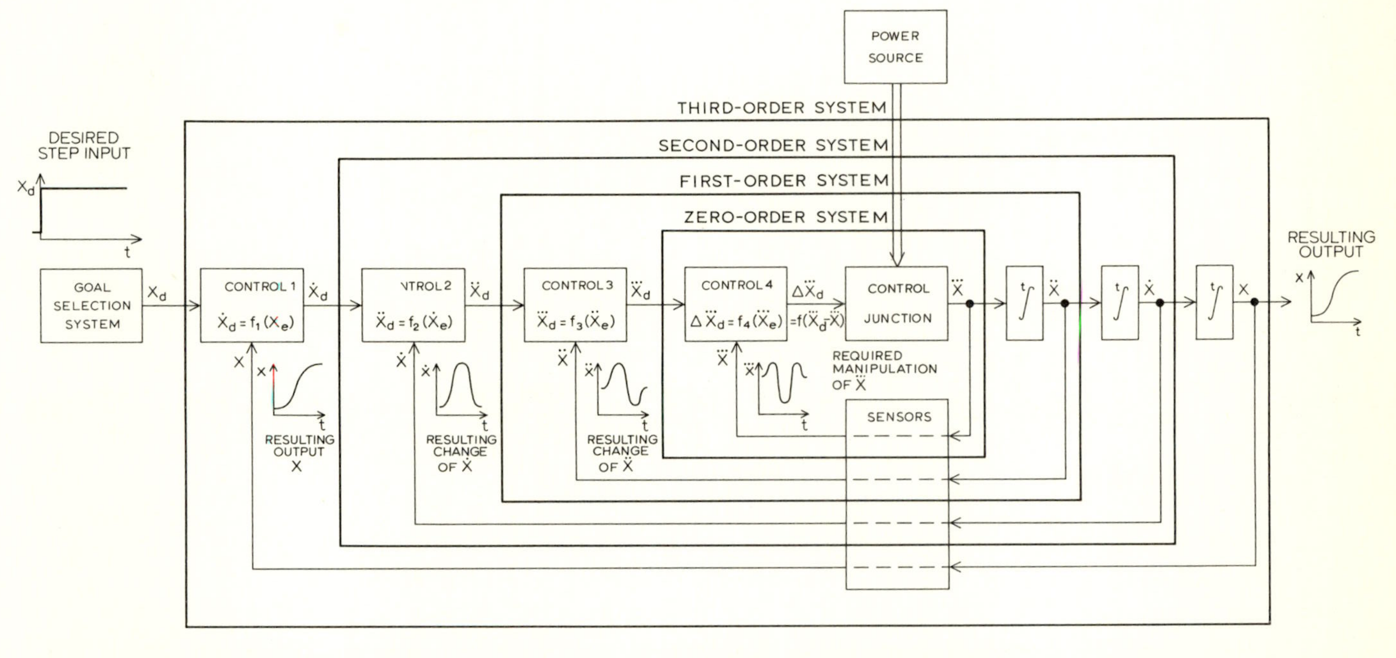

Fig. 3.13 The general structure of a complex nested feedback control hierarchical system. (After Kelley 1968.)

control of a ship by the navigator and helmsman (controllers), through the rudder and hull (control effectors) under the influence of winds and currents (environmental disturbances), and in harmony with the orders of the captain (desired goal). Fig. 3.13 depicts a similar but more complex and general instance of a nested control hierarchy showing the variations of rate of change of acceleration ($\dddot{x}$) required to be applied to the inner loop of a nested control system to bring about successive changes in acceleration ($\ddot{x}$), velocity ($\dot{x}$) and distance (x) in accordance with a stepped reference input (X_d) of desired distance in a given time. This model assumes that direction of movement is true, that position is entirely dependent on acceleration (positive or negative) involving changes in velocity, and that movement is not complicated by the introduction of external environmental disturbances. As we have seen, the latter may be envisaged simply as, for example, changes in the gravitational effects as a land vehicle goes up or down hill or in current effects acting parallel to the direction of movement of a ship. Just as these effects are randomly applied to a moving vehicle within an estimated magnitude and frequency distribution, so in nature environmental disturbance inputs occur in a wholly or partly random manner within a determinable range of magnitudes and frequencies. Fig. 3.13 shows the complexity of operation of inner loops of such a nested hierarchy which are necessary to effect relatively simple, longer-term outputs involving the outer loops.

The application of such a control model to environmental systems is in its infancy but, in theory at any rate, its potentialities are great. Almost all physico-ecological and socio-economic systems may be viewed as partaking of something of this hierarchical structure with more accessible and susceptible subsystems being embedded in more important but more remote systems. Thus, for example, one can conceive of the long-term control of erosion in a catchment being effected by short-term measures to control debris movement in the channel subsystem, or the long-term control of a regional economy by the short-term regulation of the investment subsystem. Even more interesting possibilities are explored in chapter 5 where the addition of a series of feedforward loops and batteries of environmental disturbance inputs and action outputs allow, at least in a qualitative way, the conceptualization of very complex systems.

3.4 Criteria for efficiency of control

It is usually necessary to know not only whether and how a control system can exert a regulatory influence, but the efficiency of the controller in achieving the measure of control which is desired. A number of criteria are of importance in this connection (Lange 1970):

1 Speed of control. How fast a system disturbance is eliminated and restored to equilibrium (steady state) value.

2 Precision control. How closely the regulator can achieve the desired output pattern. This is especially related to the degree of noise entering the system.
3 Reliability of control. Under what conditions the regulator will function correctly and when it will break down.
4 Stability of control. Under what conditions the controller will damp or amplify disturbances.
5 Sensitivity of control. Sensitivity of the regulator to small deviations.
6 Cost-efficiency of control. How useful the control system is in achieving control within given cost limitations.
7 Control range. Under what range of disturbances the regulator will give cost-effective control.

The overall efficiency of the control action (feedback or otherwise) on the system being controlled can be judged in two main ways. The first method is open-loop study involving the construction of *ex post* indices of regulator efficiency (coefficient of variation, relative precision, damping, etc.; see Lange 1970), and the second a closed-loop study of the transfer function properties of the combined system plus controller (the closed-loop system characteristic polynomial). If the original system is controllable then the closed loop (system plus controller) can be shifted to any assigned or arbitrary set of transfer

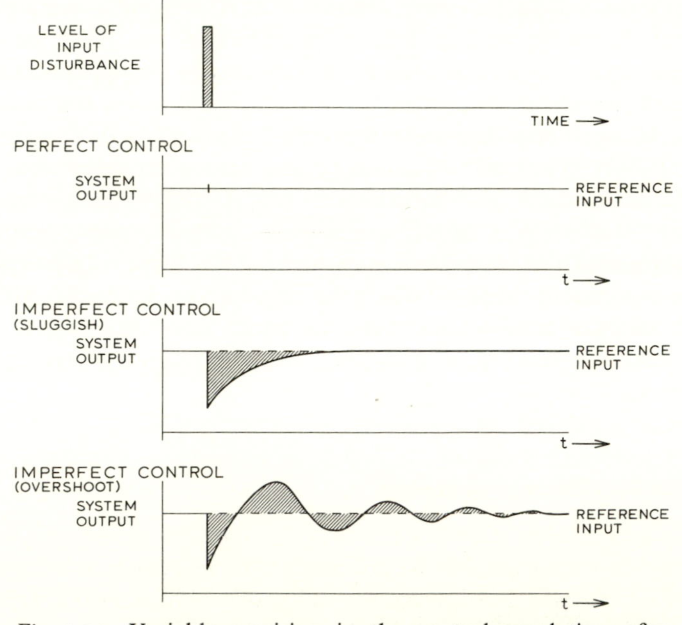

Fig. 3.14 Variable precision in the control regulation of a system in the face of a unit impulse disturbance. This may vary from perfect control to imperfect control of either the sluggish or the overshoot type.

function characteristics (Wonham 1967). It should therefore be possible to design the closed-loop system character arbitrarily closely to the desired response structure. One such method is the *Guilleman-Truxal method* which is a special case of more general state variable feedback (Schultz and Melsa 1967) using modal control. Using equation (3.1.1) this is represented by the ability to assign arbitrary eigenvalue location to the matrix $\mathbf{G} = [\mathbf{I} - \mathbf{A}]^{-1}\mathbf{B}$ by an appropriate choice of feedback gains $\mathbf{B}$. The graphical plot of the pole location of $\mathbf{G}$ under the influence of $\mathbf{B}$ gives the Bode and root-locus methods of pole and zero assignment which are especially useful in stability studies (e.g. in stabilization of economies, see chapter 8).

The closed-loop characteristic polynomial, or *ex post*, indices of control system performance can be used to examine each of the efficiency criteria mentioned above. The relative speed and precision of a number of controllers in the regulation problem of returning the system to its reference level are shown in fig. 3.14. In general, the form of the response of the control system can be evaluated in terms of a number of characteristics of the response function under control. A typical system response to control is shown in fig. 3.15. This can be compared with the response to other controllers in terms of the following parameters:

1 Delay time: time to reach half final time.
2 Dead time: time before any control response occurs.
3 Rise time: time to reach first crossing of reference level.
4 Steady-state error: tolerance level of $\pm\alpha\%$ of reference level (usually $\pm 5\%$).
5 Settling time: time to reach point at which controller is within $\pm\alpha\%$ of reference level.
6 Percentage overshoot: ratio of the difference between maximum value of controller response and the steady-state value.

If we are interested in the problem of how fast and efficiently the controller responds to changes in reference input (the servo problem), it may be necessary to invoke alternative performance criteria. The servomechanism response to changing levels of desired output is shown in fig. 3.16. The *reliability* of controllers depends upon their ability to function over the whole time range of interest, and over all ranges of disturbances possible. The reliability of control system operations can be increased by replicating the complexity of coupling and nesting of the real-world system, for example, by cross-coupling in the regulator (Von Neumann 1952). Many control instruments will be unreliable over certain ranges of system operation, for example nonlinear systems, but it is possible to construct regulators with any required degree of reliability from unreliable elements, even for the policy instruments usually available in economic regulation. This is based on the Von Neumann '*design principle*' which uses a series of *reserve elements* coupled in parallel, but with alternative coupling or switches as shown in fig. 3.17A by

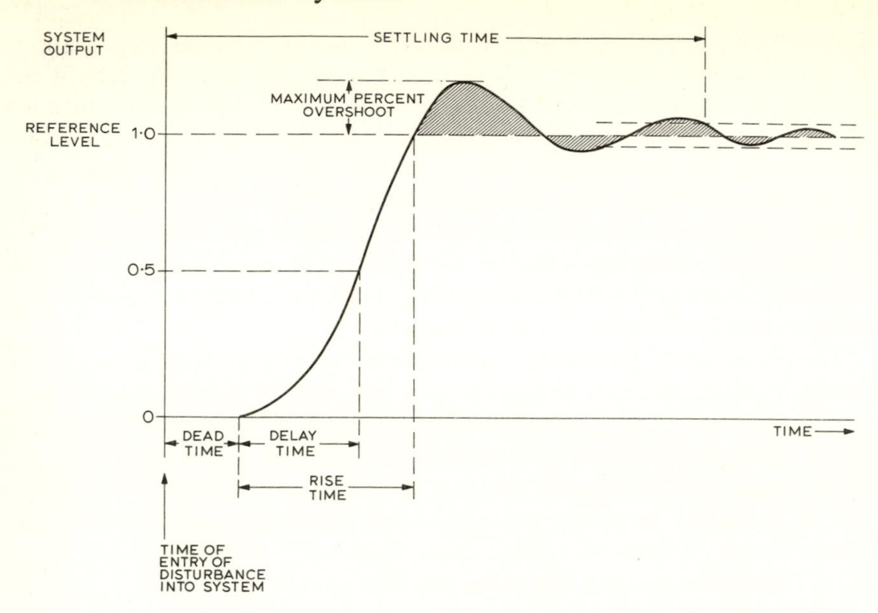

Fig. 3.15 The features of a typical response of a system to control action.

the Ⓥ symbol (see chapter 7). Examples are the replication of components in electrical systems, the many energy pathways in biological systems, and bureaucratic multiple-access systems in socio-economic control. If one pathway through the system is inoperative or fails, it is possible to switch to an alternative. This greatly reduces the possibility of failure, catastrophic collapse or blocking (see chapter 9). In such switching systems the output of the regulation elements is determined by any one of the constituent transfer functions, while the remaining system transfer function elements do not influence the control output, but are in reserve to be used if the active element fails to produce the desired control output. 'The reliability of the system (P) increases as the number of elements coupled in parallel in alternative ways increases' (Lange 1970, pp. 156–7) and is given by the formula

$$P = 1 - \prod_{i=1}^{m} (1 - S_i) \qquad (3.4.1)$$

where S_i are each of the control system m subelements. Reliability increases with the level of redundancy or switching built into the system through parallel component replication. The reliability index P for a system in series is much lower than for a replicated system, and decreases with the number of components present as S_i^n (Chorafas 1960), as shown in fig. 3.18. For example, a series of systems with $S_i = 0{\cdot}99$ mean reliability of components has an overall reliability of $P = 60\%$ for 50 components, but only $P = 19\%$ for 200 components, whereas a system in parallel will achieve 60% reliability overall, even with a mean component reliability of only 0·40 for anything upwards of

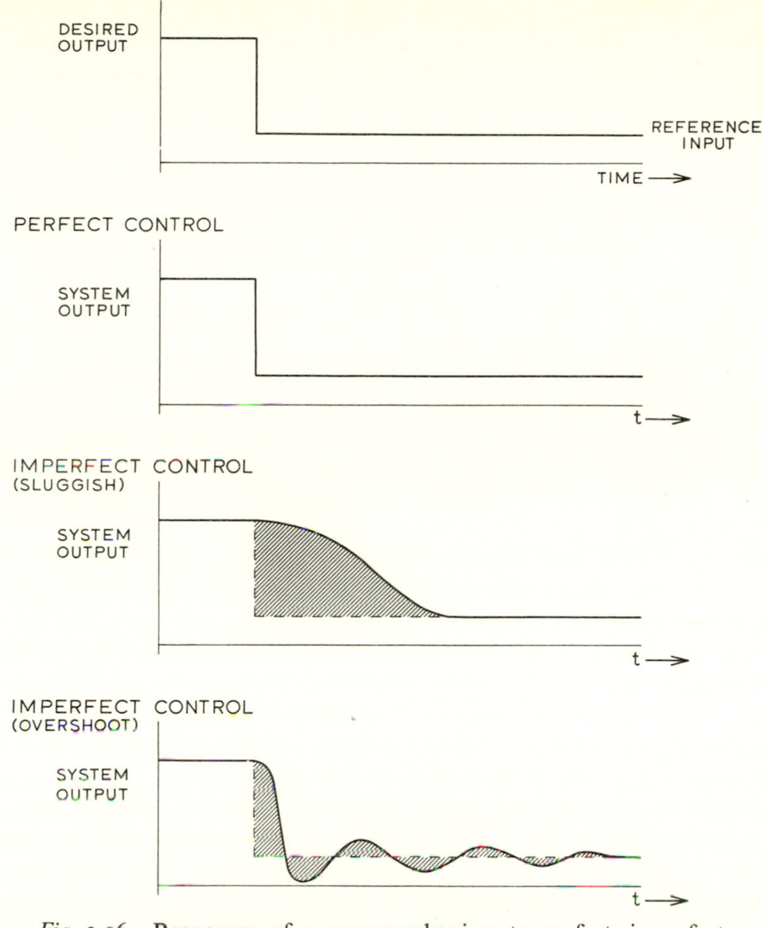

Fig. 3.16 Responses of a servomechanism to perfect, imperfect sluggish, and imperfect overshoot control action in the face of a step reference input.

two components. This has important implications for the response and control of environmental systems which are characterized by a multiplicity of nested parallel subsystem loops. In stochastic systems these concepts must be extended to encompass the relative variance around component performance targets (Chorafas 1960). This concept can also be extended to increase the reliability of complex multivariate control systems in which a whole set of regulators is connected in series, as shown in fig. 3.17B. For N regulators in series, each with m_j parallel alternative regulator elements, the reliability index is now (Lange 1970):

$$P = \prod_{j=1}^{N} \left[1 - \prod_{i=1}^{mj} (1 - S_{ij}) \right]. \tag{3.4.2}$$

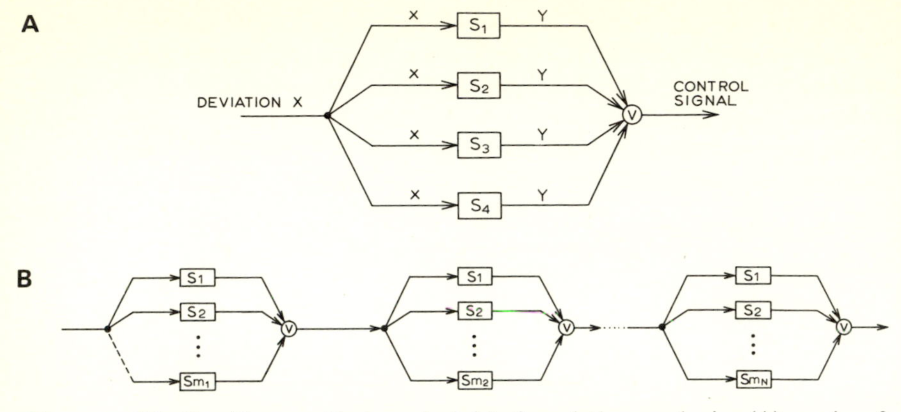

Fig. 3.17 The Von Neuman 'design principle' of regulation employing (A) a series of reserve elements coupled in parallel but with alternative switching (indicated by the symbol Ⓥ), and (B) a set of regulators in series each employing parallel alternative coupling. (Partly after Lange 1970.)

In socio-economic and physico-ecological systems, the range of system variation is very wide (chapters 7 and 8), but the switched regulator gives one method of replicating the complexity. The switching regulator also parallels the many overlapping controllers and regulators which ensure the stability of environmental systems, but when applied to the man-environment control system, switching between controls may involve 'bumps'.

In chapter 2 the differences between stable and unstable systems have been used as special cases of the effect of the configuration of the sign and magnitude of the transfer function feedback parameters. The addition of control components to the system response in the closed-loop case usually has the aim of improving stability, reducing oscillation, improving speed of response, etc. Although Wonham (1967) has proved that a system is stabilizable if all its unstable modes are controllable, it is necessary to compare and evaluate controllers to determine if stability has indeed been improved, or if, under certain conditions, the closed-loop system may have more critically unstable modes of behaviour than the original open-loop system. Stability is most critical in the feedback case and can be determined by examining the closed-loop transfer function $S/(1 - SR)$ given in equation (3.2.2) (where S is the system transfer function and R the regulator transfer function). The overall condition for stability is $|SR| < 1$, or $|R| < 1/|S|$ (Lange 1970). Hence the absolute power or gain of the controller must be less than the reciprocal of the absolute power of the system. Fig. 3.19 shows the general stability conditions for the closed-loop system for various magnitudes of the combined system plus control transfer function SR under the effect of system disturbances.

The sensitivity of the control system will be closely related to its stability (pole and zero location) and to its reliability. Most macro-systems are highly insensitive (Emery 1969, p. 97) because of the high degree of system feedback.

Component replication may aid control if a control input has a large number of pathways to achieve a desired output, but frequently the complex redundancy of environmental systems diffuses the available control instruments so that they become ineffective in achieving the desired control payoff. This certainly applies to most human systems (chapter 8) and to the complex man-environment interface discussed in chapter 9.

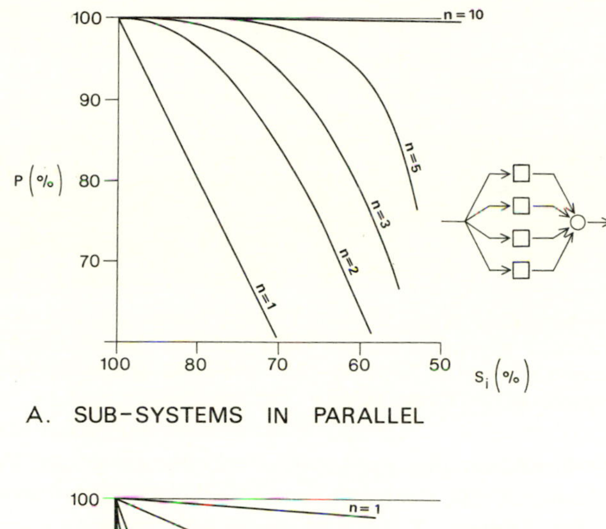

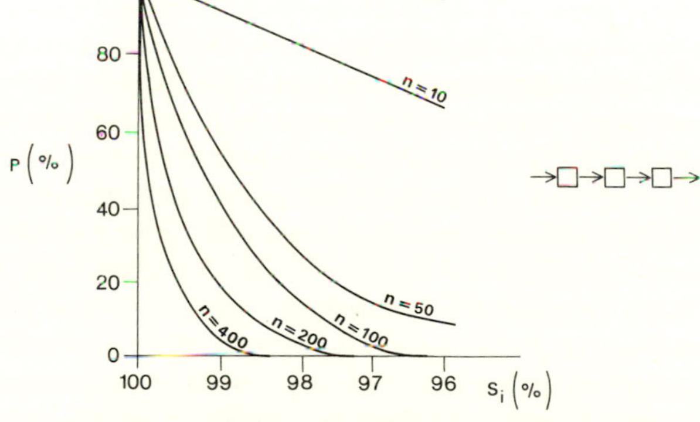

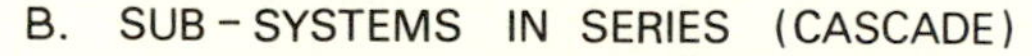

Fig. 3.18 The reliability $P(\%)$ is determined by the number (n) of subsystems (S_i) ($i = 1, 2, \ldots, n$) in a larger system, and by their configuration. Reliability is greater when the subsystems are arranged in parallel (A), rather than in a cascading series (B). For parallel systems there is a direct relation of reliability to the number of subsystems n, but for cascading systems there is an inverse relationship.

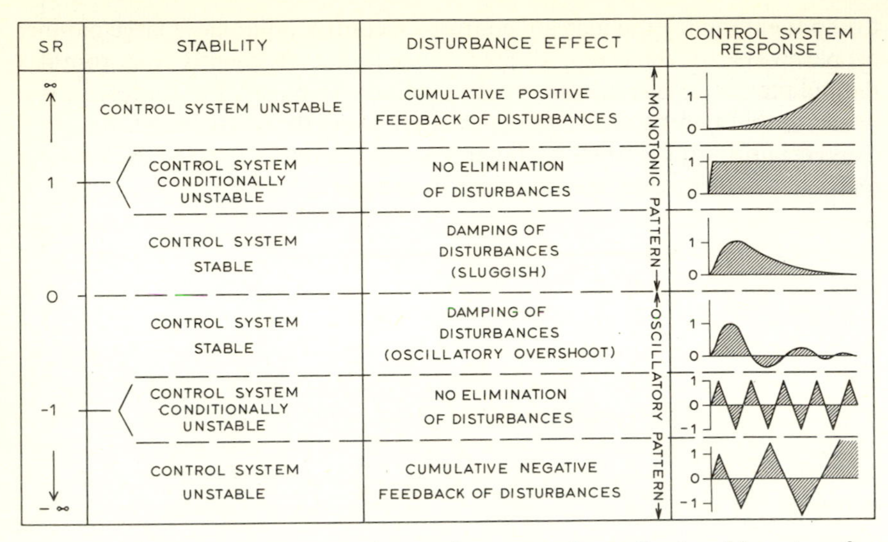

Fig. 3.19 Stability conditions and control system regimes of a closed-loop transfer function system for various magnitudes of the combined system plus regulator transfer function (*SR*) under the effects of different disturbances. (After Lange 1970.)

It is most important in all control system design to consider the cost-efficiency of control. The straightforward feedback solution which gives the control equation $U(t) = -RY(t)$ is unimportant in most socio-economic and physico-ecological systems. Such feedback control involves no consideration of the 'costs' of employing that control and it will be possible to attain any desired control target. There is no constraint on the magnitude, speed or cost of the control employed and an optimization problem of trading off deviation from target for costs does not arise. In fact, any desired system output setting can be achieved by 'piling on' unlimited control (Kalman and Koepcke 1958). In most control applications it is necessary to take into account the generalized 'costs' of control which limit the area, range and possibilities of control. 'Costs' may be in terms of economic cost, effort, political cost, energy, etc. The control problem then becomes the problem of minimizing the costs of control input, whilst keeping as close as possible to system reference levels. This overlaps the evaluation stage of decision making systems (chapter 6). In addition restrictions on the area or range of control may arise through non-linearities, time changes or spatial variations, and it may be necessary to adjust the controller to operate under different criteria, in different space-time locations, or over different system magnitudes.

3.5 Types of control strategies

The object of the controller design, as we have seen, is to set up an automated, quasi-automated or man-machine system which is capable of manipulating a set of control inputs to achieve a target setting, or to track a predetermined system course. To do this, the controller must incorporate a *strategy* which determines how the controlled system will respond to disturbances from target of any form or magnitude. The control strategy can be built into a control chart (Barnard 1959, Box and Jenkins 1970), giving a set of simple rules for controller adjustment under given disturbance conditions. This type of control necessitates a 'man in the loop' and requires decision making criteria, as discussed in chapter 6. It has been the major form of control implemented in most socio-economic systems.

Alternatively, it is possible fully to automate the control system strategy if the controller response can be expressed as a simple mathematical function, or as a set of such functions. These controllers are adjustments based on the *loss*, *error criterion*, or *maximization rule*. The rules for choosing the most appropriate criterion can be applied in one of two general forms; *suboptimal* and *optimal* control. In suboptimal control the decision rule is chosen on some informed but arbitrary basis, which can yield complete cancellation of input disturbance only under fortuitous circumstances. The optimum controller, however, provides for the possibility of complete elimination of disturbances from target.

In each case the difference revolves around the decision rule for choice of control strategy. If we define the value of the control input as $U(t)$ and the system output by $Y(t)$ and assume that the control variable is governed by the same system transfer function $S(z)$, then the choice of feedback controller of the system

$$Y(t) = S(z)U(t) \tag{3.5.1}$$

is given simply by

$$U(t) = -k(z)(Y(t) - r(t)) \tag{3.5.2}$$

where $k(z)$ will, in general, be some function of the transfer function parameters $S(z)$ and of time lag, and $r(t)$ is the reference level at time t. The *control law* given by equation (3.5.2) sets the control input $U(t)$ at a level in some sense proportional to the level of deviation of the input $Y(t)$ from its reference value. As Williams and Wilson (1978) note, the control problem most frequently encountered in environmental systems, economics and regional science may be written in the general form:

$$\max_{\mathbf{U}(t) \in V} J(\mathbf{U}) = \int_{t_0}^{t_f} \underbrace{f_1(\mathbf{Z}(t), \mathbf{U}(t), t)\, dt}_{\text{control trajectory}} + \underbrace{f_2(Z_0, t_0)}_{\text{initial conditions}}. \tag{3.5.3}$$

The maximum value of this functional over all controls

$$\mathbf{U}(t) = [U_1(t), U_2(t), \ldots, U_n(t)]$$

is subject to four conditions:

1 The system state evolution equation – as defined, for example, in equation (2.2.13):

$$\frac{d\mathbf{Z}(t)}{dt} = \dot{\mathbf{Z}}(t) = f_1(\mathbf{Z}, \mathbf{U}, t). \tag{3.5.4}$$

2 The boundary and initial condition:

$$\mathbf{Z}(t_0) = \mathbf{Z}_0. \tag{3.5.5}$$

3 Feasibility conditions, that the control trajectory $\mathbf{U}(t)$,

$$\mathbf{U}(t) \in V$$

is within the class of admissible control strategies V.

4 System regime conditions on the state vector $\mathbf{Z}(t)$ such that

$$\mathbf{Z}(t) \in \mathbf{Z}, \quad (t = t_0, t_1, \ldots, t_f)$$

is within the range of system behaviour with known equations over the time interval from the initial to the final time (t_0 to t_f).

The choice of functional, control instruments, control weighting, control interval and maximization tolerance all determine the form of control strategy adopted.

3.5.1 SUBOPTIMAL AND OPTIMAL CONTROL STRATEGIES

In suboptimal control, various feedback and feedforward strategies are built into the controller on the basis of the transfer function properties and desired control system response characteristics, but no consideration is given to control costs or to the optimality of cancellation of disturbances. A number of strategies are possible, each involving a different choice of $k(z)$ in equation (3.5.2):

1 On-off action (bang-bang control):

$$U(t) = \begin{cases} -k_1(Y(t) - r(t)), & |Y(t) - r(t)| \leqslant I \\ -k_2(Y(t) - r(t)), & |Y(t) - r(t)| > I \end{cases} \tag{3.5.6}$$

2 Proportional action, P:

$$U(t) = -k_p(Y(t) - r(t)) \tag{3.5.7}$$

3 Integral action, I:

$$U(t) = -k_I \int_{t0}^{t_f} (Y(t) - r(t))\,dt \tag{3.5.8}$$

$$\approx -k_I \sum_{t=0}^{t_f} Y_t - r_t$$

4 Derivative action, D:

$$U(t) = -k_D \frac{d(Y(t) - r(t))}{dt} \qquad (3.5.9)$$
$$\approx -k_D[(Y(t) - r(t)) - (Y(t - \delta) - r(t - \delta))]$$

On-off action is the simplest form of control which gives a switching between two values of feedback control gain depending upon the magnitude of the system deviation from target. Often the first constant k_1 will be set to zero, so that no control action will be taken if the deviation is below a certain value. This may be extended to multipositional control settings k_i ($i = 1, 2, \ldots, N$). Proportional action is a simple generalization of the multipositional and on-off cases, where the magnitude of the feedback gain is a simple linear relation to the magnitude of the deviation. Integral action gives a feedback control gain proportional to the *sum* of deviations over the previous time periods, and derivative action gives a feedback gain proportional to the rate of change of the deviations over the previous time period. Various combinations of these actions are also possible. Common examples are proportional plus integral (*PI* control, or two-term controller), and proportional plus integral plus derivative (*PID* control, or three-term controller). The frequency response of various *PID* controllers is shown in fig. 3.20. This figure should be compared with the system frequency responses of uncontrolled systems (fig. 2.45). Proportional and integral (*PI*) action induces an overall feedback (autoregressive) characteristic into a system otherwise subject only to gain (input) terms. *PD* action does not induce feedback terms, whilst *PID* action has a combination of feedback and gain at any frequency ω. Hence, the closed-loop system response parallels the character of 'free' systems in frequency terms, but with the response structure determined by the controller.

In the time domain, the closed-loop control system response using each definition of the controller can be compared with the open-loop case in fig. 3.21A for a simple first-order system, and in fig. 3.21B for a second-order system. Again the response character parallels that of the 'free' systems shown in fig. 2.42. On-off control switches continuously at an infinite rate between the values k_1 and k_2 (or on and off) and produces a spikey, oscillatory response. This control will be physically impossible to realize in almost all instances and the effect of lags and inertia in operating the on-off switch induces wild overshoots. Alternatively, the limits k_1 and k_2 may be set fairly wide, which will damp fluctuations giving hysteresis. Generally, on-off feedback action will be satisfactory when the disturbance is within a defined range and when the rate of variation of disturbances is slow (i.e. a long time between disturbances). Such 'stop-go' control of national economies, for example, has been particularly unreliable because of the amplifying effect of the control in situations subject to many and continuous disturbances (see, for example, Phillips 1954). Proportional feedback control gives a relatively sluggish response, but is particularly defective in that complete cancellation of

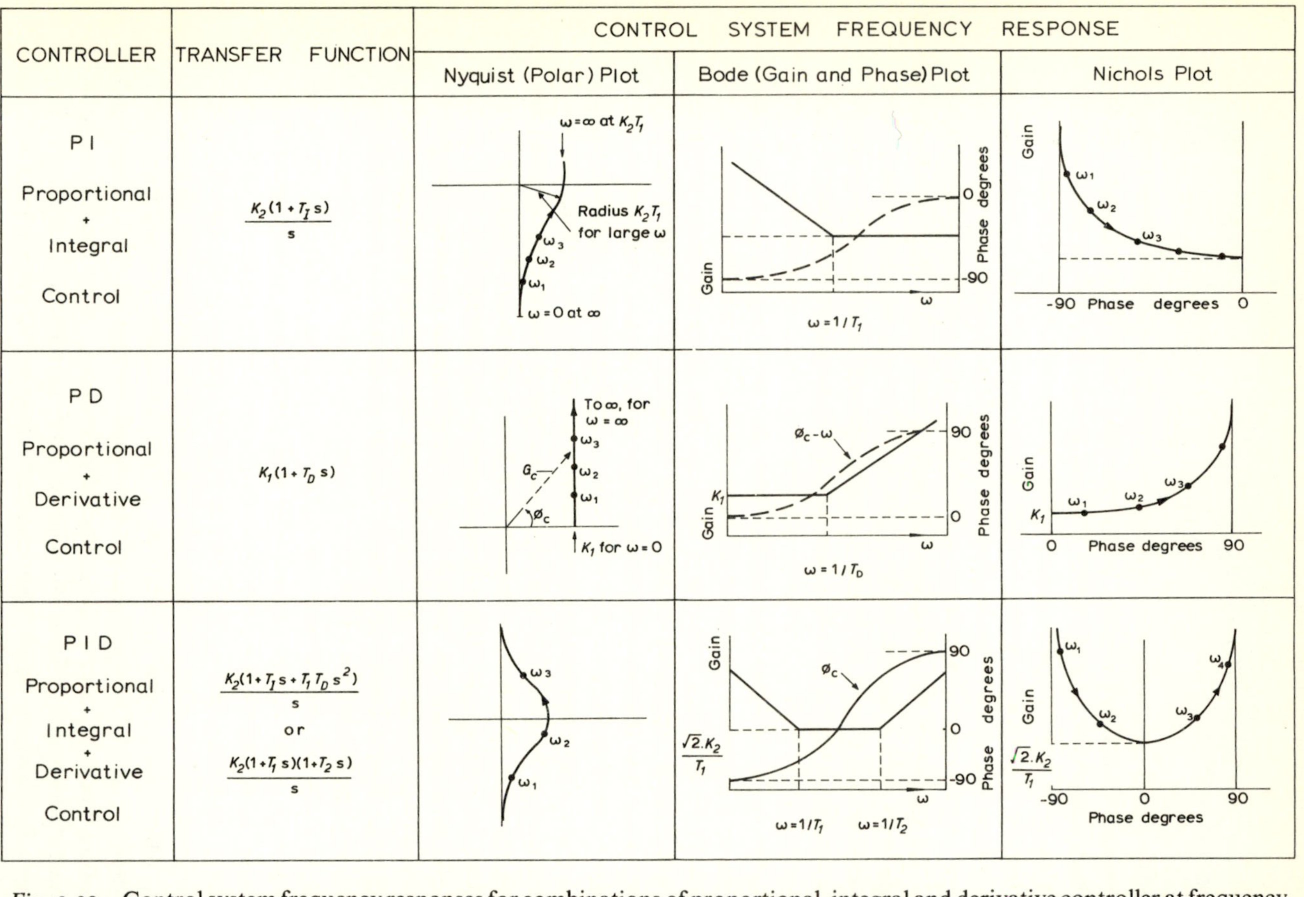

Fig. 3.20 Control system frequency responses for combinations of proportional, integral and derivative controller at frequency ω. K_1 and K_2 refer to proportional terms, T_I and T_D to integral and derivative terms, and the transfer functions given are in closed-loop form, with T_1 and T_2 as time constants.

deviations cannot be obtained. This 'offset' of the response from zero, or from the reference value, can be reduced by increasing the gain k, but higher gains introduce more instability and overshoot. In addition, very high gains may go outside the bounds V over which the system can be considered linear or stable. Integral action ignores the speed of control, can bring the system to a standstill anywhere and can achieve complete cancellation of deviations; but it accentuates oscillatory modes in the system. Derivative feedback action improves the stability of the controller and gives more immediate response, but if used for pure delay processes will have little effect, and if used in noisy stochastic systems will accentuate and amplify variations. Derivative action is never used alone since it involves no relation to the desired reference value, such that the control will be the same whether the deviation is large or small. The combination of *PI* and *PID* action gives the most desirable control action responses (fig. 3.21) with minimal overshoot, rise time and settling time, with derivative components reducing the peak height of the oscillatory effects. The deviations in *PID* controllers will again amplify noise in some stochastic systems.

The problems in applying these techniques to complex environmental systems in practice are that the system transfer function is seldom known with certainty; there are usually a few *a priori* grounds for choosing one particular controller rather than another or any particular value of k, and experimentation with the system is not possible. Much practical economic and environmental control has been rather hit and miss, by trying various policy instruments and controls in different contexts. When the transfer function is known with a reasonable degree of precision, simulation experiments may be conducted to allow the optimum choice of the feedback gain k (Naylor 1971b, Sengupta and Fox 1969, Phillips 1954, 1957). It is preferable, however, to use the available knowledge of the system transfer function to design the controller, and simulation can be avoided. This approach results in the possibility of achieving *optimum control* when the transfer function is known with complete certainty.

Given the system equation $Y_t = S(z)X_t$, as in equation (2.1.1), then the *optimum control law* can be derived by simply inverting the system transfer function operator to give:

$$\begin{aligned} U(t) &= -k(z)(Y(t) - r(t)) \\ &= -S^{-1}(z)(Y(t) - r(t)) \end{aligned} \tag{3.5.10}$$

which is a *dead beat* control strategy in deterministic systems, or a *minimum variance* controller in stochastic systems. More generally, the control action of a number of automatic control strategies can be adopted, as shown in fig. 3.22, which express the magnitude of control response in proportion to the magnitude of the deviation from target, given by W_i (the weighting or objective function). In addition, the control strategy may be a *one-stage* criterion based on the deviation of the system from its reference input at only

FIRST ORDER SYSTEM : $\frac{1}{(T_1S+1)}$

e.g. 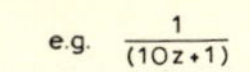

SECOND ORDER SYSTEM : $\frac{1}{(T_1S+1)(T_2S+1)}$

e.g. $Y(t) = \frac{1}{(1+10z)(1+5z)} X(t) = \frac{1}{1+15z+50z} X(t)$

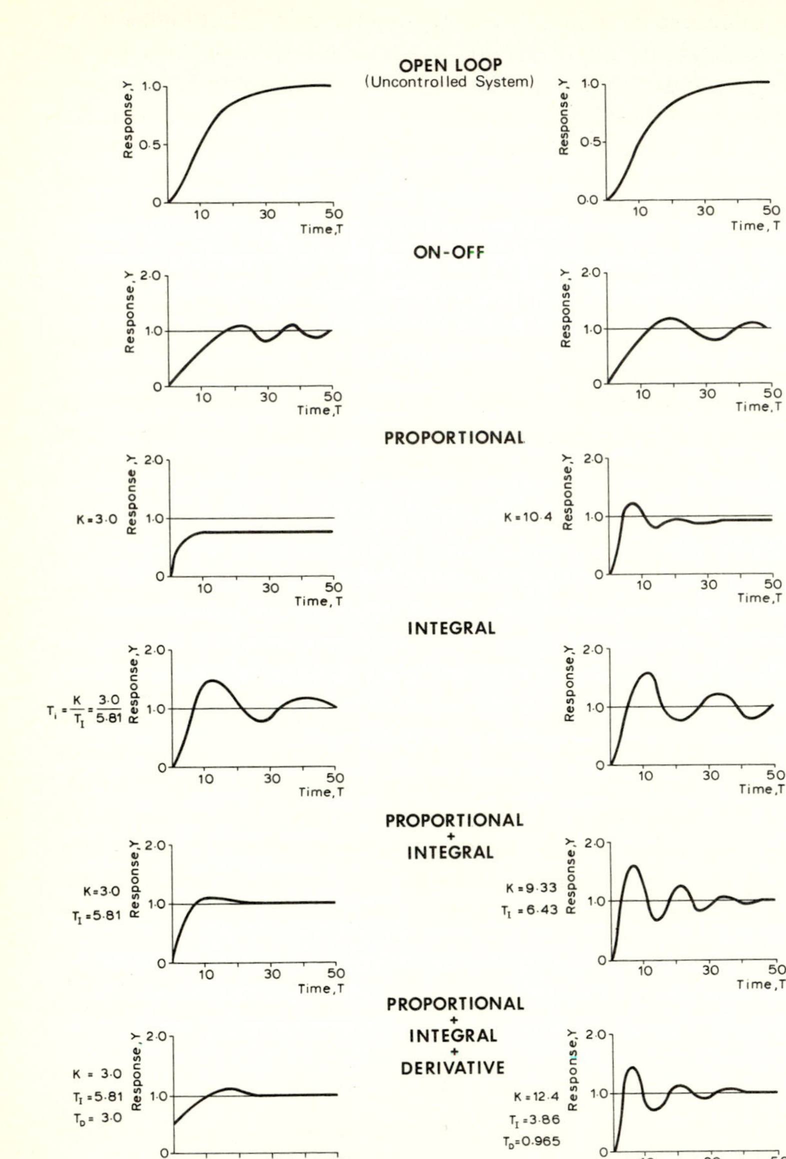

Fig. 3.21 Control system responses (Y) for (A) first-order and (B) second-order systems through time (t).

K = proportional term
T_I = integral term
T_D = derivative term
T_1 and T_2 are time constants.

one time instant, which attempts to reduce the disturbance to zero in one time period. Alternatively, the control may be an *N-stage* strategy, dependent on the sum of the deviations from target over a period of time, it being an attempt to cancel system disturbances over an extended planning time horizon. The most usual criteria adopted are those of minimum error and least-squares, but control results obtained from each may vary greatly. The criteria W_2 and W_3 based on least-squares have the useful property that the optimal control is the *dual* of the least-squares parameter estimation problem (Kalman 1958, 1960, Kalman and Bucy 1961) and is a *linear* function of the system output. The solutions using the Laplace absolute deviation criteria W_4 and W_5, however, yield nonlinear matrix Riccati equations (Bryson and Ho 1969) and the necessity to integrate both the model and control over the N steps forward in time.

In practice, most optimum controllers employ a combined *performance index* or objective function which incorporates two elements: (1) the 'cost' of deviations from targets, and (2) costs of employing a given control instrument. A common choice is the *quadratic* control criterion (objective function):

$$W_2 = E(MY_t^2 + NU_t^2) \tag{3.5.11}$$

where M and N are arbitrary weighting constants (or matrices in the multivariate case) which determine the relative 'costs' of deviations from target, and the 'costs' of control. It is especially difficult in socio-economic and physico-ecological systems to decide on appropriate values for the objective function terms M and N. Livesey (1973) suggests these should be as simple as possible, since complicated welfare functions will be no less arbitrary or more determinate. He suggests, however, that long-term research should be addressed to refining these objective functions and improving on the quadratic. Peston (1973) draws attention to the arbitrariness of treating both positive and negative deviations from target with equal weight. This will usually not be realistic in economic problems since, for example, increases or decreases from a target level of employment are not equally tolerable!

The problem is also a stochastic optimization one since most systems of practical interest are partial and, through the presence of ignored variables or modelling errors, will contain disturbances which in general will be correlated. This requires two responses by the controller. Firstly, a model of the disturbances must be built in the same manner as the noise model treated earlier in chapter 2. Secondly, optimization must be undertaken in conditions of uncertainty. This problem has been treated by Kalman (1960) and Whittle (1963b), and in an economic policy context by Holt (1960, 1962), Livesey (1973) and Chow (1973). The control solution is complicated by the necessity of considering a probability density function of control inputs. This means that it is unrealistic to attempt deterministic control, and the best the economic and environmental controller can hope to do is to control the mean and reduce the variance of a given process. Fortunately, the presence of stochastic error

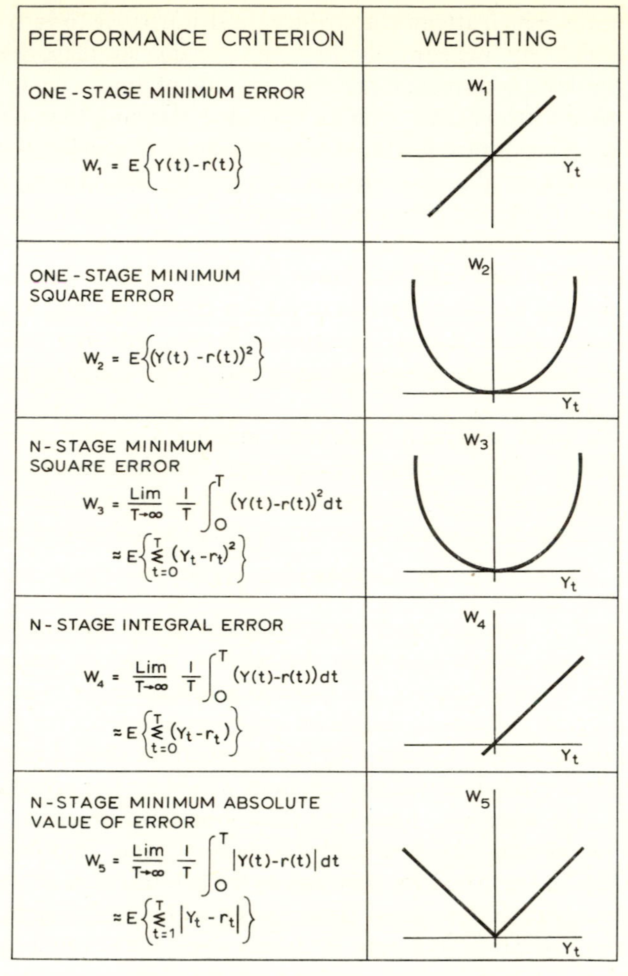

Fig. 3.22 Performance criteria (i.e. objective functions) for various one-stage and N-stage automatic control strategies. Each criterion determines the weight to be placed on deviations from set points.

processes does not complicate the control algorithm unduly since the optimum controller will have the same structure and form in the stochastic as in the deterministic case. This is due to the *certainty equivalence principle* (Theil 1957, Simon 1956), as stated by Simon: 'When a quadratic cost criterion (W_2 or W_3) is used in a control problem under uncertainty, then only the expectation and not the probability density function (*pdf*) of the controlled variables need be known.'

Hence, minimum variance mean square error forecasts can be used in the control algorithm, provided that they are unbiased, and are not consistently above or below the real values (Holt *et al.* 1960); i.e. the mean value of all system variables can be used.

3.5.2* ONE-STAGE AND N-STAGE OPTIMAL CONTROL STRATEGIES

The one-stage control problem is to shift the system from an initial point to a desired final point in one step with minimum cost. This can be stated as:

$$\text{Maximize (or minimize)} \quad W(Y) = f(U_t^1, U_t^2, \ldots, U_t^i, \ldots). \quad (3.5.12)$$

$$\text{Subject to the constraints} \quad g(U_t^1, U_t^2, \ldots, U_t^i, \ldots) \leqslant P. \quad (3.5.13)$$

This problem can be solved by two principles: linear programming and inversion of the system equation.

It is now possible to formulate and solve *linear programming* problems of very large dimension, often with as many as 10^4 equations and 10^6 variables (Lasdon 1970), but frequently large programming problems have special structures (e.g. triangularization, dimension reduction, etc.) which make them more amenable to solution, often because such problems arise from linking independent subunits in either time or space. A review of such methods is available in Dantzig (1965) and Lasdon (1970). The linear programming solution will be based on a linear objective function W_1, defined in fig. 3.22, and *constraints* $g(\cdot)$ are derived from the system equation and cost limitations.

For control strategies based on *inversion* of the system equations we may classify optimal control strategies by the cost criteria used and whether the system is deterministic or stochastic. The optimum controller for both deterministic (dead beat) and stochastic (minimum variance) systems with cost criteria W_1 and W_2 are shown in fig. 3.23. As an example of the way in which each controller is devised, consider the deterministic system (2.2.7A):

$$Y_t = A^{-1}(z)B(z)U_t. \quad (3.5.14)$$

In this case the optimum dead beat controller is given by inverting the equation:

$$U_t = -\frac{A(z)}{B(z)}(Y_t - r_t) = k(z)(Y_t - r_t). \quad (3.5.15)$$

For example, the transfer function

$$(1 - 0{\cdot}7z)Y_t = 0{\cdot}9U_t$$

becomes the controller

$$-\frac{(1 - 0{\cdot}7z)}{0{\cdot}9} Y_t = U_t.$$

The derivation of the optimum controller for stochastic systems, and for the quadratic least-squares criterion W_2, is more complex and is given in

* This section may be found difficult and can be omitted at first reading.

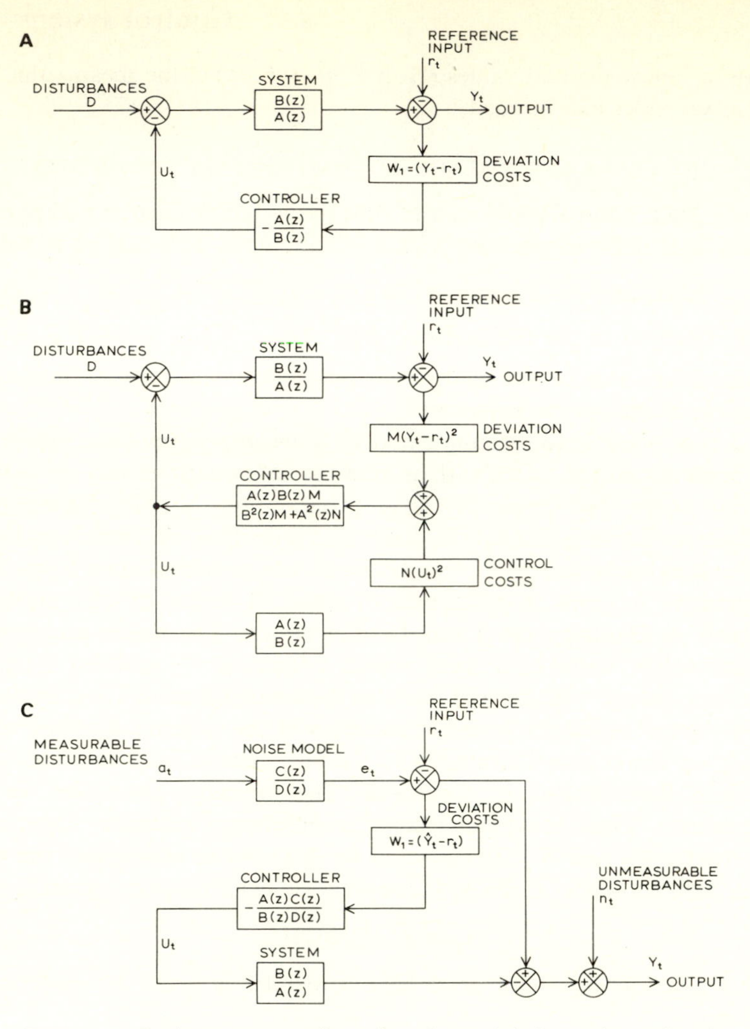

Fig. 3.23 Optimum controllers for deterministic and stochastic systems with cost criteria W_1 and W_2.

A Deterministic feedback control system:

$$k(z) = -\left[\frac{A(z)}{B(z)}\right]$$

B Deterministic quadratic feedback control system:

$$k(z) = \left[Y_t - \frac{B(z)}{A(z)} U_t\right] \frac{A(z)B(z)M}{B^2(z)M + A^2(z)N}$$

C Stochastic feedforward control system:

$$k(z) = -\left[\frac{A(z)}{B(z)} \frac{C(z)}{D(z)}\right]$$

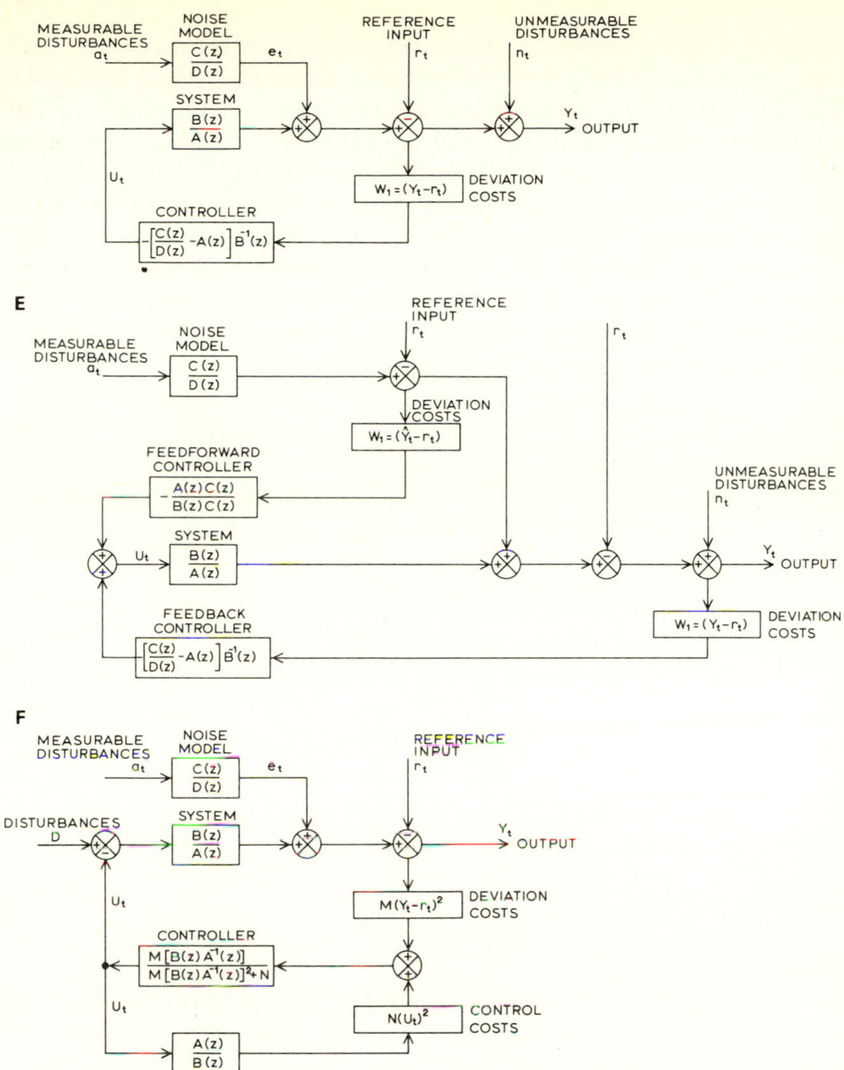

D Stochastic feedback control system:

$$k(z) = -\left[\frac{C(z)}{D(z)} - A(z)\right]B^{-1}(z)$$

E Stochastic feedback-feedforward control system:

$$k(Z) = -\frac{A(z)}{B(z)}\left[\frac{C(z)}{D(z)}\,Y_t + \frac{C(z)}{D(z)}\,e_t\right]$$

F Stochastic quadratic feedback control system: $k(z) = M\,\frac{B(z)}{A(z)}\left[M\,\frac{B^2(z)}{A^2(z)} + N\right]^{-1}$

$$\times\left[Y_t - e_t - \frac{B(z)}{A(z)}\,U_t\right]$$

Appendix 2. Further discussion of the quadratic criterion is provided by Theil (1957), Holt (1962), Sengupta and Fox (1969), Chow (1973) and Turnovsky (1973). For stochastic systems the aim of control is to keep the output $\{Y_t\}$ of the system as close as possible to some desired reference value r, or to some desired trajectory $\{r_t\}$. The problem differs from the deterministic case, however, since exact output cancellation can never be achieved because of the presence of the error disturbances. The aim must be to reduce the deviation from target to a minimum. The achievable minimum will be equal to the variance of the unknown input disturbances, so that optimal control in stochastic systems is *minimum variance* control. For example, the applicability of the feedback controller for the stochastic system:

$$U_t = -\left[\frac{C(z)}{D(z)} - A(z)\right]B^{-1}(z)Y_t = -k(z)U_t \qquad (3.5.16)$$

can be proved by substituting this control law into the system equation

$$Y_t = \frac{B(z)}{A(z)}U_t + \frac{C(z)}{A(z)D(z)}e_t.$$

Thus

$$\begin{aligned} AY_t &= -B\left(\frac{C}{D} - A\right)B^{-1}Y_t + \frac{C}{D}e_t \\ &= -B\frac{C}{BD}Y_t + \frac{AB}{B}Y_t + \frac{C}{D}e_t \\ &= -\frac{C}{D}Y_t + AY_t + \frac{C}{D}e_t. \end{aligned}$$

Therefore

$$\frac{C}{D}Y_t = \frac{C}{D}e_t$$

and the output exactly cancels the input stochastic disturbances. This controller depends solely on the character of $C(z)$ and $D(z)$, and the speed of regulation will depend upon the position of the poles of $C(z)$ and of $D(z)$, being slower the closer these are to the unit circle.

An example of the optimal stochastic control solution with the quadratic cost function W_2 is shown in fig. 3.24. The optimal control setting of instrument U_t^0 to achieve the optimal output Y_t^0 with least deviation from the desired settings U_t^* and Y_t^* is given at the point of tangency of the equi-cost ellipse with centre (U_t^*, Y_t^*) with the system equation as a linear constraint. A change in uncontrollable and unpredictable stochastic effects e_t will move the constraint line up or down in proportion to the attainable control, whilst a change in system parameters (b_0) changes the slope of the line. A change in M/N will change the shape of the ellipse (provided cov $(b_0, e_t) = 0$) (see also Appendix 2). Fig. 3.25 gives a comparison of the dead beat and minimum

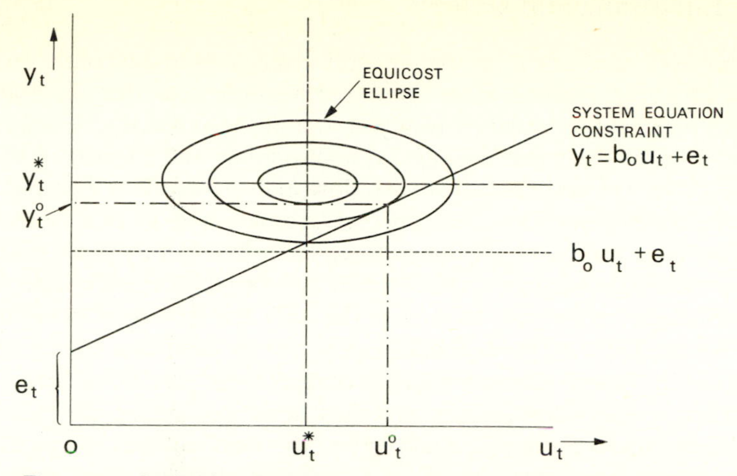

Fig. 3.24 Optimal solution to quadratic control problem with criterion

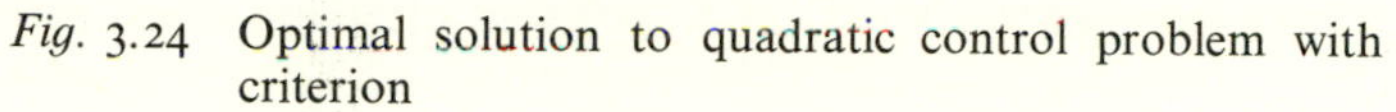

$$W_2 = M(Y_t - Y_t^*)^2 + N(X_t - X_t^*)^2$$

The system equation acts as a linear constraint on the objective function W_2. Compare this with the optimum control given in fig. 3.2.

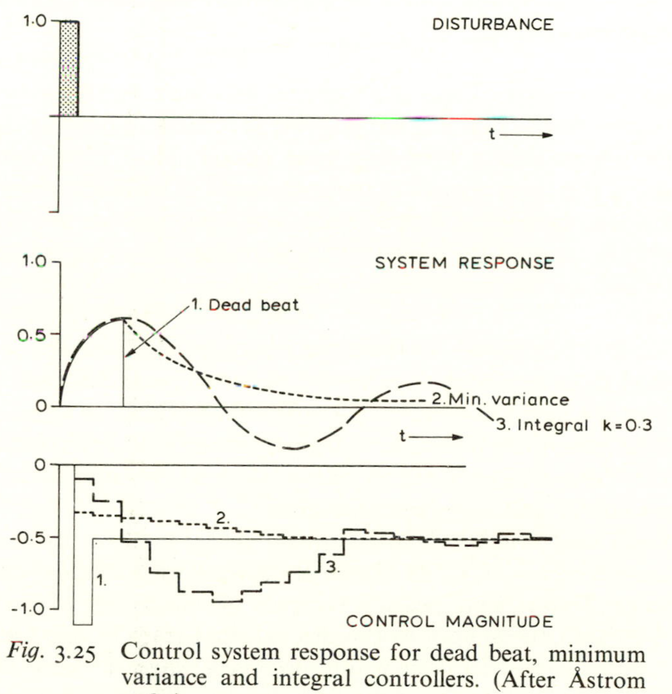

Fig. 3.25 Control system response for dead beat, minimum variance and integral controllers. (After Åstrom 1965.)

variance optimum controllers and integral strategy optimum controllers for a simple first-order system. From this figure it can be seen that the dead beat strategy produces the quickest response, but at the expense of a large initial applied control which may endanger system stability and linearity. It will also be expensive. Integral control is very slow and expensive, whilst minimum variance control performs best on cost grounds and gives a fairly rapid damping. Fig. 3.26 shows the three strategies applied to a dynamic time series exhibiting a higher degree of drift (trend) and variation in each series. There are two control problems: to remove the drift and to reduce variation. All three methods remove the drift adequately, but the dead beat strategy gives a very violent, 'spikey' response which accentuates the variations. The integral controller is also fairly spikey and has some remaining drift elements. Minimum variance control again gives the best control characteristics, although even here the variability of response is often amplified rather than damped.

In most practical problems of environmental control, one-stage controllers are of little utility, as we most often desire to derive an optimum strategy over a time horizon of variable extent (i.e. of N-steps into the future). Such controllers are termed *N-stage*, or *fixed time*, controllers which are required to produce optimum action over the whole N stages of the planning time horizon. These types of controllers are employed in the following categories of control problems:

(*i*) *Sluggish systems*. These are systems in which the response time to control is very slow so that it is not possible to achieve in practice the desired final state, except over a very long time period. Most physico-ecological systems are of this form because of the complexity of feedback loops involved which dissipate the control inputs. Socio-economic systems are also of this form, either because of the length of the learning time required or because changes which are too rapid would be socially or politically disruptive (see chapter 6).

(*ii*) *Cost-limited systems*. In these the resources available for control are smaller than those necessary to achieve a one-term controller, and hence the control strategy must be spread over a large time horizon. In socio-economic systems cost limitations reflect the marginal rate of return to control and the opportunity costs on capital. In physico-ecological systems limitations arise because of energy and resource constraints.

(*iii*) *Range and stability-limited systems*. These exist in cases in which control can be exercised, or its effects predicted with accuracy, only over a certain range. Thus it is not possible to apply the magnitude of control input necessary to shift the system to the desired final state in one time period. Instead, controls must be used within accepted ranges to keep the system inside established or tolerable behaviour bounds. This limits the choice of control $\mathbf{U}$ to the class of *admissible* controls V, ($\mathbf{U} \in V$). Most socio-economic

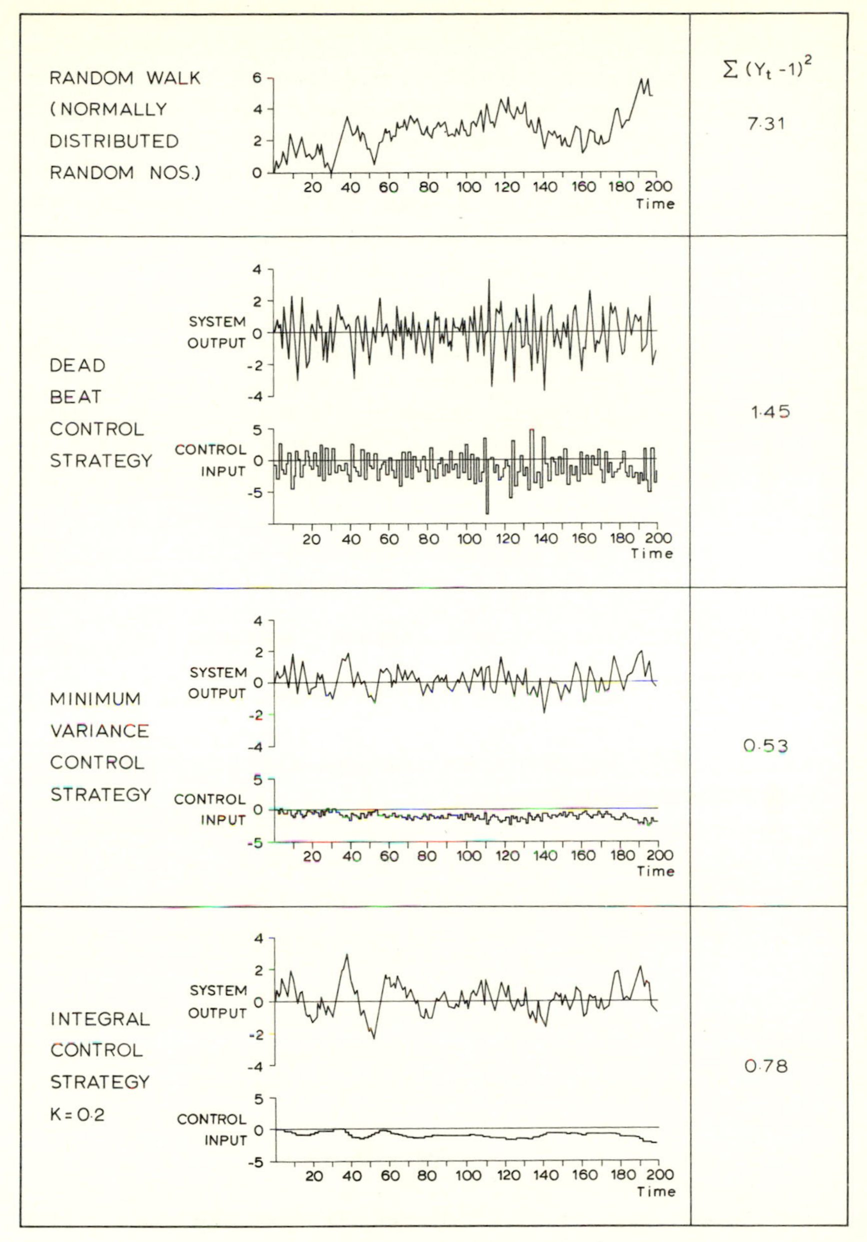

Fig. 3.26 Dead beat, minimum variance and integral control strategies applied to a dynamic (random walk) time series exhibiting a high degree of trend and variation. (After Åstrom 1965.)

and physico-ecological systems are thus limited. The *immediate* counteraction of disturbance inputs frequently requires very large magnitude control inputs which would make the system unstable, or even destroy it. The control of soil fertility by nitrate fertilizer, or of pests by pesticide and DDT are examples of control strategies with destructive and catastrophic side effects if the control used is at too great an intensity. In many feedback autostabilization mechanisms on aircraft too rapid response will lead to 'chatter' or undue stress. Similarly the excessive use of antibiotics is likely to build up bacterial resistance and destroy their regulatory potential.

In each case controls can be applied only over a time horizon often of great length from the initial time (t_0) at which it is desired to implement control to the final time (t_f) at which it is required to have brought the system to its desired level. For discrete control systems this becomes an N-stage optimization problem and in continuous systems a fixed time control problem. The most useful application of such controllers is to the quadratic cost system which is discussed below. Given the system (as in equation 2.2.19):

$$\dot{\mathbf{Z}} = \mathbf{AZ} + \mathbf{BU} = f(\mathbf{Z}, \mathbf{U}, t)$$
$$\mathbf{Y} = \mathbf{CZ} \qquad (3.5.17)$$

and a quadratic performance criterion (objective function):

$$\mathbf{J}(t, \mathbf{Z}) = \int_{t0}^{t_f} F(\mathbf{Z}, \mathbf{U}, t)\, dt = \int_{t0}^{t_f} (\mathbf{Z}'\mathbf{MZ} + \mathbf{U}'\mathbf{NU})\, dt = \mathbf{Z}'\mathbf{P}(t)\mathbf{Z} \qquad (3.5.18)$$

where $\mathbf{P}(t)$ is positive-semi-definite and represents a total cost matrix, or sum of the control and system costs.

The control problem is to find the minimum value of the performance criterion over the fixed time $t_0 - t_f$ along the optimal control trajectory $\mathbf{U}^0(t, \mathbf{Z})$. At the minimum, the objective function is given by:

$$\mathbf{J}(t, \mathbf{Z}) = \int_{t0}^{t_f} F(\mathbf{Z}(\tau), \mathbf{U}^0(\tau, \mathbf{Z}), \tau)\, d\tau. \qquad (3.5.19)$$

This is a Liapunov function (chapter 4.5) and to determine the nature of its minimum one must consider small variations away from the optimal point with time. Differentiating (3.5.19) with respect to the lower limit gives the negative term:

$$\frac{\partial \mathbf{J}(t, \mathbf{Z})}{\partial t} = -F(\mathbf{Z}, \mathbf{U}^0(t, \mathbf{Z}), t) \qquad (3.5.20)$$

or, alternatively, from (3.5.17) and (3.5.18) using the chain rule of differentiation as:

$$\frac{\partial \mathbf{J}(t, \mathbf{Z})}{\partial t} = \sum_{i=1}^{n} \frac{\partial \mathbf{J}(t, \mathbf{Z})}{\partial \mathbf{Z}_i} f_i(\mathbf{Z}, \mathbf{U}^0(t, \mathbf{Z}), t) + \frac{\partial \mathbf{J}(t, \mathbf{Z})}{\partial t}$$
$$= \nabla \mathbf{J}'(t, \mathbf{Z}) f(\mathbf{Z}, \mathbf{U}^0(t, \mathbf{Z}), t) + \frac{\partial \mathbf{J}(t, \mathbf{Z})}{\partial t}. \qquad (3.5.21)$$

This represents the change from initial conditions. Combining these two equations (3.5.20) and (3.5.21)

$$\nabla \mathbf{J}'(t, \mathbf{Z}) f(\mathbf{Z}, \mathbf{U}^0(t, \mathbf{Z}), t) + \frac{\partial \mathbf{J}(t, \mathbf{Z})}{\partial t} + F(\mathbf{Z}, \mathbf{U}^0(t, \mathbf{Z}), t) = 0 \tag{3.5.22}$$

the control problem is solved by stating the optimum control law $\mathbf{U}^0(t, \mathbf{Z})$ as a function of $\nabla \mathbf{J}$, $\mathbf{Z}$ and t; substituting this relation in (3.5.22) to give a partial differential equation and solving to give the time derivative $\mathbf{J}(t, \mathbf{Z})$. With $\mathbf{J}(t, \mathbf{Z})$ known, $\mathbf{U}^0(t, \mathbf{Z})$ can be determined. These steps can be derived as follows:

1 $\mathbf{U}^0(t, \mathbf{Z})$ is expressed in terms of $\nabla \mathbf{J}$, $\mathbf{Z}$ and t, as follows

$$\begin{aligned} \mathbf{U}^0(\mathbf{Z}, \nabla J, t) &= \min_{\mathbf{U} \in V} \nabla \mathbf{J}' f(\mathbf{Z}, \mathbf{U}^0(t, \mathbf{Z}), t) + F(\mathbf{Z}, \mathbf{U}^0(t, \mathbf{Z}), t) \\ &= \min_{\mathbf{U} \in V} H(\mathbf{Z}, \mathbf{U}, \nabla \mathbf{J}, t) \end{aligned} \tag{3.5.23}$$

for the class V of admissible controls $\mathbf{U}$. The H function states the form of optimal control function and is usually termed the *control Hamiltonian* (Schultz and Melsa 1967, Dreyfus 1965) or Pontryagin system state function (Pontryagin *et al.* 1962).

2 The expression for the optimal control law can be substituted into (3.5.22) to give the control as a function of system state $\mathbf{Z}$ and time t:

$$H(\mathbf{Z}, \mathbf{U}, \nabla \mathbf{J}, t) + \frac{\partial \mathbf{J}(t, \mathbf{Z})}{\partial t} = 0 \tag{3.5.24}$$

which is the *Hamilton-Jacobi equation* which is a nonlinear partial differential equation, and is a function of $\mathbf{Z}$ and t only.

3 The solution of the Hamilton-Jacobi equation is normally carried out by four methods: calculus of variations, Lagrangian multipliers, dynamic programming, or control charts.

The derivation of the optimal control strategy by *calculus of variation* methods (Pontryagin *et al.* 1962) is given in Appendix 2 and yields a standard form of feedback control solution for the fixed-time control problem. The optimum control law is given by

$$\mathbf{U}^0(t, \mathbf{Z}) = -\mathbf{k}'(t)\mathbf{Z} \tag{3.5.25}$$

where

$$\mathbf{k}(t) = \mathbf{P}(t)\mathbf{B}\mathbf{N}^{-1} \tag{3.5.26}$$

and $\mathbf{P}(t)$ is the solution of the *matrix Riccati equation*:

$$\frac{\partial \mathbf{P}(t)}{\partial t} = \mathbf{P}(t)\mathbf{B}\mathbf{N}^{-1}\mathbf{B}'\mathbf{P}(t) - \mathbf{P}(t)\mathbf{A} - \mathbf{A}'\mathbf{P}(t) - \mathbf{M}. \tag{3.5.27}$$

This solution allows the Hamilton-Jacobi equation to be integrated forwards from initial conditions (or backwards from final conditions) to give a value of the cost functional and control over time. The $(\partial \mathbf{P}(t))/\partial t$ term represents the functional derivative of the performance criteria with respect to changes in the system state variables, i.e. the cost gradient for the values of system behaviour under optimal regulation on the trajectory of $\mathbf{Z} + \Delta\mathbf{Z}$ integrated over the control time $t_0 - t_f$. It should be noted that, in the dual problem of linear filtering and estimation, the $\mathbf{P}(t)$ term is the covariance matrix of the optimal (minimum) error of the estimate of the system state.

Lagrangian multipliers offer a simple solution to control and optimization problems under constraints (Dreyfus 1965, Wilson and Kirkby 1975) using the property that the minimum control cost of the system equation (3.5.17) can be determined from:

$$\sum_{j=1}^{n} \lambda_j \left(\frac{\partial f_j(\mathbf{Z}, U_j, t)}{\partial U_j} \right) \tag{3.5.28}$$

i.e. the Lagrangian multipliers λ_i are those arbitrary constants which render the first derivative of the system equation with respect to the control equal to zero – a necessary condition for a minimum. Stronger conditions can also be imposed:

$$\sum_{j=1}^{h} \lambda_j \left(\frac{\partial^2 f_j(\mathbf{Z}, U_j, t)}{\partial U_j^2} \right) \geqslant 0$$

$$\sum_{j=1}^{n} \lambda_j f_j(\mathbf{Z}, U_j, t) = \text{min. along control trajectory}$$

which are analogues of the Legendre and Weierstrauss conditions (discussed in chapter 4.5) which eliminate the respective weak minima and determine the minimizing control trajectory.

As an example, consider the problem of minimizing the deviations of a system from target values over a time horizon $t_0 - t_f$:

$$J(t, \mathbf{Z}) = \int_{t0}^{t_f} F(\mathbf{Z}, \mathbf{U}, t)\, dt = g_1 \tag{3.5.29}$$

subject to the integral constraint that the total control costs should not exceed a level k:

$$k \geqslant \int_{t0}^{t_f} H(\mathbf{Z}, \mathbf{U}, t)\, dt = g_2. \tag{3.5.30}$$

Using definitions of these two functionals g_1 and g_2, the minimization problem is one of determining the values of (g_1, g_2) which minimize (3.5.29) subject to (3.5.30), and which lie on or within the boundary of the oval which describes the boundary of all possible solutions. The distance from the origin is proportional to k so that a minimum value of k is equivalent to

$$\begin{aligned} k &= ag_1 + bg_2 \\ &= a\int_{t0}^{t_f} F(\mathbf{Z}, \mathbf{U}, t)\,dt + b\int_{t0}^{t_f} H(\mathbf{Z}, \mathbf{U}, t)\,dt, \quad a \neq 0. \end{aligned} \tag{3.5.31}$$

The Lagrangian multiplier arises from dividing this equation through by a and setting $b/a = \lambda$. Bellman (1961, p. 108) proves that the resulting equation is equivalent to the combination of (3.5.29) and (3.5.30) satisfying a minimizing solution. If equation (3.5.31) cannot be solved analytically, dynamic programming usually provides a better solution since the number of computations involved is reduced from $[(n-1)n!]/[(\frac{1}{2}n)!(\frac{1}{2}n)!]$ to $(n^2/2) + n$ (Dreyfus 1965).

In many situations involving environmental systems control (especially in discrete systems) it is easier to use the *dynamic programming* solution for the choice of optimal feedback control or allocation (Bellman 1957, 1961, Dreyfus 1965). This method depends on the *principle of optimality* – namely, that the optimal control policy has the property that, whatever the initial control decision, the remaining control decisions must constitute an optimal policy with regard to control from the position of the initial decision (Bellman 1957). This is a Markovian decision rule such that the policy trajectory with gradient $\partial u/\partial t$ at the initial point, combined with the optimal trajectory over the remaining control time from the terminal point of the initial trajectory, yields Bellman's equation which converts the N-stage control problem into N single-stage problems:

$$\mathbf{J}^0(\mathbf{U}, \mathbf{Z}, t) = \min_{\mathbf{U} \in V}\left[F(\mathbf{Z}, \mathbf{U}, t) + \frac{\partial \mathbf{J}^0}{\partial \mathbf{Z}} + \frac{\partial \mathbf{J}^0}{\partial \mathbf{U}}\frac{\partial \mathbf{U}}{\partial t}\right] \tag{3.5.32}$$

or

$$\mathbf{J}_N^0(\mathbf{U}, \mathbf{Z}, t) = \min_{\mathbf{U} \in V}\left[F(\mathbf{Z}, \mathbf{U}_N, t_{N-1}) + \mathbf{J}_{N-1}^0(\mathbf{Z} + \mathbf{A}(\mathbf{Z}, \mathbf{U}_{N-1}))\right]$$

where $\mathbf{A}(\mathbf{Z}, \mathbf{U})$ satisfies $\dot{\mathbf{U}} = \mathbf{AZ} + \mathbf{U}$. The optimal control law minimizing the cost over the time interval (t_0, t_f) is the minimum of the cost functional over the initial interval (t_0, t_1), plus the minimum over the remaining interval (t_1, t_f). This is a recurrence relation. As an example of the method consider fig. 3.27. This gives the dynamic programming solution for a four-stage controller with quadratic costs which seeks to counteract deviation of magnitude Z_0 given on the left-hand side. The control costs at the final time $J_4(Z_0)$ are then given, together with the admissible control setting at each time $U_t(Z)$, $(t = 0, 1, 2, 3)$ which is restricted to be 0, 1, or -1. For example, an initial deviation of $Z_0 = 4$ will cost 34 units to counteract, with control settings given by $\{U_t(Z)\} = \{-1, -1, -1, -1\}$. Because of the limitations on the range of admissible controls to 0, 1, or -1, an initial deviation from target of $Z_0 = 5$ cannot be counteracted within the four available.

Initial deviation Z_0	$J_4^0(Z_0)$	$U_0(Z)$	$J_3^0(Z_1)$	$U_1(Z)$	$J_2^0(Z_2)$	$U_2(Z)$	$J_1^0(Z_3)$	$U_3(Z)$
5								
4	34	-1						
3	17	-1	17	-1				
2	7	-1	7	-1	7	-1		
1	2	-1	2	-1	2	-1	2	-1
0	0	0	0	0	0	0	0	0
-1	2	1	2	1	2	1	2	1
-2	7	1	7	1	7	1		
-3	17	1	17	1				
-4	34	1						
-5								

Fig. 3.27 Dynamic programming solution to quadratic control of deviations from target for a four-stage controller and admissible control limitation. (After Lee 1964.)

Control chart strategies represent the implementation of 'man in the loop' control, or man-machine systems. Such strategies are restricted to systems for which human action is sufficiently fast and responsive – i.e. systems which change slowly in time, or for which disturbance effects have long reaction and relaxation times. For example, in the operation of a chemical plant an operative may periodically take measurements, read off the required action from a control chart and carry out the action. This is especially important in environmental and economic systems. Our monitoring system informs us that the system is working correctly, or is going astray from targets. A set of control rules (including 'mixed scanning') will then determine the appropriate remedial action. In the case of large deviations, however, it may be necessary to enter into wider questions of policy principles, such that the environmental and economic controller might be involved in political as well as optimal control decisions. This is discussed further in chapter 6. For the present it is sufficient to determine how *general* operating rules and control charts can be constructed.

Fig. 3.28 shows a simple control chart for various *PID* and on-off controllers. On the chart the system output is plotted for each successive time reading by the operator who will read off the appropriate control action setting (on the right hand side) which depends upon the control strategy to be followed. For example, proportional control gives a simple linear response weighted by the proportional term k_p. On-off action here produces no control response for k_i when the system is within $\pm 20\%$ of the reference level, but outside of these bands does implement control. *PID* action produces a complex counteraction term which will have the general linear form shown in the figure. When the action axis is divided, as in this figure, the chart is normally termed a *rounded* chart (Barnard 1959, Box and Jenkins 1970, ch. 12). The action taken conforms to the midpoint of each band and gives a simplified set of control actions at the expense of larger control error.

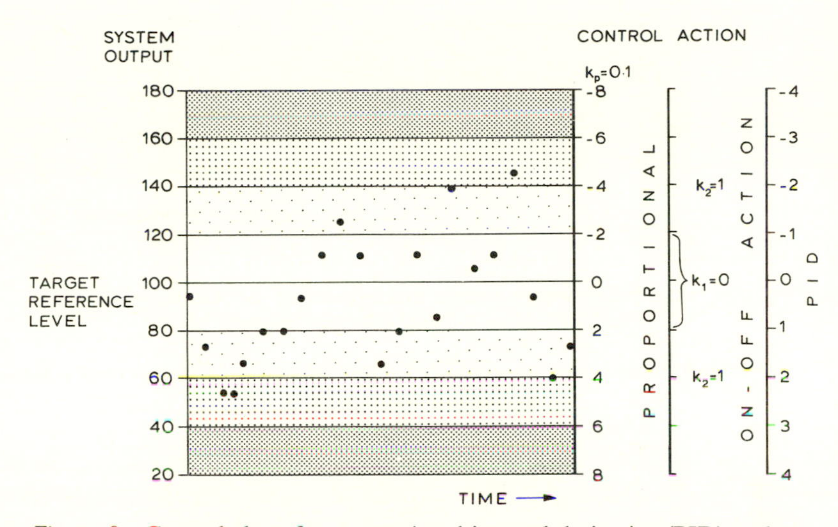

Fig. 3.28 Control chart for proportional-integral-derivative (PID) and on-off controllers. The right-hand axes give the weighting on PID and on-off feedback terms for the deviations on the control chart. The shading shows the weight given by proportional terms with deviations from a reference value of 100.

A more general set of control charts, adopting the feedback and feedforward solutions discussed above, can be implemented by the use of a control *nomogram*. In this case the optimal control law is written into a set of proportional graphs. In fig. 3.29 a feedforward control scheme of a simple second-order system is shown. The spacing of the axes is made proportional to the system parameters and the control setting U_t at the present can be read off directly for any value of present and past deviation and past control settings.

3.5.3 PROBLEMS OF CONTROL IMPLEMENTATION

In many instances the control law strategies discussed above may be inappropriate for environmental control issues and total *redesign of the system*

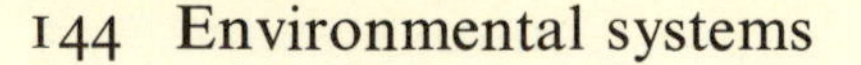

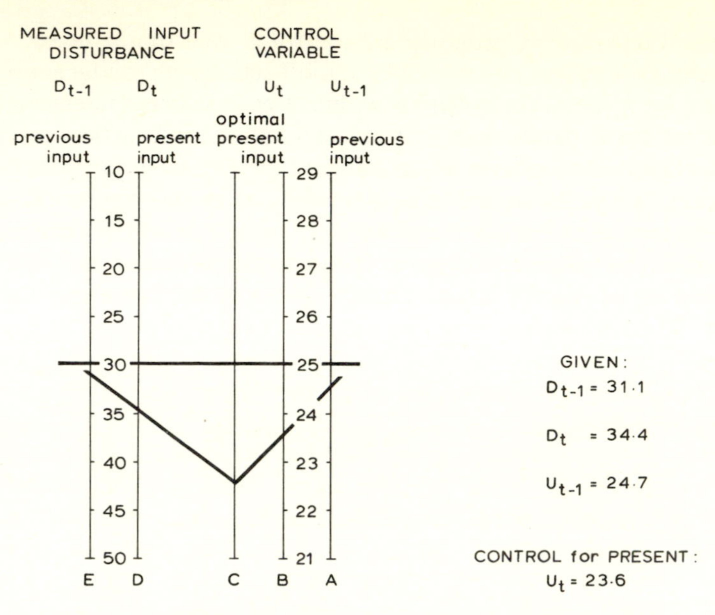

Fig. 3.29 Simple control chart nomogram for stochastic feed-forward control scheme. Control is given by:

$$(1 + \nabla)U_t = -0{\cdot}20(1 + 2{\cdot}33\nabla)D_t$$

$$(1 + \mu\nabla)U_t = U_t + \mu(U_t - U_{t-1})$$

Chart control axes satisfy:

$$\frac{BC}{AC} = 1; \qquad \frac{CD}{DE} = 2{\cdot}37; \qquad D_t = -0{\cdot}2U_t$$

for distances BC, AB, CD, DE. (After Box and Jenkins 1970.)

is required. In particular, it becomes impossible to apply the feedback–feedforward and *PID* controller when:

1. No policy instrument is available to introduce input changes.
2. No policy instrument is available with the required degree of sensitivity, selectivity, reliability, etc.
3. Any possible control is too 'expensive'.
4. The overall structure of the system is considered 'undesirable'.

The 'costs' of feedback control and the other control strategies discussed above continue over the whole operating life of the control system. If a restructuring of the system could be achieved, the costs will often be less than continuing control inputs. As an example, we might consider the costs of monitoring an old and ailing piece of machinery as against the costs of installing a new machine.

In each case it is necessary to compare the expected control costs in a feedback–feedforward, or *PID*, situation with those required for system redesign. Such comparisons, and the resulting choice of the most appropriate control system to implement, are a facet of decision making systems, project appraisal and the economics of control discussed in chapters 6 and 8. System restructuring is particularly important in socio-economic systems where some theoretical approaches, for example Marxism, advocate revolutionary changes to the structure of the social system as the only means of achieving the desired control ends. Even within mixed economies, governments frequently restructure policy instruments (tax and control legislation) so as to attain a better grip on the functioning of the socio-economic system – for example, attempts to change a regional industrial base rather than employ a subsidy (feedback) strategy.

In most practical physico-ecological and socio-economic systems, the form of the system transfer function is not known with complete certainty and a control can be implemented only after a series of system *identification and parameter estimation* stages of analysis have been undertaken (Box and Jenkins 1970, Bennett 1978). When the system variables are subject to theoretical observability and controllability restrictions (equation 3.1.2) the system state variables must be reconstructed. For deterministic systems this can be carried out using the *observer* method (Luenberger 1964), and for stochastic systems by the Kalman filter (Kalman 1960, Bryson and Ho 1969). Writing the system in state variable form (chapter 2), where **U** is a vector of control variables, the structures of the combined system model estimates and closed-loop control system are shown in fig. 3.30. The two state reconstructors are identical in structure except that in the stochastic case the statistics of the state must also be included: i.e. the expectation and variance of the state at each time instant. These estimates then allow for the presence of input and measurement noise, respectively $\{v_t\}$ and $\{w_t\}$, where it is assumed that $\text{cov}\,(x(t), w(t)) = 0$. For quadratic cost criteria the Kalman filter gives a special form of the Hamilton-Jacobi equation (3.5.24).

When the system is subject to practical observational restrictions, we have seen in chapter 2 that in identifying and estimating transfer function parameters the analyst may commit a number of blunders of unknown magnitude; mis-specification, incorrect causal hypotheses, measurement errors, etc. Additionally, our resulting parameter estimates are true only within certain statistical confidence bands which derive from the amount and quality of the data on system behaviour that are available. When the system equations have to be estimated we need to know (Wittenmark, 1969; Åstrom and Wittenmark, 1971):

1 Does a *separation theorem* exist? i.e. can estimated parameters be substituted into the control algorithms such that the estimation and control problems are independent of each other?

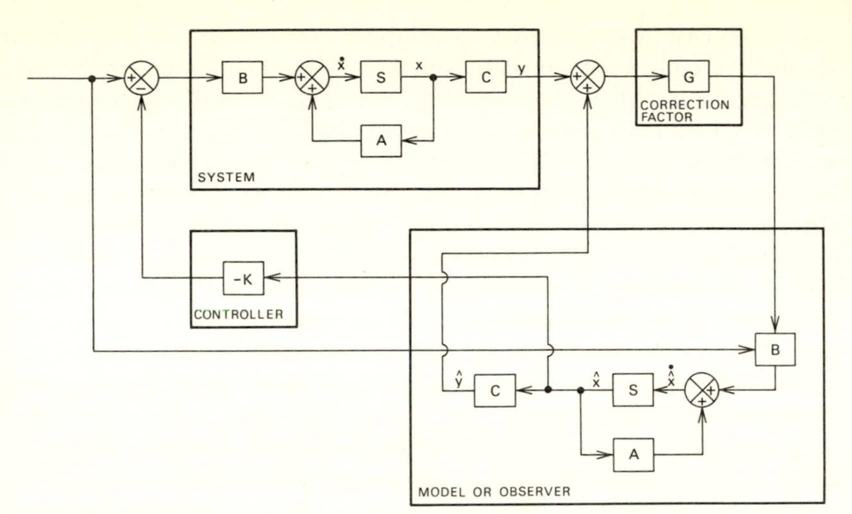

A LUENBERGER OBSERVER

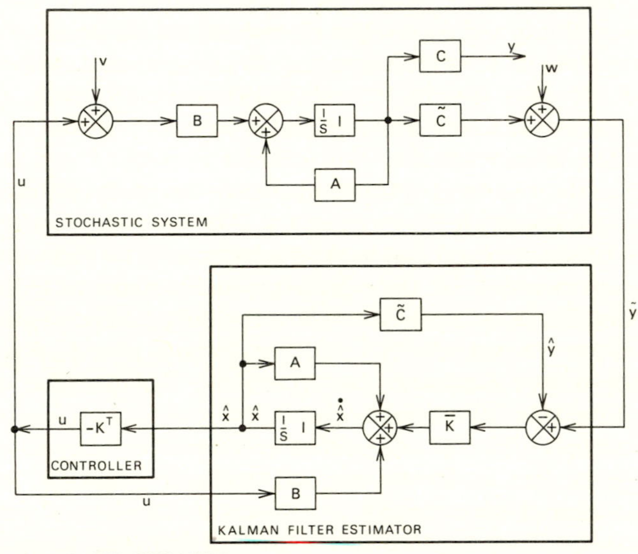

B KALMAN FILTER ESTIMATOR

Fig. 3.30 Methods for reconstructing state variables for (A) deterministic systems and (B) stochastic systems. A system model (observer or filter) is modified in the light of control system deviations (modified after Luenberger 1964, Kalman and Koepcke 1958, Schultz and Melsa 1967). The state space definition (2.2.14) is used with K the control gains and s the Laplace operator; v, w are the state and measurement noises.

2 Should parameter uncertainty be incorporated into the control regulator, i.e. does a *certainty equivalence theorem* hold for estimated systems?
3 Will the performance indicators (objective function) be satisfied in the same way as if the parameters were known?

The most important of these questions is the separation theorem. Almost all physico-ecological and socio-economic systems are understood with a limited degree of certainty and our understanding is being continually refined by hypothesis reformulation and testing. In consequence the models we can now utilize are at best approximations to the major functional relationships present and are estimated from statistical records of the activity of the system. No important environmental system is governed by completely *known* parameter relationships. In attempting to control such systems, therefore, we need to know what the effect of estimating these relationships will be. It is usually assumed that estimated system relationships can be used directly for control purposes (e.g. economic macro-economic policy models, the Forrester-Meadows' world models, etc.) with the parameter and specification uncertainties ignored. If a separation theorem does not hold, then the potentialities of achieving even poor control of environmental systems will be small. Åstrom and Wittenmark (1971) prove that *the separation theorem does not hold.* Taking the example of criterion W_2 and feedback from U_{t-1} only (a one-step controller), Åstrom and Wittenmark show that the minimum of the control objective function will be altered through the appearance of new terms which derive from the parameter uncertainty. Hence the optimum control no longer gives a global minimum for the information on system operation which is available. This has important consequences for practical control since the parameters and coefficients in environmental systems are usually derived from estimates. In addition such estimates usually have high parameter standard errors and it will be necessary to invoke and prove that a certainty equivalence theorem also holds. Both of these effects will increase the variance of control strategies and the choice of optimum policies will become a difficult task.

In response to these difficulties there have been two courses of action. An automated systems view suggests that we should rederive the control equations making allowances for parameter uncertainty and the use of estimated parameter terms. A second approach is to abandon an automated approach to control and base decisions on a wider range of 'soft' and uncertain control criteria. The automated systems view, as advocated by Åstrom and Wittenmark (1971), Chow (1975) and Bennett (1978, chapter 8), suggests the use of methods of system identification and parameter estimation which will permit control to be implemented within the range of available knowledge and certainty of the system structure. Since system control and continued system operation both generate further information as to the structure of the system, it is possible to implement joint parameter estimation, identification and control. Using least-square (O*LS*) parameter estimates, for example, Wieslander and Wittenmark (1971) and Åstrom and Wittenmark (1971) show that it is possible to design a *self-tuning regulator*. The method is:

1 Determine the parameters of the transfer function at time $t - 1$ by *OLS* assuming

$$\text{cov}\,(e_t e_{t-\tau}) = 0$$

2 Choose the control law of the stochastic problem at time $t - 1$.

3 Repeat steps (1) and (2) at time t and at each succeeding sampling interval.

The equations for this self-tuning regulator are given by a form of recursive least-squares algorithm (Plackett, 1950). Prediction of the system parameters for time $(t + 1)$ based on information at time t is given by:

$$\hat{\boldsymbol{\theta}}_{t+1} = \mathbf{F}\boldsymbol{\theta}_t + \mathbf{K}_t(\mathbf{Y}_t - \mathbf{X}_t'\boldsymbol{\theta}_t) \tag{3.5.33}$$

$$\mathbf{P}_{t+1} = [\mathbf{F} - \mathbf{K}_t\mathbf{X}_t]\mathbf{P}_t\mathbf{F}' \tag{3.5.34}$$

$$\mathbf{K}_t = \mathbf{F}\mathbf{P}_t\mathbf{X}_t[\mathbf{X}_t\mathbf{P}_t\mathbf{X}_t' + \sigma_e^2]^{-1} \tag{3.5.35}$$

Predictions of the system output are then generated from a state variable and measurement equation model (see chapter 2):

$$\hat{\mathbf{Y}}_{t+1} = \mathbf{X}_{t+1}'\hat{\boldsymbol{\theta}}_{t+1} + e_{t+1} \tag{3.5.36}$$

$$\boldsymbol{\theta}_{t+1} = \mathbf{F}\boldsymbol{\theta}_t + \mathbf{n}_t \tag{3.5.37}$$

This is a special form of the Kalman filter (Kalman 1960) as discussed earlier, and is identical in overall structure to Chow's (1975) learning algorithm for joint estimation and control. The general importance of this result is that a unified approach can be made to automated decisions for control. The Kalman filter provides an iterative 'learning' algorithm which is equally applicable to parameter estimation, parameter adaptation, and system control (Eykhoff 1974, Bennett 1978). For such an approach Åstrom and Wittenmark (1972) prove that both separability and optimality theorems hold for the optimum controller whatever the form of the closed-loop transfer functions (provided that they are stable). This automated approach to control of estimated systems is most important for environmental issues since we may estimate the control system parameters (from (3.5.33) to (3.5.37) by least-squares) and then use these estimates to implement optimum controllers, even though the separation theorem does *not* hold. In addition, the updating of the parameter estimates allows the accuracy of the control to be improved at each step (Wittenmark 1969, Åstrom and Wittenmark 1971). The method is shown in fig. 3.31A and in more detail in fig. 3.31B.

The second response to difficulties in attaining separation and certainty in control decisions is to abandon attempts at an automated approach. Such approaches recognize the difficulty that optimum control is rarely 'good' control since it is difficult to determine suitable objective functions or to make

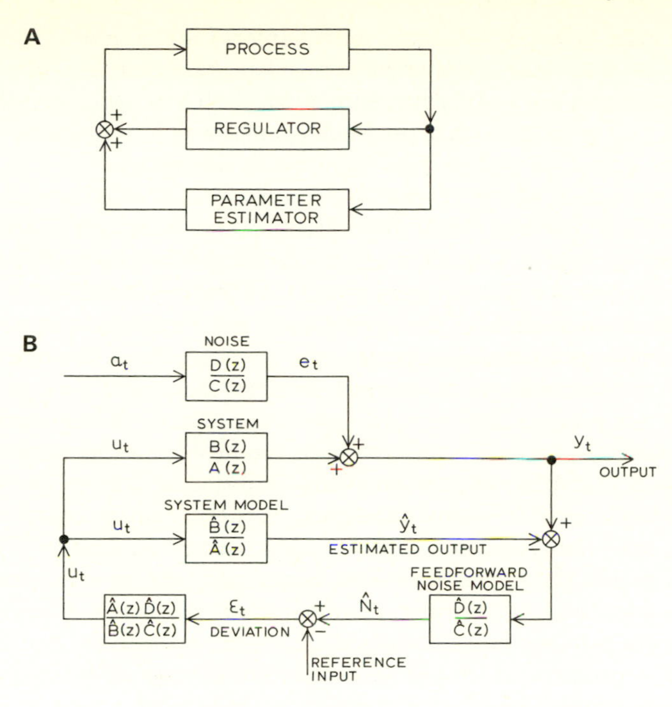

Fig. 3.31 The estimation of optimum control system parameters by least-squares regression techniques shown (A) in a general sense and (B) in more detail. A self-tuning regulator is then given by the refinements to the system model as a response to experience gained during control. (Partly after Åstrom and Wittenmark 1971.)

decisions between near-equivalent minima. Also many environmental systems are subject to severe constraints on the form of feedback or feedforward scheme that can be implemented such that optimum control may not be realizable. These constraints arise from the nonlinear and often 'chaotic' behaviour of environmental systems which makes their mathematical modelling and control very difficult. An alternative approach therefore has been to suggest a more flexible and less automated decision procedure based on close monitoring of system evolution, comparison of observed system changes against socially specified goals, and the implementation of a range of actions to counteract or anticipate these system changes if deviations from goals are likely. This approach to control encounters a wider degree of acceptance for environmental issues but is subject to considerable difficulties in the way in which we determine social goals. These require an understanding of cognitive and political issues, and the non-automated approach to control is left for more detailed discussion until chapters 5 and 6. For the present it is sufficient to note that automated controllers based on sub-optimal and optimal strategies are useful devices upon which to base some environmental decisions, and often provide a yardstick against which to judge the effect of perceptual, cognitive and political elements in decisions.

One further issue which is common to both automated and non-automated systems is the problem of identifying and estimating the system structure when it is subject to control. Remembering the earlier discussion, it will be recalled that perfect control will completely cancel the effect of system response to any inputs which cause shifts of the system away from specified targets. Similar effects are often produced by imperfect and *ad hoc* feedback strategies. In each case the implementation of control renders the operations of systems 'invisible', i.e. they are unobservable from inspection of the input and output streams alone. This is often referred to as the problem of identification in a 'closed-loop' and it can be overcome only if substantial *a priori* knowledge of system operations is available (of either the nature of the system transfer function, or of the system controller). In practice neither element is very well known in most environmental systems. Since such systems are dominated by closed loop feedbacks, either natural or man-induced, there will be substantial difficulties in gaining understanding of the system operations (Bennett, 1978, ch. 8). The problem of 'closed-loop' operations is an additional factor in generating complex and 'counter-intuitive' behaviour which is discussed further in chapter 9.

3.6 Conclusion

This chapter has been concerned mainly with the developing or understanding of automatic and semi-automatic control systems as extensions to the definition of hard systems initiated in chapter 2. Such systems have severe limitations when employed in any autonomous sense. Firstly, it is not possible to contain all possible control responses within a mathematical model. Secondly, no range of control responses (or discretion) is allowed. Thirdly, the calculations involved in any practical environmental problems preclude completely automatic solution (even though they may be theoretically feasible). The hard control systems described in this chapter must be seen instead as an algorithmic level of actions which are enmeshed in higher level man-machine systems and restructuring solutions. These latter allow the human capacities of thinking, learning and decision making to be employed, whilst leaving simple actions to autonomous units. These lead to the *minimum intervention principle* (Tomovic 1969) in multilevel control: that the importance of decisions and frequency of decisions are inversely related. This allows maximum autonomy in carrying out decisions at lower levels. This principle is based on the psychological and physiological properties of man; i.e. to minimize the effort required in making a decision. The development of the man-machine interface requires, therefore, an understanding of the cognitive and decision making capacity of man. These themes are developed in chapters 5 and 6, but first it is necessary to develop the mathematics, descriptive capacity, and control methods for systems which are distributed across space as well as dynamically evolving through time. Such space-time systems become crucial, not only in their effect on the decision making structure (chapter 6), but also in the man-environment interface discussed in chapter 9.

4 Space-time systems

He had brought a large map representing the sea,
Without the least vestige of land:
And the crew were much pleased when they found it to be
A map they could all understand.

'What's the good of Mercator's North Poles and Equators,
Tropics, Zones and Meridian Lines?'
So the Bellman would cry: and the crew would reply
'They are merely conventional signs!'

'Other maps are such shapes, with their islands and capes!
But we've got our brave Captain to thank'
(So the crew would protest) 'that's he's brought us the best –
A perfect and absolute blank!'

(Lewis Carroll: *The Hunting of the Snark*)

4.1 Man in space and time

It is often convenient to ignore the manner in which man is enmeshed within an environment which is spatially and temporally distributed. This has been the approach to much analysis of physico-ecological and socio-economic systems, and has been the approach of the preceding two chapters. Such systems are easier to understand but ignore two fundamental principles underlying all spatial relationships:

Specificity: that phenomena and systems are located at a particular spot and differ from place to place.

Relativity: that phenomena and systems interact with one another across space and over a range of time horizons such that the definition of any system must be related to its external interactions – i.e. its initial and boundary conditions which define its location in the space-time continuum.

Hägerstrand (1975, p. 12) specifies the man-environment interface in space and time as subject to eight conditions which arise from these two fundamental spatial principles of specificity and relativity:

1. The indivisibility of the human being from other living organisms and from most non-living entities.
2. The limited length of each human life and that of living organisms.
3. The limited ability of man and other indivisible entities to take part in more than one task at any time.

4 The fact that every task has a duration.
5 The fact that movement and interaction between points in space consumes time.
6 The limited packing capacity of space to absorb or contain organisms and entities.
7 The limited outer size of any terrestrial space (e.g. farm, city, region, nation, world).
8 The fact that every situation is rooted in past situations (although there may be anticipations of the future).

Lewis Carroll recognizes the analyst's inherent tendency to wish for simple systems which blank out spatial variations. However, such simplifications provide no basis for understanding and control of environmental systems in which the effect of space-time feedbacks, cascades and evolution, operating in a set of spatially nested loops, links the physical world environment with that of human relations. Space-time systems can be seen, therefore, as sitting astride the interface between automated and man-machine control systems, as discussed in the last chapter. They lie between an algorithmic and semi-automatic sensing and response to spatial location and environment, and the higher-level decision processes that allow the use, exploitation and development of society and culture by the successful symbiotic linkage of man–environment interactions.

The environment of man, therefore, should be extended to include space and time which are *resources* in the same way as are the non-renewable resources of the earth's crust (e.g. coal, oil, minerals) and the renewable resources of the physico-ecological world. As Hägerstrand (1973, pp. 77, 81) notes, 'to man time and space are not only dimensions for viewing and analysing the location of events, they are also in a very real sense *resources* – often scarce. . . . Nor can we understand how various arrangements of a political, technical and locational nature affect individuals in a population unless we can estimate how their private budget-spaces become widened or circumscribed.' The configuration of the physico-ecological and socio-economic environment of man is bounded by the specificity and relativity of his location. 'An important consequence of looking at things in this way is that one accepts spatial location to be strongly determined by the sequence of events in time' (Hägerstrand 1973, p. 81).

To concentrate on the purely spatial relationships of man and his environment and to ignore time, however, is to commit the same error as ignoring space whilst concentrating on system dynamics. A few special cases of purely spatial processes are discussed in section 4.3 which find reasonable philosophical justification outside of contexts in which time is considered. These processes are characterized by conditions of equilibrium or infinitely short relaxation times, and are *stationary* in the sense that temporal variation is zero. Such processes are of interest only philosophically, or under conditions in which observational restrictions (e.g. data availability) preclude a full space-time system study.

The development of such a space-time system study is not a simple task; nor is it a task which, despite its importance, has been tackled in any depth in most analyses of physico-ecological and socio-economic systems. For this reason, it is necessary to develop in this chapter a mathematical notation and definition of space-time systems which may be largely new to the reader. Because of the complexity of space-time systems this mathematical development may seem difficult, but we find that it has the useful properties that a state-space approach (as discussed in chapter 2) still applies, and that much of the mathematics of multivariate systems can be used to describe and solve space-time systems. The discussion is given at summary level and details of its development are given in Bennett (1978, chs 6 and 7). One of the most important conclusions is that we must treat space and time transfer functions within a system simultaneously. Because of the importance of space as well as time constraints to man-environment relations, we must analyse any system under study into its space and time components. But if we wish to say *anything* about the system as a whole, *qua* space-time system, we must look at the entire transfer function matrix of space-time interaction. It is the structure of this space-time transformation matrix, which turns out to be a matrix polynomial, to which we devote most attention in the first three sections of this chapter, whilst in sections 4.4 and 4.5 we explore the application of control concepts to space-time systems.

4.2 The structure of space-time systems

In the preceding chapters we have been concerned with the analysis of system structure within a single domain. The location of each system state variable has been assumed to be a lumped characterization of the system at a particular location. Such *lumped parameter systems* may derive from a point observation of the system at one location, or from an averaging of all the spatial input and output effects over an area or region. The *specificity* of location has been ignored. In this section we seek to show the form of system structure that results when *spatial specificity* is introduced – i.e. when the variables are considered to be *distributed* over a spatial domain or region of interest. Such systems, termed *distributed parameter systems* by the engineer, are structured according to three main elements (Butkovskiy 1969, p. 28):

1. *The connections* between the spatial elements over time: whether the connections are simple, multiple, reciprocal, disconnected, closed, bounded or unbounded.
2. *The dimensionality* of the space considered: lines, surfaces, volumes, etc. In socio-economic systems network lines and economic surfaces are the usual elements, but for physico-ecological systems volume diffusivity in solids, liquids and gases must also be considered.

3 *The form of measurement* of the space-time system which is available: whether this is a total picture of the system or only a set of sample sites; a continuously differentiable measurement function or a discrete set of measurements (e.g. a sampled data system). In this way the regularity and irregularity of the measurement of information and the degree of aggregation over the spatial regions and time periods is taken into account.

Continuous surfaces and discrete arrays in space and time may be defined as follows:

1 The spatial domain is a continuously differentiable surface. The system input and output variables are defined as $Y(t, x)$ and $X(t, x)$ at time point t, and spatial location x, where x is made up of the two Cartesian coordinates (x_1, x_2) – i.e. northings and eastings on a rectangular grid. A triangular (central place) lattice may also be adopted.

2 The spatial domain is an array of irregular regions or point samples. The system variables are defined for each location Y_{ti}, X_{ti} at time t and location i $(i = 1, 2, \ldots, N)$ for the ith sampling point or region within the whole domain of interest, divided into N regions.

Most developments of space-time models in the environmental, engineering and physical sciences have been concerned with processes defined as continuously differentiable variables, whereas most socio-economic and 'geographical' systems have been developed with N-regional, or N-sample irregular data arrays. As in the analysis of purely dynamic systems (e.g. time series streams), system response is dependent upon the form of system input, output and transfer function. In the present space-time case, however, the time stream of inputs and outputs is made up of either a surface of stimuli and responses (continuous systems) or an array of inputs and outputs (discrete systems), with dimensions equal to the number of discrete spatial locations. Impulse, step, ramp, periodic and other inputs may characterize this surface or array, and the output response may be linear or nonlinear, stationary or nonstationary, giving sets of spatial as well as temporal trends and periodicities. These may be analysed by a range of trend-surface, two-dimensional fourier and spectral techniques, and other methods. But, as in the purely dynamic system case, the most important characteristic of system response is the transfer function. This *is* the space-time system in the same way that it contains all the response information of the dynamic system. The magnitude, form, sign and behaviour of the parameters which make up the system transfer function may be constituted of space-time derivatives, integrators, lead, lag, gain and delay operators in an exactly analogous manner to purely temporal processes. In the case of space-time processes, however, the Z (lag) and Laplace operators must be considered to operate not only over time but also over the spatial region of interest.

Two approaches to the mathematical description of such transfer functions are possible. Firstly, a *space-time interaction system*, which is a mathematical description based upon defining the input and output variables over a set of N locations (point-sampled systems) or N regions (area-averaged systems). For the ith region or location in this N-location setting ($i = 1, 2, \ldots, N$) the input and output sequences can then be defined, respectively, by variables $X(t, i)$ and $Y(t, i)$. These expressions give the value of input and output at time t and location (or region) i. There will be N such inputs and outputs for each of the N regions apart from the range of n inputs and m outputs within each region which derive from any multivariate transfer function relations. A second mathematical description is based upon a series of *special space-time system* definitions. Four such special processes have been recognized (Bennett 1975b), and are shown in fig. 4.1.

1 *Barrier processes*: These occur where the interaction between zones is prevented in some way. Barrier states refer to zones with which there is a zero probability of interacting. In practice the barriers may not completely prevent interaction, and Yuill (1965) has recognized four forms of barriers: super-absorbing, absorbing, reflecting, and direct reflecting. Some of these forms have been built into the Hägerstrand (1953) diffusion model. A total barrier between two zones will result in a zero entry in the space-time transfer function matrix and the emergence of parallel systems (fig. 4.1).
2 *Hierarchy processes*: These occur when interaction between regions is controlled by the position of each region in a structural dependence based upon level in a 'hierarchy'. First noted by Ravenstein (1885), Berry (1962) has shown how such systems generate not only central place, but also shopping trip and other interactive phenomena. In another study Cliff *et al.* (1975) have shown how some stages of the measles diffusion process in Southwest England are controlled by hierarchy effects.
3 *Network processes*: The socio-economic landscape is dissected by transport networks of preferred flow, and physico-ecological systems are structured by natural energy pathways such as rivers and trophic energy chains. The nature of the networks will in each case affect the form of interaction between regions, and hence constrain the structure of the linkage matrix. Haggett and Chorley (1969) have noted the general importance of networks in spatio-temporal patterns, and Cliff *et al.* (1975) have demonstrated that some stages of measles epidemics in southwest England are governed by the layout of the transport network.
4 *Contiguity processes*: It has been frequently stated as the fundamental law of geography that everything is related to everything else but that near things are more likely to be related than distant things (Gould 1970). Hence it is natural to consider forms of linkage processes for which only lags on adjacent or near-neighbour cells are involved. Simple diffusion processes are of this form (Hägerstrand 1953, Tobler 1970). The spatial

pattern produced by each of these special processes is shown in fig. 4.1 for a simple diffusion example.

Each of the above two forms of mathematical description of space-time systems is discussed below, where it is shown that the mathematical representation of space-time systems by interaction systems contains each of the special processes detailed above as special cases.

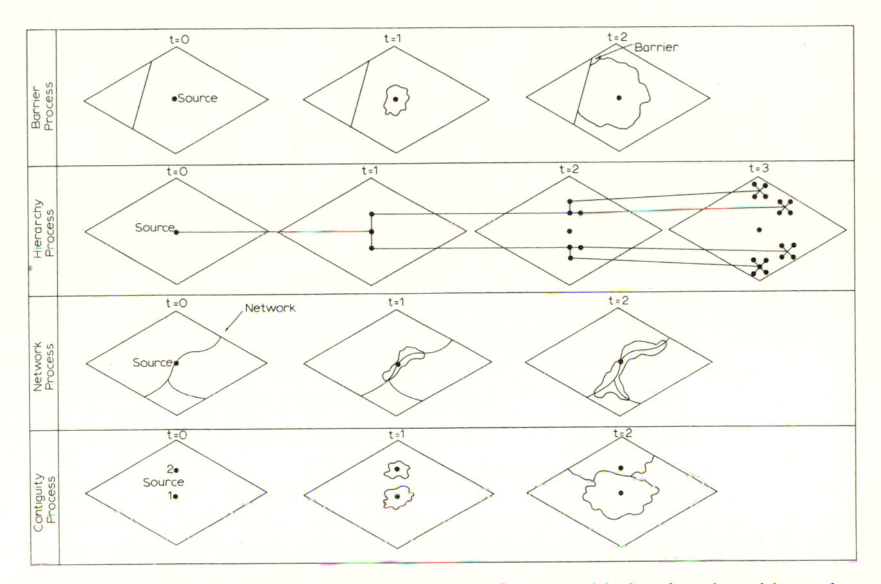

Fig. 4.1 Spatial patterns produced through time (t) by barrier, hierarchy, network and contiguity processes. (After Bennett 1978.)

4.2.1 INTERACTION SYSTEM DEFINITION OF SPACE-TIME PROCESSES

The interaction system definition of space-time processes treats each spatial region i ($i = 1, 2, \ldots, N$) in the N regions of interest as a separate entity. The complete set of cascaded inputs and outputs between locations is divided into two components. Within each region (or at each location) *specificity* effects can be modelled by dynamic system equations exactly identical to the system models discussed in chapter 2. Between locations the *relativity* effects of interaction can be modelled by systems equations which are again identical in form to those discussed in chapter 2. This combination of the two components gives a set of system equations, each individually identical to the chapter 2 dynamic systems, but in combination capable of generating the range of subtlety and complexity which typify the interaction in space-time systems. This combination of dynamic system equations can be best understood by a simple example.

To fix ideas, Bennett (1978, ch. 6) has considered the two-region system with a (2 × 2) space-time matrix transfer function operation which links the input of each region to the outputs of each other region, as in fig. 4.2. Each element in this matrix operator is defined by a separate transfer function S_{ij} which individually is identical in structure to the dynamic system of chapter 2 – i.e. it is time-indexed on a range of lag, lead, gain and delay terms. The diagonal elements S_{ii} define the *specific* locational dynamics internal to any location, and the off-diagonal terms S_{ij} $(i \neq j)$ define the *relativistic* dynamics of interaction between locations. Although each element S_{ij} is an independent transfer function made up of a combination of lag, gain, delay and memory

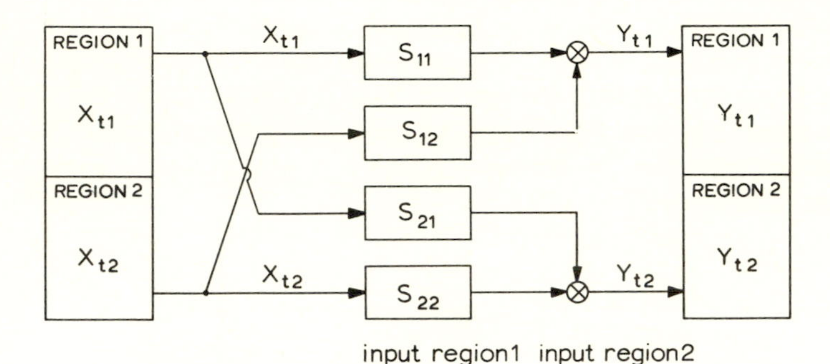

Fig. 4.2 Representation of a two-region system with 2 × 2 space-time matrix transfer function operator. (After Bennett 1978.)

terms, if we wish to say *anything* about the system as a whole (as a spatio-temporal system rather than as a single, lumped input-output relation) we have to look at the entire function matrix **S**. This is analogous to the multivariate matrix transfer function which is defined in fig. 2.17. As an example, consider the two-region transfer function matrix (given below) in which each individual transfer function element S_{ij} has been defined in terms of the z-operator:

$$\mathbf{Y}_t = \mathbf{S}\mathbf{X}_t \tag{4.2.1}$$

$$\mathbf{S} = \begin{bmatrix} S_{11} & S_{12} \\ S_{21} & S_{22} \end{bmatrix} = \begin{bmatrix} 3+z & 3+z \\ 1 & 7+z \end{bmatrix}. \tag{4.2.2}$$

In the simple dynamic system case (Chapter 2), the properties of the system transfer function **S** were examined in terms of its stability, stationarity, linearity, and response character under various input conditions (impulse, step, ramp, sinusoid, etc.) of its parameter polynomials. In the present space-time case the same approach can be adopted but we must now consider matrix properties, rather than pure polynomial properties. However, equation (4.2.2) is a matrix, each element of which is itself a polynomial. Such *matrix polynomials* yield interesting and important properties. Consider the determinant of (4.2.2) which is important in response and stability studies. This is given by

$$\det \mathbf{S} = (3 + z)(7 + z) - (3 + z)1 = 21 + 10z + z^2 - (3 + z)$$
$$= 18 + 9z + z^2 = (6 + z)(3 + z).$$

Thus the determinant of a polynomial matrix is also a polynomial, in this case expressed in terms of the lag operator z. For a more complicated example of an N-regional system, the transfer function $\mathbf{S}$ will become a $(N \times N)$ matrix polynomial operator linking the inputs and outputs of each region.

This system definition makes no assumption about the form of linkage between the N regions. It is *not* assumed that they are contiguous, joined by a particular transport network, or by any other specific structure; merely that a linkage of some form is present which allows interaction flows across space to occur. It is assumed, however, that the space-time processes are still subject to a *temporal* lag structure. Such processes have been termed N-variate interaction processes by Bennett (1975a) and possess the general $(N \times N)$ system transfer function given by:

$$\begin{bmatrix} Y_{t1} \\ Y_{t2} \\ \vdots \\ Y_{tN} \end{bmatrix} = \begin{bmatrix} S_{11} & S_{12} & \cdots & S_{1N} \\ S_{21} & S_{22} & \cdots & S_{2N} \\ \vdots & \vdots & & \vdots \\ S_{N1} & S_{N2} & \cdots & S_{NN} \end{bmatrix} \begin{bmatrix} X_{t1} \\ X_{t2} \\ \vdots \\ X_{tN} \end{bmatrix}. \tag{4.2.3}$$

In this equation there is now a set of N-outputs from each region $\{Y_{ti}\}$, a set of N-inputs into each region $\{X_{ti}\}$, and a set of $(N \times N)$ transfer functions which link the inputs into each region with the outputs of each other region in the overall system. The transfer function matrix S_{ij} will frequently be a set of gain, delay and feedback elements, such that it can be written in full in terms of a general set of polynomial distributed lags (i.e. lagged inputs of order-maximum lag, equal to q) and autoregressive feedback lags (i.e. lagged outputs of order-maximum lag, equal to p) – i.e.:

$$\begin{bmatrix} Y_{t1} \\ \\ \vdots \\ \\ Y_{tN} \end{bmatrix} = \begin{bmatrix} \dfrac{b_0^{11} + b_1^{11}z + \cdots + b_{q1}^{11}z^{q_1}}{a_0^{11} + a_1^{11}z + \cdots + a_{p1}^{11}z^{p_1}} & \cdots \\ & \ddots \\ \cdots & \dfrac{b_0^{NN} + b_1^{NN}z + \cdots + b_{qN}^{NN}z^{qN}}{a_0^{NN} + a_1^{NN}z + \cdots + a_{pN}^{NN}z^{pN}} \end{bmatrix} \begin{bmatrix} X_{t1} \\ \\ \vdots \\ \\ X_{tN} \end{bmatrix} \tag{4.2.4}$$

In this equation the $\{b_k^{ij}\}$ parameters are defined in equation (2.2.7) and the $\{a_0^{ij}\}$ parameters represent simultaneous feedback between regions and will normally be set equal to

$$a_0^{ij} = \begin{cases} 1, & i = j \\ 0, & i \neq j \end{cases}$$

where the (ij) superscript on each of the parameters represents the interaction polynomial between location i and location j, and in general the orders of each polynomial will be unequal for each (ij) combination. The rendering of equation (4.2.4) shows that S_{ij} is an $(f_i \times N)$ matrix, where f_i is the sum of q_i and p_i for region i. As in the time series case, not all elements of $\{b_k\}$, $\{a_k\}$ need be non-zero in these equations.

The structure of each transfer function component S_{ij} in the equation (4.2.4) can be looked upon as a univariate time series or dynamic system relation, as given in chapter 2 with a polynomial structure identical to that given in table 2.1, and each will also be a Padé approximant. To emphasize this symmetry between the N-regional system equations and the lumped dynamic system transfer function, we may rearrange equation (4.2.4) and re-express it in terms of arguments of the temporal lag operator z. This gives a new set of polynomial matrices in increasing temporal lags which, after separating the a_i and b_i terms, appear as:

$$\begin{bmatrix} Y_{t1} \\ Y_{t2} \\ \vdots \\ Y_{tN} \end{bmatrix} = \left\{ \begin{bmatrix} a_0^{11} & a_0^{12} & \dots & a_0^{1N} \\ a_0^{21} & a_0^{22} & \dots & a_0^{2N} \\ \vdots & \vdots & & \vdots \\ a_0^{N1} & a_0^{N2} & \dots & a_0^{NN} \end{bmatrix}, \begin{bmatrix} a_1^{11} & a_1^{12} & \dots & a_1^{1N} \\ a_1^{21} & a_1^{22} & \dots & a_1^{2N} \\ \vdots & \vdots & & \vdots \\ a_1^{N1} & a_1^{N2} & \dots & a_1^{NN} \end{bmatrix} z, \dots, \begin{bmatrix} a_p^{11} & a_p^{12} & \dots & a_p^{1N} \\ a_p^{21} & a_p^{22} & \dots & a_p^{2N} \\ \vdots & \vdots & & \vdots \\ a_p^{N1} & a_p^{N2} & \dots & a_p^{NN} \end{bmatrix} z^p \right\}^{-1}$$

$$\left\{ \begin{bmatrix} b_0^{11} & b_0^{12} & \dots & b_0^{1N} \\ b_0^{21} & b_0^{22} & \dots & b_0^{2N} \\ \vdots & \vdots & & \vdots \\ b_0^{N1} & b_0^{N2} & \dots & b_0^{NN} \end{bmatrix}, \begin{bmatrix} b_1^{11} & b_1^{12} & \dots & b_1^{1N} \\ b_1^{21} & b_1^{22} & \dots & b_1^{2N} \\ \vdots & \vdots & & \vdots \\ b_1^{N1} & b_1^{N2} & \dots & b_1^{NN} \end{bmatrix} z, \dots, \right.$$

$$\left. \begin{bmatrix} b_q^{11} & b_q^{12} & \dots & b_q^{1N} \\ b_q^{21} & b_q^{22} & \dots & b_q^{2N} \\ \vdots & \vdots & & \vdots \\ b_q^{N1} & b_q^{N2} & \dots & b_q^{NN} \end{bmatrix} z^q \right\} \begin{bmatrix} -Y_{t1} \\ -Y_{t2} \\ \vdots \\ -Y_{tN} \\ X_{t1} \\ X_{t2} \\ \vdots \\ X_{tN} \end{bmatrix} \qquad (4.2.5)$$

or more simply as

$$\mathbf{Y}_t = \mathbf{A}^{-1}(z)\mathbf{B}(z)\mathbf{X}_t = \mathbf{S}\mathbf{X}_t \tag{4.2.6A}$$

where,

$$\mathbf{X}'_t = [-Y_{t1} - Y_{t2} - \cdots - Y_{tN}X_{t1}X_{t2}\ldots X_{tN}]$$
$$\mathbf{Y}'_t = [Y_{t1}Y_{t2}\ldots Y_{tN}]$$

$\mathbf{Y}_t$ is the $(N \times 1)$ vector of regional outputs, $\mathbf{X}_t$ the $(N \times 1)$ vector of regional inputs and lagged outputs, and $\mathbf{A}(z)$ is a polynomial matrix of autoregressive parameters, and $\mathbf{B}(z)$ is a polynomial matrix of distributed lag parameters defined by:

$$\left.\begin{aligned}\mathbf{A}(z) &= \mathbf{A}_0 + \mathbf{A}_1 z + \mathbf{A}_2 z + \cdots + \mathbf{A}_p z^p \\ \mathbf{B}(z) &= \mathbf{B}_0 + \mathbf{B}_1 z + \mathbf{B}_2 z + \cdots + \mathbf{B}_q z^q\end{aligned}\right\} \tag{4.2.6B}$$

The similarity of equations (4.2.6A) and (4.2.6B) with equations (2.2.7A) and (2.2.7B) should be noted especially. For the purpose of estimating $\{a_k^{ij}\}$ and $\{b_k^{ij}\}$ parameters, the system equation can be rewritten as:

$$\mathbf{Y}_t = \boldsymbol{\theta}\mathbf{X}_t \tag{4.2.7}$$

where

$$\boldsymbol{\theta} = \mathbf{A}^{-1}(z)\mathbf{B}(z) \tag{4.2.8}$$

and $\mathbf{X}_t$ and $\mathbf{Y}_t$ are defined by equation (4.2.6). This is identical in structure to the univariate system equation (2.3.7) but with a much extended $((p + q + 1) \times N \times N)$ parameter set. Study of the structure of $\mathbf{A}(z)$ and $\mathbf{B}(z)$ yields a number of interesting mathematical properties.

1 Two autoregressive interregional feedback structures for $\mathbf{A}(z)$ are possible: (*a*) when $\mathbf{A}_0 = \mathbf{I}_N$, where $\mathbf{I}_N$ is an $(N \times N)$ identity matrix such that no simultaneous interaction between regions can take place; (*b*) when $\mathbf{A}_0 \neq \mathbf{I}_N$, and simultaneous autoregressive interaction between different spatial regions can take place.
2 When the autoregressive terms $\mathbf{A}_j(z) = 0$ for all j, there is a purely distributed lag (i.e. no feedback) space-time process.
3 When the distributed lag terms $\mathbf{B}_i(z) = 0$ for all i, there is a purely space-time autoregressive process.
4 When the input variables $\mathbf{X}_t$ into each of the N regions are not known or are not measured, there is a space-time autoregressive, moving average process. This is usually treated by assumptions similar to those employed in the case (6) example discussed in chapter 2.3, but requires the additional rigorous assumption that the input sequences treated as Gaussian random sequences $e_{ti}(i = 1, 2, \ldots N)$ into each of the N regions must be assumed independent between regions (Bennett 1975b). This is not usually realistic.

GAIN (DISTRIBUTED LAG) SPACE-TIME PROCESS:

$B_i \neq 0, (i = 0,1,2, \ldots\ldots, q)$

$A_j = 0, (j = 0,1,2, \ldots\ldots, p)$

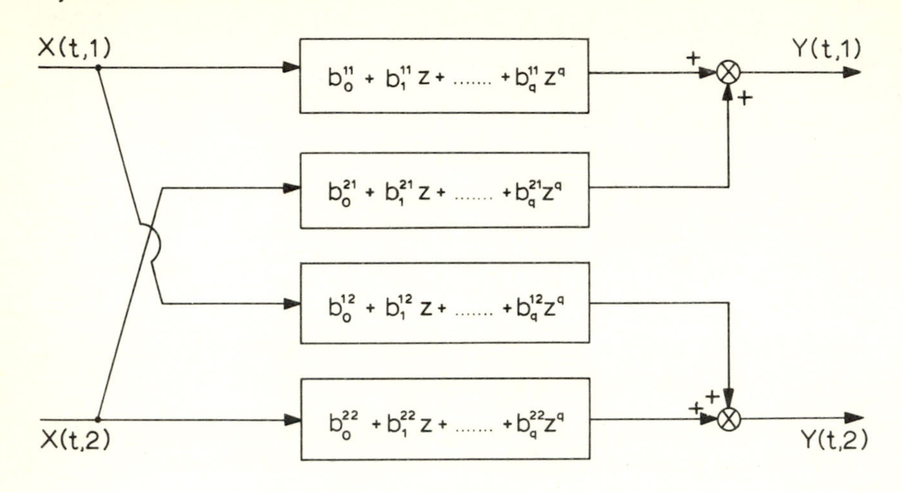

GENERAL SPACE-TIME TRANSFER FUNCTION PROCESS:

$B_i \neq 0, (i = 0,1, \ldots\ldots, q)$

$A_j \neq 0, (j = 0,1, \ldots\ldots, p)$

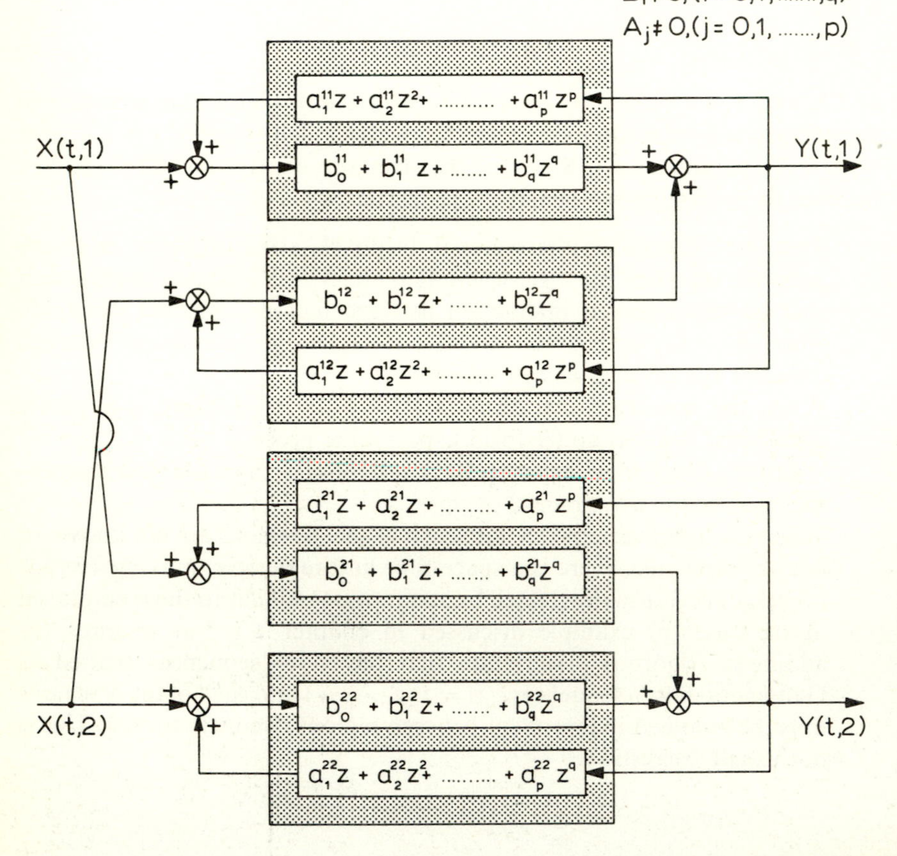

In order to see the application of these various cases, fig. 4.3 shows block diagrams of a number of space-time processes for the simple two-region case. Each S_{ij} transfer function element is now composed of a range of polynomial lag terms on the two inputs, and polynomial feedback lag terms on the two outputs. This general form may be appreciated in more detail by considering examples of low-order space-time transfer function models.

(1) *First-order space-time systems*

Fig. 4.4 gives the transfer function for three first-order systems for the two-region case where only gain terms (lags on the input X_{ti}) are present and there are no feedback (autoregressive) components. The three systems are lag, lead, and lead-lag processes:

First-order lag

$$\begin{aligned} \mathbf{Y}_t &= \mathbf{B}(z)\mathbf{X}_t \\ &= (\mathbf{B}_0 + \mathbf{B}_1(z))\mathbf{X}_t \end{aligned} \tag{4.2.9}$$

or

$$\begin{bmatrix} Y_{t1} \\ Y_{t2} \end{bmatrix} = \left(\begin{bmatrix} 3 & 3 \\ 1 & 7 \end{bmatrix} + \begin{bmatrix} z & z \\ 0 & z \end{bmatrix} \right) \begin{bmatrix} X_{t1} \\ X_{t2} \end{bmatrix}$$

First-order lead

$$\begin{aligned} \mathbf{Y}_t &= \mathbf{B}(z)\mathbf{X}_t \\ &= (\mathbf{B}_0 + \mathbf{B}_1^{+}(z^{-1}))X_t \end{aligned} \tag{4.2.10}$$

or

$$\begin{bmatrix} Y_{t1} \\ Y_{t2} \end{bmatrix} = \left(\begin{bmatrix} 3 & 3 \\ 1 & 7 \end{bmatrix} + \begin{bmatrix} z^{-1} & z^{-1} \\ 0 & z^{-1} \end{bmatrix} \right) \begin{bmatrix} X_{t1} \\ X_{t2} \end{bmatrix}$$

First-order lead-lag

$$\begin{aligned} \mathbf{Y}_t &= \mathbf{B}(z) \\ &= (\mathbf{B}_1^{+}(z^{-1}) + \mathbf{B}_0 + \mathbf{B}_1(z))\mathbf{X}_t \end{aligned} \tag{4.2.11}$$

or

$$\begin{bmatrix} Y_{t1} \\ Y_{t2} \end{bmatrix} = \left(\begin{bmatrix} z^{-1} & -5z^{-1} \\ 0 & 2z^{-1} \end{bmatrix} + \begin{bmatrix} 1 & -5 \\ 1 & 3 \end{bmatrix} + \begin{bmatrix} z & -5z \\ 0 & 2z \end{bmatrix} \right) \begin{bmatrix} x_{t1} \\ x_{t2} \end{bmatrix}$$

Fig. 4.3 (*on facing page*) Block diagrams for (A) gain (distributed lag) and (B) general transfer function space-time processes. Each transfer function component consists of a range of polynomial lag terms and is identical in structure to the system transfer functions discussed in chapter 2. (After Bennett 1978.)

The determinant of each system matrix can be used to examine the character of system behaviour. Fig. 4.4 also shows the *recursive* structure that is implicit in each polynomial matrix. The recursion diagram shows the chains of dependency of one regional output on the inputs into both the same region and the other region considered. In each first-order case the dependence structure generates an overall response which is cut off after one period forward or backward in time. This is analogous to the separate dynamic system responses shown in fig. 2.43 since, if we look at any one component in the matrix transfer function, we find identical equations to those in this figure. For example, taking the element S_{11}, we have the following three equations:

1 First-order lag:

$$Y_{t,1} = 3X_{t,1} + X_{t-1,1} + 3X_{t,2} + X_{t-1,2}$$

2 First-order lead:

$$Y_{t,1} = 3X_{t,1} + X_{t+1,1} + 3X_{t,2} + X_{t+1,2}$$

3 First-order lead-lag:

$$Y_{t,1} = X_{t+1,1} + X_{t,1} + X_{t-1,1} - 5X_{t+1,2} - 5X_{t,2} - 5X_{t-1,2}$$

These may be compared with the equations in fig. 2.42 and table 2.1.

When we consider the feedback autoregressive terms we have to deal with systems of the form

$$\mathbf{A}(z)\mathbf{Y}_t = 0 \tag{4.2.12}$$

(where here, at first, no input term X_t is considered). These can be solved only by finding the inverse matrix $\mathbf{A}^{-1}(z)$, i.e.

$$\mathbf{Y}_t = \mathbf{A}^{-1}(z).$$

The inverse of a matrix polynomial can be found by following exactly the same procedures as apply to normal matrices and given in Appendix 1. To see this, the inverse of an example system $\mathbf{A}(z)\mathbf{Y}_t = 0$ is calculated as follows:

$$\begin{aligned}\mathbf{A}(z) &= \begin{bmatrix} 2 & 5 \\ 0 & 1 \end{bmatrix} + \begin{bmatrix} 3 & 6 \\ 1 & 1 \end{bmatrix} z = \begin{bmatrix} 2+3z & 5+6z \\ z & 1+z \end{bmatrix} \\ &= \mathbf{A}_0 + \mathbf{A}_1(z).\end{aligned}$$

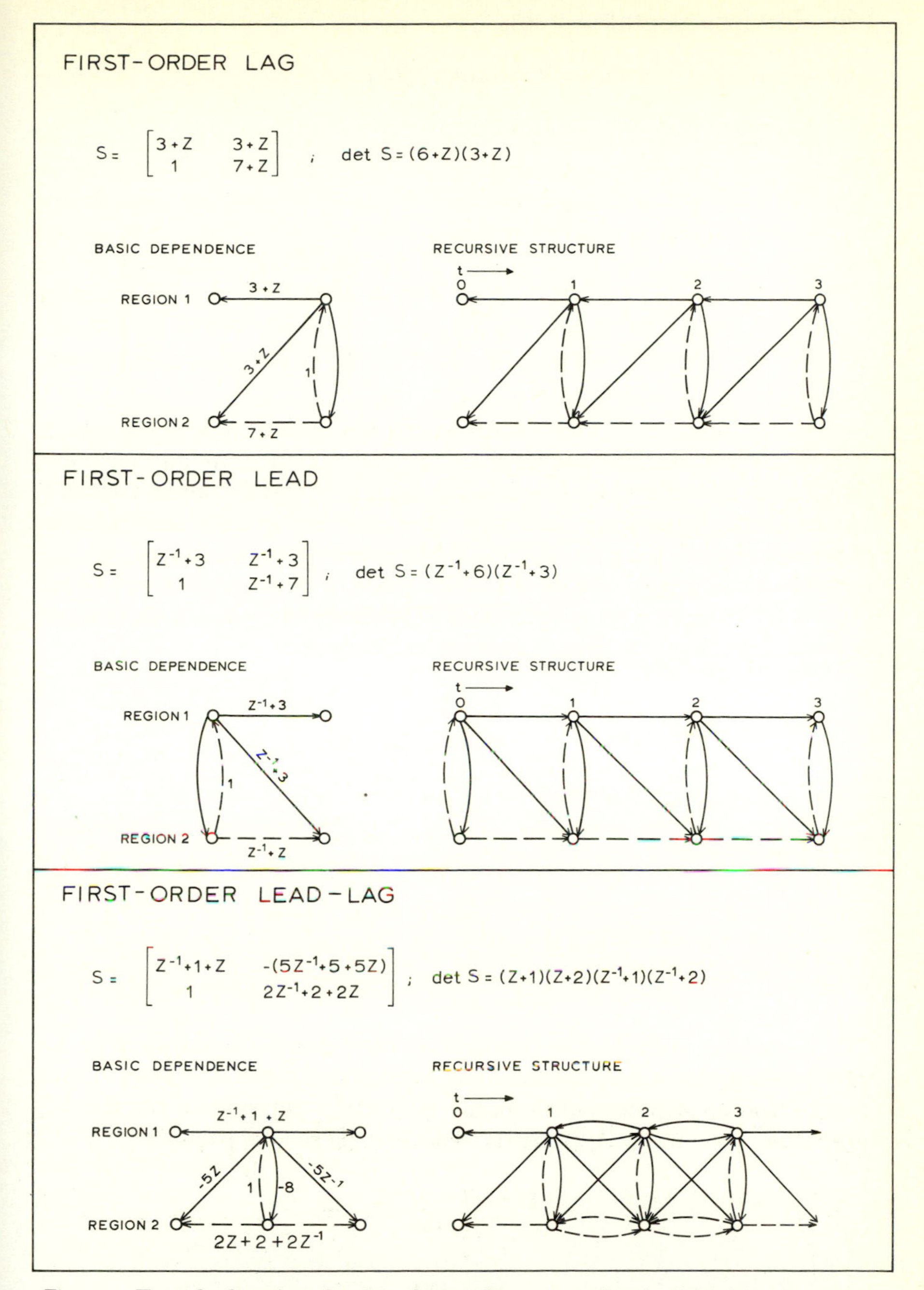

Fig. 4.4 Transfer functions for three first-order systems (lag, lead, lead-lag) in respect of a two-region case. Only gain terms are present and there are no feedback (i.e. autoregressive) components. Basic dependence and recursive structures are given. (After Bennett 1978.)

Then the inverse is given by Robinson (1967):

$$\det \mathbf{A}(z) = (2+3z)(1+z) - (5+6z)z = 2 - 3z^2$$

$$\operatorname{adj} \mathbf{A}(z) = \begin{bmatrix} 1+z & -5-6z \\ -z & 2+3z \end{bmatrix}$$

$$\mathbf{A}^{-1}(z) = \frac{1}{2-3z^2}\begin{bmatrix} 1+z & -5-6z \\ -z & 2+3z \end{bmatrix} = \begin{bmatrix} \dfrac{1+z}{2-3z^2} & \dfrac{-5-6z}{2-3z^2} \\ \dfrac{-z}{2-3z^2} & \dfrac{2+3z}{2-3z^2} \end{bmatrix}.$$

In this case the inverse of the system matrix is a ratio of polynomials, each Padé approximants. The nature of the dependence structure and recursive solution for equation (4.2.12) is shown in fig. 4.5. In this first-order feedback lag case the dependence structure generates an overall simultaneous dependence between regions and the solution cannot be determined. We are dealing with the special case $\mathbf{A}_0 \neq \mathbf{I}_N$. The nature of the solution in well-posed cases in which there is no simultaneous dependence ($\mathbf{A}_0 = \mathbf{I}_N$) of the outputs between regions can be seen from a second example:

$$\mathbf{A}(z) = \begin{bmatrix} 1-3z & z \\ 0 & 1+2z \end{bmatrix} = \begin{bmatrix} 1 & 0 \\ 0 & 1 \end{bmatrix} + \begin{bmatrix} -3 & 1 \\ 0 & 2 \end{bmatrix} z$$

$$= \mathbf{A}_0 + \mathbf{A}_1 z.$$

Then the inverse is:

$$\mathbf{A}^{-1}(z) = \begin{bmatrix} \dfrac{1}{1-3z} & \dfrac{-z}{(1-3z)(1+2z)} \\ 0 & \dfrac{1}{1+2z} \end{bmatrix}.$$

The dependence and recursive structure of this system is also shown in fig. 4.5 and generates an overall exponentially declining response with time in each region, and an exponential declining interaction dependence for the S_{12} interaction term in the $\mathbf{A}(z)$ matrix. This has the general solution:

$$\mathbf{Y}_t = (1 - e^{\mathbf{A}t})\mathbf{X}_t \tag{4.2.13}$$

or

$$\mathbf{Y}_t = e^{\mathbf{A}t}\mathbf{Y}_0$$

in terms of the matrix exponential of $\mathbf{A} = \mathbf{A}(z)$, which is exactly identical to the state-space form of equation (2.2.15). Indeed, if the two-region system equations are defined in terms of continuous variables, we can obtain exactly analogous solutions to those given in equation (2.2.18) using Laplace transforms. The strength of the polynomial matrix representation can already be seen; e.g. the structure of the mathematical solution for the system can be derived by existing mathematics of state-space equations and N-variate equation representations apply to the N-regional case. Hence, we would expect a ready extension to the higher-order system case.

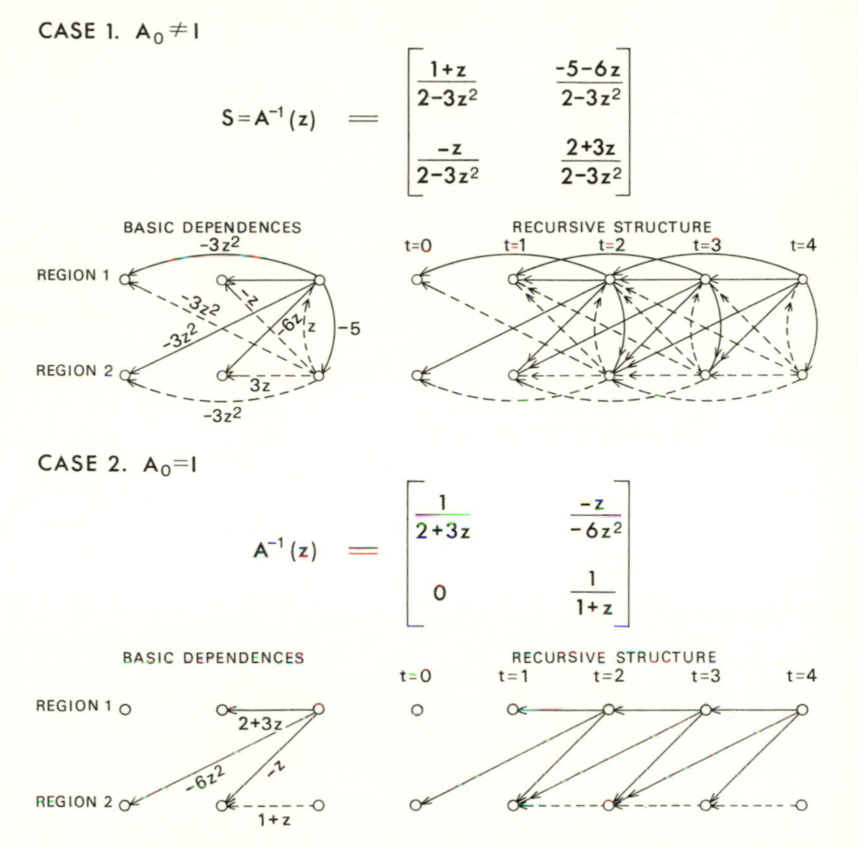

Fig. 4.5 Transfer functions for two first-order autoregressive systems ($A_0 \neq 1$ and $A_0 = 1$) in respect of a two-region case. Basic dependence and recursive structures are given. (After Bennett 1978.)

(2) *Higher-order space-time systems*

The solution to higher-order differential equations (i.e. continuous variables) and difference equations (i.e. discrete variables) for space-time systems will be more complex than for the first-order systems outlined above, but can follow essentially the same procedure. We have now merely to extend the order of lag, lead or lead-lag dependency in each independent transfer function relation S_{ij} to the full set of p and q lags given in equation (4.2.6) for the general space-time system. In each case the solution will be a sum of exponentials or complex exponentials, as in the case of the second-order systems discussed in chapter 2.2.2. The resulting matrix exponential for the purely feedback elements $\mathbf{A}(z)$ can then be defined by a block partitioned matrix in which each block represents the set of S_{ij} terms for feedback interactions between location i and location j. This block partitioned matrix is given as follows (Bennett 1975c). The system equation is first expressed in state variable form as:

$$\underset{Np \times 1}{\mathbf{Z}_{t+1}} = \underset{Np \times Np}{\mathbf{A}} \; \underset{Np \times 1}{\mathbf{Z}_t} + \underset{Nq \times Nq}{\mathbf{B}} \; \underset{Nq \times 1}{\mathbf{X}_t} \tag{4.2.14}$$

$$\underset{N \times 1}{\mathbf{Y}_t} = \underset{1 \times Np}{\mathbf{E}'} \; \underset{Np \times 1}{\mathbf{Z}_t} \tag{4.2.15}$$

where

$$\mathbf{Z}_t = [-Y_{t-1,1} - Y_{t-2,1} - \cdots - Y_{t-p1,1} - \cdots - Y_{t-1,N} - \cdots - Y_{t-pN,N}]$$

$$\mathbf{X}_t' = [X_{t,1} X_{t-1,1} \ldots X_{t-q1,1} \ldots X_{t-1,N} \ldots X_{t-qN,N}] \tag{4.2.16}$$

and the system matrix then becomes:

$$\mathbf{A} = \left[\begin{array}{ccccc|ccccc|c|ccccc}
0 & 1 & 0 & \ldots & 0 & 0 & 1 & 0 & \ldots & 0 & & & & & & \\
0 & 0 & 1 & \ldots & 0 & 0 & 0 & 1 & \ldots & 0 & & & & & & \\
 & & \ldots & & & & & \ldots & & & \vdots & & & \ldots & & \\
a_1^{11} & a_2^{11} & a_3^{11} & \ldots & a_{p1}^{11} & a_1^{12} & a_2^{12} & a_3^{12} & \ldots & a_{p1}^{12} & & & & & & \\
\hline
 & & \ldots & & & & & \ldots & & & \ddots & & & \ldots & & \\
\hline
 & & & & & & & & & & & 0 & 1 & 0 & \ldots & 0 \\
 & & & & & & & & & & & 0 & 0 & 1 & \ldots & 0 \\
 & & \ldots & & & & & \ldots & & & \vdots & & & & & \\
 & & & & & & & & & & & a_1^{NN} & a_2^{NN} & a_3^{NN} & \ldots & a_{pN}^{NN}
\end{array}\right] \tag{4.2.17}$$

For within-region dynamics, the diagonal blocks define the appropriate model structure. Between-region dynamics are defined by the off-diagonal blocks (which need not be non-zero) and there are $(N \times N)$ blocks which define the total S_{ij} interaction transfer function matrix. For each block, we may solve the system to give the time solution at any point t in the future by the same methods as in chapter 2.2.3. Hence, from equation (2.2.15) the steady state (feedback dynamic) response is given by:

$$\mathbf{Z}(t) = e^{\mathbf{A}t}\mathbf{Z}(0) \tag{4.2.18}$$

and from equation (2.2.20) the response to input stimuli X_t is given by:

$$\mathbf{Z}(t) = e^{\mathbf{A}t}\mathbf{Z}(0) + \int_0^t e^{\mathbf{A}(t-\tau)}\mathbf{BX}(\tau)\,d\tau. \tag{4.2.19}$$

The response of these general lead-lag space-time processes will be the sum of the responses of each individual interaction transfer function S_{ij} to each of the N-region inputs. For any one region the resulting impulse response will have a set of very complex superimposed properties depending upon the number, magnitude and sign (positive or negative) of the space-time transfer function parameters. A number of important cases are shown in fig. 4.6. When simple

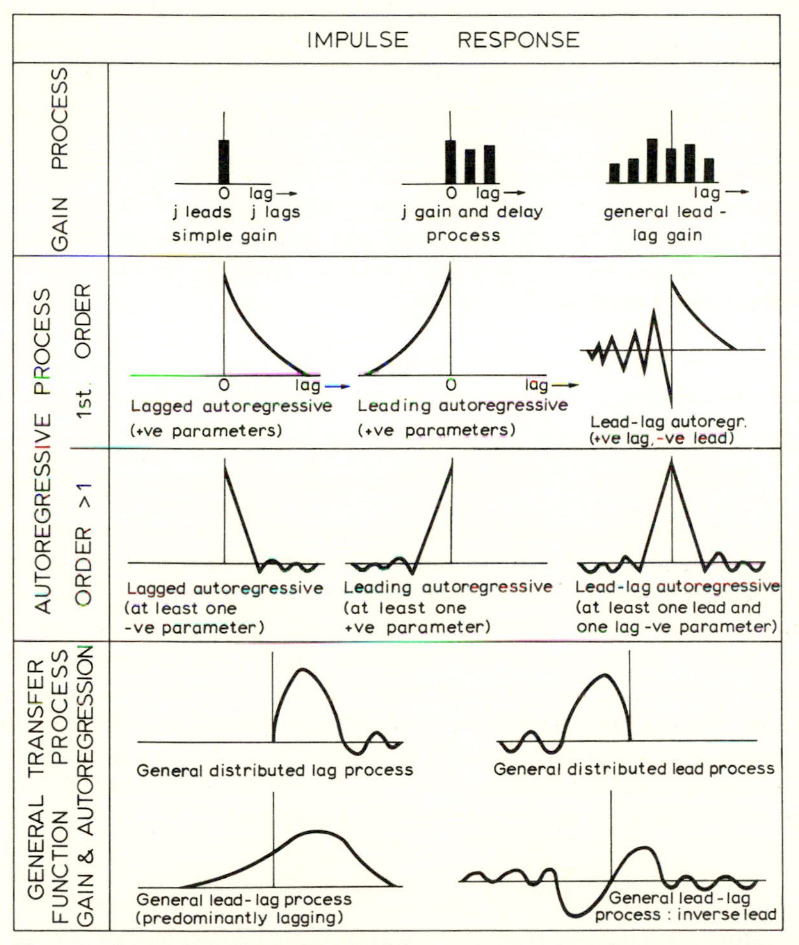

Fig. 4.6 A number of important lead-lag, space-time impulse responses, the variety of which depends on the number, magnitude and sign of the transfer function parameters. These can be compared with the cross-correlation functions from time series (see fig. 2.42). (After Bennett 1978.)

lead *or* lag processes are present, the system impulse response has the same form as the impulse response for the time series system. When a lead-lag process is present, however, the impulse response function is a combination of two time series responses, one in the negative and the other in the positive time direction. The actual form of each function is governed by the parametric structure.

The effect of these parametric modulation components is illustrated in fig. 4.7, which shows the relations between two time series for two separate interacting regions. The interzonal dependence may be confined to different linear trends, to a simple linear lead or lag, to linear amplification or damping, and to various other more complex modulations structures. Of course, any of these forms may be combined with each other, and different responses may characterize different S_{ij} linkages in the space time transfer function matrix – some sections of the matrix being made up of lead, some of lag, others of lead-lag, etc., variations.

The structure of lead-lag dependence in space-time systems is a result of the way we choose to view the system, in particular of the manner in which the system variables are defined. A pure lead process in one variable, for example, can be treated as a pure lag process in another variable. Thus the first-order distributed lead process

$$Y_t = b_1 X_{t+1} + b_0 X_t \tag{4.2.20}$$

can be transformed to a distributed lag plus autoregressive feedback process:

$$X_t = \frac{1}{b_1} Y_{t-1} - \frac{b_0}{b_1} X_{t-1}. \tag{4.2.21}$$

For more complex lead-lag relations, Bennett (1978, ch. 6) shows that an N-regional system with $(N \times N)$ transfer function matrix $\mathbf{S}$ and p-order leads for each element S_{ij} can be transformed to a pure lag with autoregressive order elements equal to $(N \times p)$. Thus a low-order space-time lead system is equivalent to a very high-order space-time lag system. Despite this large increase in the size of the parameter set that must be used to describe such systems, it is normally impossible to implement forecasting with space-time systems with lead elements since there are rarely anticipatory data available which can be used to define the $(t + 1, t + 2, \ldots)$ elements. In the following discussion, therefore, we consider only lag processes.

4.2.2 SPECIAL SPACE-TIME SYSTEMS

Of the four special space-time processes discussed earlier (barrier, hierarchy, network and contiguity), the contiguity process has received most discussion by engineers in the application of partial differential equations and by geographers searching for spatial autocorrelation. In this study a general notation has been developed for contiguity processes which can be extended

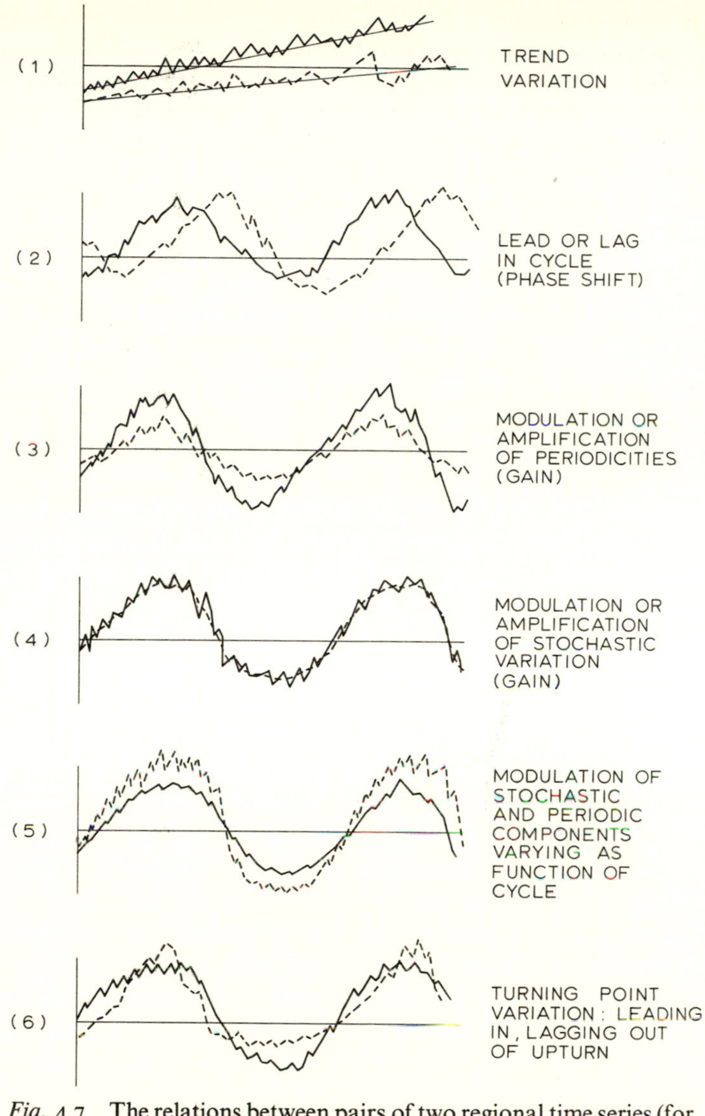

Fig. 4.7 The relations between pairs of two regional time series (for two separate, interacting regions) (i.e. space-time outputs). (After Bennett 1978.)

to the other special space-time processes. This is based on the use of a weights matrix which derives from Moran (1948), Geary (1954) and Dacey (1966). The weights matrix elements may refer to samples at a point, area integrals or region centroids. Various forms of weights matrix can be used depending upon the form of space-time interlinkage present. A number of examples of weights matrices W_{ij} are given in fig. 4.8 for a simple four-region example. The

Links matrix $\mathbf{W} = L^2$

Map	First-order links	Second-order links
1 | 2 | 3 ; 4 (below 3)	$\begin{bmatrix} 0 & 1 & 0 & 0 \\ 1 & 0 & 1 & 0 \\ 0 & 1 & 0 & 1 \\ 0 & 0 & 1 & 0 \end{bmatrix}$	$\begin{bmatrix} 0 & 0 & 1 & 0 \\ 0 & 0 & 0 & 1 \\ 1 & 0 & 0 & 0 \\ 0 & 1 & 0 & 0 \end{bmatrix}$
	Rook's case	
1 | 2 ; 3 | 4	$\begin{bmatrix} 0 & 1 & 1 & 0 \\ 1 & 0 & 0 & 1 \\ 1 & 0 & 0 & 1 \\ 0 & 1 & 1 & 0 \end{bmatrix}$	$\begin{bmatrix} 0 & 0 & 0 & 1 \\ 0 & 0 & 1 & 0 \\ 0 & 1 & 0 & 0 \\ 1 & 0 & 0 & 0 \end{bmatrix}$
	Bishop's case	
	$\begin{bmatrix} 0 & 0 & 0 & 1 \\ 0 & 0 & 1 & 0 \\ 0 & 1 & 0 & 0 \\ 1 & 0 & 0 & 0 \end{bmatrix}$	$\begin{bmatrix} 0 & 0 & 1 & 0 \\ 0 & 0 & 1 & 0 \\ 1 & 0 & 0 & 0 \\ 1 & 0 & 0 & 0 \end{bmatrix}$
	Queen's case	
	$\begin{bmatrix} 0 & 1 & 1 & 1 \\ 1 & 0 & 1 & 1 \\ 1 & 1 & 0 & 1 \\ 1 & 1 & 1 & 0 \end{bmatrix}$	

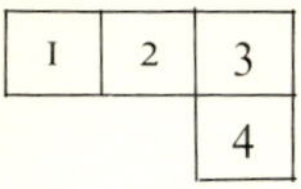

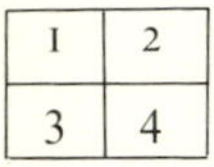

Fig. 4.8 Weights matrices for a four-region example depending on the order and nature of linkage being adopted illustrated by the moves on a chess board.

form of the weights matrix will depend upon the order of the linkage adopted, and the nature of linkage (edge to edge, or corner to corner, etc.). The weights matrix can also be extended to take into account the length of common boundary, the area of each zone, the distance between centroids, various distance transformations, etc. (see Cliff and Ord 1973). It is normal for the diagonal elements to be set to zero to constrain the sum of the elements of W_{ij} to equal unity, i.e.

$$\sum_i \sum_j W_{ij} = 1.$$

The weights matrix can also be written as an operator, L, similar to the z-operator defined in chapter 2 (Martin and Oeppen 1975). Thus,

$$L^0 Y(t, i) = Y(t, i)$$

$$L^1 Y(t, i) = \sum_{j=1}^{k} W_{ij} Y(t, j)$$

$$\vdots \qquad\qquad \vdots$$

$$L^s Y(t, i) = \sum_{j=1}^{k} \sum_{j=1}^{s} W_{ij} Y(t, j)$$

where k and s are defined either as the number of zones j which interact with zone i, or as number of zones which are s steps away from zone i (in terms of contiguities, neighbours, distance, etc.). For example, first-order nearest neighbours will be defined by L^1. k represents the sum over the k zones in the weights matrix which interact with zone i as first-order neighbours. Using this operator notation, the matrix polynomial space-time transfer function (4.2.6) can now be rewritten as

$$Y(t, i) = \sum_{k=1}^{p} \sum_{s=0}^{R} a_{ks} L^s Y(t-k, i) + \sum_{k=0}^{q} \sum_{s=0}^{R} b_{ks} L^s X(t-k, i) \quad (4.2.22)$$

where R is the appropriate order of weights matrix to allow the correct mode of interzonal linkage.

This form of representation is quite general since any form of the special space-time processes can be handled, but it is not amenable to simple mathematical solution. For contiguity systems two alternative procedures based on integral and partial differential equations are generally to be preferred. The *integral equation* case becomes (Tinline 1971, Tobler 1966) for lags T in time and S_1, S_2 in Cartesian space:

$$Y_{tij} = \sum_{k=0}^{T} \sum_{l=-s_1}^{s_1} \sum_{m=-s_2}^{s_2} W_{klm} X_{t-k,i-l,j-m} + e_{tij}. \quad (4.2.23)$$

The continuous space-time transfer function becomes:

$$Y(t, ij) = \int_0^t \iint_{-\infty}^{\infty} W(k, l, m) X(t-k, i-l, j-m)\, dk\, dl\, dm + e(t, i, j). \quad (4.2.24)$$

The *partial differential equation* case is given by:

$$\frac{\partial Y(t, x)}{\partial t} + \cdots + \frac{\partial^n Y(t, x)}{\partial t^n} + \frac{\partial Y(t, x)}{\partial x} + \cdots + \frac{\partial^n Y(t, x)}{\partial x^n} + \frac{\partial^2 Y(t, x)}{\partial t \partial x} + \cdots + \frac{\partial^n Y(t, x)}{\partial t^{n/2} \partial x^{n/2}} \quad (4.2.25)$$

for general spatial coordinate x. In discrete systems this equation is subject to spatial lag operators analogous to the z-shift operator. Such difference operators are of most utility in deriving discretization schemes for environmental systems and industrial processes where physical diffusion, conduction and heat transfer equations operate (e.g. river flow, pollution, atmospheric and other physical processes). Consider the simple one-dimensional diffusion equation:

$$\frac{\partial Y(t, x)}{\partial t} = \frac{\partial^2 Y(t, x)}{\partial x^2} + U(t, x).$$

This can be approximated using the central difference operator:

$$\frac{Y_{t+1,j} - Y_{t,j}}{\Delta t} = \frac{Y_{t,j+1} - 2Y_{t,j} + Y_{t,j-1}}{(\Delta x)^2} + U_{tj}.$$

This is equivalent to cutting a time-space slab into slices ($j = 1, 2, \ldots, N$) for finite increments Δx and Δt. This yields the solution recursively at any time t and location j, and can be reduced to state space form:

$$Y_{t+1,j} = Y_{t,j} + \frac{\Delta t}{(\Delta x)^2}(Y_{t,j+1} - 2Y_{t,j} + Y_{t,j-1}) + \Delta t U_{tj}.$$

Thus, for each location we have

$$\begin{aligned} Y_{t+1,1} &= Y_{t,1} + \frac{\Delta t}{(\Delta x)^2}[Y_{t,2} - 2Y_{t,1} + Y_{t,0}] + \Delta t U_{t,1} \\ &\vdots \\ Y_{t+1,N} &= Y_{t,N} + \frac{\Delta t}{(\Delta x)^2}[Y_{t,N+1} - 2Y_{t,N} + Y_{t,N-1}] + \Delta t U_{t,1} \end{aligned}$$

or

$$\mathbf{Z}_{t+1} = \mathbf{A}\mathbf{Z}_t + \mathbf{B}\mathbf{U}_t \tag{4.2.26}$$

and where,

$$\mathbf{Z}_t' = [Y_{t,1}, Y_{t,2}, \ldots, Y_{t,N}]$$

$$\mathbf{A} = \begin{bmatrix} 1-\mathbf{\Delta} & 1 & 0 & 0 & \ldots & & & \\ 1 & 1-2\mathbf{\Delta} & 1 & 0 & \ldots & & & \\ 0 & 1 & 1-2\mathbf{\Delta} & 1 & \ldots & & & \\ \cdot & \cdot & \cdot & \cdot & \cdot & \cdot & \cdot & \cdot \\ 0 & 0 & 0 & 0 & \ldots & 0 & 1 & 1-\mathbf{\Delta} \end{bmatrix}$$

$$\mathbf{B} = \Delta t\mathbf{I}; \quad \mathbf{\Delta} = \frac{\Delta t}{(\Delta x)^2}.$$

This describes the evolution of the system state through time and along the linear one-dimensional spatial region governed by the discretization distances Δt in time and Δx in space. Both the integral and partial differentiation equation solutions are based on assuming that the space-time system has a continuous spatial, as well as temporal, lag structure. This is restrictive but does help us to make clear the assumptions involved in such representations. In particular the spatial analogues of time stationarity are required. This involves not only distance across space, but also direction. Thus a stationary space-time process must satisfy:

1 *Homogeneity*: (i.e. distance invariance) the variable $Y(t, x_1, x_2)$ being invariant under translations: $Y(t, x_1 \pm K_1, x_2)$, etc. for arbitrary K.
2 *Isotropy*: (e.g. direction invariance) the variable $Y(t, x)$ being invariant under permutation or rotation of the spatial coordinate axes $Y(t, x_1 \pm K, x_2 \pm s)$ for any K, s.
3 *Stationarity*: (i.e. time invariance) the variable $Y(t, x)$ being invariant under permutations and rotation of both the space and time coordinate axes $Y(t \pm p, x_1, x_2)$ for any p.

Processes satisfying these three conditions simultaneously are space-time stationary, or *solenoidal* (Bass 1954), in that the vector derivative has zero divergence, as shown in fig. 4.9. If the space-time system does not have vanishing divergence then there is a source or sink (Preston 1974), or there is more coming out of the small unit cube δt, δx_1, δx_2 in fig. 4.9 than is going in, or *vice versa*. This characterizes many pollution processes, fluid dispersal in river channels, and biological dispersal mechanisms. It is almost always the case in socio-economic systems.

The assumption of contiguity processes (i.e. spatial lag structures) and other special processes allows a considerable degree of simplification to the resulting space-time process and has been adopted in most engineering applications and also in geography. It has the particularly useful properties that the resulting partial differential equations yield the possibility of analytical solutions (usually in terms of Bessel functions; Whittle 1963a). However, the application of the mathematical representation is limited by the fact that few environmental systems follow pure spatial contiguity or lag solutions. This is in contrast to time series dynamic processes which may frequently be considered to operate such that an impulsive input into the system propagates in a unidirectional functional manner:

$$Y(t) = f_1(Y(t-1)) = f_2(Y(t-2)) = \cdots \qquad (4.2.27)$$

This has the especially important Markov property noted by Whittle (1963a). Because of this relation it is reasonable that dynamic systems should be represented by the order of time lag involved. 'In the spatio-temporal case, however, such sequences of dependence need not necessarily exist. Commonly, barriers, hierarchies, networks and distortions of the economic and physical space occur such that the structure of the spatial and temporal transfer function cannot be characterized by simple lagged dependence operators' (Bennett 1978, ch. 6). Although the L operator method allows these developments, the $(N \times N)$ matrix transfer function developed for interactive systems in the previous section permits more general mathematical solutions. The L operator form discussed here is useful only in certain purely spatial processes discussed in section 4.3 or when it is possible to reduce the order of the $(N \times N)$ transfer function matrix by introducing a degree of previous

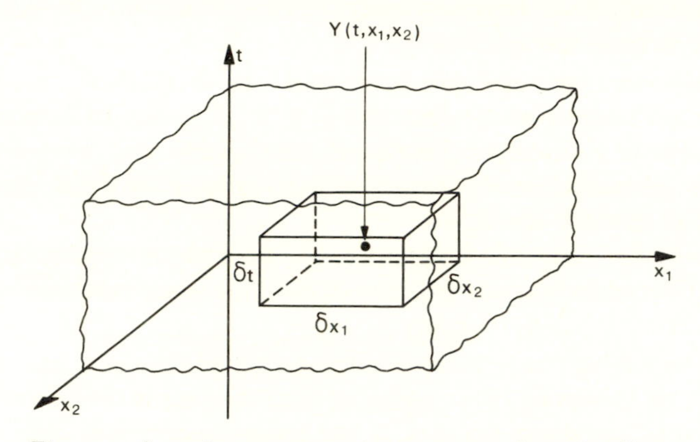

Fig. 4.9 A stationary space-time system, where the divergence is defined by:

$$\nabla Y(t, x_1, x_2) = \alpha \frac{\delta Y(t, x_1, x_2)}{\delta t} + \beta \frac{\delta Y(t, x_1, x_2)}{\delta x_1} + \gamma \frac{\delta Y(t, x_1, x_2)}{\delta x_2} = 0.$$

This expression for the divergence will equal zero only for the stationary space-time process in which the flow into the infinitesimally small box $(\delta x_1, \delta x_2, \delta t)$ is equalled by the outflow.

knowledge, e.g. that space-time interaction can be expressed as some simple function of the linkage between regions (e.g. adjacency, distance, network configuration, etc.). This requires a degree of system knowledge which is not usually available.

4.2.3 IDENTIFICATION AND ESTIMATION OF SPACE-TIME SYSTEMS

In the analysis of space-time systems a similar set of observational and theoretical restrictions arise to those discussed in the context of dynamic systems (ch. 2.3), i.e. from observability, controllability and measurement deficiencies, from the presence of stochastic errors, and from the incremental way in which we acquire knowledge.

For the case of theoretical restrictions, fig. 4.10 shows for a (2×2) regional system transfer function matrix that the effect of zero and non-zero elements in the system matrix (**S**) leads to various special forms of space-time interaction. When all elements of $S_{ij} \neq 0$, then there will be four transfer function terms and in general $(N \times N)$ terms, each of which represents the *cascade* of the input into one region affecting the output in another region. There may be cascaded outputs as well. When the leading diagonal of **S** contains the only non-zero components, the two regions have independent temporal evolution, do not interact and are *parallel* systems. Various other special forms are possible. When S_{ij} contains a row of zeros, the space-time system is only partially observable since some of the output effects cannot be observed. When S_{ij} contains a column of zeros, the space-time system is only partially controllable, since some of the input effects cannot be modelled.

To be valid the space-time system transfer function of the $(N \times N)$ operator between subregions must generally satisfy the four properties (Weissbrod 1974, p. 23):

1. $f(Y_{ti}, Y_{tj}) > 0, \quad i = j$ (non-zero diagonal property)
2. $f(Y_{ti}, Y_{tj}) > 0, \quad i \neq j$ ($(N \times N)$ interaction property)
3. $f_1(Y_{ti}, Y_{tj}) = f_2(Y_{tj}, Y_{ti})$ (asymmetry property)
4. $f_1(Y_{ti}, Y_{tk}) = f(Y_{tk}, Y_{tj}) + f(Y_{tj} Y_{tk})$ (additivity and superimposition properties)

which give the important superposition property between zones. When a row-column combination has only one non-zero item, the space-time system is partially uncontrollable and partially unobservable, since only one output effect can be measured and only one input term registers on the system output.

Exploiting the analogy between N-variate processes and N-regional space-time processes, the necessary and sufficient conditions for a space-time system to be observable and controllable can be stated as follows. For the system defined in equation (4.2.14) and (4.2.15) with $\mathbf{Z}_t$ a vector of space-time state variables and $\mathbf{X}_t$ vector of input variables, then:

$$\begin{bmatrix} Y_0 \\ Y_1 \\ \vdots \\ Y_{Np} \end{bmatrix} = \begin{bmatrix} \mathbf{E}' \\ \mathbf{E}'\mathbf{A} \\ \vdots \\ \mathbf{E}'\mathbf{A}^{Np-1} \end{bmatrix} \mathbf{Z}_0 = \mathbf{F}\mathbf{Z}_0 \tag{4.2.28}$$

is observable if **F** has rank Np. It is controllable if the matrix **G** in

$$[\mathbf{B}, \mathbf{AB}, \ldots, \mathbf{A}^{Np-1}\mathbf{B}] \begin{bmatrix} X_{Np-1} \\ X_{Np-2} \\ \vdots \\ X_0 \end{bmatrix} = \mathbf{GX} \tag{4.2.29}$$

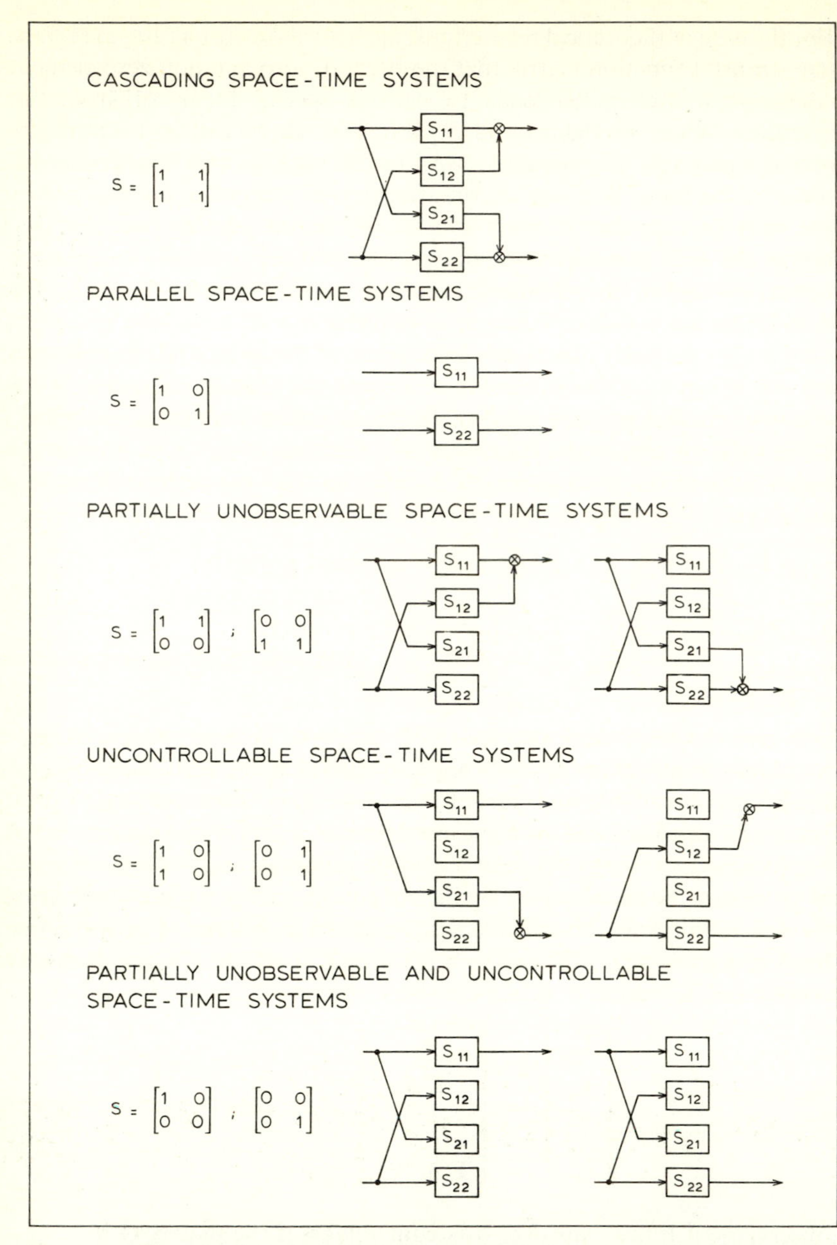

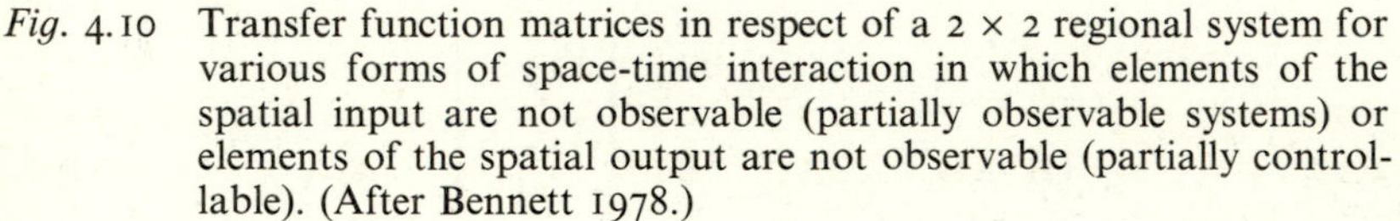

Fig. 4.10 Transfer function matrices in respect of a 2×2 regional system for various forms of space-time interaction in which elements of the spatial input are not observable (partially observable systems) or elements of the spatial output are not observable (partially controllable). (After Bennett 1978.)

is also of rank Np, where the rank of a matrix is as defined in chapter 2 (the common value of its column and row ranks, or the order of the highest non-vanishing minor, see Appendix 1). These conditions, together with those given by Weissbrod above, determine the conditions under which it is theoretically possible to observe and/or control various components of the time-space system.

The effect of practical observational restrictions is identical to that discussed in chapter 2, the effect being to introduce uncertainty into the structure of the space-time input-output relations such that any system model must be identified and estimated and checked for accuracy (the same stages which are summarized in fig. 2.40). Identification in this context requires the orders p and q of the lag operators in the system transfer function to be determined for each element of interaction within and between each region. We restrict attention here to the general interactive systems discussed in section 4.2.1. For these systems there is a massive identification problem for which $(N \times N \times p)$ transfer functions orders must be determined. For the identification of multivariate time series Whittle (1963b), Quenouille (1957) and Hannan (1969) have discussed time domain schemes which employ the system autocorrelation, cross-correlation and partial correlation functions, together with associated standard errors, in analogous manner to the univariate case discussed in section 2.3.3. In the frequency domain, multivariate frequency response functions have been discussed by Granger and Hatanaka (1964) and by Jenkins and Watts (1968). These methods can all be applied to the space-time identification problem using figs 2.41 to 2.44 as diagnostics of model order. But the size of most regional systems and the degree of dependence between the variables in each region makes the use of such methods far from simple. A special difficulty is that the inputs into each of the N regions (and within any one region) are normally highly linearly dependent. In addition the output from each of the N regions which is fed back will also be highly linearly dependent between regions, as well as within any one region. These linear dependencies, termed *multicollinearity* (Johnston 1972) lead to indeterminate results after applying most identification techniques. More specifically, multicollinearity arises for four reasons:

1 The dependence *between* the input into region i and that into region j.
2 The dependence *between* the output from region i and the output from region j.
3 The dependence between the inputs *within* region i (due either to the presence of lag terms or collinearity between regressors).
4 The dependence between the outputs *within* region i.

To overcome this problem attempts can be made to construct *orthogonal* sets of input and output variables. Orthogonality in this case requires that each input is uncorrelated with each other input in each of the N regions and within each region, and similarly for each output. Such an approach leads to a *canonical factorization* of the system transfer function in which each orthogonal input or output variable (i.e. canonical variable or factor) will be a linear sum of the various original input and output variables. Techniques for accomplishing this orthogonal transformation of the system are based on canonical correlation analysis (Akaike 1973) and factor analysis and are summarized in Bennett (1978, ch. 7).

The application of canonical factorization leads to the definition of simple and unique system equations, but this is generally at the expense of the ease with which the resulting orthogonal variables and parameters can be interpreted. As these are linear sums of the original system variables, the orthogonal (canonical) variables are a confused sum of partial effects of each of the original variables weighted by an additional set of estimated parameters (the canonical loadings or factors).

Because of this interpretation problem, many practical applications of space-time systems identification have ignored the multicollinearity issue (Bennett and Haining 1976) and have instead adopted the univariate techniques discussed in chapter 2. Thus, in an exactly analogous manner to the time series analysis methods discussed in chapter 2, the time domain identification of the lead-lag dependencies between each region i and each other region j $(i, j = 1, 2, \ldots, N)$ can use the time series diagnostic of the cross-correlation function. This will have precisely the same form as the cross-covariance between regions given in chapter 2, provided that lead relationships have been eliminated. In the frequency domain, cross-spectral analysis performs a comparison of the decompositions for the time series of each two-region comparison with respect to the frequency components (see Granger 1969). In the Bode plot, the phase shift represents the time lag between one frequency component in one region and the corresponding frequency in the other region, and the gain indicates the relative size or amplification of the corresponding frequencies in the two regions. A value of one indicates that the cyclical components at the given frequency have identical amplitudes in the two regional series; a value greater than one indicates that the cyclical component in the first regional series (at the given frequency) has a greater amplitude than the cyclical component at the same frequency in the second regional series. For example, Haggett (1971) and King *et al.* (1969) have applied simple (1×1) identification to the relation between unemployment time series. Cliff, Haggett *et al.* (1975) have extended this analysis to cover measles and other epidemiological systems. Bennett (1975a), in a regional economic model of Northwest England, identified a five-output, 69-region, space-time system using univariate techniques. Bennett notes that the use of univariate techniques in this space-time context will *not* give the correct identification diagnosis since the partial effects of other endogenous (autoregressive feedback) variables are not included and controlled in the conditional expectation of any individual equation for any one region. Univariate techniques will not eliminate interaction and degeneracy effects between variables so that, for example, it will not be possible to differentiate an autoregressive moving average process in N variables (or regions) from an N-order process with correlated residuals in one variable.

If the linearity and additivity of all inputs and outputs in the system can be assumed, however, then univariate techniques may give a good approximation:

> Assume that all input series are held zero for all time, except for the one input series j, which is in no way restricted. Furthermore, let us assume that there is no output noise in the system. We then measure the variations of output series Y_i, from its steady state level, and ignore the other output series.... We consider ourselves to be in a univariate situation which can be modelled independently of the 'ignored' systems. (G. T. Wilson 1970, p. 40)

This theoretical experiment is valid only if it can be assumed that the jth output is unaffected by other outputs i at the same sampling instant. This, in turn, requires that there is no simultaneous feedback between regions; i.e. that $\mathbf{A}_0 = \mathbf{I}_N$ in equation (4.2.6). In practical instances, univariate identification may be used as a simple and rapid technique to generate first hypotheses, perhaps preliminary to canonical factorization (Bennett 1978).

For unbiased and efficient estimation of the parameters of space-time systems the identification of a non-degenerate set of system equations is a prerequisite. Special conditions specifying identifiability criteria which permit efficient parameter estimation are given in Bennett (1978, ch. 7). It is sufficient to note here that if the space-time system is specified in terms of mutually independent inputs and outputs (i.e. no multicollinearity), then multivariate least-squares can be used for parameter estimation and yields an estimator identical in form to the univariate OLS estimator (2.3.9). Ord (1975), for example, gives the multivariate least-squares estimator:

$$\hat{\boldsymbol{\theta}} = [\mathbf{X}'\mathbf{X}]^{-1}\mathbf{X}'\mathbf{Y} \tag{4.2.30}$$

where

$$\mathbf{X}' = [\mathbf{X}_1, \mathbf{X}_2, \ldots, \mathbf{X}_T]$$

$$\mathbf{X}'_t = [-Y_{t-1,1} - Y_{t-21} - \cdots - Y_{t-p1,1} - \cdots - Y_{t-1,N} - \cdots - Y_{t-p_N,N} X_{t,1} X_{t-1,1} \ldots X_{t-q1,1} \ldots X_{t,N} \ldots X_{t-q_N,N}]$$

$$\mathbf{Y}' = [Y_{t1} Y_{t2} \ldots Y_{tN}]$$

$$\boldsymbol{\theta}' = \mathbf{A}^{-1}(z)B(z). \tag{4.2.31}$$

This estimator differs from (2.3.9) only in as far as the parameter and variable vectors now include all ($N \times N$) terms, rather than the single univariate series for any one location. The use of multivariate OLS (as given in equation (4.2.30)) to estimate the parameters of the N-regional system has a number of advantages (Bennett and Haining 1976). First, the application is extremely simply accomplished, as shown by Cliff and Ord (1975) and Bennett (1975a). Second, the N-variate sampling theory, significance tests and other properties of the multivariate estimators can be assumed to apply to the $(N \times N)$ transfer function input-output matrix of the N-regional system. Hence, geographic applications can go forward using these estimators as a basis. However, as pointed out above, a considerable degree of collinearity will usually be present, and there will be marked loss of degrees of freedom. Considerable simplification can be achieved if *a priori* restrictions are introduced, such as contiguity constraints, distance decay, etc. But most progress will be made by the derivation of practical small sample estimators (including estimators on the boundary) specifically designed for space-time estimation problems. Hence, there are at present considerable research needs in this area. For example, multivariate OLS will be unbiased and efficient only if the elements within $\mathbf{X}$ and within $\mathbf{Y}$ are each linearly independent of each other, *and* if the estimation errors are also independent in time (i.e. lack autocorrelation) and are independent between equations. If any of these conditions is not met, alternative estimators to least-squares must be sought, for example, based on multivariate maximum likelihood, generalized least-squares and other methods (Johnston 1972, Anderson 1958). These can be combined with nonlinear and non-stationary (adaptive) transfer function estimators as discussed in later chapters.

4.3 Purely spatial processes

We discuss in this section the special case of purely spatial processes in which simultaneous dependence occurs between the output in each of the N regions in the system of interest. This is equivalent to the case when $\mathbf{A}_0 \neq \mathbf{I}_N$ in the system equations (4.2.6A) and (4.2.6B). For the spatial distribution of unemployment, for example, this allows the level of unemployment in one region to depend upon the level of unemployment in another region *at the same time instant*, there being no allowance for any diffusion time for the effects of unemployment outputs in one region to reach the other region. This type of process has very important implications since it states that causality exists without any measurable response time being observed. With such processes time need no longer be considered.

The discussion of purely spatial systems is a difficult field, different from that which arises with the space-time systems discussed above. It requires different mathematical assumptions and manipulations, and draws on different fields of interest and classes of problems since the concept of time is normally considered to be fundamental to the understanding of *processes* as sequences of causal or functional ordering. This lies behind each of the philosophical system structures discussed in chapter 1 and is also crucial to the representation of spatio-temporal systems as discussed in this chapter. It is philosophically problematic how spatial relationships can be appreciated out of their temporal context, although geographers have been notably neglectful in realizing this obstacle in the 1960s. Bunge (1966) and Haggett (1965), for example, place great emphasis on purely spatial geometry. More recent studies of spatial autocorrelation (Cliff and Ord 1973 1975) have hypothesized purely spatial generating processes drawing on statistical models given by Whittle (1954, 1963a). Hägerstrand (1973, p. 81) notes that 'how far momentarily observed relations, measured in terms of distances, configurations and densities can reveal the counterpoint structure over time must remain a matter for investigation.'

Although purely spatial processes are philosophically highly problematic, Bennett (1978) notes that their study may be of interest for three reasons: first, they may, under special conditions, actually exist; second, they may seem to exist; and third, there are a number of important purely spatial problems. Spatial processes do actually exist when the system under study has reached equilibrium. A static equilibrium pattern, such as a stable crystal lattice, potential energy field, etc., or a dynamic equilibrium (steady-state) pattern such as characterizes thermodynamic gas laws, the Ising model of ferromagnetism, and other physical phenomena (Whittle 1963a), may arise in many situations in the physical sciences. Alternatively, some processes might be considered to be governed by infinitely short reaction and relaxation times: an impulsive input entering the system has immediate effect which decays to zero in infinitely short time. Equilibrium and zero time attenuation are conditions which seldom if ever arise in socio-economic or environmental systems, and indeed characterize physical systems only under exceptional conditions which are usually only theoretically realizable.

More usually, purely spatial processes may *seem* to exist to the observer or analyst. This will occur in some of the situations listed in table 2.2 and when the space-time data are available with only very wide time separation which does not allow the recognition of reaction, relaxation and lagged dynamics, as shown in fig. 4.11. This problem arises with decennial or even quinquennial censuses, and renders them of doubtful utility for many systems analysis problems. The necessity to consider purely spatial processes most commonly arises, however, from the availability of only a single cross-section of data – a single map, one-off sample surveys or censuses. This is really a special case of the widely separated temporal sampling internal case. Again no information on short-term adjustment dynamics will be available.

In other cases, the lack of adequate temporal data may be no disadvantage if a realistic purely spatial problem exists. For example, spatial extrapolation, interpolation of missing data points, tests of residuals from regression analysis, and selected problems of statistical inference, are meaningful spatial studies. For spatial extrapolation models, Bunge (1966) suggests that the spatial model of a known map can be used to predict spatial patterns in adjacent maps. Spatial forecasting can be distinguished as the determination of the values of functions within the map – especially at missing data points.

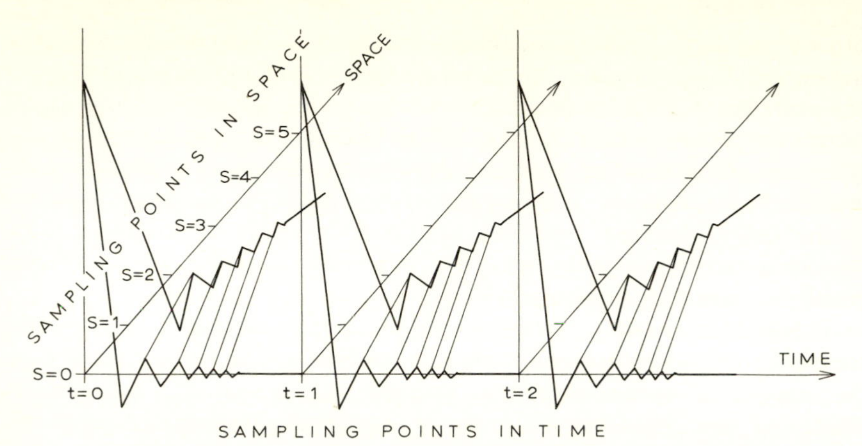

Fig. 4.11 Space-time dependence decaying to zero before the next observation in time. From the widely spaced observations the lag adjustment dynamics of the system cannot be observed and the spatial responses *appear* to be simultaneous across the map.

An important subset of the latter techniques is the Kriging method (Krige 1966, Agterburg 1970). Spatial models may also give indications of the likely form of space-time generating processes, but the mapping from space to space-time models is seldom unique (Whittle 1963a), in that many space-time models can generate the same spatial pattern. Models of purely spatial processes have been justified by Besag as attempting to describe the here and now of a wider process.

> In many practical situations this is a remarkable standpoint, since we can only observe the variables at a single point in time.... Ideally, even when dealing with a process at a single instant of time, we should first set up an intuitively plausible spatial-temporal model and then derive the resulting instantaneous spatial structure. This can sometimes be done if we are prepared to assume stationarity in both time and space but, unfortunately, such an assumption is unlikely to be realistic.... However when this approach is justifiable, it is of course helpful to check that our spatial models are consistent with it.... Otherwise, regarding the transient spatial structure of a spatial-temporal process, this (stationarity assumption) is almost always intractable and hence there exists a need to set up and examine purely spatial schemes without recourse to temporal considerations. (Besag 1974, p. 193)

A further problem in the analysis of spatial data is the effect of stationarity (in this context, homogeneity and isotropy). This is a very restrictive assumption; as Granger (1969) notes, this implies that the relationship between variates measured at Oxford and London will be the same as between variates measured at two Lincolnshire villages 55 miles apart, or indeed any other locations at the same spatial distance which are considered to be within the same stationary region. This requires that we deal with restricted areas of considerable internal homogeneity at least with respect to covariance properties, since trends in the mean can be handled using spatial differencing (Bennett and Haining 1976). The correlation of purely spatial processes can be modelled by defining $\mathbf{A}_0 \neq \mathbf{I}$ in equation (4.2.6) which can be written:

$$Y_i = A_0^{-1} B_0 U_j + e_j \qquad (4.3.1)$$

which is a purely spatial transfer function expressing dependence in region i upon zones ($j = 1, 2, \ldots, N$) in the spatial region. However, systems of the form of (4.3.1) are of little utility when parameter estimation is required, since the number of parameters in the matrices, which are each of ($N \times N$) dimension, will generally exceed the number of observations available. To overcome this loss of degrees of freedom, the weights matrix proposed by Cliff and Ord (1973) specifies the spatial transfer function as:

$$Y_j = \sum_{s=0}^{R} a_s L^s Y_j + \sum_{s=0}^{R} b_s L^s U_j + e_i \qquad (4.3.2)$$

where R is the maximum order of spatial weighting involved, and s is the degree of the weights matrix (compare equation (4.2.22)).

Two particularly important processes, based on spatial lags, are the *unilateral* and *multilateral* processes defined, respectively, as:

$$Y_{ij} = a(Y_{tij-k} + Y_{ti-lj}) + e_{tij} \quad (k, l = 1, 2, \ldots) \qquad (4.3.3)$$

and

$$Y_{tij} = a(Y_{ti-kj} + Y_{ti+kj} + Y_{tij-l} + Y_{tij+l}) + e_{tij} \quad (k, l = 1, 2, \ldots) \qquad (4.3.4)$$

where a Cartesian lattice (i, j) has been assumed. The parameter (a) must satisfy $|a| < \frac{1}{2}$ for the spatial process to be stationary in the unilateral case, and $|a| < \frac{1}{4}$ in the multilateral case (Whittle 1954, 1963a, Haining 1977). In the special case $k = l = 1$ we have first order unilateral and multilateral processes respectively. For $k = l = 1$ in equation (4.3.4) the term quadrilateral process is frequently used. In one dimension, the multilateral process reduces to a bilateral representation. Multilateral processes are generated in many models of interplant competition in ecology (Matern 1947, 1960, Mead 1971), whilst bilateral processes are especially important to describe network processes (Haining 1977). Unilateral models are characterized by at least one spatial axis being time-like (Whittle 1954, Heine 1955) or having a unidirectional dependence structure (for example, the diffusion of foot and mouth disease under the influence of a strong prevailing wind (Tinline 1971). Fig. 4.12 shows a characteristic spatial sample of each process, where the time cross-sections show the evolution pattern that would occur if a time series were available.

The philosophical difficulties in handling purely spatial processes also affect the statistical methods of system specification and estimation that can be employed. The essence of the spatial definition problem is that Y_{ti} is included in the conditional *pdf* of Y_{tj} when the value of Y_{ti} itself depends upon Y_{tj}; the Jacobian of the transformation from the system errors to the system output is not unity (Whittle 1954, 1963a). Hence if two entities A and B are found located together across a spatial field or map, on the basis of that single map

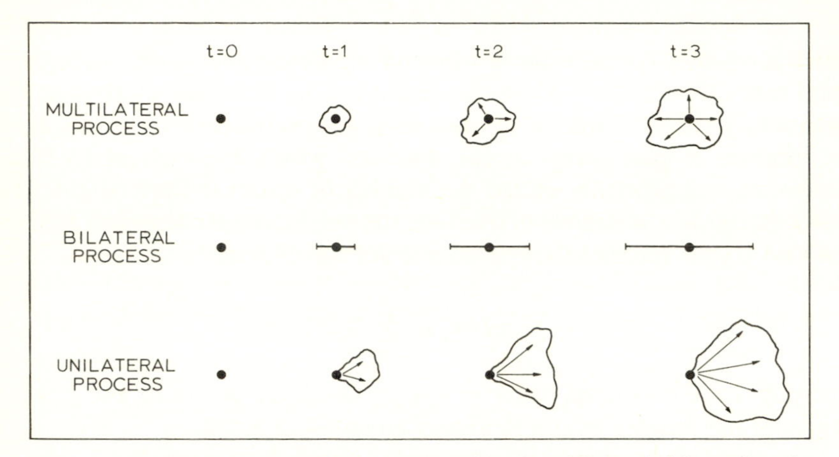

Fig. 4.12 Spatial examples through time (t) of multilateral, bilateral and unilateral processes.

realization, or indeed upon even an infinite number of realizations, we cannot differentiate between

1. $A = f(B)$
2. $B = f(A)$
3. $A, B = f(C)$

or a number of other generating hypotheses. The only methods of resolving the question of the chain of dependence is to invoke either exogenous *a priori* behavioural information, i.e. we know *ex hypothesi* that A causes B or B causes A; or to invoke the Markov property for equilibrium spatial fields. In the time series dynamic systems of chapter 2, the conditional probability function is

$$p(Y_t \mid Y_o, Y_1, \ldots, Y_{t-1}) = p(Y_t \mid Y_{t-1}) \tag{4.3.5}$$

and is equivalent to the *Markov property* – that the value of Y_t is dependent only upon the value at Y_{t-1}. This can be expressed as *equivalent* to the joint *pdf* relation

$$p(Y_1, Y_2, \ldots, Y_t \mid Y_0) = \prod_{i=1}^{t} Q_i(Y_i, Y_{i-1}) \tag{4.3.6}$$

giving the product of the joint probabilities from the initial conditions. In the purely spatial case, however, it is not possible to identify the conditional probability definition with the joint *pdf* in the form given in equation (4.3.6). Instead it is necessary to hypothesize a Markov random field structure for the internal sites $\mathbf{Y}_I$ and boundary sites $\mathbf{Y}_B$ of the spatial system. Bennett (1978, ch. 7) notes that it would be desirable to specify this structure as $p(\mathbf{Y}_I \mid \mathbf{Y}_B)$, that is the probability of the values of Y_{ij} for initial data points $(ij) \in D$, conditional upon the boundary data $Y_{ij} \in D$. This cannot be achieved directly because no ordering is available and the conditional probability involves dependence upon itself. More especially, Besag (1972) notes that it will not be possible to define the joint probabilities by the multiplication relation:

$$p(Y_{MN}, \mathbf{Y}_{I^*} \mid \mathbf{Y}_B) = p(Y_{MN} \mid \mathbf{Y}_{I^*}, \mathbf{Y}_B)p(\mathbf{Y}_{I^*} \mid \mathbf{Y}_B) \tag{4.3.7}$$

where $\mathbf{Y}_{I^*}$ is the set of internal (MN sites excluding the MNth, when D is the domain of interest with boundary ∂D). Instead, it is possible to apply the Hammersley and Clifford (1971) theorem to the spatial Markov field (see also Preston 1974, Moussouris 1974). This yields a ratio of probabilities:

$$p(\mathbf{Y}_I \mid \mathbf{Y}_B)/P(\mathbf{Z}_I \mid \mathbf{Y}_B) \tag{4.3.8}$$

where Z_I is a new variable for which $Z_{ij} = Y_{ij}$ (for $ij \in I$) and may take any value for $(ij \in I)$. This approach expresses the ratio probabilities of the given map realization for sites internal to the region $\mathbf{Y}_I$ conditional upon the boundary data $\mathbf{Y}_B$, to the realization of a map $\mathbf{Z}_I$, again conditional upon the boundary data $\mathbf{Y}_B$ (Moussouris 1974, Besag 1974). By defining $\mathbf{Z}_I = 0$ for all $(ij \in D)$ we obtain the conditional Markov definition of equation (4.3.7) as the ratio of the probabilities of the actual realization to that of the probability of a map of all zeros, i.e. the joint *pdf* becomes:

$$Q_{ij} = \ln\{p(Y_{ij}) \mid p(0)\}. \tag{4.3.9}$$

The spatial Markov property applies only to equilibrium situations and is hence highly restrictive since it involves the assumption of *time-reversibility*, i.e.

$$\Pi(A)G(A, B) = \Pi(B)G(B, A) \quad \text{for all } A, B \in D. \tag{4.3.10}$$

Preston (1974, ch. 2) shows that any state Π satisfying this condition must be an equilibrium state, and thus must be *the* equilibrium state of the spatial system. Time-reversibility requires that 'if a film were made of the time evolution of the model, with the model started in an equilibrium state, then it would not be possible to distinguish (in statistical terms) between the film run forwards and the film run backwards' (Preston 1974, pp. 11–12), and no additional information is gained from observing the temporal evolution of the system. The use of the spatial Markov property allows the identification and estimation of models of purely spatial processes and may often be very artificial. It does, however, allow the purely spatial problem to be handled.

Other special assumptions which allow treatment of spatial models are discussed by Bennett (1978, ch. 7).

Purely spatial processes are abstract constructs which describe either very special systems or unique measurement conditions. They will be useful in certain contexts, e.g. patterns of wheat yield or voting behaviour (Haining 1978), but they permit functional interpretation only by the use of *a priori* hypotheses – for example, maximum entropy assumptions and constraints on interaction between locations (Wilson 1974, Bennett 1975a). Also purely spatial models do not allow the definition of the optimal control systems discussed in chapter 3. As such, purely spatial processes are special cases of more general space-time interaction systems, and are treated in the following chapters mainly in those contexts where the ergodic hypothesis is employed to generate forecast futures in the absence of temporal data (for example, in chapter 9).

4.4. Space-time control systems

In the previous sections of this chapter system analysis representations have been developed to deal with problems involving the specificity and relativity of location. In the control of space-time systems the impact of space must again be incorporated. 'The planning of spatial location will not work well until we have a better understanding of how projects as wholes must be accommodated in a space-time budget space' (Hägerstrand 1973, p. 81). The theoretical developments which allow the control of space-time systems are very restricted. In engineering most of the mathematics is restricted to the special contiguity case in which the system is governed by partial differential operators such as equation (4.2.25). In economics and geography the main developments have been of either once-and-for-all decisions on the location allocation of projects, or have dealt with the control of purely dynamic systems, as discussed in chapter 3. For example, Hägerstrand (1973, pp. 81–2) notes that:

> Most quantitative techniques so far applied . . . seem to be best fitted to deal with the real-world situation of an old-fashioned, stable rural environment where friction of distance is immensely high and projects related to human action are on the whole repetitive and restricted to compact space-time 'bubbles' which are elongated in time, but very narrow in space.

In this section we develop an integrated approach to space-time systems, *qua* space-time systems, governed by the transfer function matrix S_{ij} of elements of interaction between region i and region j.

Space-time control systems can be seen as a more general case of the control systems discussed in chapter 3. They can be manipulated by feedback and other regulators, by man-machine interaction, and changes in system structure (reorganization). In each case, however, the controller, like the system itself, must take account of the specificity and relativity of location and will be space-time dependent. There are five important categories of space-time controllers:

1 *Dynamic control*: This reduces the space-time system to a single aggregated *lumped* control system in which no spatial dependence is taken into account. This corresponds to the purely time series case discussed in chapter 2, but ignores spatial linkages. There are two subcategories:
(*a*) Area-integrated control, in which the applied dynamic control is uniform over the entire spatial region of interest – e.g. national macro-economic policy, interregional legal regulations, etc.
(*b*) Boundary control, which is applied at the margin of the spatial region and relies upon the internal diffusivity of the system – for example, tariff barriers and frontiers in economic trade and migration plans, heating of the outside of a metal object, etc.
2 *Spatial control*: This is a second reduction of system control to a simplified form in which spatial variation is taken account of, but no variation in control over time is possible. Again, there are two subcategories:
(*a*) Static spatial control, involving a once-and-for-all decision on spatial layout or structure. For example: the spatial airframe of an aircraft or rocket, building design (Butkovskiy 1969); the arrangement of wells in an oil field; the layout of services in a city (hospitals, schools, police stations, shopping centres, etc.); and administrative zoning (Scott 1971, Massam 1975).
(*b*) Dynamic spatial control, involving a once-and-for-all decision on spatial control strategy in a dynamic equilibrium situation. A continuous but unchanging input in time is required – for example: static tax structures; regional subsidies; allocation of economic growth between industrial sectors in economic policy (Mennes *et al.* 1969); continuous steam and temperature feeds in industrial processes; and continuous flows of nutrients and energy in ecosystems.
3 *Space-time regulation and servomechanisms*: If full account is taken of spatial dependence and temporal dynamics in the control system, it will be possible to adjust control strategies in space and time to keep system variations as close as possible either to a fixed reference point (regulator), or to track a changing reference value (servomechanism). There will be important differences in the form of these control policies depending upon the nature of the space-time system.
(*a*) Space-time interaction systems (including barrier, hierarchy and network processes) will require treatment by control algorithms developed from equations (4.2.6A) and (4.2.6B) – for example: regional economic policy; federalist control; differential heating in furnaces; controlled extraction from oil strata; and transport scheduling.

(*b*) Space-time contiguity systems in which the dependence of the spatial system on explicit functions of lag and distance allows some simplified control algorithms to be developed using solutions to partial differential equation systems (Butkovskiy and Lerner 1960, Butkovskiy 1961, 1969, Wang 1964). For example: control of heat diffusion in metals, solids, liquids and gases; atmospheric and water diffusivity; conduction; magnetism; pollution (see chapters 7 and 9); weather modification; etc.

4 *Space-time redesign control*: In many cases it may be most appropriate to change the spatial and temporal behaviour modes of the system in order to achieve different response characteristics. In particular, it is often possible to redesign or restructure the economic and administrative system boundaries and behaviour levels, or to change the nature of interzonal linkage – for example: noise level attenuation; dams and other physical barriers; adjustments to transport rates and networks; new liquid, metal or crystal structures altering diffusivity and conductivity properties; and location-allocation shifts.

5 *Control by separation of space-time variables*: In certain special cases noted by Butkovskiy (1969, pp. 142–5) it is possible to solve the system specification and control problems of space-time systems as in mathematical physics, by forming an infinite series of products of two functions, one depending only on time, the other only on space. For example, Fourier expansion and separation of variables method may be used. In this case, system control for the *i*th distribution in time *t* and space *x* can be written:

$$Y(t, x) = \sum_{k=1}^{\infty} g_k(t)h_k(x) \tag{4.4.1}$$

giving a set of dynamic (parallel) control systems (case 1 above) and a set of spatial (simultaneous) control systems (case 2 above). A special case of the separation of variable technique is the use of the Galerkin method of weighted residuals which may be useful in approximating space-time partial differential equations. The existence of systems satisfying (4.4.1) is rare in practical environmental control, except as an approximation, and this method of control is not discussed in detail. Each of the first four categories of space-time controllers is discussed in turn below and overall constraints on space-time control strategies developed in section 4.5.

4.4.1 DYNAMIC CONTROL

Pure dynamic control of *lumped systems*, as discussed in chapter 3, takes no account of spatial distributed effects which are treated as integrated over the whole area of interest of the system. In many cases, however, it is the spatial distribution effects that generate the specific lagged, lead and gain response characteristics of the lumped system; wherein each spatial subcategory presents different possibilities of storage, delay and anticipation. Control of such systems assumes that both the ignored spatial relationships and the dynamic relationships remain constant over the planning time horizon. Hence, control must be administered on a *uniform* basis across the entire spatial domain. In socio-economic systems most national macro-economic applications of control use this approach (as discussed in chapter 8), whilst many world-based control systems also impose uniformity between separatist distributed systems (see chapter 9). All such control systems are based upon the suppression or attenuation of internal variety and impose upon the system of interest an unnatural uniformity (Beer 1974). This has important implications for the imposition of uniform goals and system behaviour, and in socio-economic systems for the suppression of freedom of the individual, the social group, or the region (Hayek 1944). The effect of such inappropriate lumped controls has been to lead to increasing demands for spatial disaggregation of sociopolitical and economic administration, as discussed in chapters 6 and 8. In physico-ecological systems, the suppression of variety in attempts to simplify control can be no less disastrous. The attenuation of species diversity in ecosystems, the constriction of channel flows and catchment storage in hydrologic systems, and monoculture agricultural management systems, all reduce the variety of environmental response to system inputs and dynamics. The loss of variety is especially important in weakening the negative feedback damping of environmental disturbances and of extreme events (as discussed in chapter 7) and increases the number of singularity points (or points of natural catastrophe) discussed in section 4.5 of this chapter. The treatment of the world ecosystem by lumped controllers will be shown in chapter 9 to lead to an increased illusion of catastrophic collapse. Hence it is often the *representation* and *treatment* of control systems, rather than the structure of the systems themselves, which lead to overshoot and collapse.

An alternative method of dynamic control of spatial systems is to impose control on the boundary. The internal spatially distributed effects which induce diffusion of control influences across the entire spatial system are allowed for and constrain the control law, but it is assumed that it is impossible to influence system response internally. This constraint will operate, for example, in heating of a metal or gas; in affecting cognitive norms of an individual or social group; in affecting economic response by manipulation of frontier and tariff barriers; in treatment of river, lake and air pollution (see chapter 9); and in control of health without surgery. Two approaches to boundary control of distributed systems are possible, which depend upon the degree of knowledge about internal spatial structure.

(*i*) *Spatial dynamics unknown*: Here the system is treated as a black box with controls chosen on the basis of operating experience with the distributed system under a range of conditions. The resulting lumped system is then treated as a time series. Examples are the influence of traders in stock market situations (Zeeman 1973, Box and Jenkins 1970) and bank support in face of currency speculation, for which the spatial (and other) dynamics are extremely complex.

(*ii*) *Spatial dynamics known*: In this case the system is treated as a white box (or a grey box with partial information on spatial dynamics) but it is assumed that physical, environmental or sociopolitical constraints limit the possibility of internal intervention. The resulting boundary control is then of a canonical lumped form. Tzafestas and Nightingale (1968, 1969) give the resulting lumped controlled solution using algorithms based on dynamic programming and the Hamilton-Jacobi approach yielding the Riccati matrix differential equation for boundary control.

In case (*i*) we have a partially unobservable system and in both cases partially uncontrollable systems, since all of the states cannot be accessed and manipulated. In each instance control solutions are obtainable by transforming the multidimensional space-time model into an equivalent lumped single dimensional model. Solutions that are derived for the lumped system can then be transformed back into space-time solutions in order to determine the distributed effects (Tzafestas and Nightingale 1968). In diffusion (contiguity) systems this is relatively simple, as shown in section 4.4.3 (see also Butkovskiy 1969), because of the triple diagonal structure of the finite difference matrix and assumed constant diffusivity over unobservable states. But in general space-time interaction systems, or in systems with unknown spatial structures, it will be difficult to determine the equivalence transformation of time series models from given space-time characteristics. Again, the treatment of space-time systems by lumped controllers is, at best, a partial solution and, at worst, endangers system stability.

4.4.2 SPATIAL CONTROL

Once-and-for-all static spatial control decisions on allocation and location of networks and facilities depend upon the construction of an optimal site selection table (Scott 1971, Massam 1975) which is ranked in terms of the utility of each payoff on the control objective function. Typical examples are: the assignment of location to a school, retail or servicing facility (Scott 1971); determining a minimum cost location for an administrative centre (Massam 1975); the design of flow systems and networks to minimize interaction costs (Haggett and Chorley 1969); or the design of built-form structures (March and March 1972). In each case the control problem is written as one case of a general programming problem:

$$\max \text{ (or min) } J = \sum_{j}^{N} \sum_{j}^{N} d_{ij} Y_{ij} \tag{4.4.2}$$

where d_{ij} is a cost of impedence for movement between zone i and zone j, and Y_{ij} is the element assigned or allocated from i to j. The specific form of the combinatorial problem and solution will depend upon the system itself, which defines the character of each variable, and the constraints under which the optimization takes place. Scott (1971, chs 4–9) summarizes the resulting spatial control problem as follows:

(*i*) *Assignment problem* (zero-one programming problem): optimal assignment of n elements Y_{ij} to n categories or regions subject to,

$$\sum_{j=1}^{N} Y_{ij} = 1; \qquad \sum_{i=1}^{N} Y_{ij} = 1; \qquad Y_{ij} \geqslant 0. \tag{4.4.3}$$

Each element is assigned to only one region, and each region receives only one element under equal unitary demand.

(*ii*) *Transportation problem*: optimal (minimum transportation cost) assignment of flows Y_{ij} between N regions without exceeding productive capacity of any region subject to

$$\begin{aligned} &\sum_{j=1}^{N} Y_{ij} \leqslant s_i, \quad (i = 1, 2, \ldots, N), \quad \text{maximize supply capacity } s_i \\ &\sum_{i=1}^{N} Y_{ij} \geqslant r_j, \quad (j = 1, 2, \ldots, N), \quad \text{minimize demand } r_j \\ &\quad Y_{ij} \geqslant 0 \end{aligned} \tag{4.4.4}$$

(*iii*) *Trans-shipment problem*: optimal assignment of flows Y_{ij} between N regions subject to production and supply constraints, as in (*ii*), but where flows may be routed through any intermediate region, subject to

$$\begin{aligned} \sum_{i=1}^{N} Y_{ij} - \sum_{\substack{i=1 \\ i \neq j}}^{N} Y_{ji} = r_j - s_i, \quad (i, j = 1, 2, \ldots, N) \\ Y_{ij} \geqslant 0 \end{aligned} \tag{4.4.5}$$

(*iv*) *Network problem*: optimal assignment of expenditure on network linkage improvement or creation to yield equilibrium of arc capacities q_{ij} and flow problems Y_{ij} subject to

$$\begin{aligned} &\sum_{j=1}^{N} Y_{ij} \leqslant s_i \quad (i = 1, 2, \ldots, N) \\ &\sum_{i=1}^{N} Y_{ij} \geqslant r_j \quad (j = 1, 2, \ldots, N) \\ &Y_{ij} - \Delta q_{ij} \leqslant q_{ij}, \quad (\text{all}, i, j) \\ &\sum_{i=1}^{N} \sum_{j=1}^{N} \varepsilon_{ij}\, \Delta q_{ij} \leqslant Q \\ &Y_{ij}, \Delta q_{ij} \geqslant 0 \end{aligned} \tag{4.4.6}$$

where Δq_{ij} is the increment in arc capacity costing c_{ij} to improve, and Q is an overall global expenditure limitation. The total flow over arc (ij) cannot exceed the existing plus new capacity of that arc.

(*v*) *Location-allocation problem* (optimal partitioning): optimal assignment of N points or regions in continuous or discrete space to m service centres. This is composed of two components: optimal allocation of regions and optimal location of centres. Optimization is over a modified form of equation (4.4.2) given by

$$J = \sum_{i=1}^{m} \sum_{j=1}^{N} Y_{ij} r_j \, d_{ij}$$

subject to

$$\sum^{m} Y_{ij} = 1, \quad (j = 1, 2, \ldots, N)$$
$$Y_{ij} = \begin{cases} 1 \\ 0 \end{cases} \qquad (4.4.7)$$

where Y_{ij} determines whether region j is assigned to centre i, d_{ij} is distance or cost as before, and r_j is the level of demand in region j upon the function which is being located. Special cases of the location-allocation problem are discussed by Scott (1971, ch. 7) under capacity constraints on supply at centre i, index fixed costs of location at i, and index network constraints.

At least six species of algorithms exist for the solution of these control-algorithm problems based on mechanical, geometrical, heuristic, numeric-analytic, simulation and intuition methods (Scott 1971, Massam 1975).

A special form of spatial control involves once-and-for-all decisions to allocate resources to a particular strategy but requires in addition a continuous input of control at a constant level to maintain the optimal solution. An example is the quantity of permanent subsidies to industry in problem areas. This form of spatial control can also be conveniently formulated as a programming problem (Scott 1971). Mennes *et al.* (1969, ch. 3) discuss a model for cost minimization in development planning involving optimal allocation of investment with respect to spatial cost differentials in production and demand. Their model can be applied to general once-and-for-all allocations of investment, production or subsidies as a Hitchcock linear programming problem:

$$\min J(Y) = \sum_{i=1}^{N} c_i(Y_i(f) - Y_i(0)) \qquad (4.4.8)$$

subject to

$$Y_i(f) - Y_i(0) = d_i(Y_i(f) - Y_i(0)), \quad \text{given change in demand in region } i.$$

$$\sum_{i=1}^{N} (Y_i(f) - Y_i(0)) = (Y(f) - Y(0)) = b, \quad \text{given change in output in region } i.$$

$$Y_i(f) - Y_i(0) \geqslant 0$$

where c_i are the costs of control in region i in the N-regional system per unit change in $Y_i(t)$, the controlled variable. The planning time horizon is $(0-f)$ and d_i is the demand parameter giving a preference weighting of deviations of the system output from target in each region i. Boundary control effects can be introduced via flow-balance equations which give the net flux to or from the system when optimal solution to the above is desired. Mennes *et al.* (1969, chs 3 and 5) also give the solution, discussed further in chapter 8, to optimal allocation for multiple input-output systems with varying spatial constraints (e.g. mobility) on the inputs and outputs.

4.4.3 SPACE-TIME CONTROL

If full account is taken of space-time interdependence over the system of interest, a set of regulators and servomechanism control problems result which are special cases of the automated, algorithmic and man-machine controllers discussed in chapter 3. We discuss in this section two cases of control of such systems; first, control of $(N \times N)$ space-time interaction systems, and, second, control of space-time contiguity systems which are amenable to simplified mathematical solutions which are of special relevance in physical diffusion and heat conduction.

For space-time interaction systems which are partitioned into N interacting regions (based on spatial separation of location, and political, economic or physical boundaries) it is necessary to solve the optimization problem by consideration of simultaneous control of all spatial regions to independent targets by independent controllers. Indeed, the system equation (4.2.6) is not fully controllable for equation (4.2.29) unless an independent instrument is available in each region. Moreover, the optimization problem must be solved for the N regions jointly since interaction between the regions must be taken into account. Given these constraints, the disaggregated space-time system must be optimized subject to separate regional targets and also overall system goals. This gives a form of hierarchical or nested control problem. Spatial control thus normally necessitates the use of hierarchical solutions. We may approach such hierarchical control problems by three main methods (Schoeffler 1971) dependent upon *structural*, *influence*, and *control decomposition*.

(1) *Structural decomposition* involves partitioning the system equations into separate subsets, one set for each spatial region, each with separate target levels of reference input and a separate set of controllers. The overall control system must then be coordinated as a form of hierarchical control design. Two examples of control of space-time interactive systems by feedback and feedback-feedforward controllers are shown in fig. 4.13. In each case the transducers are control elements individual to each subsystem (region) but involving a level of coordination in controller response. For example, a

negative feedback to counteract a positive disturbance in region 1 will have an output also in region 2 which might accentuate disturbances in that region rather than damping them. Because of this 'leakage' between regions it is not possible to design the controller response in one subsystem without regard to the consequences, or trade-offs, which affect other regions and subsystems (i.e. there is a joint optimization problem). The multilevel coordination systems required to integrate space-time interaction in socio-economic systems are discussed in chapter 8 and for physico-ecological systems in chapter 7. In each case a joint simultaneous solution is required. One example for a linear spatial problem is given by Butkovskiy (1969) of the control of a billet heating furnace and rolling mill in metal manufacture. The control problem is to keep

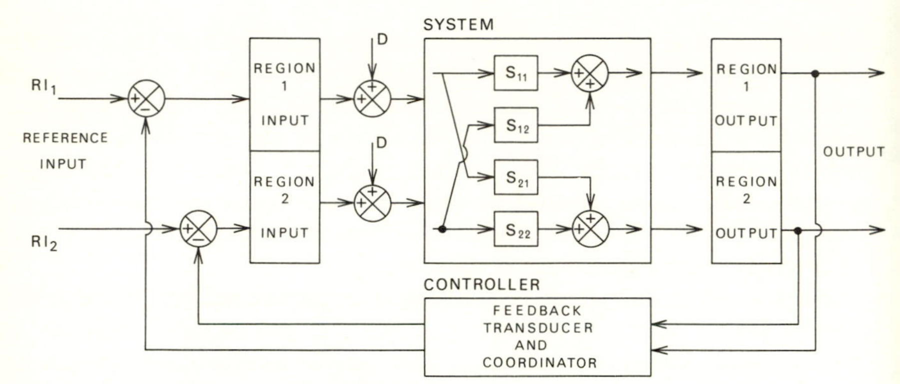

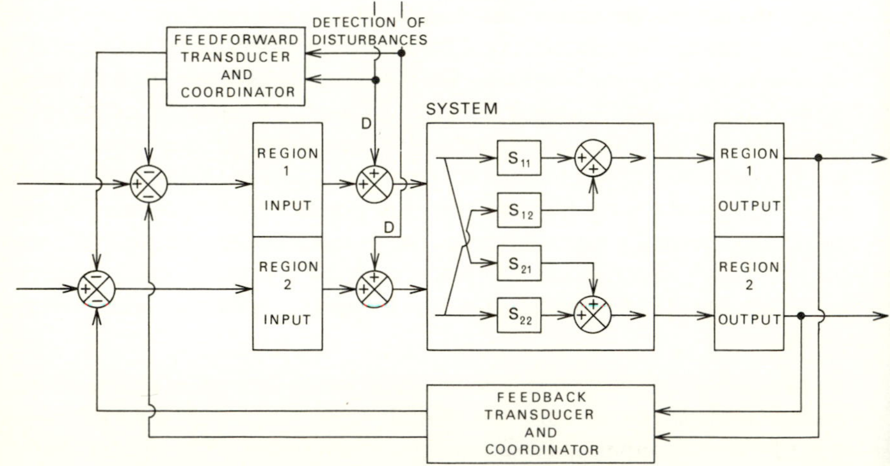

B SPACE-TIME FEEDBACK-FEEDFORWARD CLOSED-LOOP CONTROLLER

Fig. 4.13 The control of space-time systems by feedback and feedback-feedforward closed-loop controllers. The output from each region is compared with a reference set point and deviations counteracted. In the feedforward case this is combined with anticipation of system deviations.

the furnace and rolling unit at particular temperature targets using feedback heat devices, as shown in fig. 4.14. This exemplifies the particular problems of controllability in space-time systems, since control inputs can be effected at only three locations, whilst variations in billet temperatures can occur with much greater rapidity at many locations. This is only one example of many physical processes in which the control is aided by knowledge of the physical model that governs the process – in this instance the heat conduction equation.

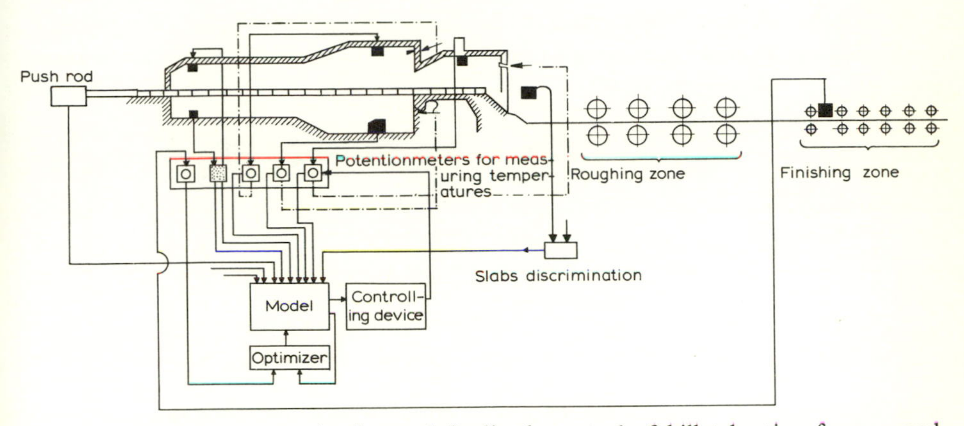

Fig. 4.14 Temperature monitoring and feedback control of billet heating furnace and rolling mill. The U_i elements are monitoring devices and control is fed to furnace inputs along the dashed-dotted lines (i.e. temperature and speed controls). (After Butkovskiy 1969.)

(2) *Influence decomposition* involves a partitioning of the control system itself into levels of control between subsystems or regions. More generally, this implies the use of a control design based upon the hierarchy of control policy. It introduces the question: Who controls? If it can be determined that one region or subsystem has an overriding influence on other sets of regions and subsystems, then action taken within that region will determine, or at least influence, system response in the other 'dependent' regions. This represents the partitioning of the control system between priority levels akin to management control systems, and hence the partitioning of the spatial domain into executive (i.e. policy determining) and subordinate regions. Thus, once again, the space-time controller is hierarchical. As an example, fig. 4.15 shows a feedback-feedforward controller for a two-region system in which region 1 consistently leads region 2 in registering changes, evolution, the effect of disturbances, etc. This is an especially important case of the general space-time transfer function process of equation (4.2.6) and fig. 4.3 in which, in addition to the normal distributed lag and autoregressive elements, a lead relation between region 1 and region 2 can be used as a realization of

disturbances which will occur in region 2. This information is employed in the feedforward transducer, which, together with feedback information, will permit very accurate control. Since stimuli in region 1 affect region 2 without the latter being able to achieve any control over them, it is natural that region 1 would become the executive controller of a dependent region 2. In socio-economic systems the element of foreknowledge in the leading region may also be a great aid to control and decision making in the dependent region (chapter 6).

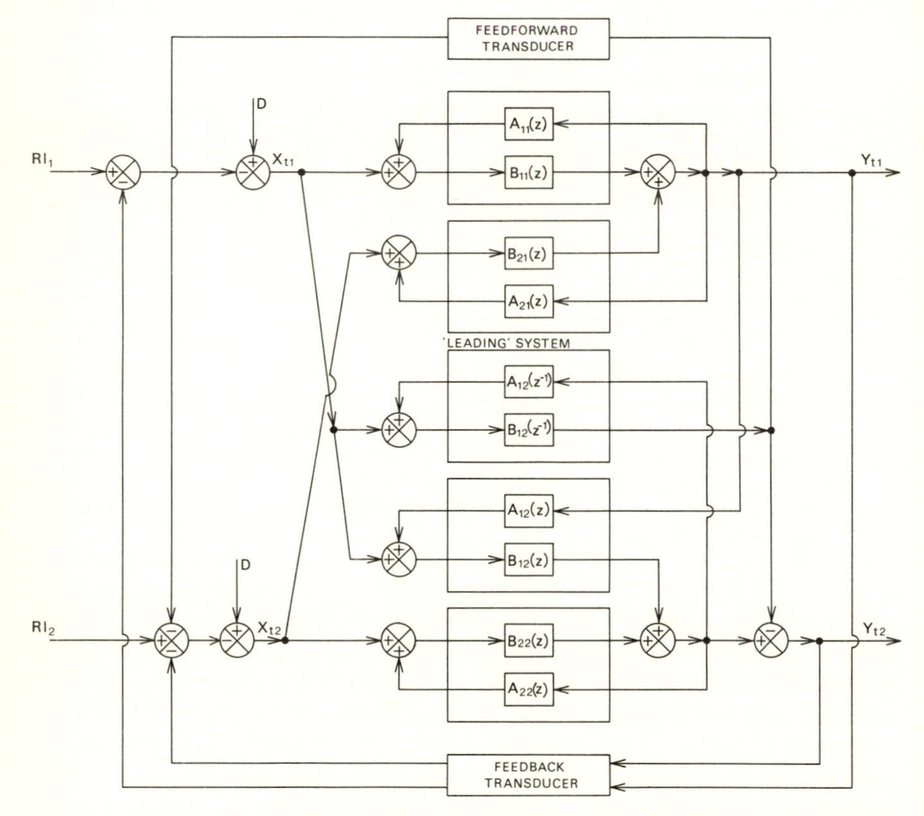

Fig. 4.15 Space-time feedback-feedforward controller for a two-region system with region 1 leading. RI_1 and RI_2 are the reference inputs into regions 1 and 2 respectively.

More generally, the executive action of executive regions in the space-time domain may set three levels of control over the dependent regions.

1 *Direct intervention*: This involves the overriding of lower-level decisions by administrative action backed by legislative and executive power in socio-economic systems. In environmental systems this corresponds to the direct control of system inputs into the subsystem by the executive

region – e.g. the control of discharge input into one reach of a river by the next reach upstream, control by adjacent aquifer storages or by adjacent gene pools, etc.

2 *Control intervention*: This aims at limiting the control actions at lower levels. In socio-economic systems this represents the freedom of dependent regions to make administrative 'implementation' decisions, but within overall controls and laws by which executive and legislative action is retained in the executive region. In environmental systems many examples exist of the relative loosening of control bonds between regions based upon increased spatial separation ('islands') or differences of system type: e.g. stable ecological niches resisting invasion, direct linkages affected by random disturbances (pollution and wind direction), etc.

3 *Goal intervention*: This involves controlling the goals and objectives of lower levels, or at least the parameters associated with these goals. Only executive policy making action is operated from the leading region, such that the dependent regions have autonomy in administrative and legal controls. In physico-ecological systems this corresponds to the overall governing of system levels or ecotypes by high-level controls within which many individual subsystems can interact to produce quite distinctively different steady states in different areas – e.g. overall climatic controls on vegetation character determined in each zone by microclimate, species availability, competitive equilibrium, etc.

With physico-ecological systems the levels of control correspond to tighter nesting of executive loops often to the degree of spatial contiguity, and hence to the scale of the system and its strength of coupling (see chapter 7). In socio-economic systems, the tightness of control depends upon the level of administrative, political and legal control within each region. This will affect the form of mixed scanning response and possibilities of economic and governmental structure discussed in chapters 6 and 8.

(3) *Control decomposition* is a third method of resolving space-time control problems which is determined not by the structure of the system nor by control equations, as in the case of structural and influence decomposition, but by the mathematics of the optimization solution itself. This partitioning may have no meaningful counterparts in terms of control system structure but is convenient for mathematical programming solution and allows simpler coordination based upon the relative frequency of disturbances entering the system (see table 3.1). Within this control decomposition method Schoeffler (1971) recognizes two sets of algorithms for achieving practical solution: goal coordination and model coordination.

The *goal-coordination* (or dual-feasible) method partitions the system objective function into N parts for each of the ($i = 1, 2, \ldots, N$) regions in the spatial system. The interactions between these N systems are then taken into a set of **Z** variables which act as arbitrary manipulated variables (Schoeffler 1971). This allows complete decoupling by the interaction-balance principle – i.e. the independently selected between-region interaction variables $\{Y_{ti}\}$ and $\{Z_{ti}\}$ are set equal. The solution path is shown in fig. 4.16 for a two-region system for control variables U_1, U_2. At level 1 each region has the optimum control solved independently, then at level 2 the first level solution is forced to satisfy the interaction-balance principle by modifying the goals of level 1. The

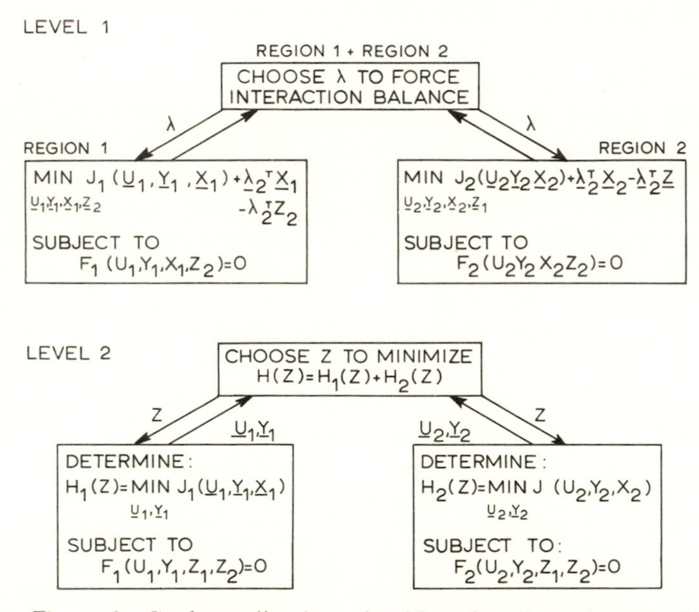

Fig. 4.16 Goal-coordination algorithm for the solution of optimal control of space-time interaction systems in respect of two regions. There is two-level optimization with the optimum from level 1 forced to conform with the separate optima for each region in level 2, and so on interactively. (After Schoeffler 1971.)

difference between true system interactions $\{X_{ti}\}$ and slack variables $\{Z_{ti}\}$ is used to define a set of Lagrangian multipliers $\boldsymbol{\lambda}(\mathbf{Y} - \mathbf{Z})$. A necessary condition for a minimum is that the derivation of this term equals zero (Schoeffler 1971). Using this function maximization is carried out with the cost function:

$$J(\mathbf{U}, \mathbf{Y}, \mathbf{X}, \mathbf{Z}, \boldsymbol{\lambda}) = J_1(\mathbf{U}_1, \mathbf{Y}_1, \mathbf{X}_1) + J_2(\mathbf{U}_2, \mathbf{Y}_2, \mathbf{X}_2) + \boldsymbol{\lambda}'(\mathbf{X} - \mathbf{Z}). \tag{4.4.9}$$

The *model coordination* method of control decomposition (or feasible method) is based on a disaggregation of the system equation or model to give another two-level optimization problem for the two region system as:

Level 1: Determine

$$H(z) = \min_{\mathbf{U}, \mathbf{Y}} J(\mathbf{U}, \mathbf{Y}, \mathbf{Z})$$
$$= J_1(\mathbf{U}_1, \mathbf{Y}_1, \mathbf{X}_1) + J_2(\mathbf{U}_2, \mathbf{Y}_2, \mathbf{X}_2) \qquad (4.4.10)$$

Subject to: $F(\mathbf{U}, \mathbf{Y}, \mathbf{Z}) = 0$.

Level 2: Minimize $H(Z)$.

Subject to: $\mathbf{X} = \mathbf{Z}$.

Again the interactions are taken into the slack variable **Z**. At level 1 the optimal value of the objective function is given by choosing that **U** and **Y** to minimize $J(.)$ *after* level 2 has provided an estimate **Z** of **X** at its optimal level of interaction. The new values of **U** and **Y** determined at level 1 are then used in level 2 to produce a better estimate of **Z**, and so on until convergence. Hence, the first level assigns the control variable, whilst the second level fixes the interaction variables (Wismer 1971).

A number of space-time control problems involving *space-time contiguity* systems, discussed in section 4.2.2, occur in practice. In industrial processes Butkovskiy (1969) gives examples of metal heating for rolling and tempering, drying and finishing in furnaces, sintering, distillation, crystal growth, induction heating, electrical circuits, etc. In physico-ecological systems contiguity control arises in atmospheric and water circulation, disease, plant and animal diffusion. In socio-economic systems contiguity control problems are not common. They may characterize some trade flows and information diffusion examples, but are most important in man-environment systems: air and water pollution, and the extraction of oil or water from underground deposits.

The optimal control problem is now treated in terms of partial differential equations defined on time and space (equation 4.2.25). The control problem is to choose the optimal distributed and boundary feedback and feedforward elements to render output deviations to a minimum. The optimization can adopt each of the control algorithms discussed in chapter 3. Butkovskiy (1960, 1961, 1969) has developed distributed optimization of contiguity systems based on the calculus of variations, which has been reduced by Wiberg (1967) to a Lagrangian multiplier problem. Tzafestas and Nightingale (1968, 1969) and Meditch (1971) give the alternative dynamic programming solution. The latter is the most useful in space-time optimization usually characterized by short time series and irregularly discretized spatial series. For each method Butkovskiy (1969, p. 38) and Tzafestas and Nightingale (1968, 1969) state the means of control of spatial contiguity processes analogous to the dynamic control solution discussed in chapter 3. For a dynamic programming solution the Pontryagin state function (control Hamiltonian) H is analogous to that described in equation (3.5.23) but must now be expressed as a function of time and space derivatives of the cost function. The optimal control law is given by

$$\mathbf{U}^0(\mathbf{Z}, \nabla J, t) = \min_{\mathbf{U} \in V} H(\mathbf{Z}, \mathbf{U}, \nabla J, t) \tag{4.4.11}$$

in which J is a cost functional and $\mathbf{Z}$ is the vector of space-time state variables. Equation (4.4.11) is the space-time equivalent of the Hamilton-Jacobi equation (Bellman's equation) analogous to equation (3.5.24), and equation (3.5.32) is given by Butkovskiy (1969) and Tzafestas and Nightingale (1969) and is identical to equation (3.5.24).

As an example, consider a simple diffusion system such as the flux of a pollutant in a lake. The control problem is to derive an N-stage quadratic controller with costs given by equation (3.5.11) which can check the spread of the pollutant with minimum cost. The control is effected by the use of the control variable (instrument) $U(t, x)$ which can be manipulated freely at any point in space x and time t. The N-stage recursive controller equivalent to that derived in chapter 3 is based on the system discretized approximation

$$\mathbf{Z}_{t+1} = \mathbf{A}\mathbf{Z}_t + \mathbf{B}\mathbf{U}_t$$

given in equation (4.2.26). The control problem is then stated as

$$\max_{\mathbf{U} \in V} J = \tfrac{1}{2} \int_0^{t_f} \int_0^D [\mathbf{M}\mathbf{Z}^2(t, x) + \mathbf{N}\mathbf{U}^2(t, x)]\, dt\, dx$$

$$\approx \tfrac{1}{2} \Delta t \Delta x \sum_{t=0}^{N-1} [Z_t' \ \mathbf{M}\mathbf{Z}_t + \mathbf{U}_t' \mathbf{N} U_t] \tag{4.4.12}$$

which gives the integral over the time horizon t_0 to t_f with final time t_f. In discrete terms this becomes the N-stage optimization with each state variable Z_{ti} being defined at location i in a Cartesian lattice, and at time t.

The Hamiltonian is given by the maximum principle as the sum of the cost function and the system constant equation (see fig. 3.24), i.e.

$$H(\mathbf{Z}_t, \mathbf{U}_t, \lambda_t, t) = \tfrac{1}{2} \Delta t\, \Delta x [\mathbf{Z}_t' \mathbf{M}\mathbf{Z}_t + \mathbf{U}_t' \mathbf{N}\mathbf{U}_t] + \lambda_{t+1} [\mathbf{A}\mathbf{Z}_t + \mathbf{B}\mathbf{U}_t]. \tag{4.4.13}$$

At an optimum this yields:

$$\mathbf{Z}_{t+1} = \mathbf{A}\mathbf{Z}_t + \mathbf{B}\mathbf{U}_t$$

$$-\lambda_t = \Delta t\, \Delta x\, \mathbf{M}\mathbf{Z}_t + \mathbf{A}'\, \lambda_{t+1}, \quad \lambda_t = 0$$

$$\mathbf{U}_t = \frac{1}{\Delta t\, \Delta x} \mathbf{N}^{-1} \mathbf{B}\, \lambda_{t+1}$$

which can be solved for each time and space location by summing the discrete Riccati equation formed in time, i.e.

$$\mathbf{P}_t = \Delta t\,\Delta x\mathbf{M} + \mathbf{A}'\mathbf{P}_{t+1}\mathbf{A} + \mathbf{A}'\mathbf{P}_{t+1}\left[\frac{1}{\Delta t\,\Delta x}\mathbf{A}^{-1}\mathbf{B}\mathbf{N}^{-1}\mathbf{B}'\mathbf{P}_{t+1}\right]^{-1} \quad (4.4.14)$$

$$\mathbf{Z}_t = \left[\mathbf{I} + \frac{1}{\Delta t\,\Delta x}\mathbf{B}\mathbf{N}^{-1}\mathbf{B}'\mathbf{P}_t\right]^{-1}\mathbf{A}\mathbf{Z}_{t+1} \quad (4.4.15)$$

$$\mathbf{U}_t = -\frac{1}{\Delta t\,\Delta x}\mathbf{N}^{-1}\mathbf{B}'\mathbf{P}_t\mathbf{Z}_{t+1} = -\frac{1}{\Delta t\,\Delta x}\mathbf{N}^{-1}\mathbf{B}'\mathbf{A}^{-1}[\mathbf{P}_t - \mathbf{M}]\mathbf{Z}_t \quad (4.4.16)$$

which is a Kalman controller. Sage (1968, pp. 160–1) gives the solution to this space-time control problem for the example system, as shown in fig. 4.17. The control magnitude $U(t, x)$ is greatest nearest the origin (both in space and time) of the pollutant (bottom left of the figures). The size of counteracting control that is necessary to remove the pollutant decreases with time and with distance away from its original location. The figure also shows the effect of two different initial conditions. In fig. 4.17A an initial state of the pollution variable $Z(0, x)$ is set equal to $[1 + Z(t_f, x)]/N$, whilst in fig. 4.17B the initial state (i.e. concentration of the pollutant) is twenty times as large. The effect of this greatly increased system disturbance is to require a much greater magnitude of system control with greater fineness across space and time.

4.4.4 SPACE-TIME REDESIGN CONTROL

The restructuring of space-time systems is mainly achieved by adjusting the form of linkage between zone i and zone j in the N-regional system. There are three main forms of this.

1. Creating a new linkage, e.g. a road crossing, new market or port of call; or removing an old linkage, e.g. political boundary, removal of bridge, construction of road or motorway impeding interaction of social groups, etc.
2. Modifying the form of linkage; by adjusting speed, cost, volume and capacity constraints on interzonal linkage, especially through changes in network structure or layout.
3. Creating a new demand or supply (input and output) node (i.e. the dynamic location-allocation problem).

The input-output relation of the space-time system can be written in terms of explicit spatial dependence by disaggregating the output of any N-regional system into the responses to each of the N-regional stimuli that contribute to

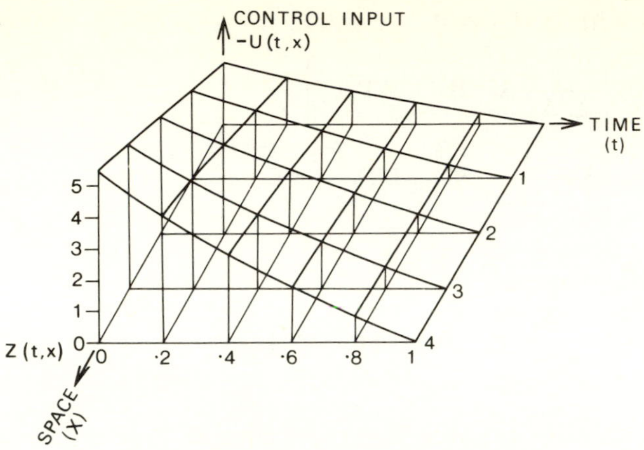

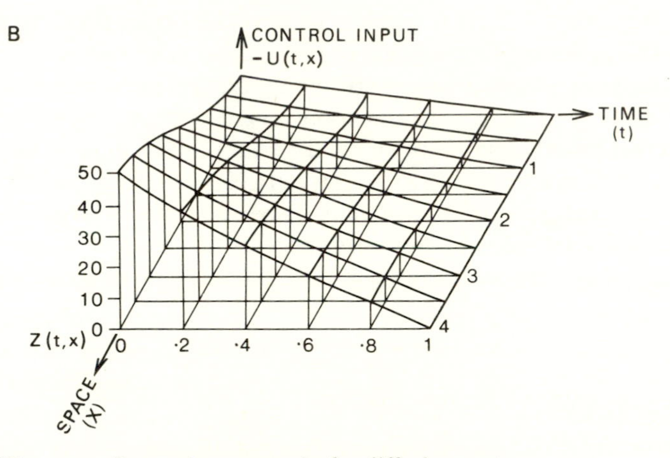

Fig. 4.17 Space-time control of a diffusion system.
A $t_f = 1{\cdot}0$; $\Delta t = 0{\cdot}01$; $B = 0{\cdot}01I$
$Z_f = 4{\cdot}0$; $\Delta x = 0{\cdot}5$; $Q = R = 1$
Initial conditions:

$$Z(0, x) = \frac{1 + xZ(t_f, x)}{N}$$

B As in A, but with initial conditions:

$$Z(0, x) = \frac{1 + 20xZ(t_f, x)}{N}$$

The control input varies over space and over time but is subject to smooth continuity relations. (After Sage 1968.)

that output. Fig. 4.18 shows an entire $N \times N$ set of spatial flows. The transfer function equations used in earlier sections have been dealing with the row and column sums – i.e. the total input $X_j(t)$ from region j into all other regions with which j interacts, and the total output $Y_i(t)$ from region i in response to all input stimuli received. The input-output matrix gives the set of impulse response elements $\{Y_{ij}(t)\}$ between each region at time t, of which some elements will be zero. Node and linkage creation and removal are subsets of network control problems discussed in section 4.4.2(*iv*). The dynamic location-allocation of nodal structures within a network requires the assumption that the centres are located in stages at each time interval. Scott (1971, ch. 8) gives the following solution to this nodal allocation. The objective function derives from equation (4.4.2) and is given by:

$$\min J = \sum_{i=1}^{m} \sum_{j=1}^{N} Y_{ij}(t) d_{ij}(t) + \sum_{j=1}^{N} Y_{kj}(t+1) d_{kj}, \quad (k = m+1, \ldots, N) \tag{4.4.17}$$

subject to

$$\sum_{i=1}^{m} Y_{ij}(t) + Y_{kj}(t+1) = 1, \quad (j = 1, 2, \ldots, N)$$

$$Y_{ij}(t),\ Y_{kj}(t+1) = \begin{cases} 1 \\ 0 \end{cases}.$$

The constraints derive from the equation (4.4.6) but with temporal dependence. At any time m nodes are already in existence, and the problem is to assign optimal locations to the new $(M\text{-}m)$ nodes. For the restructuring of system network linkages between nodes we can rewrite the system input-output matrix as:

$$Y_{ij}(t) = f(T_{ij}(t))$$

where $Y_{ij}(t)$ are interregional input-output flows (fig. 4.18) and $T_{ij}(t)$ are the network constraints (e.g. capacity, costs, etc.) of the ij links and $f(\,.\,)$ is a function of these constraints. The redesign control problem is to adjust the network constraints $T_{ij}(t)$ to shift the input-output interaction between region i and region j from an initial state $Y_{ij}(o)$ to a prespecified target final state $Y_{ij}(f)$ under minimum cost criteria. Adopting quadratic costs, Garg (1973) gives the solution to the problem as follows:

$$\min_{T_{ij} \in V} J = \sum_{i=1}^{m} \sum_{j=1}^{N} (Y_{ij}(f) - Y_{ij}(0))^2. \tag{4.4.18}$$

The Hamiltonian is given by

$$H = J + \sum_{i=1}^{m} \sum_{j=1}^{N} \lambda_{ij} Y_{ij}$$

where

$$\lambda_{ij} = -\frac{\partial J}{\partial Y_{ij}} - \frac{\partial \lambda_{ij}}{\partial Y_{ij}}$$

and the optimal control is given by

$$T_{ij}^{\circ}(t) = T_{\max} - \sum_{i=1}^{m} \sum_{j=1}^{N} \lambda_{ij} T_{ij}(t)$$

where $T_{\max}$ is the upper limit of costs of control permissible. Garg (1973) gives a particular example of control of trade by manipulating transport costs. The

$$\text{Inputs} \left\{ \begin{bmatrix} Y_{11}(t) & Y_{12}(t) & Y_{13}(t) & \ldots & Y_{1N}(t) \\ Y_{21}(t) & Y_{22}(t) & Y_{23}(t) & \ldots & Y_{2N}(t) \\ Y_{31}(t) & Y_{32}(t) & Y_{33}(t) & \ldots & Y_{3N}(t) \\ \cdot & \cdot & \cdot & \cdot & \cdot \\ Y_{N1}(t) & Y_{N2}(t) & Y_{N3}(t) & \ldots & Y_{NN}(t) \end{bmatrix} \right. \quad \begin{matrix} \sum_j Y_{1j}(t) \\ \sum_j Y_{2j}(t) \\ \sum_j Y_{3j}(t) \\ \\ \sum_j Y_{Nj}(t) \end{matrix}$$

Total input from region j at time t

Total output in region i at time t

Fig. 4.18 Space–time input–output flows in an N-regional system. $Y_{ij}(t)$ is the input into region j and output in region i at time t.

change in flow pattern and transport cost structure is shown in fig. 4.19. The details of market price and supply cost effects in this model are discussed in chapter 8. It is sufficient to note here that the desired final state of interaction $Y_{ij}(f)$ is achieved, but at the expense of negative costs of control (subsidies) for T_{11} and T_{22}. The total costs of control will be the difference between the subsidy outlay and any transport tax income.

An analogous example of the effect of changes in linkage structure is given for island ecosystems (MacArthur and Wilson 1967) in which the nature of the linkage to gene and animal pools becomes modified. Lidicker (1966), in a study of a feral house mouse population on a harbour breakwater showed how this population declined to extinction when the linkage of the breakwater to the mainland (and to the maintaining population of house mice) was severed. Analogous examples for gene pools are also common.

4.5 Additional topics in space-time control systems

We have been able in this chapter to define the full environment of physico-ecological and socio-economic systems in their space-time context. It now becomes possible to develop further some of the concepts that underlie the

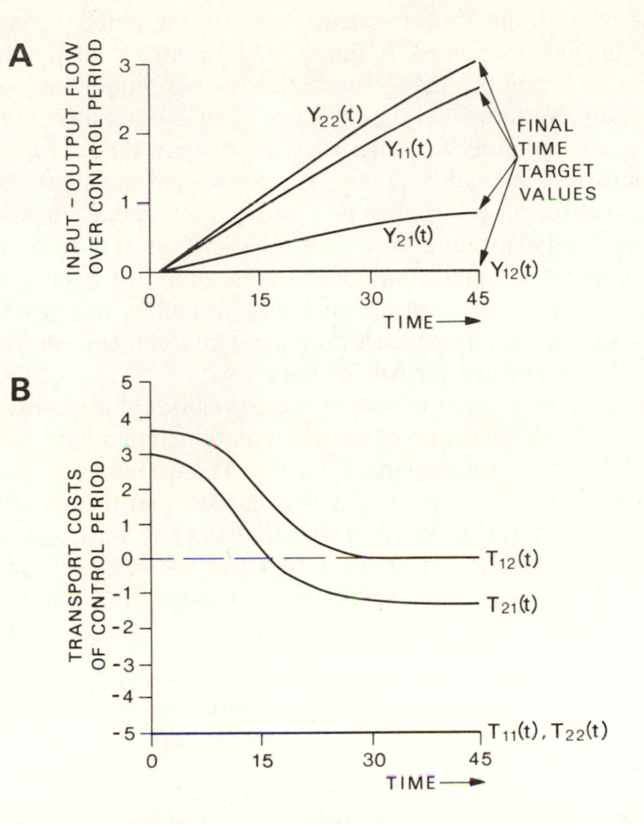

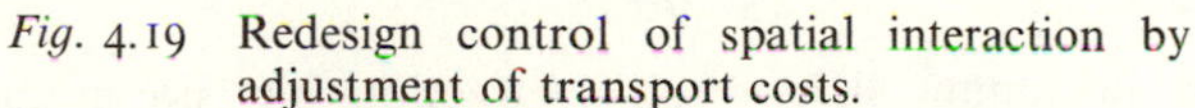

Fig. 4.19 Redesign control of spatial interaction by adjustment of transport costs.

$$Y_{ij}(0) = \begin{bmatrix} 0 & 0 \\ 0 & 0 \end{bmatrix}; \qquad Y_{ij}(f) = \begin{bmatrix} 2{\cdot}5 & 0 \\ 0{\cdot}7 & 2{\cdot}8 \end{bmatrix}$$

$$t_f = 45; \qquad T_{\max} = 5{\cdot}0; \qquad f(\cdot) = q = 100$$

$$\min_{T_{ij} \in w} J = \sum_{i=1}^{2} \sum_{j=1}^{2} (Y_{ij}(f) - Y_{ij}(0))^2$$

Market price:

$$PM_1 = 5{\cdot}0 - \sum_{i=1}^{2} Y_{i1}; \qquad PM_2 = 5{\cdot}0 - \sum_{i=1}^{2} Y_{i2}$$

Supply cost:

$$PS_1 = 1{\cdot}0 + \sum_{j=1}^{2} Y_{1j}; \qquad PS_2 = 1{\cdot}0 + 0{\cdot}5 \sum_{j=1}^{2} Y_{2j}$$

The interaction between the two regions $Y_{ij}(t)$ is controlled by optimum adjustment to the transport costs $T_{ij}(t)$. (After Garg 1973.)

problems of controlling these systems and also the purely dynamic systems discussed in chapters 2 and 3. In this section we discuss, firstly, the necessary and sufficient conditions that characterize the optimum in a control system and, secondly, we show that the existence of such optima occurs only under special system operating characteristics. In many instances apparent optima may be only ridge and saddle points. As a consequence of this recognition, we show that control in environmental systems (automated, man-machine, or restructured) may encounter discontinuities and thresholds. At these points the system may 'flip' rapidly and often catastrophically from one behaviour regime to another. The presence of such discontinuities in system behaviour is shown to have great consequences for man-environment interrelations and these points are pursued further in chapter 9.

All control strategies rely upon one or more of a set of diagnostic conditions which allow the identification of an optimum (maximum or minimum) of the objective function or performance criterion. These conditions determine that a minimum of the cost criterion $J(\mathbf{X}, \mathbf{U}, t)$ exists, that the minimum value is unique, and how it is characterized (Pontryagin *et al.* 1962, Arrow and Kurz 1970) and are usually divided into two sets of necessary and sufficient conditions. For a single-input single-output (univariate) system these can be stated as:

Necessary condition: at a maximum (or minimum) of the cost function $J(\mathbf{X}, \mathbf{U}, t) = J$, J is optimal relative to neighbouring points a and b ($a < U_0 < b$), if

$$\frac{\partial J}{\partial U} = 0 \tag{4.5.1}$$

i.e. at the maximum the objective function has no gradient (i.e. is horizontal) with respect to the control U.

Sufficient condition: that

$$\frac{\partial^2 J}{\partial U^2} \begin{cases} < 0 & \text{(maximum)} \\ > 0 & \text{(minimum)} \\ = 0 & \text{(no maximum or minimum: saddle point)} \end{cases} \tag{4.5.2}$$

i.e. there is non-zero curvature of the surface away from U_0.

In the multivariate case (multicriterion objective functions) the conditions are more complex. For the function $J(U_1, U_2, \ldots, U_N)$ defined on the N control objectives $\mathbf{U}$ we have:

Necessary condition:

$$\frac{\partial J}{\partial U_1} = \frac{\partial J}{\partial U_2} = \cdots = \frac{\partial J}{\partial U_N} = 0. \tag{4.5.3}$$

Sufficient condition:
(*i*) *Two-criterion objective function*:

$$\frac{\partial^2 J}{\partial U_1^2} < 0 \quad \text{and} \quad \frac{\partial^2 J}{\partial U_1^2}\frac{\partial^2 J}{\partial U_2^2} - \left(\frac{\partial^2 J}{\partial U_1 \partial U_2}\right)^2 < 0 \quad \text{(maximum)} \tag{4.5.4}$$

$$\frac{\partial^2 J}{\partial U_1^2} > 0 \quad \text{and} \quad \frac{\partial^2 J}{\partial U_1^2}\frac{\partial^2 J}{\partial U_2^2} - \left(\frac{\partial^2 J}{\partial U_1 \partial U_2}\right)^2 > 0 \quad \text{(minimum)} \tag{4.5.5}$$

$$\frac{\partial^2 J}{\partial U_1^2} = 0 \quad \text{and} \quad \frac{\partial^2 J}{\partial U_1^2}\frac{\partial^2 J}{\partial U_2^2} - \left(\frac{\partial^2 J}{\partial U_1 \partial U_2}\right)^2 = 0 \quad \text{(indeterminate)} \tag{4.5.6}$$

$$\frac{\partial^2 J}{\partial U_1^2} = 0 \quad \text{and} \quad \frac{\partial^2 J}{\partial U_1^2}\frac{\partial^2 J}{\partial U_2^2} - \left(\frac{\partial^2 J}{\partial U_1 \partial U_2}\right)^2 < 0 \quad \text{(no maximum or minimum, saddle point with multiple branches)} \tag{4.5.7}$$

(*ii*) *N-criterion objective function*: First rewrite the system in matrix form (Appendix 1)

$$\begin{aligned} J(U_1, U_2, \ldots, U_N) = {} & a_{11}U_1^2 + a_{12}U_1U_2 + \cdots + a_{1N}U_1U_N \\ & + a_{21}U_2U_1 + a_{22}U_2^2 + \cdots + a_{2N}U_2U_N \\ & \vdots \\ & + a_{N1}U_NU_1 + a_{N2}U_NU_2 + \cdots a_{NN}U_N^2 \end{aligned}$$

Note the symmetry $a_{ij} = a_{ji}$ and define

$$\mathbf{A} = \begin{bmatrix} a_{11} & a_{12} & & \\ a_{21} & a_{22} & & \\ & & \ddots & \\ & & & a_{Nm} \end{bmatrix}; \quad \mathbf{U}' = [U_1 U_2 \ldots U_N]$$

then $J(.)$ becomes a quadratic form

$$J(U_1, U_2, \ldots, U_N) = \mathbf{U}'\mathbf{A}\mathbf{U}. \tag{4.5.8}$$

Using the quadratic structure the conditions for a maximum and minimum can be stated in terms of the *definiteness* conditions of matrices. These can be stated in terms of Sylvester's theorem (Pontryagin *et al.* 1962) that the maximum or minimum of a quadratic form

$$\sum_{i=1}^{m} \sum_{j=1}^{N} a_{ij}U_iU_j$$

is *positive definite* if all its determinants (principal first minors) are positive. These minors are given (Appendix 1) by:

$$\det \mathbf{A}_1 = |a_{11}|, \det \mathbf{A}_2 = \begin{vmatrix} a_{11} & a_{12} \\ a_{21} & a_{22} \end{vmatrix}, \ldots, \det \mathbf{A}_n = \begin{vmatrix} a_{11} & a_{12} & \ldots & a_{1N} \\ a_{21} & a_{22} & \ldots & a_{2N} \\ \vdots & \vdots & & \vdots \\ a_{N1} & a_{N2} & \ldots & a_{NN} \end{vmatrix}$$

The quadratic **A** is *negative definite* if $-\mathbf{A}$ is positive definite, and it is *indefinite* if it is neither positive nor negative definite. Hence the sufficient condition for a maximum in the n-criterion control system can be stated as:

A is positive definite: $J(U_1, U_2, \ldots, U_n)$ has a minimum.
A is negative definite: $J(U_1, U_2, \ldots, U_n)$ has a maximum.
A is indefinite: $J(U_1, U_2, \ldots, U_n)$ has no maximum or minimum provided that the matrix

$$\begin{vmatrix} \frac{\partial^2 J}{\partial U_1^2} & \frac{\partial^2 J}{\partial U_1 \partial U_2} & \vdots & \\ \frac{\partial^2 J}{\partial U_2 \partial U_1} & \frac{\partial^2 J}{\partial U_2^2} & & \\ & & \ddots & \\ & \ldots & & \frac{\partial^2 J}{\partial U_N^2} \end{vmatrix} \neq 0. \qquad (4.5.9)$$

Alternative conditions are the Kuhn-Tucker conditions, Weierstrass, Legendre, Jacobi and various other conditions (Dreyfus 1965).

The control problems discussed in this and the preceding chapters have all assumed that an optimum of the system objective function does indeed exist (i.e. the $J(U_1, U_2, \ldots, U_n) = \mathbf{A}$ is positive or negative definite). However, many physical and economic systems that are encountered in practice have indefinite derivative matrices **A**. These cases correspond to discontinuities in the system objective function and may also characterize system behaviour in ordinary (uncontrolled) conditions as well.

It has been assumed throughout the discussion of dynamic systems in chapter 2, and implicitly in many of the philosophical models discussed in chapter 1, that systems behave in an ordered functional manner which is describable by simple mathematical functions. Indeed Stewart (1975) notes that the very success of the Leibniz and Newton systems models which apply to phenomena exhibiting 'smooth' behaviour, has to some extent limited the sort of questions that the analyst may seek to ask. When the system under study exhibits sudden discontinuities in space or time it is necessary to extend the range of transfer function description available. To do this Thom (1975) has adopted a highly generalized topology of *catastrophe theory* which permits the description and prediction of a number of general kinds of discontinuous processes – for example, shock waves, turbulent flow, phase transition from liquid to gas, biological cell division and embryonic development, and many human relations (love-discord, war-peace) (see Thom and Zeeman 1975, Zeeman 1973, Stewart 1975, Thom 1969).

The basic nature of catastrophic system behaviour has been discussed for simple environmental systems by Chorley and Kennedy (1971, p. 239) who recognize the presence of 'discontinuities in the phase space separating it into zones of differing system economies or as deformities in the manifold. These can lead to catastrophic transition ("flips") of the system state from one domain of operation into another, perhaps irreversible one'. The typical behaviour of such space-time 'flipping' is illustrated for a time series of system output observations in fig. 4.20. At the points t_1, t_2 and t_3 the magnitude of the system input has caused the system behaviour to cross some threshold and produce different response magnitude and form; in this case, higher magnitude with greater degree of stochastic variation.

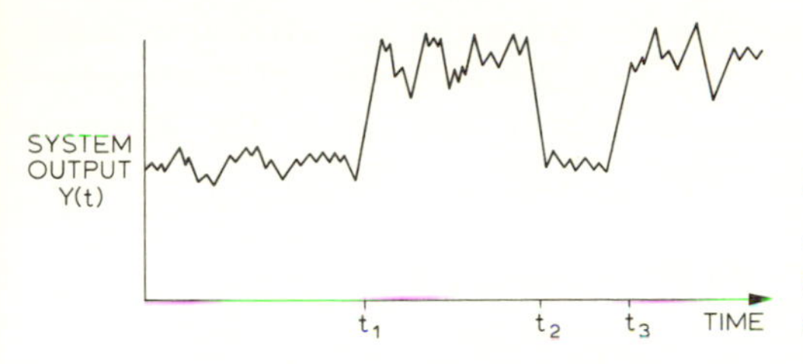

Fig. 4.20
'Flipping' illustrated with respect to a time series output; the output shifts rapidly between two steady state regimes.

To understand the structure of such behaviour, any system must be characterized in terms of a *system state function* or *behaviour function*, say $V(\mathbf{Z})$, which is some measure of the overall system state. In this case $\mathbf{Z}$ is the vector potential and is identical to the Pontryagin state function H of equation (3.5.23) with space-time state variables as defined in equations (2.2.13) and (2.2.14). Since each of the state variables is assumed to be mutually orthogonal, we may plot the state space for two state variables as in fig. 4.21, which shows contours for one particular state description with $V(\mathbf{Z}) = \mathbf{Z}'\mathbf{M}\mathbf{Z}$, a quadratic form. On to the contours of $V(\mathbf{Z}) = k_i$, for various values of k_i, is superimposed an example system trajectory over time t. The passage of the system through time follows a 'downhill" course of decreasing value of the state function $V(\mathbf{Z})$, and exhibits so-called 'stability in the sense of Liapunov'. This behaviour is important to the definition of discontinuities (i.e. catastrophes) and can be defined as follows (Schultz and Melsa 1967, p. 164). The origin of the state space is said to be *stable in the sense of Liapunov* if corresponding to each region bounded by $V(\mathbf{Z})$ there is a region $V(\mathbf{Z}^i)$ such that system solutions starting in $V(\mathbf{Z}^i)$ do not leave $V(\mathbf{Z})$ as $t \rightarrow \infty$, and for any particular value $V(\mathbf{Z})$ of the state variable $\mathbf{Z}$. If, in addition, the system solution approaches zero as $t \rightarrow \infty$, then the system is asymptoti-

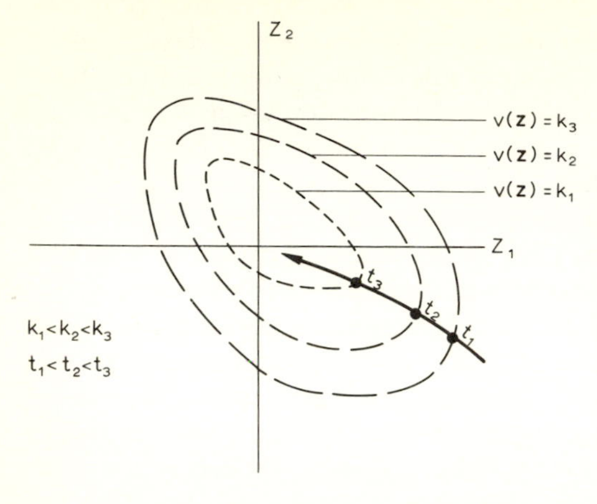

Fig. 4.21
System trajectory over time (t) as a geometric representation of the system state measured by the state function $V(\mathbf{Z})$ (with quadratic form $V(\mathbf{Z}) = \mathbf{Z}'\mathbf{M}\mathbf{Z}$). The path of evolution is towards a global equilibrium downhill on the state function. The contours plot equal value of the state for the two state variables $Z_1(t)$ and $Z_2(t)$.

cally stable. Asymptotic stability requires that the system returns to rest as time passes. Fig. 4.22 shows the behaviour of three systems with asymptotic stability, Liapunov stability, and unstable behaviour for deterministic and stochastic cases. In general, the stability conditions for the global asymptotically stable system and the unstable system will satisfy the following properties (Schultz and Melsa 1967, pp. 170–2).

Global asymptotic stability

(i) $V(0) = 0$

(ii) $V(\mathbf{Z}) > 0 \quad \text{for } \mathbf{Z} \neq 0$

(iii) $V(\mathbf{Z}) \to \infty \quad \text{as } |\mathbf{Z}| \to \infty$

(iv) $\dfrac{\partial V(\mathbf{Z})}{\partial t} \leqslant 0$

Instability

(i) $\dfrac{\partial V(\mathbf{Z})}{\partial t}$ exists and is continuous

(ii) $V(\mathbf{Z}) = 0$ on boundary $V(\mathbf{B})$

(iii) $V(\mathbf{B})$ is on boundary

(iv) $Y(\mathbf{Z})$ and $\dfrac{\partial V(\mathbf{Z})}{\partial t} > 0$

These stability properties enable the definition of the unstable-stable system boundary which is fundamental to defining catastrophes in control systems. Consider the state function $V(\mathbf{Z})$ defined on one state variable and one control variable $\{U_t\}$ in fig. 4.23. At the point P the system can be considered in equilibrium or to be globally stable. If changes in U_t induce a shift of the system state $V(\mathbf{Z})$ along the line PQ to point Q, the system becomes progressively less stable until at the point Q itself the system is conditionally unstable. To search out a stable equilibrium position, it may return to point P or may move along a new course towards a new stable position shown by the arrow. At this point of inflexion (Q) the system is in a *critical equilibrium state* with condition $\partial^2 V(\mathbf{Z})/\partial \mathbf{Z}^2 = 0$ (remember the sufficient condition for a maximum equation (4.5.2)). We may now restate the global stability condition in simplified form.

Global stability:

$$\frac{\partial V(\mathbf{Z})}{\partial \mathbf{Z}} = 0; \qquad \frac{\partial^2 V(\mathbf{Z})}{\partial \mathbf{Z}^2} > 0; \qquad \frac{\partial^2 V(\mathbf{Z})}{\partial \mathbf{Z} \partial U} > 0 \tag{4.5.10}$$

Critical equilibrium state:

$$\frac{\partial V(\mathbf{Z})}{\partial \mathbf{Z}} = 0; \qquad \frac{\partial^2 V(\mathbf{Z})}{\partial \mathbf{Z}^2} = 0; \qquad \frac{\partial^2 V(\mathbf{Z})}{\partial \mathbf{Z}\, \partial U} > 0 \tag{4.5.11}$$

Since it is movement along the U control axis which brings the system to this critical point, it is the single variable or *control parameter* U which is necessary

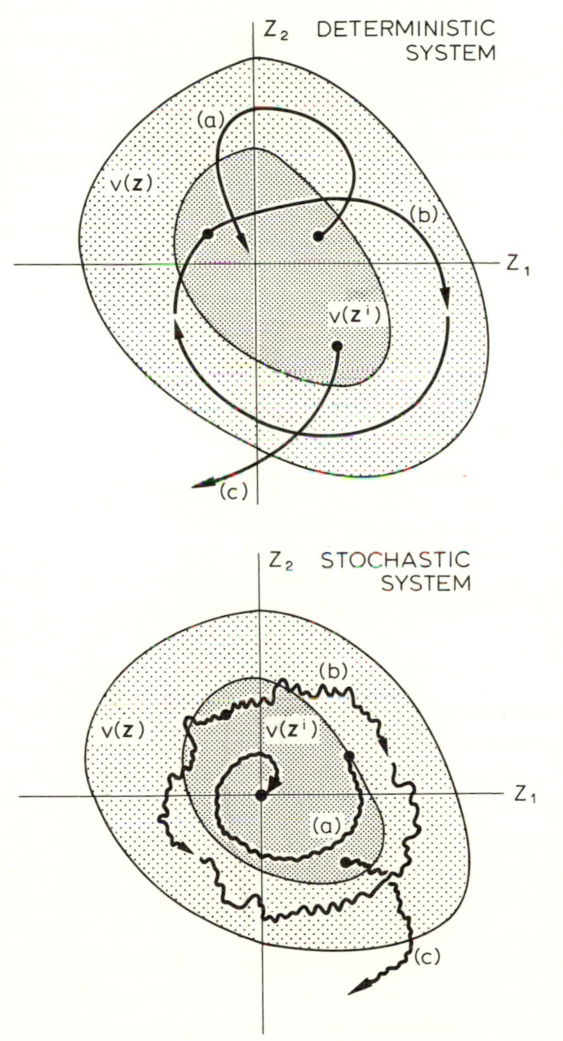

Fig. 4.22
Stability trajectories in state space from varying origins. Three system trajectories are shown for deterministic and stochastic systems:

(*a*) Asymptotically stable system.
(*b*) Stable system in the sense of Liapunov.
(*c*) Unstable system.

The heavy shading defines the area of initial conditions for which the system is asymptotically stable, and the lighter shading the bounds of Liapunov stability. Systems with their origin at initial conditions in this outer region will be unstable. (Partly after Schultz and Melsa 1967.)

to *unfold* the singularity at point Q, or to make the system state shift from a minimum at P to a minimum not at P. For any such singular point satisfying the critical equilibrium condition, there will exist a K-dimensional family of deformations of $V(\mathbf{Z})$:

$$V = V(\mathbf{Z}) + U_1 g_1(\mathbf{Z}) + U_2 g_2(\mathbf{Z}) + U_3 g_3(\mathbf{Z}) + \cdots + U_k g_k(\mathbf{Z}) \tag{4.5.12}$$

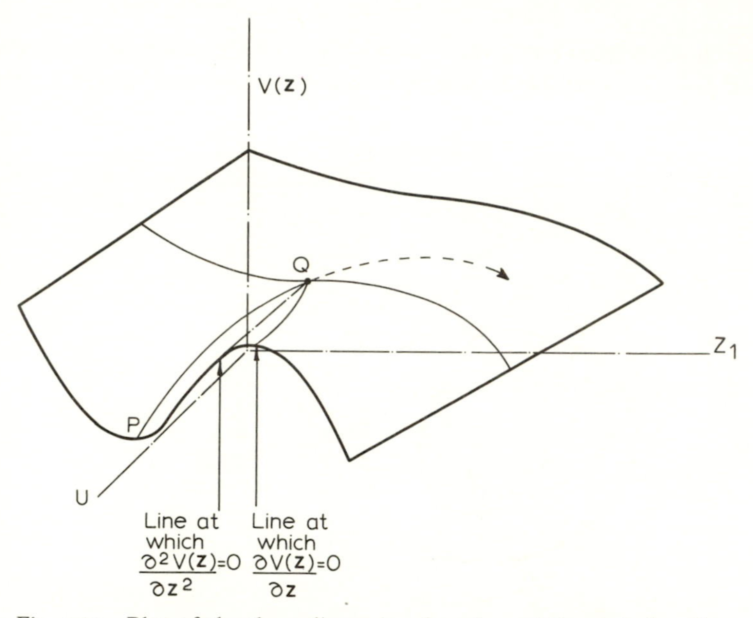

Fig. 4.23 Plot of the three-dimensional surface of the state function $V(Z)$. At the point of critical equilibrium or universal unfolding (Q) the system becomes conditionally unstable after moving along a trajectory from a point of global stability (P). At Q the lines $\partial^2 V(Z)/\partial Z^2 = 0$ and $\partial V(Z)/\partial Z = 0$ converge.

corresponding to the set of control parameters U_i. This family of transformations of the system state is the *universal unfolding* or the *organizing centre* of the family (Thom 1969). The dimension K of the universal unfolding is the *finite codimension* of the singularity of $V(\mathbf{Z})$, or the number of ways in which an elementary catastrophe can occur for that system. These correspond to the saddles in fig. 2.33. The state-space system evolution graphs given in this figure, and by Lotka (1956) and Hirsch and Smale (1974), should be compared with fig. 4.22.

Various elementary catastrophes can be constructed on the basis of the state or behaviour variables $Z_1, Z_2, Z_3, \ldots = \mathbf{Z}$ and the control variables or parameters $U_1, U_2, U_3, \ldots = \mathbf{U}$. Thom (1975) gives seven *elementary*

State Variables	Codimension	Catastrophe name	Point of singularity or organizing centre	System state function $V(\mathbf{Z})$ or universal unfolding	Spatial interpretation of catastrophe	Temporal interpretation of catastrophe
1	0	simple minimum	$V(\mathbf{Z}) = Z_1^2$	Z_1^2	a being, an object	to be, to last
	1	fold	$V(\mathbf{Z}) = \frac{1}{3}Z_1^3$	$\frac{1}{3}Z_1^3 + u_1Z_1$	the boundary, the end	to end, to start
	2	cusp	$V(\mathbf{Z}) = \frac{1}{4}Z_1^4$	$\frac{1}{4}Z_1^4 + \frac{1}{2}u_1Z_1^2 + u_2Z_1$	a pleat, a fault	to separate, unite, change
	3	swallowtail	$V(\mathbf{Z}) = \frac{1}{5}Z_1^5$	$\frac{1}{5}Z_1^5 + \frac{1}{3}u_1Z_1^3 + \frac{1}{2}u_2Z_1^2 + u_3Z_1$	a split, a furrow	to split, tear, saw
	4	butterfly	$V(\mathbf{Z}) = \frac{1}{6}Z_1^6$	$\frac{1}{6}Z_1^6 + \frac{1}{4}u_1Z_1^4 + \frac{1}{3}u_2Z_1^3 + \frac{1}{2}u_3Z_1^2 + u_4Z_1$	a flake, pocket or fish scale	to fill or empty, to give or receive
2	3	hyperbolic umbilic	$V(\mathbf{Z}) = Z_1^3 + Z_2^3$	$Z_1^3 + Z_2^3 + u_1Z_1 + u_2Z_2 + u_3Z_1Z_2$	crest of wave, arch	break of wave, collapse
	3	elliptic umbilic	$V(\mathbf{Z}) = Z_1^3 - 3Z_1Z_2^2$	$Z_1^3 - 3Z_1Z_2^2 + u_1Z_1 + u_2Z_2 + u_3(Z_1^2 + Z_2^2)$	needle, spike, hair	drill, fill, indent
	4	parabolic umbilic	$V(\mathbf{Z}) = Z_1^2Z_2 + Z_2^4$	$Z_1^2Z_2 + Z_2^4 + u_1Z_1 + u_2Z_2 + u_3Z_1^2 + u_4Z_2^2$	jet, mushroom, mouth	break, open, close, pierce, cut, pinch

Fig. 4.24 The seven elementary catastrophes. (After Thom 1975.)

catastrophes, as shown in fig. 4.24 (Stewart 1975). Each of these elementary catastrophes expresses the behaviour of a system in the neighbourhood of the point of singularity in the state space, on the universal unfolding at the 'flip point' or singularity in $V(\mathbf{Z})$ – the organizing centre. The graphical form of each of these catastrophes is shown in fig. 4.25. It is perhaps easiest to follow

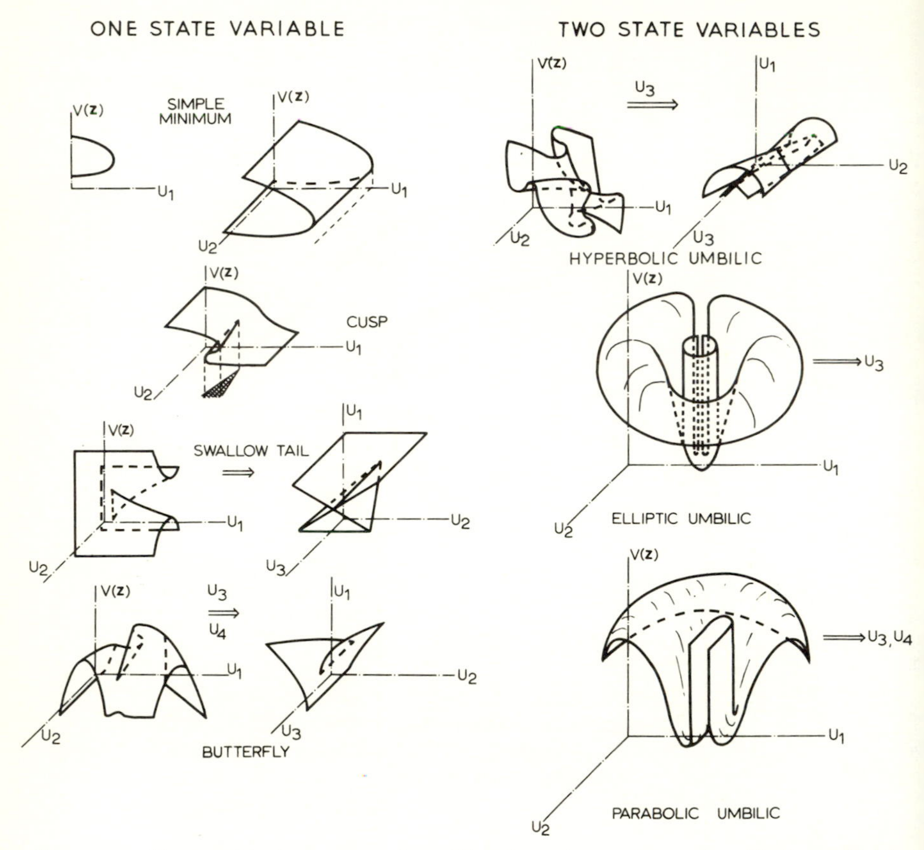

Fig. 4.25 The graphical representation of the elementary catastrophes. Each manifold is plotted in state space $V(\mathbf{Z})$ and the control space defined by the parameters of the control variables U_i. For the higher order catastrophes it is only possible to depict the state space at single locations and the control space is preferable. (After Woodcock and Poston 1974.)

the action of the catastrophe at the point of singularity by considering two systems paths on the surface of a cusp as shown in fig. 4.26. For the system A, changes in variable U_1 produce a smooth system response, but for the system B, starting at a higher magnitude on variable U_2, a marked flip occurs with changes in U_1. For space-time systems the set of spatiotemporal variables and lags can be depicted by U_1 for time, and U_2 for space. In this case the unfolding of the singularity introduces a discontinuity in space and time for the *B* system, but for the *A* system there is a smooth pattern of temporal evolution. The true overall constraint on these forms of system behaviour is space-time location.

The definition of space and time for the control variables U_1 and U_2, respectively, occurs frequently in practice. Thom's original theory of morphogenesis was developed to deal with temporal development and/or evolution in which the growth or spatial spread of an organism is modelled as a series of gradual changes triggered by, and in turn triggering, catastrophic

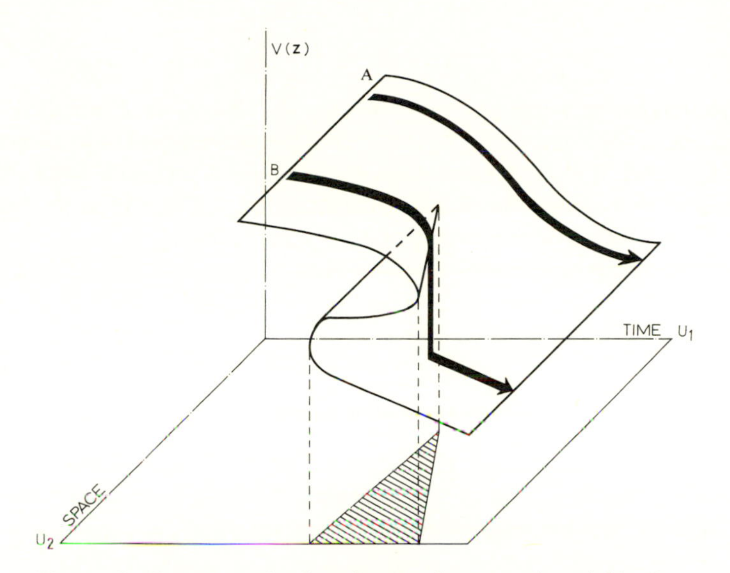

Fig. 4.26 System paths for changes in control variable U_1 (time) at two different values of the second control variable U_2 (space). Path *A* traces a smooth pattern of evolution, whilst path *B* shows a flip over the point of the cusp. (Partly after Stewart 1975.)

jumps in biochemistry linked to large-scale changes in the form of the organism (Stewart 1975). It is possible to depict discontinuities in many of the space-time transfer function processes discussed in previous sections in the same manner. Many types of nonlinear and non-stationary parameter behaviour also conform to elementary catastrophes (Bennett 1978, ch. 5, Goldfeld and Quandt 1973), whilst the action of many control variables may shift system behaviour suddenly from one regime of behaviour to another. If the control is spatially variable, then there will be a space-time flip. In each case, the system state vector **Z** is 'pushed over' a catastrophic threshold by the action of input stimuli. For the *N*-region system (equation 4.2.3) there are two space-time model regimes. The switching between these two models is controlled by the magnitude of the θ_{ti} parameter:

$$\left.\begin{aligned} Y_{ti} &= \mathbf{S}_1 Z_{ti}, \quad \theta_{ti} \in I_1 \\ Y_{ti} &= \mathbf{S}_2 Z_{ti}, \quad \theta_{ti} \in I_2 \end{aligned}\right\}. \tag{4.5.13}$$

This is a form of on-off control (equation 3.5.6) in which the switching may be dependent upon three conditions:

(1) *Input magnitude* (quasi-nonlinear catastrophe):

$$\begin{gathered} \theta_{ti} = Z_{ti} \\ (k_1 \leqslant Z_{ti} \leqslant k_2) \in I_1 \\ (k_3 \leqslant Z_{ti} \leqslant k_4) \in I_2 . \end{gathered} \tag{4.5.14}$$

The switching between the two regimes I_1 and I_2 is controlled by the magnitude of the input term $\{Z_{ti}\}$. If Z_{ti} is less than or equal to k_1 then the first regime is used, if Z_{ti} is greater than k_3 the second regime is used. Note, in general, $k_2 \neq k_3$ since the point of catastrophe will be different depending upon from which side the catastrophe is approached.

(2) *Time value* (irreversible catastrophe):

$$\begin{gathered} \theta_{ti} = t \\ (k_1 \leqslant t \leqslant k_2) \in I_1 \\ (k_2 \leqslant t \leqslant k_4) \in I_2 . \end{gathered} \tag{4.5.15}$$

The switching is controlled by the value of time, and if this is assumed irreversible, then the process is irreversible. Note $k_3 = k_2$ in this case.

(3) *Spatial location* (space-time catastrophe):

$$\begin{gathered} \theta_{ti} = Z_{ti} \\ (k_1 \leqslant t_i \leqslant k_2) \in I_1 \\ (k_3 \leqslant t_j \leqslant k_4) \in I_2 . \end{gathered} \tag{4.5.16}$$

The switching is controlled by the location (i, j) in space and also the location in time. Some spatial locations will obey one regime at one time while others obey another; at other times the locational models may be reversed.

In an example of a spatial discontinuity problem Butkovskiy (1969, p. 118) notes that at a singularity there are 'singular segments' of any optimal control trajectory and the Hamiltonian H of equation (3.5.23) will be independent of these segments. The space-time control problem will be complex since sudden changes cannot be perfectly compensated. For example, sudden changes in thickness of billets in Butkovskiy's heating kiln (shown in fig. 4.14) require discontinuities in control input with infinite (or very high) temperature gradients which cannot be realized in practice. The theoretical control required with space and time discontinuities is shown in fig. 4.27. In practice,

Butkovskiy and Lerner (1960) adopt the assumption that the control at each location can be treated as independent of the spatial location in each heating zone, i.e. as an N-regional system ignoring conductivity effects of cross-region linkage.

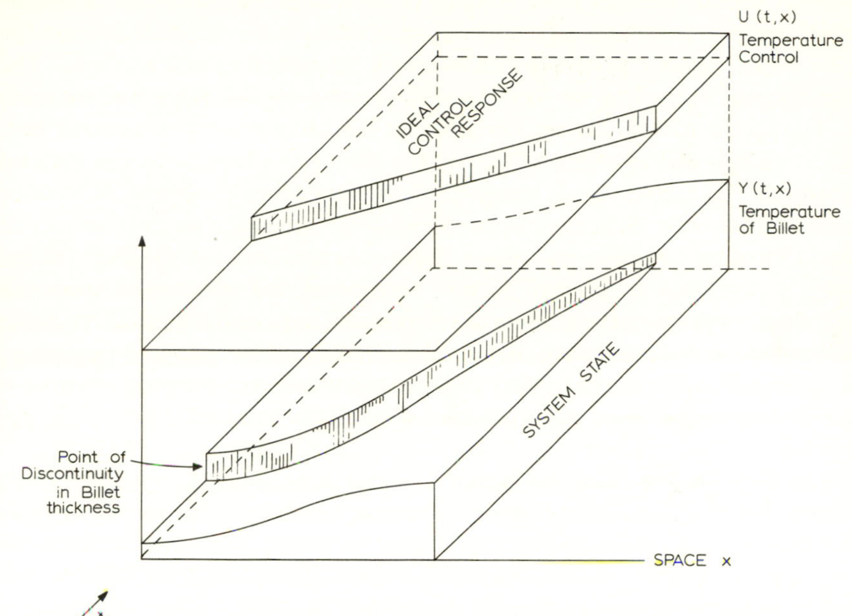

Fig. 4.27 The theoretical control of temperature $\{U(t, x)\}$ required to control the temperature of a rolled metal billet when sudden changes of thickness of the billet occur (see fig. 4.14). The instantaneous change in temperature gradient required within the heating furnace is not physically realizable. (Partly after Butkovskiy 1969.)

4.6 Conclusion

The description of space-time systems, together with the mathematics of spatiotemporal control, which has been developed in this chapter allows the application of systems analysis techniques to a wide range of practical environmental problems. The nature of man-machine decision making systems at the interface between man and the physical environment necessitates the consideration of space in almost all facets of the behaviour of environmental systems. Models which are capable of describing and adjusting the counterpoint interplay of spatial specificity and relativity are crucial to the understanding of such systems. However, in the definition of space-time systems as developed in this chapter we must not attempt a spatial reduction and classification of regions and locations which ignores evolution and shifts within the system. 'Timing and spacing, temporization and regionalization are coordinates. Planning in time is not linked to definite periods. There are no such things as temporal "regions" with "natural" or scientifically defined limits. In the same way the spatial entities have no precise and relatively fixed boundaries' (Groenman 1972, p. 104). In particular, the space and time over which we have to define and explore system behaviour is dependent upon the view of the spatial system which we seek to adopt in any analysis. The consequence of the way we view the system is the way we define and measure it. The mathematical description given in this chapter is quite general and, based on sets of Padé approximants, has the same property as the systems discussed in chapter 2 of being able to approximate system behaviour arbitrarily closely. 'Space is one of the abstract schemata we impose on our world in order to make experience more coherent and meaningful' (Miller 1966, p. 132) and it has two forms: perceptual and conceptual space. In perceptual space we have a frame of reference as sensed which is used to guide human action and movement. In conceptual space we have an extended *image* of space built up by fitting smaller regions into larger ones forming a matrix of spatial knowledge based on a system of abstract coordinates (Miller 1966, pp. 133–5).

The hard systems discussed in part I of this book allow the specification in precise and analytical form of man's *conceptual space* – i.e. his view of the way in which physical, social and economic systems behave. The interface between these hard conceptual systems and man as a perceiver, appraiser and decision maker requires the development of mixed man-machine systems of cognition and decision making in perceptual space. It is to these soft systems that we turn in chapters 5 and 6. With reference to the space-time systems discussed specifically in this chapter, Groenman (1972, p. 105) notes that 'the proximity-collectivity or the community *sec* in space is to be called region . . . the region is within us'. The specificity and relativity of space as perceived and used by man is a further development of our maps of system behaviour as important as maps based on space and time as coordinates. These 'mental maps' and models of spatial environmental cognition are the subject of part II.

Part II
Soft Systems

5 Cognitive systems

> I incline – though with hesitation – to the view that there will ultimately be a science embracing both physics and psychology, though distinct from either as at present developed. The technique of physics was developed under the influence of a belief in the metaphysical reality of 'matter' which now no longer exists. . . . The technique of psychology, to some extent, was developed under a belief in the metaphysical reality of the 'mind'.
>
> It seems possible that, when physics and psychology have both been completely freed from these lingering errors, they will both develop into one science dealing neither with mind nor with matter, but with events, which will not be labelled either 'physical' or mental.
>
> (Bertrand Russell: *Religion and Science*)

In the foregoing three chapters the general structural and behavioural characteristics of what have been termed 'hard systems' have been described. Hard systems are those which are susceptible to rigorous specification, quantification and mathematical prediction in terms of their responses. With cognitive systems we encounter a 'soft system', one which, although representing 'a portion of the world that is perceived as a unit and that is able to maintain its identity in spite of changes going on in it' (Van Gigch 1974, p. 51), is not tractable by mathematical methods. Whether or not such systems will ever acquire such tractability, the foregoing treatment of hard systems is essential to the understanding of the structural features of soft systems and to their possible behaviour in the face of the various inputs to which they are subjected.

5.1 Mental functions and psychological models

Human behaviour is extremely complex, both individually and especially as a group. Models of human behaviour thus all too often appear as crude caricatures, each stressing and distorting one or more mental functions. At the beginning of the present century two classes of models stood in clear opposition, the psychoanalytical and the behavioural. The former, largely associated with the work of Freud and Jung (which will be treated in more detail later), were non-material in the sense that they were based on the assumption that adult behaviour is determined in large part by the, probably unconscious, 'resolution of psychological conflicts experienced earlier in life' or influenced by 'biologically transmitted traces of earlier experiences in human behaviour' (e.g. as subsumed by such terms as the 'collective unconscious' and 'racial memory') (Downs and Stea 1973, p. 3). In contrast to such models in which immediate social influences were assumed to play only a limited, and environmental disturbances a negligible, role in adult patterns of behaviour was the behavioural model of J. B. Watson (1913) which assumed that all behaviour would ultimately be explainable, at least in principle, given full knowledge of the environmental effects on an organism (Powers 1974, pp. 1–3). These genetic and functional, or heredity and environmental, explanatory models have been subsequently rejected in their simplest terms because the former dismissed the mental effects of the feedback on human action of the surrounding environment; and because the latter could neither satisfactorily explain why the same environmental stimulus does not always produce the same psychological reaction in a given person or group, nor account for such manifestations as reflexes, deep psychological actions and memory (Koffka 1935, p. 50). As a consequence, much work during the present century has been directed towards the generation of hybrid behavioural models, such as the 'psychophysical field' concept of Koffka (1935, pp. 66–7) which views behaviour in terms of its links with various psychological environments (see next section) or the 'image' of Boulding (1956, pp. 5–6), a 'cognitive structure' of what one believes to be true, together with the valuations placed on it, which may be triggered off in response to an immediate stimulus. As Downs and Stea (1973, p. viii) point out, the behavioural, or stimulus-response, model takes one only a short way towards the understanding of the cognitive-valuation structure.

It is already clear that some confusion exists between behavioural models and models of mental processes, because it is often only by means of observing behavioural patterns that mental processes can be assessed. A later section of this chapter will examine this relationship between 'image' and 'action'. A further semantic difficulty is presented by the use of the terms 'perception' and 'cognition'. Perception is sometimes used in the widest sense to encompass the mental processes of interpretation – namely, the consensus of knowledge of what is and what happens around us (Powers 1974, p. 35) or the fitting of information into a coherent picture which is internally consistent and involving the following mechanisms (fig. 5.1) (Litterer 1973, pp. 106–7):

1 *Selectivity*: the identification of relevant information.
2 *Closure*: the structuring of information into a meaningful whole.
3 *Interpretation*: the judging of information in the light of past experience.

In terms of information processing operations, the term perception may sometimes be relegated to a relatively primitive step in a mental ladder linking information reception and physical implementation (Fogel 1967, p. 61):

1 *Reception*: the process of accepting an energy stimulus from the real world and translating this signal into a form which can then excite –
2 *Perception*: the encoding of the received information into a neural

message which is then transferred to the appropriate brain centre which can achieve –

3 *Cognition*: the assimilation of information with respect to data extracted from memory, thus achieving identification based on past experience; this process may, or may not, be accompanied by –

4 *Apperception*: the conscious awareness of the result of cognition; information, categorized by cognition, may receive –

5 *Mediation*: the transformation of information from immediate and past cognitive experience and inherent structure into directives in the form of neural messages which are then subjected to –

6 *Implementation*: the decoding of the neural signals into a physical energy output.

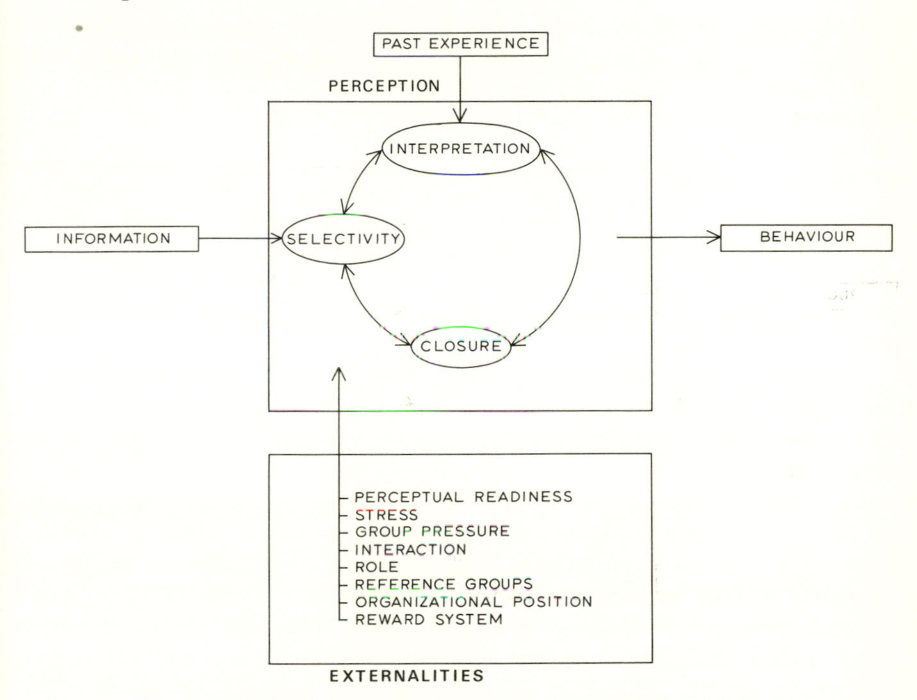

Fig. 5.1 The operation of the perception system in translating information inputs into behaviour outputs, under the influence of past experience and externalities. (After Litterer 1973.)

This hierarchical view of mental processes will be elaborated in the later section on 'images' in which perception does play a relatively restricted role involving 'the process that occurs because of the presence of an object' (Downs and Stea 1973, p. 14) and which results from the selective apprehension of that object.

In this chapter the concept of 'cognition' is employed in a much broader sense than by Fogel, and more in line with Neisser's (1967) view of it as a study of the vicissitudes of information *en route* from an exterior impingement to the final uses to which it is put. Following the usage of Downs and Stea (1973, p. 14), cognition is recognized as a very much more general term than perception, not necessarily linked with immediate behaviour but also with the past and the future, and including thinking, problem solving, organization of information and ideas, memory, and perception. Cognition includes the processing of information too extensive to be classed as perceptive and relies heavily on long-term, as well as short-term, memory. One of the objects of this chapter is therefore to suggest some of the salient features for a general cognitive system (fig. 5.7) linking environmental disturbances and human action by the mental processes involved in:

1 *Information filtering*: the conscious or unconscious acceptance or rejection of environmental disturbance information by means of specification thresholds.
2 *Shunting*: the activation, inactivation or inhibition of certain mental processes (Gagné 1966, p. 38).
3 *Memory*: the capacity to retain information, models, rules, action sequences, etc., against which environmental disturbances are assessed and by which they are modified; memory may be instinctive, conscious or subconscious, as well as short-term or long-term, the latter (sometimes termed 'perceptual constancy') being particularly well developed in man as distinct from machines (Gagné 1966).
4 *Image building*: the development of a hierarchy of mental functions in which cognitive structures are constructed from filtered environmental information which has been subjected to the processes of shunting and memory.
5 *Feedback*: the effect of human action on the input of environmental information.

It will perhaps be of some assistance to try to link the foregoing with the more familiar psychological models of the individual as proposed by Freud and Jung. Freud's view was that our image of reality which conditions our behaviour (or the 'output' of our psychological system) is the result of the intersection of the conscious and unconscious subsystems, and it was one of his major achievements to suggest means by which the latter could be investigated and resistance to its revelation broken down. The conscious subsystem Freud divided into the *ego*, the 'organized and realistic' part, and the *superego*, which is concerned with the critical and moralizing faculties. The unconscious (which, in later life, Freud came to recognize as a highly ambiguous term), termed the *id*, is concerned with desires or wishes that derive their energy from primary physical instincts of a largely sexual and destructive nature, aimed at obtaining immediate satisfaction. These instinctive desires derive largely from phenomena associated with early childhood, having no organization or coordination, and explain why illogical contradictions can occur within a personality. The id may therefore be out of step with the conscious subsystem of the mind which is concerned with such practical and logical matters as adaptation to reality, avoidance of external dangers and forward planning.

In systems terms, the psychology of Jung (Hall and Nordby 1973, pp. 32–70, on which valuable work much of the following section is based) is more complex, being divisible into the linked subsystems or psychic levels (fig. 5.2):

(1) *The conscious.* This involves that part of the mind known directly by the individual and which develops as a result of the mental functions of thinking, feeling, sensing and intuiting. The two basic attitudes of extroversion and introversion orient the mind towards the external objective world and the inner subjective world, respectively. External environmental information (the disturbance inputs) are operated on by a strong executive (see chapter 1), the ego, which controls the organization of the conscious psychic subsystem which is composed of perceptions, thoughts, feelings and memories. The information stream of ideas, feelings, and the like can only be brought into awareness if acknowledged by the ego, and the latter is responsible for the suppression of much external information, much of which passes into the personal unconscious subsystem.

(2) *The personal unconscious.* This is the repository for the perceptions, thoughts, feelings and memories which have been suppressed by the ego, either as the result of distress, conflict, morality, or simply assumed irrelevance. In this subsystem this information is regrouped into *complexes* (e.g. the mother complex) which Freud would ascribe to traumatic experiences in early childhood, but which Jung believed originated in processes operating in the deeper level of the psyche, the collective unconscious subsystem.

(3) *The collective unconscious.* This was one of Jung's most revolutionary suggestions and one which makes his work, at one and the same time, of immense interest to the student of environmental systems and very much out of keeping with so-called progressive ideas of egalitarianism and equality of opportunity. The collective unconscious subsystem of the mind is a reservoir of latent *primordial images* (e.g. the fear of snakes in the dark) which do not derive from the personal experience of the individual or from phenomena which have ever been conscious to the individual. The images, or *archetypes* (i.e. original models), are made up of evolutionary and hereditary blueprints linking the mind with the distant past, just as the body is so linked by genetic information. Jung likened these archetypes or predispositions not to photographs, but to photographic negatives which may be developed and find expression in ways which depend on the experience of the individual. The collective unconscious was for Jung 'the unwritten history of mankind from time unrecorded':

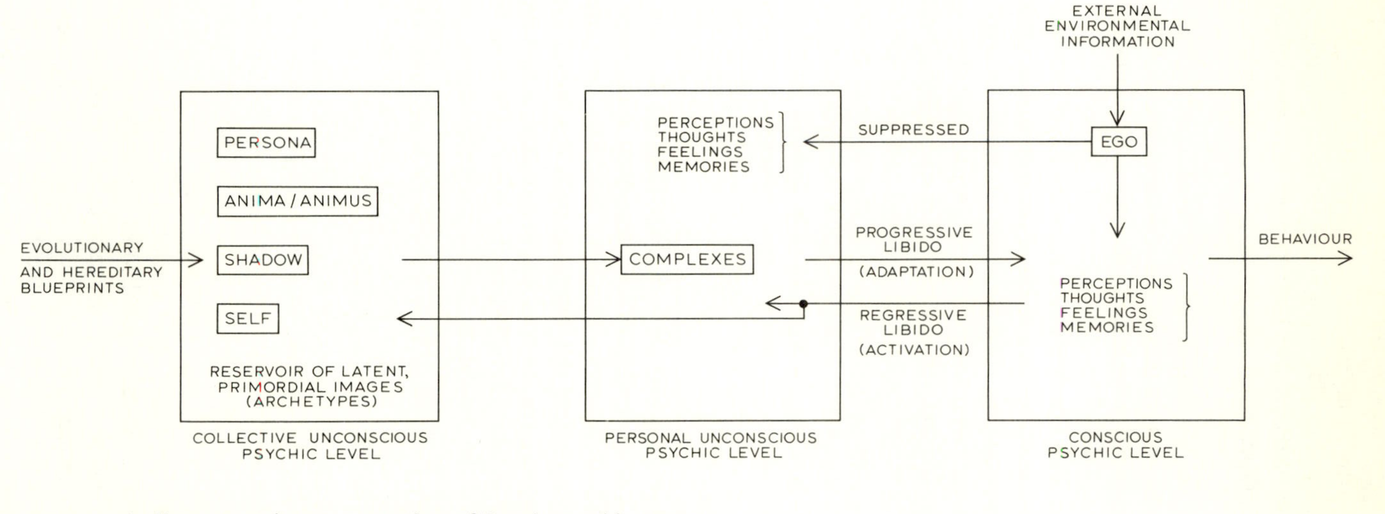

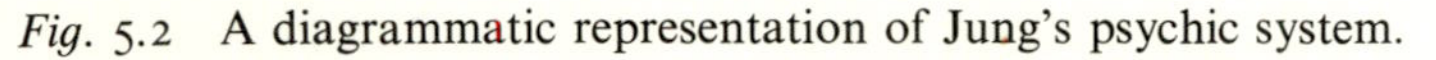

Fig. 5.2 A diagrammatic representation of Jung's psychic system.

> . . . every civilized human being, however high his conscious development, is still an archaic man at the deeper levels of his psyche. Just as the human body connects us with the mammals and displays numerous vestiges of earlier evolutionary stages going back even to the reptilian age, so the human psyche is a product of evolution which, when followed back to its origin, shows countless archaic traits (Jung 1931, para. 105).
>
> Just as a man has a body which is no different from that of an animal, so also his psychology has a whole series of lower storeys in which the spectres from humanity's past epochs still dwell, then the animal souls from the age of Pithecanthropus and the hominids, then the 'psyche' of the cold-blooded saurians, and, deepest down of all, the transcendental mystery and paradox of the sympathetic and parasympathetic psychoid systems (Jung 1955–6, para. 279).
>
> Archetypes are like riverbeds which dry up when the water deserts them, but which it can find again at any time. An archetype is like an old watercourse along which the water of life has flowed for centuries, digging a deep channel for itself. The longer it has flowed in this channel the more likely it is that sooner or later the water will return to its old bed (Jung 1936, para. 395).
>
> All the most powerful ideas in history go back to archetypes. This is particularly true of religious ideas, but the central concepts of science, philosophy and ethics are no exception to this rule. In their present form they are variants of archetypal ideas, created by consciously applying and adapting these ideas to reality. For it is the function of consciousness not only to recognize and assimilate the external world through the gateway of the senses, but to translate into visible reality the world within us (Jung 1927–31, para. 342).

The most important archetypes are the persona, the anima/animus, the shadow and the self.

Persona is the conformity archetype, the mask or facade which one assumes to portray a public character. If the persona becomes too important, the ego identifies with it (i.e. there is *inflation*) and psychological tension develops.
Anima is the feminine side of the male psyche, *animus* the male side of the female psyche. These acquired characteristics of the opposite sex govern our relationships with it.
Shadow contains much of one's basic animal nature (good and bad) and is the most powerful and potentially dangerous of the archetypes, which influences relationships with one's own sex. It is associated with the motive power for spontaneity, creativity, strong emotions, deep insights and wisdom of an instinctual nature. It is most beneficial to the individual when the ego accepts and works in harmony with the shadow, but 'civilization' presents problems by applying certain suppressions to the latter.

Self is the organizational archetype of the personality and is the central archetype of the collective unconscious. It is concerned with order, organization and unification, develops best after middle age and is the basis of religious feelings. It imparts a sense of *oneness* with the universe and is thus related to the universe subconscious in fig. 5.7 (p. 248).

Jung was very much concerned with the balance of the psychic system maintained by flows of psychic energy (*libido*), and he called these processes *psychodynamics*. As is shown in fig. 5.2, the libido can flow progressively (i.e. feedforward) or regressively (i.e. feedback), the former operating towards adapting the personal unconscious to external stimuli (in so far as it is allowed to do so by the ego) and the latter towards activating the unconscious. Psychodynamics is based on two principles:

Equivalence: The concept that if psychic energy in one psychic subsystem decreases, that in another increases. For example, if energy seems to disappear from the conscious psychic level, it is merely being transferred to the personal or collective unconscious.

Entropy: The principle that the flow of psychic energy occurs from areas of high to low energy, bringing about a tendency for a balance among the subsystems. Where there is imbalance, this is manifested by tension, stress and conflict.

Thus the model of Jung contained many attributes of a true system, not least in the stress which he placed on the importance and dynamism of the feedforward and feedback flows of energy.

When one turns to group behaviour, which has always been the predominant mainspring around which environmental systems have been fashioned, the basis provided by formal systems theory becomes even more apparent. Most relevant in this context is the work of the behavioural school of social anthropology deriving largely from the ideas of Lévi-Strauss which rest on the premise that 'society is an articulated system which exists in its own right independently of the individuals who make it up' (Leach 1973, p. 38). Much modern social anthropology is based on the concept of *structuralism* which developed in linguistics some half century ago, and Lévi-Strauss drew the analogy between language and culture in that both can be viewed as mechanisms of communication between conscious humans. Social actions and cultural objects, thus conceived, represent structurally organized systems in space and time which convey meaning and which can only be understood as a whole, like language or music (Leach 1973, pp. 40, 43). Piaget (1971, p. 5) viewed such a structure as

> a system of transformations . . ., these transformations involve laws; the structure is preserved or enriched by the interplay of its transformation laws, which never yield results external to the system nor employ elements

> that are external to it. In short, the notion of structure is comprised of three key ideas: the idea of wholeness, the idea of transformation, and the idea of self-regulation.

Piaget's structures are thus systems composed of organized state variables, connected by transfer functions and subject to feedback. If one sets aside the all-embracing definition of system boundaries adopted by Piaget, then the systems analogy may be further exploited. According to Leach (1973, pp. 37–9) behaviour is made up of two basic components, a practical one which 'alters the state of the world' and a ritual or symbolic one which 'says something' about the social situation. Classical anthropology sought to rationalize behaviour either in terms of *cultural universals* which transcend individual or group idiosyncracies (Malinowski) or of the group *collective consciousness* characteristic of each particular society (Durkheim). By comparing culture with language Lévi-Strauss unites these two concepts in terms of influences operating at three distinct levels:

1. The way culture is manipulated is the outcome of individual self-interest.
2. Superficial cultural patterns are peculiar to particular social systems.
3. The ultimate structural underpinnings of culture are 'deep level' human universals.

This attempt by Lévi-Strauss to synthesize together human universals and cultural specifics may help to resolve the question posed in chapter 1, of whether environmental systems exist independent of the observer. The abstract level is distinguished from the cultural, but both are combined as an explanation of system survival. Social systems are represented as a structure of cultural objects, or artefacts, and social actions, linked by complex transfer functions defined in terms of the dynamism of social groups. The output of such systems is made up of the information communicated by the system operating as a structural unity, resulting from inputs of information provided at each of Lévi-Strauss's three levels. To some degree, therefore, social systems represent grey boxes, the output of which one examines in an attempt to identify the inputs at one or more levels. Information inputs from individuals are strongly stochastic disturbances which are of major interest to the historian. A much less random stream of disturbance inputs, partly resulting from environmental stimuli, produces Durkheim's collective consciousness which it is the aim of many social anthropologists to identify from the output stream of social and cultural communication. Of increasing interest to social anthropologists are attempts to isolate the cultural 'deep level' universal structures generated by human brains, which underlie and transcend cultural diversity (Leach 1973, p. 40). Piaget believed that the search for deep structures – or, in systems terms, basic reference inputs – derives from an interest in transformation laws (i.e. transfer functions); and Lévi-Strauss held that society reflects the formal intellectual properties of man, an unconscious conceptual structure which is discoverable by the elaboration of abstract structural models (Piaget 1971, pp. 98, 107). As Tuan (1971, p. 183) has pointed out, human energy is directed by means which are either rational or irrational, but the ends to which it is directed are neither rational nor irrational, lying in the existentialist realm of the will and of search for meaning. Thus disciplines like anthropology and geography, it is argued, can be used to illuminate human nature at a more general level by the phenomenological assumption that objects and events are invested with, and communicate, a symbolism and significance deriving from a deep structure of mental, cultural and environmental underpinnings. By examining structurally organized systems *phenomenology* attempts to transcend the 'what' of experience to explain the 'why'.

A self-consciousness of the 'why' is a stimulus to social action and, like Marxism and critical social theory, structuralism has been an inspiration to those who wish to induce changes in the structure of social values and cultural patterns at least at Lévi-Strauss's level 1, and also often at level 2. Structural organization also, like the Marxian and critical approaches, can encompass historical perspective in which each society is seen as a particular instance of the way in which man comes to grips with nature and maintains his cultural patterns. Each approach also rejects the epistemology of positivist scientific method and hence replaces Darwinian views of biological evolution by the alternative emphasis on Lamarckian views which underline the importance of transmitted cultural artifacts, language and accepted behaviour patterns. The critical approach, for example, attempts to explain social processes not just in segmented terms, but as wholes emerging from the form of production practised in different societies. Horkheimer (1972) and Habermas (1973) extend this conception by recognizing that from such structural knowledge comes prediction, and from prediction comes control: that 'natural laws' of social behaviour can be turned backwards to change social systems like the control systems discussed in chapters 3 and 4. Structuralism suggests this same alternative to functional modes of explanation by favouring the study of system wholes, but also by encompassing 'natural laws' based on human universals it is also a major stimulus to the collection and maintenance of information bases as discussed in chapter 6. Structuralism tends to play down historical causes and is closely related to the realist basis for theory construction outlined in chapter 1.2. It differs from the functional approach in that, like Marxism with which it has been compared (Althusser 1969, Glucksmann 1974), it distinguishes between the form of artifacts and actions, on the one hand, and their causes, on the other – holding that the external appearance of things is not an unambiguous guide to the processes which produce, sustain and transform them, at least on any superficial level.

The structuralist approach has important implications for the way in which we view spatial systems as discussed in chapter 4. In particular, structuralism forces the analyst to treat spatial distributions and interactions as the output of the socio-economic, and hence the cognitive processes that have created them. This forces rejection of geometric analyses of spatial structure, as evidenced by Bunge (1966) and others, since not only do such analyses represent a description of the *status quo* of spatial pattern, or at best a functionalist description of interrelationships, but they also tell us nothing of the underlying determining relations to which the pattern owes its genesis. The reinterpretation of spatial analysis which structuralism requires, links closely with the recognition of spatially specific and spatially relative components of space-time systems discussed in chapter 4. Structuralist analysis forces us to recognize that on the one hand many system inputs and outputs are spatially specific and independent of inputs and outputs at other locations, whilst on the other hand some system inputs will have relativistic interactions with each other (through various transformation operators) across space. In the context of any one location, therefore, some system inputs are partly endogenous and some partly exogenous. Functional explanations, it is argued, concentrate on the specifics of internal transformations alone, whilst structuralism allows the study of wider exogenous interactions (Georgescu-Roegen 1971, Sayer 1976). Obviously this is an argument partly about the degree of system closure accepted in any cognitive system, for which structuralism offers the wider critical awareness not only of 'self' but also of interrelations with other entities and individuals. From this awareness, it is argued, we have a greater facility of predicting future social and spatial evolution which arises from the readjustments of social values (e.g. dialectic). We are drawn away from social and spatial equilibria towards methods of resolving what Piaget (1971) terms the central problem of structuralism – that of reconciling structure and genesis. The structuralist reconciliation is in terms either of gradual genetic change and social culturization and learning, or of sudden (less frequent) transitions between social phase states caused by 'flipping' to new social regimes (as in the catastrophe manifolds discussed in chapter 4.5).

The tasks of structuralist social anthropology – to identify the inputs of collective social consciousness and of deep level cultural universals which sustain and are revealed by the social system states, transfer functions and communication outputs – seem daunting. Nevertheless they are of great interest to the environmental systems analyst, although his needs require something of a reverse approach as the result of his differing aims. It is clear that social and cultural mechanisms operate at all levels when socio-economic systems interact with physico-ecological ones, and that no adequate understanding of integrated environmental systems can be achieved without an appreciation of these mechanisms. To recognize the possible importance of the latter is one thing, however, but the demonstration of their specific contribution in the production of existing or planned environmental systems is quite another. Suffice it to say that environmental scientists are becoming increasingly aware of the importance of the work of social anthropologists and of the potential value of their contribution to the understanding of environmental systems.

5.2 Models of man and environment

The complexity of human behaviour means that the world of man is not merely a stage, but a simultaneous plethora of mental environments within which he assumes many roles at any one time. These roles, or 'models of man' (Simon 1957), and mental environments are, of course, merely simplifying artifices which act as frameworks for the partial rationalization of human behaviour, each accentuating a part of the total cognitive structure. There are four important models of man which are of most interest to those concerned with environmental systems:

1. *Psychoanalytic man*: Although Freud's theory involving the id, ego and superego is regarded by some as merely a pseudomodel in that it is concerned with observed behavioural symptoms rather than their underlying properties (i.e. it is 'functional' rather than 'realistic': see chapter 1) (Powers 1974, p. 15), the model of the non-rational adult whose behaviour is largely determined by the resolution of psychological conflicts has formed a powerful model for man in the present century (Downs and Stea 1973, p. 3).
2. *Computer man*: The logical theorist who objectively screens and possesses information without inhibition (see later section on 'shunting') and with reference to long- and short-term memory stores is clearly opposed to the foregoing. Like all the other models of man he is an abstraction, as attempts to model brain-like computers attest.
3. *Economic man*: In terms of ideal exchange, man has been viewed by classical economists in a predictive sense as totally material and influenced in his behaviour only by objective factors external to himself of which he has total knowledge. Economic man as a consumer is dominated by the wish to maximize utility and as an entrepreneur to maximize his profit (Downs and Stea 1973, p. 3).
4. *Bounded rational man*: The failure of the foregoing model of man wholly to predict human economic behaviour either in time or space (Wolpert 1964) has led to a model proposed by Simon (1957, pp. 196–200) of 'bounded rational' or 'satisficing' man (Golledge, Brown and Williamson 1972). The limitation of human sources, storage and processing of information, relative to the magnitude of environmental problems, implies that man's decisions and actions are taken in response to restricted and fragmentary information and within very real time constraints. Thus, rather than optimizing, man is usually committed to satisficing, a course of action which is 'good enough' for the situation as he comprehends it (Downs and Stea 1973, p. 4). This model is one of several which depict man as operating intermediately between his respective psychological, behavioural and logical roots.

Each man, or group of men, operates in a variety of environments, most of which are unique to himself. Kuhn (1974, p. 52) describes the impinging of environment on man as a result of merely correlating devices (as in historical explanation). For him the relation between sensation and possible explanation is due to conditioning of the cognitive response correlating sensation patterns with environmental patterns. This is continually strengthened by reinforcement of this correlation, related to the frequency, intensity and recency of reinforcement. Environments impinge upon man in terms of a common set of attributes (Ittelson 1973, pp. 13–15):

1 They envelop man, such that he does not observe them, but explores them as a participant.
2 They contain a variety of stimulation and information sources which simultaneously impinge on man.
3 Some of the sources are immediate but others are peripheral and background.
4 They contain an excess of information, some of which is redundant, inadequate, ambiguous, conflicting, and contradictory, and requires filtering and reconciliation.
5 Their observation by man cannot be passive and always involves purposeful action.
6 For each person or group they possess special symbolic meanings and motivational messages.
7 They possess an 'ambiance', an 'atmosphere' which is difficult to define, but which interacts strongly with social attitudes and aesthetic patterns of thought.

The development of gestalt psychology has been very much concerned with the identification of the different types of environments within which man mentally operates, and with the relations between them. Koffka (1935, pp. 31, 40, 66–7), for example, opposed the behaviourists in proposing a 'psychophysical field' theory in which he viewed the behaviour of an organism as embedded both in its own behavioural environment (producing 'phenomenal' or 'experienced behaviour') and in those of other organisms (producing 'apparent behaviour'), which in turn lie within organic environments enveloped by a general geographical environment which is capable of alteration by the behaviour of the organism (fig. 5.3). This hierarchical nesting of environments underlay the concept of topological psychology (Lewin 1936) which assumed a topological structuring of the environment such that no single connected space exists which encompasses physical reality and that each life space is dynamically unique (Downs and Stea 1973, p. 5). Although Koffka (1935, p. 49) saw problems in the specification of interaction between the behavioural and geographic environments and questioned how a cause in one universe of discourse can produce an effect in another, his proposition connected behaviour with remote, as well as immediate, causes. Sonnenfeld

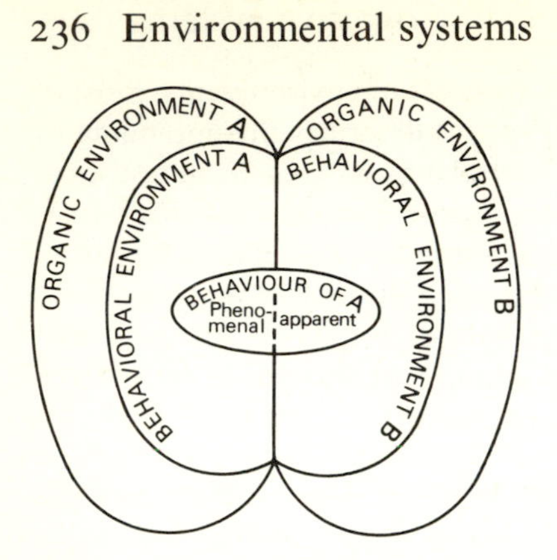

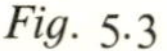

Fig. 5·3
A diagrammatic representation of Koffka's psychophysical field. (Adapted from Koffka 1935.)

(1972, pp. 247–9) has proposed the following nested hierarchy of environments:

1 *Geographical environment*: encompassing the whole external environment and approximating with Lewin's (1936, p. 214) physical space, which includes the totality of all physical facts that exist at a certain time.
2 *Operational environment*: 'the functional portion of the geographical environment, that which impinges on man as an individual or a group, influencing behaviour in some way or another' (English and Mayfield 1972, p. 247).
3 *Perceptual environment*: that part of the operational environment of which man is conscious, either through physical sensitivity or due to a sensitivity deriving from man's learning and experience. Spatially this environment is expressed in terms of spatial perception and often in terms of preference by means of mental mapping (Gould 1970). Lewin's (1936, pp. 53–5) distinctions between topological and metrical spaces are central to such notions.
4 *Behavioural environment*: that part of the perceptual environment which 'elicits a behavioural response or towards which a behaviour is directed, such as results in a conscious utilization or transformation of environment' (English and Mayfield 1972, p. 249). This is akin to Lewin's (1936, p. 216) psychological space as 'the totality of facts which determine the behaviour of an individual at a certain moment'.

The interaction between these four views of the environment is well evidenced in the diverse lines of geographic research into perception and cognition. Gould (1976, pp. 15–20) identifies six overlapping approaches. Firstly, there is that of anthropological and cultural conditioning based on conditioning both at the matter-energy level (artefacts and behaviour) and at the information level (concepts, syntax, dialectic, religion, art, etc.) (Kuhn 1974, pp. 156–7), which gives rise to the variety of ethno-scientific systems. A second approach is based on analysis of the perceptions which result from different social and physical environments (Lowenthal 1975, Tuan 1976).

Gould identifies the 'Chicago School' as a third approach, based on studies of human response to various frequencies and magnitudes of environmental hazards (White 1945, Burton, Kates and Snead 1969) involving various measures of impact assessment. A fourth approach is especially concerned with evaluation of urban and transport plans using assessments of 'environmental desirability' which is an extension of revealed space preferences (Golledge, Briggs and Demko 1969, Johnston 1971). A fifth approach parallels the Piagetian developmental psychologists and seeks to determine how spatial perception evolves in children. Finally, Gould's (1965, 1976) own approach is focused on the construction of aggregate perception surfaces of spatial evaluation, so-called 'mental maps' which often exhibit national, regional and local components in which revealed space preferences appear as local domes of desirability. The resulting information space offers the tempting prospect of returning to an older determinism, substituting for the physical environment a wholly human-created environment of aggregate information flows (Gould 1976, p. 145). In perceptual space the specific and relative components of spatial system models, described by the technology of transfer function polynomials in chapter 4, are replaced by soft information flows themselves constrained by space-time 'bubbles' of interaction possibilities.

5·3 Images

The building blocks of the cognitive system are the beliefs or images, changes in which express the meaning of the flow of information messages to the brain (Boulding 1956, p. 7). The levels of information transfer depend upon the degree of coding of messages and the mechanisms of transfer (direct, mediated, or social communication). Images are arranged in a control hierarchy of adjustable goal perception purposes, involving negative feedback loops which, in turn, influence behaviour and performance (Powers 1974, pp. 52–4). Although any categorizing of these building blocks must be an arbitrary one, it is convenient here to recognize six, the mutual operations of which together constitute the 'cognitive image' (see fig. 5.7).

Sensing is the most reflexive and automatic image, but nevertheless subsumes an extremely complex set of mental processes. Sensing involves, first, the primitive and immediate acceptance of information signals from the environment (Fogel 1967, p. 61) by means of somatic (skin), chemical (taste and smell), auditory and visual recognition (Deutsch 1967, p. 256). Such processes require short and instinctive memory reference, and range from the binary ability to distinguish presence or absence, for which man often has a more limited capacity than a machine (Gagné 1966), to feelings of basic emotions – such as joy or fear (Kelley 1968, pp. 37–40).

According to the terminology adopted here, *perception* involves the selection, identification and coding of relevant information within a remembered context of past information, separated by thresholds, so that considerations of form, configuration, change, transition and sequence apply (Kelley 1968, Litterer 1973, p. 107, Fogel 1967, and Powers 1974, pp. 115, 129, 137).

The distinction between perception and *programming* is an example of the difficulty of separating distinct images of the cognitive system. Programming thus partly involves the concept of taxonomy, the associating of what has been sensed and perceived with a known class of circumstances such that the operator can distinguish between different classes of stimulation and make different responses to them (Gagné 1966). Programming, however, is also concerned with the structural sequence of operations between events (Powers 1974, pp. 154, 160) and the association of information with respect to data from memory to achieve an identification between present and past experience, translated into consciousness by the process of apperception (Fogel 1967, p. 61). The effect of programming is the pregnant association of information or 'invention by serendipity' (Deutsch 1967, pp. 257–8).

The next higher level of image is that of *principles* which involves 'closure', the structuring of information into a meaningful whole and its weighting as to importance, as well as 'interpretation', the judging of the significance of inputs in terms of past experience as to their effects on the system, rather than their intrinsic appearances (Gagné 1966, Litterer 1973, pp. 107–12). This process draws heavily on memory to develop rules, often probabilistic, and through heuristics to create models of the real world (Powers 1974, p. 168). It is thus more than a simple sum of sensing, perceiving and programming.

The highest levels of images draw increasingly on the deeper and subconscious levels of the memory, and indeed upon 'folk memory'. *Belonging* is an instinctive desire to conform to the group and to perpetuate the well-being of the species. *Oneness* subsumes the concept of individual and corporate unity with the total environment which has been most popularly expressed in eastern philosophies (Powers 1974, pp. 171, 174). The western philosopher William James termed this 'cosmic elation', when the ego expands so that it merges with the whole cosmos.

5.4 Environmental disturbances, shunting and memory

Within the cognitive system the flow of information, leading to the production of images from which action results, is broadly the result of what may be termed 'environmental disturbances'. These disturbances cover an extremely wide spectrum of stimulation, ranging from the instinctive release of *genetic* information from within the body itself to the immediate sensing of physical changes in the environment. The genetic information may be the prime stimulus for the development of a philosophical potential. Another class of environmental disturbances is concerned with information relating to *society and group attitudes*, or to competitive behavioural systems, and leads to a potential attitude to other human beings which feeds the belonging image. Culture defines truth since messages are accepted as truthful if they conform to the images of the society (Kuhn 1974, p. 157), and, although perceptions are improved by communication of ideas about reality which eventually achieves consensus, the accordance with reality need not be close. As will be seen later, social information is particularly liable to inhibit the mental processes, but once the belonging image is established it is a powerful motivator of human action in terms of environmental planning which, in turn, has a strong feedback effect on the environmental disturbances. Other classes of environmental disturbances are the *structural* and the *physical*. Structural information is that which affects the organizational frameworks within which information is structured, and which results in the development of potentialities for model building and taxonomy. This information is thus responsible for the stimulation of the most analytical parts of the cognitive system, which lie midway between the areas of genetic instincts and sense feelings. Physical information is that which relates more immediately to perception and the senses, producing the potentials for feeling and for short-term selection of immediate environmental information. The whole question of selection or 'filtering' of information – the specification of the type of information to be handled by the cognitive system, the cutting out of 'noise', etc. – is a very complex one. Filtering clearly operates to a varying degree across the whole spectrum of environmental disturbance information inputs and is also much bound up with the mental process of 'shunting'.

One of the basic functions of the mind is to process the flood of incoming information which is being fed forward through the cognitive system, and which we have arbitrarily divided into that relating to feeling, selection, taxonomy, modelling, belonging and philosophy. This processing is effected by associated operations which may be described as *shunting* and *memory*, which operate to produce image construction and belief.

It has been suggested that one of the major functions of the brain is to be eliminative, in order to protect us from this incoming flood of environmental information (Huxley 1954, p. 16). Viewed another way, man possesses certain cognitive functions which may be active, inhibited or inactive (i.e. 'shunted out'), and the operations of which it is one of the tasks of the science of psychology to investigate (Gagné 1966, p. 38). Such shunting of certain functions is believed to enhance some or all of the remainder, both relatively and possibly absolutely. Experimental work relating to shunting has been carried out in the simple context of operators reacting to visual displays with respect to which they were required to respond at the different levels of *sensing* the presence or absence of a visual image, its more precise *identification* within different possible classes of images, and the interpretation of the image (e.g. Is there a radar 'blip'? Is it an aircraft? Is the aircraft friendly or hostile?) (fig. 5.4). At each level shunting interacts in a complex manner with both long-term and short-term memory. The stimuli which produce shunting, under

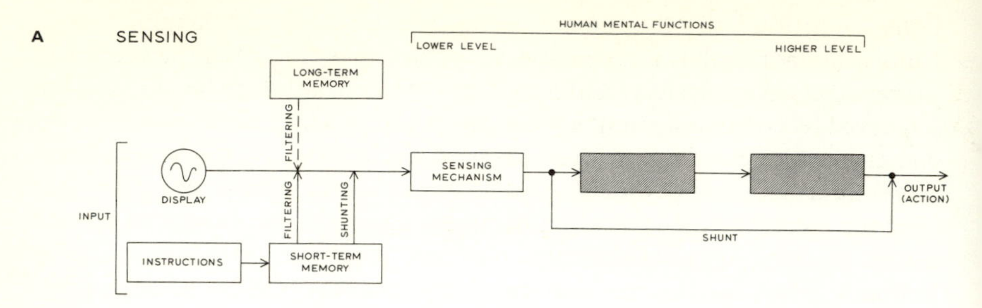

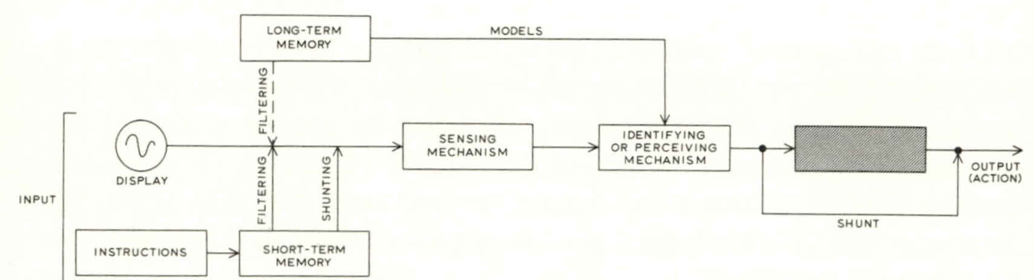

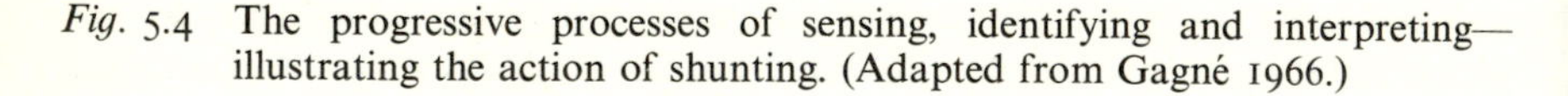

Fig. 5.4 The progressive processes of sensing, identifying and interpreting—illustrating the action of shunting. (Adapted from Gagné 1966.)

uncontrolled real-world conditions of information reception are clearly very complex but, in terms of environmental control systems, five classes may be suggested here:

(1) *Social stimuli*: Litterer (1973, pp. 117–30) (fig. 5.1) has suggested a number of social factors which may influence the shunting process:

Perceptual readiness – the clarification of goals both personal and of a wider nature.
Stress – it is known that social stress leads to quicker, less tolerant, less discriminating decisions, but stress results from many causes and is important enough to form a stimulus class of its own.

Group pressure – the effect of the pressure to conform depends on a complex of variables including individual personality, the existence of allies and the setting of the group pressure.
Interaction – the interaction of individuals or groups.
Role – the fulfilment of the roles one expects, or is expected, to play.
Reference groups – the association with organized groups (political, religious, social, occupational etc.) the approval of which one values, together with the sense of belonging.
Organizational position – one's position in a hierarchy and in the information network.
Reward systems – the significance of social rewards.

(2) *Stress stimuli*: Image construction under stress and its effects on resulting action is a classic example of shunting, and one which has been the subject of considerable psychological testing. It has been shown that the appraisal of potentially threatening events is, firstly, concerned with the evaluation of their magnitude, imminence, duration and likelihood, and that the appraisal depends not only on the intrinsic qualities of the threat itself (e.g. the magnitude and frequency of events inputs to physico-ecological systems) but on the characteristics of the evaluator, such as his past history of dealing with similar events. The second stage of appraisal is concerned with the developing of images or 'coping strategies' and it is in this respect that shunting may be particularly important. Shunting under stress has been shown to result in a characteristic pattern of 'dysfunctional' reactions (Lazarus 1966, Wolpert 1972) involving a generally less adaptive behaviour and a belief in a narrowing range of available options; a greater than usual tendency for aggression; a fixation on untested hypotheses; an increased rate of error in the performance of standard operations; stereotyped responses; disorganized activity; an increased rigidity in problem solving; a reduction in time and space perception, i.e. more concern for the immediate rather than the distant; an increased impression of the pressure of time; less belief in a benign environment; and an increased reliance upon *ad hoc*, extraordinary or improvised communication channels through which stereotyped messages are passed.

(3) *Physiological stimuli*: These are more difficult to isolate in the sense that all stress stimuli produce physiological effects, for example the secretion of adrenalin. However, some Yoga practices have been used in cognitive shunting, for example '*pranayama*', the control of breathing to influence mental processes (Koestler 1966, p. 89).

(4) *Drug stimuli*: Drugs have long been known to inhibit certain mental functions and to heighten others. Huxley (1954) has described experiments with the drug mescalin to change the quality of his consciousness and has shown its affinity with naturally occurring adrenalin. He described the results of his experience in terms of a change of perception of the environment: 'The

other world to which mescalin admitted me was not the world of visions; it existed out there, in what I could see with my eyes open. The great change was in the realm of objective fact . . . spatial relationships had ceased to matter very much. Place and distance cease[d] to be of much interest. The mind [did] its perceiving in terms of intensity of existence, profundity of significance, relationship within a pattern' (Huxley 1954, pp. 11, 14). He concluded that the drug did not significantly affect his power of reasoning (i.e. programming), intensified his visual processes (i.e. feeling), and inhibited the will (i.e. adherence to rules).

(5) *Religious stimuli*: Although it is clear that religious contemplation exercises, through what some theologians term 'gratuitous grace', a powerful shunting effect, it is felt of limited value to pursue the matter further in the present context. Nevertheless it is clear that the image of oneness is considerably enhanced by religious experience, for example as expressed by Teilhard de Chardin who wrote, 'Throughout my whole life, during every moment I have lived, the world has gradually been taking on light and fire for me, until it has come to envelop me in one mass of luminosity, glowing from within. . . . The purple flush of matter fading imperceptibly into the gold of spirit, to be lost finally in the incandescence of a personal universe' (quoted in Leroy 1960, p. 13) (note the analogy with James's 'cosmic elation').

Environmental information input into the cognitive system seeks memory trace patterns in thought recognition, or memory, storage. Usually this process acts 'like a deleterious mutation', but occasionally it may lead to image construction by 'invention by serendipity' (Deutsch 1967, pp. 257–8). Memory is concerned with the sifting, classification, coding, matching, storage, retrieval and association of acquired information, resulting in image construction, together with the provision of information sets which influence the processes of filtering and shunting. In systems analysis it is convenient to distinguish between short- and long-term memory, although the two are clearly not distinct and are interactive. Short-term memory has a photographic quality and is subject to rapid decay, but it results in man's perceptual experience acquiring a form, structure and continuity over time whereby 'discrete and erratic sensory input emerges as continuous and ordered perception' (Ittelson 1973, p. 11). Man possesses a very flexible short-term memory, primarily concerned with the processing of sense and perceptual information, but, compared with machines, he has a very limited capacity for reacting to rapidly changing frameworks of analysis (Gagné 1966). In contrast, long-term memory is inherently symbolic in character, involving the retrieval and processing of information in a form related to learning. In its extreme form, long-term memory becomes involved with such subconscious processes as are termed 'folk memory', or the collective unconscious of Jung. However, long-term memory also relates to the retention and development of models, rules and action sequences against which the transient reality of environmental disturbances is assessed (Gagné 1966). Man has an especially well-developed capacity to store such information patterns, and this is termed 'perceptual constancy'.

5.5 Belief and action

The relationship between belief, or image, and action, or behaviour, is an extremely close one. Boulding (1956, p. 6) points out that behaviour or action is dependent upon image, but others would argue that in some respects image and action are synonymous. The Marxist attitude is of this nature (see chapter 1). This viewpoint is most easy to understand in respect of sensory action. Arbib (1972, p. 7) points out that millions of receptors in the human body continually monitor the environment, both external and internal, and generate signals which are combined with those encoding memory in a network of billions of neuron cells to barrage the motoneutrons which control muscle movement and gland secretion and which implement adaptive interactions with the environment. At the other end of the action spectrum the unconscious sensory reflex is paralleled by the subconscious association of the oneness image with deep psychological actions, such as occur during the act of meditation. Between these extremes, however, lies an area of conscious action, both individual and with respect to groups, which appears more clearly divorced from image or belief, and at the core of which lies analytical decision making which forms the subject of chapter 6.

One of the main obstacles to the development of a viable environmentally oriented cognitive system is that the environment (both man-influenced and external to him) both influences and is influenced by behaviour (Walmsley 1973, p. 52). An important theme of the volume on *Environmental Psychology* by Proshansky *et al.* (1970) is that psychological processes manifest themselves only in specific environmental contexts and are thus not only dependent on environmental influences, but inextricably bound up with them. In developing a social psychological environment-behaviour system, Studer (1970, p. 58) has viewed the designed or man-influenced system in a 'prosthetic' sense (i.e. as, in some sense, taking over human functions), as it both configures specific behavioural topographies and gives psychological support to behavioural goals. The designed environment is modelled as being interposed between a collection of human participants (H) and the 'antithetical forces' of an 'impinging milieu' making up the existing physical system (S_1^{de}). Most important to the resulting interaction between these two forces are the social, technological and altruistic constraints which inspire and limit man-environment interaction at the cultural level by the emergence of man as an entity different from other organisms: by the emergence of what Polanyi (1958) terms ontogenetic maturation to phylogenetic heuristics.

Studer (1970, pp. 59–65) has given in fig. 5.5 a sample iterative strategy for the design of a behaviour system (R) to accommodate the goals of the human

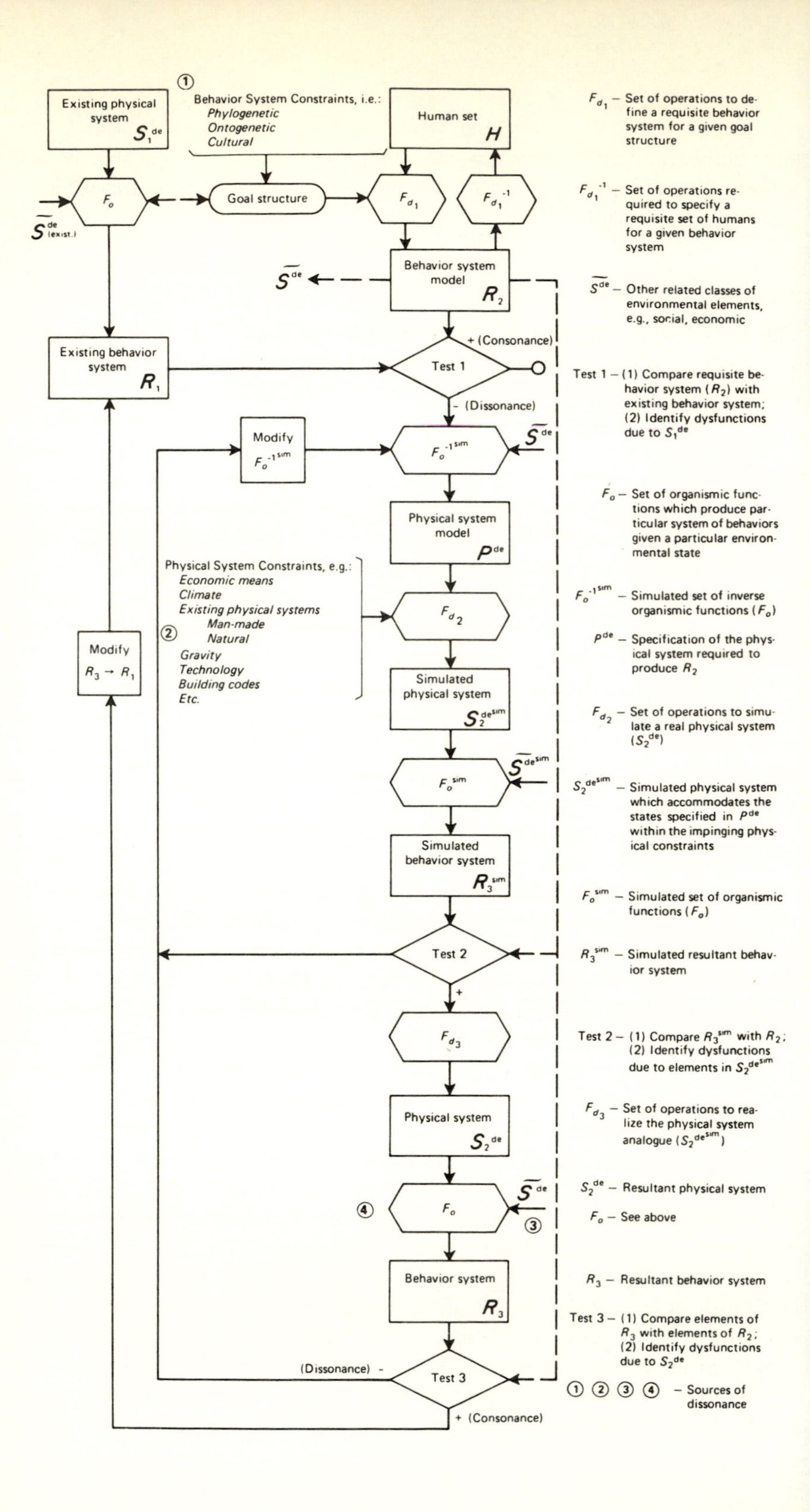

Existing physical system
Behavior System Constraints, i.e.:
Phylogenetic
Ontogenetic
Cultural
Human set
Goal structure
Behavior system model
Existing behavior system
Test 1
+ (Consonance)
- (Dissonance)
Modify
Physical system model
Physical System Constraints, e.g.:
Economic means
Climate
Existing physical systems
Man-made
Natural
Gravity
Technology
Building codes
Etc.
Modify
Simulated physical system
Simulated behavior system
Test 2
Physical system
Behavior system
(Dissonance) -
Test 3
+ (Consonance)
F_{d_1} – Set of operations to define a requisite behavior system for a given goal structure
$F_{d_1}^{-1}$ – Set of operations required to specify a requisite set of humans for a given behavior system
$\tilde{S}^{de}$ – Other related classes of environmental elements, e.g., social, economic
Test 1 – (1) Compare requisite behavior system (R_2) with existing behavior system; (2) Identify dysfunctions due to S_1^{de}
F_o – Set of organismic functions which produce particular system of behaviors given a particular environmental state
$F_o^{-1 sim}$ – Simulated set of inverse organismic functions (F_o)
P^{de} – Specification of the physical system required to produce R_2
F_{d_2} – Set of operations to simulate a real physical system (S_2^{de})
$S_2^{de^{sim}}$ – Simulated physical system which accommodates the states specified in P^{de} within the impinging physical constraints
F_o^{sim} – Simulated set of organismic functions (F_o)
R_3^{sim} – Simulated resultant behavior system
Test 2 – (1) Compare R_3^{sim} with R_2; (2) Identify dysfunctions due to elements in $S_2^{de^{sim}}$
F_{d_3} – Set of operations to realize the physical system analogue ($S_2^{de^{sim}}$)
S_2^{de} – Resultant physical system
F_o – See above
R_3 – Resultant behavior system
Test 3 – (1) Compare elements of R_3 with elements of R_2; (2) Identify dysfunctions due to S_2^{de}
① ② ③ ④ – Sources of dissonance

set (H) under the influence of sources of dissonance involved in (1) behaviour, (2) physical and (3) socio-economic system constraints. The modelling of an initial behaviour system (R_2) allows the identification of the relevant features of the existing physical system (S^{de}_{exist}) which influence the existing behaviour system (R_1), and R_1 and R_2 can be tested for dissonance, and the latter related to elements of the existing physical system (S^{de}_1) or to other social, economic, educational, etc., aspects of the environment (S^{de}). The recognition of dissonance and its possible causes leads to the successive development of an alternative physical system model (P^{de}), a simulated physical system ($S_2^{de^{\text{sim}}}$) and a simulated behaviour system (R_3^{sim}) which is tested and modified for dissonance. The modified simulated physical system is then made operational, and, under the influence of other external environmental elements (S^{de}), gives rise to a new behaviour system (R_3) which is again tested for dissonance, and, when consonant, it replaces R_1. The value of this highly conceptualized and theoretical scheme in the present context is that it draws attention to: the close connection between the human social set; the existing man-influenced environment; the behavioural, physical and socio-economic constraints; the planning process; modelling, simulation and reality; good structures; and iterative learning operations.

A rather more sophisticated behaviour-environment cognitive system has been developed in fig. 5.6, largely based on the work of Chen (1954), Emery and Trist (1965) and Walmsley (1973). As always, the fundamental inputs are the most difficult to express, and suffice it to say that there is at present a considerable body of qualitative work dealing with that vision of the environment associated by Tuan (1973, p. 256) with dreams of Utopia. This vision, tempered by a feedback loop of rethought objectives based on experience of a changing man-influenced environment, produces environment needs which are balanced against a feedback loop of perceived opportunities inherent in the environment. The relation between needs and perceived opportunities produces a learning situation which inputs *stress* into the behaviour system causing a *strain*, or a tendency for a change of behaviour (Herbst 1961, p. 71, Walmsley 1973, p. 50). Depending on the individual or group *stress-strain conversion function*, which varies with the individual or group behavioural tolerance levels, an *action potential* is generated, the intensity of which determines the equilibrium level of the behavioural system (e.g. the number of group members, their level of functioning with reference to group activities, and the interrelationship maintained between individuals and on-going group activities) or the extent to which existing behavioural patterns are disrupted (Herbst 1961, p. 72). Action potential produces a behavioural output of decision making and planning (see chapter 6) which results in a combination of an adjustment of objectives, a change in perception of the

Fig. 5.5 (*opposite*) An iterative strategy for the design of a behaviour system. (From Studer 1970.)

environment and its opportunities, and a change of operation on the components of the man-influenced environmental system (Walmsley 1973, p. 50, after Herbst 1961). The whole area of decision making and its consequences is very difficult to specify, not least because of the problem of identifying the ever-changing and arbitrary boundary between the behavioural and environmental systems. Two important components of the man-influenced environmental system are, firstly, the occurrence and distributional relationship of *goals and noxiants* and, secondly, the *causal texture of the environment*. Chen (1954) defined goals as objects or situations

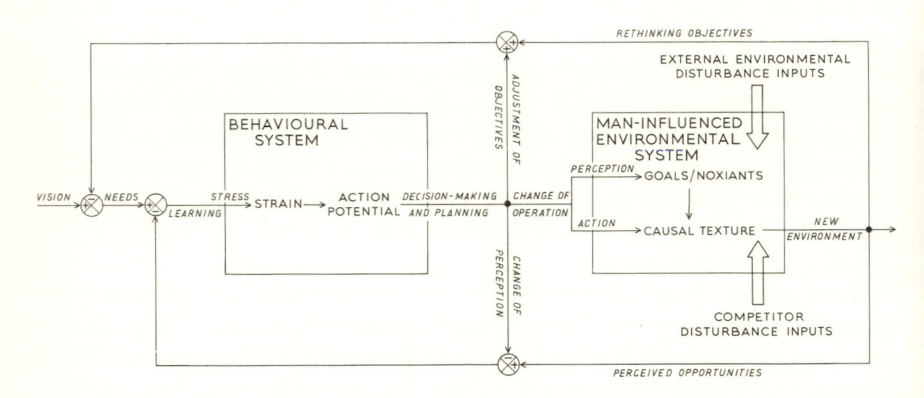

Fig. 5.6 The basic structure of a behaviour-environment cognitive system. (Partly after Emery and Trist 1965, Walmsley 1973.)

which can serve as need satisfiers, and noxiants as those which can produce pain or unpleasantness. Clearly their definition is a matter of perception but, once defined, they are capable of specification in terms of their occurrence, relationships or interconnections, and their stability. Emery and Trist (1965; see Walmsley 1973, p. 52) introduced the concept of the causal texture of the environment as the process of interdependence between the parts of the man-made environment which determines to a large extent its interaction with the behavioural system. Walmsley (1973, pp. 53–4, following Emery and Trist 1965, pp. 24–5) has recognized four levels or 'ideal types' of environmental causal texture, depending on the goals/noxiants characteristics and on disturbance inputs from the environment external to the man-made and from competitive behavioural systems:

(1) *The functional*: This characterizes a *placid-randomized environment* wherein the goals and noxiants are relatively unchanging and randomly distributed with little interconnection (i.e. clustering). Such a simple situation operates in the absence of competitor disturbance and results, by the previously outlined feedback processes, in a response by the behavioural system characterized by primitive trial and error learning or choice situations

(i.e. similar to the simple learning process in laboratory animals) with no distinction between tactics and strategy.

(2) *The goal-directed*: This characterizes a *placid-clustered environment* wherein the goals and noxiants are, although relatively unchanging, related in some comprehensible way and are therefore susceptible to a more complex learning process on both tactical and strategic levels. Such problems involve optimization based on hierarchical, information-gathering organizations which direct the system towards longer-term states of potential richness. Emery and Trist (1965, p. 24) included within this behavioural response level the 'imperfect competition' situation of the economist, and Walmsley (1973, p. 54) the conventional geographical location problems assuming the existence of economic man.

(3) *The purposive*: This characterizes a *disturbed-reactive environment* with similar goals/noxiants characteristics as in (2) but with the presence of competitor disturbance inputs from a competing behavioural system. This gives rise to a problem-solving situation in which actions must be selected which will not only advance one's own long-term strategy but will impede or neutralize that of one's competitors.

(4) *The ideal seeking*: This characterizes the most complex and common environmental states wherein the causal texture shows an ever-changing dynamism resulting from a variability and lack of clear distinction between goals and noxiants, from the unpredicted actions of competitors, from unpredicted external environmental disturbance inputs, from the action of complex linkages of the components of the environmental system (see Forrester's 'counter-intuitive behaviour', ch. 9), as well as from the interaction of the environment with the behavioural system. Such an environmental state, as Walmsley (1973) points out, constantly confronts the behavioural system with unplanned consequences leading to adaptive planning outputs sometimes of a primitive trial and error kind.

5.6 A general cognitive system

Powers (1974, pp. 50–4) has suggested that there is a hierarchy of cognitive control involving different levels of organization, whereby the higher (i.e. outer) levels operate progressively by negative feedback to prevent the images from deviating significantly from certain reference states, which may be viewed as the goals of the cognitive system. Fig. 5.7, to which reference has been copiously made throughout the chapter, depicts such a system, drawing on the principles of nested hierarchical control introduced in chapter 3. In this system the throughput between the image levels is dominated by a hierarchy of nested positive feedforward loops subjected to shunting and memory storage, which are circularly linked to less impeded, negative feedback loops in which the consequences of actions resulting from the images are compared with the appropriate parts of the spectrum of environmental disturbances. As Bruner

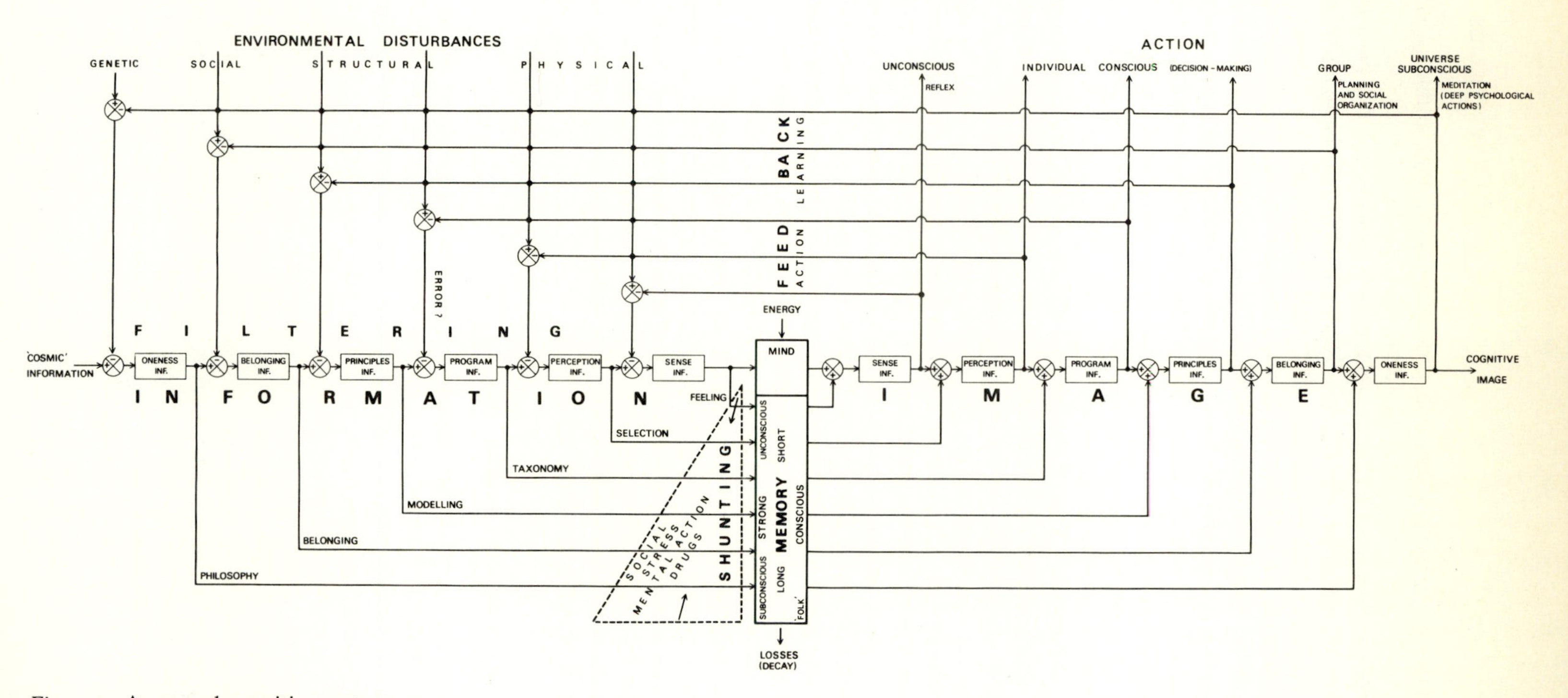

Fig. 5.7 A general cognitive system.

(1970, p. 83) put it, 'what holds together the structure of sensory organization and the structure of action is, in effect, the signalling of intended action, against which the feedback of the senses during action is compared in order to provide discrepance information and the basis of correction.' The negative feedback loops themselves form a spectrum (Annett 1969) between 'action feedback' which is internal to the individual and virtually instantaneous, and the results of deep psychological activity on the part of the individual. Between these lies the area of 'learning feedback', the delayed consequences of individual or group planned decision making which is based on a conscious knowledge of its possible results. It is this type of decision making that is treated in chapter 6.

6 Decision making systems

Men make virtues out of various acts thrown together by mere chance.

(La Rochefoucauld: *Maxims*)

6.1 The decision making process

The increasing interlinkage between the physico-ecological and socio-economic sectors of society has placed a variety of pressures upon decision makers concerned in government, administration and the economy. When man is incorporated into systems as a controller or decision maker (i.e. man-machine control) the need for full mathematical description is complemented by the human capacities of anticipation, learning, innovation, adaptation and thought. Early applications of systems analysis to decision making were extremely simplistic (fig. 6.1) in that they were applied to small units on the scale of the individual firm; goals and objectives were thought to emerge clearly and unambiguously from needs alone; a finite number of separate alternatives appeared to present themselves; and actions were viewed as dominantly deterministic rather than stochastic. However, the possibilities of producing potentially disastrous side effects from any control action, together with the demand upon governments to satisfy an increasing proportion of human needs, desires and aspirations, have stimulated the creation of large-scale decision making systems through which administrators can operate in determining actions which are in some sense optimal and which commonly involve stochastic outcomes. These demands have arisen in environmental control (Meadows *et al.* 1971, Forrester 1971, Clark *et al.* 1975) through the realization that catastrophic, irreversible effects can result from very small and seemingly unimportant decisions, and in socio-economic control through an increasing awareness that the unfettered 'free economy' can generate a number of violently inequable distributions, sectorally, socially and spatially (Harvey 1973, McCloughlin 1973).

Techniques for aiding the decision maker have been developed in management science following the seminal work by Taylor (1911), Fayol (1925) and Urwick (1933); in systems engineering following Jenkins and Youle (1971) and Reira and Jackson (1972); in economics (Tinbergen 1937, Theil 1966), and elsewhere. Eddison (1973) has summarized the role of such management control systems as an attempt to rationalize or optimize the decision makers' actions in a stochastic environment governed by risk and uncertainty. This uncertainty exists since the shifts in societal goals, realizations, aspirations and events cannot be predicted with any great accuracy. Any control policy which is constructed may be rapidly superseded. Hence a continuous planning, 'monitoring', or control approach is required in which the decision maker can compare the rapidly changing socio-economic environment with more slowly changing societal objectives and rationalize

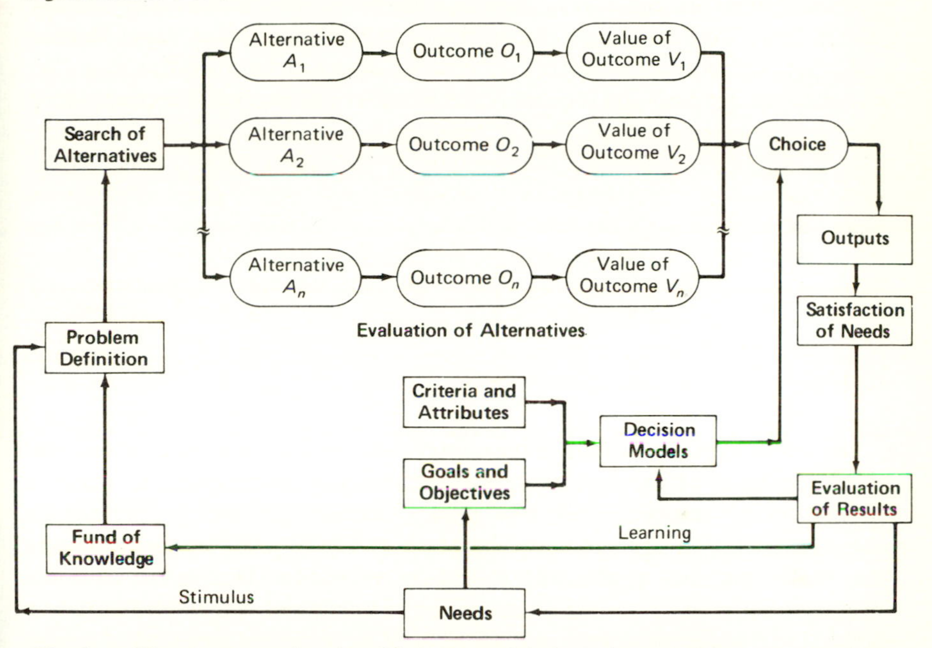

Fig. 6.1 The structure of a simplified deterministic decision making system. (After Cyert and Marsh, from Van Gigch 1974.)

appropriate responses. From these considerations various 'management by objectives' approaches have been proposed (Friend and Jessop 1969, Glendinning and Bullock 1973, Gillis *et al.* 1974, etc.). Marschak (1971), following Carnap (1966, pp. 253–4, 257), states the aim of such control that 'once we see clearly which features of prediction are desirable, then we may say that one given theory is preferable to another or if the predictions yielded by the first theory possess on average more of the desirable features than the predictions based on the other theory. . . . We should choose that action for which the expectation of the utility of the outcome is a maximum'. The aim of the decision maker is to regulate unwanted disturbances in order to achieve a greater degree of personal or group satisfaction.

The role of decision making systems is part of the 'man-machine' interface which can be fundamentally distinguished from that of automated regulatory systems. In the latter decisions are made on a routine basis, often called implementation in planning (McCloughlin 1973). The aim of a regulator is to match the disturbances in the system environment and to operate counteracting strategies: to match the variety of the input disturbances with the variety of the control system (Ashby 1956) (see chapter 3). Decision and control systems govern the choice of implementation or response strategies available to the regulator, as shown in fig. 6.2. But the regulator *R* is fundamentally constrained in that 'the Law of Requisite Variety says that *R*'s capacity as a regulator cannot exceed *R*'s capacity as a channel of communication' (Ashby 1956, p. 211). The constraints of communication channels in socio-economic and other environmental control systems are mainly administrative, but also involve legal, political and societal limits which constrain both the channels of communication and the degrees of freedom for adoption of administrative system responses, and hence the capacity of the controller to control (Beer 1966).

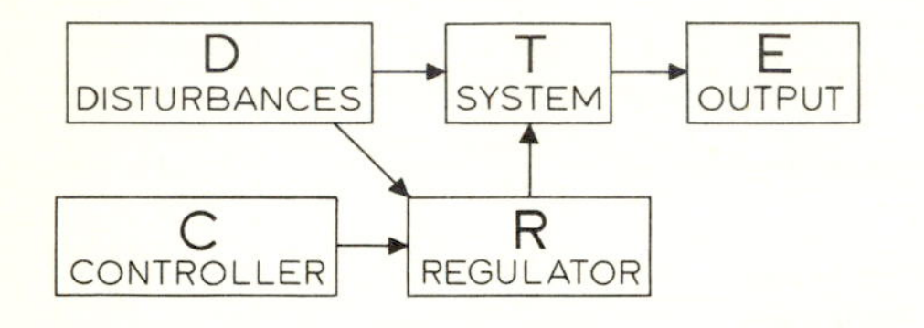

Fig. 6.2
The structure of regulation and control. The controller represents a decision maker who defines rules for the regulation process. (After Ashby 1956.)

The role of decision making in the context of environmental control systems is, on the basis of the best available information and theory, to make a disposition of resources between policy instruments and other system variables such that the control policy objectives are achieved as closely as possible within cost, probability or other constraints. Hence, the decision making system is a resource allocation mechanism which attempts to rationalize and optimize system performance to achieve a given set of goals. As such, the decision maker performs one essential link between the control targets, the control costs and the individuals, groups and components which make up the system being controlled. For this reason the decision maker and the decision making process must usually be viewed as political.

The position of the decision maker within the overall control system is shown in fig. 6.3. The general control issue or problem which is being studied leads to the need for the definition of a number of policy objectives with specific operational goals and targets. These objectives, which give the overall desired system configuration, must be compared with the existing real-world situation. This requires the assembly of a general background information system to determine what is happening, and, especially, the continuous monitoring of variables which are especially important in determining how far the control objectives are being achieved: i.e. 'what is going on, and what is going wrong' (Thorburn 1975). These crucial variables, often termed performance indicators, allow the detection of deviations from targets. On the basis of the nature, magnitude and persistence of these deviations a series of switches are triggered within the decision making system. This switching process, frequently termed 'mixed scanning' after Etzioni (1967), determines what level of action is required from the control system in response to deviations. There are a range of possible policy responses ranging from complete reappraisal of policy strategy and objectives through impact assessment to 'no action' control. Finally, the result of the decision maker's

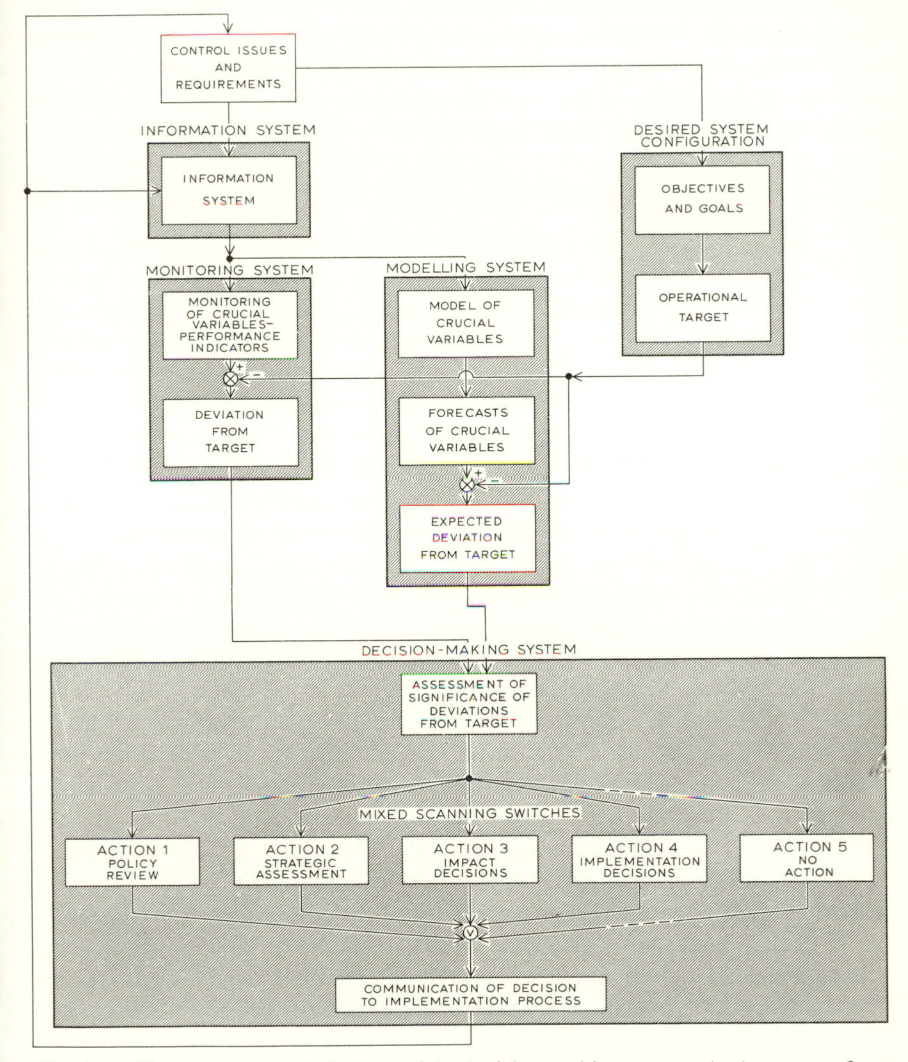

Fig. 6.3 The structure and relations of the decision making system. Actions range from null response (5), through routine regulation (4), decisions on limited impacts (3), reappraisal of decision strategies (2), to reassessment of policy goals and control issues (1) (see pp. 260–5).

response must be communicated to those individuals or groups who are responsible for the actual implementation of the control process. Additionally, in many cases it is essential to construct a model of the real-world system and, on the basis of this model and of the existing information system, to generate forecasts of future system behaviour. From these forecasts it is then possible to anticipate future system deviations before they occur. This anticipation or feedforward control is of greatest utility in systems with long lag, reaction and relaxation times, and where there are significant lags in the provision of information on system behaviour and performance. Without a modelling system to allow forecasts, the decision making process can act only as a feedback control system (chapter 3) which is capable of counteracting deviations from target only *after* they have occurred.

6.1.1 CONTROL ISSUES, GOALS AND OBJECTIVES

The overall decision making process is fundamentally determined and constrained by the components and information upon which the decisions must be based – i.e. for what, for whom, and *by* whom is control intended? The overall goals and objectives of system response govern the general trajectory or time path of system evolution. These goals may be of two forms: exogenously set, or self-regulatory. Where the system goals are determined outside the system of interest the controller attempts to keep the system at the exogenously determined reference input set point by continuous feedback of deviations, or by more complex feedback-feedforward systems, as discussed in chapter 3. In environmental systems this corresponds to subsystem regulation within nested control loops back in chain to socio-economic controls. Within the socio-economic system exogenous goals are set under idealist design only in totalitarian regimes. These contrast with self-regulating systems in which goals are determined by consensus as a sum of component interaction involving political, social, economic and environmental feedbacks. Such are the social goals in democracies and anarchies which maintain an internal *homeostasis* or constancy of operation in the presence of external fluctuations. It will be difficult to forecast the consensus goal derived, since this would depend upon knowing each of the individual democratic forces – i.e. the sociopolitical power of each component. The resulting effect is easily recognized, however, as a dynamic equilibrium maintained, often in the face of many lagged dynamic responses, according to *Le Chatelier's principle*. As shown in chapter 2, such critical damping corresponds strictly only to systems with roots inside the unit circle, but can be extended also to dynamic systems with evolving state transition matrices (Smale 1967).

6.1.2 INFORMATION SYSTEMS

The issues and goals of the overall control system determine the nature of the information system – i.e. we cannot collect information until we know on what information is required. Dunn (1974) has noted that increasing emphasis on the generation and use of information systems arises, firstly, from the increasing complexity and interlinkage of modern society such that there is the increasing possibility of producing catastrophic side effects from any control action. Alternatively, there is the increased possibility of satisfying a greater variety of human aspirations: economic, artistic, social and recreational.

Information collection is based upon the ideal of reducing system uncertainty and complexity. Hermansen (1968) has classified information systems elements into system and operative information, to which we may add a third category of distributive information:

1 *System information*: This determines the nature of the processes being controlled, their dynamic properties, relationships and possibilities for control. It also includes the system goals and methods by which control can be achieved, especially which policy instruments can be adopted.
2 *Operative information*: This determines the present state of the system controlled with respect to objectives, likely trends and deviations. It should also indicate the state of the control system itself, such as present policies, costs, benefits and effectiveness of each instrument, and the effect of other (interrelated) decision making areas. Information on the effect of decisions should be maintained so that the cumulative commitment of many small decisions, and creeping commitment decisions can be monitored (Reira and Jackson 1972, Langefors 1973).
3 *Distributive information*: It is especially important to disaggregate the effect of system responses and policy decisions into categories and sectors affected. There are three major categories. Firstly, the *client category*, composed of socio-economic groups and types – e.g. age, class, sex, race, sociocultural groups. Secondly, the *sector category*, composed of industrial and employment sectors – e.g. basic–non-basic, manufacturing, services, primary industry, white collar, blue collar, managerial and professional. Thirdly, the *area category* involving the spatial zone, country or region affected – e.g. rural-urban, federal states, developed–underdeveloped, eastern-western block, etc. The importance of categorical disaggregation is that the *distributive* effects of control policies can be assessed. This becomes crucial in assuring fair working of administrative, political and legal components of decision making systems (section 6.5).

It will not be possible to maintain information systems in each of these categories. Reira and Jackson (1972) have recognized the importance of maintaining updated information only on crucial *performance indicators* of policy effects and decision inputs, which are temporally based to allow the significance of trends to be assessed, and spatially, socially and sectorally disaggregated to allow the individual, intergroup, intersector and interzone distribution effects to be gauged. This is especially important for social policy (Harvey 1973). The performance indicators suggested by Gillis *et al.* (1974) for subregional planning control fall into four categories: (1) long-term forecasts to determine strategy shifts; (2) short-term targets, e.g. annual required levels of house-building; (3) strategic balances, e.g. sign of changes (net migration positive), or relationships (quality and growth targets); and (4) indices and descriptive indicators such as unemployment rate. To these should be added information on the decision making process itself. The information system is analogous to the nervous system in an animal involving sensory classification and cognition, transmission, storage, retrieval, transformation and display (Lotka 1956, Emery 1969). Hence it is a special form of the cognitive systems discussed in chapter 5, the special features of which should concern information compression and filtering which allow storage only of significant information. In the presence of stochastic uncertainty this will correspond, firstly, to extraction of signal from noise (as a trend plus seasonal plus stochastic persistence as discussed in chapter 2, for example), secondly, to the use of mean and variance information (certainty absorbtion), and, thirdly, to use of the mean only (certainty equivalence) (see chapter 4, and Emery 1969). In each case the cost of the information system is critical. The value of information can be judged by the effect of new information on the controller's view of the world, the effect on his decisions, and the utility increase of any modified decision. This can be measured as the information gain (i.e. entropy loss) at any stage according to Shannon's information formula and laws of redundancy (Brillouin 1956, Shannon 1948). Alternatively, Emery (1969, p. 67) has associated an information matrix of messages for actions A_i and states of nature X_j with a payoff matrix u_{ij} associated with the value of each message. This gives a Bayesian utility structure which values new information relative to the prior information available. The value of the new information is given by the reduction of uncertainty upon receipt, or the difference of the prior and posterior probabilities (see pp. 270–4).

It is also crucial that information systems have a spatial capacity whenever diffusion and social, economic or physical linkage across space are involved. Control actions should be ranked in terms of payoffs of different spatial policy configurations which should involve the existing and desired spatial pattern of contacts between individuals in the production of goods, allocation of investment and disposition of consumption (as discussed in chapter 4) (Hägerstrand 1972). On the basis of these patterns the executive will wish to know how any control decision will modify or constrain the individual and group time-space budget cascades and hence limit or encourage one type of production, consumption or information exchange at the expense of another (Pred 1974). In physico-ecological systems the decision maker requires to quantify the energy and input-output flows between each level in the climatic-ecologic-hydrologic cascade to enable the determination of the effects of any change in the inputs or outputs at any one level on the behaviour of system components at subsequent levels (Chorley and Kennedy 1971). In each case it is necessary to: quantify the responses in subsystem A, usually lagged, to changes in any subsystem variable, or constraint, B; quantify the magnitude and nature of any feedback effects of the induced changes in B upon the original subsystem A; and quantify the distributional effects of these changes upon dependent outcomes in all other subsystems (i.e. across space, between group and sector, and over time). The effects of autonomous system changes and induced changes arising from policy shifts must therefore be analysed into each of the sections shown in fig. 6.4. Attempts to design information systems

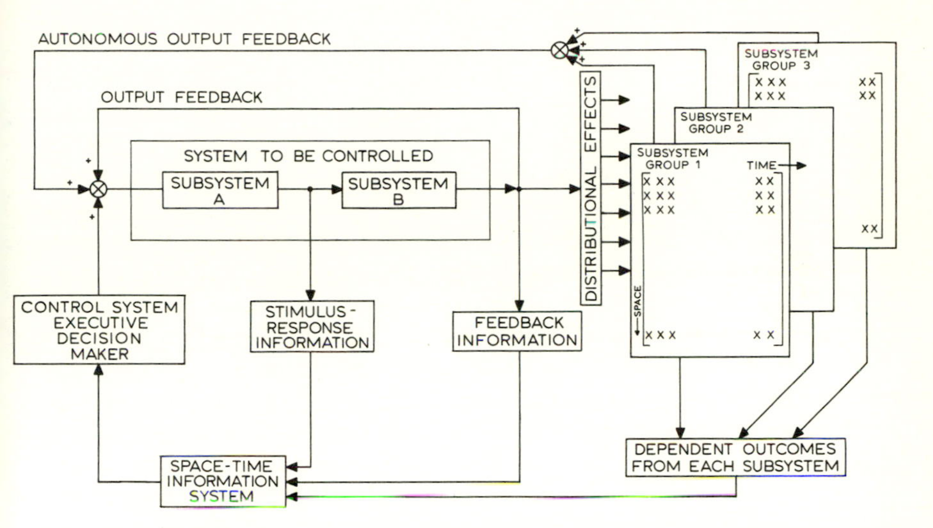

Fig. 6.4 An information system of space-time dependent outcomes disaggregated by the groups and spatial regions involved.

structures are particularly constrained by the necessity of providing information at the appropriate level to allow policy response. Fig. 6.5 shows various mismatches of information messages and action areas (i.e. policy sectors, spatial or other economic-environmental units). To overcome these mismatching problems with spatial data Cripps (1970), DOE (1972), OECD (1973) and Thomas (1971) have suggested systems such as the General Information System for Planners (GISP) based upon a set of elemental spatial units which can be aggregated to any arbitrary spatial policy area. More general coordinate-referenced geographical information systems have been discussed by Thomlinson (1972), DOE (1973), OECD (1973), IGU (1975), Ronnberg *et al.* (1973, 1974) and Aldred (1974). These techniques allow arbitrary aggregation of inner decision subsystems to outer policy boundaries. In hierarchical systems, these boundaries should be chosen 'from decisions made for the outer boundary ... [and] are not to be determined independently by the subsystem management itself' (Langefors 1973, pp. 56–7). This has important consequences for goal setting and *who* controls (chapter 8). In the United Kingdom the reception of the GISP study recommendation of a Basic Spatial Unit linked through a National Gazetteer shows the inherent difficulties of designing, creating and using such systems. Whilst the general aim of 'integrating the extremely wide range of interconnected data needed for planning and public administration' (DOE 1972, p. XV) is generally agreed, the costs are often prohibitive, and there is little immediate return until the changeover to the new system is complete (OECD 1974). There are also objections to socio-economic system information based on

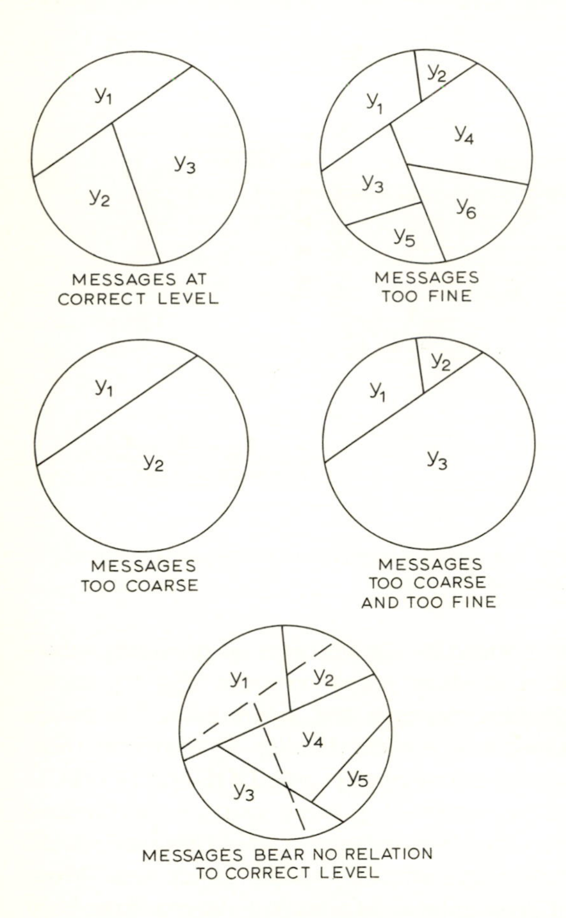

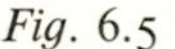
Fig. 6.5
Matches and mismatches of information and action decision areas. Top left: Correct matching. Top right: Aggregation required to achieve correct matching. Mid left: No disaggregation information for areas y_2 and y_3. Mid right and bottom: No ready matching of information and action areas. (Partly after Emery 1969.)

grounds of confidentiality (Niblett 1971) and who has access to files of data on individuals. Niblett notes that the law in most countries gives little protection on grounds of privacy (see also section 6.5.3). Practical difficulties of changes in Basic Spatial Units over time also arise (e.g. the disaggregation of hereditaments). Indeed it would seem that present spatial information systems all assume spatial stationarity and/or Markovity (if they make any pretence of describing relationships at all) and are inadequately based to deal with the control issues involved in the theory of dynamic systems discussed in chapters 2,

3 and 4 and exampled in chapters 7, 8 and 9. Future information system design must be based upon space-time formulations recognizing dynamic adjustment mechanisms and control lags.

6.1.3 MONITORING SYSTEMS

The aim of the monitoring process is to use space-time information systems in order to detect deviations from target values of each performance indicator. On the basis of this monitored deviation, it is then hoped to assess the significance for the achievement or non-achievement of control objectives. The essential roles of monitoring are fourfold: firstly, to determine how far a given policy is succeeding in meeting objectives; secondly, to detect important environmental and economic problems at an early stage and to determine if the best use is being made of available opportunities; thirdly, to determine how far the control implemented is meeting the assumed costs and benefits; and, fourthly, to determine what modifications, if any, are necessary to improve control system performance (Harris and Scott 1974, Cowling and Steeley 1973). The structure of the resulting monitoring process has been characterized by Harris and Scott (1974), as shown in fig. 6.6. This process allows the adoption of plans in the face of dynamic changes and recognizes uncertainty within a *continuous planning* process. The assessment of the significance of deviations of the control system from the target levels of the various indices is one of the more problematic steps in monitoring decision systems. Not only is a subjective judgement required as to the nature of the information system assembled and as to the crucial variables as performance

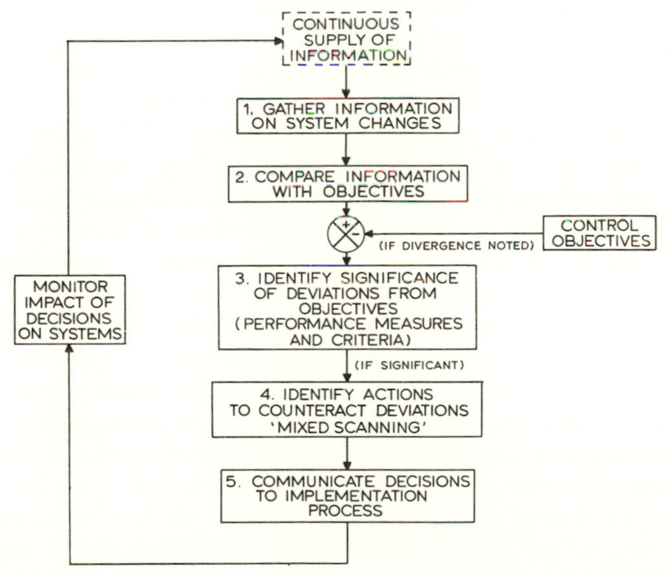

Fig. 6.6 The structure of the monitoring process. (After Harris and Scott 1974.)

indicators monitored, but the weighting placed on the magnitude, nature and direction of deviations cannot, in general, be automated. The degree of impact of any deviation can be assessed on the basis of forecasts, extrapolations and simulations of the future impact of the deviation (Naylor 1971a, Theil 1966, Berry 1971), or by analysis of the details of the underlying causes. The detailed analysis of the generating process is usually a complicated, and hence expensive, procedure which will often require resampling of the real-world system to obtain more specific information (Gillis *et al.* 1974). Hence, in most cases, simulation and forecasting models are of greatest utility in impact assessments – i.e. decisions relating to major once-and-for-all projects (action 3 of fig. 6.3).

In general, deviations will be weighted in significance by a set of tolerance levels associated with each performance indicator. Because of the politicization of these tolerance levels, it is usually thought preferable that they are defined in the original control system (Gillis *et al.* 1974), since attempts at assessing tolerance levels when particular individuals and groups are affected becomes difficult, but this will require agreement on overall social goals. The procedure will be further complicated since each performance indicator will not usually be independent of the other indicators used, and any collinearity will normally affect the estimation properties of any forecasting models adopted (Bennett 1978). Tolerance levels will also be related to the accuracy of the original data base, to criteria of model accuracy and to the sensitivity of the real-world system to changes and policy action (e.g. BOD in rivers (see chapter 9), minimum acceptable streamflow, etc.). These factors will determine if a given deviation from expectation could be the result of purely chance observational or modelling errors, and if the deviation has any significance in terms of repeatability.

Monitoring is central to control in general management and advisory systems. Hence, Jenkins and Youle (1971), Reira and Jackson (1972) and Eddison (1973) have incorporated monitoring systems into management by objectives and systems engineering frameworks. Cowling and Steeley (1973) and Gillis *et al.* (1974) have applied monitoring and advisory systems to United Kingdom subregional and structure planning, whilst Wedgwood-Oppenheim *et al.* (1975) have advocated a Performance Evaluation and Policy Review Unit (PEPR) for monitoring the United Kingdom Strategic Plan for the North West. Hence, monitoring has usually been seen as interlocked with the decision making stage in management information systems (MIS) (McKinsey 1975), although the spatial element has often been suppressed. Some indeterminateness surrounds the exact location of the monitoring role within the MIS. For example, experience with the Nottinghamshire/Derbyshire study (Cowling and Steeley 1973) and the Strategic Plan for the South East in the United Kingdom (DOE (SPSE) 1973, 1975) suggests that a monitoring unit separated from the decision making centre will be ignored if the information and advice delivered becomes politically unacceptable.

Hence, the location of monitoring systems as the essential link between information and decision is crucial in the bureaucratic and political structure of decision making systems discussed later.

The concept of MIS by objectives is also open to criticism. Whilst a set of objectives and goals must underlie any control system approach, in many cases, especially where 'soft' socio-economic systems are involved, a wider framework may be of more utility. The detection of important environmental and socio-economic problems, unconstrained by the performance indicators and information system chosen, and the more widespread use of soft information, which is a necessary part of the decision making process, allow a more flexible approach to monitoring. 'Soft' information, especially that involving social value judgements, is not excluded from the systems engineering approach and its role has been emphasized by Gillis *et al.* (1974) and by Harris and Scott (1974). In practice, however, soft data tend to become ignored in both socio-economic and physico-ecological monitoring systems, and the primary emphasis is often placed upon automation of the decision process as a control system. Burke and Heaney (1975, p. 203) note, for example, in water resources planning that it is often the underlying model of planning behaviour that limits the inclusion of social values, and they attempt to include such soft value information after other decisions. It is preferable, however, to enmesh the planning and political decision processes from the start, but the realization of such 'total planning' leads to the conflict between individual freedoms and administrative discretionary powers in socio-economic systems (section 6.5.3). The United Nations Environment Programme, Scientific Committee on Problems of the Environment (SCOPE), offers an example of this loose monitoring role in which the effect of important environmental and economic impacts cannot usually be predicted and may be counter-intuitive (see chapter 9) (Meadows *et al.* 1971). SCOPE is based on:

1. Assessment of the extent to which existing knowledge gives a sound basis for prediction and decisions on environmental policy and management.
2. Identification of deficiencies in existing knowledge.
3. Definition of programmes to remedy information deficiencies.
4. Providing a source of objective advice upon which government and intergovernment agencies can base environmental decisions.

Three major environmental monitoring issues can be identified. Firstly, for biogeochemical cycles it is necessary to know the size of environmental reservoirs and rates of transfer within each cycle – such as carbon, oxygen, nitrogen, phosphorous, sulphur and calcium, and those of toxic element cycles especially lead, mercury, cadmium, plutonium, DDT and PCB. This will allow for environmental control and regulation through steady state and transient models. Secondly, contingent upon, but somewhat independent of, the biogeochemical cycles are the human impacts on renewable natural resources which often lead to deterioration in the present ecosystem. These require monitoring of desertification, salinization and waterlogging, and of other human uses in excess of replacement. Thirdly, it is necessary to monitor the man-environmental interface, especially the impact of human settlements – for example, spatial redistribution giving rise to areas of rapid growth or decline. The environmental impact of growth areas puts pressure upon building materials and aggregates, water supply, air quality, noise levels and sanitation. Declining areas have the problems of derelict land, abandoned industrial and residential property, and of a generally degraded environment (see DOE (SPNW) 1973).

In practice, world-based monitoring systems concentrate on five different types of variables:

1 Levels of important chemicals in air, water, soil and biotic media.
2 Physical attributes of each medium (e.g. solar radiation, soil properties, etc.).
3 Magnitude and frequency analysis of effects on the ecotome and resources of changes in 1 and 2.
4 Inventories of climatic change and human impact.
5 Source strength of trace substances.

To be of utility in control, the monitoring system evolved to assess changes in each of these variables should exert pressure at three points in the environmental-economic system. As shown in fig. 6.7, these should be within

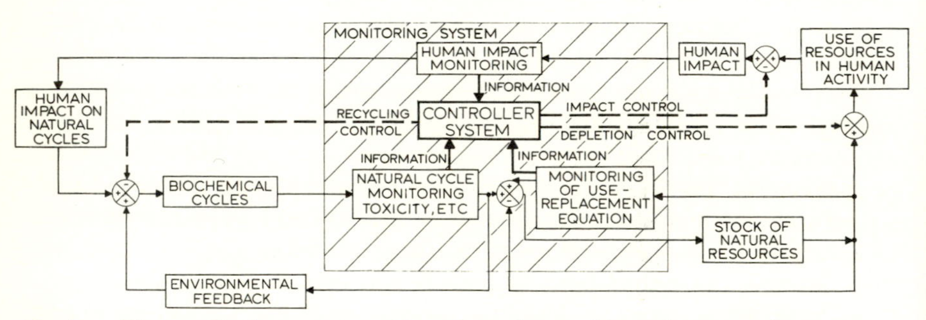

Fig. 6.7 The monitoring system and its points of control in environmental systems.

the biogeochemical cycles to regulate the recycling and accumulation of toxic elements, within the resource-regeneration-use equation to maintain a balance of the two factors, and within the human impact system to minimize adverse environmental consequences.

6.1.4 MODELLING/FORECASTING SYSTEMS

The importance of modelling and forecasting systems is both to simplify the real-world system in order to aid decision making, and to allow the anticipation of changes and deviations within system behaviour which will require counteracting control to be implemented. The modelling of the spatiotemporal structure of environmental and socio-economic systems has been discussed extensively in previous chapters. The most important result, however, is the idea deriving from Kalman (1960) and Kalman and Bucy (1961) that the model estimation and control problems are duals of one another. Hence for optimal control an optimal model forecast is required. This approach has been developed by Theil (1966), in space-time systems by Berry (1971), and in many urban and regional planning studies. Monitoring and decision systems which exclude the modelling and forecasting roles act as pure feedback control, as discussed in chapter 3. As such they are unable to respond to deviations until *after* they have occurred, they tend to over-react and amplify disturbances, and they are subject to retardation by system lag and delay dynamics. Recent criticism of the use of models in urban planning (e.g. Broadbent 1975) seems to ignore this important point. Whilst present forecasting models in almost all sectors are not fully adequate, they at least expose the assumptions on which extrapolations are based. When combined with Bayes forecasting, models allow a completely flexible approach to whatever information the decision maker has available prior to a decision, which can be combined with the information received posterior to the outcome. Moreover, modelling is not limited, as some claim (Harvey 1973, p. 129), to positivistic modes of development and functionalist maintenance of societal *status quo*, since any political, scientific or ideological assumption can be built into priors. Forecasts can also be used to determine the degree of accordance with historicist or even ontological trajectories (e.g. the Marxist development path to an ideal communist state).

It is often objected that forecasts made upon the basis of models are highly constrained by those models, or have not wide enough applicability. However, all forecasts imply some form of model, however vague. As Theil (1966, pp. 2–3) states, 'There is always some *theory*, however naive it may be . . . such a theory always amounts to the assumption that something remains *constant*. [Also] there are always some *observations*, however few and untrustworthy these may be . . . each forecast has two inputs (theory and observations).' Thus even a 'no change' forecast implies a model in which present relationships will remain constant; a linear extrapolation implies a model in which the present rate of increase will remain constant in the future, and so on. In the use of distributed lag space-time models for forecasting, as discussed in chapter 4, it is assumed that the dynamics of memory, lag, delays, leads and gains across space will all remain constant over the forecasting period. But nonlinear and nonstationary models deriving from the Kalman filter and Bayes formulations remove most of these limitations (Bennett 1978, ch. 5).

> If it is the function of a forecast to reduce (the decision maker's) uncertainty, and if each forecast has two inputs (theory and observations),

> the marginal revenues of these inputs with respect to uncertainty reduction should be equal. Assuming that neither input is 'inferior', we may conclude that the use of *both* should increase when the total resources available for the uncertainty reduction increase (Theil 1966, p. 3).

Hence models deriving from theory, from observation of the present system based upon information sources, and from ontological specifications, should all be used to the fullest extent in order to reduce the uncertainty of the future. Three basic classes of control uncertainty have been recognized by Friend and Jessop (1969, ch. 5):

1 *Class UE*: Uncertainties in knowledge of the external planning *environment*, including all uncertainties relating to the structure of the world external to the decision making system, and also all uncertainties relating to expected patterns of future change in this environment, and to its expected responses to any possible future intervention by the decision making system.
2 *Class UR*: Uncertainties as to the future intentions in *related fields* of choice including all uncertainties . . . within the decision making system itself, in respect of other fields of discretion beyond the limited problem which is currently under consideration. [This will include ontological trajectories.]
3 *Class UV*: Uncertainties as to appropriate value *judgements* including all uncertainties [of] . . . consequences . . . which cannot be related to each other through an unambiguous common scale.

Hence, forecasting systems should anticipate not only 'natural' system behaviour (i.e. behaviour with no control imposed), but also the responses of perceptual subsystems which anticipate the control action and the development trajectory to ontological goals. However, since the controller is attempting to anticipate the system reactions to control, the perceptual subsystem will attempt to anticipate the anticipatory control and take action away from trajectories with which there is no agreement. This process proceeds in the iterative fashion, shown in fig. 6.8, until an equilibrium control and system behaviour results. This equilibrium represents the greatest measure of control that can be imposed upon the anticipatory system, given disagreement with some goals and sufficient freedom to operate away from the desired trajectory. Greater control can be achieved only by the reduction of freedom, for example in socio-economic control in totalitarian states (chapter 8).

On the basis of anticipated economic and environmental impacts, or upon the registering by the information and monitoring systems of deviations of the control system from desired trajectories or set points, a control policy decision must be resolved and communicated to those parts of the system where some counteracting process can be activated. Alternatively, the system is redesigned to ontological ends. Depending upon the importance of the control deviation observed, or the future problem anticipated, it is normal for the control response to be geared in level to the significance of the undesired element. This mixed scanning process (Etzioni 1967) leads to a range of possible control responses ranging from 'no action' for control issues of little significance that are unlikely to be repeated, through a range of routine implementation

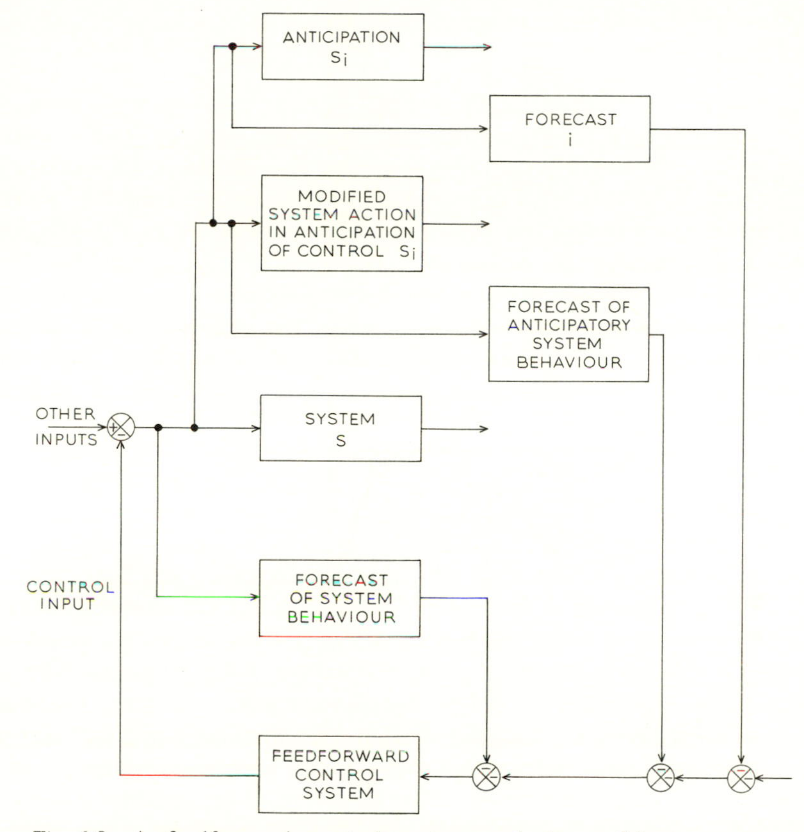

Fig. 6.8 A feedforward control system and the anticipatory system iteration.

decisions, to major decisions. The latter may involve impact control decisions to counteract important changes or strategic decisions (Anthony 1965) in which the decision maker has to cope with decisions in 'a dynamic, imperfectly understood, and imperfectly controlled environment, encompassing an unlimited field of interest' (Wedgwood-Oppenheim 1975). At this level, policy review may need to be undertaken with 'continuous planning' or 'continuous control'. In the presence of very major impacts which cannot be regulated from the system, the control system must undertake a complete reappraisal or action accommodating to the environmental shift. This may lead to an adjustment of objectives, but will certainly lead to an adjustment of policy targets. The effect of each level in the mixed scanning process in this continuous planning approach is shown in fig. 6.9, where a major deviation leads to the definition of a new policy target at later stages. The range of decision making responses in mixed scanning is largely one in which

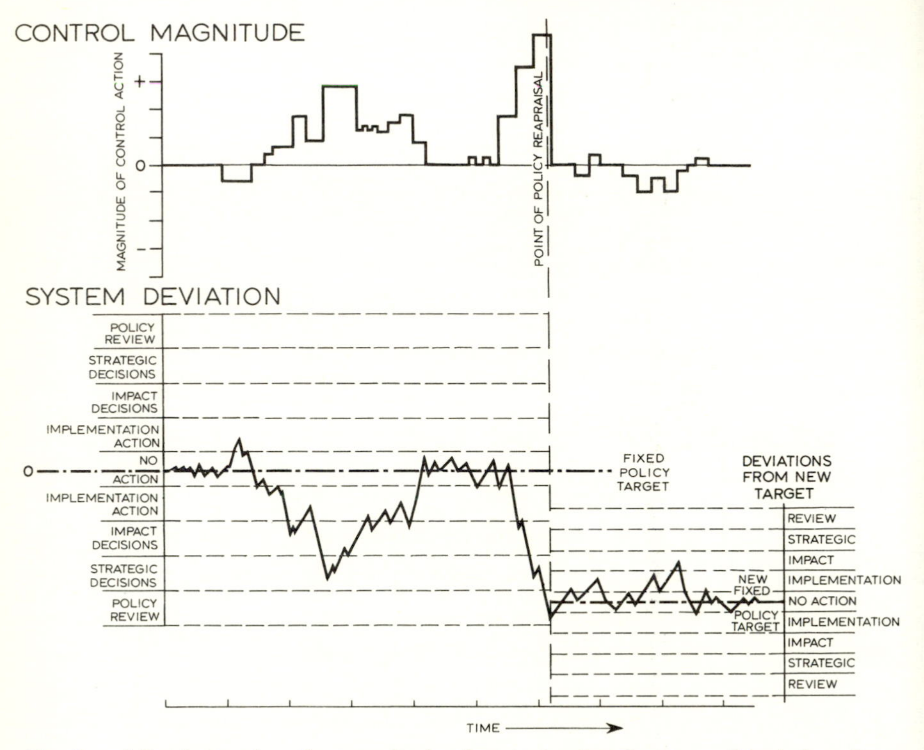

Fig. 6.9 Mixed scanning, the magnitude of control action (in terms of deviations from target) and the adjustment of mixed scanning after policy review.

administrative triggers are switched in order to achieve control. This, in turn, usually determines who implements the control action. The decision making process then follows a number of similar stages:

1 Definition of the decision making *environment* (See section 6.2).
2 Definition of the *economics* of the control decision composed of the costs of control, the marginal rate of return on control, and the marginal rate of substitution between control strategies (See section 6.3).
3 The evaluation of the 'cost' and 'benefits' of each possible control strategy (See section 6.4).

The overall decision process is what Mannheim (1966) has termed a search and selection procedure. Search generates the population or set of possible actions; whereas selection gives the range of choices from this population, a prediction of the effects of a given action, an evaluation of costs attached to each, and a decision on final desirability relative to objective and costs. Each decision is, in effect, experimental and multilevel. 'An experiment e_{ij} consists of the application of a single-level operator i to the non-elemental action j. The results of an experiment are a new action and a cost associated with that action. If the action is non-elemental, then the "cost" is some parameter of the distribution costs of those elemental actions included within the non-elemental action' (Mannheim 1966, p. 43). If p_i is the prior probability of state X_i occurring over the decision time t (representing information on X_i given by messages Y_j), then

$$\text{prob}\,(Y_j) = \sum_i \text{prob}\,(Y_j \mid X_i)\,\text{prob}\,(X_i) = \sum_i q_{ij}p_i \qquad (6.1.1)$$

where q_{ij} is the information structure of the conditional probabilities (i.e. that message Y_j will be received if state X_i occurs). The receipt of this message causes changes in the estimate of the probability of X_i occurring (posterior probabilities). Following Bayes theorem the resulting posterior probabilities are given by

$$\text{prob}\,(X_i \mid Y_j)\,\text{prob}\,(Y_j) = \text{prob}\,(X_i, Y_j) = \text{prob}\,(Y_j \mid X_i)\,\text{prob}\,(X_i).$$

Therefore,

$$\text{prob}\,(X_i \mid Y_j) = \frac{\text{prob}\,(Y_j \mid X_i)\,\text{prob}\,(X_i)}{\text{prob}\,(Y_j)} = \frac{q_{ij}p_i}{p(Y_j)}. \qquad (6.1.2)$$

The choice of control action will be based on the utility or payoff of each action on each state u_{ij}. The maximum utility is given by

$$\max_j U = \sum_i p(X_i \mid Y_j)u_{ij}. \qquad (6.1.3)$$

The value of the information, given by the reduction of uncertainty upon receipt, is expressed by the difference of prior to posterior probabilities,

i.e.

$$\max_j \sum_i q_{ij}p_i u_{ij} - \max_j \sum_i p_i u_{ij}. \qquad (6.1.4)$$

Hence, each decision is an experiment which depends upon the outcome of the last experiment. The decision making system is a 'learning' system which adapts and modifies its response dependent upon the nature and utility of the outcomes of previous decisions. Boulding (1966) extends this learning concept to include the way in which changes in the 'soft' information value systems modify the decision process. This especially affects ideological decisions towards ontological goals. Decisions are value-laden. The value elements enter in a number of ways which Burke and Heaney (1975) classify as a nine-dimensional typology shown in table 6.1. The intersection of these nine possibilities allow decision models to be characterized and different value environments to be assessed.

Table 6.1 Typology of decision situations (after Burke and Heaney 1975, p. 64)

Decision component	Bipolar distinction
1 Nature of problem	Structured vs unstructured
2 Character of analysis	Normative vs behavioural Prescriptive vs descriptive
3 Perspective	Individual vs aggregate Atomistic vs organismic
4 Decision maker(s)	Unitary vs collective
5 Nature of Resources	Private vs public
6 Problem attributes	Single vs multiple
7 Model-of-Man	Rational vs nonrational
8 Goals	Given vs fluid
9 Influence of time	Static vs sequential

Overlying the value-laden element of decision is the problem of how to achieve decision by collective action. Collective action should permit the 'resolution of diverse value systems or the aggregation of individual preferences into a single expression of collective choice' (Burke and Heaney 1975, p. 74). For example, Talcott Parsons' (1966) integrative subsystem permits questions of equity, fairness and distribution (between groups and over space) to be resolved. Various mechanisms of collective action and conflict resolution are possible and are summarized in table 6.2 for both normative and descriptive approaches to decision making. In the following sections we discuss the application of these concepts to the decision making environment. In section 6.5 the resolution of collective action alternatives is examined in the context of administrative, political and legal systems.

6.2 The decision making environment

The decision making environment is governed largely by the extent of knowledge about the effects of the decision, and the *degree of interaction* with other decision makers. If the decision is independent then the effect of other decision makers is absorbed into the environment or state of nature. If other decision makers are taken into account, then the resulting interdependent decision process is more complex and often involves sociopolitical bargaining. Depending upon the nature of the information surrounding a decision (i.e.

Table 6.2 Typology of collective action mechanisms in normative and descriptive situations (After Burke and Heaney 1975, pp. 76–7)

1 NORMATIVE MODELS (how we *should* decide)
  (*a*) Individually owned resources
    1 Welfare economics
    2 *N*-person game theory
    3 Normative legal structure
  (*b*) Collectively owned resources
    1 Normative political and policy theories
    2 Voting rules
    3 Bargaining theories
    4 Normative organization theories
    5 Normative legal structure
    6 *N*-person game theory

2 DESCRIPTIVE MODELS (how we *do* decide)
  (*a*) Individually owned resources
    1 Theories of social cohesion
      (*a*) Exchange behaviour
      (*b*) Shared norms and values
      (*c*) Conflict and conflict resolution
    2 Theories of group behaviour
    3 Operational games
  (*b*) Collectively owned resources
    1 Descriptive political behaviour
      (*a*) Voting, elections
      (*b*) Models of policy making process
      (*c*) Power structure and decision making structure
    2 Models of organization and administration
    3 Studies of bargaining and coalition behaviour
    4 Operational games

whether it is independent or interdependent), the decision environment may be governed by certainty, risk or uncertainty. A certainty environment allows a choice between policies, the outcomes of which are all completely *known* (fig. 6.10). It is then possible to make a choice based on the perceived utility u_{ij} of each outcome for each possible state of nature under any policy action. A risk environment allows a choice between actions for which the state of nature and their outcomes will be known only to a level of probability p_i of occurrence. Decisions are based upon considering the complete range and probability of outcomes for each alternative. When uncertainty conditions govern decision making the range of *probabilities* of outcomes is completely unknown and decisions are then based upon choosing between optimization resulting from given alternative criteria; such as maximin, regret, Laplace, maximum entropy and Bayes. The differences between each of these decision making environments and between the various choice criteria under uncertainty can be best considered through an example.

Assume that a policy maker can assign four policy instruments U_i, each of which can be effective (E), or ineffective (I) to achieve a given policy. Assuming that it is possible for each policy instrument to be either E or I, there are 2^4 (i.e. 16) different possibilities of combination of effective and ineffective

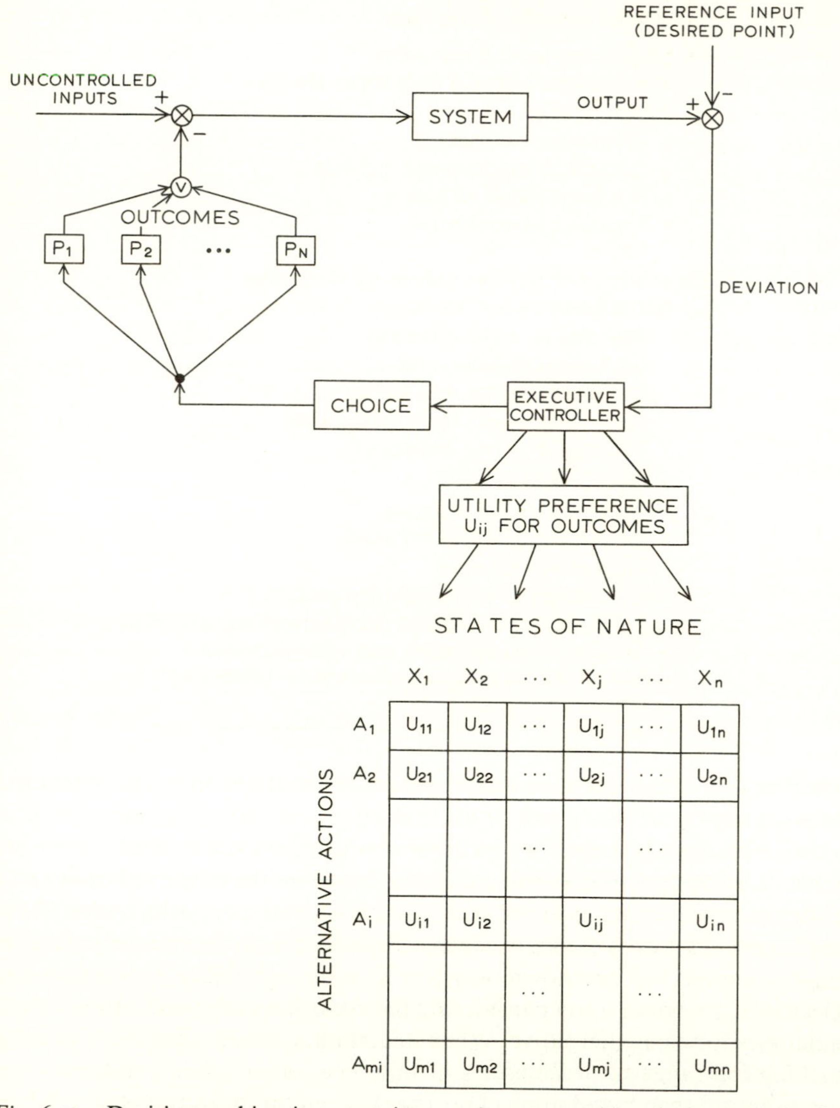

Fig. 6.10 Decision making in a certainty environment with a known range of outcomes. A decision is made on the basis of the highest utility u_{ij} in utility/preference table.

policies (fig. 6.11). Each of these possibilities is termed a *state of nature* X_i, and each state is one configuration of effective and ineffective instruments. If we further assume that the four instruments can be assigned to two policies, there will be 2^4 ways of assigning instruments to policies giving an action matrix A_i, as in fig. 6.11. Each such assignment is termed an action. For example, we may consider the two alternative policies of investing in pollution removal devices, or changes in taxation on emitters of pollutants (which is a frequent policy choice in respect of river and lake pollution). If we associate a utility u_i to the policy options of $u_1 = 40$, $u_2 = 10$, then a 16×16 payoff matrix can be constructed for the interaction of the state of nature (ineffective or effective instruments) with the policy action with these instruments (remove pollution, or tax), as shown in fig. 6.11. In response to this payoff matrix the decision maker will make his choice of policy between removal of pollution or taxation, and his choice of instrument to achieve that policy. The choice of instrument and the choice of policy are interlocked through the choices of being effective and the utility of the outcome. From the result of constructing the payoff matrix in fig. 6.11, and certain knowledge of the state of the matrix X_i, the decision maker would choose policies in the order of their utilities, A_1 then A_2. However, in most situations the decision maker has imperfect knowledge as to the probability of particular instruments being effective, and in this case there are a range of strategies open to him (De Neufville and Stafford 1971, Tribus 1969):

(1) *Certainty: strategy with complete information on outcome.* When the decision maker has complete knowledge of all outcomes ($P_i = 1$) of effectiveness of instruments, the only problem is one of choice which will be based on the maximum utility in terms of costs, preferences, etc., as discussed above. The choice may still be difficult, however, since generally each alternative strategy is characterized by several attributes which are not entirely comparable. The problem is to find weights (e.g. prices) which can be associated with each attribute as a measure of effectiveness. The product of the weights and effectiveness then gives a measure of utility and ranking to each strategy.

(2) *Risk: strategy with complete information on probabilities.* If the decision maker knows the q_{ij} probabilities of ineffectiveness then the best policy is to assign policies in the order A_1 and A_2. The probability of any policy being ineffective determines the value of the perfect information (Emery 1969, pp. 79–80) and is binomial. In general, preferences are often ordered so as to minimize loss rather than to maximize returns, which yields game theory solutions and parallels the Bayesian solution under uncertainty for which probabilities are only estimated.

(3) *Uncertainty: strategy in the absence of prior probabilities.* Under uncertainty nothing is known as to the likelihood of particular outcomes. In response to this problem five forms of inference have been proposed:

(A) SYSTEM STATUS (q_{ij})

	X_1	X_2	X_3	X_4	X_5	X_6	X_7	X_8	X_9	X_{10}	X_{11}	X_{12}	X_{13}	X_{14}	X_{15}	X_{16}
u_1	E	E	E	E	I	E	E	I	I	I	E	E	I	I	I	I
u_2	E	E	E	I	E	I	I	I	E	E	E	I	I	I	E	I
u_3	E	E	I	E	E	E	I	E	E	I	I	I	I	E	I	I
u_4	E	I	E	E	E	I	E	E	I	E	I	I	E	I	I	I

(B) POLICY ACTIONS WITH INSTRUMENTS

	a_1	a_2	a_3	a_4	a_5	a_6	a_7	a_8	a_9	a_{10}	a_{11}	a_{12}	a_{13}	a_{14}	a_{15}	a_{16}
u_1	A_1	A_1	A_1	A_1	A_2	A_2	A_2	A_2	A_1	A_1	A_1	A_2	A_2	A_2	A_1	A_2
u_2	A_1	A_1	A_1	A_2	A_1	A_2	A_1	A_1	A_2	A_2	A_1	A_2	A_2	A_1	A_2	A_2
u_3	A_1	A_1	A_2	A_1	A_1	A_1	A_2	A_1	A_2	A_1	A_2	A_2	A_1	A_2	A_2	A_2
u_4	A_1	A_2	A_1	A_1	A_1	A_1	A_1	A_2	A_1	A_2	A_2	A_1	A_2	A_2	A_2	A_2

(C) PAYOFF MATRIX FOR EACH ACTION

Actions (policy instrument assignments)	States of nature (effectiveness of instruments)											
	X_1	X_2	X_3	X_4	X_5	X_6		X_{12}	X_{13}	X_{14}	X_{15}	X_{16}
a_1	40	40	40	40	40		...	40	40	40	40	0
a_2	50	40	50	50	50		...	40	10	40	40	0
a_3	50	50	40	50	50		...	40	40	10	40	0
a_4	50	50	50	40	50		...	40	40	40	10	0
a_5	50	50	50	50	40		...	10	40	40	40	0
a_6	50	50	50	50	50		...	10	40	40	10	0
a_7	50	50	50	50	50		...	10	40	10	40	0
a_8												
a_9												
a_{10}												
a_{11}												
a_{12}												
a_{13}												
a_{14}												
a_{15}												
a_{16}												

Fig. 6.11 Choice criteria under uncertainty.

A The application of policy instruments (U_i) to various system states (X_i) may effective (E) or ineffective (I). Each of the sixteen columns represents a state of natu (X_i).

B The application of the policy instruments (U_i) to sixteen policy actions (a_i) in terms two alternative policies (A_1, A_2).

C The payoff matrix of policy actions (a_i) versus states of nature (X_i) where the utiliti for the two policy options are 40 and 10 respectively. (Partly after Emery 1969.)

(*i*) *Maximum criterion*: This gives the maximum value of the minimum utility, regardless of the state of nature X_i. The decision maker expects the worst to happen and is very risk averse (i.e. he has an extreme utility function), which gives a reasonable payoff for action in most cases but only rarely a high payoff. In fig. 6.11 the maximum policy yields a choice of policy A_1 (e.g. games against nature; Gould 1963). This gives a value of 40 for all states except for X_{16}, *which has four ineffectives and which will be zero for all strategies. Hence a payoff is nearly always achieved.*
(*ii*) *Regret criterion*: This is defined as equal to the difference between the value of an outcome under the policy instruments employed and the maximum value of that outcome possible for any combination of events and instruments (Luce and Raiffa 1957), and thus additionally weights the best outcome. From the payoff matrix (C) in fig. 6.11 it is possible to derive a regret matrix. The optimal policy decision will be that action which minimizes the maximum regret (minimax). Again the instrument A_1 is optimal, which yields the same result as in the direct minimax case. The problem with this procedure is that the regret values are weightings of the payoff values in the matrix (C) in fig. 6.11 and are thus relative to the alternatives considered and will be seriously affected by the introduction of irrelevant alternatives.
(*iii*) *Laplace criterion* (principle of insufficient reason): If the probabilities of the events are not known then they should be assumed equal. Although this has intuitive appeal it is highly affected by low probability events, and by the range of alternatives considered. This criterion is, however, correct for the example given in fig. 6.11.
(*iv*) *Maximum entropy criterion*: The least biased assumption with unknown prior probabilities on the basis of past frequencies. The definition of all possible system states gives an entropy function – e.g. by a multinomial probability function (Tribus 1969). This function is then maximized subject to whatever information is available incorporated as constraints. This generates an expectation for the most probable (i.e. maximum entropy) distribution of future states. The application of this criterion to decision making in urban and regional planning is discussed by Wilson (1970, 1974) and to more general decisions under uncertainty by Tribus (1969). In Wilson's application to planning, the decision maker asks, 'What is the least biased conclusion that I can draw from a given data set?' The least biased assumptions define a prior probability distribution which is maximized, in the trip distribution case (e.g. journey to work, journey to shop) subject to knowledge of 'trip ends' (origin and destination sums of trips) and costs of travel.
(*v*) *Bayesian criterion*: Bayesian analysis gives a more general framework than entropy maximizing by which to combine information that is known about the likelihood of outcomes with new information, as this becomes available. It is thus ideally suited to decision making in dynamic systems

(Mannheim 1966). For example, consider the case in fig. 6.11 as defining a set of prior probabilities for the effectiveness of instrument A_1:

$$P_{A_1}(\text{effective, ineffective}) = P_{A_1}(0{\cdot}5, 0{\cdot}5).$$

This can then be applied to another decision making situation in which only policy A_2 has been applied. The percentage of times that A_2 was effective in this other situation when it was also effective in the old situation is,

$$P_{A_2}(\text{effective in new/effective in old}) = 0{\cdot}3.$$

This can be used to update our expectation of effectiveness of A_1 in the new situation. Using Bayes's theorem (equation 6.1.2):

$$(P_{A_1/A_2} \text{ effective in new})$$
$$= \frac{P_{A_1}(\text{effective in old})P_{A_2}(\text{effective in new/effective in old})}{P_{A_1}(\text{effective in old})P_{A_2}(\text{effective in new/effective in old}) + P_{A_1}(\text{ineffective in old})P_{A_2}(\text{effective in old})}$$
$$= \frac{(0{\cdot}5)(0{\cdot}3)}{(0{\cdot}5)(0{\cdot}3) + (0{\cdot}5)(0{\cdot}7)} = 0{\cdot}3.$$

Since A_1 and A_2 perform similarly in the old situation the best prior judgement is that they will behave similarly in the new situation with a probability of effectiveness of 0·3.

6.3 The economics of decision making for control

The fundamental problem of decision making for control is one of allocation of scarce resources. Resources that are not scarce can be ignored from most decision making processes (Lerner 1956). Scarcity arises when the supply of resources is insufficient to meet the derived demand. When the supply is sufficient to meet the derived demand there is no allocation problem between users unless the resource is non-renewable (see Connelly and Perlman 1975, and chapter 9 of this volume). The fundamental decision making problem is one of choice and the optimal division of a resource or a factor among different demands is based upon Lerner's Rule (1946): *The value of the marginal product derived by employing a given factor should not be less than the value of any alternative marginal product derived by employing any other factor.* Applied to the control situation this rule requires that the instrument employed to achieve a desired target or tracking value should yield the optimum return to cost of control.

The influence of a particular control variable as an instrument upon a system is akin to the production function in economics. A typical form is shown in fig. 6.12 where continuous increases of the value of the control instrument eventually yield a negative return, which corresponds to a negative marginal return for each increment in applied control. With respect to this figure, the same value of the controlled variable can be achieved by an instrument setting at either A or B, but A will use less resources. The character of the rate of return on control for various instrument settings is described by *marginal rate of return*:

$$MR_i = \frac{\partial Y}{\partial U_i}. \tag{6.3.1}$$

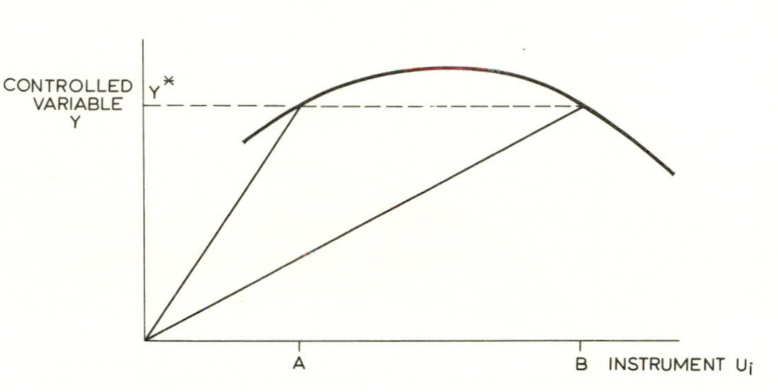

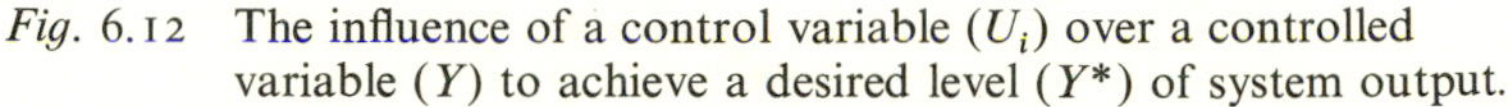

Fig. 6.12 The influence of a control variable (U_i) over a controlled variable (Y) to achieve a desired level (Y^*) of system output.

If two or more instruments can be employed each will have its own marginal rate of return given by an equation such as (6.3.1). Fig. 6.13 shows two return functions for the control of a system output Y by instruments U_1 and U_2. All possible combinations of instruments which allow the achievement of a desired level Y^* of the controlled variable are given for any level of Y by an *isoquant* (indifference curve): a locus joining all possible control configurations which will yield a desired value of Y^* for the controlled variable. It is defined by a plane at a height of Y^* over the U_1, U_2 plane of instruments. Any combination of U_1 and U_2 along this line will achieve the same level of Y^*, although each point on the isoquant will be associated with a different combination and hence different costs of the U_1 and U_2 controls.

In order to determine the best or 'cheapest' combination of instruments it is usual to consider the rate of change of the isoquant at any point with changes in the combination of any of the instruments each taken two at a time. This gives the *marginal rate of substitution* between instruments and defines the relative usefulness for control of one instrument in comparison to another independent of their magnitude. This is given mathematically by Euler's theorem as:

$$\partial Y = \sum \frac{\partial Y}{\partial U_i} \partial U_i = \sum MR_i \, \partial U_i \tag{6.3.2}$$

with MR_i being the marginal return for the ith instrument. The magnitude of each instrument setting multiplied by the marginal products summed for all instruments gives the total product or control setting. Along the isoquant $\partial Y = 0$, and all $\partial U_i = 0$. Hence,

$$\partial Y = MR_1 \, \partial U_1 + MR_2 \, \partial U_2 = 0 \tag{6.3.3}$$

and marginal rate of substitution

$$(U_1, U_2) = \frac{\partial U_1}{\partial U_2} = -\frac{MR_2}{MR_1} = -\frac{c_2}{c_1}. \tag{6.3.4}$$

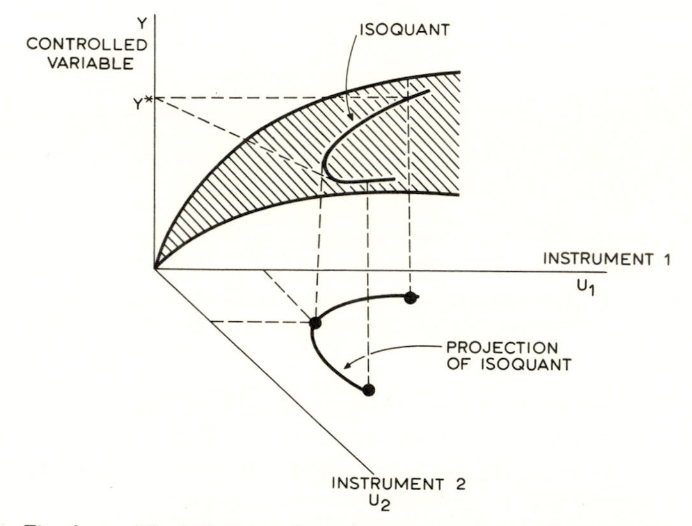

Fig. 6.13 The influence of two instruments (U_1 and U_2) over a controlled variable (Y), where the desired level of Y is represented by an isoquant (indifference curve).

The *marginal rate of substitution* of one instrument for another is equal to the negative reciprocal of their marginal returns for each increment in control magnitude or cost (c_i). Equation (6.3.4) is important in two ways. Firstly, it can be used to define a criterion for selecting the optimum combination of instruments which achieves a given target setting. The optimality criterion is usually stated as:

$$\frac{MR_i}{c_i} = \frac{MR_j}{c_j}. \tag{6.3.5}$$

On an isoquant, for a given target value, the ratios of the marginal costs of control and the marginal returns will be equal for all pairwise comparisons between instruments. Secondly, equation (6.3.4) can be used in the optimization decision process of determining an optimum under cost minimization criteria. The optimum configuration of controls can be defined by the intersection of the plane formed by the technically efficient solutions, and the plane formed by the relative values or costs of the instruments employed (the mechanical and economic factors). The relative costs of each instrument define a line of constant cost or budget curve in the instrument plane. The 'cost' associated with each instrument can be equal and constant for all magnitudes of input. In this case the optimum control configuration is given at the point of tangency of either of the cost lines to the isoquant defining the technical combinations of instruments (fig. 6.14A) which will yield the desired value of the controlled variable Y^*. It is more usual for the unit cost of the instruments to change at varying magnitudes. For example, the frequently observed convex cost curve produced by cost increases with higher levels of instrument settings, shown in fig. 6.14B. In both cases the point of tangency A is the point at which the desired reference value Y^* (the controlled variable) is achieved with minimization of costs (i.e. the minimax solution).

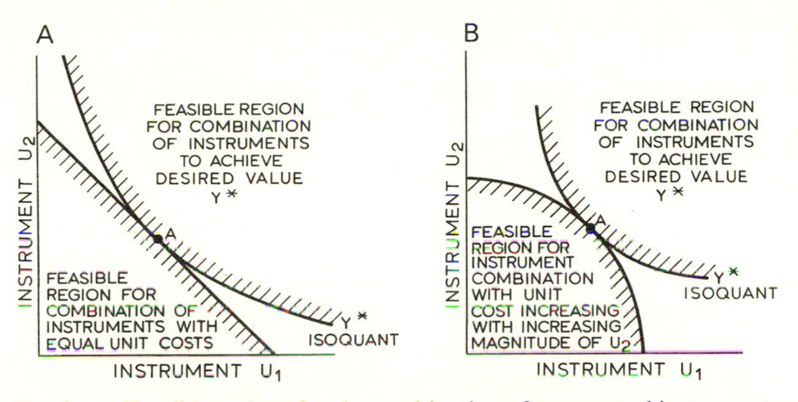

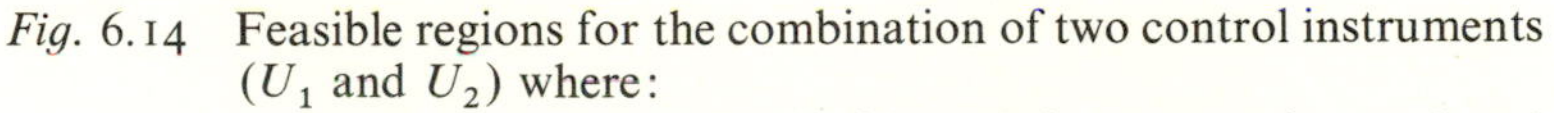

Fig. 6.14 Feasible regions for the combination of two control instruments (U_1 and U_2) where:

A The unit cost associated with each instrument is equal and constant for all magnitudes of input (straight budget line).

B The unit cost associated with each instrument changes at varying magnitudes as a function of magnitude (curved budget line).

The optimum combination of instruments is achieved at that point (A) at which the budget lines cut the highest indifference curve (isoquant). (After de Neufville and Stafford 1971.)

If a whole range of possible desired values of the controlled variable Y_i^* is considered, there is a set of minimax solutions each with tangential points. The locus of these optimum points defines the expansion path (de Neufville and Stafford 1971) of minimum cost combinations of instruments required to obtain any given level of control as shown (fig. 6.15A). Each of these points satisfies the optimality criterion (equation 6.3.5). The sum of the costs associated with each instrument U_i varies along this expansion path locus so that to obtain the final control solution two further functions are often considered: the total cost function and the cost effectiveness function. The total cost function, shown in fig. 6.15B, is usually a monotonically increasing function of the magnitude of the desired control configuration. The cost-effectiveness function $J(Y)$ (fig. 6.15C) is a form of utility function (use value) representing the subjective trade-offs between different instruments (or policies). Generally successively higher desired levels of control will cost increasingly more and be decreasingly effective. In other situations, Burke and Heaney (1975, p. 87) suggest that it may be fruitless to increase control inputs further when acceptable levels of control (or agreement) have been achieved,

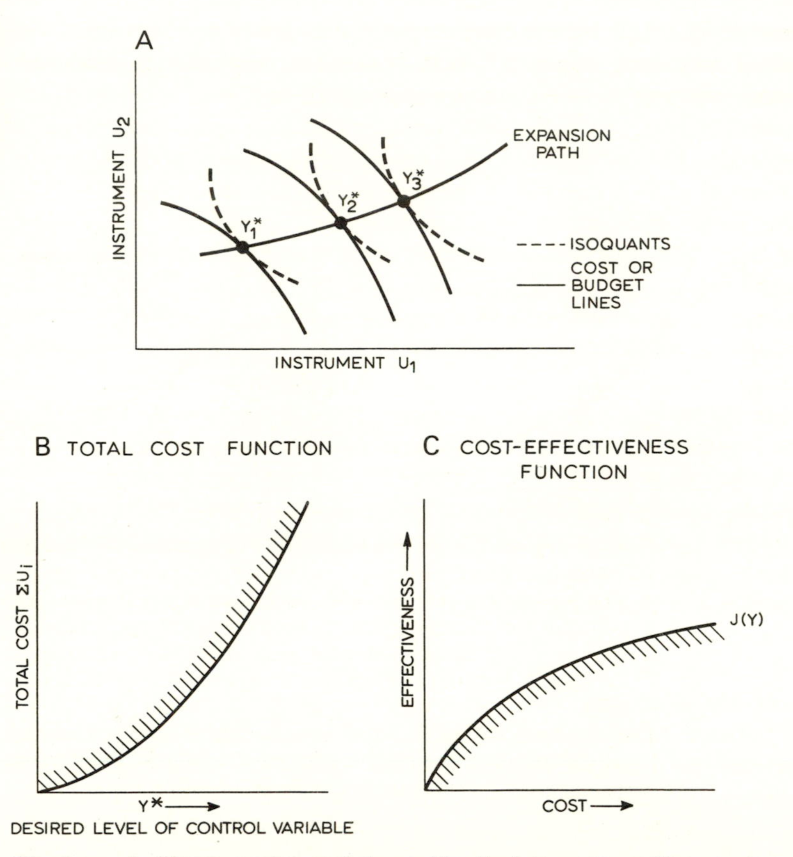

Fig. 6.15 A The expansion path formed by the intersection of isoquants and budget lines for various desired levels of system output (Y_i^*).
B The total cost function which is an accelerating function of the desired level of control.
C The cost-effectiveness function which expresses the return from control at various levels of expenditure.
(After de Neufville and Stafford 1971.)

when impasse has developed, or when the control resources of the decision maker have been exhausted.

The desired optimization problem of decision making can now be stated as maximizing the effectiveness of the system subject to cost constraints, i.e.

Maximize (or minimize):

$$J(Y) = f(U_1, U_2, \ldots, U_i, \ldots) \tag{6.3.6}$$

Subject to the constraints:

$$g(U_1, U_2, \ldots, U_i, \ldots) \tag{6.3.7}$$

where $J(Y)$ is the cost effectiveness (utility) function or some other *objective function* describing the rate of return or preference for each magnitude of each control instrument. The constraints $g(\cdot)$ will be either cost constraints, as discussed above, or system behaviour constraints of space-time dynamics (see chapter 4). The optimization problem posed by equations (6.3.6) and (6.3.7) can be solved by a number of the methods discussed in chapter 3. If $J(Y)$ is linear, linear and dynamic programming can be used. If $J(Y)$ is nonlinear (e.g. quadratic) dynamic programmes and matrix Riccati solutions must be sought. It is possible to formulate and solve computer programming problems of very large dimension, often with as many as 10^4 equations and 10^6 variables (Lasdon 1970). Frequently, however, large programming problems have special structures (e.g. triangularization, dimension reduction, etc.) which makes them more amenable to solution, frequently because large problems arise from linking independent subunits in either time or space. A review of such methods is available in Dantzig (1965) and Lasdon (1970).

An alternative application of Lerner's Rule, as stated at the outset of this section, is to individual decisions and to the aggregation of individual decision making actions. This aggregation yields the individualist approach to decision by *collective action* (Buchanan and Tullock 1969, Burke and Heaney 1975). At the level of the individual, each decision maker, when confronted by real choice, is considered to execute decisions and actions based on Lerner's Rule. Hence, the alternative marginal products of each action will be weighed and the optimum choice made of that which yields the greatest return. For example, the industrialist will scale the value of various alternative factor inputs, the consumer the value of alternative good-purchasing opportunities, and so on. If each individual acts rationally and conforms to the fundamental economic postulate of always choosing more rather than less, then each individual will seek to maximize his own utility function in any decision. Each decision maker is assumed to have his own personal utility function composed of a set of ranked choices, which is further assumed to have a fixed and transitive ranking, extendable also to public goods. Transitivity is a specially important assumption and requires that if an individual prefers A to B and B to C, then he will prefer A to C. Hence, all preferences can be ordered along a single scale of utility. The final optimum decision will be based on that action or choice which yields the greatest utility (rate of return) after considering all possible substitutions between alternatives – i.e. it is that individual action which satisfies the optimality criterion (6.3.5) at which the marginal rate of substitution between alternatives (goods, actions) is equal to the reciprocal of the ratio of their prices (marginal returns) (Boulding 1955, Lerner 1946).

Collective decision arises when individuals choose to accomplish purposes collectively rather than individually (Buchanan and Tullock 1969). The collective aggregation of preferences gives rise to collective group machines of action constructed to carry out collective actions – i.e. government. Collective organizations will arise when there is an expectation of increased utility to be derived from such organization. This will be at the point at which the external costs to be paid by the individual are outweighed by the external costs received by the individual, i.e. where the costs to the individual resulting from the actions of others and from his participation in collective actions give a greater return than can be derived by individual action (Buchanan and Tullock 1969, pp. 30–60). In the general case of N persons trying to make collective decisions, macro-economics resolves the problem of choice of actions within the collective by converting all decisions to marginal decisions. This is made possible by the divisibility of goods and the availability of alternatives to both buyers and sellers. In political bargaining the assumption of divisibility and alternatives may be invalid since usually no alternative to a national government decision is available to the individual (except migration). Buchanan and Tullock's calculus of consent acts as a necessary and sufficient condition on collective action which is based upon unanimous agreement on any action before it can be carried out. Hence, the lack of alternatives in the politico-economic arena is an incentive to bargain and cooperate, but the failure of individual A to bargain prevents B from carrying out any action.

In individual action, in collective action, and in control system decisions, the most crucial and difficult stage of the decision process is the determination of the objective (utility) function. There are special difficulties with public goods, facilities and services where competitive market conditions do not apply (Lerner 1946) and where environmental costs are unvalued. There are problems in how public or welfare benefits are measured, and how renewable and non-renewable national resources are costed (Connelly and Perlman 1975). When the utility function is a collective aggregate of individual preference functions, we have the problems associated with Arrow's impossibility theorem discussed in chapter 8 – i.e. that collective preferences have no uniquely ordered or ranked form when certain reasonable conditions are placed on individual preference functions. Because of each of these problems which constrain the direct implementation of optimization on a defined objective function, the decision maker must usually employ a weaker decision structure which is more open to alternative information and options. As de Neufville and Stafford (1971, pp. 111–12) note:

> The selection of a system is an art which requires the analyst to balance all the many consequences of an alternative. . . . It is an art supported by a considerable amount of theory . . . [but] this decision theory has principally defined important issues in the analysis of decisions without yet providing techniques which can be categorically applied to the problems of system selection.

Hence in any decision situation the decision maker should be aware of the *range* and *probability* of each possible outcome of a choice, and should also give a rational weighting between each outcome so that an objective choice can be made. Thus decision making is most frequently approached through consideration of the merits of a range of alternatives and is hence an *evaluation* problem.

6.4 Evaluation in decision making

Evaluation can be loosely defined as the consideration of the advantages and disadvantages of a particular control policy against one or more possible alternative courses of action (Lichfield *et al.* 1975), and against a global objective function of control objectives (Tinbergen 1937). Evaluation then provides the factual basis upon which a decision can be based. This is determined by some measure of how far each alternative control strategy is successful in achieving the desired target values within generalized 'cost' (e.g. opportunity cost, disutility cost) constraints. This approach to decision making depends upon subjecting each decision to a comparison of achievement against objectives, as a 'welfare test' (Lichfield *et al.* 1975). Five main forms of such welfare or achievement tests have been adopted (Lichfield *et al.* 1975):

(1) *Financial investment appraisal* is based upon estimates of the future cash flow of a given investment project. This involves projections of operating costs, income and taxation by the decision maker. For the controller of investment the cash flow over the life-time of the project, or more usually over the life-time required for amortization of the original capital investment, must be appraised to determine whether an adequate return for control can be achieved. The adequacy of the control is usually evaluated either by the rate of return or by the net present value of the control. The rate of return is a crude test to determine if the monetary returns on capital exceed the costs of the capital invested in the control policy. The net present value is based on discounting the cost flows received from control in the future to the value they would have if received at the present time (i.e. the time of initial investment). This is a more useful criterion which takes into account the 'opportunity costs' on capital, the property that money received now is more useful than money received in the future since it can be invested and will yield an interest return. The use of financial appraisal in social, community and planning contexts has

been criticized by Lichfield *et al.* (1975) on the grounds that many of the returns of control to society have no easily identified cash value. However, this method may be of great utility in rendering a judgement between control strategies with similar social spinoffs but different cash flows. For example, NEDO (1971) has considered the comparative efficiency of government tax concessions as spatially differentiated controllers of the relative efficiency of investment projects between development areas and the rest of the United Kingdom.

(2) *Check list of criteria.* This method seeks to rank various control strategies on a subjective basis with regard to how well each policy performs when compared against a number of specified or desired criteria. The final choice between these ordinal rankings is very difficult in practice. Nevertheless it is argued that the method helps to clarify the issues and tradeoffs involved in adopting any one strategy, and the approach may also be combined with financial ranking of strategies as an additional criterion. This approach has been adopted in a number of planning studies.

(3) *The goals-achievement matrix* (Hill 1968) approach to control strategy evaluation seeks to place the check list method into a more explicit comparison with control target values, rather than relying upon the exogenous application of subjective processes to resolve the strategic choice issue. The extent to which each control strategy achieves each of the prescribed set of control goals or targets is given some weight or ranking. On the basis of the rankings of each strategy upon each of the goals a final choice is made, again mainly on a subjective basis.

(4) *Cost-benefit analysis* is an attempt to incorporate the concepts of social welfare and public goods into the financial appraisal of control strategies so that the Lerner-type substitution analysis can be applied. It is analogous to profit and loss accounts in business management and is an attempt to quantify the externalities of public investment involved in any control strategy (Prest and Turvey 1965), mainly in terms of the benefits accruing to the affected individuals, groups or society as a whole. Given a cost (production) function of any control policy $f(U, Y)$ as a function of control $U(t)$ and system output $Y(t)$, and a net benefit function $J(U, Y)$, the cost-benefit problem is to maximize $J(U, Y)$ subject to the constraint $f(U, Y) = 0$. Using Lagrangian multipliers the resulting equation solution is given by the results of chapter 3 as the zero point (maximum) of the partial derivatives:

$$\frac{\partial J}{\partial U} = -\lambda \frac{\partial f}{\partial U}, \qquad \frac{\partial U}{\partial Y} = -\lambda \frac{\partial f}{\partial Y} \tag{6.4.1}$$

derived from the Lagrangian

$$L(U, Y) = J(U, Y) + \lambda f(U, Y). \tag{6.4.2}$$

Great controversy has surrounded the values given to the benefits, expressed as physical or social output. The most common criterion is based on a measure of benefit equivalent to 'the quantity of alternative goods and services which would give the same amount of satisfaction to the beneficiaries. Conversely, a cost is measured by the goods and services which would provide sufficient compensation to the losers' (Lichfield *et al.* 1975, p. 59). It is this and other associated definitions of externalities or public goods that are the most difficult stage to justify in the cost-effectiveness process. One variant of the cost-benefit criterion is to estimate the value of the benefits and costs to the community of the resources used. This resource cost assessment method is critically dependent in the non-renewable resource case upon whether the resources are treated as capital or income (see chapter 9). A second important variant of the cost-benefit approach is planning balance sheet analysis. This attempts to identify the groups affected by each controller and by each strategy. The planning balance sheet is a two-way tabulation of linkages between 'producers/operators' and individuals concerning their goals and services. Each link represents a 'transaction' which can be quantified in terms of costs and benefits to each group (Lichfield *et al.* 1975). The approach is especially useful for planning problems where multisector allocations are involved and where distributional effects between groups are of importance. In any instance, the evaluation method chosen by the executive controller will depend upon the nature of the problem, upon the nature of the criteria which it is to satisfy and the objective it is sought to achieve, and upon the nature of the decision making environment.

(5) *Operational analysis.* This final form of policy and control evaluation is made up of an assemblage of methods based more upon pragmatic considerations than the rationalizations of cost benefit, financial appraisal, etc. It is not desired explicitly to cost or quantify the effects of policy but merely to determine how far given sets of objectives have been achieved. Operational analysis has been distinguished from the other forms of evaluation (loosely termed systems analysis) by Rosenbloom and Russell (1971) as developing efficiency of government and control within existing institutions, i.e. an incremental approach. Within this category fall the organization and methods (O and M) approach which seeks to give advice upon the structure of organization, management and control with the aim of advising, eliminating, reducing, simplifying and speeding control actions (Sherman 1969). To some extent, management by objectives also enters into this category (Glendinning and Bullock 1973), as do any approaches which seek to classify and improve control without explicit consideration of cost criteria.

Each of these five evaluation methods give some measure of the costs of control and the achievement of targets. None gives an automated procedure by which the decision maker resolves conflicts and preferences between strategies. Each evaluation method relies upon subjective judgement in the definition of benefit criteria and in the choice between benefits of alternatives.

6.5 Components of decision making systems

Each stage in the decision making process proceeding through information gathering, monitoring, forecasting, and evaluation with respect to objectives has presupposed that the manager or decision maker has had the power to act out the control decisions reached. In practice, however, control by objectives, and economic control evaluation by rate of return and by substitution analysis are highly constrained by the social, administrative, legal, and political environment in which they are embedded. In most complex control situations in which the executive is a manager of resources, or of society itself, management must involve a consensus over a range of conflict issues. There will usually be multiple control cost minima, or the cost minima will be separated by irrelevantly small cost differentials. Moreover, the achievable control will normally be only partial since a limited range of instruments and levers will be available for manipulation.

The overall constraints on control reduce the number of executive degrees of freedom in three ways. Firstly, he will be tied into an administrative hierarchy (i.e. a nested control system) in which he has only limited responsibilities and limited certainty. Secondly, he will be governed by horizontal side rules that limit his bounds of control through sociopolitical and legal acceptability. Thirdly, he cannot usually regulate the disturbances that enter his area of responsibility but only try to ameliorate them. He is a decision maker in a multi-organization constrained by uncertainty. Hence any decision making system cannot be considered outside the context of the multi-organization and of multigroup society and environment of which it is a part. The form of decision structuring depends upon the degree of spatial interlinkage and interdependence. This has been structured into five categories by Laudon (1974) (fig. 6.16):

(*i*) Pooled interdependence. Decision of one unit affects the whole system (i.e. a homogeneous decision system).
(*ii*) Sequential interdependence. Decisions of one unit affect other parts of the system in sequence in one direction only.
(*iii*) Reciprocal interdependence. Decisions of one unit affect the inputs into that unit via feedback from other decision making systems.
(*iv*) Sector interdependence. Multiple input-output dependence of one decision unit upon many, and effect upon many.
(*v*) Market interdependence. A few decision units control the whole decision hierarchy by competition or monopolistic collusion.

The general interrelation between decision areas given by these categories of active and passive system components requires what Burke and Heaney (1975) term a behavioural collective action approach to conflict resolution and decision. Hence political, administrative and legal conflict resolution methods may be employed, depending upon the societal structure and nature of the control problem. The use of such behavioural models removes the decision problem from creating desirable forms (ontology) to the description of the interactive relations between decision areas (the groups, sectors or locations affected by a decision). The study of decision making systems in this context requires knowledge of the study of organizations (March and Simon 1966), of community groups and of identifiable regions, and their interaction one with each other. Hence each category of interdependence between

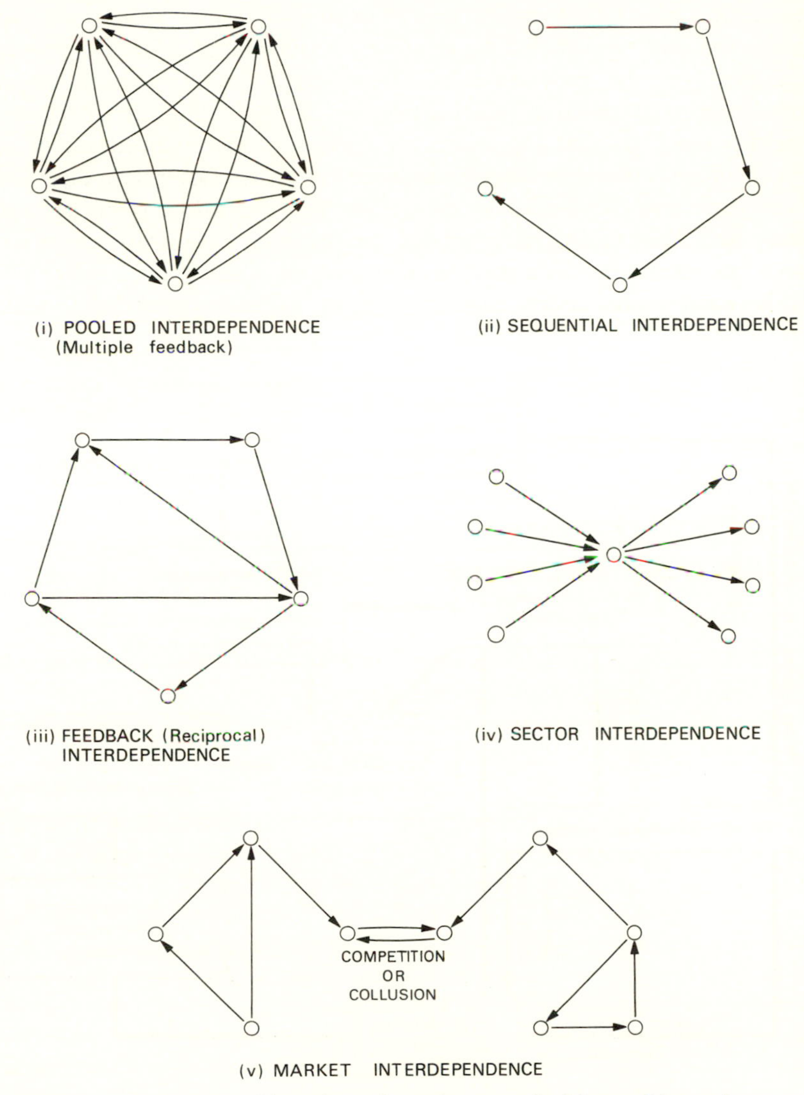

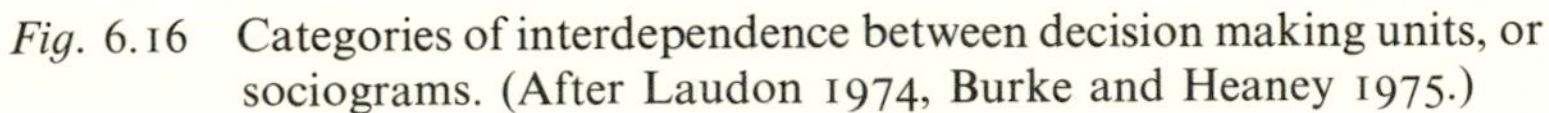

Fig. 6.16 Categories of interdependence between decision making units, or sociograms. (After Laudon 1974, Burke and Heaney 1975.)

decision areas requires a means of resolving conflicts and overlaps of aim and objectives. Such action has three major components – administrative, sociopolitical, and legal – which are discussed below.

6.5.1 THE ADMINISTRATIVE COMPONENT

The administrative component of decision making systems is clearly a very variable one and dependent to some degree on sociopolitical considerations (see section 6.5.2). The institutions of modern Western society are partly maintained and rejuvenated by relatively unstructured collective processes dependent on the generation of stress, but containing both negative (stabilizing) and positive (elaborating or disorganizing) feedback loops (Buckley 1967, p. 137) (fig. 6.17). A feature of the elaboration of such organizations is that their interdependence has grown faster than their ability to integrate and coordinate, to create the 'new pluralism' (Drucker 1969). Emery and Trist (1965) have termed the resulting organizational environments 'turbulent', in which there is an escalation of stress and conflict within and between organizations, together with an increasing inability to select a

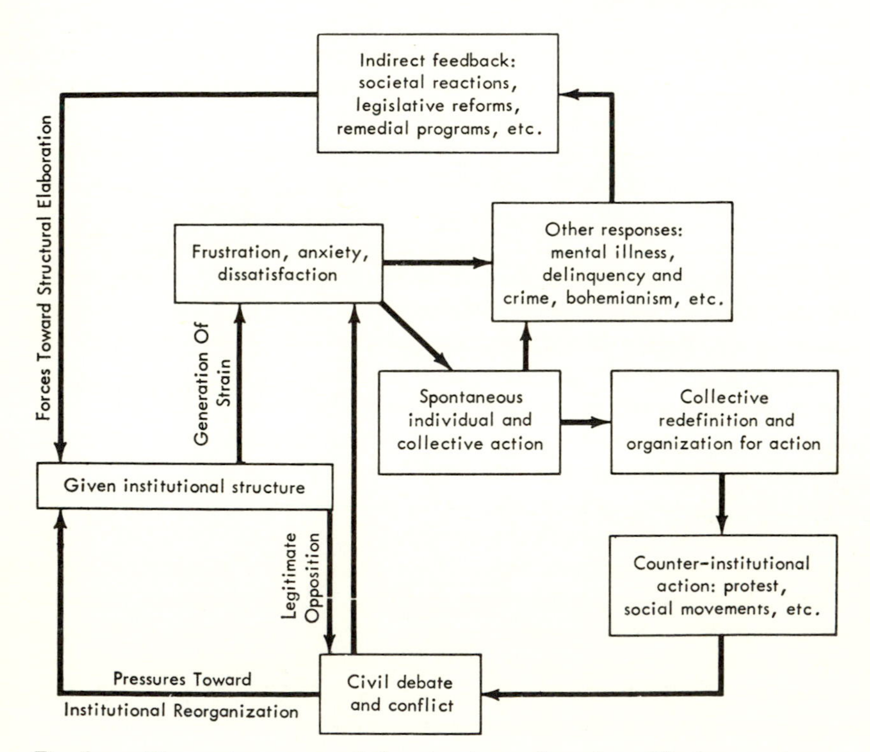

Fig. 6.17 The maintenance of the structure of modern Western society. (From Buckley 1967.)

viable course of action (Metcalfe 1974). However, any type of administrative structure contributes to a common set of decision making problems and the atrophy in Western organizations may not be greatly different from that described by Kafka in eastern Europe, despite (or perhaps because of) the alleged importance of negative feedback loops in the Marxist evolutionary society (Sztompka 1974, pp. 162–78).

The day-to-day implementation of control decisions within environmental control is managed by bureaucratic machines. In environmental systems analogous structures are built up out of nested governor systems (see chapter 3). The degree of organization in any system is dependent upon its scale. With n elementary tasks, a decision making system will have $(2^n - 1)$ relationships between tasks (Emery 1969). All practical socio-economic and physico-ecological systems are large scale and involve many nested loops. Their administration is thus the implementation of multi-organizational decision making. The multi-organization is defined as 'the union of parts of several organizations, each part being a subject of the interests of its own organization' (Stringer 1967, p. 107) and has three main aspects. Firstly, members are held in uncertainty and have little idea of a common definition of improvement; secondly, it is difficult to represent the interests of the people in the tasks being performed; and thirdly, the elements are intercorporate – there is no management from a centre, processes are taking place between internal subunits. These internal processes have the attributes of power, independence, progressiveness in disaggregation, overlapping and distinct decision areas (Stringer 1967, Ureña 1975). Because of the complexity of multi-organizational decision systems, bureaucratic administration is normally oriented towards simplifying the decision environment of each component in order to reduce the number of relations that need to be considered in coping with higher level tasks. Emery (1969, p. 11) suggests that the only way to proceed is to fragment or *decouple* decision areas (groups, sectors or locations affected) and then recoordinate them (fig. 6.18). This is related to goal and model decomposition discussed in chapter 4. Thus if a subset of only s of the n tasks are involved in any decision, $(2^s - 1)$ relationships need be considered. Coordination of these subsets in the face of interactions depends upon organizing a hierarchy of decision areas. In a non-hierarchical system there are $\frac{1}{2}n(n - 1)$ pairs of direct links between the n decision areas. In a hierarchical system there are $(n - 1)(s/(s - 1))$ direct links between the subset decision areas. Hence, hierarchical organization greatly reduces the complexity of decision making to a proportional, rather than power, function of the number of tasks. This is important in the way we approach the man-environment interface (Mesarovic and Pestel 1975, and see chapter 9 of this volume). The manner in which the number of relationships is modified by different linkages is shown in fig. 6.18. A simple hierarchical system of direct single relationships gives the simplest relationship structure.

Decoupling of decision areas can be undertaken in a number of ways

(Emery 1969, pp. 24–6). Firstly, subsets of decision areas with limited interfaces can be identified, or where interfaces are dependent upon certain thresholds being exceeded. Secondly, buffers can be created, as in stock control subsystems. This reduces the need to coordinate inputs and outputs of a decision area since the system contains a built-in time lag (or integrator); for example, creation of a land bank to control building development, water reservoir systems, and gene pools. In socio-economic systems a third method of

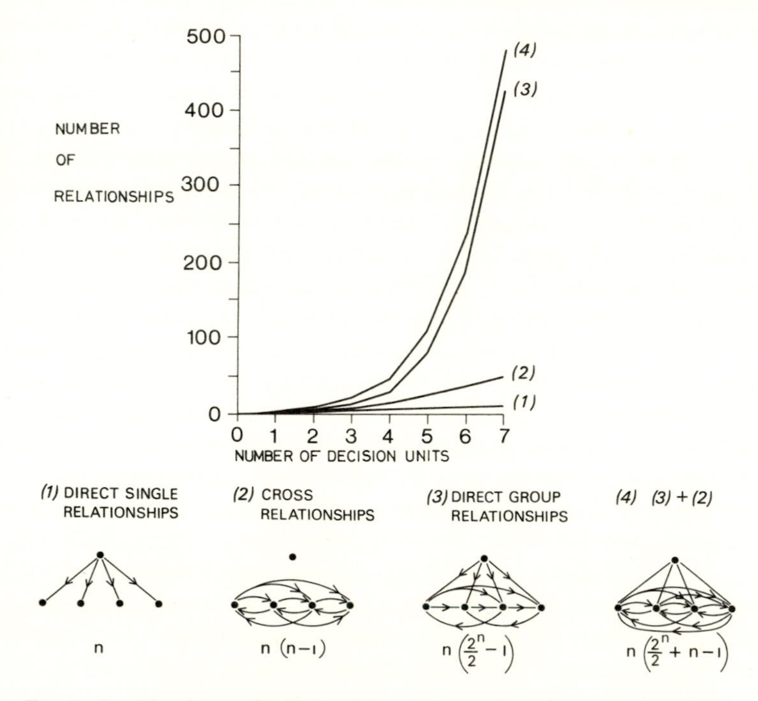

Fig. 6.18 Numbers of relationships (Y) as a function of number of decision units (X) for four classes of linkages (1–4). n is the number of independent decision making units which may be linked in one direction (simple), or reciprocally (cross), or in sets of semi-independent groups. (Partly after Gulick and Urwick 1937.)

decoupling is to increase the range of discretion in mixed scanning to allow the decision maker to improve his ability to deal with random shocks. Various mathematical methods of achieving optimal decoupling have been discussed by Yore (1968), Tyler and Tuteur (1966) and Gilbert (1969). These attempt by the appropriate choice of feedback and feedforward gains to cancel the effects of control inputs into one subsystem while driving another subsystem to an arbitrary reference point or goal. This corresponds to the seizure of power by one group at the expense of others. An example from Yore is given in fig. 6.19.

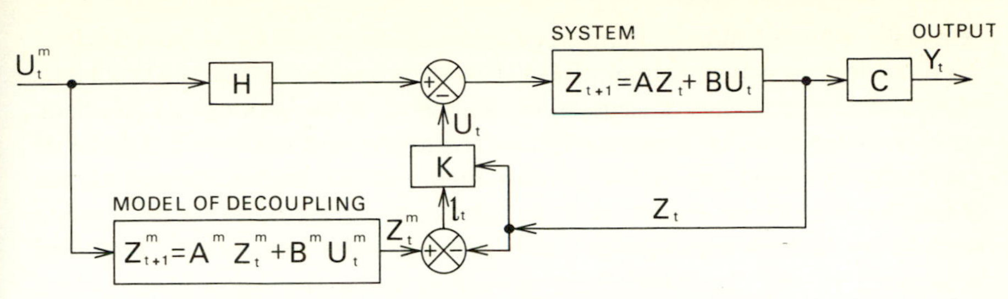

Fig. 6.19 Decoupling of real-world system (equations 6.5.1 and 6.5.2) with complex interaction to achieve optimal correspondence with a desired pattern of decoupling given by equation (6.5.3). The result (equations 6.5.7 to 6.5.11) is a modified form of the matrix Riccati equation for quadratic control discussed in equation (3.5.27). (After Yore 1968.)

System equations requiring decoupling (cf. equations 2.2.13 and 2.2.14):

$$\mathbf{Z}_{t+1} = \mathbf{AZ}_t + \mathbf{BU}_t \tag{6.5.1}$$

$$Y_t = \mathbf{EZ}_t \tag{6.5.2}$$

Model of decoupling desired:

$$\mathbf{Z}^m_{t+11} = \mathbf{A}^m\mathbf{Z}^m_{t+1} + \mathbf{B}^m\mathbf{U}^m_t \tag{6.5.3}$$

Combined system (equations 6.5.1 + 6.5.2 + 6.5.3):

$$\mathbf{Z}_{t+1} = \begin{bmatrix} \mathbf{A} & 0 \\ 0 & \mathbf{A}^m \end{bmatrix}\mathbf{Z}_t + \begin{bmatrix} \mathbf{B} \\ 0 \end{bmatrix}\mathbf{U}_t + \begin{bmatrix} \mathbf{B} & \mathbf{H} \\ & \mathbf{B}^m \end{bmatrix}\mathbf{U}^m_t \tag{6.5.4}$$

$$\mathbf{U}_t = \mathbf{U}_t + \mathbf{HU}^m_t \tag{6.5.5}$$

Control problem: to minimize the differences of the real system from the observed form of decoupling:

minimizing
$$J = \sum_{t=0}^{T} (\mathbf{Z}_t'\mathbf{MZ}_t + \mathbf{U}_t'\mathbf{NU}_t) \tag{6.5.6}$$

subject to (6.5.1) to (6.5.5) and

$$\mathbf{M} = \begin{bmatrix} \mathbf{E'ME} & -\mathbf{E'M} \\ -\mathbf{ME} & \mathbf{M} \end{bmatrix}$$

Solution (optimal control decoupling): feedback-feedforward control system gains required for optimal decoupling in real system.

(1) Feedback gains: optimal control input $\{U_t\}$ after deviations:

$$\mathbf{U}_t = \mathbf{KZ}_t = \tfrac{1}{2}\mathbf{H}^{-1}[\mathbf{B}'0]\mathbf{P} \tag{6.5.7}$$

$$= \mathbf{P}_1\mathbf{Z}_t + \mathbf{P}_2\mathbf{Z}^m_t \tag{6.5.8}$$

with $\mathbf{P}_i$ given by solution of matrix Riccati equation (3.5.27).

$$\begin{bmatrix} \mathbf{P}_1 & \mathbf{P}_2 \\ \mathbf{P}_2 & \mathbf{P}_3 \end{bmatrix}\begin{bmatrix} \mathbf{A} & 0 \\ 0 & \mathbf{A}^m \end{bmatrix} - \begin{bmatrix} \mathbf{A}' & 0 \\ 0 & \mathbf{A}^m \end{bmatrix}\begin{bmatrix} \mathbf{P}_1 & \mathbf{P}_2 \\ \mathbf{P}_2 & \mathbf{P}_3 \end{bmatrix} + \frac{1}{\lambda}\begin{bmatrix} \mathbf{P}_1 & \mathbf{P}_2 \\ \mathbf{P}_2 & \mathbf{P}_3 \end{bmatrix}\begin{bmatrix} \mathbf{B} \\ 0 \end{bmatrix}$$

$$\times [\mathbf{B}' \quad 0]\begin{bmatrix} \mathbf{P}_1 & \mathbf{P}_2 \\ \mathbf{P}_2 & \mathbf{P}_3 \end{bmatrix} - \begin{bmatrix} \mathbf{E'ME} & -\mathbf{EM} \\ -\mathbf{ME} & \mathbf{M} \end{bmatrix} = 0 \tag{6.5.9}$$

(2) Feedforward gains: optimal control input $\{\mathbf{U}_t\}$ in anticipation of deviations:

$$\mathbf{U}_t = \mathbf{KZ}_t + \mathbf{HU}^m_t \tag{6.5.10}$$

$$\mathbf{H} = [\mathbf{EB}]^\dagger\mathbf{B}^m \tag{6.5.11}$$

where (†) denotes the pseudo-inverse (see Appendix 1).

A model of the ideal decoupling required is designed which is used as a set point or desired system configuration. It is then possible to use the quadratic minimization procedure (discussed in chapter 3) to derive optimal feedback gains **K** by solution of a matrix Riccati equation and feedforward gains **H** (normally by pseudo-inversion). Adjustments to the Jordan normal matrix (equation 2.2.13) are the basis of an alternative state variable decoupling derived by Gilbert (1969), which is based upon the recognition of certain invariants of the feedback gains **K** which leave the open loop system unchanged. The choice of gains are those which yield a canonical diagonal structure similar to the Jordan matrix (the Falb-Wolovich matrix) which allows placement of system poles within this set of invariants such that the system outputs $\mathbf{Y}_t$ form an orthogonal (decoupled) set. Optimal decoupling provides a useful approach to the problem of how to deal with the complex man-environment systems discussed in chapter 9, in which the number of interacting elements is on a scale of millions of subsystems. In these situations conflict resolution and control by iterative processes of collective action would be so time consuming that potential collapse modes would occur before any countervailing action could be agreed upon and implemented. As Mesarovic and Pestel (1975) note, decision making in complex systems can be undertaken only by recognizing the *organic* interrelations between system components and this requires practical ordering of decoupled system patterns into a coordinated hierarchy (fig. 6.20A).

The price paid for the simplification offered by decoupling is that decision areas become divorced from each other in administrative hierarchies. This may be overcome in two ways (Emery 1969): firstly, by the creation of supplementary direct links between interconnected decision areas (coordination) shown in fig. 6.20B, and secondly, by the creation of a common information pool (i.e. information interlinkage) (fig. 6.20C) which allows all decision areas to be aware of relevant changes in other decision sectors. A special form of information interlinkage is the creation of a unit to maintain information and coordination to each of the divorced decision areas. This is to some extent the role of monitoring units (see section 6.1.3). Decoupling of decision areas has been strangely unsuccessful in practice – for example, as employed by the Department of Economic Affairs, and Regional Economic Planning Boards, in the United Kingdom. Whilst the theory of decoupling and hierarchical organization works well in physico-ecological systems of the environment, socio-economic applications show that jealousy and 'infighting' between decision units often make decoupling counterproductive. Despite this drawback, decoupling must be *made* to work since in complex man-environment systems (see chapter 9) this may be the only method of averting collapse modes. Emery (1969, p. 31) suggests that the optimum degree of decoupling can be derived as a rationalization of conflict between coordination and independence of subsystems as a cost-minimization problem, as shown in fig. 6.21. Beer (1972, 1974) argues that the failure of decoupling in socio-economic administration can be rectified by designing the administrative structure of decision making systems with reference to Ashby's Law of Requisite Variety. Only with regard to this law can efficient system operation be achieved with maximum individual freedom whilst reducing disturbances to the societal structure.

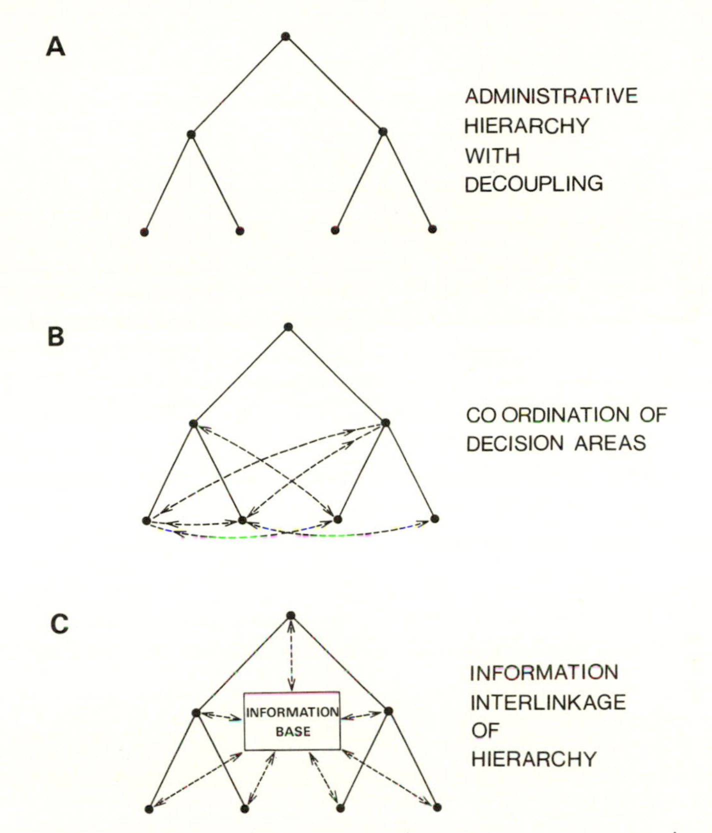

Fig. 6.20 The interrelations between complex system components in decision making. (Partly after Emery 1969.)

All our major societary institutions are high-variety systems; all of them need to have a finite relaxation time; but all of them are subject to constant perturbation. . . . What is it that controls variety? The answer is dead simple: variety. Variety absorbs variety, and nothing else can . . . when varieties are disbalanced, as they usually are, we structure our organization to cope. Fundamentally there are two ways, and only two ways of doing this . . . [first] to reduce the variety generated by the system so that it matches the available supply of regulatory variety. . . . The alternative is to amplify the variety of the regulatory part of the total system . . . in societal

> systems, this is the preferable way to proceed, because it helps to preserve individual freedom. (Beer 1974, pp. 11, 21, 23, 24, 25)

Administration is the variety controller in the decision making machine. The control achieved is determined by the relative position and balance of variety amplification and variety attenuation.

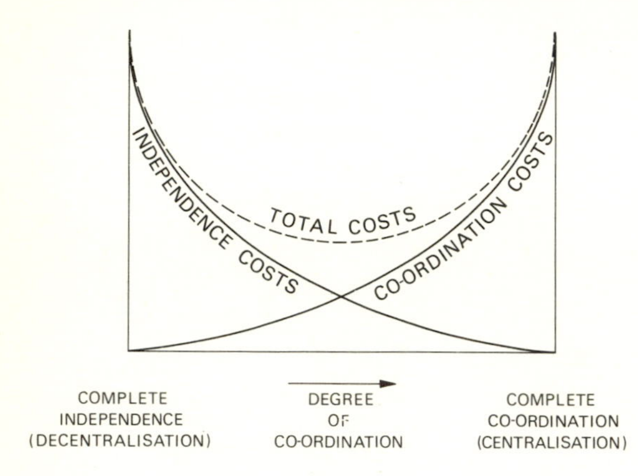

Fig. 6.21
The optimum degree of decoupling as a rationalization of the conflict between coordination and independence of subsystems as a cost minimization problem. (Partly after Emery 1969.)

The practical implementation of variety-amplifying administration is difficult to realize. Gulick and Urwick (1937) and Self (1972) recognize four principles of departmentalization:

1. *Client principle*: This involves a separate decision making entity for each group of people or problem (e.g. a ministry for the old, for the unemployed, etc.) which represents sectional groups. The United States Bureau of Indian Affairs is an example in practice.
2. *Purpose principle*: Under this a separate decision is defined for each control goal based on the output of each subsystem. This is the usual governmental department structure, favoured in the United Kingdom since Haldane in 1918.
3. *Process principle*: Here different decision systems are defined for each type of expenditure or skill (e.g. engineers, architects, departments of local authorities). This is the normal method of division of labour and breakdown of decision making units in industrial control, but it has the effect of blurring decision goals.
4. *Areal principle*: This allows decision systems to be evolved for each territorial unit with total devolution to lower geographical levels. Self (1972, p. 57) notes that this does not usually solve the departmentalization problem since functions must then be allocated at the new spatial level.

Practical variety amplifying systems will possess multiple switches between each of these principles, as shown in fig. 6.22. Beer (1974) argues that at

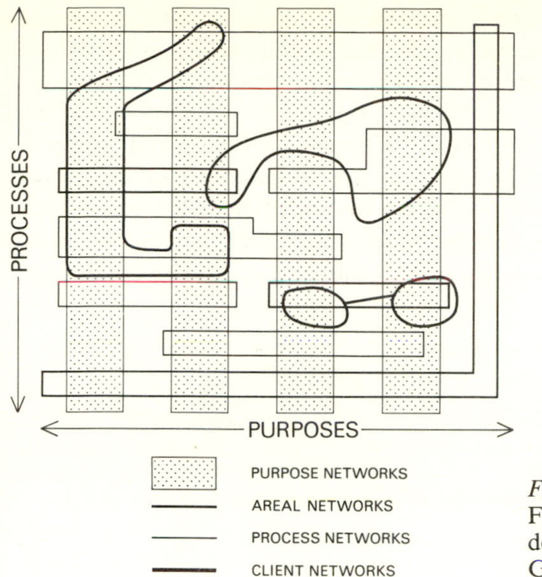

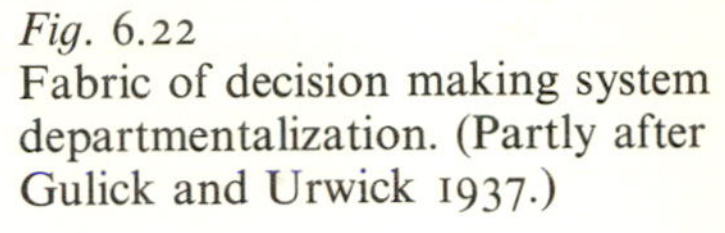

Fig. 6.22
Fabric of decision making system departmentalization. (Partly after Gulick and Urwick 1937.)

present the tools of computers, teleprocessing and organization through cybernetics are being used on the wrong side of the variety equation – in other words, to attenuate and reduce rather than increase variety, as shown in fig. 6.23. Government administration reduces the variety of human and group responses: every government department constructs a model of the country; every department builds a model of the component enterprises for which it is responsible; each of the departmental models is then aggregated; all models are subject to delays in collection and use of information. Variety is reduced at each stage by imperfectly understood partial models derived from the vertical distribution of control in parallel sectors with few horizontal links. In place of this, Beer proposes a recursive dynamic system in which each level fits within the previous level with continuous information flows both up and down. This is a symbiotic relationship based on integrated information systems. In Beer's conception, the decision maker is a cybernetic regulator who stands in his control room with all information registering in real-time computer storage systems. Various levels of filtering produce pictures of how the system is operating at various scales of complexity. The decision maker then pulls the appropriate policy lever or makes the most appropriate decision. This is communicated back to the information sources and acted on, all in real time. The failure of this system in Chile can be attributed largely to the failure of individuals to supply the appropriate information. Symbiotic decisions must rely on political consensus for collective action (Burke and Heaney 1975) since then conflict between coordination and independence (centralization–decentralization) does not arise. True symbiosis allows each individual community to pursue its own interests and ideals consistent with leaving other individuals and groups unaffected (i.e. horizontal constraint) and consistent with the freedom and aims of groups at higher or lower levels in the recursion (i.e. vertical constraint). Only by taking into account the individual and continuously changing roles and goals of each individual and group by amplifying the variety of the regulator can societal administration become stable, i.e. be capable of adapting smoothly to unpredicted changes. Because of the integrated information base the environment and the control system become the same entity. The basis of this model is that 'the freedom we

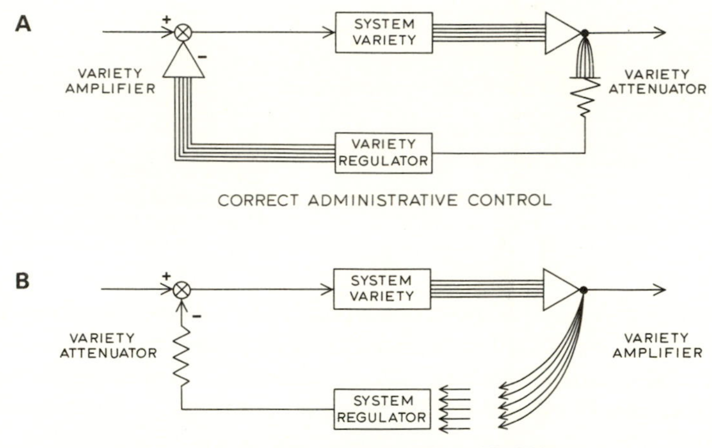

Fig. 6.23 Administrative variety control by placement of amplifiers and attenuators. (Partly after Beer 1974.)

embrace must yet be "in control". That means that people must endorse the regulatory model at the heart of the viable system in which they partake, at every level of recursion' (Beer 1974, p. 88).

The necessity of forging stronger links between administrative units has frequently been recognized in multi-organizations (Stringer 1967) and intercorporate environments. Examples include problems of combining the essentially interconnected decision roles of local, regional and central government planning (as shown in fig. 6.24), the coordination of disparate decision making units within the firm (Beer 1972), and coordination within bureaucracies and between the fragmented agencies responsible for economic, environmental and man-environment management (as discussed in chapter 9). Ureña (1975) has shown how regional decision making systems can be built along Ashby's cybernetic lines (fig. 6.25), the politico-administrative system being the regulator constraint through its capacity to transmit information

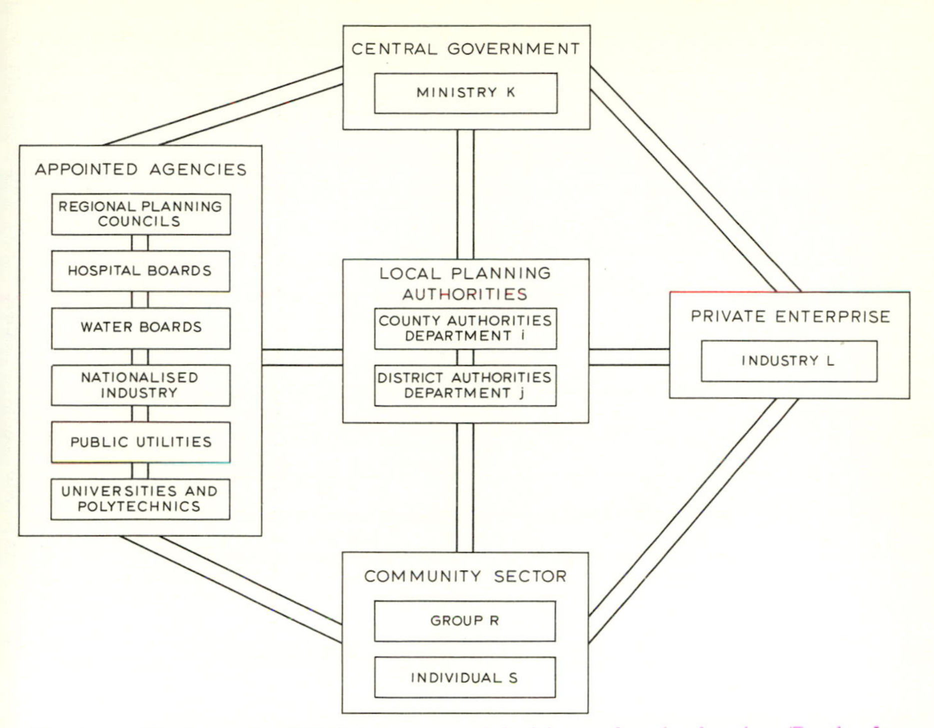

Fig. 6.24 The interrelated decision areas and decision makers in planning. (Partly after Friend, Power and Yewlett 1974.)

and amplify variety. In environmental control a fragmentation of authority and information lines is common within sociopolitical systems which are ill adjusted to the modern scale of possible human impacts. Parkes (in progress) has demonstrated, for example, how management of the foreshore in the United Kingdom is split between at least six decision making bodies (fig. 6.26).

As an aid in determining the relations between each decision making sector, Friend and Jessop (1969) and Friend *et al.* (1974) have suggested an 'Analysis of Interconnected Decision Areas' technique (AIDA) based on Stringer's (1967) study of multi-organizations. The AIDA approach employs the intersection of a strategy graph and an option graph for each decision area, as shown in fig. 6.27. This can be resolved into an associated set of matrices (Wedgwood-Oppenheim 1970) for a local government land allocation problem. Given a range of decisions that must be made, AIDA facilitates the resolution of the various possible alternative solutions consistent with satisfying the demands in each decision area based on the structuring of the problem into two graphs of direct and related decision areas (**D** and **R** networks) (fig. 6.27). Analysis of decision systems allows short-term coordination and improvement of decision; but it does not define the long-term ideals or trajectories which systems should follow, nor the optimal

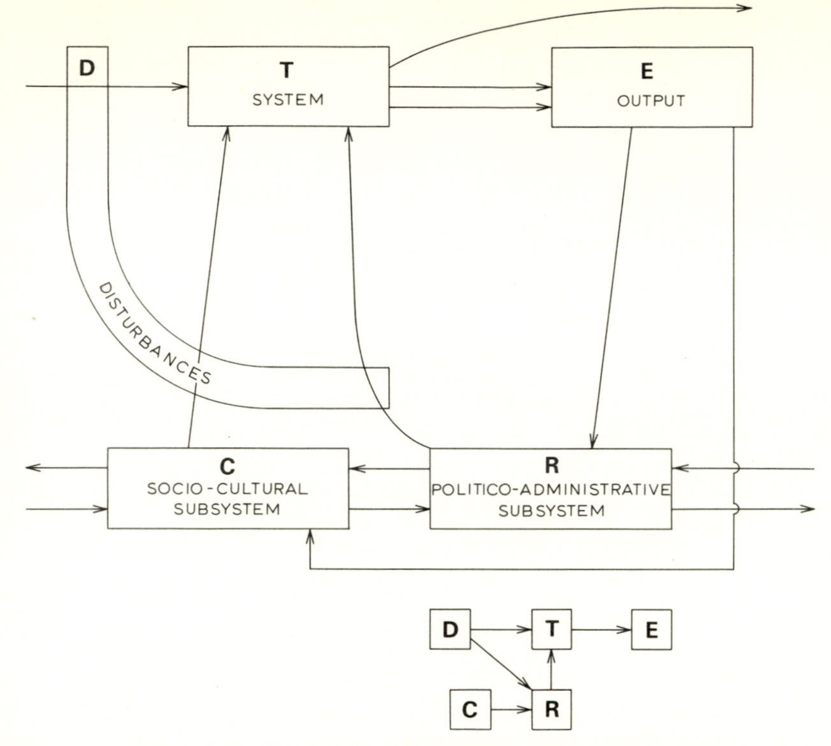

Fig. 6.25 The regional controller-regulation system on Ashby's principle (compare with fig. 6.2). (Partly after Ureña 1975.)

decision system to achieve these ideals. The socio-economic institutions (political and legal) necessary to such trajectories are discussed in the next section and in chapter 8.

6.5.2 THE SOCIOPOLITICAL COMPONENT

Planning and decision making can be viewed as societal guidance systems at the core of which is an image of the environment acted upon by the social will, or its sociopolitical expression (Faludi 1973). Cohen and Rosenthal (1971, pp. 12–15) have examined the manner in which societal forms operate through political and administrative structures to produce interactions of political processes and geographical space which underlie landscape planning. Sociopolitical systems can be seen as a search for justice and the good life (Plato and Aristotle); as a means of securing society against life which is poor, nasty, brutish and short (Hobbes); as a means of promoting private and collective interests to the greatest happiness by 'felicific calculus' (Bentham and the Utilitarians); as an instrument by which a class of exploiters maintains

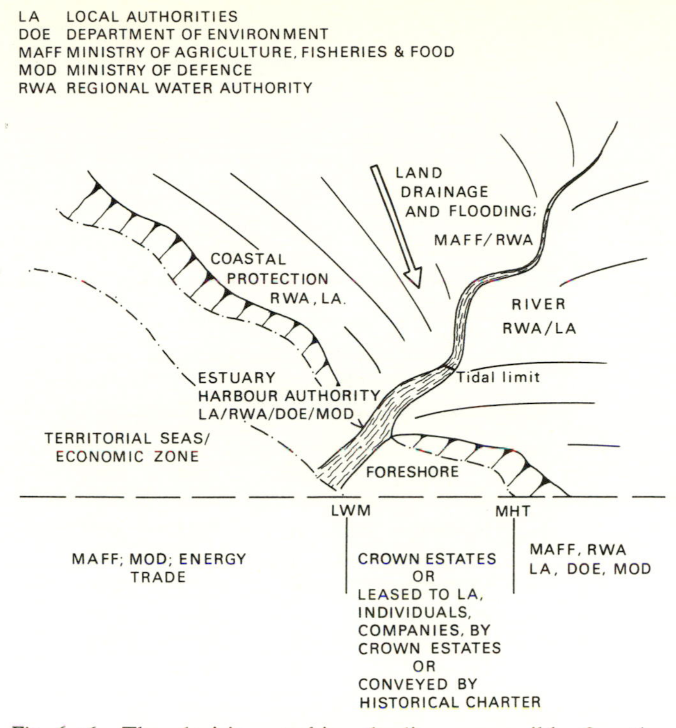

Fig. 6.26 The decision making bodies responsible for the management of the foreshore in the United Kingdom. (After Parkes, in preparation.)

its monopoly and exploitation (Marx); or, as the modern scientific school suggests, as a pragmatic regulation-distribution decision mechanism (chapter 1 and Eulau and March 1969). Whichever ideological position is adopted, the sociopolitical system acts as a resolvant between four interlinked processes; the structure of socio-economic and environmental processes, the public and private agencies and other bodies executing activities, the decision makers who decide what activities are carried out, and the political representation and power of groups in the society (Easton 1965a, 1965b) (fig. 6.28).

The subjugation of decision making to political factors is an antithesis to the Machiavellian view that the decision maker ought always to take counsel but only when he wishes, not when others wish – on the contrary he should discourage absolutely attempts to advise him unless he asks it. Because of the complexity and scale of modern society the modern 'Prince' has increasing dependence on the advice of experts and delegated administrative structures. Within multi-person systems and multi-organizations, decisions can be resolved in two ways depending upon the nature of the control goals (March and Simon 1966):

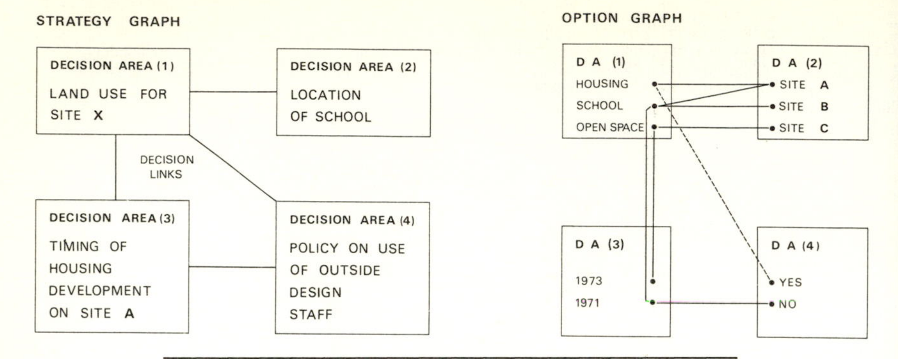

Solution	Decision area: 1 Use	2 Site	3 Year	4 Staff
1	School	*X*	1971	Yes
2	Open space	*Y*	1971	Yes
3	Open space	*Z*	1971	Yes
4	Housing	*Y*	1971	Yes
5	Housing	*Y*	1973	Yes
6	Housing	*Y*	1973	No

(After Wedgwood-Oppenheim 1970.)

Fig. 6.27 AIDA resolution of solutions to strategic choice problem. Strategy graph shows policy choices; option graph shows land use, site, timing and staffing choices; matrix shows intersection of strategy and option graphs as an interplay of contingent choices.

1 *Shared goals*: Here differences in opinion and courses of action can be resolved by analysis of the different options available to give an optimal set of decisions to achieve the goals.
2 *Goals not shared*: Here decisions cannot be made by analysis but must rely upon bargaining power, the strength of each interest group, and the possibilities of alternative courses of action.

In large-scale socio-economic systems few goals are shared by any wide group (Hayek 1944) and attempts are made in democratic organizations to achieve shared goals by various forms of representative machinery. Self (1972, pp. 281–2) recognizes three major mechanisms of achieving goal agreement: representative democracy, pluralist democracy, and populist democracy. In representative democracies decision making is assigned to elected representatives of the society as a whole with accountability through periodic elections. Considerable freedom is available to the decision makers between elections who will rarely be truly representative of the electorate as a whole. Pluralist democracies are based upon groups organized to press particular

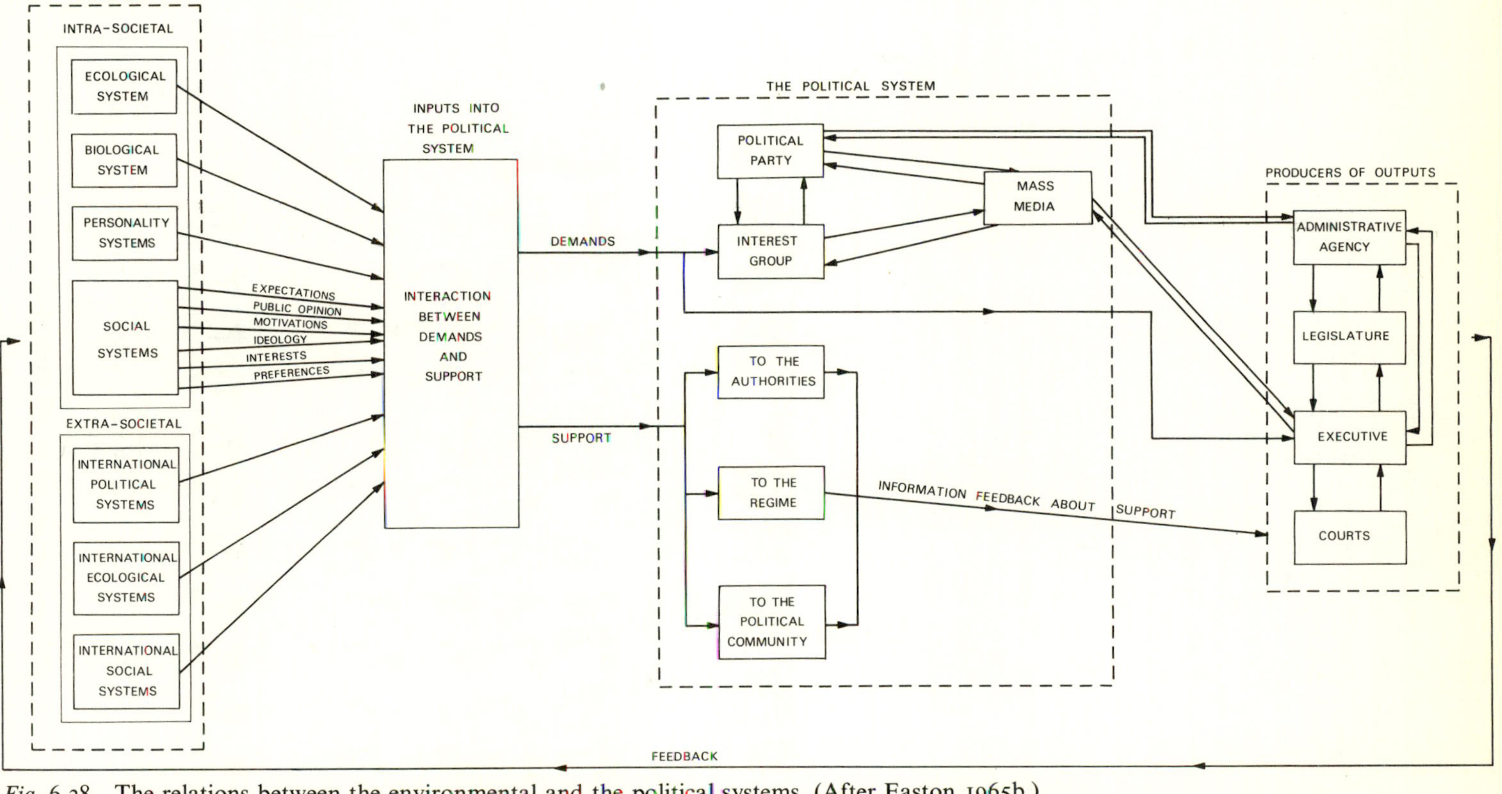

Fig. 6.28 The relations between the environmental and the political systems. (After Easton 1965b.)

interests and viewpoints. The success of these will depend upon the size of the group, its social powers and ability to make coalitions. Decision making then becomes a bargaining process not based upon shared goals. In populist democracies decision making is in the hands of the individual himself, either through collective consultation devices such as referenda, or through the sovereignty of the market. Such democracies are based upon minimal control by objectives since few goals can be agreed as shared (Hayek 1944). In practice, most democratic societies are organized as a mixture of all these principles with enhanced scope for individual safeguards through public accountability, legal obligation and public participation.

The sociopolitical component of decision making in systems in which goals are not shared is based on social power; in fact the absence of sufficient social power is one of the generators of uncertainty for the controller and decision maker. Social power is the ability of persons to carry out their will, even when opposed by others (Weber 1924, p. 51). Strong political actors do not need the advice of control systems for there is no need to optimize under uncertainty, since little can interfere with the achievement of their desires or chances of

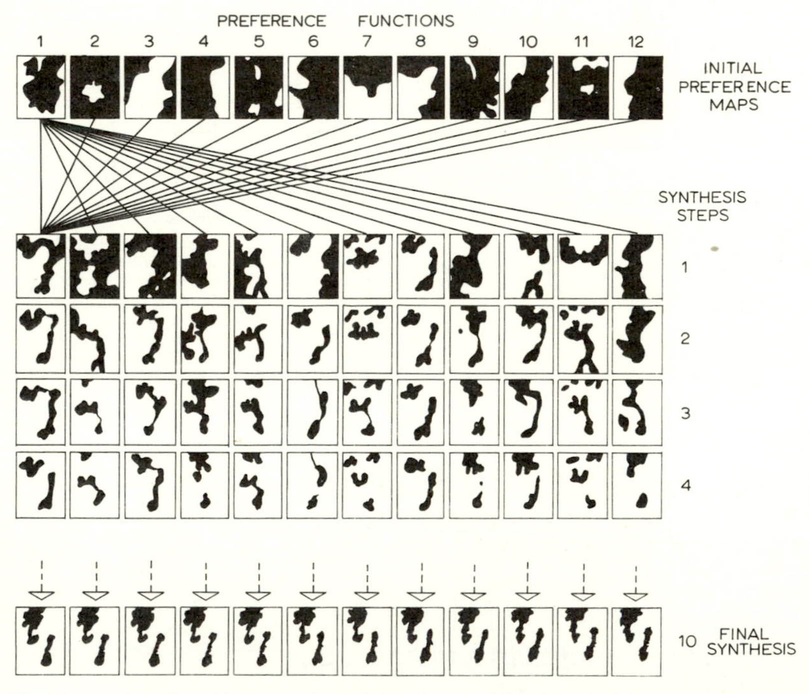

Fig. 6.29 Areal preference functions for residential development in the Macclesfield area showing examples of the averaging process for the contribution of initial preference to final syntheses; the first synthesis results for group one as the average of the 12 initial preference maps and so on. (After Batty 1971.)

success. The degree of uncertainty entering the control system is then dependent upon the degree of power that is vested in the decision maker, and in each of the actors affected by the decision. The cybernetic regulator proposed by Beer (1972, 1974) as a symbiotic decision maker aided by real-time information systems eliminates uncertainty by enmeshing the regulator and the environment. In a decision situation based on social power uncertainty is reduced by expressing the relationships between power groups or individuals in terms of their power to influence. Alexander (1964), Lindblom (1968) and Batty (1971, 1974) have derived solution methods to the resulting graph of decision areas which yield the compromise solution, given different initial response functions of a set of actors and different power linkages. Fig. 6.29 shows the resultant preference function of a decision making system derived from the initial preference function of each actor. The various actors attempt to convert their neighbours in their contact system, the resultant being a 'conflict resolution which relates to their power in the political arena' (Batty 1974, p. 292) which is an average of the previous preferences of each actor. The solution is derived from the Markov assumption and can be obtained by stage-wise algorithms such as dynamic programming. This gives the basis for the *theorem of social power*, that 'in a strongly connected symmetric power structure, the influence which each actor has on the final solution is in exact proportion to the number of channels the actor has to transmit or receive messages to or from other actors' (Batty 1974, p. 301). Various categories of social power result from different configurations

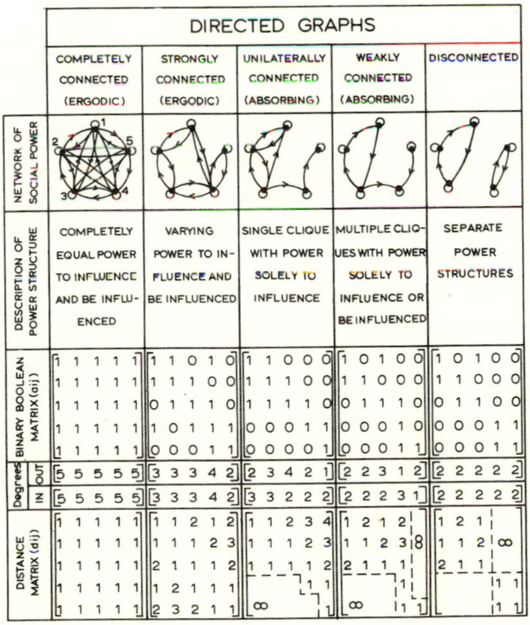

	DIRECTED GRAPHS				
	COMPLETELY CONNECTED (ERGODIC)	STRONGLY CONNECTED (ERGODIC)	UNILATERALLY CONNECTED (ABSORBING)	WEAKLY CONNECTED (ABSORBING)	DISCONNECTED
NETWORK OF SOCIAL POWER					
DESCRIPTION OF POWER STRUCTURE	COMPLETELY EQUAL POWER TO INFLUENCE AND BE INFLUENCED	VARYING POWER TO INFLUENCE AND BE INFLUENCED	SINGLE CLIQUE WITH POWER SOLELY TO INFLUENCE	MULTIPLE CLIQUES WITH POWER SOLELY TO INFLUENCE OR BE INFLUENCED	SEPARATE POWER STRUCTURES
BINARY BOOLEAN MATRIX (aij)	1 1 1 1 1 1 1 1 1 1 1 1 1 1 1 1 1 1 1 1 1 1 1 1 1	1 1 0 1 0 1 1 1 0 0 0 1 1 1 0 1 0 1 1 1 0 0 0 1 1	1 1 0 0 0 1 1 1 0 0 1 1 1 1 0 0 0 0 1 1 0 0 0 0 1	1 0 1 0 0 1 1 0 0 0 0 1 1 1 0 0 0 0 1 0 0 0 0 1 1	1 0 1 0 0 1 1 0 0 0 0 1 1 0 0 0 0 0 1 1 0 0 0 1 1
Degrees OUT	5 5 5 5 5	3 3 3 4 2	2 3 4 2 1	2 2 3 1 2	2 2 2 2 2
Degrees IN	5 5 5 5 5	3 3 3 4 2	3 3 2 2 2	2 2 2 3 1	2 2 2 2 2
DISTANCE MATRIX (dij)	1 1 1 1 1 1 1 1 1 1 1 1 1 1 1 1 1 1 1 1 1 1 1 1 1	1 1 2 1 2 1 1 1 2 3 2 1 1 1 2 1 2 1 1 1 2 3 2 1 1	1 1 2 3 4 1 1 1 2 3 1 1 1 1 2 1 1 ∞ 1	1 2 1 2 1 1 2 3 ∞ 2 1 1 1 1 ∞ 1 1	1 2 1 1 1 2 ∞ 2 1 1 1 1 1 1

Fig. 6.30
Classification of social power networks. (After Batty 1974.)

of the connection matrix in fig. 6.30 and are analogous to ergodic, absorbing and disconnected Markov chains. In the completely connected graph each actor interacts with all other actors and a compromise is reached immediately. In a disconnected graph local compromises are possible, but a global compromise can never be achieved. The unilateral and weakly connected graphs represent the usual practical cases of unidirectional influence or domination of one group by another.

There appear to be at least five methods of defining the weights given to each power group (Batty 1974):

1 Equal weights.
2 Distance weights (e.g. number of links).
3 Preference conflict between weights a_{ij} with range zero to one,

$$a_{ij} = \frac{m_{ij}}{m_{ij} + u_{ij}} \tag{6.5.12}$$

where m_{ij} is the number of matching cells between two factors and u_{ij} is the number of non-matching cells.
4 Statistical calibration.
5 *A priori* definition.

The Markov decision machine is at best a simplified view of the compromise processes at work in the political arena. Although it incorporates something of the view of the satisficer, it does not allow for the effects of collusion and cooperation between power groups. Laudon (1974) has analyzed this in terms of four models dependent upon the formality of contact and group-adoptive mechanisms, recognizing the self-adoptive nature of control systems in the sociopolitical dimension (see p. 312). More generally, Von Neumann and Morgenstern (1944) have attempted to rationalize political bargaining in terms of game theory: when does it pay to collude? Collusion is adopted when two groups (or individuals) perceive the opportunity of obtaining a higher political (or economic) return from their joint monopoly than is available from oligopoly (the competitive Cournot-Nash equilibrium). This is the prisoners' dilemma (Luce and Raiffa 1957, sec. 5.4), under which two prisoners suspected of a crime are questioned separately and each told that if one confesses, and the other does not, then the non-confessor will be punished. If neither confesses, both must go free. In political bargaining the return r_i of each (competitive) decision maker is a function of his social power s_i, and the competitive solution is given at the point where for two bargainers

$$\frac{\partial r_1(s_1, s_2)}{\partial s_1} = \frac{\partial r_2(s_1, s_2)}{\partial s_2} = 0. \tag{6.5.13}$$

This solution is depicted in fig. 6.31 where it is assumed that two equal levels of social power s'_1 and s'_2 are available. For these levels a competitive equilibrium at Q^0 gives a return to both after the expenditure of a given level of political

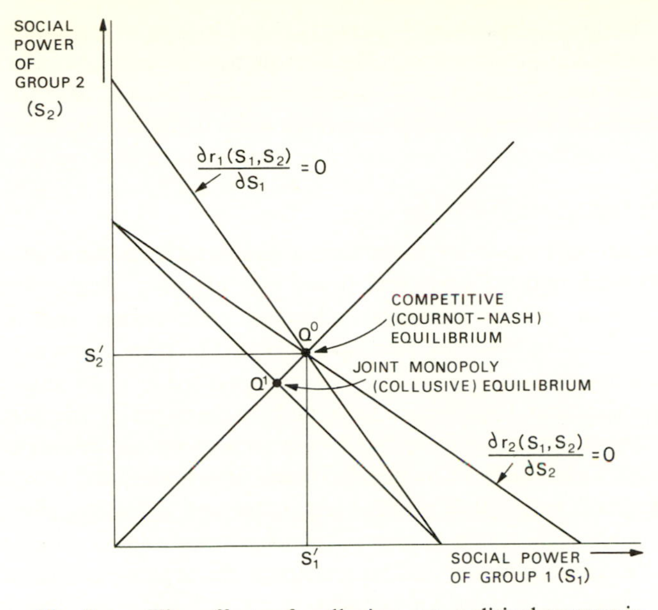

Fig. 6.31 The effect of collusion on political-economic returns to social groups. (After Von Neumann and Morgenstern 1944.)

bargaining. However, a collusion between the two groups yields a higher rate of return for monopoly at Q^1 but the solution cannot be enforced. If the two groups agree to collude but one group cheats, then the cheater gains a temporarily larger return than his share of the maximum joint monopoly return. Noton *et al.* (1974) have developed a primitive systems model involving two conflicting communities living under one jurisdiction and with a single economic system (e.g. the South African situation). For each community there was developed:

1 A sociopolitical model, involving variables expressing community size, political power, intracommunity cohesion, aggregate discontent, hostility towards the other community, conflict behaviour and external support.
2 An economic model, involving such variables as the basic growth rate of wages, real wage rates, level of investments, etc. This system model was used to simulate changes in conflict behaviour and hostility, associated with increases in the gross national product, the population of one community and a sustained increase of external support for one community.

A final, but increasingly important, area of sociopolitical decision making to which systems analysis is beginning to be applied is that relating to the anticipation of, and response to, the occurrence of natural hazards and disasters. Milete, Drabek and Haas (1975) have adopted such an approach, and the embryonic information/belief/action flow chart shown in fig. 6.32 embodies the main sequence of operations and, above all, stresses the importance of the feedback loop involving prior hazard experience.

6.5.3 THE LEGAL COMPONENT

Legal systems exist in order to ensure a predictable pattern of response between decision making units and hence operate as a variety attenuating system to reduce the uncertainties in human intervention and control of socio-economic and physico-ecological systems. The principal ways in which legal systems regulate society are to make certain conduct criminal, and hence subject to punishment, to provide civil remedy in damages for the effects of the actions of others, to require licensing before an activity can be carried out, to define the rights and duties of ownership (state or private), and to regularize a variety of indirect methods of control (e.g. taxes and subsidies) (Swan 1976). Additionally, the legal system is seen as an *independent* adjudicating controller regulating the decision maker and representing 'the organized force of society brought to bear on the individual who does not conform' (Swan 1976). Ideally, the check operates so that administrative decision making units: 'have no separate authority of their own which would enable them to respond to the public in a less inhibited way. . . . Scope for the individual approach may be limited, and officials for the most part have to apply and defend laws and rules for which they are not responsible and which in fact they may not like' (Royal Commission on the Constitution 1973, pp. 370–1). This ensures that the law is 'reasonably certain and predictable; where the law confers wide discretionary powers there should be adequate safeguards against their abuse; like should be treated as like, and unfair discrimination must not be sanctioned' (de Smith 1971, p. 40). Montesquieu, following Aristotle and Locke, and the American Constitution have developed the view of the legal system as the judicial component in the separation of powers to balance the strength of the executive and legislature in government. Whereas the policy and rules are executive or legislative functions made outside of the legal system, the judiciary adjudicates on breaches in the rules. The legal framework is, however, a system adaptive to changes in executive policy which necessitates the formulation of new laws to overcome inadequacies of legal performance, and to assimilate changes in social values which invalidate or make inappropriate present practice. The result is that the legal process is iteratively evolving both as a simple double feedback loop system (fig. 6.33) and as an adaptive feedback control system for the United States, as shown in fig. 6.34.

Loucks (1972) has reviewed the techniques by which environmental evidence may be developed into a powerful systems format for purposes of litigation. However, the legal system is not only adaptive to changes in social

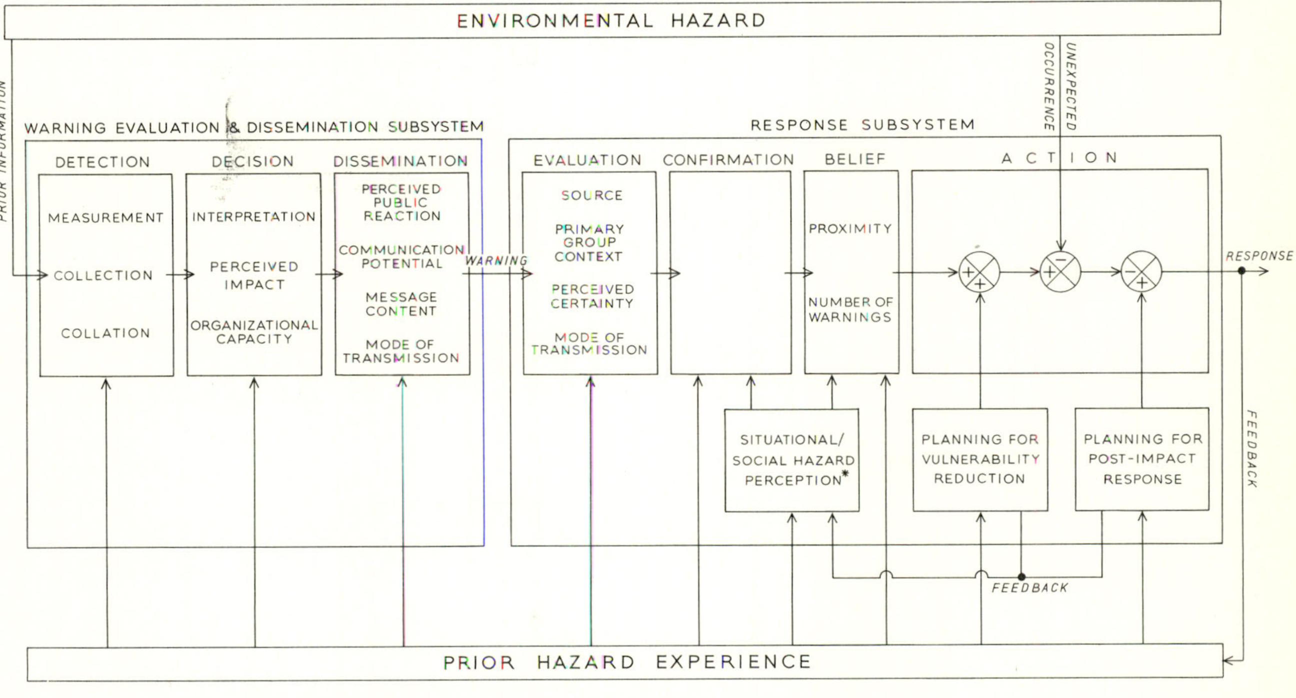

Fig. 6.32 Information-belief-action flow chart for natural hazards and disasters.

* Situational/social hazard perception includes a mass of factors which are difficult to define, including: interpretation of the environmental clues; observed action of others; experience of others; amount of breathing space; socio-economic status; urban/rural residence; race; ethnicity; primary group context; role conflict; perceived escape alternatives; etc. (Partly after Milete, Drabek and Haas 1975.)

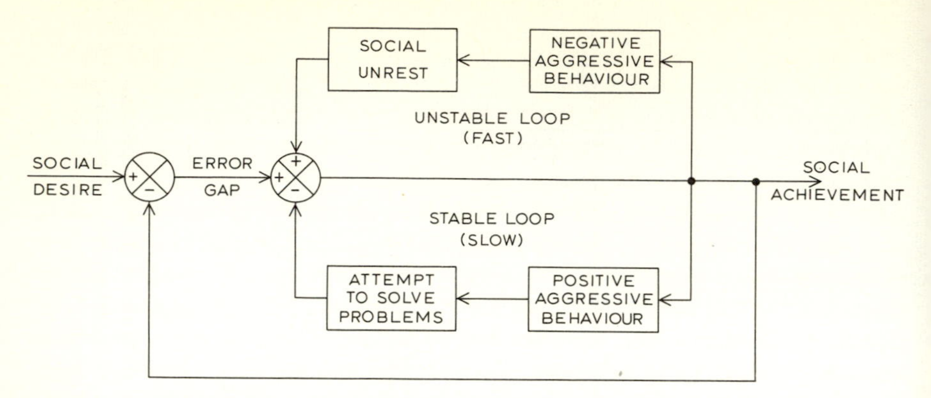

Fig. 6.33 The legal process as a double feedback loop system. (Partly after Attinger 1970.)

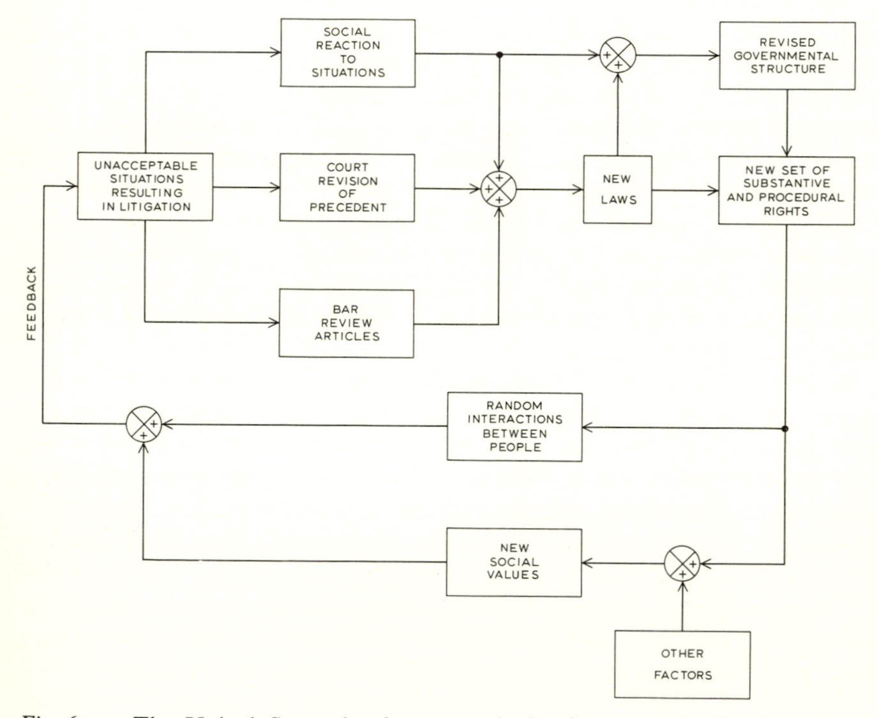

Fig. 6.34 The United States legal process depicted as an adaptive feedback control system. (After Vanyo 1971.)

and executive values over time, but is constrained at any one time within defined limits of discretion. As a system, the law attempts to regulate the loss of adaptive qualities (i.e. freedom) of one subsystem when it is integrated into higher-level control (e.g. the individual in society). This system becomes very complex and potentially indeterminate when the higher-level system is adaptive (i.e. discretionary). Thus, constraints operate at any one time to limit the range of discretionary power and hence limit also the range of adaptation of the social and executive control system to environmental changes. Apart from legal restraints on confidentiality of information which are poorly developed (Niblett 1971), there are two major constraints which derive from the complex interplay between governmental separation of powers and socially acceptable solutions. Firstly, there is a constraint to operate within legitimate bounds of authority, which may to some extent be discretionary to the decision maker, but is always limited by *vires*. Secondly, there is a constraint to observe certain principles of natural justice. These constraints operate irrespective of the type of sociopolitical system. For example, communist concepts of 'socialist legality' are symmetric with the liberal view of the 'rule of law', although the legal structure in the two cases will differ (de Smith 1971, p. 40).

Adjudication on the *vires* of decision making systems is a determination of whether or not decisions are being made within the bounds of what the law permits (*intra vires*). The decision making system acts *ultra vires* in substance when an action is without legal basis, such as the prevention of United States state subsidies to railways or United Kingdom local authority subsidies to the rents of private tenants or gifts of money to the aged (even when that action may be laudable). Decisions may also be *ultra vires* in procedure or form (de Smith 1971, pp. 548–9) if a mandatory requirement is not observed – for example, a period of notice or procedure on a compulsory purchase order. The definition of *vires* determines, in legal terms, the hierarchy of decision making discretion and thus represents absolute limits on decoupling control within hierarchical administration.

The constraint on decision making observing minimal principles of natural justice falls into three categories (de Smith 1971, pp. 557–68). Firstly, the principle of *nemo judex in causa sua* (no one should be the judge in his own case) provides for the removal of bias. This becomes difficult to uphold in a highly totalitarian state or where the controller has monopolistic powers – for example, in Department of Environment transport inquiries (see chapter 8). There may also be conflicts between public control systems and the individual. If a local authority is adjudicator on a land allocation decision, as in the United Kingdom Community Land Act, its attempts to save itself money help the community as a whole but may be to the detriment of individuals – for example, in compensation for compulsory purchase. The second principle of natural justice of *audi alteram partem* (no party should be condemned unheard), provides for notice of action to be given and opportunity to put a case opposing the action. This applies to much appeal machinery in United Kingdom environmental and structural planning. A third principle, that every party is entitled to know the reason for a decision, is less rigidly applied, especially to discretionary bodies. De Smith (1971, p. 568) notes that the principles of natural justice are very weak and ill-defined. They impose 'no more than a duty to observe minimum standards of procedural fairness. [They do] not require the decision to be right or even just; or that reasons be given for decisions; or the proceedings be conducted in public; or that a record of proceedings be maintained'.

There is a dilemma surrounding the legal constraints on decision making systems. Decision implies the discretionary power to choose between alternative courses of action. The legal limits on discretionary power seek to control executive and legislative action but in practice are limited to the questioning of very few decisions. De Smith (1971, pp. 271–5) summarizes the areas where questioning occurs in the U.K. as: inadequate reason, absence of reason, taking irrelevant factors into account, and reasonableness. The decision making system must be free to act, but a compromise must be struck with principles of natural justice. Discretion taken to its extreme 'means that some authority is given power to make with the force of law what to all intents and purposes are arbitrary decisions usually described as judging the case on its merits"'. This leads to the indeterminacy 'that many of the laws affect people's lives so closely that elasticity is essential! What does it mean if not conferment of arbitrary power, power limited by no fixed principles?' (Hayek 1944, p. 49). This has led to a series of ground-rules for the exercise of discretion within decision making systems – for example, in the United Kingdom by the Donoughmore Committee Report (1932), the Franks Report (1957) and attempts at increased electoral representation (Reorganization of Local Government 1974, Devolution proposals 1975–6).

One approach to the definition of optimal legal control levels subject to social, political and economic constraints is via linear programming. If the cost of employing two (or more) control policies can be defined, together with maximum (or minimum) constraints derived from socio-economic and political feasibility, then the legal maximal or minimal setting of a control policy can be defined as the solution to a simple linear optimization under constraints. An example of an optimal setting of acceptable air pollution levels is given by Vanyo (1971) using two policy options; investment in smog control devices and rapid transit system taxes (fig. 6.35). Alternative legal solutions in terms of dynamic programming and feedback-feedforward control systems under various 'cost' criteria can be developed for more complex examples, using the methods of chapter 3.

Legal systems represent, or attempt, a codification of the values, goals and ethics of society. They may often be outdated and slow to change, but represent attempts to balance conflicts in society. Increasingly, however, legal systems have to balance not only conflicts in society (i.e. man-man relations), but also conflicts of man's actions and the environmental consequences (i.e. man-environment relations). Hence, in Western societies there has been a tendency to move away from the legal institution of private property towards common property (Swan 1976) which recognizes both the increasing complexity of society's needs which can be provided only in common, and the increasing externalities of human actions in one sector which affect other sectors and the environment. This is especially important in limiting the private rights of the individual in terms of the nuisance, the air, water and noise pollution, and the environmental degradation that should be tolerated. As the complexity of society has grown, so the need has increased for

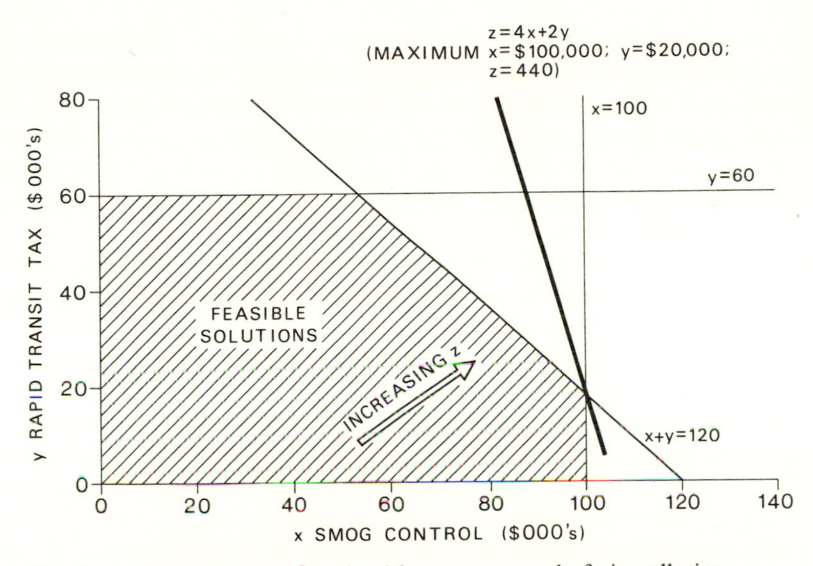

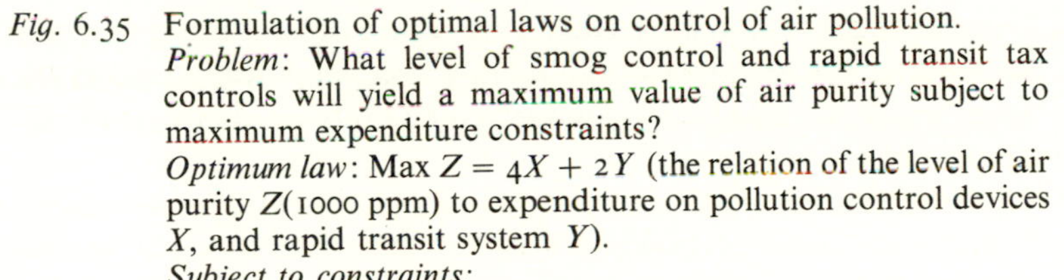

Fig. 6.35 Formulation of optimal laws on control of air pollution.
Problem: What level of smog control and rapid transit tax controls will yield a maximum value of air purity subject to maximum expenditure constraints?
Optimum law: Max $Z = 4X + 2Y$ (the relation of the level of air purity Z(1000 ppm) to expenditure on pollution control devices X, and rapid transit system Y).
Subject to constraints:

$X \leqslant$ 100,000 (max. cost of smog control \$100,000)
$Y \leqslant$ 60,000 (max. tax for rapid transit \$60,000)
$X + Y \leqslant$ 120,000 (max. cost of both controls \$120,000)

(over unit planning horizon)
Solution: Subject to the stated assumptions, the legal limit on the air pollution considered should be 660 ppm, which is obtainable by spending \$100,000 in smog control devices and \$20,000 on rapid transit. Any other expenditure pattern gives a lower level of air purity. (After Vanyo 1971.)

regulation of the rights of the individual and their control to improve the general common good or social benefit.

6.6 Spatial structure of decision making systems

The administrative, sociopolitical and legal components of decision making systems have important spatial implications which are often crucial in determining organizational, representational or *vires* structures. These depend upon the degree of spatial interlinkage and interdependence. Three forms of structuring of decision units attempt to link spatial separation and hierarchy with maximum coordination, dependent upon the degree and character of interaction between the subcategories as shown in fig. 6.36:

1. *Separatist decision units*: Here complete sovereignty in all matters resides in each decision making system. Each will have its own representation and legal framework and have the ability to create and break inter-unit linkages as desired, but subject to its bargaining position. In government each unit corresponds to the nation-state linked by international alliances, trade agreements, etc., with separate internal economic and environmental regulators.
2. *Federalist decision units*: Here complete sovereignty is divided between the higher and lower spatial level decision making systems according to purpose or function. Each has complete autonomy in its sectors of responsibility and neither is subordinate to the other. This is a mixture of the purpose and areal principles (Gulick and Urwick 1937) with legally entrenched demarcation of discretion. In government the higher-level spatial units correspond with federal congresses or parliaments, and the lower-level units with state, provincial or regional assemblies, as in the case of the West German länder, the Swiss cantons, the Canadian provinces and the states of Australia and the United States. The split between legislative and executive discretion differs in each example. In the United States and Australia mainly residual power is held at the state level, whilst in Canada and Switzerland mainly residual powers are held at the federal level.
3. *Devolved decision units*: Here sovereignty in all matters is retained in the higher-level decision making systems, but discretionary powers are delegated to lower-level spatial units. A multitude of different decision making interlinkage is possible depending upon which powers (legislative, executive, judicial or administrative) are developed and what organization of linkage is imposed. Legislative devolution gives virtually individual responsibility to each lower-level unit with only residual powers left at the higher level and is akin to federation. Executive devolution allows discretion on how higher policy is implemented with 'participation and variation consistent with the general policy aims' of the higher level unit (Royal Commission on Constitution 1973, p. 279).

Administrative devolution allows for only the operation of implementation discretion at lower levels with control policy, goal definition and decision response all retained at the higher level.

The various levels of spatial structuring of decisions are a response to different possibilities of control system deviation. Mixed scanning of response to

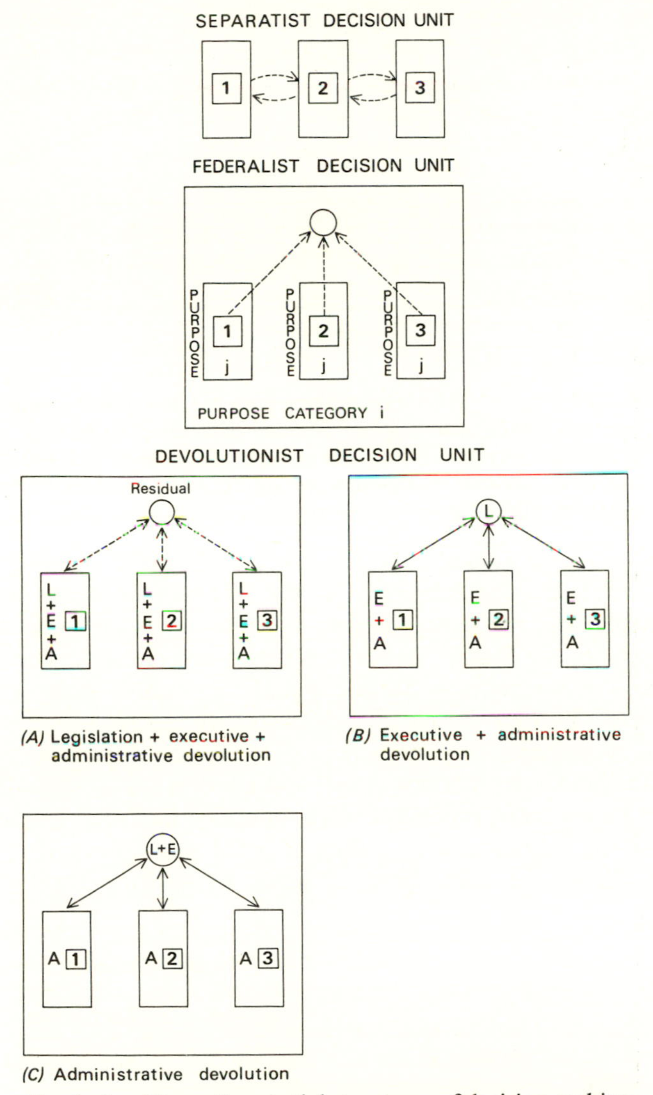

Fig. 6.36 Alternative spatial structures of decision making units (E = executive, L = legislature, A = administration) and three different decision units are considered.

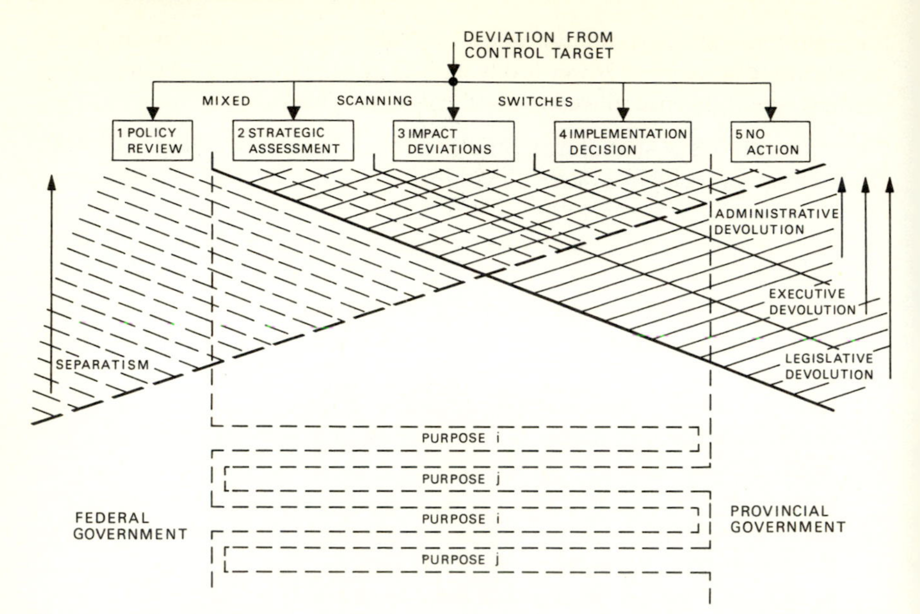

Fig. 6.37 Spatial structure of decision making units to allow mixed scanning. Various levels of mixed scanning are possible depending upon what powers are available at any spatial level. The greater the degree of discretion allowed, the higher is the level of mixed scanning decisions possible within one unit (compare fig. 6.3).

deviation from target values involves a structure of spatial scanning within the spatial decision making hierarchy at the interface between the decision making system and the spatial discretion structure, as shown in fig. 6.37.

In addition to the formal spatial structure based on successive degrees of devolution, a number of more flexible spatial organization structures can also be employed. Various forms of administrative decentralization and institutional coordinates have been incorporated into a general typology of bureaucratic models by Selznick (1949) and Laudon (1974). This depends on the spatial bureaucratic process at work (cooption and recruitment) and the content of the administrative bodies (formal and informal) (fig. 6.38). Cooption allows the adjustment of goals to environmental differences

		Adaptive processes	
		Co-option	Recruitment
Context	Formal	Notables	Reputational
	Informal	Pluralist	Collegial

Fig. 6.38 A general model of bureaucratic models. (From Laudon 1974.)

through coalition; recruitment reinforces existing individuals, élites and institutions. Formal content requires the sharing of executive and administrative burdens, goals and methods; informal content requires accommodation of local centres of influence, interest and power. The content/process interface gives rise to four bureaucratic models (fig. 6.39) (Laudon 1974):

1. *Pluralist model* (*informal cooption*): The level 1 system coopts at level 2 those local power groups which can decide the success or failure of the decision making system. This is appropriate with large numbers of heterogeneous units with strong facilitative interdependence and does least to modify existing level 2 élites, for example the TVA land grant colleges.
2. *Collegial model* (*informal recruitment*): The level 1 decision maker encourages level 2 groups to take over control aims often using level 1 'seed' money. It is least élitist and minimizes the chance of decisions taking on sectional interests. It is most appropriate when the level 2 decision units have high degree of homogeneity and are strongly interdependent with each other. Examples are the regulation of professional bodies or industries and community action programmes.
3. *Notables model* (*formal cooption*): The level 1 decision units select visible important individuals, groups or components of the level 2 system. This gives a highly élitist decision system but is useful in environments in which the social milieu is hostile and competitive at level 2. It can claim to be representative but is usually short lived. An example is the colonization process.
4. *Reputational model* (*formal recruitment*): This involves level 1 decision units selecting level 2 decision bases which eventually lose their identity as spatially disaggregated systems. Examples are provincial commissioners and agents of government. The model is most appropriate with small or competitive interdependence between level 2 decision units which are highly internally homogeneous. As time progresses the decision units become more élitist and depart further from level 2 group norms as they become more enmeshed in the level 1 decision making process.

Further issues of the size and shape of spatial administrative decision making systems are discussed by Bunge (1966) and the optimal allocation of electoral districts according to spatial control algorithms by Scott (1971).

6.7 Conclusion

The early part of this chapter has been concerned largely with normative models of decision making in which there are prespecified objectives, decision criteria and models of system behaviour. This has developed from the semi-automatic controllers discussed in chapters 3 and 4. The later part of the chapter has contrasted these normative models with descriptive forms where overlaps between decision areas, relations between components, groups, or

1 PLURALIST

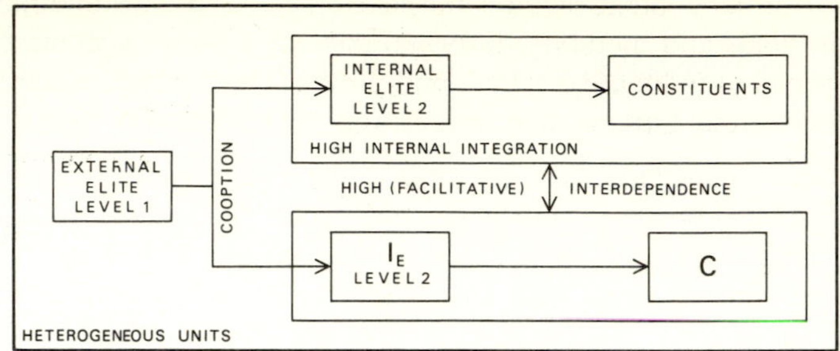

2 COLLEGIAL

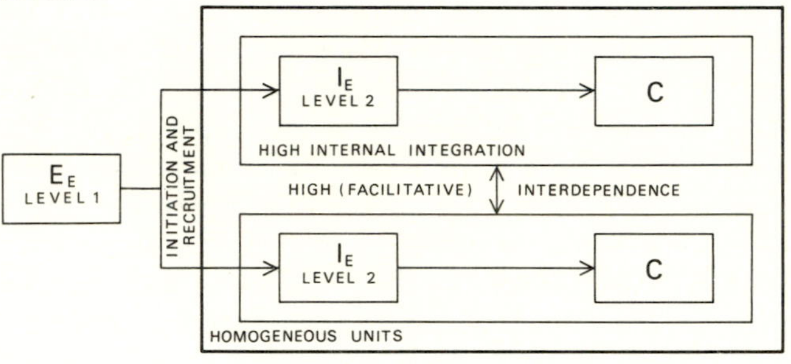

3 NOTABLES

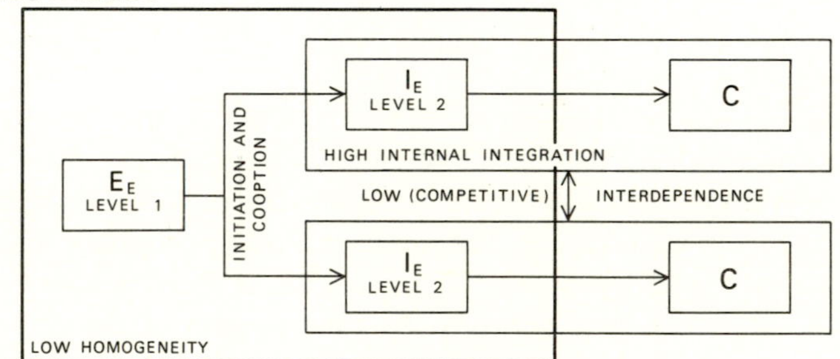

4 REPUTATIONAL

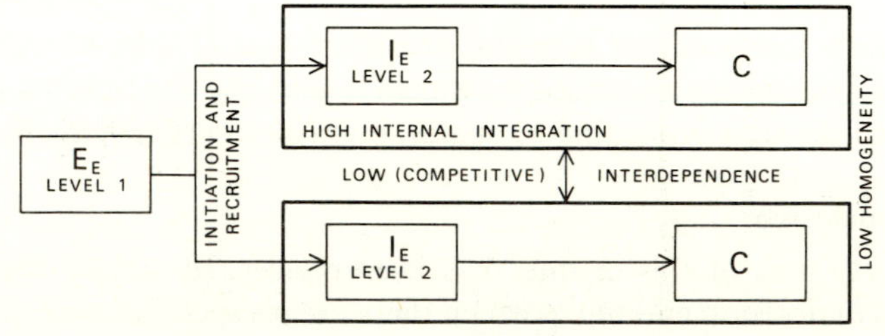

Fig. 6.39 The comparative structure of the four types of bureaucratic models. (After Laudon 1974.)

locations of part of the system are resolved by interaction, bargaining and collective action (i.e. as 'soft' control systems emphasizing man-machine interaction). Decision making based on descriptive models can easily become normative or prescriptive. This has little consequence in the physico-ecological systems discussed in chapter 7, but with socio-economic systems (see chapter 8) and man-environment systems (see chapter 9) the transition from decision making with descriptive models to prescriptive control creates important ethical problems. Burke and Heaney (1975, p. 134) note, for example, that the extension of Simon's satisficer principle to justify prescriptive forms of behaviour was straightforward: 'after all, Simon did say that people in actuality "satisfice" rather than optimize. Then why should people strive for any more?' In these terms any form of administrative or bureaucratic error can be justified. More generally, normative models indicate that it should be possible to aggregate individual system component preference functions into a single expression of collective choice which can be applied to whole systems. Moreover, decisions should be implemented with computerized information and monitoring devices since 'human beings do not have sufficient means for storing information in memory to enable them to apply the efficient strategy unless the presentation of stimuli is greatly slowed down, or unless subjects are permitted external memory aids, or both' (Simon 1969, p. 33) (see chapter 5).

The ethical dilemmas of such a course of action (developed in chapter 8) show that decision making must proceed by a mixture of normative and collective models of action. Decisions are made and cannot be avoided; often they must be made on an intuitive basis with only partial information. Fully developed monitoring and information systems are not always possible. Such decisions, and indeed decisions based upon collective action, may become normative. A decision once made may not be reversible. If a decision is not reversible subsequent actions must accommodate to it, and it is questionable if such irreversible decisions can ever be termed 'wrong'. In all real decisions normative models of control based on reason and rational design are of limited applicability on their own. In the extreme case we may take the view that a scientific approach to decision making is of limited utility since individuals are interested in the effects of a decision on their desires. 'Logical argument deals with the surface not the substance of politics. The very nature of politics is that people *don't* heed fact and logic. . . . Why? Because, the great issues arise from the stormy clash of incommensurable values, backed by the hunger for justice and self advantage . . . action, not analysis is its culmination' (Roszak 1972, p. 243). Alternatively, some actions and some environmental futures may be seen to be of such complexity that scientific analysis and rational design are unavoidable (see chapter 9). Political debate and intergroup problem-resolution by collective action may on occasion be either too slow, or may yield solutions which, although being politically and socially acceptable, are environmentally disastrous.

Part III
Complex systems

7 Physico-ecological systems

> Our experience hitherto justifies us in believing that Nature is the realization of the simplest conceivable mathematical ideas. I am convinced that we can discover, by means of purely mathematical constructions, the concepts and laws connecting them with each other, which furnish the key of the understanding of natural phenomena. Experience may *suggest* the appropriate mathematical concepts, but they most certainly *cannot* be deduced from it. *Experience remains, of course, the sole criterion of the physical utility of a mathematical construction.* But the creative principle resides in mathematics. In a certain sense, therefore, I hold it true that pure thought can grasp reality, as the ancients dreamed.
>
> It seems that the human mind has first to construct forms independently before we can find them in Nature. Kepler's marvellous achievement is a particularly fine example of the truth that knowledge cannot spring from experience alone but only from the comparison of the inventions of the intellect with observed facts.
>
> (Albert Einstein)

7.1 Mathematics and nature

The belief by Einstein in the underlying mathematical rationality of natural phenomena is not shared by most contemporary environmental systems scientists. While some environmental sciences (e.g. hydrology) are viewed, in the first analysis at any rate, as exact sciences, in general it is believed that natural processes are immensely complex and that the role of mathematics, however indispensable, resides in providing either elegant, but blind, working approximations of observed input/output relationships, or a basis for the creation of individual system linkages, the material interaction of which can only be idiosyncratically treated by some computer-based accounting procedure. There can be no doubt that our attitudes towards the control of environmental systems are strongly conditioned by our view of the essential structure of these systems. On the one hand, the input/output preoccupation of systems analysis prompts one to predict output from input such that it is possible to forecast system behaviour in a manner which encourages the planning of corrective or regulatory measures to accommodate the output after it has occurred. On the other hand, where the synthetic representation of the system has been modelled one is encouraged to identify those parts of the system where prior intervention might affect the system performance before it occurs. Obviously the above distinction between the use of mathematics to model physico-ecological systems in an overall, analytical mathematical or in an empirical synthetic manner is certainly not clear cut. In any event, complete reliance on mathematical models is based on the underlying belief that the real world operates with the same logic that human beings think in (Kowal 1971, pp. 124–5), and this, of course, is the most striking assumption made by Einstein, and one which united him more securely with nineteenth century scientific philosophy than with that of the twentieth.

There are three, related, classes of attributes of control systems which present problems in their modelling; nonstationarity (or time variance), uncertainty (or their stochastic nature) and nonlinearity (Amorocho 1967, p. 862, 1969, p. 420 and 1973, p. 204). The nonstationarity of physico-ecological systems arises from their inherent nature as cascading subsystems containing lagged storages, which may vary significantly with time. In the basin hydrological system soil moisture is a critical storage (see later in this chapter) and, in this respect, an unvegetated shale badland catchment with a thin soil cover is easier to model than a vegetated humid one. An analogous storage in the ecosystem is the nutrient pool represented by the soil and litter cover. In tropical rainforests this particular nutrient pool is small, compared with that of most other ecosystems, because the majority of the nutrients are recycling in other parts of food webs at a rate three to four times that in other ecosystems. The varying state of these storages, together with other varying system conditions, means that the form of the output depends, to a greater or lesser extent, on the particular moment in time that the input is applied, as well as on the input itself (Amorocho and Orlob 1961, pp. 15–16). Besides the time variance deriving from the varying processing of inputs in respect of the storage states, nonstationarity of physico-ecological systems can result from other controls operating in both the short and long term. Man-made changes, particularly in system storages (e.g. channel, sediment, soil and vegetational), are a prime mechanism of such invariance in that they cause systems to change with time relative to their past performances and, thus, human action itself is a major impediment to attempts at the simple mathematical modelling of such systems. Another set of factors disturbing the time invariance of systems comprises the slower natural changes of the structural state of systems, of which examples are climatic changes, the progressive weathering of surface material and natural plant succession. Since nearly all environmental parameters shift constantly but at variable rates, magnitudes and frequencies, ecosystems and individual organisms must track and adapt to these changes in order to survive. The frequency-magnitude range of biological and sociological responses to the spectrum of environmental disturbances and parameter shifts can be depicted as a multilevel tracking problem (fig. 7.1) (Wilson 1975) Organismic responses are stimulated by changes within one life span, and are developmental and stabilizing control systems. At higher levels and at faster

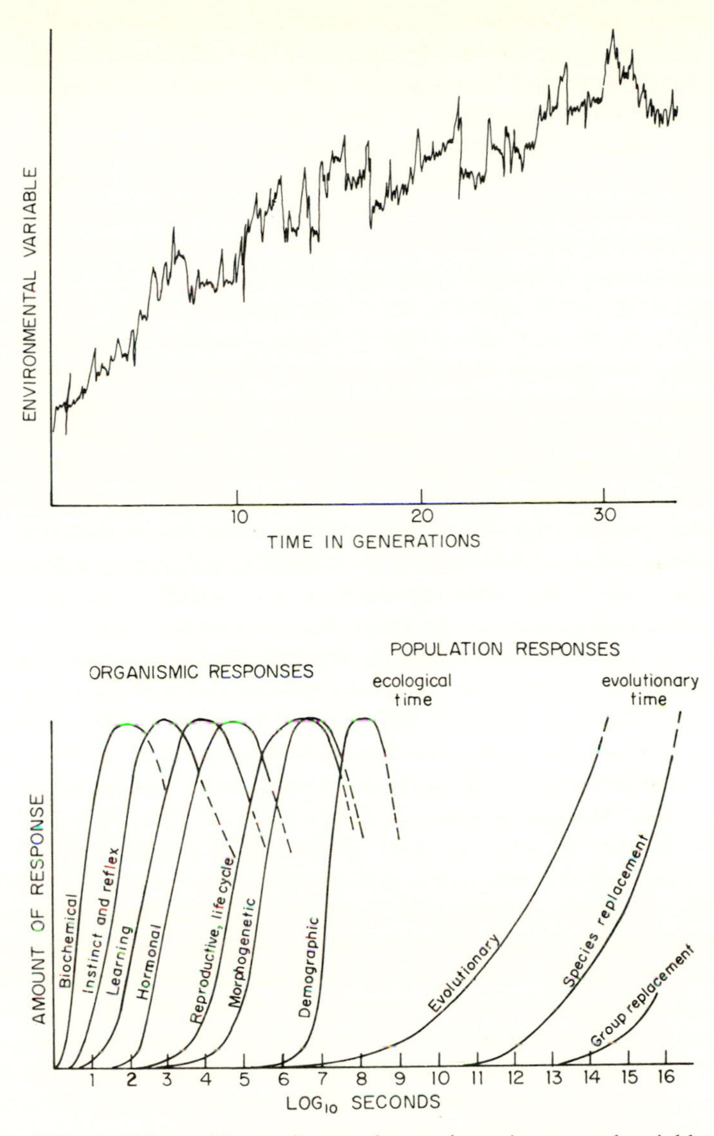

Fig. 7.1 Above: Nonstationary changes in environmental variables involving changes to the steady state level of response of ecosystem organization. Below: Multilevel hierarchy of biological and sociological responses to environmental changes at various intensities with geometrically increasing response time for changes of increasing magnitude. (From Wilson 1975.)

rates of change in system storages, sociological, territorial and evolutionary changes are necessary for the maintenance of a population; whereas high magnitude or rapid environmental changes will lead to the annihilation of the population and its replacement by other species (Wilson 1975, pp. 144–75). Operating within a wide range of temporal scales are disturbances to do with the regime of inputs – for example, those associated with climatic changes involving changes of solar energy inputs into ecosystems and of precipitation into hydrological systems. The complex interrelation between temporal scale and probability of such variations causes them to overlap with considerations of uncertainty. Some systems are more protected from these last non-stationarity effects than others and, for example, tropical forest ecosystems (particularly those of the forest floor) are protected by the buffering effect of their vegetational layering from certain effects of climatic changes (see chapter 9). All models which depend on historical records for determining the values of their parameters assume time invariance or stationarity (Amorocho and Hart 1964, p. 309) but, in reality, only relatively small systems, considered over short periods, may be viewed as time invariant in that they can be expected to produce approximately the same output functions when subjected to the same input functions, regardless of time (Jacoby 1966, p. 4811).

A second impediment to the successful modelling of physico-ecological systems derives from their uncertainty. Environmental scientists broadly have two attitudes to uncertainty of natural processes, either that the world is basically stochastic in character and must be modelled in terms of stochastic systems, or that stochastic modelling is a necessary evil in the absence of present understanding of the underlying order, albeit complex, but that, as understanding increases, the success of dominantly deterministic methods will improve (Eagleson 1969, p. 400). Uncertainty arises in terms of the magnitudes and the space/time (i.e. areal location/temporal frequency) attributes of inputs, outputs and states of the subsystems (e.g. storages) (Amorocho 1973, p. 204). Uncertainty of the magnitudes and temporal frequency of natural processes increases as the variability of the subsidiary inputs, throughputs and outputs (e.g. evaporation in the hydrological cycle) increases relative to that of the main input (e.g. precipitation). All physico-ecological systems have areal extent and thus uncertainty also resides in the spatial location of inputs and outputs. The normal reaction to uncertainty is, firstly, to assume the existence of both temporal invariance (which also subsumes nonstationarity) and spatial invariance by the 'lumping' of components (Jacoby 1966, p. 4811, Amorocho 1973, pp. 242–3). In this way temporal variations may be rationalized by averaging, smoothing or trend removal in some manner (e.g. by polynomial detrending, harmonic and Fourier analysis). Each method assumes the constancy of the input-output relation over time (i.e. the system parameters are not adaptive, as discussed in chapter 2). Spatial inputs may be similarly treated by spatial averaging (e.g. as in precipitation by means of Thiessen polygons or by attempts to relate storm sizes to average rainfall intensities to obtain spatial mean intensities) (Amorocho 1973, pp. 242–3), or by spatial smoothing (e.g. by trend surface, spatial moving-averages or double Fourier methods). Spatial lumping subsumes specific and unique locational effects and averages cross-spatial interactions as overall system attributes (see chapter 4). Temporal and spatial lumping remove the system model at least one stage from the system of interest. Thus it becomes difficult, if not impossible, to interpret the resulting structure and/or parameters in terms of physical attributes. Hence we are often forced back upon justifying the model as a mathematical approximant of reality (as, for example, with the Padé approximants discussed in chapter 2). The second reaction to uncertainty is to construct entirely stochastic systems which are geared to treat temporal and spatial variations by Markov, Monte Carlo, or similar autoregressive and moving-average techniques. If the state of the system at any point in time or space is affected by its state at some previous time or proximate space, the system is said to be dynamic and to be state determined with a finite memory (for example, the state space equations discussed in chapter 2). With simple linear systems relationships between lumped components can be modelled by ordinary differential- or difference-equations (depending on whether the temporal or spatial processes are being continuously or discretely monitored), whereas those relationships between unlumped components can be treated with difficulty by the use of partial differential- or difference-equations (Kowal 1971, p. 129) or by matrix polynomials. However, when the systems are nonlinear other approaches are necessary, and, unfortunately, the majority of physico-ecological systems fall into this category.

The energy and mass transfers in most natural systems are complex and nonlinear (Amorocho 1967, p. 862, 1969, p. 421, 1973, p. 204). Nonlinearity implies that some or all of the mathematical parameters employed in the description of the system operation are not identical over all magnitudes and regimes of system operation – in other words, that they are free, within certain limits, to change their values in harmony with changes in the system variables. As a consequence, nonlinear systems cannot be accurately described by linear algebraic or linear differential-equations. Thus, in most situations nonlinear systems will not satisfy the principle of superimposition (i.e. the addition or subtraction of subsystem outputs to produce a total system response), will not exhibit simple proportionality between the amplitudes of input and output, and will generate in the output not only the frequencies contained in the input but new frequencies related to these (Kisiel 1969, p. 8). Whereas linear systems possess single, unique steady states, nonlinear systems exhibit multiple steady states which makes their behaviour indeterminate and, to a degree, unpredictable. Each steady state regime corresponds to a different response to system initial and boundary conditions (the stimuli for system change) and must be approached from global considerations which permit the manifold of variable system dynamics and thresholds to be encompassed (see chapter 4).

Langbein and Leopold (1964) have shown this to be the case in respect of the hydraulic geometry system of natural rivers (Chorley and Kennedy 1971, ch. 6). Also important from the point of view of possible control, especially of ecosystems, is that, whereas linear systems are unambiguously either stable or unstable, nonlinear systems are extremely ambiguous in their behaviour (Patten *et al.* 1975, p. 217). The nonlinearity of physico-ecological systems derives from the time variance of the system state, the spatial variance of inputs and transfers, and from the thresholds and discontinuities which produce sharp disjunctions in the behaviour of the subsystems (Amorocho 1973, pp. 209–10, Patten *et al.* 1975, pp. 215–16). However, the most obvious source of nonlinearity is the nonstationarity of the storages, in that they operate relatively slowly and their rate of intake depends not only on the rate of supply but also on the state of the storage (Dooge 1968, p. 67). In hydrological systems, for example, a major source of nonlinearity is the behaviour of soil moisture storage and the decay of infiltration capacity with moisture content. In ecosystems, too, nonlinearity results from the fact that rates of mass or energy transfer (fluxes, defined by non-dynamic state variables) between subsystems (storage compartments – e.g. biomass – defined by dynamic state variables) are dependent upon the simultaneous effects of both the 'donor control' (e.g. resources or food limited) and of 'recipient control' (e.g. predator limited) (Kowal 1971, p. 140, Patten *et al.* 1975, p. 218). Fig. 7.2 illustrates the nonlinearity of simple hydrological systems. Fig. 7.2A shows the apparently simple pattern of total runoff from small experimental plots subjected to an artificial rainfall pulse sequence of 1·55 inches/hour for 64 minutes, 3·30 inches/hour for 34 minutes and 1·55 inches/hour for 32 minutes. However, a detailed examination of an even simpler situation in fig. 7.2B reveals the basic nonlinear behaviour of runoff from a 9·9 × 120 cm metal box containing a 10 mm layer of crushed diorite gravel (Amorocho and Orlob 1961, pp. 55–7). The threefold artificial rainfall pulse sequence (1, 2, 3) produced curve *a*; curve *b* was generated by a single rainfall pulse of 3·46 cc/sec, applied for 161 seconds (a volumetric input equal to *a*), which overestimated the peak discharge and gave poor shape reproduction; curve *c* involved the superimposition of the upper part of curve *b* at the point where pulse a_2 began; and curve *d* is the linear superimposition of pulses 1, 2 and 3. The nonlinear nature of this extremely simple system is clearly demonstrated by the discrepancy between curves *d* and *a*, the former representing output shape but grossly underestimating peak discharge. Ironically it is in the estimation of maxima that systems methods often find their greatest potential utility.

The complexity of natural systems has led some environmental scientists to adopt two distinct approaches to their modelling, systems analysis and systems synthesis (e.g. Amorocho and Hart 1964).

Systems analysis involves the development of expedient mathematical functions which approximate to the overall operation of the system without

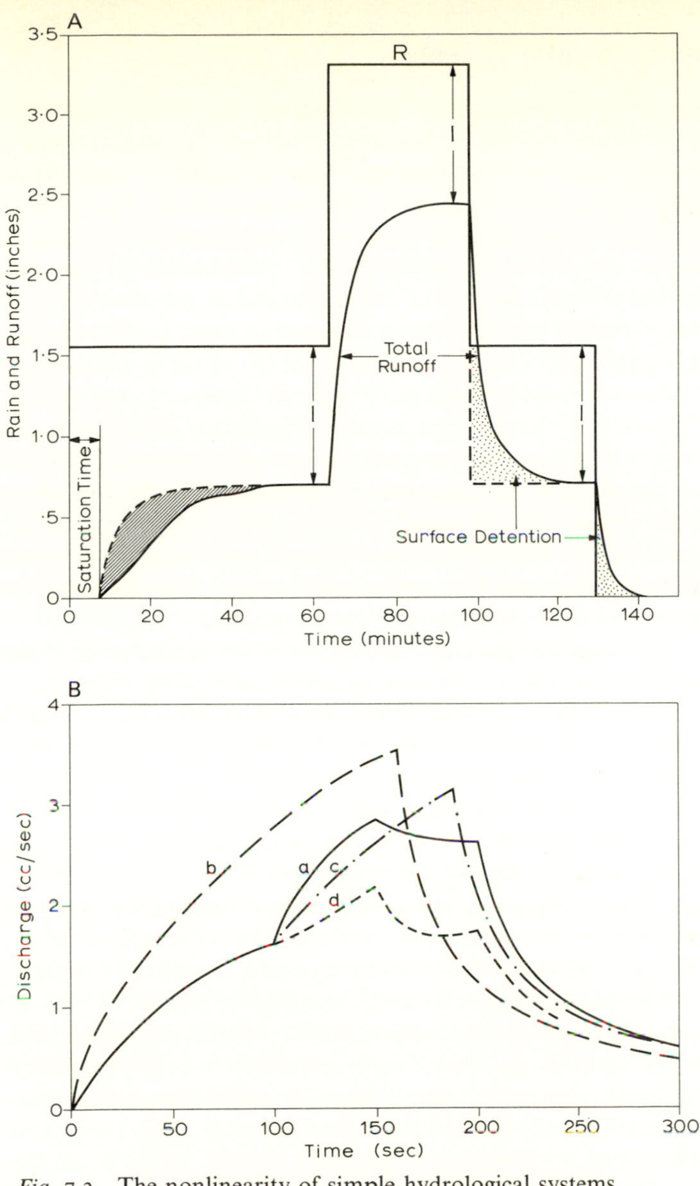

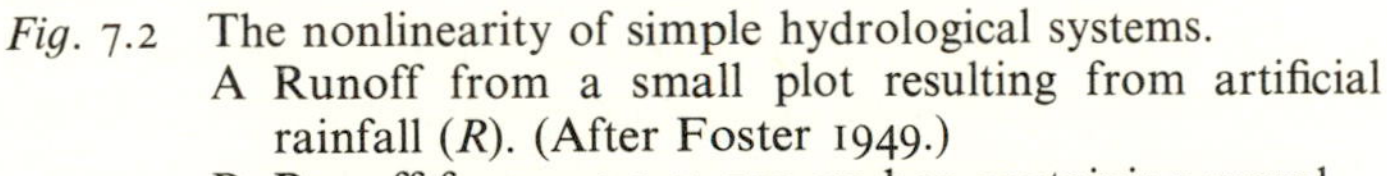

Fig. 7.2 The nonlinearity of simple hydrological systems.

A Runoff from a small plot resulting from artificial rainfall (*R*). (After Foster 1949.)

B Runoff from a 9·9 × 120 cm box containing gravel.
Pulse sequence: 1 + 2 + 3.
1 = 100 sec at 2·56 cc/sec
2 = 50 sec at 3·46 cc/sec
3 = 50 sec at 2·56 cc/sec
Curve *a* = Actual measured response.
Curve *b* = Response to single pulse at 3·46 cc/sec.
Curve *c* = Upper curve *b* shifted.
Curve *d* = Linear superposition of single pulses.
(After Amorocho and Orlob 1961.)

the explicit use of the physical laws involved. It is thus termed the 'black box' approach. The aim of the analytical system is to produce a satisfactory reproduction of the overall system operation by means of a model which is dictated by mathematical expediency, rather than one which models in detail the operations of the real system. Thus no correspondence can be assumed to exist between the individual mathematical components of the analytical system and the physical elements of the real world (Amorocho 1973, pp. 207–8, 213). Such systems relate real-world output to input by the application of expedient, established mathematical models (e.g. those based on time-series analysis, Markov methods, Monte Carlo modelling, curve fitting, etc.), usually involving the parallel arrangement of arbitrary subsystems of varying order and linearity, which have no concern for the internal relationships of the system, no physical meaning in terms of the processes involved, and do not fulfil dimensional consistency (Amorocho and Hart 1964, p. 311). The aim of the systems analyst is to take observed data and produce efficient estimates of the spectral representations (i.e. variance spectra) or of the moment functions (correlograms) of the process (Kisiel 1969, pp. 9–10).

Systems synthesis, on the other hand, relies on inductive and deductive reasoning, based largely on a study of recorded data, to develop a conceptual model of the operation of natural systems in terms of a linkage or combination of components whose presence is assumed to exist in the system and whose functions are known and predictable. In this way, a 'white box' system is constructed in terms of empirical mathematical relationships linking the dynamic state variables expressing the storage condition of the subsystems or compartments and the non-dynamic state variables, or flows between the subsystems, to produce a mathematical model of the total system based on differential- or difference-equations, the parameters of which are estimated to yield satisfactory relationships between measured inputs and outputs (non-dynamic state variables) or between acceptable values of the storages (dynamic state variables) (Amorocho and Hart 1964, p. 311, Amorocho 1973, pp. 207–8, Patten 1971, pp. 32–3). Such a system can only be handled on an accounting basis by means of high speed electronic computers, and can be used as a simulation model to predict the results of changing inputs, storages and transfers. These systems have proved extremely popular in the earth, biological and social sciences, but they possess a number of obvious disadvantages. Synthetic systems are not unique but are highly idiosyncratic; they are based on recorded time-series data commonly of a lumped variety (especially spatially), which may be short, subject to errors and unrepresentative of extreme values; their empirical character makes their application suspect outside the range of these original data; the empirical mathematical relationships are usually complex and often intractable; the model structure, as conceived, may be imperfect; and the complex and often suspect web of empirical mathematical relationships on which they rest may produce widely fluctuating predictions involving a collapse or an explosion of variables. Such catastrophic behaviour of synthetic systems may often be dependent not so much on the essential nature of the real-world system but upon the choice of parameters and other purely mathematical considerations (Amorocho and Hart 1964, pp. 311, 314, Kisiel 1969, p. 9). Later in this chapter this problem will appear in connection with ecosystem modelling and, in a more extreme form, in chapter 9 when symbiotic predictive models are treated.

7.2 Systems analysis

The most immediate reaction to the apparent complexity of real-world systems is to assume that they are simple and to treat them as if they were linear, despite the fact that the resulting systems are crude and often valid for only a limited range of the variables, including time (Kowal 1971, p. 140). A most obvious example of this has been the development of functional relationships between hydrologic input (X = precipitation) and output (Y = stream runoff), as follows (Amorocho 1973, pp. 209–10):

$$X = Y + \frac{dV}{dt} \tag{7.1}$$

where V is the instantaneous single storage volume. Let $V = KY^m$, where K is the 'routing constant', the storage delay time for a single linear reservoir and $m = 1$ (for linear systems). Therefore:

$$X = Y + K\frac{dY}{dt} \tag{7.2}$$

or

$$X = [1 + K(1/s - 1)]Y$$

where $1/s - 1$ = the Laplace transform of the differential operator (see chapter 2.2). Let us assume initial conditions of $Y = 0$ and $X = 0$ at time $t = 0$, then we may employ the methods discussed in chapter 2 (equation 2.1.5) deriving the analytical solution to the rainfall-runoff equation for any time t in the future. This solution has the general structure of a negative exponential integrated over all time from the initial conditions to the present. Therefore:

$$Y(t) = \int_0^t K^{-1} \exp[-(t - \tau)/K] \, . \, X(\tau) \, d\tau, \tag{7.3}$$

where:

$Y(t)$ = the value of the output (i.e. storm runoff volume) at time t,
τ = the time variable of integration over range from initial conditions $\tau = 0$ to present $\tau = t$.

Let

$$K^{-1} \exp[-(t-\tau)/K] = h(t-\tau)$$

Therefore

$$Y(t) = \int_0^t h(t-\tau) \cdot X(\tau)\, d\tau \tag{7.4}$$

(Dooge 1968, p. 63, Kisiel 1969, p. 57, Amorocho 1973, pp. 209–10). This is the integral function or convolution integral for a time invariant, constant-parameter system, which embodies the principle of antecedence in that past values can influence future states, where:

$h(t-\tau)$ = the unit impulse response function, or 'kernel', lagged in time by τ,
$X(t)$ = the lagged input (i.e. effective precipitation) that is acted upon by the kernel.

It is clear that, once the kernel function is known for any watershed, the direct runoff (Y) can be calculated from any rainfall excess input (X) (Delleur and Rao 1971, p. 116).

For the instantaneous unit hydrograph (i.e. the runoff related to a rainfall input approaching zero duration) the kernel function is $h(t)(=K^{-1} \exp[-t/K])$. Where more than one linear storage is assumed (Nash 1957, see also Dooge 1959):

$$h(t) = [K\Gamma(n)]^{-1} . e^{-t/K} . (t/K)^{n-1}, \tag{7.5}$$

where:

$\Gamma(n)$ = the gamma function,
K = the average storage delay time for each single linear reservoir,
and n = the number of linear reservoirs.

This deterministic function can be translated into stochastic terms either by introducing a random operator to vary K and n in accordance with some probability law (Kisiel 1969, p. 59) or by considering only the moments (means, variances, autocovariance, etc.) of the system input and output. In the case in which K and n are stochastically varying we have an adaptive system model with stochastic parameters which reproduces variable initial and throughput storages, but when we describe system variables by their moments we employ the principle of 'certainty equivalence' of setting the system outputs equal to their mean expectation (chapters 2 and 3) which ignores variable initial and throughput effects, so concentrating attention on the general rather than the variable character of hydrological response.

The unit hydrograph model (i.e. the theoretical discharge resulting from a uniform rainfall input into a small catchment) is based on the assumption that storm discharge from a stream basin is linear and that its predictive power depends on the antecedent input. If this assumption of linearity between rainfall input and runoff output were true, however, time of flow from a given point on the catchment to the outflow point should remain constant for all discharges and under all conditions. Clearly the impulse response functions will vary throughout protracted inputs (Amorocho 1963, p. 2248), and fig. 7.3 shows the highly variable relation of travel time to discharge of labelled water between the stream head and the gauging station (2500 feet apart) in a small catchment (Pilgrim 1966, p. 319). The limited success which the unit hydrograph has achieved is due to the fact that under certain conditions its

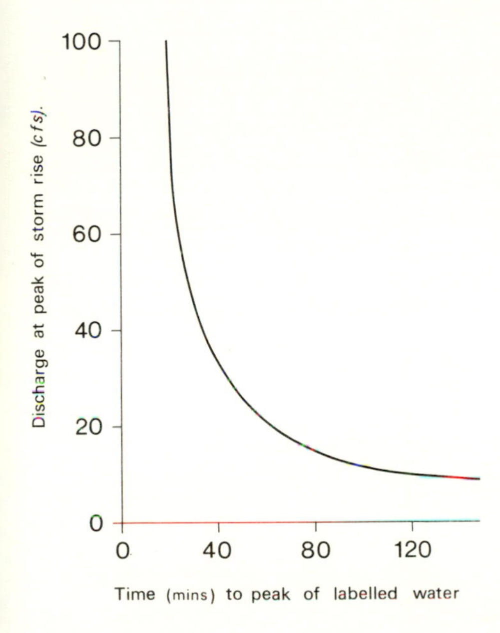

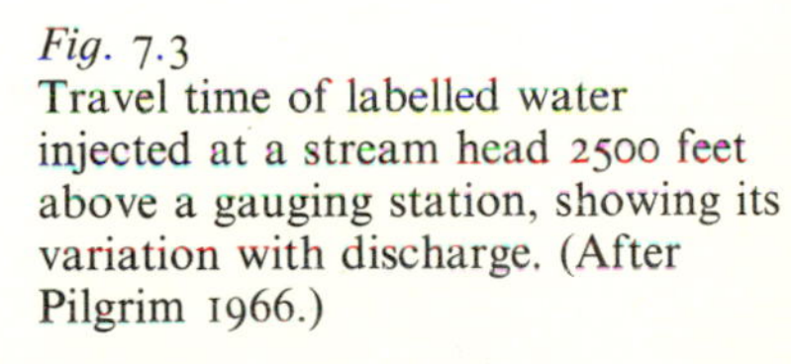
Fig. 7.3
Travel time of labelled water injected at a stream head 2500 feet above a gauging station, showing its variation with discharge. (After Pilgrim 1966.)

linear assumptions of lumping, time invariance, consistency of storage operation, etc., do approximately hold (Jacoby 1966, p. 4811) and fig. 7.2 indicates that watershed behaviour is more linear at high discharges, to which the unit hydrograph has been particularly applied (Pilgrim 1966, p. 317, see also Pilgrim 1976).

The nonlinearity of environmental systems, for example where $m \neq 1$ in the assumptions leading to the equation (7.2) above, results in nonlinear integral functions which do not have a closed solution which can be derived from simple integration. In his quest for greater predictive power, the systems analyst has a number of courses open to him which were explored by electronic engineers during the period 1942–59. These generally involve the development of different kernels (Amorocho 1973, pp. 209, 211) usually by decomposing the natural system into a parallel number of linear subsystems of

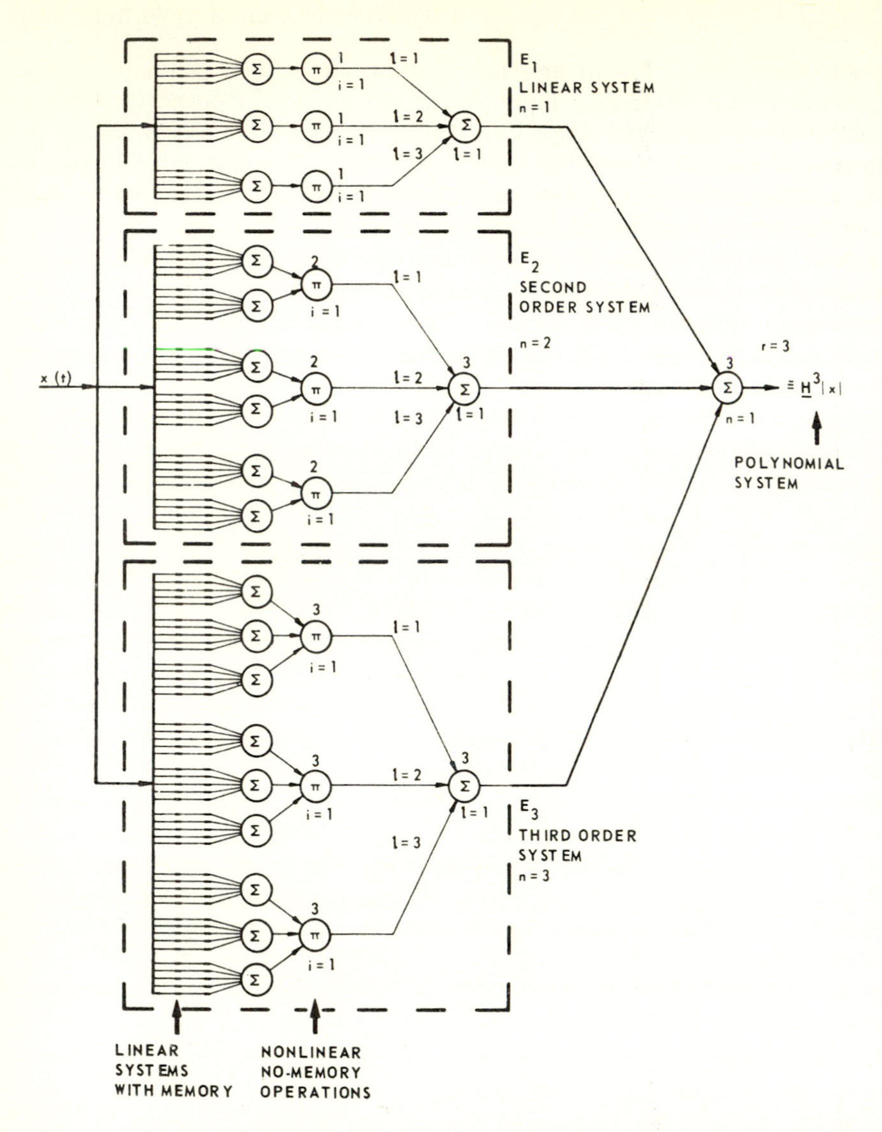

Fig. 7.4 A general model for systems with unrestricted arbitrary inputs, developed by Jacoby (1966), based on the decomposition of the system into basic linear (time lag) systems with memory and nonlinear (no time lag) no-memory systems. The figure expresses a complex cascade equivalence of a third-order polynomial system, of the following relationship:

$$H[x] \simeq \sum_{n=1}^{r} \left[\sum_{l=1}^{m_n} \prod_{i=1}^{n} \left(\sum_{m=0}^{\infty} b^*_{l,i,m} l_m[x'] \right) \right]$$

where l_m denotes the Laguerre filter (i.e. a linear system with memory of a special class).

Networks of Laguerre systems cascade into first-order, no-memory progressively higher degree or order (i.e. polynomial (Padé) series or Taylor series and other series expressions) cascading into a zero-memory nonlinear output (Amorocho 1967, Kisiel 1969, pp. 384–5). Thus the output of a nonlinear system can be approximated to any desired accuracy by combining a finite number (n) of linear subsystems forming a convergent series in which the contribution of each term (subsystem) to the total output gets progressively smaller and yields a more progressively nonlinear result (Amorocho and Orlob 1961, p. 8). A simple example of the nonlinear integral equation kernels necessary to reproduce hydrological response has been given in chapter 2 (fig. 2.11, p. 40). The approximation of nonlinear time lag hydrological systems by a mathematical decomposition model in which the properties of nonlinearity and time lag are separated by the use of such a parallel arrangement (summation) of n polynomial subsystems of order 1 to n (i.e. a polynomial nth-order system) is shown for a simple example in fig. 7.4 (Amorocho 1967, 1969, pp. 448–61, 1973, p. 212, Jacoby 1966). The nonlinear cascade shown in this figure is a Laguerre expansion which is a form related to the Padé series discussed in chapter 2. In such a series the first term is a linear, first-order system analogous to the ordinary convolution integral (equation 7.4) which expresses the unit hydrograph, the second term is a quadratic term, and so on (Amorocho 1973, p. 212). The linearity assumption is tantamount to truncating such an infinite series after the first term, although the whole series has terms involving the input which converge on the true output. Nonlinear hydrological systems can be approximated by such a truncated series but more than one term must be retained (Jacoby 1966, p. 4811). Similar linear approximations have been proposed for many other environmental systems such as air pollution and river channel morphometry (Bennett *et al.* 1976, Bennett 1976), BOD levels in rivers (Wastler 1969) and climatic regimes (Bryson 1974). A short memory can be built into such models by cascading the output into another parallel set of linear systems of increasing order, each possessing a time lag (Jacoby 1966, p. 4814). Such approximations of nonlinear systems naturally assume (Amorocho 1973, pp. 209–10):

1 That inputs and outputs are defined *a priori* in a lumped manner. For example, that inputs, although in reality varying in space as well as in time, are equivalent to inputs which are functions of time only.
2 That the system is, has been, and will be time invariant.
3 That, relative to the time variability of the main input (e.g. precipitation), the variability of the other natural inputs and outputs (e.g. transpiration) is small.

---

systems (a description of this model is given by Amorocho (1969, pp. 459–61) and a discussion of the role of such series expansions in system approximation is given in chapter 2). (From Amorocho 1969.)

Stochastic methods can be adapted to the construction of such analytical systems. We then have a problem in inference: from system knowns to estimates of system unknowns on the basis of statistical confidence criteria (see chapter 2, table 2.2). In most cases it is the system transfer function and storage terms that are not known. Stochastic methods of estimating these elements require either the comparison of statistical estimates of impulse and frequency response functions against theoretical structures (as in figs 2.42 to 2.45) by some form of eyeballing or curve fitting, or the use of sequential estimation methods (e.g.: stepwise regression; canonical, principal component and factor analysis; iterative estimation; and spectral factorization) (Box and Jenkins 1970, Jenkins and Watts 1968, Bartlett 1966, Bennett 1978). These statistical diagnostics and estimators can incorporate nonstationary and nonlinear responses, but most models of such learning and adaptation in physical systems are almost invariably black box since the historical record is so limited. Hence Lamb (1966) and Bryson (1974) have constructed models of long-term climatic shifts which are essentially trend extrapolations for which there is little basis in explanatory mechanisms (but see Rosenberg and Coleman 1974, who attribute changes to latitudinal oscillations in the jet stream). On the basis of these models Lamb (1966) and Winstanley (1973) have argued that some zones are undergoing long-term shifts in climate, e.g. the Sahel. Hence the human response in terms of land use should also be modified. In a detailed study in Arizona, Cooke and Reeves (1976) have shown significant shifts to occur in rainfall frequency at different magnitudes even when overall rainfall totals do not change significantly (fig. 7.5) and these magnitude-frequency shifts may have had important effects on arroyo initiation and development. For example, the slight decrease in summer precipitation, the greater decrease in winter precipitation, together with slight increases in temperature, and the relatively low frequency of both summer and non-summer light rains, in the late nineteenth century point to a period of greater aridity with less vegetation at that time.

Attempts to introduce information on the dynamics controlling shifts in environmental system response for both deterministic and stochastic models can be based on the Kalman filter and Luenberger observer (Kalman 1960, Luenberger 1964), as discussed in chapters 2 and 3. These provide the most flexible method of approximating the high-variety response of physico-ecological systems. Although each estimation is constrained by its efficiency, consistency and bias properties (chapter 2) the interfacing of these adaptive estimators with global response functions over the infinite range of system state initial conditions allows very flexible and elegant mathematical solutions (Thom 1975, Smale 1974) which permit continuous evolution or discontinuous 'flipping' between different regimes of system behaviour. It seems likely that it is in this area that future advances in systems analysis will be made (Harrison and Stevens 1971, Bennett 1978, ch. 5).

Application of global models to the evaluation of variable regimes of

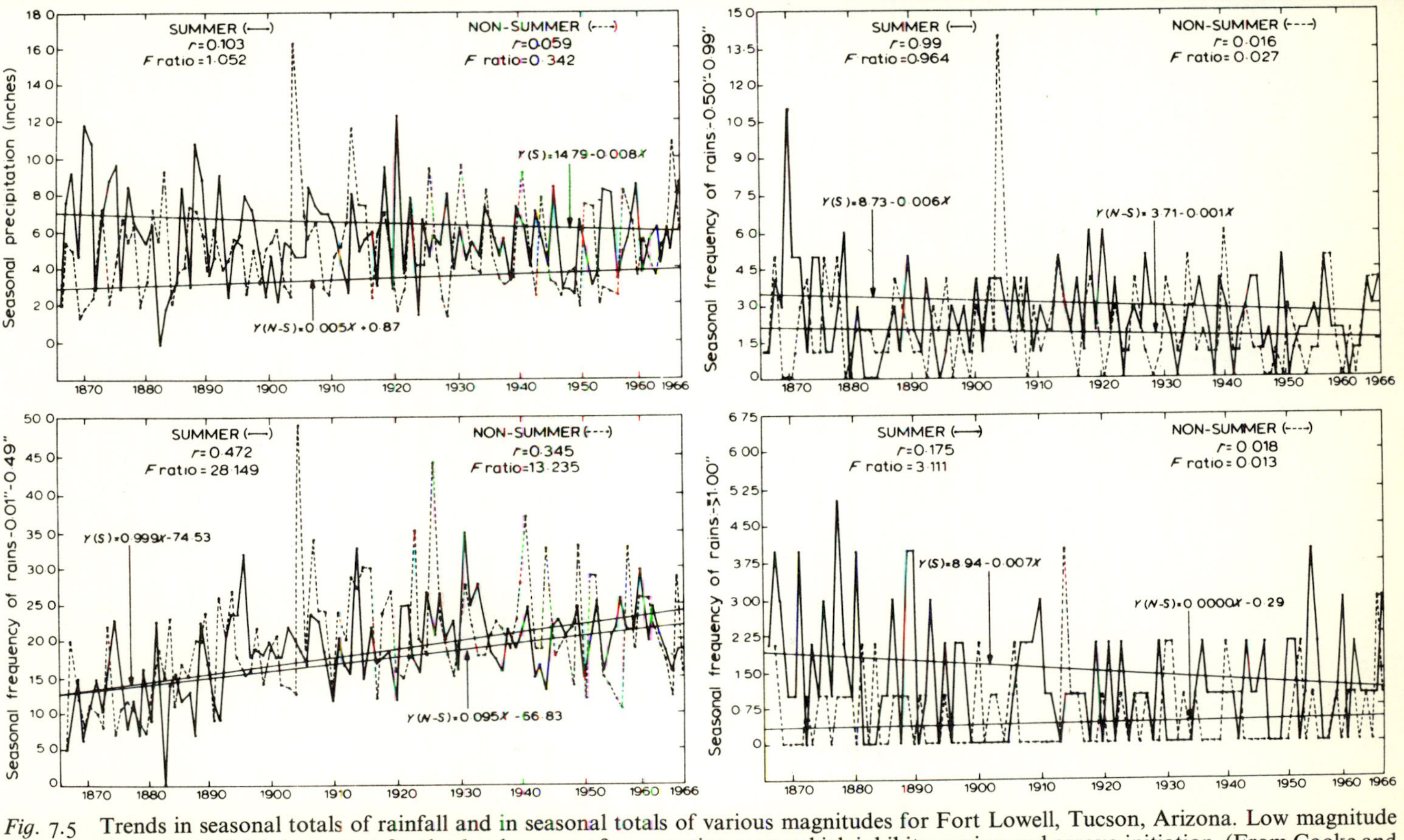

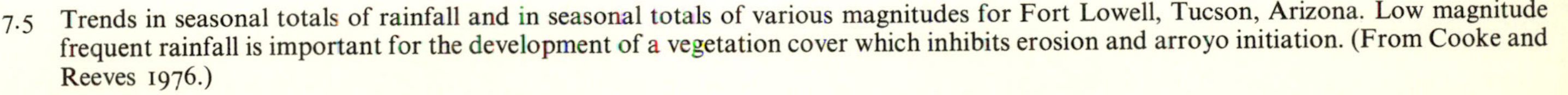

Fig. 7·5 Trends in seasonal totals of rainfall and in seasonal totals of various magnitudes for Fort Lowell, Tucson, Arizona. Low magnitude frequent rainfall is important for the development of a vegetation cover which inhibits erosion and arroyo initiation. (From Cooke and Reeves 1976.)

environmental system behaviour using the Kalman filter and other parameter evolution models has been limited. Young and Beck (1974) have applied an extended Kalman filter model to water-quality control in the Cam, and Bennett *et al.* (1976) have analysed the changing persistence pattern in pollution concentration of SO_2 and smoke in London. For SO_2 this study (fig. 7.6) demonstrates that reduced levels of pollution emission may be linked in nonlinear fashion through complex photochemical processes, for example reactions of SO_2 with metallic oxides giving oxidation to SO_3, with water to give an H_2SO_4 aerosol mist and with the NO_2-olefin mixture of car exhausts to further increase aerosol formation. The nonlinear response characterizes different time periods in fig. 7.6B and is expressed in modified autocorrelation and spectral functions with increased annual and monthly cyclic components. Fitting a black box second-order autoregression model to the SO_2 time series

$$Y_t = a_1 Y_{t-1} + a_2 Y_{t-2} + e_t \tag{7.6}$$

and tracking changes in the two autoregressive parameters a_1 and a_2 over time results in the identification of two components of non-linear variation. These components are, firstly, random and step variation in the feedback terms at different times which cannot be accounted for within the black box structure, and, secondly, a slow drift in parameter values to larger negative values. Hence fig. 7.6C represents a parameter map of variable pollution response under a range of nonlinear, nonstationary and stochastic regimes.

A similar study by Bennett (1976) of the control of sediment transport equations on river channel morphology used Yatsu's (1955) spatial series of height loss along the course of a number of Japanese rivers. Yatsu found that sediment in the rivers was divided into two regimes; an upper reach with predominant gravel fractions and a lower reach with sand fractions, and that this influenced overall bed slope. Estimates of the autocorrelation, partial autocorrelation and spectrum of height loss along one of these rivers (figs 7.7A and 7.7B) show clearly a first-order moving-average process to be an appropriate black box description (compare fig. 7.7B with fig. 2.42, p. 82), i.e.

$$\nabla H_s = d_1 e_{s-1} + e_s. \tag{7.7}$$

Kalman filter estimates of the moving-average parameter d_1 (fig. 7.7C) confirmed Yatsu's findings that changes in sediment fraction from predominantly gravel to sands (i.e. from -6ϕ to $+1\phi$; where $\phi = \log_2$ particle diameter) are linked with important changes in river long profile and bed morphology (sinuosity).

As has been emphasized, it can be appreciated that the use of systems analysis in modelling natural systems results in a certain inevitability in one's attitude to input-output relationships and that, in these terms, environmental control consists mainly of attempting to predict outputs from possible inputs and in the taking of steps to respond to these outputs – i.e. by containing,

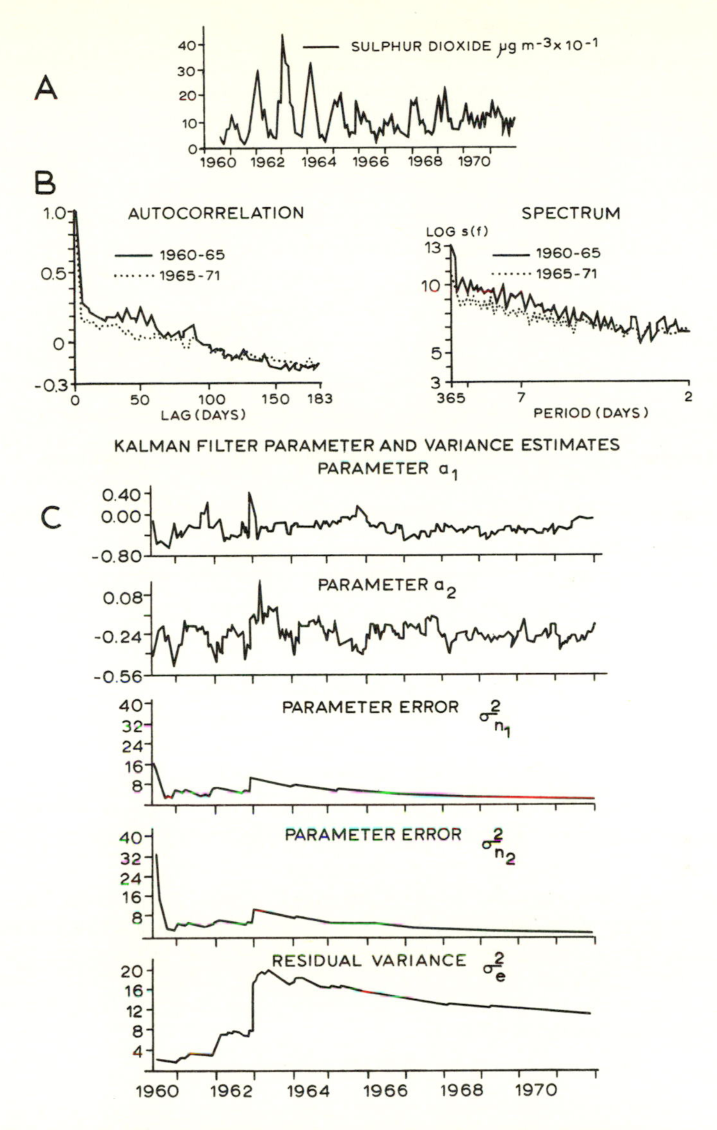

Fig. 7.6 Analysis of daily SO_2 concentrations at Kew (London) for the period 1960–71 using Kalman filter estimates. Estimates of the autoregressive model are derived from least-squares regression; estimates of the nonlinear parameter evolution model derive from the use of a Sage-Husa filter (Sage and Husa 1969, Bennett 1978), which gives the value of each parameter a_i^t at time t as

$$a_1^t = a_1^{t-1} + n_t^1 \qquad a_2^t = a_2^{t-1} + n_t^2$$

(From Bennett, Campbell and Maughan 1976.)

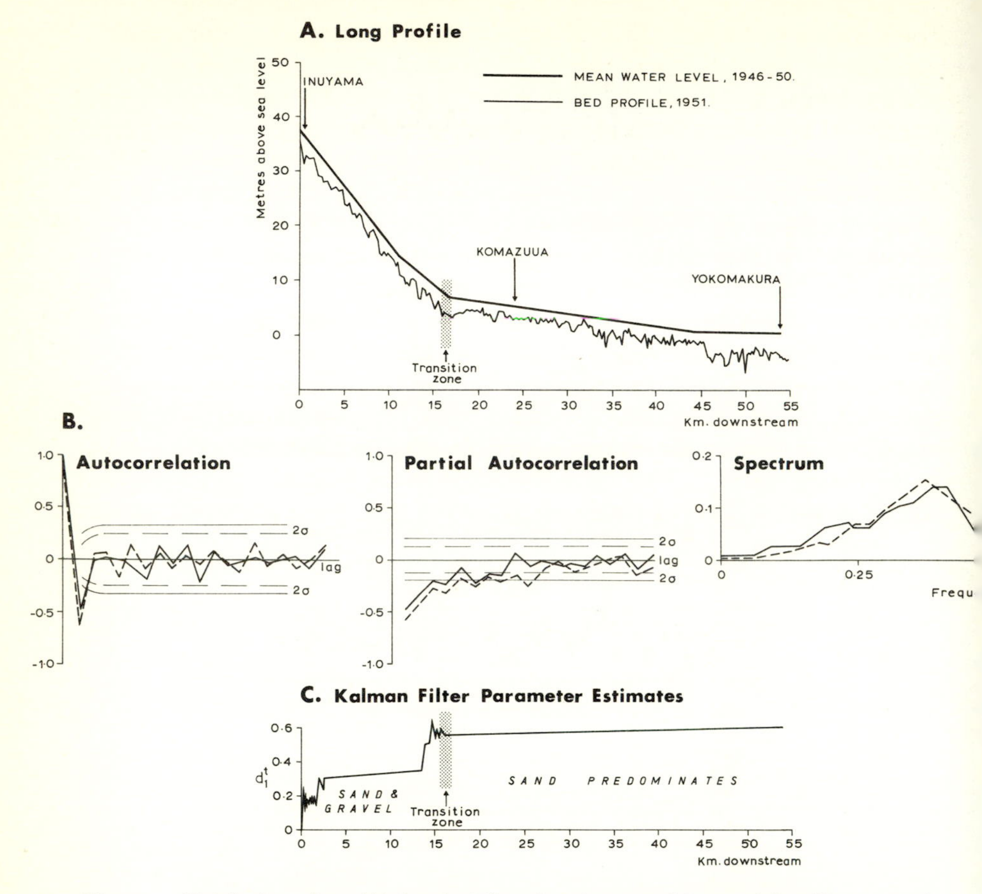

Fig. 7.7 Height-loss data (A) for the Kiso river in central Japan (after Yatsu 1955), together with (B) estimates of system autocorrelation and spectrum, solid upstream and dashed downstream sections; and (C) Kalman filter parameter estimates showing the point of change in bed oscillation from rough features to more attenuated bed forms (after Bennett 1976). The moving-average model is given by

$$d_1^t = d_1^{t-1} + n_t.$$

channelling or dissipating them. Such attitudes are thus not only conditioned by the nature of the real-world system itself but also by the type of modelling applied to it.

7.3 Systems synthesis

The more complex a segment of the real world, the greater the likelihood that it will be modelled as a synthetic system. Thus it is not surprising that the ecosystem provides a prime example of systems synthesis in the environmental sciences (see valuable discussion by Stoddart, 1967). Ecosystems can be represented as a complex of feedforward cascades of energy or mass between a hierarchy of storage compartments of biomass (e.g. different trophic, or feeding, levels), possessing feedback which may be either negative (i.e. self-regulatory) or positive (self-forcing), nested within grand positive-feedback mineral nutrient recycling loops (Van Dyne 1969, Kormondy 1969, p. 4, Patten *et al.* 1975, p. 218). The complexity of such a system resides in the variety of nonlinear flows between the many possible storage compartments, the complex linkages of the network of storages and the fact that the system focuses attention on the state of the energy/mass storages (as distinct from the cascades or flows between them, which are highlighted by hydrological systems) which can only be rationalized by reference to the chemical fluxes of the nutrient cycling. It is common, therefore, for ecosystems to be modelled predominantly as energy/mass cascades, with the nutrient recycling being introduced in terms of the effects which the storage and flux of nutrients have on the maintenance, stability and thresholds of the biomass compartments. Fig. 7.8 shows the energy cascade and nutrient cycling for a simple ecosystem

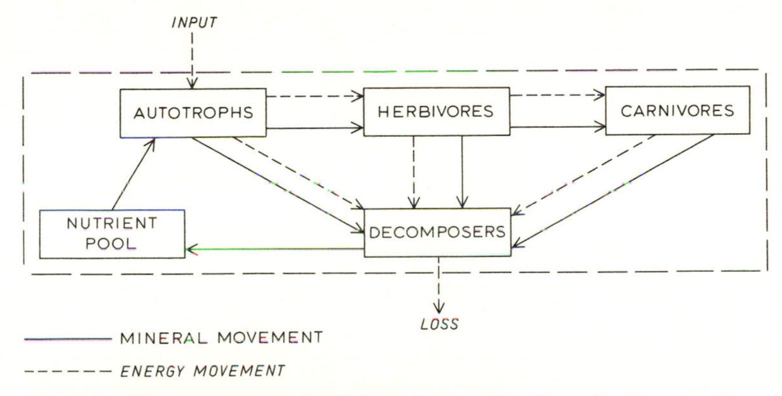

Fig. 7.8 The energy cascade and nutrient cycling for a simple ecosystem. (After Kormondy 1969.)

in which the solar energy produces autotrophs (i.e. 'self-feeders', like vegetation), which are consumed by herbivores at the next trophic level, which in turn are consumed by carnivores, and all three storage compartments cascade energy into a decomposers compartment of organic matter. The nutrients produced by respiration and decomposition are stored in various nutrient pools (i.e. the soil) to be recycled, after a certain delay, to the trophic cascade by means of the autotrophs (Kormondy 1969). Nutrient cycles involve a variety of chemicals, chiefly carbon, nitrogen, sulphur, phosphorus, potassium and calcium; some operate rapidly, others slowly; and some operate on a world scale, whereas others are extremely local. It is convenient to divide nutrient flows into open cycles in which the nutrients (of which carbon and nitrogen are the most important) have direct interchange with the worldwide atmospheric pool (figs 7.9A and 7.9B show the structure and relations of these two open cycles and the nested energy/mass cascade), and into small-scale closed cycles (e.g. potassium) which depend on the local decompositional weathering of parent material without significant external inputs, except for minor ones associated with rainfall, etc. (Usher 1973). The dominant features of ecosystems, and those which bear most on their control,

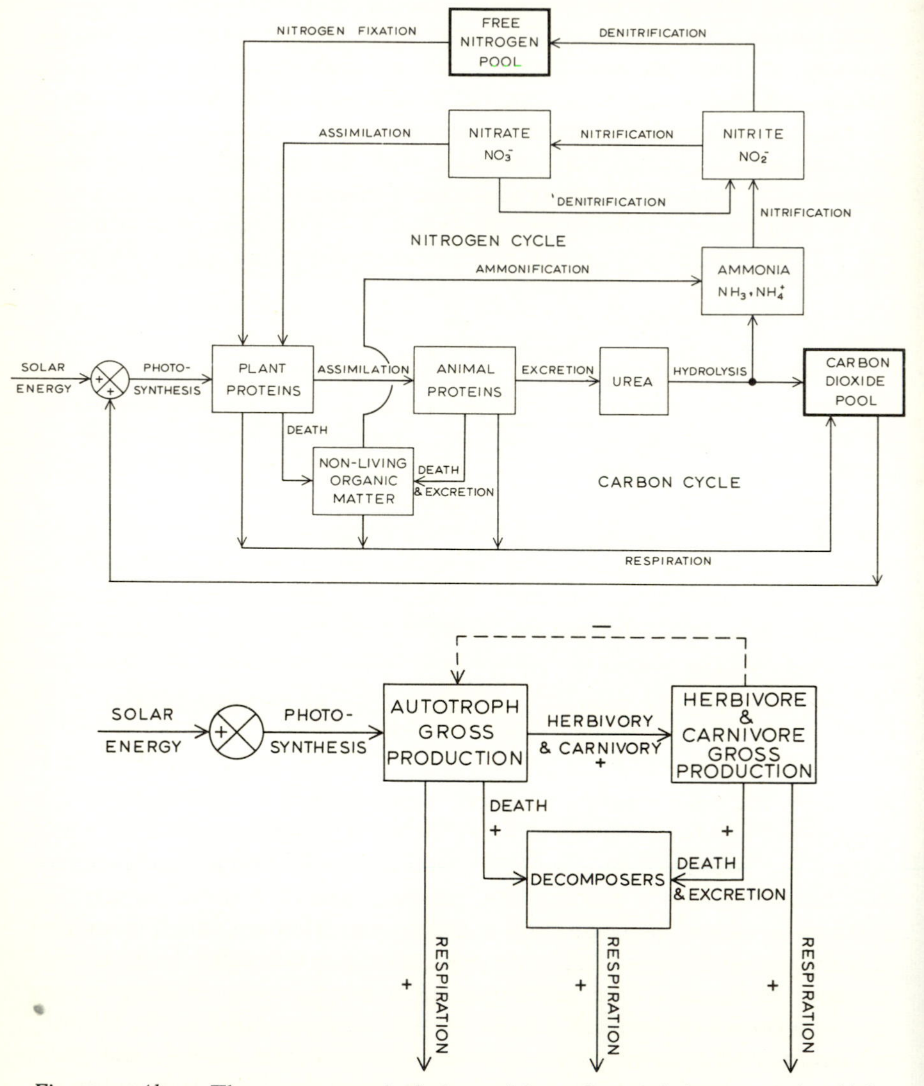

Fig. 7.9 *Above*: The structure and relations of the carbon and nitrogen cycles. (After Rich 1973.)
Below: The detail of the nested ecosystem.

are thus the self-regulation and stability which may exist within the complicated causal cascades of energy and mass, together with the nutrient recycling loop (Kowalik 1974, p. 165). Stability in ecosystems is thus maintained by:

1 The operation of self-regulatory feedback in the energy/mass cascades (e.g. size of prey related to size of the predator population).
2 The operation of nutrient cycling, which may be relatively slow because of the large size of many nutrient storages (e.g. carbon dioxide in the atmosphere) and because of the time taken for nutrients in some organic matter to be released into the soil and recycled back into plants. In general, the more diverse the ecosystem (e.g. tropical rainforest, as distinct from tundra) the more stable the production of recycled nutrients (see chapter 9).

For example, both tropical rainforest and coral reef biomes (i.e. ecosystems characteristic of a broad environment) possess stable physical environments, but they have no large storage pools or major inputs of essential elements and are sustained by tightly closed and rapid nutrient cycling supported by quick regeneration (e.g. by evapotranspiration and by symbiotic pairs, respectively). Consequently they are both slow to recover after disturbances of nutrient cycling (Pomeroy 1970, pp. 183–4). Two particularly contrasting biomes are the tundra and the saltmarsh which are both relatively simple and possess a low diversity of species, but whereas the former is very unstable and subject to oscillations and collapse, the latter is highly stable (Pomeroy 1970, pp. 183–4). The reason for this contrast is that the tundra has no large nutrient inputs, most of the nutrients being tied up in the lemming and litter biomass compartments, the lack of nutrient reserve being due to lemming grazing combined with the small root compartment providing little nutrient storage (i.e. 'buffering') to supply and sustain the surface vegetation compartment (e.g. the root storage can buffer the leaf compartment against a loss of phosphorus for only one year: Jordan and Kleine 1972, p. 43). The saltmarsh, on the other hand, possesses no dominant grazer, is a system which is relatively open to tidal nutrient supply, and has little nutrient shortage. Considerations of stability of the storages (represented by dynamic state variables) are thus paramount in ecological modelling (Kowal 1971, p. 176), which usually involves:

1 The assumption of a steady state for the whole system. This often presents difficulties in that a model can behave much like the real-world ecosystem even if some of the basic assumptions are very wrong (Pomeroy 1970, p. 186).
2 Each storage compartment is assumed to maintain a steady state.
3 Intercompartment fluxes may be modelled as linear and controlled by the donor compartment and yield unrealistically stable models, or they may be complex, nonlinear and unpredictable.

4 The assumption that the system possesses no 'dead' time between changes of input and the output response (Child and Shugart 1972, p. 107).

The purpose of such modelling is to examine the behaviour of the system under conditions of control to determine its sensitivity and under what conditions it may be expected to be stable, decay or explode (Kowal 1971, pp. 175–6).

A synthetic ecosystem is defined by a set of coupled differential-equations, one for each of the storage components (i.e. dynamic state variables $-X$) which is expressed in terms of energy/mass fluxes (i.e. non-dynamic state variables $-F$). It is assumed that the system is continuous, dynamic and spatially lumped, and the equations are state space equations of the form:

$$\dot{X}_1 = f(X_1, X_2, X_3, \ldots, X_n, t)$$

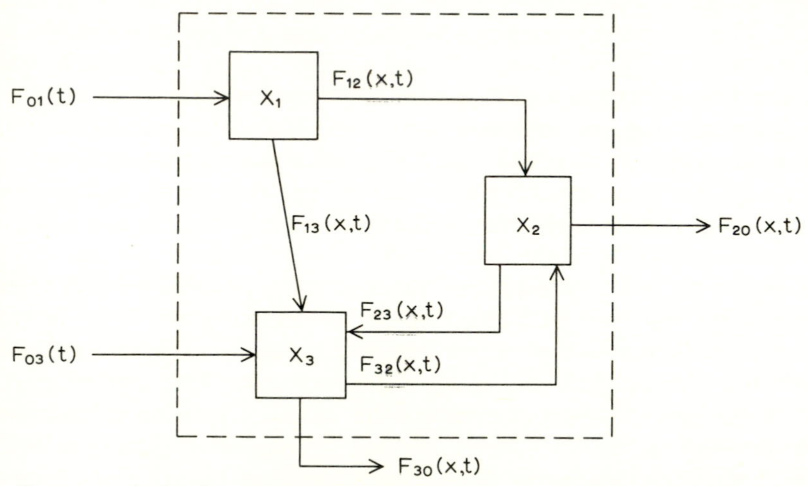

Fig. 7.10 A simple ecosystem structure.
X_1, X_2 and X_3 = storage components
F_{ij} = rate of flow of mass or energy from i to j
X_i = concentration of mass or energy in the donor compartment
X_j = concentration of mass or energy in the recipient compartment

$$\frac{dX_1}{dt} = F_{01} - F_{12} - F_{13}$$

$$\frac{dX_2}{dt} = F_{12} + F_{32} - F_{20} - F_{23}$$

$$\frac{dX_3}{dt} = F_{03} + F_{13} + F_{23} - F_{30} - F_{32}$$

(After Patten 1971.)

where the function maps the real l-space where elements are at time t, and the real n-space or vector space where elements are vectors X (Kowal 1971, pp. 127–30). Fig. 7.10 shows such a simple ecosystem structure with three components. Flow rates between these components may take a number of forms (Patten 1971, pp. 34–5), the first three being linear, the last three non-linear:

1 Constant ($F_{ij} = k$).
2 Proportional (gain) to the state of the donor storage ($F_{ij} = \phi_{ij}X_i$).
3 Proportional (gain) to the state of the recipient storage ($F_{ij} = \phi_{ij}X_j$).
4 Proportional to the joint states of donor and recipient storage (multiplicative gain) ($F_{ij} = \phi_{ij}X_iX_j$).
5 Proportional to the state of the donor and to the joint states of donor and recipient storage

$$[F_{ij} = \phi_{ij}X_i(1 - \alpha_{ij}X_j)]$$
$$[F_{ij} = \phi_{ij}X_j(1 - \alpha_{ij}X_i)]$$

(i.e. first term being linear, positive feedback and the second term non-linear and negative feedback).
6 Lotka-Volterra type (two negative feedback loops)

$$[F_{ij} = \phi_{ij}X_j(1 - \alpha_{ij}X_i - \beta_{ij}X_j)].$$

The interaction of the various ecosystem components represents a special form of space-time interaction and cascading between components, and in fig. 7.11 are shown the relations between simple cascades and interactive networks for forced (i.e. subject to external inputs) and unforced linear and nonlinear networks. Each cascade is a special form of linear or nonlinear transfer function, as discussed in chapter 2. The steps in the construction of synthetic systems are given below, and involve initially defining the system in terms of both algebraic and differential-equations and then eliminating the former (Kowal 1971, pp. 130–46; and see chapter 2 of this volume, fig. 2.39):

1 Specification of the state variables of interest, involving the storages and fluxes.
2 The construction of a block diagram specifying the immediate cause and effect relationships between the variables.
3 The classification, operational definition and specification of units of the variables.

Input variables (F_{01}, etc.), being external causes ('forcings') commonly representing input fluxes or environmental controls (e.g. solar energy, precipitation, etc.) (units: gm/unit area/day, etc.).

State variables, which represent the state of the system at any given time. These are divisible into:

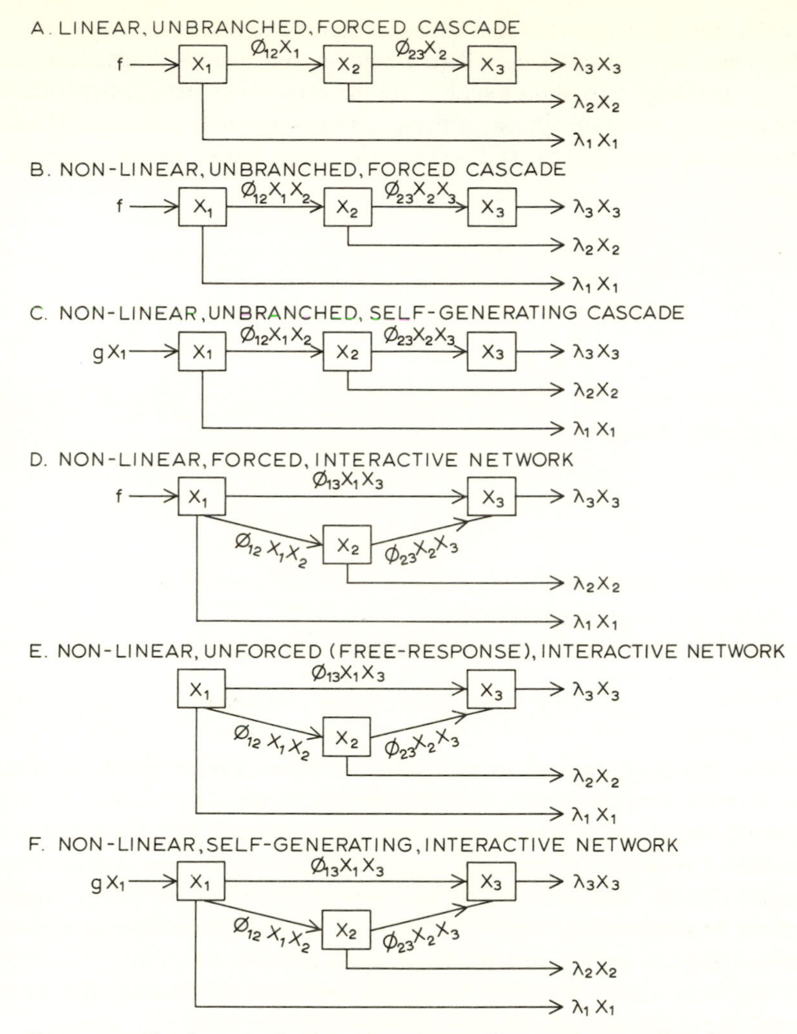

Fig. 7.11 Six forms of space-time interaction and cascading between ecosystem components. (Partly after Williams 1972.)

(a) Non-dynamic state variables (F_{ij}), representing the rates of transfer or flows between the compartments; they are thus instantaneous, zero energy, non-energy-storing variables. The values of state variables do not depend on their previous values and they may be defined therefore in terms of algebraic equations, later to be eliminated. They include herbivory, evapotranspiration, etc., and are expressed in the same units as the input variables.

(b) Dynamic state variables (X_n), representing the state of the storage compartments (e.g. biomass). They are of finite memory or energy storing, depend on previous states and must be defined by differential-equations, for example $\dot{X}_n = dX/dt =$ gain fluxes – loss fluxes (units: gm/unit area, etc.).

4 Specification of the form of the linking equations, which involves expressing the state variables as a function of input variables and other state variables. These equations are the axioms of the mathematical structure and are commonly of the form of power series, statistical regressions, etc., being based on mathematical, physical, chemical and biological principles, as well as on the literature, personal knowledge and intelligent guesses. The set of equations thus represents a complex hypothesis in which each equation is an independent subhypothesis capable of being independently tested and altered. Time variance and time lags impose special problems in the modelling of synthetic ecosystems. The former assumption means that mathematical solutions cannot be obtained in an exact literal form and that behaviour must be described without an attempt to specify its causes completely. This is commonly achieved by converting time-variant dynamic state variables into input variables. Where there are significant time lags in the system either more state variables are necessary (i.e. to specify state as a function of prior states) or modified differential-equations are required (Kowal 1971).

5 Parameter estimation. The values of the constants are obtained by measurement and analysis (using statistical methods, by hypothesis and from the literature). A special problem with the equations specifying synthetic systems is that, as both sides of these must in theory have the same units, the empirical constants are often not pure numbers but possess unusual units (i.e. $°C^{-1}$, etc.).

6 The testing and modification of the synthetic systems under simulated forcing conditions (Rykiel and Kuenzel 1971, p. 527). These conditions may be:
   (a) Free response (zero forcing) – where a system in an equilibrium state is deprived of inputs and permitted to run down.
   (b) Forced response – where the initial compartment conditions (i.e. dynamic state variables) are zero and the system is forced by a step input, in a constant manner (i.e. leading to a constant input into each of the storages equal to their annual gross productivity), by a sine input (e.g. representing seasonal changes), or by a self-generating forcing involving positive feedback by at least one of the storage components. If the feedback equals the losses, a steady state results similar to constant forcing.

A number of examples will serve to show something of the structure, complexity and possibilities for the simulative exploitation of synthetic systems. Fig. 7.12 and table 7.1 show the energy flow, notional block diagram,

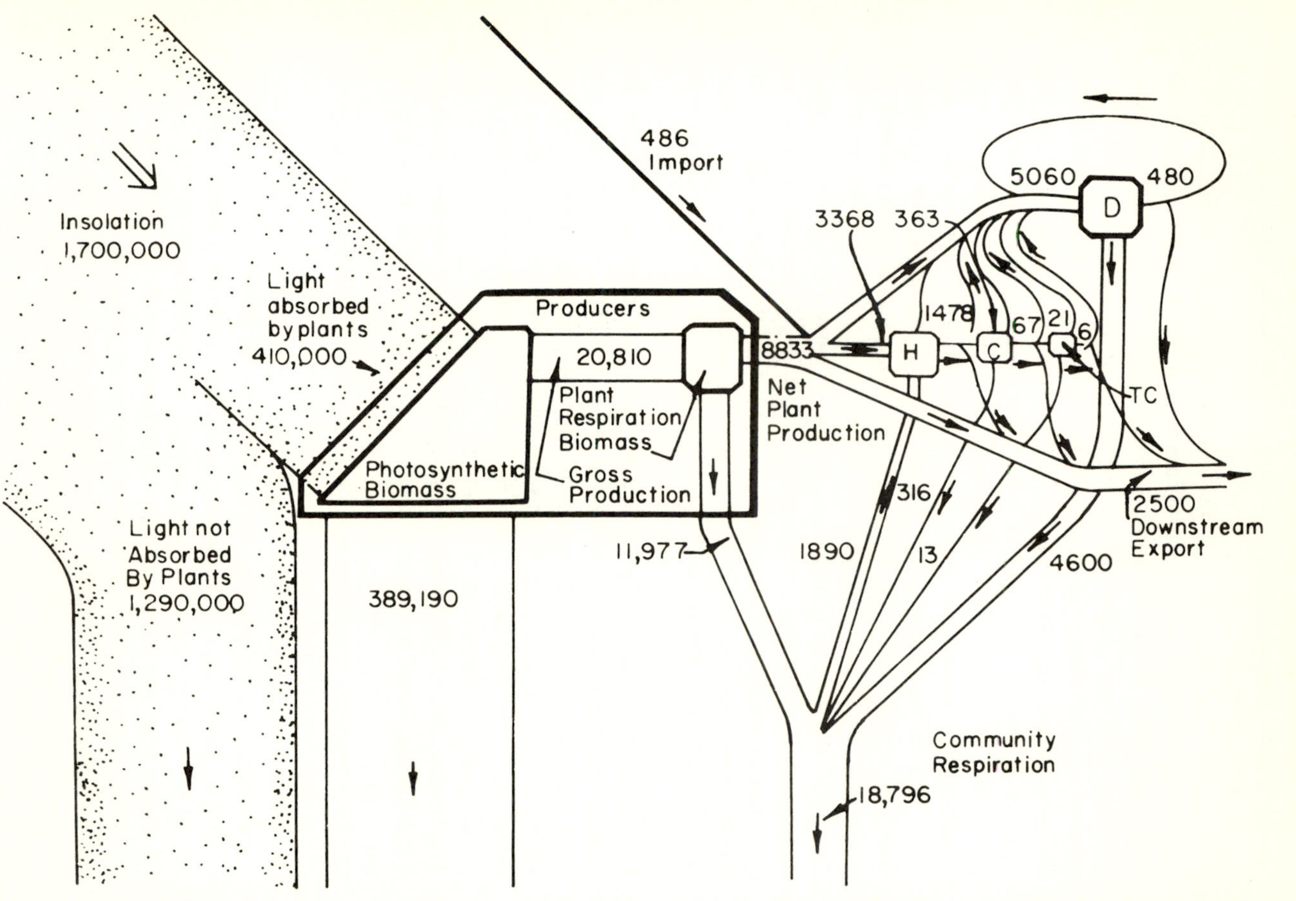

Fig. 7.12 (*Above*) The energy flow (in Kcal/m^2/year) and (*Opposite*) the notional block diagram for the Silver Springs ecosystem. (From Odum 1957, and Patten 1971.)

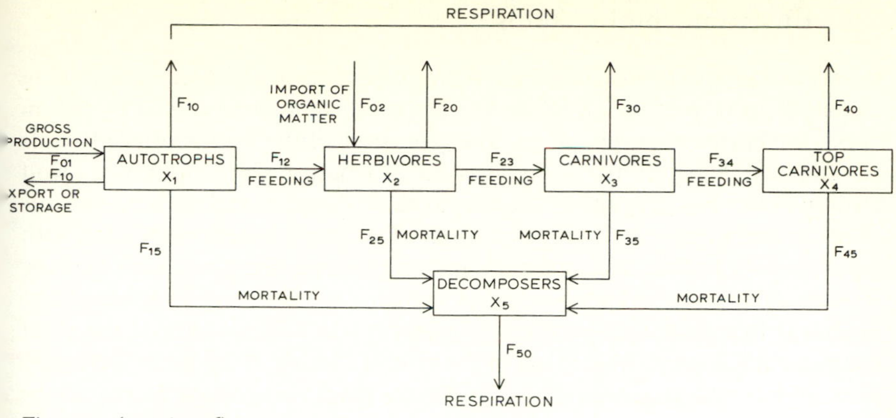

Fig. 7.12 (*continued*)

Table 7.1 Data for Silver Springs ecosystem (see fig. 7.12A) (after Patten 1971, table 1, pp. 38–9)

VARIABLES AND CONSTANTS

X_1 = plant (respiration) biomass
X_2 = herbivore biomass
X_3 = carnivore biomass
X_4 = top carnivore biomass
X_5 = decomposers biomass

τ = trophic level feeding rates
ρ = respiration
μ = natural mortality
λ = losses downstream

NON-FORCING ENERGY FLOWS (*kcal/m²/year*)

Feeding	Mortality	Respiration	Export
F_{12} = 2,874	F_{15} = 3,455	F_{10} = 11,974 (11,977)*	F'_{10} = 2,498 (2,500)
F_{23} = 382	F_{25} = 1,095	F_{20} = 1,891 (1,890)	
F_{34} = 21	F_{35} = 46	F_{30} = 317 (316)	
	F_{45} = 6	F_{40} = 13	
		F_{50} = 4,598 (4,600)	

* *Figures in parenthesis are those given in fig.* 7.12*A*.

LINEAR SYSTEM EQUATIONS

$$\dot{X}_1 = F_{01} - \tau_{12}X_1 - \mu_{15}X_1 - \lambda_{10}X_1 - \rho_{10}X_1$$
$$\dot{X}_2 = F_{02} + \tau_{12}X_1 - \tau_{23}X_2 - \mu_{25}X_2 - \rho_{20}X_2$$
$$\dot{X}_3 = \tau_{23}X_2 - \tau_{34}X_3 - \mu_{35}X_3 - \rho_{30}X_3$$
$$\dot{X}_4 = \tau_{34}X_3 - \mu_{45}X_4 - \rho 40X_4$$
$$\dot{X}_5 = \mu_{15}X_1 + \mu_{25}X_2 + \mu_{35}X_3 + \mu_{45}X_4 - \rho_{50}X_5$$

NONLINEAR SYSTEM EQUATIONS

$$\dot{X}_1 = F_{01} - \tau'_{12}X_2X_1 - \mu_{15}X_1 - \lambda_{10}X_1 - \rho_{10}X_1$$
$$\dot{X}_2 = F_{02} + \tau'_{12}X_2X_1 - \tau'_{23}X_3X_2 - \mu_{25}X_2 - \rho_{20}X_2$$
$$\dot{X}_3 = \tau'_{23}X_3X_2 - \tau'_{34}X_4X_3 - \mu_{35}X_3 - \rho_{30}X_3$$
$$\dot{X}_4 = \tau'_{34}X_4X_3 - \mu_{45}X_4 - \rho_{40}X_4$$
$$\dot{X}_5 = \mu_{15}X_1 + \mu_{25}X_2 + \mu_{35}X_3 + \mu_{45}X_4 - \rho_{50}X_5$$

the tabulated energy flows and the form of the linear and nonlinear linking equations in respect of the Silver Springs ecosystem made classic by Odum (1957, see Patten 1971, pp. 38–9). The ecosystem of the wolves of Isle Royale (Rykiel and Kuenzel 1971) involves an instructive plant → moose → wolf feeding chain. Fig. 7.13 shows the simple block diagram and table 7.2 the sets of linear and nonlinear linking equations. The linear system responded with great stability to free, step, constant and sine forcing (fig. 7.14), with a steady state being approached in about eight years for the first two. In contrast, the nonlinear model showed poor stability and it was difficult to hold the plant component constant. Regardless of the type of forcing, if the latter increased, even slightly, the system exploded and if the plant component decreased the

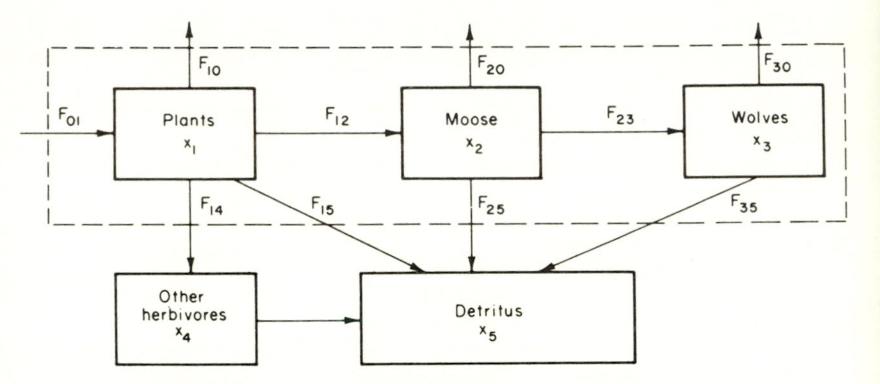

Fig. 7.13 Block diagram of the Isle Royale ecosystem. (From Rykiel and Kuenzel 1971.)

Table 7.2 Data for the wolves of Isle Royale ecosystem (see fig. 7.13) (after Rykiel and Kuenzel 1971, pp. 522, 533)

VARIABLES AND CONSTANTS

X_1 = plant biomass (kcal/m^2)
X_2 = moose biomass
X_3 = wolves biomass

τ = feeding rates (kcal/m^2/year)
ρ = respiration
μ = mortality
λ = other energy losses

LINEAR SYSTEM EQUATIONS

$$\dot{X} = F_{01} - (\rho_{10} - \tau_{12} + \tau_{14} + \mu_{15})X_1$$
$$\dot{X}_2 = \tau_{12}X_1 - (\rho_{20} - \tau_{23} - \lambda_{25})X_2$$
$$\dot{X}_3 = \tau_{23}X_2 - (\rho_{30} + \lambda_{35})X_3$$

NON-LINEAR SYSTEM EQUATIONS

$$\dot{X}_1 = F_{01} - (\rho_{10} + \tau_{14} + \mu_{15})X_1 - \tau_{12}X_2X_1$$
$$\dot{X}_2 = \tau_{12}X_2X_1 - (\rho_{20} + \lambda_{25})X_2 - \tau_{23}X_3X_2$$
$$\dot{X}_3 = \tau_{23}X_3X_2 - (\rho_{30} + \lambda_{35})X_3$$

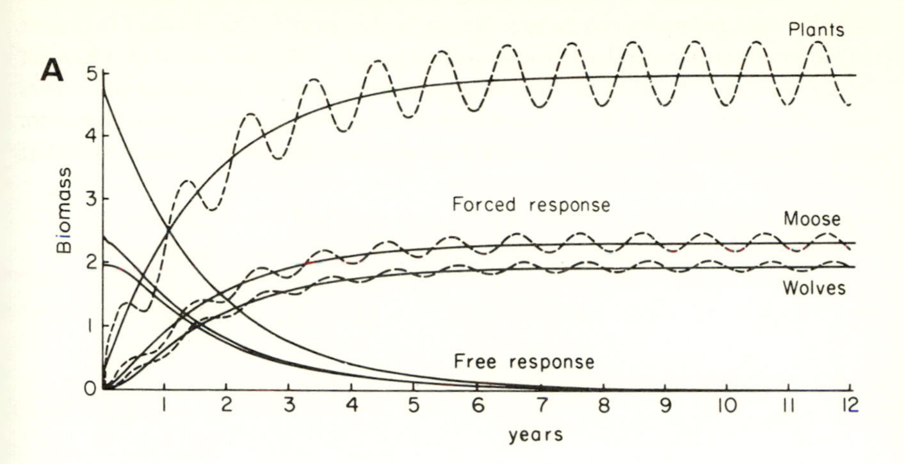

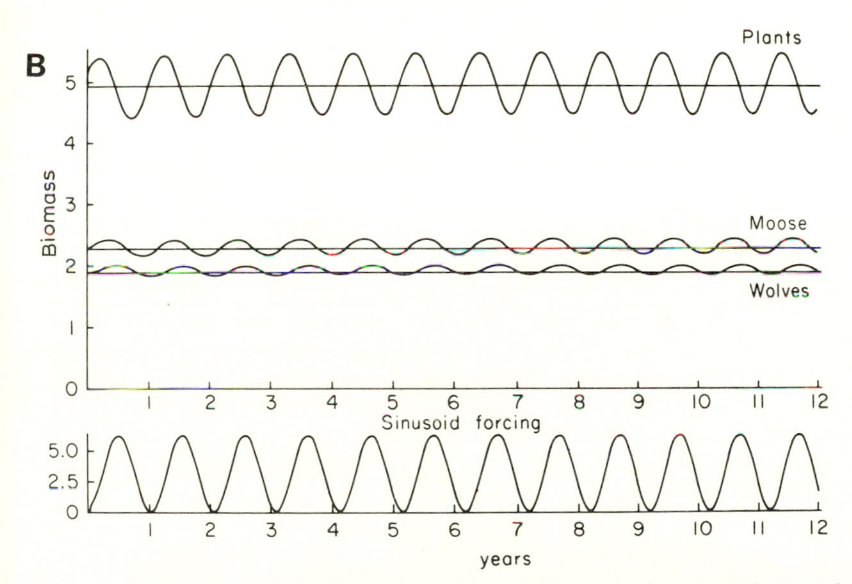

Fig. 7.14 The results of the linear model simulation experiments with the Isle Royale ecosystem.

A Responses of the linear model. Free (unforced) and forced responses go through a period of transient behaviour and approach a steady state when input balances output in about eight years.

B Steady states of the linear model with constant and sine forcing. (From Rykiel and Kuenzel 1971.)

system collapsed (fig. 7.15). Actual observation, on the other hand, showed that the wolf peaks lagged behind those of the moose, that wolves fluctuate with smaller amplitude than moose and that a unique steady state should exist for a given set of conditions – i.e. the cyclic variation in number of predators was out of phase with cyclic variation in number of prey. It was therefore concluded that the nonlinear model was unacceptable in behaviour, but that

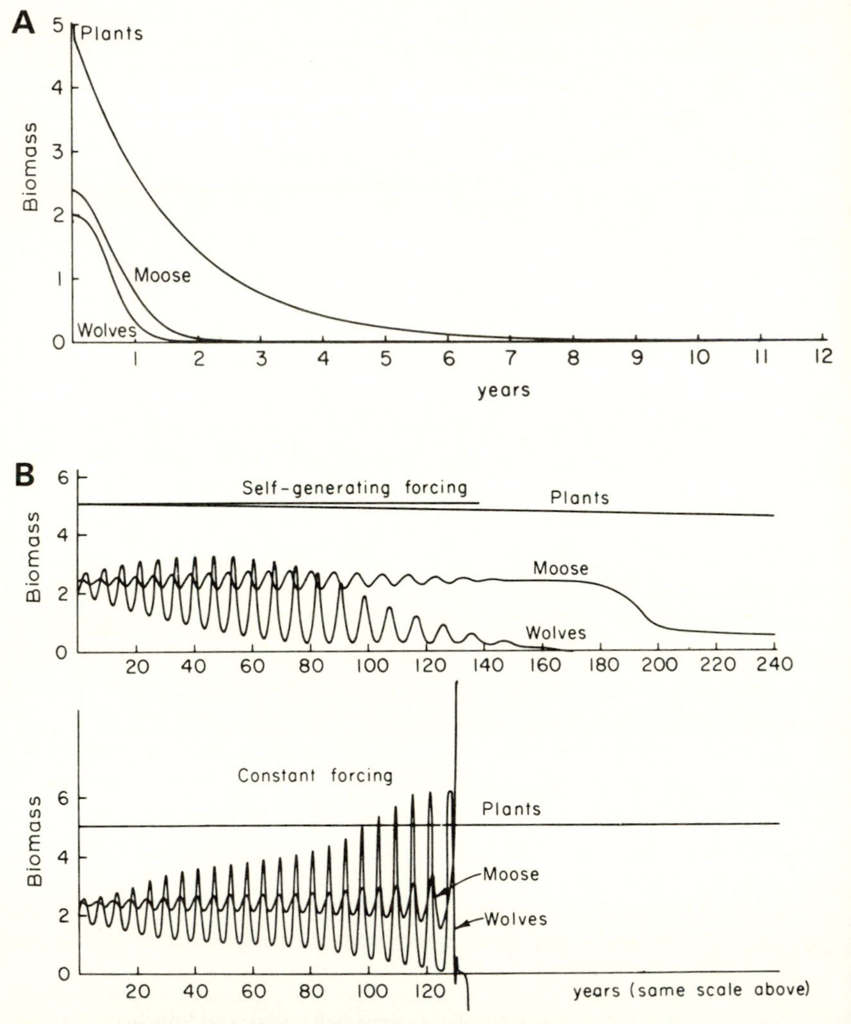

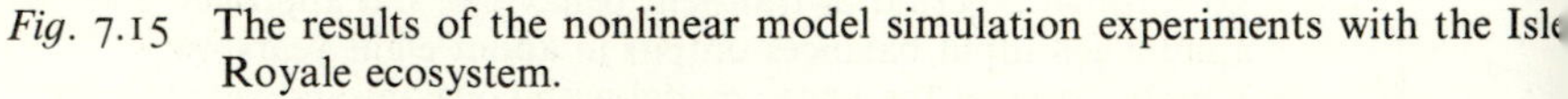

Fig. 7.15 The results of the nonlinear model simulation experiments with the Isle Royale ecosystem.
A Free (unforced) response.
B Long-term behaviour resulting from self-generating and constant forcing.
(From Rykiel and Kuenzel 1971.)

the simpler and less realistic model satisfied all three of the empirical observations. Moreover the balance between the moose and wolf population is governed by a form of homeostatic feedback control system in which exponential growth in the number of moose is arrested by a rise in the number of predatory wolves, and exponential decline in the wolf population is arrested by a rise in the number of prey (moose) (Gazis *et al.* 1973). Note that exponential decline or growth are the two possible modes of behaviour resulting from the state equations (table 7.2) and their interaction is responsible for homeostasis. A similar study has been made of the Cedar Bog Lake ecosystem (Williams 1971, pp. 560–4), the subject of Lindemann's classic pioneer work on ecosystem dynamics (fig. 7.16). A ten-component

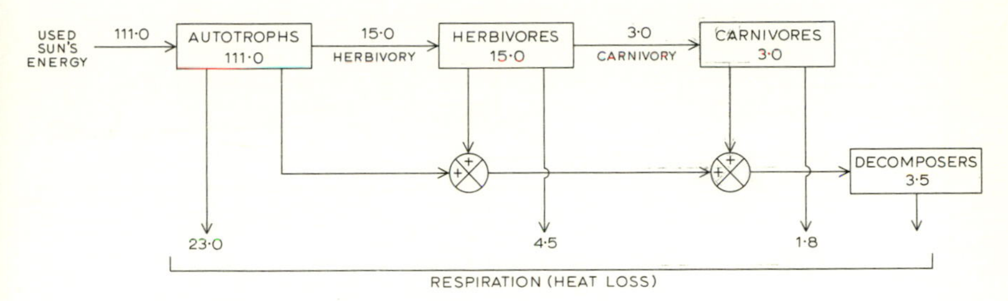

Fig. 7.16 Energy flow diagram of the Cedar Bog Lake ecosystem showing gross production and flows in gcal/m^2/year. (Data from Lindemann, after Kormondy 1969.)

nonlinear model was developed (fig. 7.17 and table 7.3) and tested with constant solar energy (X_1) input, the ooze component (X_{10}) beginning at its average value, nannoplankton (X_3) at five times its average value, and the remaining components at half their average values. This resulted in much of the model being sluggish and unstable (fig. 7.18) with, for example, X_2, X_5 and X_7 oscillating widely and slowly, X_6 increasing markedly at first and then declining to below its initial value, X_8 increasing markedly and then being almost exterminated, and X_9 increasing irregularly during the four-year simulation period. The behaviour of the model was improved by varying the flux into the swimming predator (X_9) compartment and by adjusting the parameters a and ϕ. A similarly complex nine-compartment, multitrophic level model has been constructed for a grassland ecosystem (fig. 7.19) and seasonal biomass and respiration changes simulated with respect to the seasonal variation in solar energy input (fig. 7.20) (Van Dyne 1974, pp. 93–4).

The above examples of ecosystem models bring one back to the question which was posed at the beginning of the chapter – namely, to what extent is it justified to proceed with the construction of multicompartment, nonlinear synthetic systems in the belief that an increasing 'reality' and complexity of detailed system operations will lead to the construction of models which behave as a whole in ever more realistic a manner? Ecosystems are not the only environmental systems which have been modelled in this way, and similar but more complex symbiotic world models will be treated in chapter 9. In hydrology the Stanford IV basin model (Crawford and Linsley 1966, Chorley and Kennedy 1971, pp. 116–22) has been similarly constructed to simulate flows between storage compartments (fig. 7.21), but even with tight control

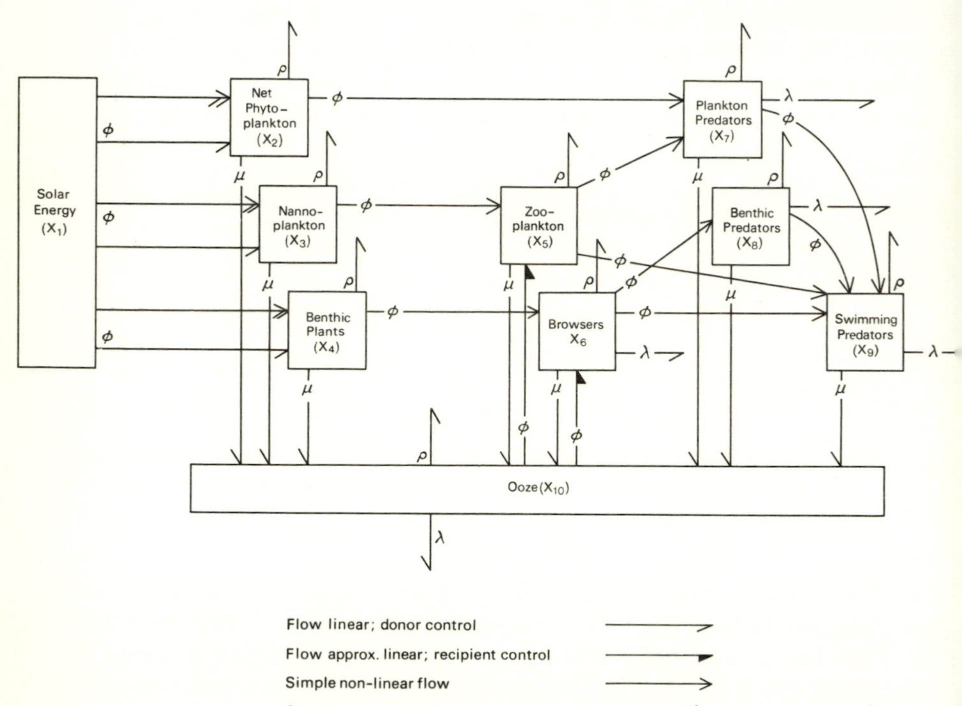

Fig. 7.17 A ten-compartment nonlinear energy flow model for the Cedar Bog Lake ecosystem. (After Williams 1971.)

(e.g. transfer of water between storage compartments every simulated 15 minutes, and removal of water from interception storage and routing of streamflow every hour) the model does not predict peak basin discharges with complete reliability (fig. 7.22). Fig. 7.23 shows a simplified synthetic basin hydrological system in a form compatible with earlier figures in this book, and illustrating the troublesome soil moisture negative feedback effect.

Another group of similar synthetic physical systems have to do with sediment movement. Indeed such pressing environmental issues as the erosion of soil from agricultural and building land (Graf 1975), the silting of reservoirs, the denudation of slopes (Kirkby 1976), erosion of channels, and beach transportation and deposition, promise to be controlled most

Table 7.3 (A) Flow rates, self-inhibitions, and residuums for (B) a ten-compartment nonlinear model of energy flow in the Cedar Bog Lake ecosystem (rates cal/cm²/year). (From Williams 1971.)

A		Transfer (ϕ)		Respiration (ρ)	Loss to ooze (μ)	Loss to outside (λ)	Self-inhibition (a)	Residuum (b)
Compartment		Path	Rate (yr^{-1})	Rate (yr^{-1})	Rate (yr^{-1})	Rate (yr^{-1})	Rate (yr^{-1})	Rate (yr^{-1})
Net phytoplankton	X_2	1, 2	0·000474	12·3	32·73		0·520	0·000178
Nannoplankton	X_3	1, 3	0·000549	12·3	34·94		0·572	0·000154
Benthic plants	X_4	1, 4	0·0000315	0·66	1·99		0·0126	0·0794
Zooplankton	X_5	3, 5	6·55	16·40	21·52			
		10, 5	42·4					
Browsers	X_6	4, 6	0·0228	1·25	0·05	0·05		
		10, 6	3·25					
Plankton predators	X_7	2, 7	14·1	4·16	1·13	1·16		
		5, 7	14·1					
Benthic predators	X_8	6, 8	12·0	2·80	0·70	0·60		
Swimming predators	X_9	5, 9	2·37	1·37	0·50	0·50		
		6, 9	2·37					
		7, 9	2·37					
		8, 9	2·37					
Ooze	X_{10}			1·27		0·416		

B

$$X_1 = 118{,}625 \quad (1)$$
$$\dot{X}_2 = X_1\phi_{12}[b_2 + X_2(1 - a_2X_2)] - X_2(\rho_2 + \mu_2 + \phi_{27}X_7) \quad (2)$$
$$\dot{X}_3 = X_1\phi_{13}[b_3 + X_3(1 - a_3X_3)] - X_3(\rho_3 + \mu_3 + \phi_{35}X_5) \quad (3)$$
$$\dot{X}_4 = X_1\phi_{14}[b_4 + X_4(1 - a_4X_4)] - X_4(\rho_4 + \mu_4 + \phi_{46}X_6) \quad (4)$$
$$\dot{X}_5 = X_5[\phi_{35}X_3 + (\phi_{10,5} - \phi_{35}X_3) - (\rho_5 + \mu_5 + \phi_{57}X_7 + \phi_{59}X_9)] \quad (5)$$
$$\dot{X}_6 = X_6[\phi_{46}X_4 + (\phi_{10,4} - \phi_{46}X_4) - (\rho_6 + \mu_6 + \lambda_6 + \phi_{68}X_8 + \phi_{69}X_9)] \quad (6)$$
$$\dot{X}_7 = X_7[\phi_{27}X_2 + \phi_{57}X_5 - (\rho_7 + \mu_7 + \lambda_7 + \phi_{79}X_9)] \quad (7)$$
$$\dot{X}_8 = X_8[\phi_{68}X_6 - (\rho_8 + \mu_8 + \lambda_8 + \phi_{89}X_9)] \quad (8)$$
$$\dot{X}_9 = X_9[\phi_{59}X_5 + \phi_{69}X_6 + \phi_{79}X_7 + \phi_{89}X_8 - (\rho_9 + \mu_9 + \lambda_9]) \quad (9)$$
$$\dot{X}_{10} = \mu_2X_2 + \mu_3X_3 + \mu_4X_4 + \mu_5X_5 + \mu_6X_6 + \mu_7X_7 + \mu_8X_8 + \mu_9X_9 - [X_0(\rho_{10} + \lambda_{10}) + X_5(\phi_{10,5} - \phi_{35}X_3) + X_6(\phi_{10,6} - \phi_{46}X_4)] \quad (10)$$

immediately and effectively by means of synthetic systems modelling. Both the potentialities and the problems of sediment systems modelling have been reviewed by J. P. Bennett (1974) and have been well illustrated by the computer simulation work of Negev (1967), associated with that on the Stanford Watershed Model IV. This sediment model (fig. 7.24) considered two erosional contributions of sediment to basin discharge.

1 Sheet or overland erosion. This is made up, firstly, of raindrop impact soil splash which detaches surface soil and makes it available for surface transport. This sediment contribution was estimated by hourly rainfall

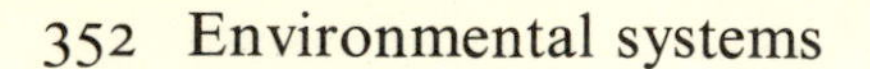

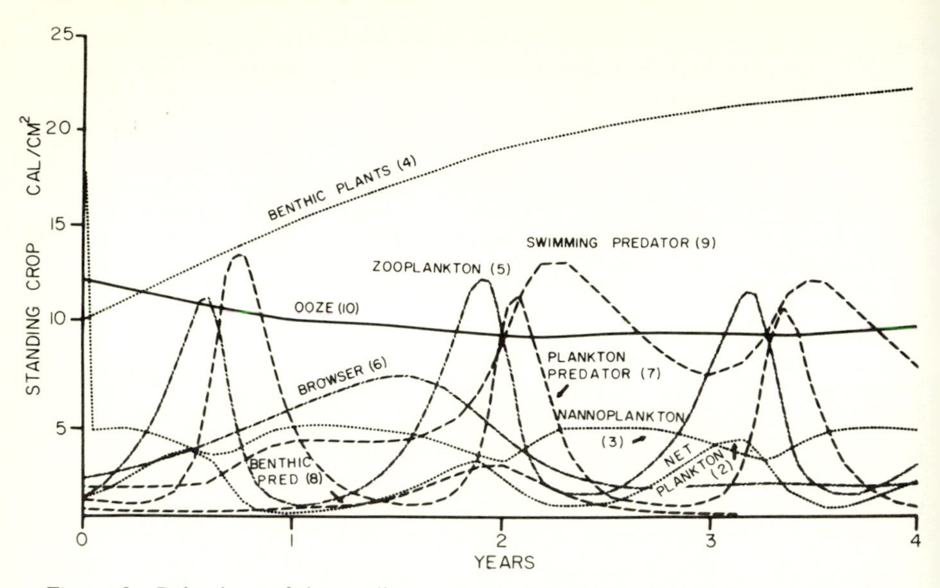

Fig. 7.18 Behaviour of the nonlinear model of the Cedar Bog Lake ecosystem (fig. 7.17) based on the equations given in table 7.3B. The simulation was begun with the ooze compartment (X_{10}) at its average value; nannoplankton (X_3) at five times its average value, and the remaining compartments at half their average values. Values on the ordinate for ooze are multiplied by 0·1, for benthic plants (X_4) by 1·0, and for the remaining compartments by 10·0. (From Williams 1971.)

data and by means of a parameter varying with soil type and surface cover. Secondly, overland flow was assumed to be available to erode the surface soil and to wash away detached soil. This contribution is derived from assumed soil splash pickup, hourly overland flow data, estimates of availability of fine particles, etc.

2 Channel erosion. This is made up, firstly, of a bank-caving component which is generally ignored, together with sediment supplied by rill and gully erosion estimated from the observed grain-size characteristics of the channel material. Secondly, stream bed erosion is calculated from the amount of sediment in storage and from daily discharges. To circumvent the perennial problem of the definition of bedload a new component, 'interload', is introduced (a stream sediment load component having characteristics in common with both suspended and bedload and comprising material finer than about 95% of the material by weight).

Inputs into the system are rainfall data, streamflow data (actual or simulated from the Stanford Watershed IV model) and simulated overland flow data (fig. 7.24). It is clear that, because sediment data are very difficult to derive empirically, the flows of sediment in the system are based either on very generalized estimates of sediment availability or on hydrological surrogates of

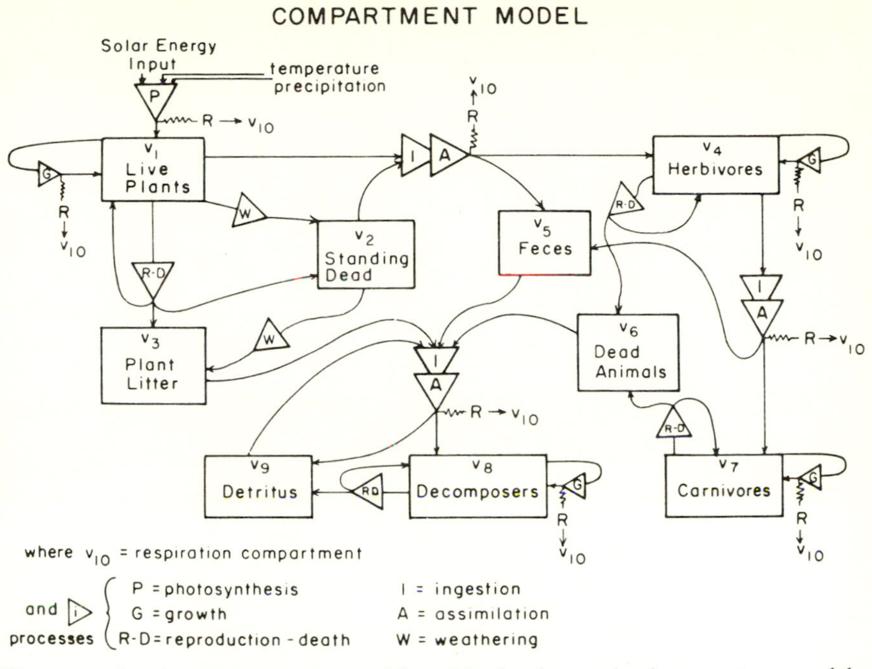

Fig. 7.19 A nine-compartment multitrophic level grassland ecosystem model. Energy flow is shown by arrows. Compartments (rectangles) are connected by processes (triangles). At many steps there is a respiratory energy loss (*R*) to the environment (V_{10}). (From Van Dyne 1974.)

sediment transport (e.g. stream discharge). Fig. 7.25 shows the relation between recorded and simulated suspended sediment discharged by the Napa River near St Helena, California, during the period December 1957 to March 1958. The model produced a surprisingly good correspondence of annual and monthly sediment discharges, with the errors in computed annual loads being less for years with high sediment yields. It is clear that the detailed predictive errors, particularly in short-term sediment budgets, are due to estimates of sediment availability and characteristics, to the use of hydrological surrogates for erosion, to the assumption of spatial rainfall lumping and to the lack of specific data relating to bank caving and to gully erosion. Allen (1974) has outlined the manner in which fluvial and tidal sedimentary systems may be recognized, as well as examining their reactions, relaxation and lags (see also Sunamura 1976), and it may well be that much of future process-response geomorphology may be reformulated in terms of sediment systems. Trudgill (1976, pp. 90–2) has suggested a systems-based relationship between variations in process intensity due to climatic change and the variously lagged and damped erosional outputs affecting micro-, meso- and macroscale landform systems (fig. 7.26) depending on their *lability* (i.e. the relationship of

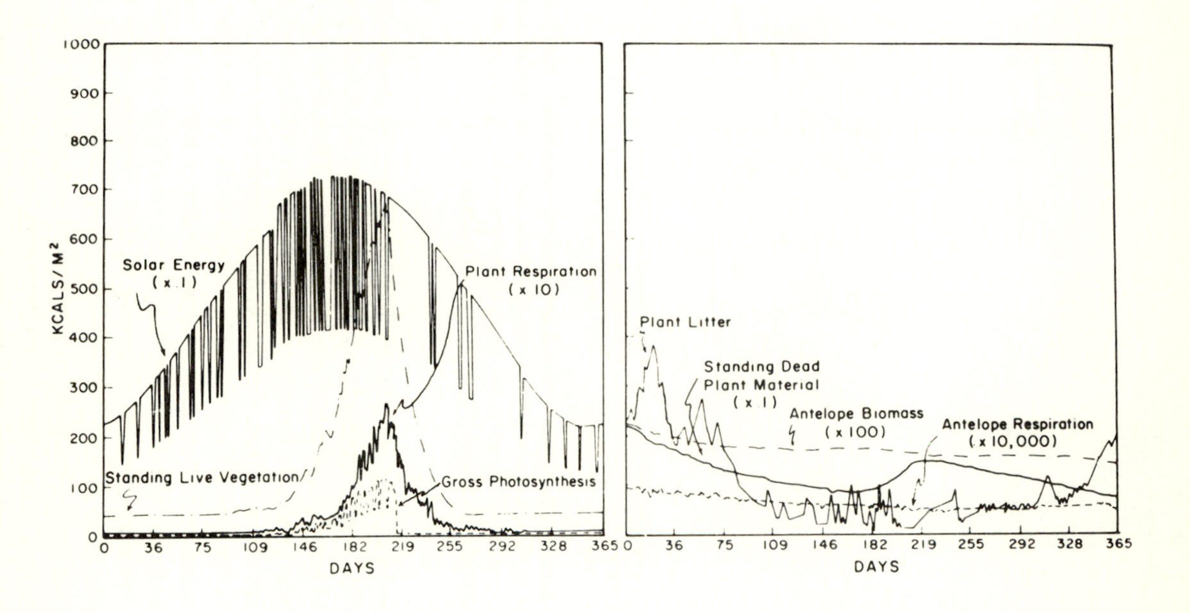
KCALS/ M²
1000
900
800
700
600
500
400
300
200
100
0
Solar Energy
(x 1)
Plant Respiration
(x 10)
Standing Live Vegetation
Gross Photosynthesis
0
36
75
109
146
182
219
255
292
328
365
DAYS
Plant Litter
Standing Dead
Plant Material
(x 1)
Antelope Biomass
(x 100)
Antelope Respiration
(x 10,000)
DAYS

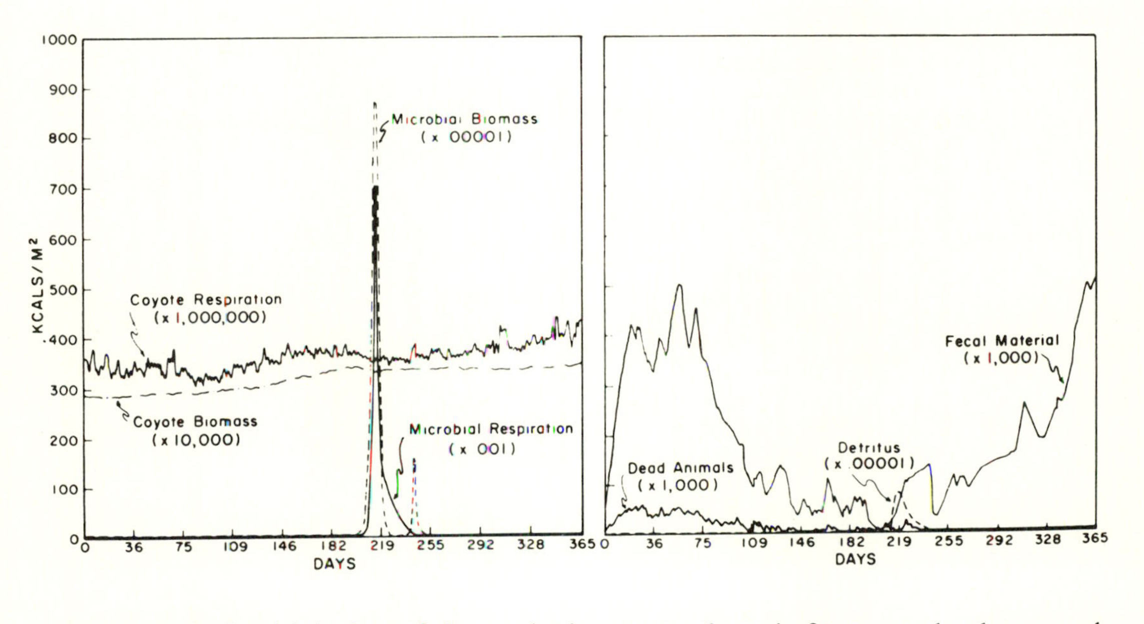

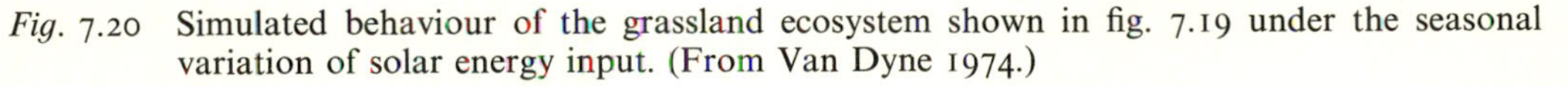

Fig. 7.20 Simulated behaviour of the grassland ecosystem shown in fig. 7.19 under the seasonal variation of solar energy input. (From Van Dyne 1974.)

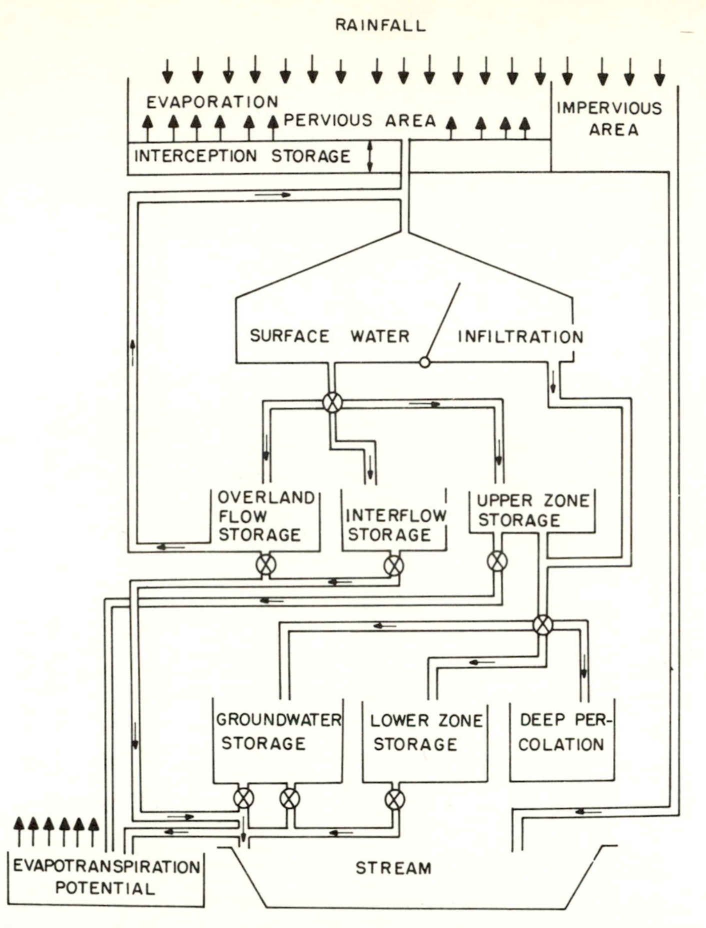

Fig. 7.21 Schematic diagram of the Stanford Watershed Model IV. (From Moore and Claborn 1971.)

process intensity to system resistance). This accords well with the view that many landform assemblages can be conceived of as a palimpsest of cascading space-time sediment systems (Chorley and Kennedy 1971, p. 248). Meteorological and climatic processes are difficult to specify in systems terms. It is only recently that specific climatological use is being made of the systems model in other than extremely small spatial units (Terjung 1976) (see chapter 9).

It is increasingly apparent that the detailed synthetic modelling of environmental systems, when carried to its logical extremes, does not always result in a realistic simulation of the real-world system operation, and some workers are being thrown back to a more pragmatic modelling approach akin to systems analysis wherein, if the systems model seems to work, it is

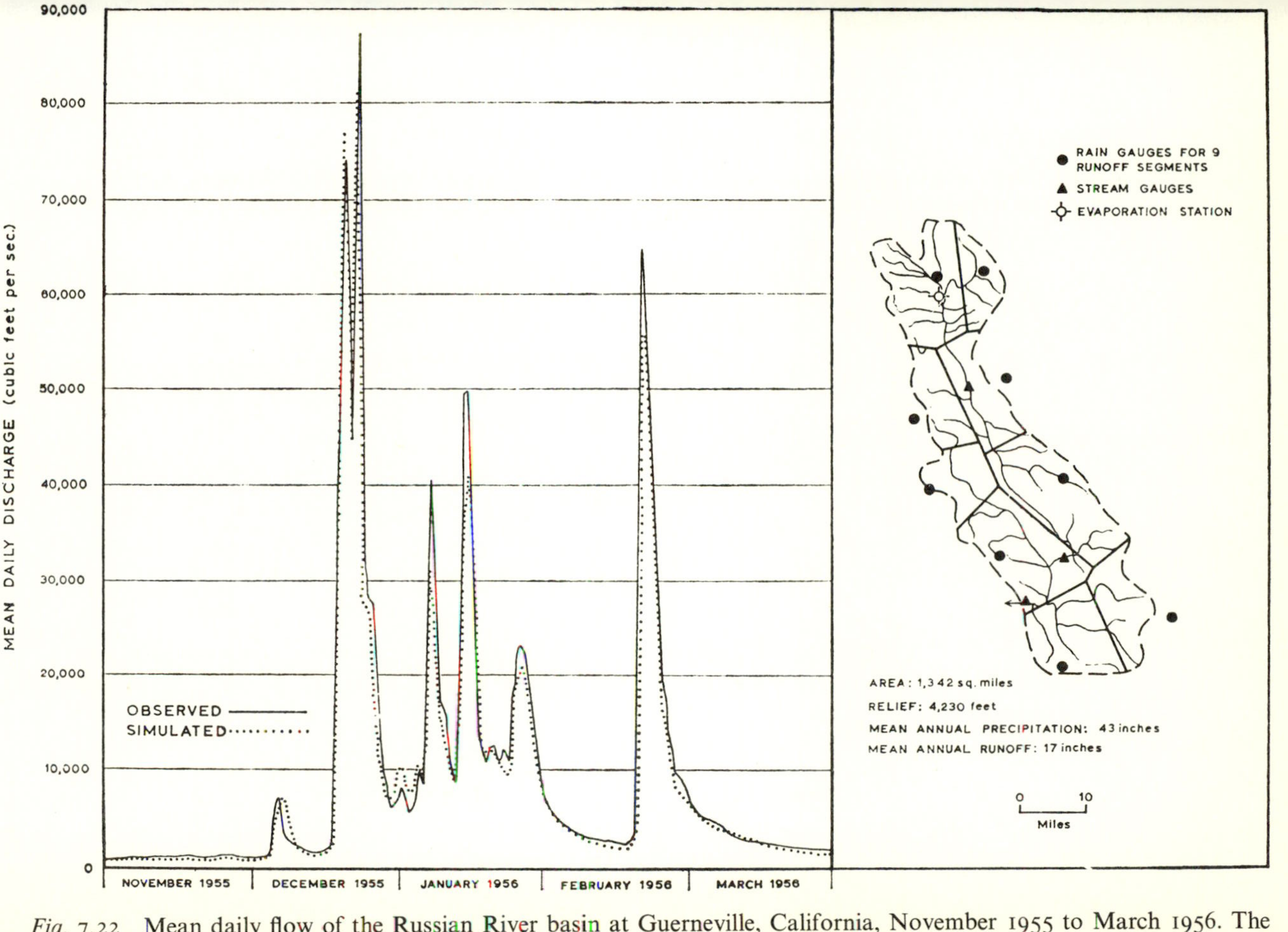

Fig. 7.22 Mean daily flow of the Russian River basin at Guerneville, California, November 1955 to March 1956. The observed flows were those recorded by the gauging station and the simulated ones were obtained from the Stanford Watershed Model IV. (From More 1969, after Crawford and Linsley 1966.)

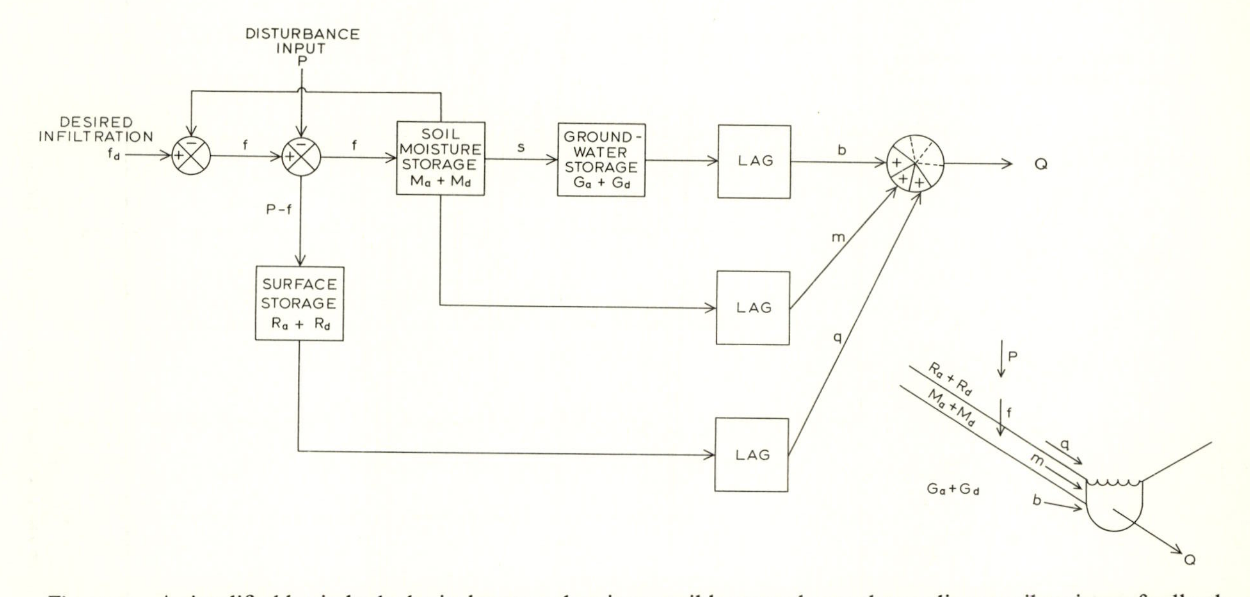

Fig. 7.23 A simplified basin hydrological system showing possible control over the nonlinear soil moisture feedback loop. Subscripts *a* and *d* refer to 'active' and 'dead' storage, respectively. M_d is thus equivalent to field moisture capacity.

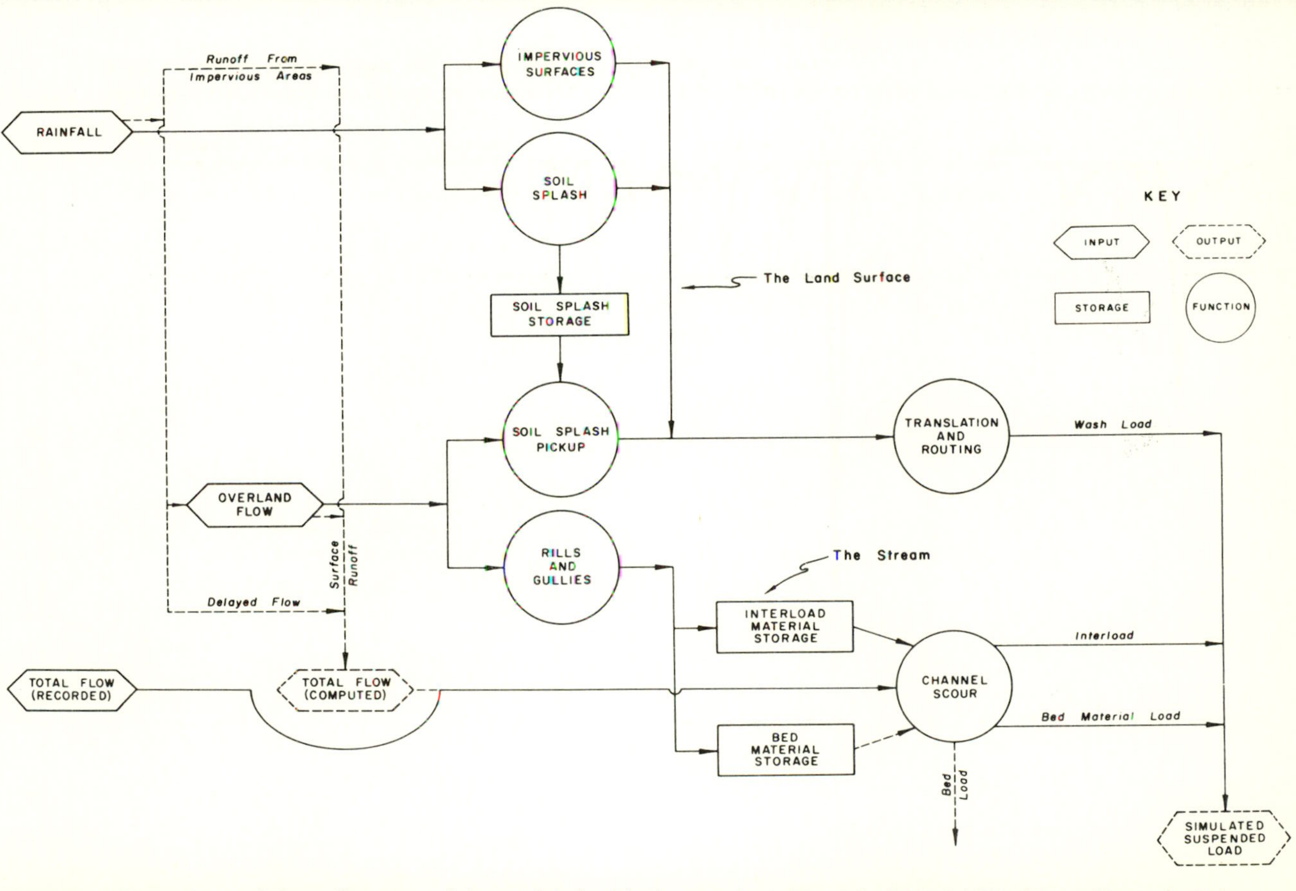

Fig. 7.24 Flow chart of the sediment model associated with the Stanford Watershed Model IV. (From Negev 1967.)

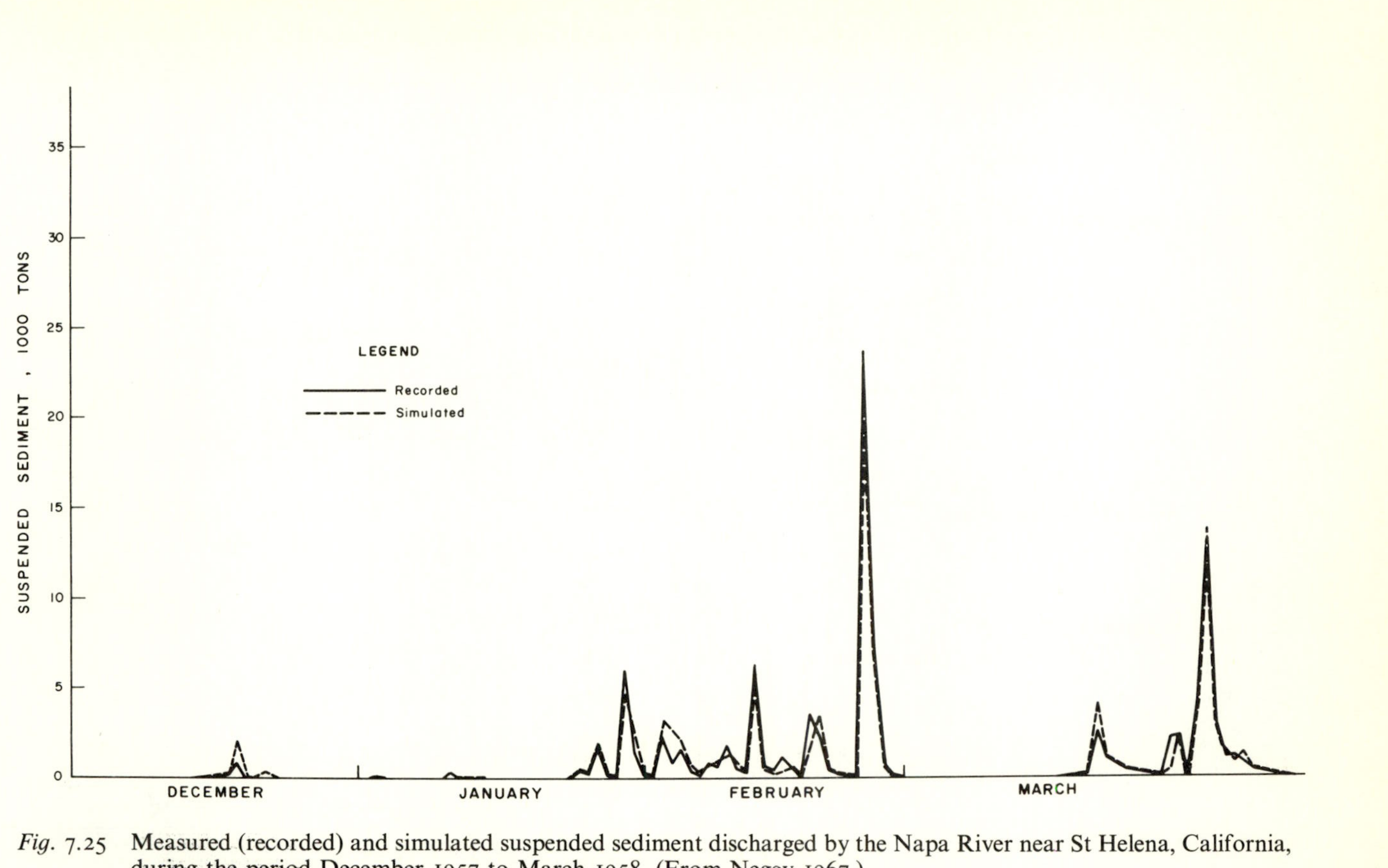

Fig. 7.25 Measured (recorded) and simulated suspended sediment discharged by the Napa River near St Helena, California, during the period December 1957 to March 1958. (From Negev 1967.)

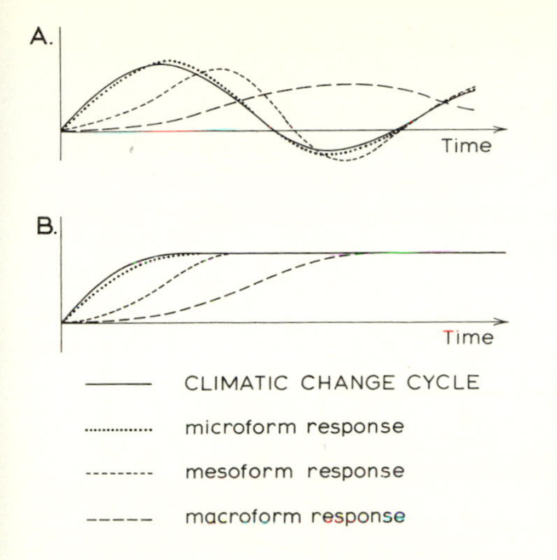

Fig. 7.26
The relations between (A) cyclic and (B) quasi-stepped inputs of climatically induced processes and micro-, meso- and macro-scale erosional outputs. (After Trudgill 1976.)

acceptable even if it does not faithfully reflect the assumed internal complexity of the system. Fig. 7.27 shows a grey box approach to drainage basin hydrology, lying mid way between the white and black box approaches (Amorocho and Hart 1964). Similarly, Patten *et al.* (1975, pp. 215–22) have argued forcibly that the undeniably nonlinear behaviour of individual ecosystem components and of the detailed dynamics is damped by the very complexity of the system such that the macroscale dynamics are regular, uniform and predictable, largely because superposition of inputs and outputs occurs at this level. It is suggested that the following mechanisms make ecosystem stability as a whole different from that of the individual organisms:

1 Species replacement provides a buffering mechanism.
2 Feedback control of population dynamics is based on population size.
3 Communities themselves exert a control *over* the environment, so damping the microclimatic and biochemical irregularities.
4 The steady state stabilization of internal conditions leads to balanced inflows and outflows.

Such mechanisms (sometimes collectively termed the 'law of large systems') imply that ecosystems 'adaptively evolve an internal control structure that maintains linear behaviour in the large through natural selection against component configurations that give untenable (poorly behaved) nonlinear dynamics' (Patten *et al.* 1975, pp. 217–18). This is held to be supported by the following observed nonlinear behaviour features of ecosystems.

1 They tend to have single steady states (e.g. involving ecological phenomena of convergent succession), whereas nonlinear models have multiple steady states 'usually too different in magnitude to depict the mosaic of communities which occur over the landscape' (Patten *et al.* 1975, p. 217).

2 Sinusoidal (i.e. seasonal, etc.) inputs result in outputs of the same angular frequency (see chapter 2).
3 They are essentially stable. Linear models are either unambiguously stable or unstable, whereas nonlinear ones are much more ambiguous in their behaviour.
4 If external inputs (i.e. forcings) are removed the system decays rapidly to zero, like a stable linear model.

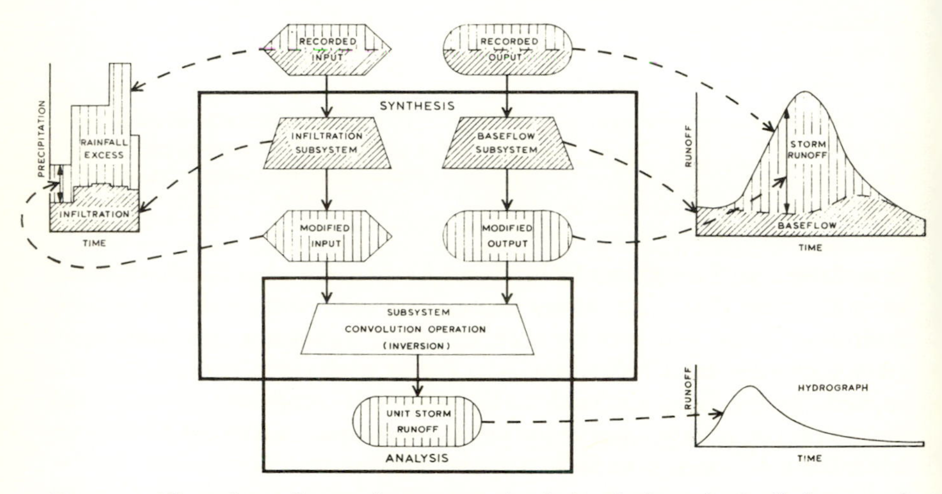

Fig. 7.27 Flow chart of a grey box approach relating drainage basin discharge and the unit hydrograph to rainfall input by use of partial system synthesis with linear analysis. (From Amorocho and Hart 1964.)

Two further theoretical considerations are adduced to explain the macroscale linear behaviour of complex ecosystems: firstly, that despite complex feedback connections the control proceeds 'downstream' from the input of limited resources so that donor-controlled flows can be used in modelling compartment interactions; and, secondly, that deviation of system dynamics is restricted because, unlike social systems, the ecosystem is capable of capturing, mobilizing and controlling only a small and geographically restricted portion of the energy and matter with which it comes into contact. Patten *et al.* (1975, p. 222) conclude that if the ecosystem is modelled as an interconnection of realistically modelled component mechanisms this does not usually result in adequate simulations of the whole ecosystem, and that linear models are preferable at this scale. This view raises a most important issue in both the modelling of environmental systems and in our attitudes to the environment – namely, to what extent are our views of reality conditioned, not by reality itself, but by the type of systems models we choose to adopt? The synthesis of complex systems may involve dozens of equations and detailed behavioural modelling, but it may result in a synthetic system which is unreal *in toto*, over-sensitive and subject to explosion or collapse due to the initial assumptions regarding the models themselves, rather than to any actual features of the real-world systems (see the discussion of 'world models' in chapter 9). For example, some ecological fears may have been generated by the behaviour of the man-made models rather than by nature herself (see chapter 9) and Patten *et al.* (1975, pp. 215–16) consider that ecosystems are not generally fragile but can withstand 'brutal' disturbances (except those massive enough to inhibit the operation of natural selection), being stable against perturbations as the result of evolutionary processes in which homeostatic feedback mechanisms are developed. As in the case of systems analysis, we require models which permit understanding of the system of interest under the global range of initial conditions and disturbances to which the system will be subject. Only in this way will the sudden 'unpredictable' and often catastrophic elements of natural system behaviour be amenable to understanding and prediction. For example, the pattern of earthquakes, volcanic eruptions, slope failure, snow and ice failure and phase transitions, flash floods, and atmospheric thunderstorms and tornadoes (Scheidegger 1975) are all amenable to treatment as singular points on a manifold of system dynamics defined on a space of fundamental control parameters (Thom 1975; and see chapter 4). However, even in systems synthesis, we must rely upon probabilistic predictions in most cases, which are often based on recurrence periods for rare events (Gumbel 1941, Scheidegger 1975) which we know from our restricted historical record with very limited accuracy.

7.4 Spatial systems

In chapters 2 and 4 the concept of *confined spatial cascades* was introduced where it was shown that an impulse cascaded through a series of first-order linear systems would result in progressively damped and attenuated outputs (see fig. 2.27 and Dmowski 1974). Despite the simplicity of the assumptions many natural phenomena (for example, the mass movement of waves in confined channels or confined shallow aquifers) conform with this pattern of behaviour. The concept of flood routing (Leopold and Maddock 1954, pp. 37–40) assumes that a rectangular rainfall input (impulse) into a channel is transmitted downstream as if the channel system were acting as a series of linear storages. Fig. 7.28 shows the effect of rainfall excess (i.e. rainfall minus infiltration) of 1 inch in 4 hours over a headwater fingertip tributary in terms of measured discharges at four points downstream; fig. 7.29 illustrates the case of the passage of an actual flood peak along the Chattooga River resulting from a rainstorm in the headwaters (Strahler 1965, pp. 276, 278); and fig. 7.30 shows synthetic results of a simulated storm over the whole basin of the Saline River above Tescott, Kansas (7300 square km.) where input occurred simultaneously into four spatial units (Linsley, Kohler and Paulhus 1949, pp. 534–5). Constricted subsurface flow produces something of the same

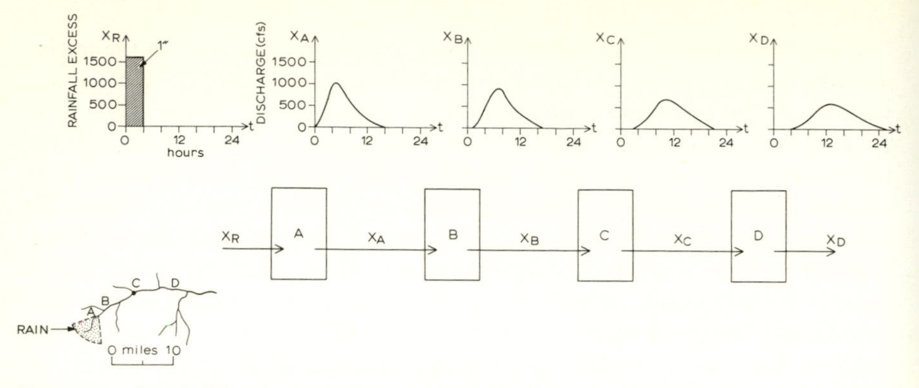

Fig. 7.28 A rainfall excess (input) (X_R) of one inch in four hours over a headwater catchment cascaded through four channel reaches (A–D) giving four outputs of discharge (X_A–X_D). (After Leopold and Maddock 1954.)

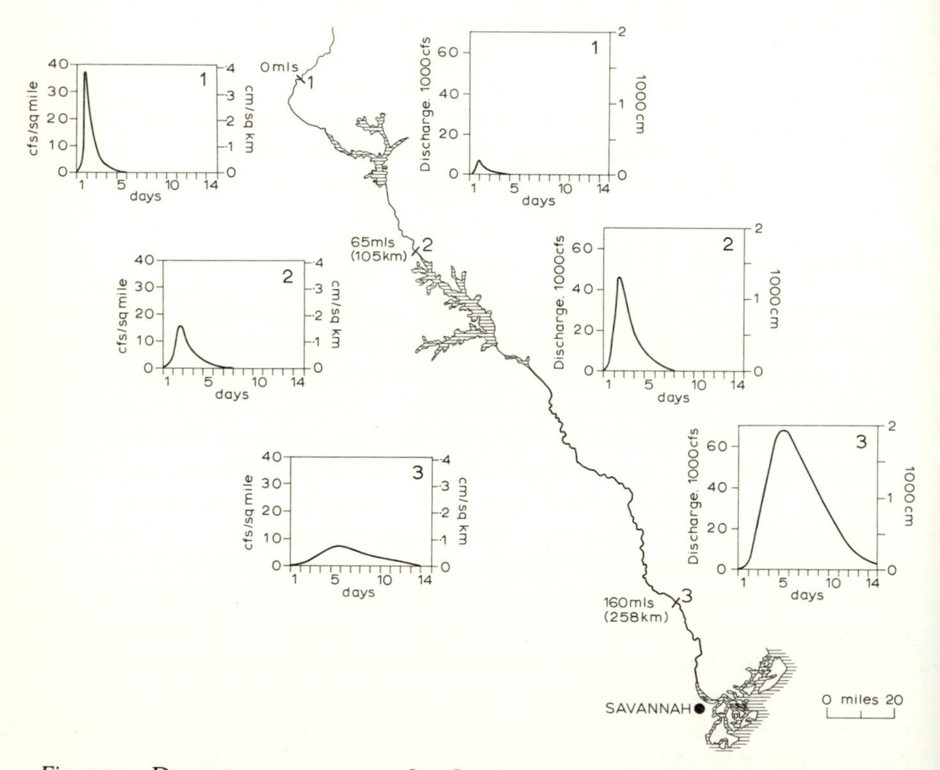

Fig. 7.29 Downstream progress of a flood wave on the Chattooga River, South Carolina and Georgia, resulting from heavy rainfall inputs in the headwaters. (Data from Hoyt and Langbein, partly after Strahler 1965.)

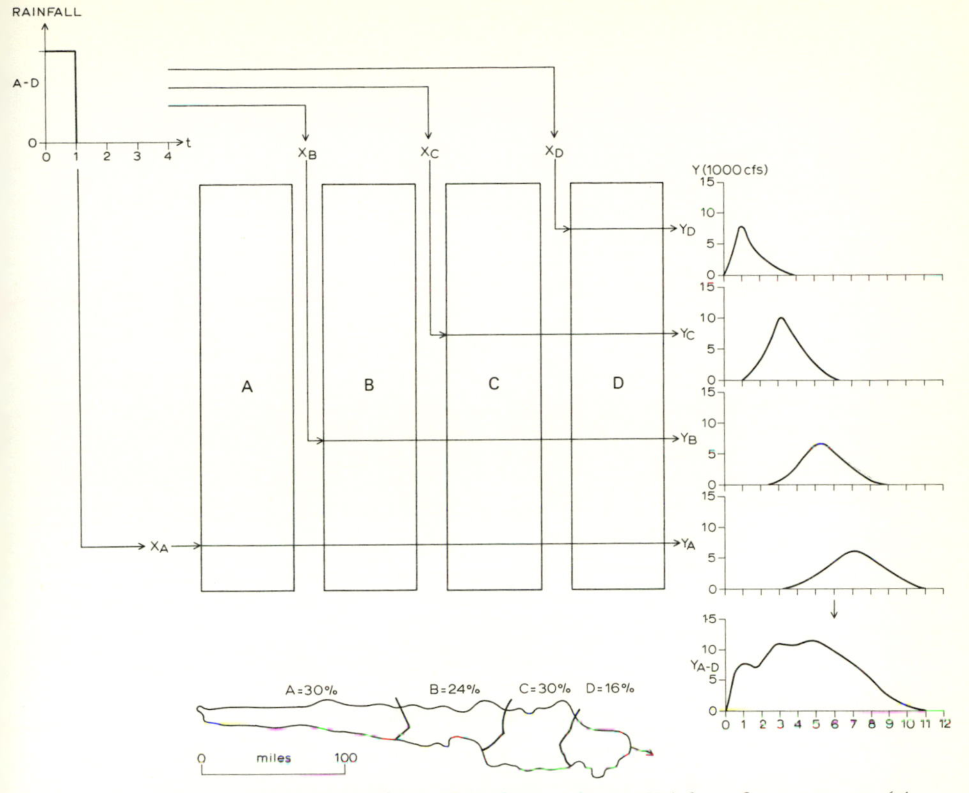

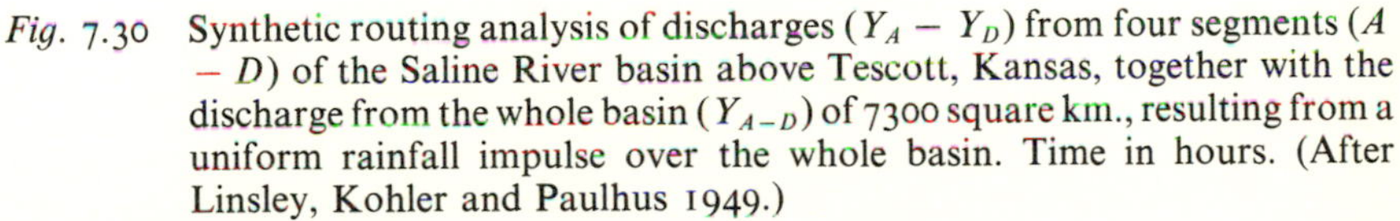

Fig. 7.30 Synthetic routing analysis of discharges ($Y_A - Y_D$) from four segments ($A - D$) of the Saline River basin above Tescott, Kansas, together with the discharge from the whole basin (Y_{A-D}) of 7300 square km., resulting from a uniform rainfall impulse over the whole basin. Time in hours. (After Linsley, Kohler and Paulhus 1949.)

result when the effects of vertical infiltration and lateral throughflow are combined. A natural plot (17 m long and 2·44 m wide), sloping 28 %, containing four soil layers (0–56 cm, 8 % clay; 56–90 cm, 15 % clay; 90–120 cm, 20 % clay; 120–150 cm, 29 % clay) produced the downslope discharges shown in fig. 7.31 at the various soil levels when subjected to artificial rainfall. Fig. 7.31A gives the discharges which resulted from an artificial storm of 5·1 cm/hour for 120 minutes when the soil condition was dry, and fig. 7.31B for a storm of 1·81 cm/hour for 100 minutes with wet antecedent conditions (Whipkey 1965).

Returning to confined channels, the transport of tracers introduced into streamflow (fig. 7.32) takes place by the operation of two processes, advection and diffusion, and it is necessary to understand something of this type of spatial cascade as a prelude to the references to river pollution in chapter 9. The following equation expresses the substantial time (t) rate of change in

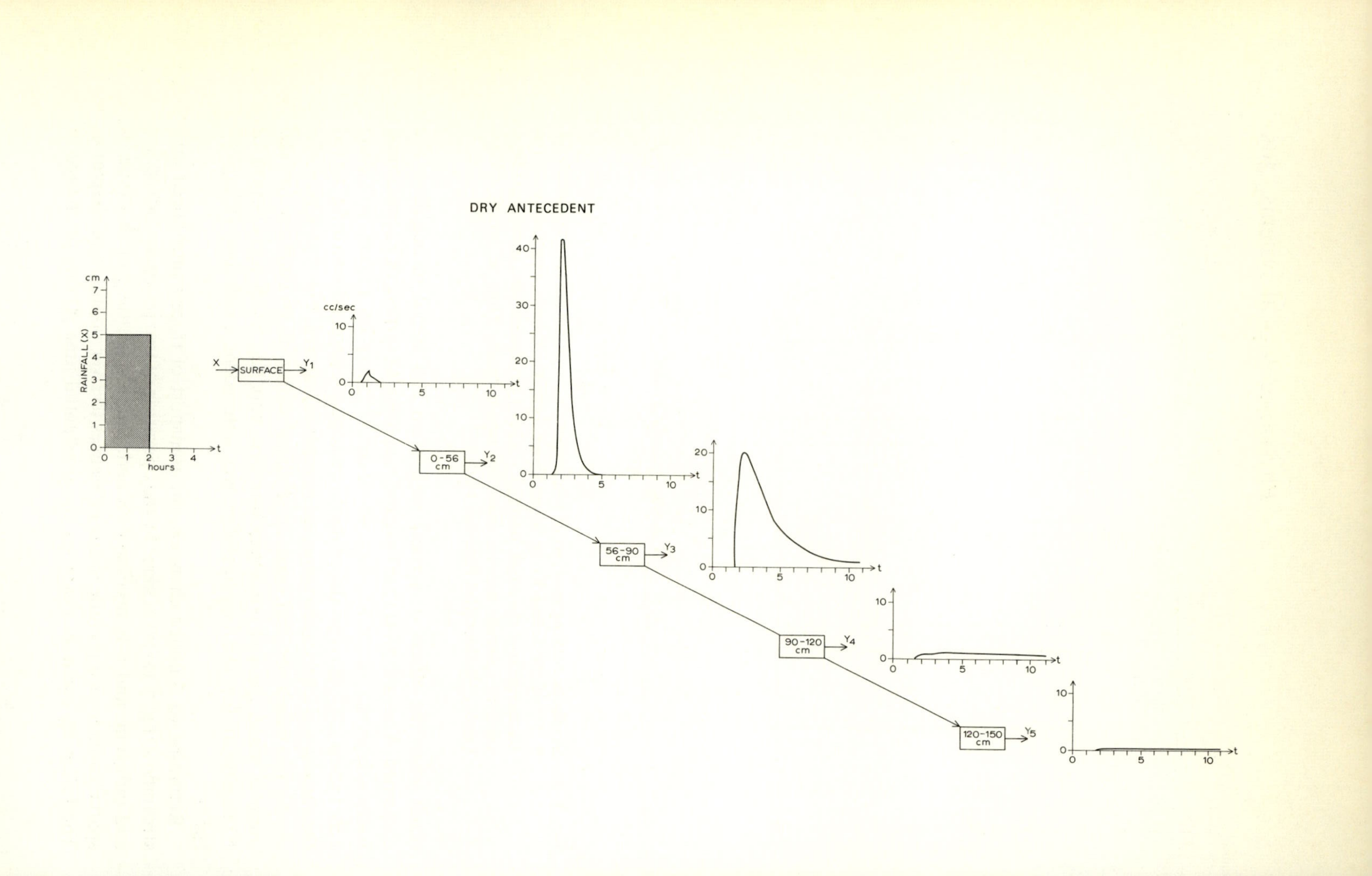

DRY ANTECEDENT
cm
RAINFALL (X)
hours
t
X
SURFACE
Y1
0-56 cm
Y2
56-90 cm
Y3
90-120 cm
Y4
120-150 cm
Y5
cc/sec

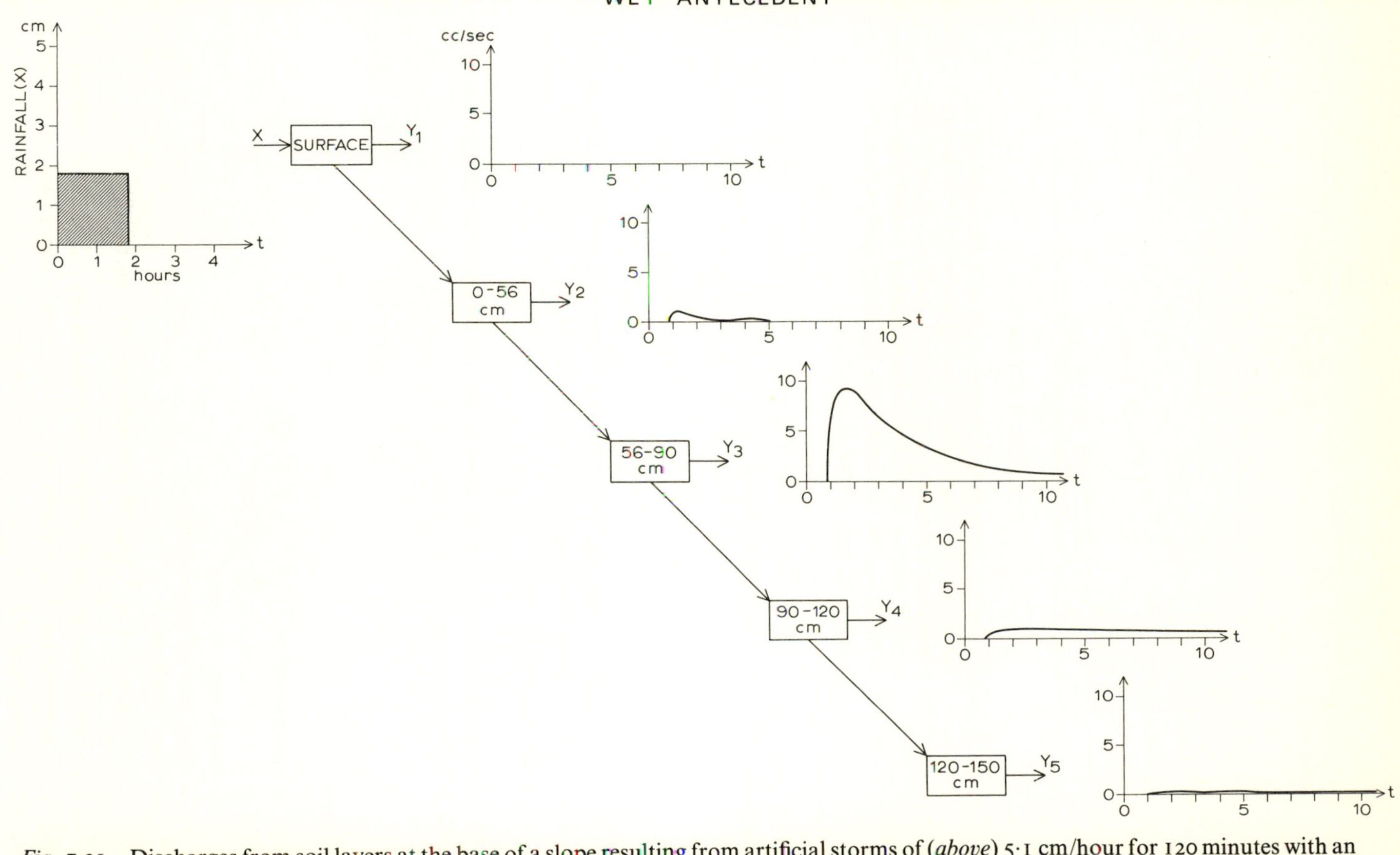

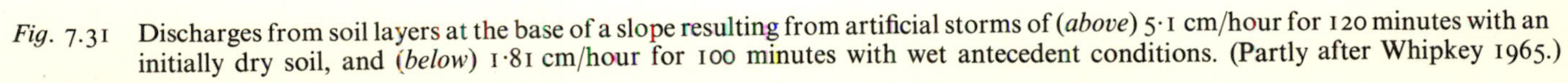

Fig. 7.31 Discharges from soil layers at the base of a slope resulting from artificial storms of (*above*) 5·1 cm/hour for 120 minutes with an initially dry soil, and (*below*) 1·81 cm/hour for 100 minutes with wet antecedent conditions. (Partly after Whipkey 1965.)

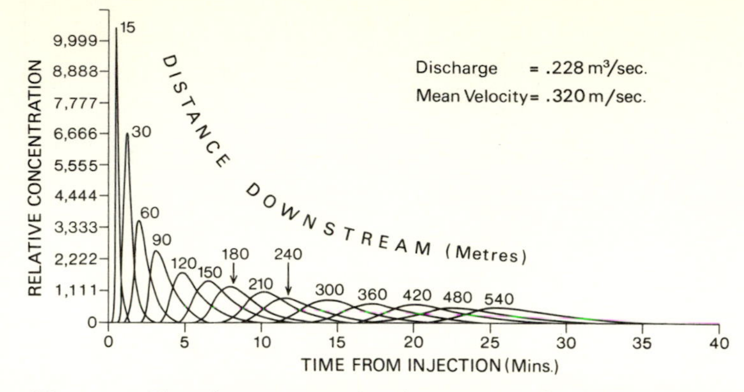

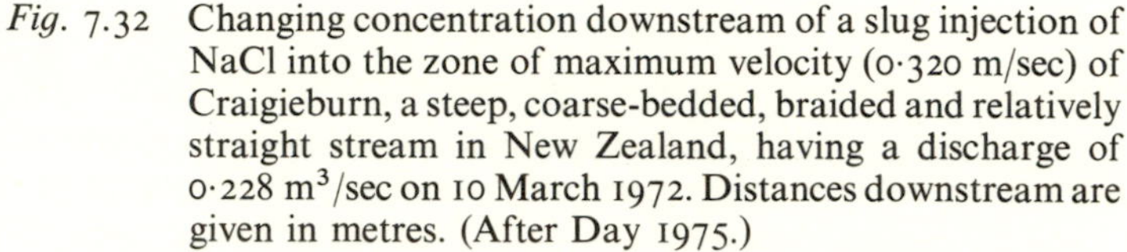

Fig. 7.32 Changing concentration downstream of a slug injection of NaCl into the zone of maximum velocity (0·320 m/sec) of Craigieburn, a steep, coarse-bedded, braided and relatively straight stream in New Zealand, having a discharge of 0·228 m^3/sec on 10 March 1972. Distances downstream are given in metres. (After Day 1975.)

tracer concentration (dc/dt) as it moves advectively in the flow direction (X) at the same velocity as the fluid (Rich 1973, p. 2):

$$\frac{dc}{dt} = \frac{\partial c}{\partial t} + U_x \frac{\partial c}{\partial x} + U_y \frac{\partial c}{\partial y} + U_z \frac{\partial c}{\partial z} \qquad (7.8)$$

where:

C = cross-sectional average of tracer concentration ($\triangleq FL^{-3}$: e.g. mg/l),
U_x, U_y, U_z = velocity components of the fluid (fig. 7.33).
$\triangleq$ = dimensionally equal to.

The advective model for a single, non-conservative variable can be otherwise stated (Thomann 1973, p. 357) as:

$$\frac{\partial c}{\partial t} = -U\frac{\partial c}{\partial x} - KC \qquad (7.9)$$

where:

C = concentration [$C(0, t) = W(t)/Q$]
K = first-order decay coefficient*
U = mean river velocity*
W = mass rate of waste input
Q = river discharge
(*assumed to be spatially and temporally constant).

This model leads to oscillatory pollution levels downstream, the amplitude of which is independent of the frequency of the waste input and the phase is linearly related to Kx/U.

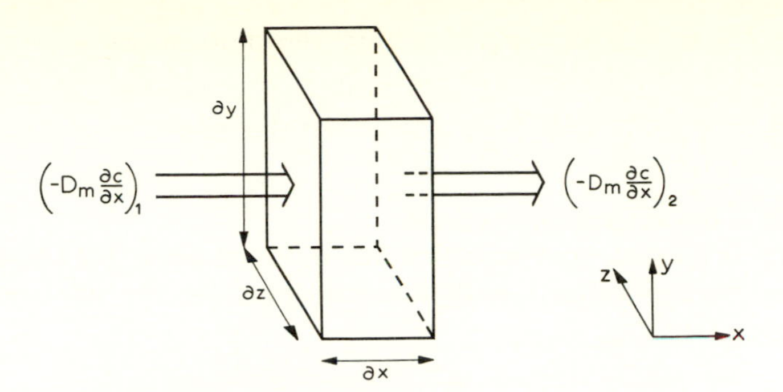

Fig. 7.33 Diffusion into and out of an incremental volume. (After Rich 1973.)

In contrast, diffusion involves the migration of a substance in solution or suspension through another substance in response to a concentration gradient. If the system is stationary in space and time (i.e. is homogeneous and isotropic; see chapter 4), then Fick's laws of diffusion (Rich 1973, pp. 3–5) can be used to express the rate of mass transport by diffusion (N_X) across an element of area A normal to the diffusion direction (X) as proportional to the concentration gradient of the diffusing substance (fig. 7.33):

$$N_X = -D_m \frac{\partial c}{\partial x} \tag{7.10}$$

where:

D_m = molecular diffusion coefficient ($\hat{=} L^2 T^{-1}$), which is directly proportional to absolute temperature, and inversely proportional to the molecular weight of the diffusing material and to the viscosity of the dispersing medium through which it is diffusing,

$\partial c / \partial x$ = concentration gradient of diffusing phase ($\hat{=} F L^{-4}$).

For the unsteady state in one dimension:

Accumulation = mass in − mass out.

Therefore

$$\frac{1}{\partial A} \cdot \frac{\partial m}{\partial t} = \left(-D_m \frac{\partial c}{\partial x}\right)_1 - \left(-D_m \frac{\partial c}{\partial x}\right)_2,$$

or

$$\frac{1}{\partial A} \cdot \frac{1}{\partial x} \cdot \frac{\partial m}{\partial t} = \frac{\partial c}{\partial t} = D_m \frac{\partial^2 c}{\partial x^2}, \tag{7.11}$$

which can be written for the general three-dimensional case as:

$$\frac{\partial c}{\partial t} = D_m\left(\frac{\partial^2 c}{\partial x^2} + \frac{\partial^2 c}{\partial y^2} + \frac{\partial^2 c}{\partial z^2}\right) = D_m \nabla^2 c \tag{7.12}$$

where ∇^2 = the differential operator.

Where, as is usual, both advection and diffusion occur, the above can be employed by combining it with the expression for the substantial time derivative:

$$\frac{dc}{dt} = D_m \nabla^2 c. \tag{7.13}$$

To introduce turbulence, the instantaneous velocity terms (U_x, U_y, U_z) can be replaced by temporal mean velocity terms ($\overline{U}_x$, $\overline{U}_y$, $\overline{U}_z$) plus the turbulent velocity fluctuations (U'_x, U'_y, U'_z). For the one-dimensional case:

$$\frac{\partial c}{\partial t} + (\overline{U}_x + U'_x)\frac{\partial c}{\partial x} = D_m \frac{\partial^2 c}{\partial x^2}. \tag{7.14}$$

Sometimes it is better to replace the turbulent velocity fluctuations by a turbulent diffusion coefficient (e), so that:

$$\frac{\partial c}{\partial t} + \overline{U}_x \frac{\partial c}{\partial x} = (D_m + e)\frac{\partial^2 c}{\partial x^2}. \tag{7.15}$$

For a typical environmental situation this may be further modified by replacing the advective term ($\overline{U}_x$) by a cross-sectional average of the velocity (U) and the diffusion term ($D_m + e$) by a longitudinal diffusion coefficient (E) ($\triangleq L^2T^{-1}$) such that

$$\frac{\partial c}{\partial t} + U\frac{\partial c}{\partial x} = E\frac{\partial^2 c}{\partial x^2}. \tag{7.16}$$

Thomann (1973, pp. 357–9) has expanded equation (7.9) to give the following mass-balance equation as a general dispersive model for a single non-conservative variable:

$$\frac{\partial c}{\partial t} = E\frac{\partial^2 c}{\partial x^2} - U\frac{\partial c}{\partial x} - Kc + \frac{W(t).\delta(x)}{A}. \tag{7.17}$$

where:

A = the cross-sectional area of the river
$\delta(x)$ = the distance delta function $1/L$.

If $P = KE/U^2$, then values of P under natural conditions are:

Upstream feeder streams	$<0{\cdot}01$
Main drainage rivers	0·01–0·5
Large rivers	0·5–1·0
Tidal rivers	1·0–10·0
Estuaries	$>10{\cdot}0$

Fig. 7.34 gives calculated values of C at distances of 10 and 30 miles downstream of a waste discharge input varying with a period of seven days; where $K = 0{\cdot}1$/day (i.e. organic matter first-order decay rate), $U = 0{\cdot}2$ fps (3·2 miles/day), and $E = 1$ miles^2/day – therefore $P \approx 0{\cdot}01$ and $K/U = 0{\cdot}031$/mile.

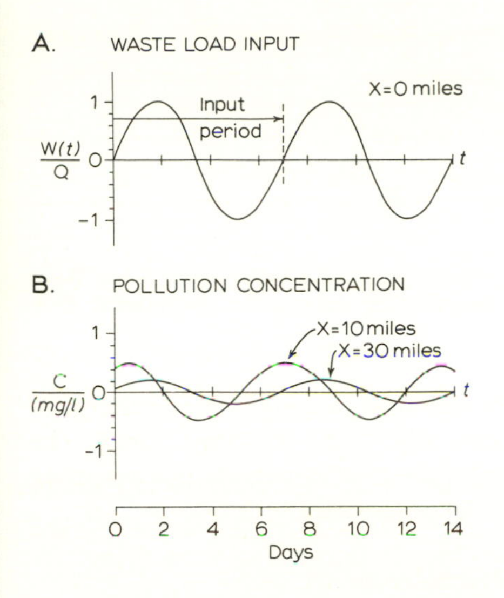

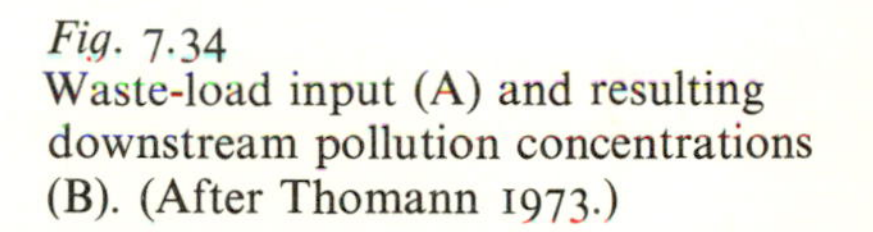
Fig. 7.34
Waste-load input (A) and resulting downstream pollution concentrations (B). (After Thomann 1973.)

Where one is concerned with the output resulting from an input of advecting and diffusing tracer into a constricted spatial cascade it is convenient to examine the limiting gross cases of continuous flow models: plug flow and mixed flow (fig. 7.35) (Rich 1973, pp. 31–2). *Plug flow* involves increments of the influent moving through the system without mixing, and being discharged in the same sequence as they are input. With *mixed flow* increments of influent are dispersed uniformly throughout the system such that the properties of the output are identical with the contents of the system. Fig. 7.35 shows the patterns of output for step and impulse inputs into systems having plug and mixed flow, together with the realistic streamflow case of intermediate mixed flow, which partakes of features of both plug and mixed flow. It is interesting to compare the output patterns of intermediate mixed flow with the measured output resulting from injections of a salt solution (rectangular and sudden or 'gulp' – i.e. a pulse larger than instantaneous)

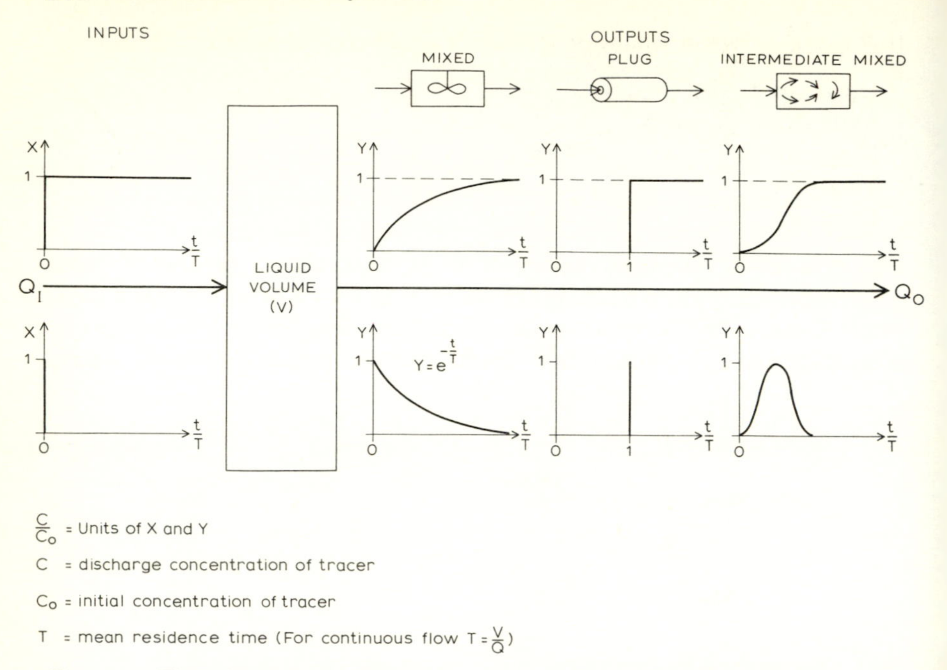

Fig. 7.35 Time-domain, residence-time distribution curves for continuous flow models. (After Rich 1973.)

which were made in connection with the chemical gauging of the discharge of a small stream having 9·94 cusecs flow (fig. 7.36) (Collinge 1964). The following equations relate tracer concentration to stream discharge, assuming there to be no initial tracer concentration in the river:

$$C_s = C_I \frac{q}{Q} \quad \text{(rectangular input)}$$

$$\int_0^T C_s \, dt = \frac{VC_I}{Q} \quad \text{(sudden or 'gulp' input)} \qquad (7.18)$$

where:

C_s = concentration of tracer in stream at sampling point (e.g. mg/l Na^+)
C_I = concentration of tracer in dosing solution (e.g. mg/l Na^+)
q = rate of injection of tracer (e.g. cusecs)
Q = stream discharge
V = volume of injected tracer solution.

Similar spatial smoothing and cascading effects can be observed in estuarine flow and pollution patterns (see, for example, Gameson 1973) where heat gradients from power stations, or changes in biochemical oxygen demand materials (BOD) and dissolved oxygen (DO) concentrations are usually considered. Low levels of DO induce death of fish and other aquatic life, a critical level being less than about 3 mg/l in temperate latitudes. High levels of BOD result from inputs of wastes, dissolved and suspended sludge, organic carbon and nutrients which increase the oxygen absorbing potential of the water. The interaction of DO and BOD can be reduced to first-order differential equations with transport delays which allow for transport lags between spatial locations (Young and Beck 1974). However, in a study of the

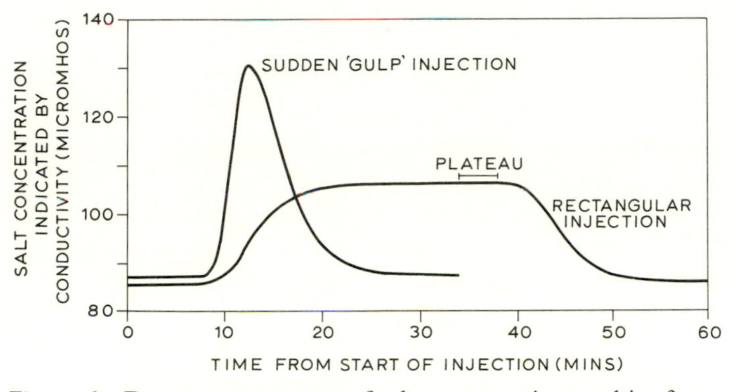

Fig. 7.36 Downstream outputs of salt concentration resulting from gulp (impulse) and rectangular inputs into a small stream (discharge = 9·94 cusecs) in connection with chemical discharge gauging (mho = 1/ohm). (After Collinge 1964.)

effect of tidal mixing from one waste source down the Potomac estuary, Wastler (1973) was able to isolate spatial and temporal cascades in three separate frequency regimes using spectral analysis, as discussed in chapter 2 (fig. 7.37): long term, diurnal and semidiurnal. Spatial modulation effects can be seen in the original time series, with dissolved oxygen levels decreasing down-estuary due to overall flow regimes and to tidal mixing effects for all spectral frequency components except the diurnal. Whereas at the annual and semidiurnal (single tide) frequencies tidal flow is important in reaeration of the estuary, at the diurnal level reaeration arises mainly from photosynthesis of plankton and the diurnal cycle of waste inputs. For BOD the inverse effect is shown but with a concentration of mixing by semi-diurnal tides through downstream drag in the immediate vicinity of the waste source below the outlet. A depth/down-estuary study of the effects of changes in DO and H_2S (fig. 7.38) for a Norwegian fjord shows this eutrophication effect very clearly and also the resulting decrease in numbers of aquatic invertebrates. Other estuarine studies show similar effects in the United Kingdom for the Thames, Clyde and Tay estuaries (Gameson 1973).

Atmospheric pollution is also governed by a set of space-time partial differential diffusion equations which describe the rate of transfer or flux of

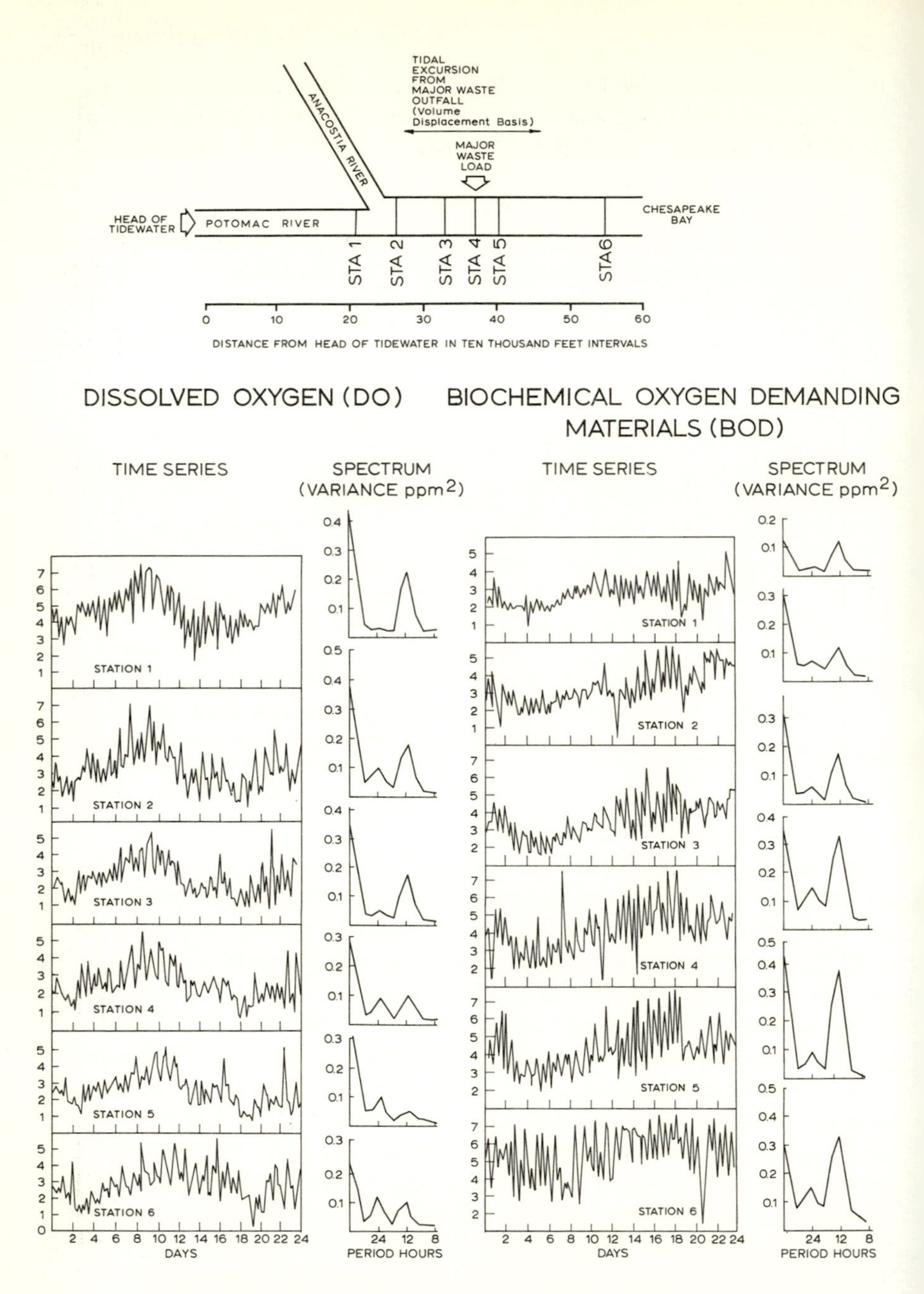

Fig. 7.37 A Location of six monitoring stations within the tide-water range of the Potomac Estuary.

B Modulation of the time series of measured dissolved oxygen (DO) and biochemical oxygen demand (BOD) at the monitoring stations and their spectra.

(After Wastler 1969.)

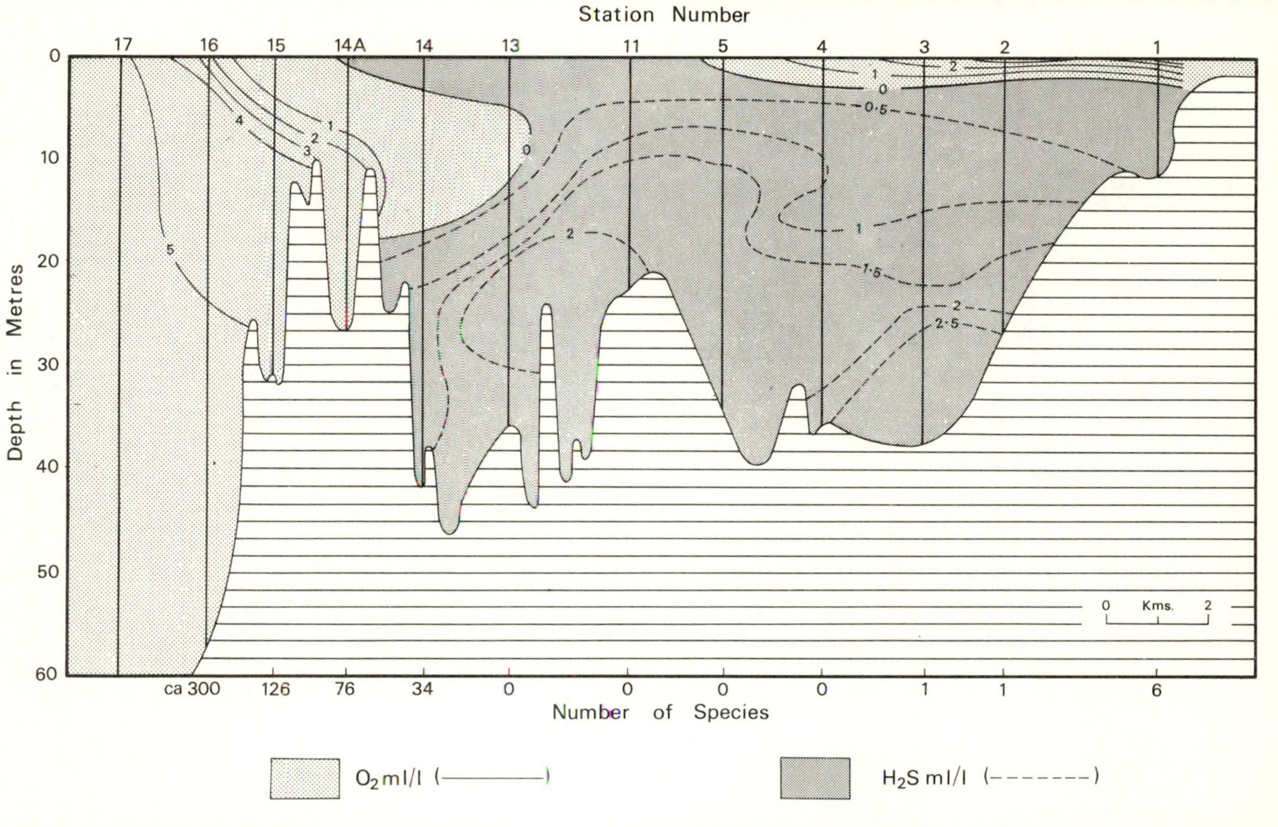

Fig. 7.38 Oxygen/hydrogen sulphide balance and its effect on the number of species of macroscopic invertebrates (excluding *Nemertini*) in heavily polluted Idefjord on the Swedish-Norwegian border. (After Dybern 1974.)

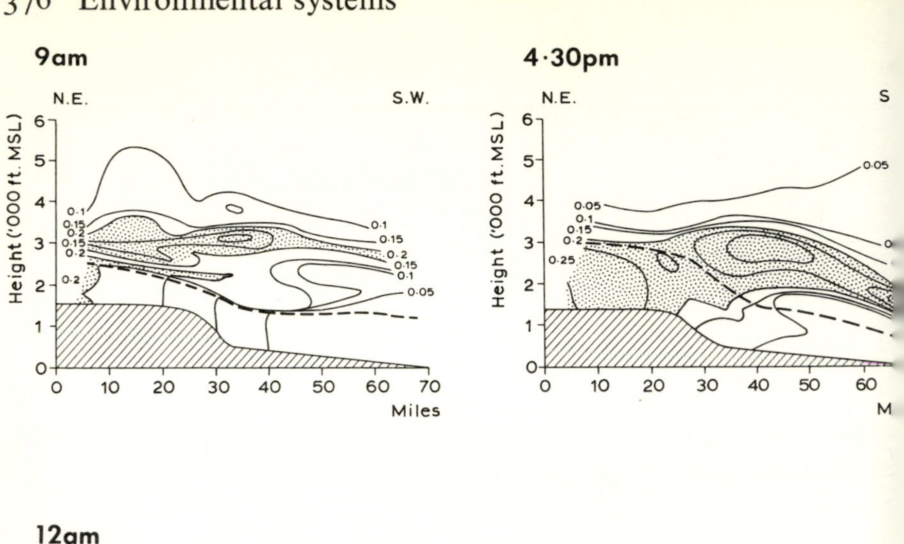

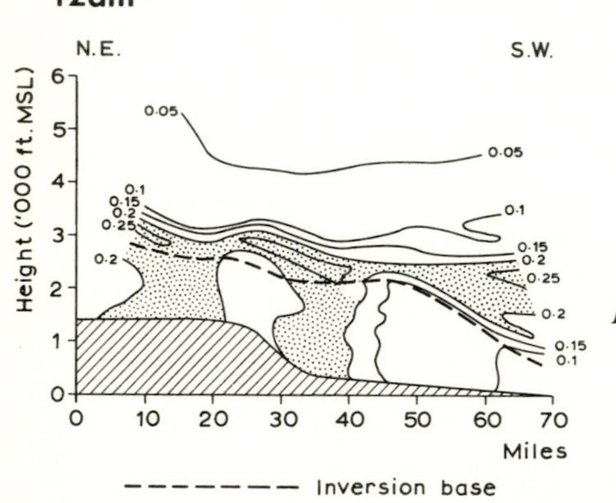

Fig. 7.39 Vertical oxidant concentration (ppm) in the Los Angeles basin (Rialto-Miro) for 20 June 1970. (After Edinger 1973; see Liu and Goodin 1976.)

pollutants from one spatial zone to another. An urban airshed is thus a dynamic space-time system with uncontrollable meteorological inputs and controllable inputs of pollution source and location in spatial cells, and one of whose principal time units of analysis is diurnal, because of the repetition of pollution patterns (Seinfeld and Kyan 1971, p. 174). Pollution control may be either open-loop, based on the initial state of the system and a sequence of predicted inputs, or closed-loop, based on the continuous manipulation of pollution inputs as the result of a monitoring procedure (see also chapter 9; for a more general and theoretical approach to airshed pollution control see Malone 1972). The systems approach to the study of atmospheric pollution must take into account the two major classes of pollutants. The primary pollutants (e.g. CO and SO_2), which do not undergo transformation in the atmosphere, are probably linearly related to total emissions, and in

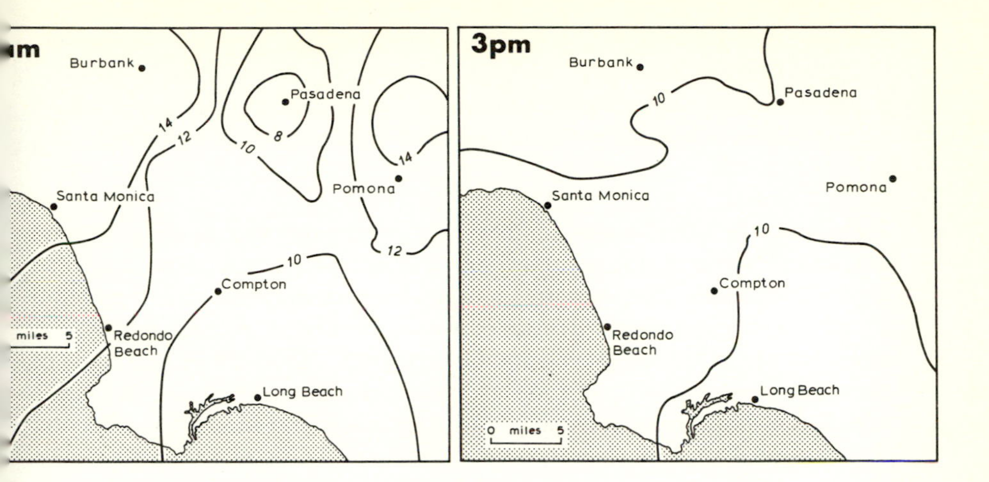

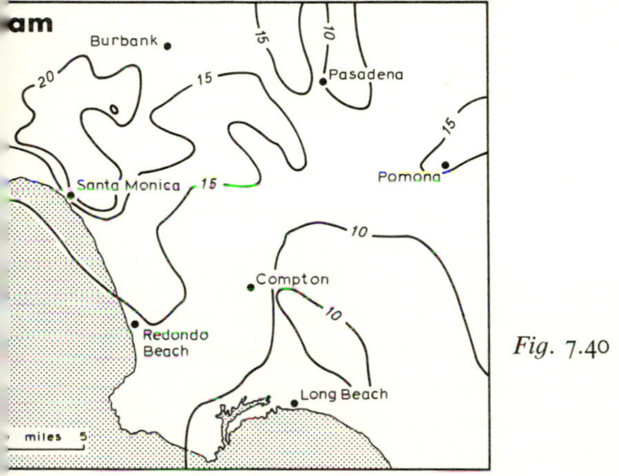

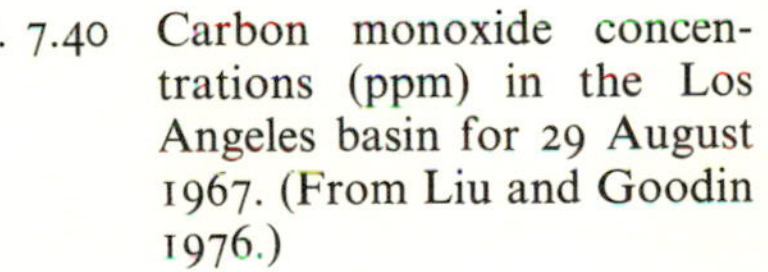
Fig. 7.40 Carbon monoxide concentrations (ppm) in the Los Angeles basin for 29 August 1967. (From Liu and Goodin 1976.)

consequence are relatively easy to predict by means of diffusion modelling. The secondary pollutants (e.g. photochemical smog) are produced by atmospheric chemical reactions and are more difficult to predict, even with simulation models (Seinfeld and Kyan 1971, p. 177). An urban airshed space-time system has three major classes of components:

1 Transport and diffusion transfer functions, concerned with atmospheric transport and dispersive processes between spatial cell units.
2 Reaction kinetics, involving meteorological inputs and photochemical transfer functions, describing rates of reactions occurring in the atmos-

phere as a function of pollution concentration, intensity of radiation, temperature, etc.

3 Emission inputs, involving a complete source inventory of mass pollutant emissions as a function of time and cell location (Seinfeld and Kyan 1971, p. 181).

Emissions are often point sources (e.g. chimneys and cooling towers) distributed across space with variable height, rate, temperature and time regime of emission. They most frequently form periodic inputs into the overall climatic-ecological system due to seasonal, weekly or diurnal patterns of domestic and industrial heating and of industrial activity. However, many of the most important pollutants are also widely distributed and mobile sources of pollution (e.g. cars, aircraft, ships and other transport), in addition to having a cyclical intensity. Whatever the source and form of air pollution inputs, once they are emitted they are controlled by the air medium in which they are diffused. Hence meteorological conditions of vertical temperature gradient, eddy forms, wind speed and direction control the spatial diffusion rate and direction. For example, Edinger (1973) shows the variation of oxidant concentrations in Los Angeles over a typical summer day as a function of the level of the inversion base, height above sea level, topography and distance inland (i.e. downwind) (fig. 7.39). The zone below the inversion base has rapid vertical mixing resulting in vertical centres of pollutant concentration, and the zone below the inversion base has rapid horizontal mixing and advection controlled by prevailing wind conditions. The spatial variation will also differ during the day based on the source of emission and the components of horizontal and vertical mixing (as shown for the Los Angeles basin in fig. 7.40). The two preceding figures show the control of the diurnal meteorological cycle on pollution concentration. Longer-term variations will depend on whether or not the synoptic situation is conducive to pollution stagnation (as with anticyclones during the British winter and stable inversions in California). In addition, when the pollutant is photochemically active, its rate of reaction with other pollutants from other sources must be taken into account. Also pollutants become absorbed into the overall physico-ecological system and are cycled through various hydrological and biotic subsystems. The form of this environmental spatial interaction for the sulphur cycle (fig. 7.41) gives an understanding of two forms of pollutant-biosphere interaction: *synergism* resulting from the simultaneous effects of two pollutants (e.g. SO_2 and ozone, or SO_2 and nitrogen oxides) to produce harmful effects on vegetation, and *antagonism* resulting from non-harmful interaction between two pollutants (e.g. SO_2 and ammonia to give ammonium sulphate). One of the effects of SO_2, oxidants and fluorides is to discolour and kill vegetation (Air Pollution Control Association 1970).

Meteorological conditions are also very important in their effects on disease diffusion across space. Tinline (1971, 1972), in a study of the 1967–8 foot and mouth epidemic in Britain, found wind direction and intensity to be the most important factor in generating long-distance diffusion (especially through the meteorological effect of lee waves) (fig. 7.42A), whilst short-distance diffusion effects resulted mainly from local contiguous contamination which could be modelled by simple spatial Markov (lag) operators (fig. 7.42B). Disease diffusion is governed by the contact patterns of the carriers so that in most cases it is the structure of human contact systems which is most important.

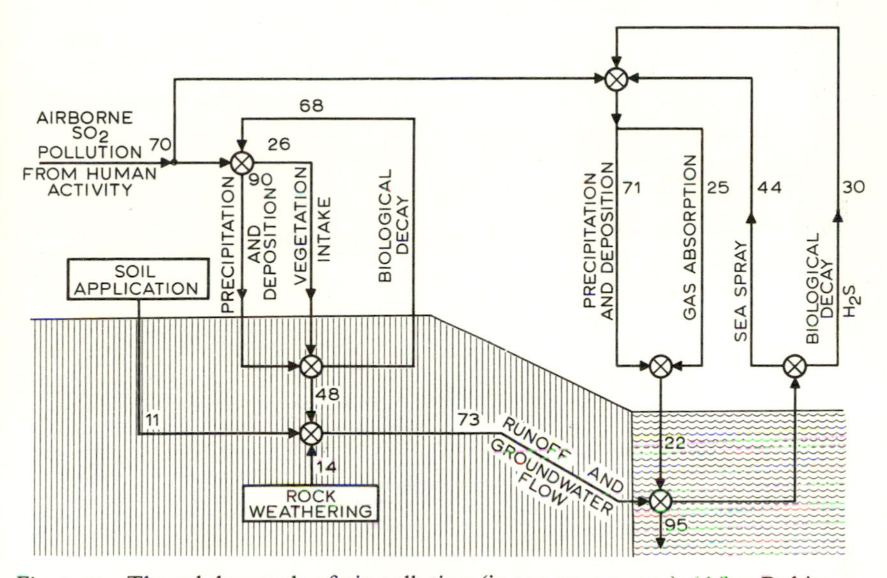

Fig. 7.41 The sulphur cycle of air pollution (in tonnes per year). (After Robinson and Robbins 1968, from Garland 1975.)

Measles epidemics, for example, can be explained as a result of socio-economic contacts (fig. 7.43, inset) based upon the structure of transport networks and urban hierarchies (Haggett 1972, Cliff *et al.* 1975) which can be recognized by tests of spatial autocorrelation and modelled by space-time lag operators incorporating network and hierarchic effects into the weights matrix (see chapter 4). More sophisticated forms of cross-correlation effects between measles notification time series (fig. 7.43) and the fitting of stochastic models of diffusion have been pursued by Cliff *et al.* (1975).

The pattern of climatic and meteorological shifts is itself a spatially evolving system and is perhaps, through its effects on all other physico-ecological systems, the most important overall control of the magnitude and frequency of physical system response. Short-term and long-period secular changes in climate have been recognized in many studies (Lamb 1966, Bryson 1974, Winstanley 1973) and have important impacts on resources in man-

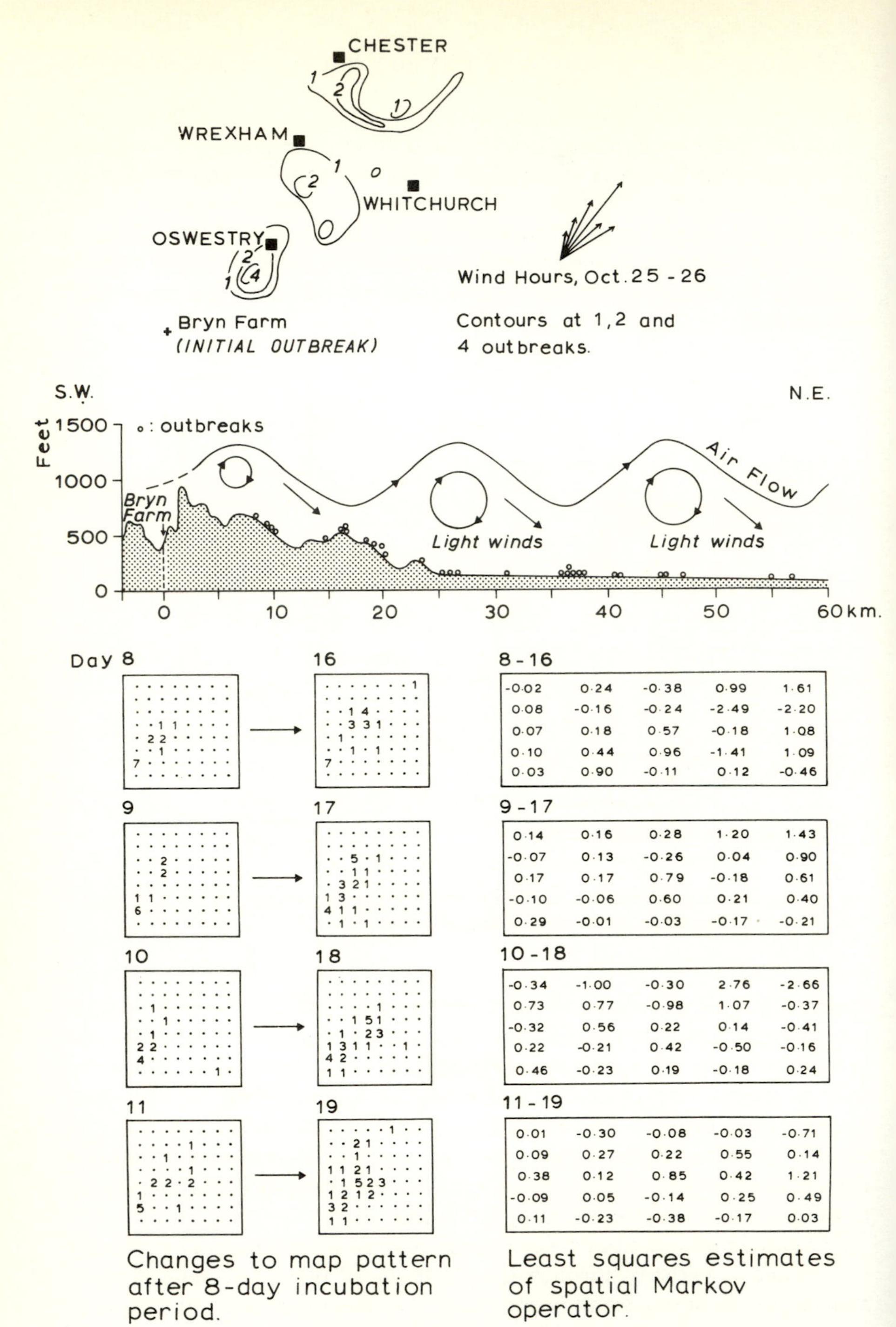

Fig. 7.42 Spread of foot and mouth disease in the English Midlands, 1967–8.
Above: Spatial and distance distribution as a function of wind direction and lee waves.
Below: Short-distance diffusion effects modelled by spatial Markov (lag) operators.
(After Tinline 1971, 1972.)

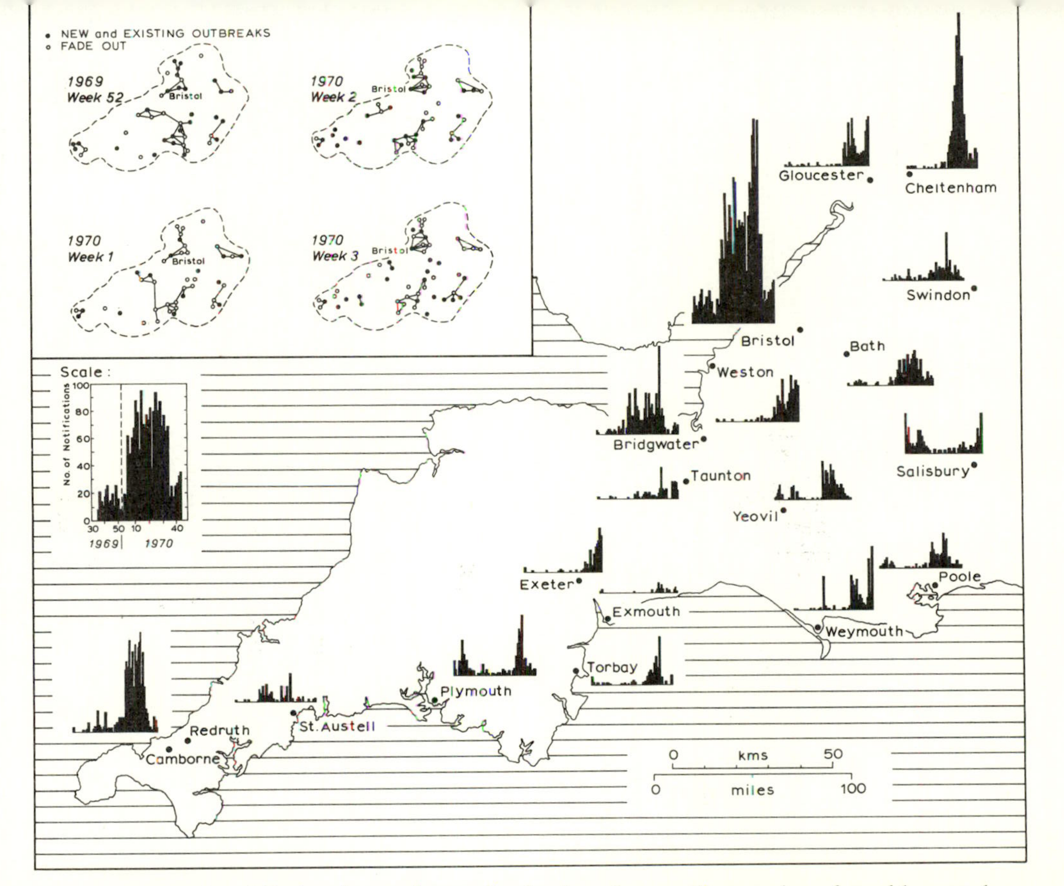

Fig. 7.43 Measles diffusion in southwest England, 1969–70. Time series of weekly measles notifications for a number of General Register Office areas. Inset shows the spatial pattern of measles notification for a four-week period represented as a planar graph with infected General Register Office areas linked. (Partly after Haggett 1972; see also Cliff *et al.* 1975.)

environment interfacing (see chapter 9). The spatial pattern of the changes controlling this variation seems to result from latitudinal oscillation of the jet stream on a world scale governed by the seasonal and latitudinal distribution of solar radiation. The most convincing explanation of these changes seems to be variation in the Earth's orbit around the Sun. Milankovitch (1920) hypothesizes three components of this variation: eccentricity (stretch of the size of the Earth's orbit), wobble (of the Earth's axis), and roll of the perihelion (shifting longitude). Each component has cyclic periods of 90,000–100,000 years (stretch), 41,000 years (wobble), and 20,000 years (roll) (Vernekar 1971, Calder 1974). The superimposition of these components provides some convincing correlation of this theoretical model with available fossil evidence: paleobotany and oxygen isotope changes in marine fossils. Using deductive formulae for these orbital changes and the latitudinal equations of Brouwer and van Woerkom, and Sharaf and Budnikova, world-scale shifts in climatic belts derived from variations in solar radiation receipt have the space-time pattern shown in fig. 7.44 (derived from Vernekar 1971) from 1 million years ago to 100,000 years into the future. A new ice age, which is just beginning, corresponds to the coincidence of small wobble and large stretch and is forecast to intensify considerably before ending in 120,000 years time (Calder 1974). Although synthetic in their content, such world-scale models are still essentially black box since the linkages to levels of radiation receipt and to overall mechanisms which control the spatially shifting patterns are remote and tenuous. Nevertheless, models of the past, and especially of the future, evolution of climate systems become crucial in considerations of man-environment interfacing since they govern the level of input of the major source of renewable energy and also the overall level of activity in the biomass which is essential to the maintenance of man's existence.

In contrast with the spatial black box models adopted for many spatial diffusion type studies (e.g. disease diffusion), ecologists are adapting systems synthesis techniques to the spatial analysis of ecosystems based on evolution of ecological communities as organisms following ideas of Clements, Tansley and Cooper. This is well illustrated by a recent complex study of phytoplankton biomass in western Lake Erie (Di Toro *et al.* 1975), where the microscopic plant life could reach densities detrimental to water use by affecting benthic organisms, using dissolved oxygen, releasing organic carbon and nitrogen, etc., affecting water transparency and the solar energy economy. The biomass model involved some eighteen variables (some exogenous – e.g. water temperature, quality of tributary streams; some endogenous – e.g. nitrogen production) relating to seven spatial segments of western Lake Erie for seven months of the year (April–October), giving forty-nine space-time compartments. Fig. 7.45 shows the causal interrelationships between the endogenous variables and fig. 7.46 the steady state exchange and advective water flows between the spatial cells. The model is based, in part, on mass-conservation equations of the form:

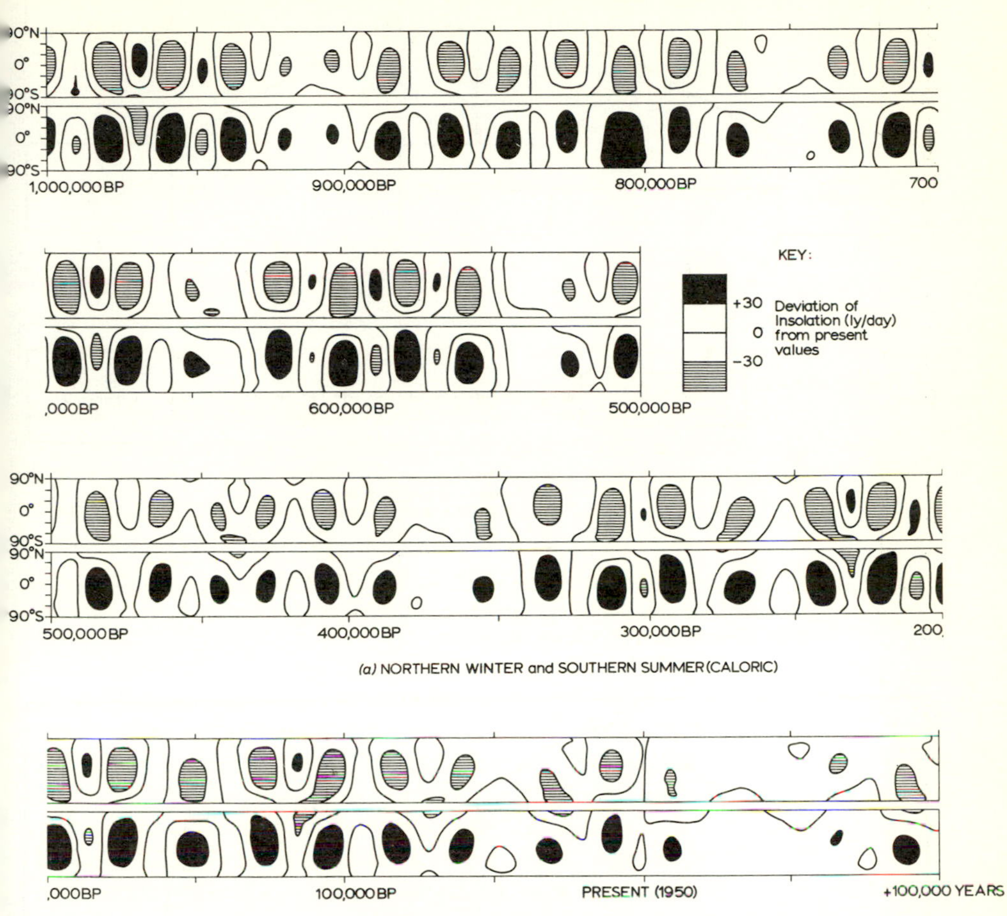

Fig. 7.44 World-scale climatic shifts of solar radiation based on three superimposed periodic components of variation in the Earth's orbit: latitude variation on *Y* axis, time on *X* axis (thousands of years before and after the present: 1950). (After Vernekar 1971.)

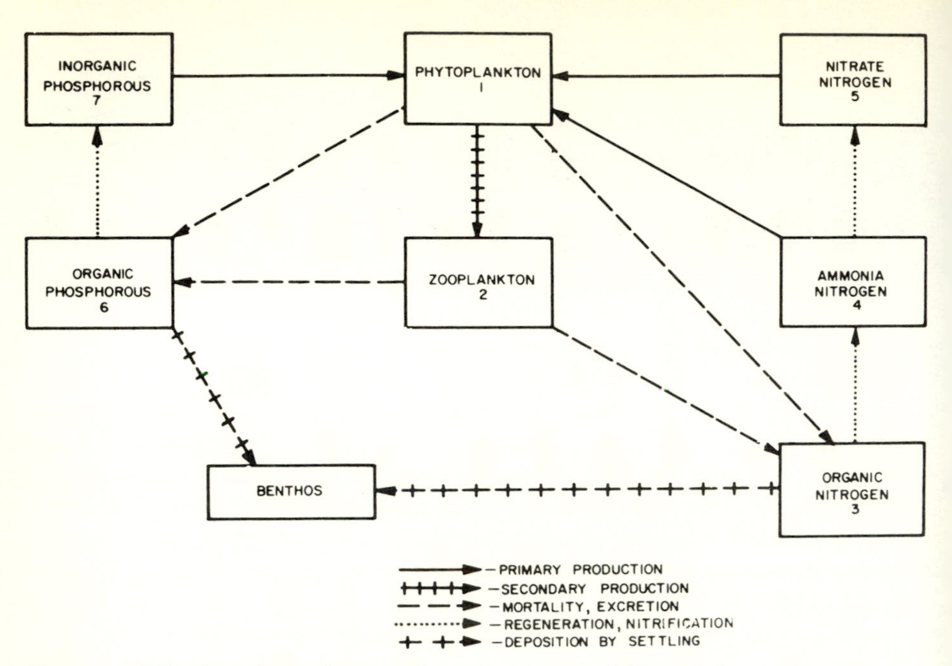

Fig. 7.45 Kinetic pathways between the endogenous variables relating to the simulation model of the space-time system of phytoplankton production in western Lake Erie. (From Di Toro *et al.* 1975.)

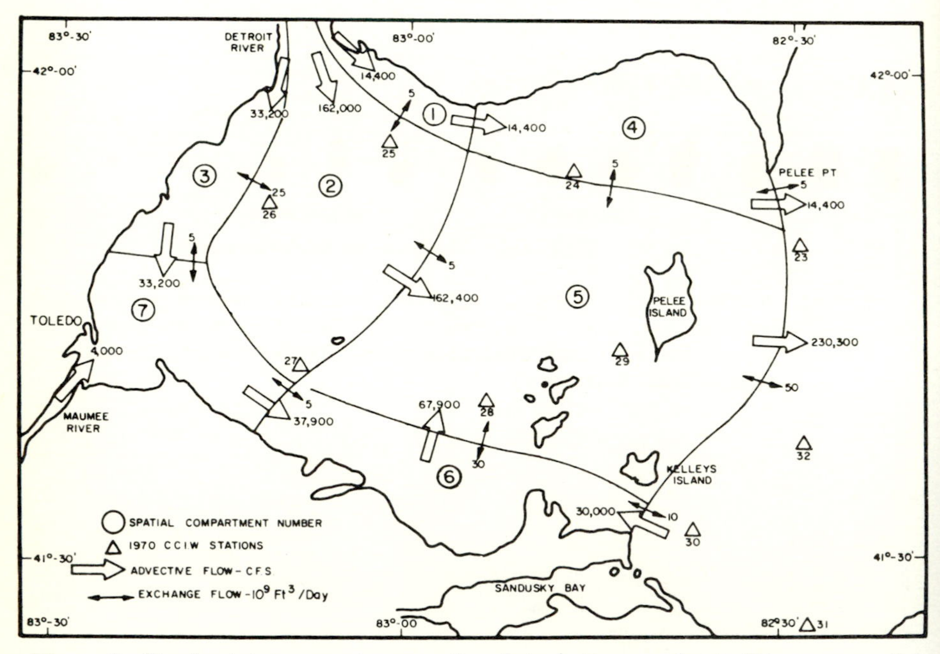

Fig. 7.46 Steady state transport assumptions into, between and out of the seven spatial cells of western Lake Erie. (From Di Toro *et al.* 1975.)

$$V_j \frac{dc_{ij}}{dt} = \sum_k Q_{kj} C_{ik} + \sum_k E'_{kj}(C_{ik} - C_{ij}) + \sum_k S_{ijk} \qquad (7.19)$$

where:

V_j = segment volume,
S_{ijk} = the kth source (+) or sink (−) of substance i in segment j,
E_{kj} = the bulk rate of transport of C_{ik} into, and C_{ij} out of, segment j for all segments k adjacent to segment j,
Q_{kj} = the net advective flow rate between segments k and j.

Fig. 7.47 gives examples of input variations of two of the exogenous variables (water temperature and tributary stream organic nitrogen concentration) and fig. 7.48 compares the seasonal variation of phytoplankton chlorophyll concentration in the lake segments predicted by the model with measured amounts. Of more general interest than the relations between predicted and measured amounts of phytoplankton is the apparent stability of this complex spatial model.

Similar diffusion patterns also characterize macroscopic plant and animal

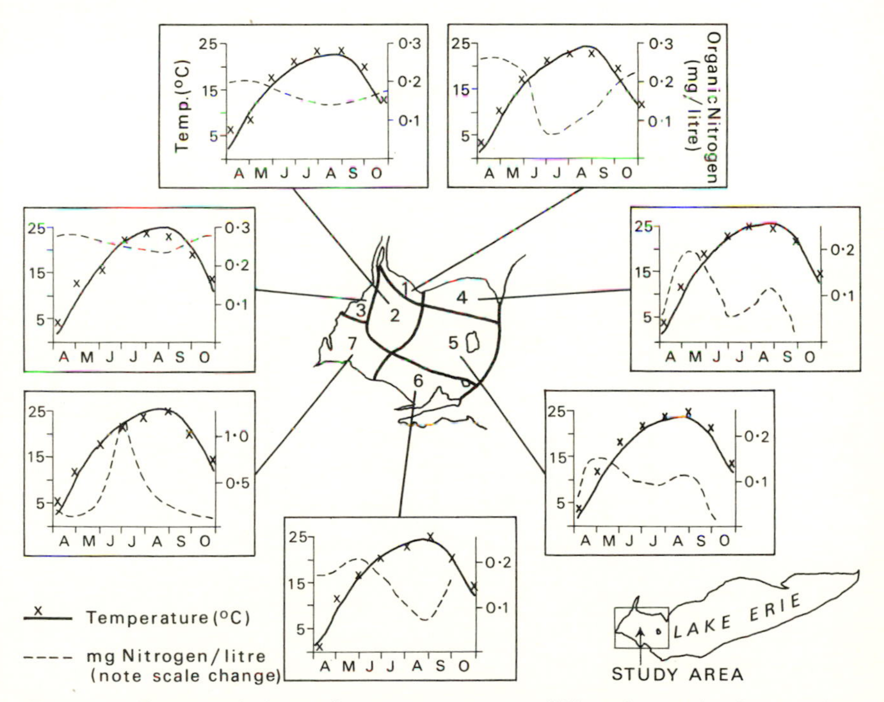

Fig. 7.47 Input variations of water temperature (°C) and organic nitrogen (mg nitrogen/litre; dominantly from tributary streams) into western Lake Erie during the period April–October 1970. (After Di Toro *et al.* 1975.)

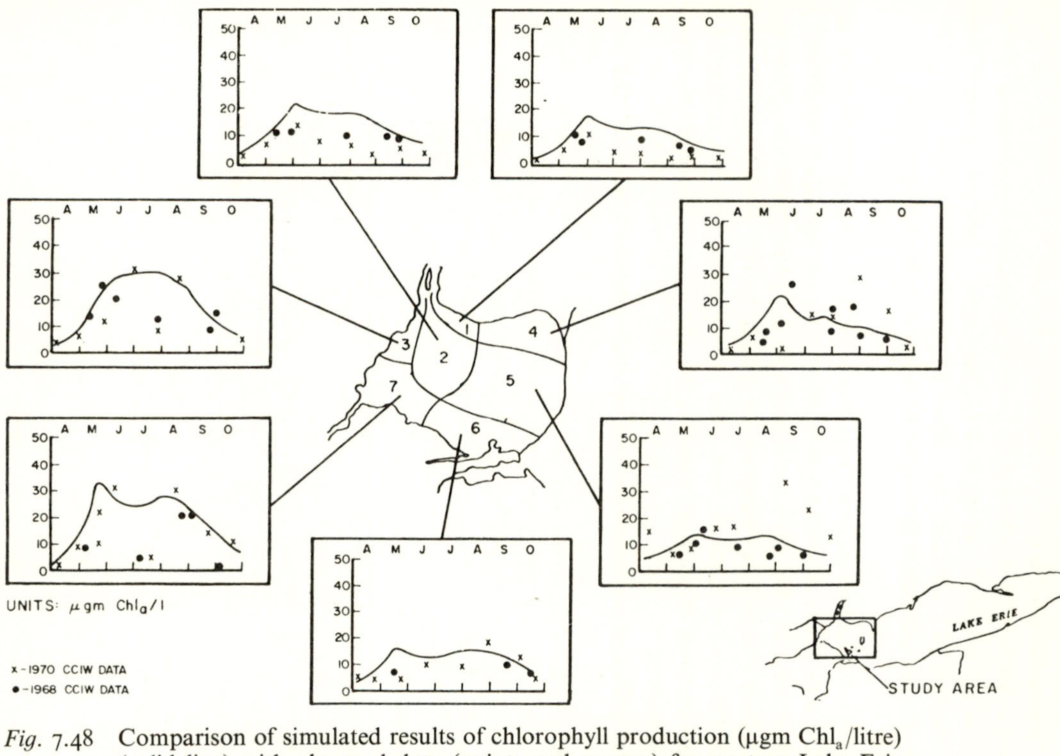

Fig. 7.48 Comparison of simulated results of chlorophyll production (μgm Chl_a/litre) (solid line) with observed data (points and crosses) for western Lake Erie during the period April–October. (From Di Toro *et al.* 1975.)

populations. In the spatial pattern of these, the overall controls are the uneven distribution of environmental resources together with physiological mechanisms (mainly in plants, e.g. dispersal by wind, water and ballistic action) and sociological forces (mainly in animals, e.g. shoal behaviour, herding and society with division of labour, role playing and age structure) (Odum 1963). In each case aggregation, herding or clumping gives better chance of survival by confusing predators, encouraging cross-pollenation and cross-breeding. This gives rise to close-packing distributions which can be described by Thiessen polygons resulting in patterns similar to the central place packing of retail market areas; each zone being governed by individual or collective territoriality. Territoriality is a function of the kind and degree of sociality in the population; group size, age and sex pattern, cohesion of connections in the herd, tribe or flock, the permeability to other sociological formations, the degree of internal compartmentalization and differentiation of behaviour, and patterns of information flow (Wilson 1975, pp. 16–18). The resulting patterns of sociobiological groups are variable (some examples are shown in fig. 7.49) but resemble the multilevel structure of decision making units (see chapter 6). Changes to these territorial areas occur through changes to the social networks and by diffusion and migration. Changes in social networks between species represent changes in interspecific territoriality which is common in birds and is also present in ants, crayfish, some lizards, squirrels, chipmunks, gophers and gibbons (Wilson 1975). The several pathways of interspecific development, shown for two habitats and three species in fig. 7.50, can be described by stochastic models of the various stages of the diffusion process:

1 Initial implanting of colonists.
2 Diffusion around new location.
3 Condensation, density rising to competition and displacement of existing population,
4 Saturation with a slow increase to the filling of all available ecological niches.

Monte Carlo methods can be used to simulate these stages (Hägerstrand 1953 1967) (fig. 7.51). An example of diffusion of the ant species *Anoploleris longipes* in Tanzania at the expense of competitors (fig. 7.52) shows how the diffusion process moves as a 'wave' to cover all areas which can support the species (i.e. sandy, thinly vegetated soil) (MacArthur 1972). Various probability density models can be used to describe the resulting point pattern of locations (i.e. Neyman type A, Poisson, negative binomial, Pearson curves, etc.) (Bartholomew 1973, Cliff *et al.* 1975). Over long periods there may be changes to the whole biomass associated with shifts in climatic belts and with ice ages: for example, the paleobotany studies of fauna and flora changes since the Pleistocene (Lamb 1966, Deevey 1949, Hultén 1937, Vita-Finzi 1973) especially using pollen analysis, diatom counts, carbon 14 and isotope dating.

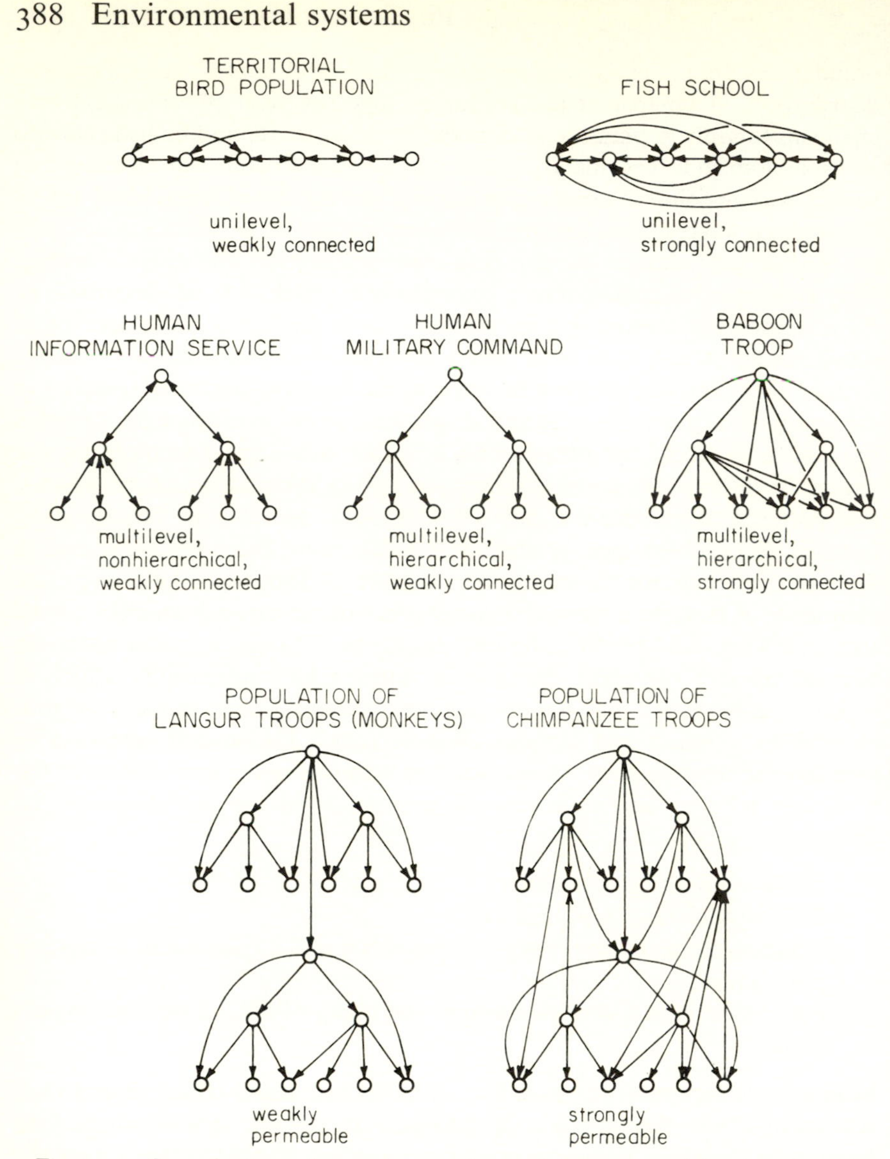

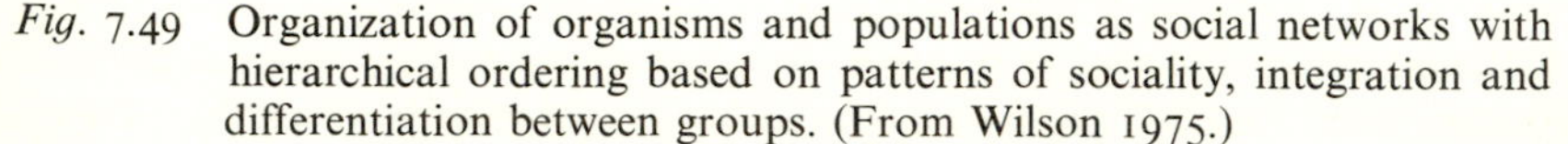

Fig. 7.49 Organization of organisms and populations as social networks with hierarchical ordering based on patterns of sociality, integration and differentiation between groups. (From Wilson 1975.)

Shifts in population concentrations and distribution over time may be affected by various biological, environmental, physical and man-made barriers to movement. The interaction of these barriers with the genetic, physiological and sociological forces which determine territoriality lead to the progressive creation and annihilation of differentiation between spatial areas. This results either in the generation of distinct regional communities, or in

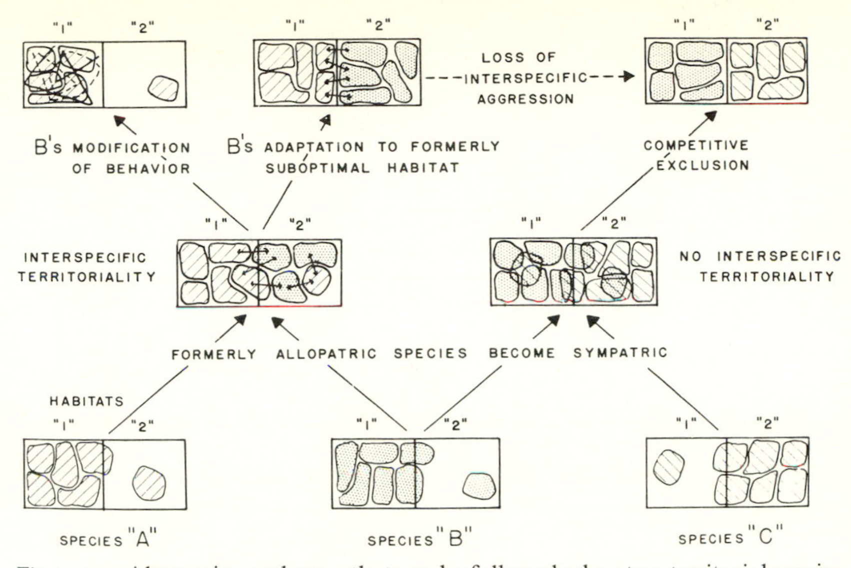

Fig. 7.50 Alternative pathways that can be followed when two territorial species become sympatric (overlap in range) after evolving in separate (allopatric) range. Species *A* and species *B* are best adapted to habitat 1; species *C* is best adapted to habitat 2. If species *B* and *C* become sympatric (compete for some resource and are not interspecifically territorial) their co-occupancy of habitats 1 and 2 (centre right) would not persist indefinitely before competitive exclusion resulted in habitat segregation (upper right). If species *A* and *B* become sympatric (do not compete for resources – for example, food, nesting sites – other than territorial space, and are interspecifically territorial) then species *B*, if subordinate to *A*, will be forced out of its optimal habitat (centre left). Species *B* may either modify its territorial behaviour (upper left) or become adapted to its suboptimal habitat (upper centre). If species *B* subsequently loses its interspecific territoriality through intraspecific selection then habitat segregation not distinguishably different from that resulting from competitive exclusion could occur. (From Murray 1971, in Wilson 1975.)

homogeneous spreads of population. Any one such distribution can be seen as a 'climax' distribution interpreted as a steady state or dynamic equilibrium of the population under prevailing internal and external (environmental) control mechanisms (Whittaker 1953). An interesting example is that of diffusion through breeding and allelomorph selection of genes (one of an alternate pair of characteristics is inheritable) which under uniform mixing and selection results in various gene distributions (fig. 7.53). Brues (1974) suggests that such processes, controlled by partial barriers, govern racial differentiation of skin colour in Europe and Africa across the Saharan barrier (and not the traditional explanation of migration and invasion). Simulation of diffusion of three gene types (fig. 7.53) shows that their diffusion depends upon the effect of genetic barriers. The resulting overall diversity of species is a result of

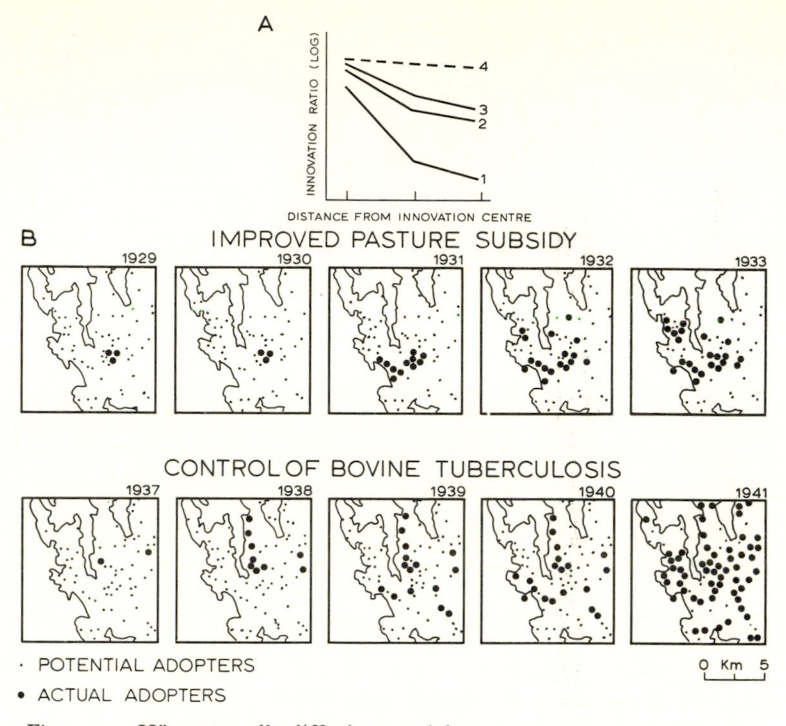

Fig. 7.51 Hägerstrand's diffusion models.
A Diffusion waves in stages of initiation (I), diffusion (II), condensation (III) and saturation (IV).
B Two diffusion examples of innovation adoption in Sweden by a rural farming community of (above) subsidy for improved pasture on small farms and (below) control of bovine tuberculosis. Small points are potential adopters and large dots adopters.
(After Hägerstrand 1967.)

breeding between individuals. Interesting cases occur when barriers reduce the interaction of an area or cut it off completely, as in the case of islands. Each distinct community of any type is an island, but 'true islands' show a greater attenuation of species for a given size (figs 7.54C and 7.54D) (MacArthur and Wilson 1967, MacArthur 1972). The total number of any species in such an isolated community is the result of the balance between the rate of immigration and rate of extinction, plus the rate of internal mutation (figs 7.54A and 7.54B). Islands along a chain can act as stepping stones with decreasing probability of occurrence of a given species with distance down the chain (fig. 7.54B) (Wilson 1971), as observed by Greenslade (1968) for insects in the Solomon Islands. Spatial effects become compounded with territoriality principles and can often result in a 'checkerboard' effect in which two or more closely related species occupy similar areas to the exclusion of each other in an

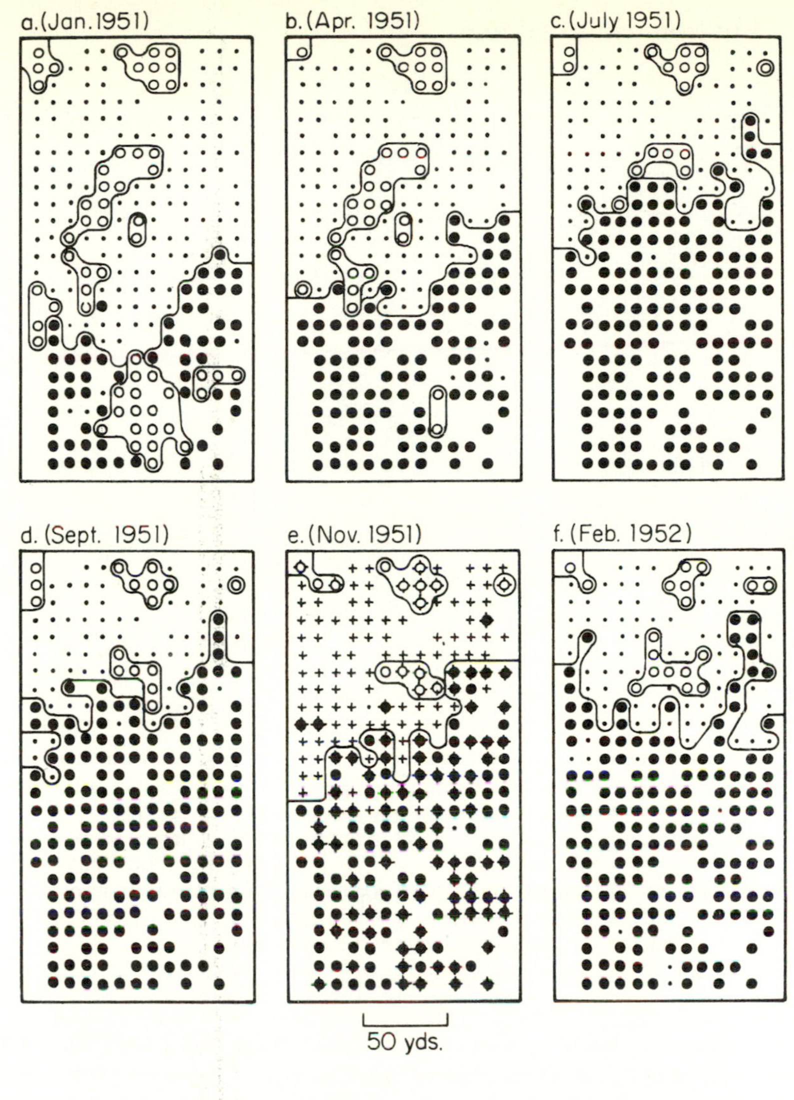

Fig. 7.52 Diffusion of the ant *Anoploleris longipes* and the exclusion of its competitor *Oecophylla longinoda* in a coconut plantation in Tanzania. Exclusion occurs through fighting at the colony level: in areas of sandy soil with sparse vegetation, *Anoploleris* replaces *Oecophylla*, but where the vegetation is thicker and the soil less open and sandy, the reverse often occurs. A third species, *Pheidole punctulata*, is occasionally abundant. (From Way 1953.)

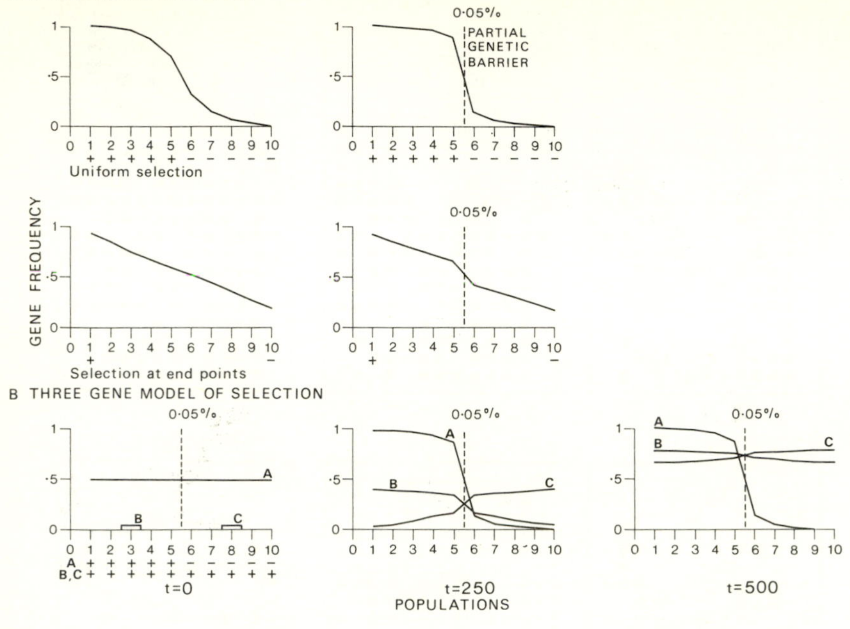

Fig. 7·53

A Differentiation of communities (1–10) resulting from allelomorph selection of genes, giving different proportions of genes with distance plus the effect of a partially permeable genetic barrier.

B Simulation diffusion of three genotypes with the effect of a partial genetic barrier. Type *A* is initially uniformly distributed but is subject to allelomorphic selection. Types *B* and *C* are initially concentrated but diffuse towards uniform distributions over time since there is no selectivity between the communities (1–10).

irregular pattern. This is especially common for bird populations (MacArthur 1972), for example the checkerboard distribution of humming birds in the West Indies discussed by Lack (1971).

Similar patterns for human populations have also been observed, and Gheorghiu and Theodorescu (1956) have shown that the rural/urban spatial differentiation of human populations can be described by Volterra equations similar to those for interactions of plants and animals of two species (Lotka 1956), i.e.

$$\frac{dR(t)}{dt} = rR(t) - gR(t) \quad \text{change in rural population } R(t)$$

$$\frac{dU(t)}{dt} = uU(t) + gR(t) \quad \text{change in urban population } U(t)$$

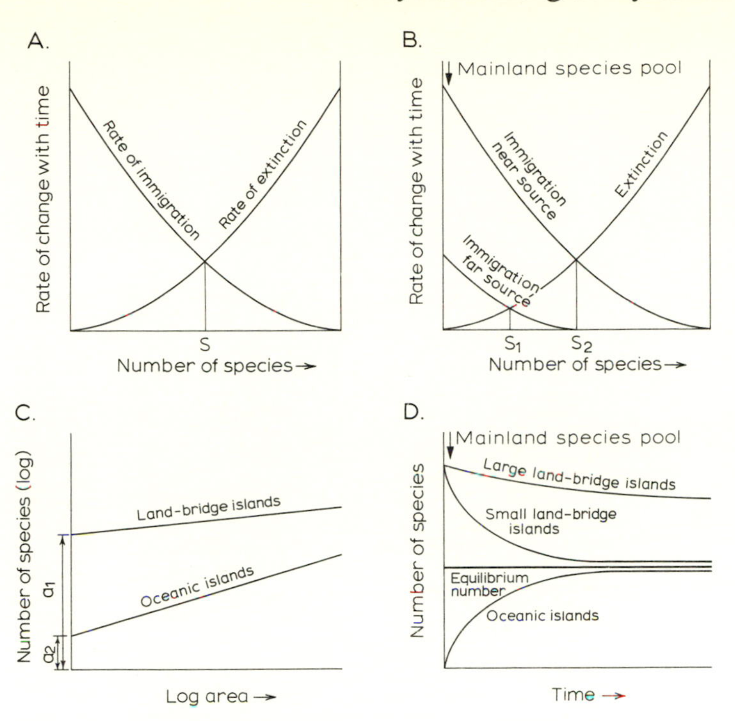

Fig. 7.54 Island bird migration.
A Number of species as the result of interaction between immigration and extinction rates.
B Number of species resulting from the effect of spatial separation of islands at various distances from the mainland species pool.
C Regression results of species/area relations (i.e. species/abundance curve) for oceanic and land-bridge islands.
D Convergence of numbers of island species over time. (After MacArthur 1972, see also Wilson 1971.)

for which g is a coefficient of rural-urban migration, and r and u are coefficients of respective rural and urban natural increase. More general interregional population migration models depend upon the age-specific characteristics, birth, death and fertility rates in each region (Keyfitz 1968, Rogers 1968), and have been placed on a rigorous accounting basis by Rees and Wilson (1977). The resulting system equations are complex and are a special form of state space representation (see chapters 2 and 4). However, an understanding of the manifolds of system dynamics traced by such state spaces under global ranges of initial conditions can be generated from the simpler case of parasitic invasion of a spatially isolated insect species (Thompson 1922, Lotka 1956). For this example (fig. 7.55) the manifold of

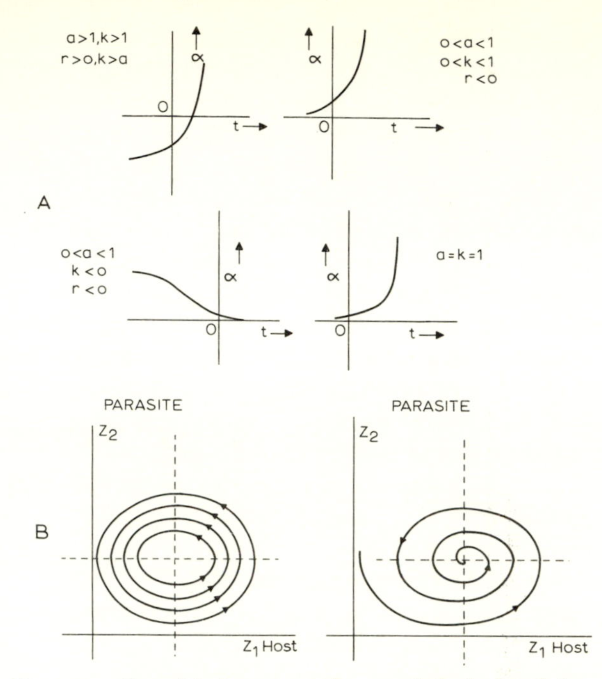

Fig. 7.55 Parasitic invasion of a spatially isolated insect species.

A Time (t) response of host population attacked. (Partly after Thompson 1922, Lotka 1956.)

Fraction of host population attacked (α)

$$= \frac{pa^t}{n - p(a^t - a)/(a - 1)} = \frac{a - 1}{Ke^{-rt} - 1}$$

where n = number of insect hosts
p = number of parasites
a = ratio of reproductive powers of parasite to host

$$K = \frac{n(a + 1)}{p} + a = 1 + (a - 1)\frac{(n + p)}{p}$$

$$a^t = e^{rt}$$

B Global trajectories in state space of numbers of host (Z_1) and numbers of parasites (Z_2) for two values of the parameters K and a. (Partly after Lotka 1956.)

dynamics is controlled by the two control parameters K and a (the respective time constant and reproductive power ratio). Depending upon the value of these two parameters we obtain four types of system trajectory (response over time) for changes in the host population over time, as in fig. 7.55A. Incorporating birth and death rates of host and parasite and assuming every host who is attacked is killed, there are two state space trajectories (fig. 7.55B). In the first case there is an increase in host followed by increases in parasites which overtake it and reduce the host to a new level from which it again develops (as in the case of moose and wolves: figs 7.14 and 7.15). In the second case the expansion of the parasite increases continuously resulting in the decrease of the host population until it annihilates the host, but as the host population decreases so the parasite population decays through lack of food (Thompson 1922, Lotka 1956). Disease diffusion can be described by similar global trajectories, and the spread of malaria (fig. 7.56A) can be defined by integral curves (deriving from the Ross equations; Lotka 1956) which provide a topographic map of dynamic development of the disease. There are two singular points. Unstable equilibrium obtains at G where malaria spread may suddenly increase or suddenly decrease. The singular point F describes the stable equilibrium situation of constant malaria presence. The representation of this manifold as a 'landscape' provides a purely qualitative interpretation of system development with point F singularities representing a col and point G singularities a pit (fig. 7.56B).

The importance of constructing global maps of system evolution is reinforced by recent discussions by May (1975) and Kolata (1975) of fish, insect and cicada populations in which infinitely small initial differences in population sizes permit growth trajectories which differ markedly, and can become totally unpredictable. The population model analyzed by May (1975) is given by the nonlinear equation $P_{t+1} = \lambda P_t \exp(-\gamma P_t)$, where λ is the population growth rate and γ is a density term. This expression is equivalent to a third order difference equation with properties: $P_{t+3} \leqslant P_t$, and $P_t < P_{t+1} < P_{t+2}$. Thus there is a pattern of growth, then catastrophic decrease as the population reaches the limit γ. The structure of the solution of this equation depends upon the value of λ (the population growth rate), and there are three possibilities:

(i) $7{\cdot}34 > \lambda > 0$	Stable equilibrium level of population.
(ii) $14{\cdot}77 > \lambda \geqslant 7{\cdot}34$	Stable oscillations in level of population.
(iii) $\lambda = 14{\cdot}77$	Infinite number of different oscillatory population trajectories: Chaotic solution.

As a consequence of this behaviour, even when the parameters of the system are constant and known perfectly, probability theory and stochastic models allow the best description of system response for the chaotic case. Thus stochastic models can arise not from sampling errors and ignored variables, but from known and constant nonlinear equations with chaotic solutions. Since such models characterize any population which grows at low densities and crashes at high densities, the implications of these models are very disturbing for the system analyst. As we shall see in chapter 9, the growth and crash model has been hypothesized for world population levels, but from the discussion of May and Kolata, we cannot expect to reach any conclusions

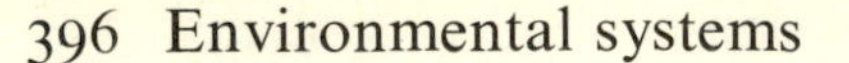

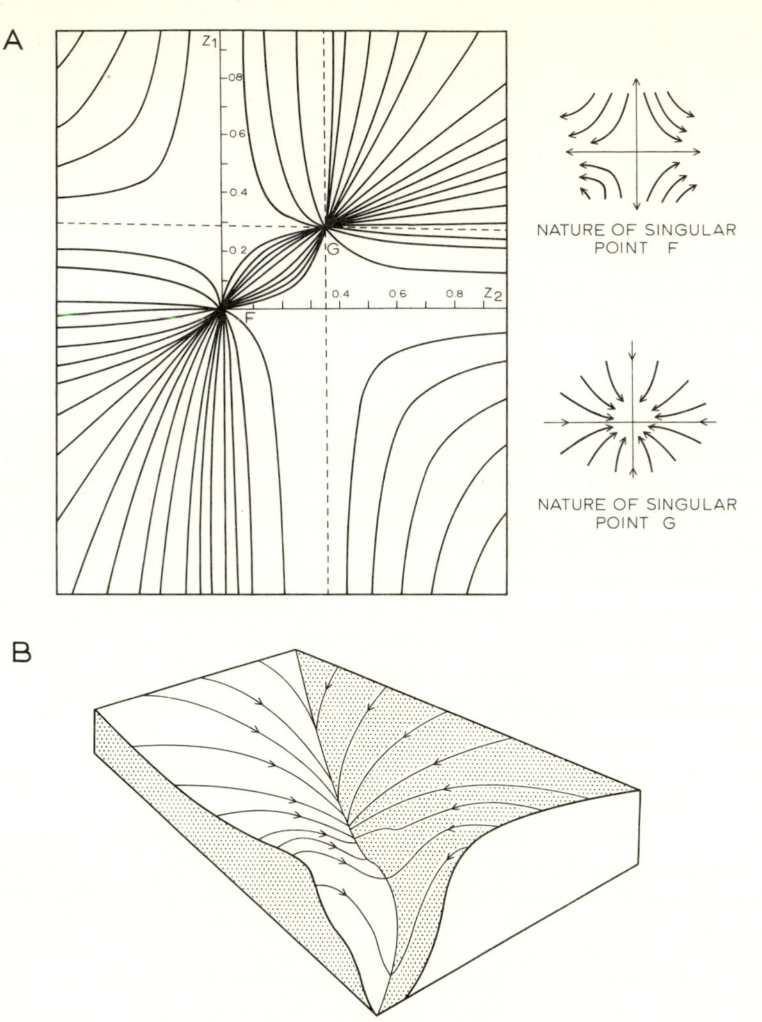

Fig. 7.56 The spread of malaria with two state variables Z_1 and Z_2.
A The differential equation solution to the Ross equations for the spread of malaria, showing a stable equilibrium at point *G* and an unstable equilibrium at point *F*.
B The qualitative description of the topology of malaria spread shown as a 'landscape'.
(After Lotka 1956.)

about the validity of this hypothesis without stochastic methods and global analysis.

The use of global spatial models of physico-ecological systems allows the description of system change over time as a shift on a topological manifold (or landscape) defined in a space with dimension equal to the number of overall control parameters. As changes, learning and evolution shift and modulate the character of system response parameters, the analyst can determine how the shifts affect the overall time location of the physico-ecological system on stable, unstable and chaotic pathways of development. The interaction of purely physical evolution with man-induced changes is discussed as a result of intervention and symbiosis in chapter 9. For the present we may conclude that whatever the cause of shifts in system initial conditions and parameters (natural or man-made) it is important to evaluate these on the landscape map which describes the location of the new response pattern relative to the range of alternative patterns of system operation conceivable.

7.5 Conclusion

Looking back to the treatment of control systems in chapter 3 and forward to the interaction of physico-ecological and socio-economic systems in chapter 9, it is clear that the role which man plays by intervening in many terrestrial physico-ecological systems is limited but potentially decisive. For example, attributes of soil and vegetational covers, accessible to and commonly easily manipulated by man, have been shown to operate as key regulators of the ecosystem nutrient pool, of the water supply within the hydrological system and of sediment availability by rainsplash, overland flow and rill/gully flow in surface-sediment systems. As we have seen, the case study by Cooke and Reeves (1976) and by Cooke (1974) which attempts a systems-based explanation of the 1865–1900 period of arroyo cutting in the southwestern United States shows the possible delicate interaction of human land-use changes with random magnitude-frequency variations and secular climatic changes based on time series analysis of historical climatic data. Other studies have emphasized the interaction of man with physico-ecological systems to encourage investment and creation of resources or degradation (e.g. desertification, pollution and erosion). As will appear in chapter 9, the distinction between the creation or degradation of environmental resources depends upon the ratio of the storages and relaxation time in the physical world to the storages and relaxation times in the socio-economic world.

This chapter began with a philosophical speculation regarding the possibly simplistic behaviour of apparently complex systems; it will end with another. Pattee (1973, pp. 73–4) has suggested that the existence of hierarchical organization depends on functional simplicity and that there exists a paradox in that hierarchical controls, even apparently arbitrary ones, both limit freedom and, at the same time, confer more freedom. Thus legal restraints are necessary to establish a free society, and the constraints of spelling and syntax are necessary for the free expression of thought. 'Igor Stravinski writes in *Poetics of Music*, "The more constraints one imposes, the more one frees one's self of the chains that shackle the spirit . . . the arbitrariness of the constraint serves only to obtain precision of execution"' (Pattee 1973, p. 74). Followers of ballet and cricket would undoubtedly agree. Pattee (1973, pp. 90–3) refers to the views of Goodwin and Cohen (1969) that biological growth occurs as if every cell need only read a clock and a map, but that each higher level of hierarchical organization needs a different clock and map which progressively eliminate the need for temporal and spatial detail. In social hierarchical control, too, details can be selectively ignored, as in the umbrella concept of 'equality under the law', and the secret of the successful interpretation and control of hierarchical systems is to ignore just the correct amount of detail at the relevant scale of space or time. Pattee (1973, p. 83) suggests a *principle of optimum constraint* involving a set of constraints which hold between certain degrees of freedom but which do not result in rigidity. In social systems the optimum lies between anarchy and bureaucratic 'freezing up'. Pattee (1973, p. 93) concludes:

> Many hierarchical *structures* will arise from the detailed dynamics of the elements . . . but the optimum degree of constraint for hierarchical *control* is not determined by the detailed dynamics of the elements. The dynamics of control is determined by how these details are ignored. In other words, *hierarchical controls arise from a degree of internal constraint that forces the elements into a collective, simplified behaviour that is independent of selected details of the dynamic behaviour of its elements.*

8 Socio-economic systems

It has been agreed that every state is not a single entity, but all are composed of many parts. Now if our chosen subject were not the forms of constitution but the forms of animal life, we should first have to answer the question 'What is it essential for every animal to have in order to live?' And among those essentials we should have to include such things as organs of sense perception, organs for the acquisition and assimilation of nourishment, mouth and stomach, and in addition parts of the body which enable the animal to move about. If these were all that we had to consider and there were differences between them, different kinds of mouth, of stomach, of sense organs, and of locomotion, then the number of ways of combining these will necessarily make a number of different kinds of animals. For it is biologically impossible for one and the same species of animals to have different kinds of mouth or ears. So when you have included them all, all the possible combinations of these will produce forms of living creatures and the number of forms of animal life will be equal to the number of collocations of essential parts. We may apply this to the constitutions mentioned; for states too have many parts. . . .

(Aristotle: *The Politics*, Book IV)

8.1 Introduction

The socio-economic system of the state is a composite of a large number of interacting individuals. Aristotle's view, that there will be as many constitutions of society as there are permutations of individuals within it, accords well with the combinatorial statement of the law of requisite variety. From this we deduce that any individual, group, or society as a whole, has a capacity of organization limited to its capacity as a channel of communication (Ashby 1956). The organization of societal units at various scales articulates through the structure of their interlinkages (internal and external) which operate as absolute constraints. Each level of organization gives an increase in the certainty and predictability of the societal system to the individuals of which it is composed, and corresponds to an increase in negentropy (Emery 1969, Kuhn 1974). Negentropy of information is the metabolism whereby society draws continually from the physical environment and the social milieu, so avoiding decay to equilibrium and maximum uncertainty. As the complexity of modern societal systems increases so the demands made upon the environment become greater, thus increasing the pressure to control both the physical and social environment. In the previous chapter it was concluded that a *principle of optimum constraint* operates within hierarchical environmental systems to constrain individual subsystem behaviour within collective norms, goals or behaviour modes (Pattee 1973). The operation of analogous constraints within societal systems leads to a conflict of the individual *qua* individual and the individual within the collective of society. The conflict can be rationalized within any of Aristotle's combinatorial constitutions and leads to the question of the most appropriate level at which the individual's negentropy gain (societal organization) is offset by his entropy loss (freedom of action). We have already seen in chapter 6 how we may apply optimal decision rules at any level from the individual through to the collective action of large groups of individuals. In each case, optimal decisions are made on the basis of Lerner's rule: that choice is made which yields the highest rate of return (i.e. it maximizes the individual, collective or control objective function). We shall see below that as the size of the collection of individuals increases, the scope for unanimity of agreement and collective action decreases (Buchanan and Tullock 1969). The gain to the individual in participating in large collective systems is a greater certainty as to the environment of his independent decision making. The industrialist or economic actor gains most certainty if his external environment is totally predictable, but this gain is made at the expense of his own actions becoming predictable to others within the collective. He must conform to decrease their uncertainty and entropy. The participation of the individual in the collective will be determined by each decision maker from his utility scaling of the costs and benefits derived from his involvement, which is discussed in section 8.2.

The interplay of the individual and the collective accords with the $\sqrt{n}$ law (i.e. the law of large numbers) as it applies to the *explanation* of socio-economic structures:

> An organism must have a comparatively gross structure in order to enjoy the benefit of fairly accurate laws, both for its internal life and for its interplay with the external world. For otherwise the number of cooperating particles would be too small, the 'law' too inaccurate. The particularly exigent demand is the square root. For though a million is a reasonably large number, an accuracy of just 1 in 1000 is not overwhelmingly good. (Schrödinger 1967, p. 19)

This property of explanation has led the analysts to approach socio-economic systems at a number of spatial levels from the individual to the macroscopic and, between these two extremes, at various levels of aggregation of the collective.

(*i*) At the scale of the *individual*, socio-economic behaviour is a function of cognitive-perceptual responses and physico-chemical processes (see chapter 5). However, the answer to the question 'How can events in *space and time* which take place within the boundary of a living organism be accounted for by physics and chemistry?' (Schrödinger 1967, p. 3) has seldom been clear. Schrödinger (1967, p. 4) has the opinion that 'the obvious inability of present-day physics and chemistry to account for such events is no reason at all for doubting that they can be accounted for by those sciences'. Alternatively, Campbell (1961) suggests that the individual organism be treated as a hierarchy of organizational concepts ranging from genetic mutation, through learning and perception to social decision mechanisms. Hence, rather than studying the physico-chemical decision rules, most systems studies have followed Simon (1957) who sought individual and group situations that appear to be governed by rules, and then tried to discover the nature of these rules. Simon suggested that decisions at the individual level can be divided into two components, intra- and intersystem linkages which specify the respective system behaviour at a point and between locations (i.e. the individual, group and sector at different spatial points). Each of these components can be further subdivided into detector, selector and effector functions (Kuhn 1974, p. 9), as shown in fig. 8.1. The interaction of these three components determines the nature and direction of behaviour in a stimulus-response sequence: communication of information (detector), transfers of values (selector), and organization of effects and decisions (effector). Hence, Simon sees the structure of social systems as similar to that of physical control systems, except that no deterministic relations are possible. Environmental inputs are detected and monitored. Detection also involves a correlative process (Lotka 1956) of inputs and expected effects. This stimulates selection which involves a value judgement based on choice of possible

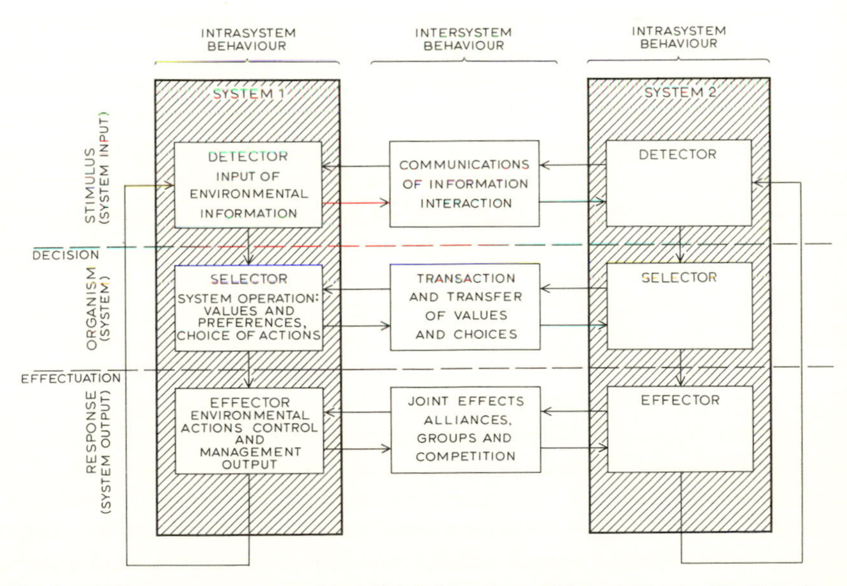

Fig. 8.1 System response and spatial interaction of two systems. (Partly after Kuhn 1974.)

alternatives (including the null action) and parallels the decision making systems discussed in chapter 6 (including, however, the effect of personal preference functions which may have no logic structure or explanation, nor may they have any confidence of repetition). Out of the detection-selection process arises a set of rankings of alternative actions which may be relatively stable in replications of choice situations over time. This stable set of rankings of alternatives yields the economists' preference or utility function of each individual. Arrow (1951) suggests that ordering of utilities is based on two axioms:

1 Comparability of alternatives yielding order: e.g. alternative *A* is preferred to *B*.
2 Transitivity of alternatives: e.g. if *A* is preferred to *B*, and *B* to *C*, then *A* is preferred to *C*.

Aggregation of these preference orderings, or of the more general detection-selection-effection process, is usually undertaken to obtain more widely applicable socio-economic models.

(*ii*) Various *aggregative* levels of scale amplification of societal system analysis are possible above the individual level, each corresponding to a different scale of generalization (Boulding 1956). In general, Feibleman (1954 p. 64) notes that 'for an organization at any level, its mechanism lies at the level below, and its purpose at the level above'. One approach to the construction of aggregate generalizations derives from psychological behaviourism and conceptualizes societal responses as the result of learning and decision processes. Various levels are possible from interpersonal influence, to group and societal interaction (Simon 1957), each with its own activity system in space (Chapin and Hightower 1965). More pragmatic procedures have been developed by Blalock (1967), who grouped variables into morphological and cascading systems structures by the use of correlators, cluster analysis, canonical correlation, path analysis and factor analysis methods and the basis of 'causal' relations built up from specified system elements. Each method involves *a posteriori* search of linkage matrices in order to find 'minimal' structures which may be based upon intergroup or interspatial components (Bennett 1978). This often introduces sophisticated spatial search procedures, as evidenced in consumer behaviour studies by Kotler (1968) and Rushton (1969). The economic basis of intergroup and societal response to decision making involves 'a comparison and evaluation of alternative spatial and non-spatial opportunities against a personal preference function'; from this, 'it is possible, from any consistent statement of preferences . . . to derive a unique ranking of objects' (Rushton 1969, pp. 391, 394). This utility analysis can be used to derive spatial indifference curves extending Huff's (1963) behavioural framework (Niedercorn and Bechdolt 1969), or general indifference functions between the whole set of spatial and non-spatial alternatives. Spatial group response must be seen against the background of spatial opportunities; spatial societal processes are controlled by spatial structure. In consumer behaviour Berry (1962) uses the range and threshold of a good as concepts to connect hierarchical spatial behaviour to central place theory, whilst Curry (1962, 1967) suggests that the varying frequency of good purchases gives a stochastic basis for the behaviour-location linkage.

Aggregative choice has also been approached from physical analogy with gravity and diffusion processes (Carey 1858, Dodd 1950). More acceptable justification is given by Ravenstein (1885), Reilly (1929) and Stewart (1947), but all such formulations derive to some extent from a mapping from physical system laws to social systems. Such laws of 'social physics' must always be treated with care, but may be useful in hypothesis creation (A. G. Wilson 1970). The gravity analogy and maximum-entropy derivation of the gravity interaction model due to Wilson (1970) have both been criticized on the grounds of ignoring the particular spatial configuration of opportunities which constrain the individual choice mechanism (Curry 1967, Rushton 1969). However, the wider criticism is that, like all aggregative models, they can be interpreted in the control situation as the two-step process:

Explanation of human action $= f$ (statistical maximum likelihood)

(8.1.1)

Desirable human action $= f$ (statistical maximum likelihood)

which is surrounded by severe ethical problems (Olsson 1975, p. 470 and ch. 14), i.e. the extension of descriptive social science into prescriptive social engineering discussed further in section 8.5.

(*iii*) At a *macroscopic* scale synthetic system explanation of societal behaviour, like the aggregative approach, recognizes a series of levels of explanation, but unlike aggregative models incorporates the idea of a conceptual discontinuity in the hierarchy. This is not suggested in aggregative social models. The discontinuity arises through system indeterminacy which itself is a product of taking a holistic approach:

> The way not to tackle such a system is by analysis, for this process gives only a vast number of separate parts or items of information, the results of whose interaction no-one can predict. If we take such a system apart, we find we cannot reassemble it! (Ashby 1956, p. 36)

Global analysis of such systems (Smale 1974), as discussed in chapter 4, takes as its core the complexity of societal systems. This is the key to the indeterminacy concept which can be viewed at two levels. Firstly, when systems are large it becomes impossible to specify the exact causal chains which relate variables together (i.e. there is indeterminacy of state). Secondly, measurement is inexact or impossible (i.e. there is indeterminacy of measurement). The chief characteristic of the aggregative approach is that all variables are recognized and incorporated into the resulting socio-economic model. Deutsch (1969) notes that this is a feasible approach only if:

1 All variables are recognizable and can be defined.
2 The variables are uniform with small variance.
3 Correlations between variables are exhaustive and explain most of the variance.
4 Correlations are stable over time periods which are long relative to planning lead times.

A limit will always be reached at which it is impossible to include further parameters. This indeterminate limit (Reichenbach 1944, p. 4) generates demand for parameter attenuation such as parsimony (Tukey 1961) or minimality (Kalman 1960) in the resulting model.

> Where a large number of interacting factors are involved in a large number of individual cases or examples, the possibilities of combination are so great that the physical laws governing forces and motions are not sufficient to determine the outcome of these interactions in the individual case. (Leopold and Langbein 1963, p. 189)

System complexity, of *necessity*, leads many analysts to describe socio-economic systems at a level of generality by statistical laws and stochastic processes. This requires invoking a societal process of indeterminacy:

> Correctly understood the [indeterminacy] principle correlates the standard deviation residing in [any property] ϕ *before* measurement . . . [the location in state space] p and q, and the deviations are only meaningful for ensembles. (Margenau 1963, p. 142)

This approach to systems analysis requires that we treat the outcome of any given trial as random and describe social systems by a frequency distribution of outcomes. The values of social system variables can then be given as their means, which can be used as expected value linear operators in statistical laws, as distinct from deterministic laws. Indeterminacy of measurement arises, as discussed in chapter 2, from pure measurement error (observational or instrumental), from slight variations in variables assumed held constant in the system environment, from inherent variations in behaviour and personal preference functions, from changes in thinking and preferences imposed by measurement (e.g. questionnaire analysis; Mackay 1957, p. 96), and from perceptual indeterminacy caused by inability to perceive small spatial or utility differences. Harvey (1969, pp. 240–3, 270–2, 321) justifies the use of stochastic models for space-time processes largely in terms of measurement indeterminacy, but it is wrong to concentrate unduly on this form of indeterminacy. All forms of explanation of socio-economic systems are subject to measurement error; but only macroscopic systems, because of their

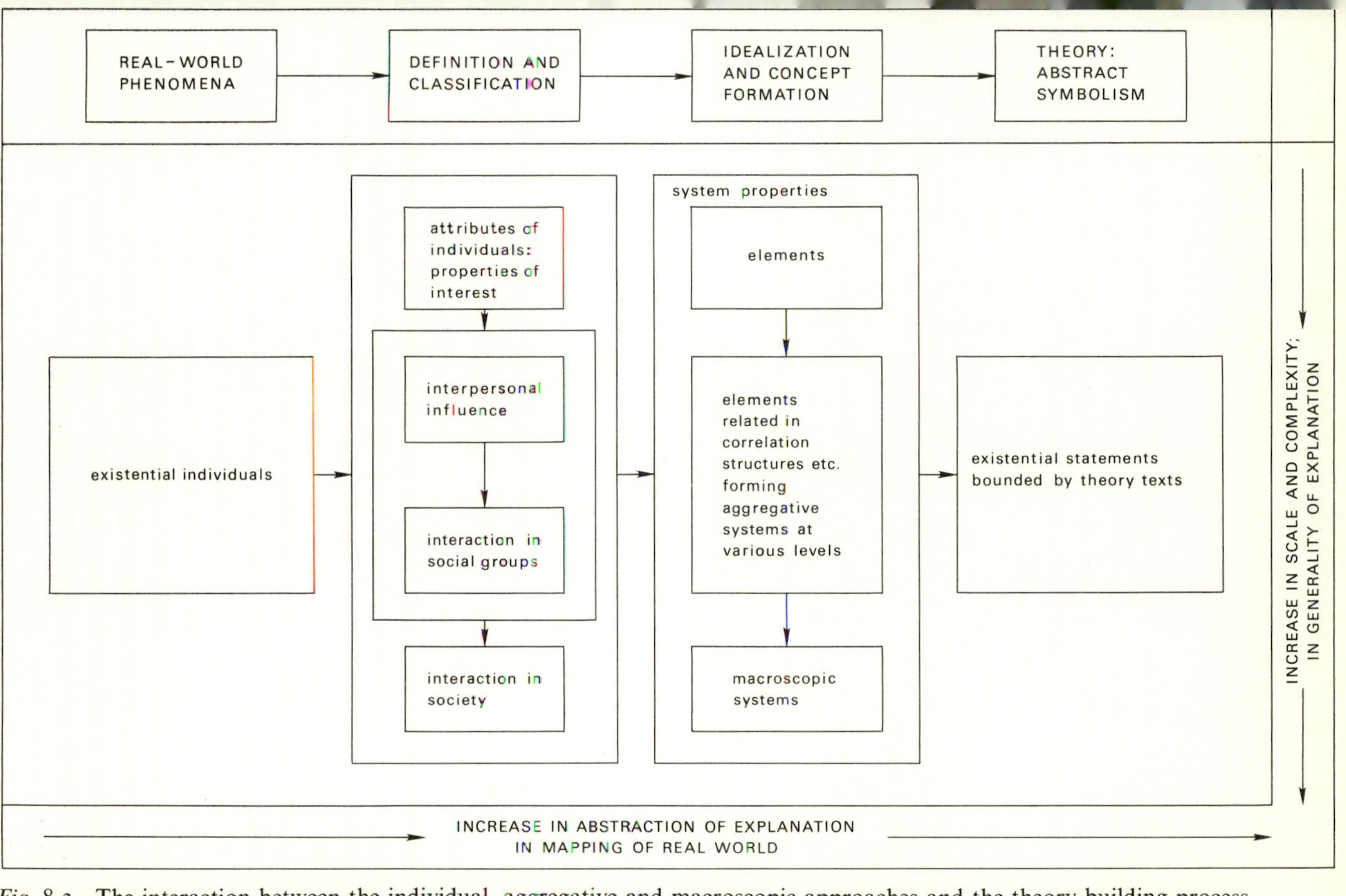

Fig. 8.2 The interaction between the individual, aggregative and macroscopic approaches and the theory-building process.

large numbers of degrees of freedom, are necessarily indeterminate, and hence must be described by stochastic models.

Individual and aggregate approaches of systems analysis are inherently more satisfying explanations of societal systems. As Schlesinger (1959, pp. 248–50) notes, to

> be able to see 'how things really work' we feel we must break up the system under consideration into its smaller elements, the properties of which give rise to the collective properties of the whole. . . . This however is an illusion. With regard to scientific theories the questions we have to think about are questions of their generality, truth, validity and usefulness. As long as the premises of one particular explanation are no more true than that of another, the logic employed no more rigorous, the predictions on it no more numerous, there is no good reason to prefer it to any other type of explanation.

The interaction between the individual aggregative and macroscopic approaches and the theory-building process itself is shown in fig. 8.2. The transition of explanation modes with scale is akin to the systems analysis/systems synthesis duality recognized in chapter 7. In 'soft' socio-economic systems this division results largely from the interface with explanation paradigms. Explanation of hermeneutic response at the individual level may be possible for simple individual cognitive systems, as discussed in chapter 5, but 'on the macro-level, it seems reasonable to assert that one must always make some assumptions about individual motivations in order to develop meaningful theory appropriate to groups as units of analysis' (Blalock 1967, p. 27).

8.2 Systems synthesis

In common with the approach to system synthesis discussed in chapter 7, synthesis in socio-economic systems corresponds to building a conceptual model as close as possible to the structure of the real world. This white or grey box system contains the empirical mathematical relationships linking state variables and subsystem components. As in the case of physico-ecological systems, synthetic socio-economic models may have a number of disadvantages of idiosyncracy, indeterminate response to extreme events, poor spatial capacity, empirical reinforcement of the *status quo*, and often depend as much on the techniques of model construction, both perceptual and mathematical, as on the real social processes represented. In this section various models of individual behaviour are analysed, but because of the extensive treatment of these models in chapter 5, most attention is directed here to the aggregation of models of individual behaviour.

Sociologists have tried to recognize patterns in man's activities that can be subsumed under role playing. This has been criticized by others as a form of reification: reducing man to a 'thingified' entity and erecting concepts to an artificial objectivism. Positivism has been criticized as the main cause of these difficulties and Habermas (1971) calls for deeper understanding based on two higher levels of (1) historical-hermeneutic and (2) self-reflective understanding. These permit the development of 'self knowledge' which emancipates decisions and emphasizes the responsibility of the individual (Keat and Urry 1975, pp. 222–7). This unifies structural understanding of what goes on inside the black box of society and subjective (ideological) meaning, which is unavailable to positivism and reflects much of Freud's cognitive system.

An alternative approach is based on a concentration on systems components. This requires the definition and abstraction of societal elements which may be individual people, firms, entrepreneurs, households or other simple group units. Each of these units is defined under specified boundary conditions which allow the simulation of the response of each unit as a decision making entity responding to various forms of environmental stimulus. By successively complicating the stimuli influencing the societal elements, it is possible to construct a picture of responses under various degrees of interaction and complexity. The decision of any firm, entrepreneur or individual is a response selection under conditions of complexity and uncertainty (Kuhn 1974, p. 106) being governed by a combination of the risk-aversion criteria discussed in chapter 6. The utility of any action will be based upon the perceived satisfaction to be procured deriving from ranked preference orderings. Subject to specified constraints, similar preference ranks and decisions may be expected to occur under similar stimuli (Simon 1957). This is the state at which the system variables move to a position of equilibrium and hence from an improbable to a probable and predictable configuration. Pure equilibrium will occur when some property (X) of the societal unit remains unchanged over a period of time, i.e.

$$\lim_{t \to \infty} \frac{dX}{dt} = 0.$$

The concept of pure dynamic equilibrium is not apposite to social systems because 'chiefly in the world of equilibrium theory there is no gravity, no evolution, no sudden changes, no efficient prediction of the consequences of friction over time' (Deutsch 1951, p. 198). The consequences of time friction are the lag, lead, gain, delay and memory components of system response discussed in chapter 2. Societal units treated as cybernetic entities operate under feedback control, goal direction and purposive behaviour (Rosenblueth *et al.* 1943, Buckley 1967). 'Process adaptive' mechanisms will produce 'stable spatial and/or temporal relations between distinguishable elements' (Buckley 1967, p. 62) such that at any time the system will fluctuate around an 'equilibrium' position at which all the elements are adjusted to each other in a minimum variance relation (fig. 8.3) (Scheidegger and Langbein 1966). The overall system will also be subject to changes over long periods of time arising from changes to overall goals (i.e. preference functions) and adaptive behaviour (i.e. learning). Brown (1969, p. 407) conceives of cities as a typical societal servomechanism unit oriented and maintained by people to achieve psychologically oriented goals (Meier 1962, Webber 1972).

The analysis of societal systems as servomechanisms reacting to minimize deviation from perceived personal and intrapersonal targets is frequently oriented around the reaction of groups to cost differentials. Costs may be measured in the most general terms: opportunity cost, costs foregone, disutility costs (i.e. costs of dissatisfaction accepted in an action) or pure economic costs and benefits. A rational societal decision then depends upon the selection of that alternative action where future 'benefit' exceeds future 'cost'. This is based upon selection of alternatives from each individual's or

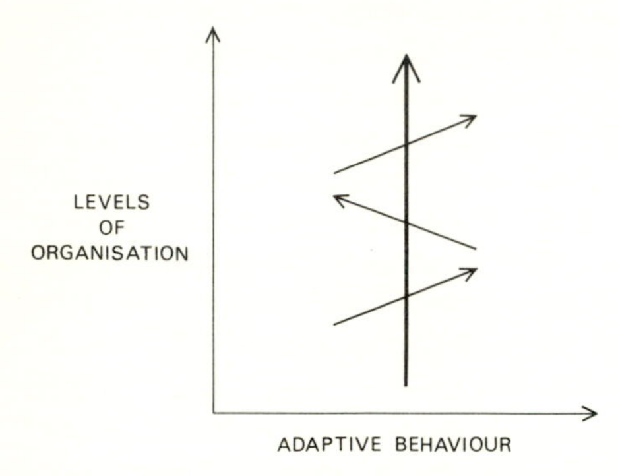

Fig. 8.3
The fluctuation of systems behaviour around an equilibrium position at different levels of organization. This form of systems synthesis envisages evolution around an evolving but fairly stable set of social modes. (After Buckley 1967.)

group's opportunity function (i.e. range of possible actions) based on pure economic appraisal and the marginal costs and utility of benefits. Various of the evaluation methods discussed in chapter 6 can be adopted, depending upon the strength of the data available and the certainty with which the individual or control objective function can be defined. Thus individual and group economic and environmental projects can be evaluated according to their rate of return, cost-benefit ratio, concordance with biological and conservation goals, and so on. In each case the individual or group decision maker is trying to maximize his return under uncertainty. In the simplest case this may be a game against the environment (Gould 1963) which leads to some optimal pattern of action, for example an agricultural system which minimizes climatic risk. This may evolve to give complex collective societal responses to environmental risk (Harris 1969, O'Riordan 1971). A good example of the latter is provided by large projects requiring changes of established group practices aimed at achieving increased return but involving a risk associated with shifting from a practice which is known to work. In such cases Bayesian or other decision trees may help to show the minimum risk development programme, as discussed in chapter 6. Such decision trees can be examined for risk among decision alternatives for irrigation improvement and agricultural development in peasant communities (Gates and Gates 1972), industrial location and investment decisions (Webber 1972) and social actions (O'Riordan 1971). Increasingly, however, the interplay between the socio-economic decisions of man and his environmental impact lead us to adopt macroscopic approaches to explanation and to policy which are outlined in the next section and in chapter 9.

Under conditions of interaction between individuals or groups, rational societal decision procedures will depend upon the nature of the interaction (i.e. cooperation, conflict, tension) and the nature of the possible action (i.e. the opportunity function). Under selfish-indifferent assumptions (i.e. when each party tries to maximize his own benefit, giving as little and receiving as much as possible, and is indifferent to the position of the others), Kuhn (1974, p. 175) develops a model of interaction (fig. 8.4) in which each party neither

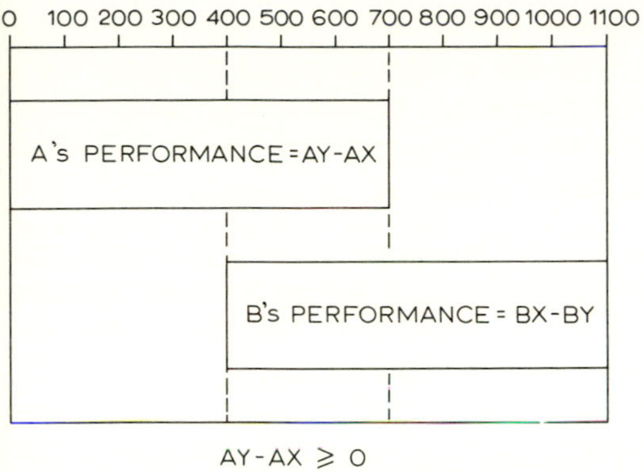

Fig. 8.4
Transaction diagram for good X, divisible (e.g. money), and good Y. A is the buyer and B the seller. The axis is measured in money units and the area between the dashed lines shows the scope for a bargaining solution. (Partly after Kuhn 1974.)

helps nor hinders the other such that any 'transaction' breaks down into two parts:

1. One party's ability to provide another's need (supply sector-opportunity function).
2. The other party's choice of response (demand sector – preference function).

For example, a transaction between individual A and individual B in which A has good X and B has good Y will occur only when AX, AY, BX, BY (i.e. A's and B's desire for X and Y) satisfy

$$X \text{ and } Y \text{ are indivisible} \tag{8.2.1}$$

$$AY - AX \geqslant 0 \qquad BX - BY \geqslant 0. \tag{8.2.2}$$

This corresponds to the situation in which A and B have different preference orderings; A prefers Y to X, and B prefers X to Y (Kuhn 1974, p. 177) and is analogous to a positive or zero-sum game. The first terms of the transaction are the interaction of the two contingent decisions and can be determined from the transaction diagram (Boulding 1955). *A* is a buyer and *B* is a seller. A transaction will occur up to a 700 purchase price and 400 sale price, which represents the area of overlap between *A*'s and *B*'s desire for goods *X* and *Y*. This is the area of feasible or bargaining solutions of which the midpoint will normally be the equilibrium value. However, under conditions of monopoly power held by individual classes or social groups controlling supply or demand, dominance-submission, and aggression solutions will yield alternative equilibrium points, such that a good might be so attractive as to make it profitable for a stronger power to go to war.

The interlinkage of buyers Y_i and sellers X_i over society has been explained by some analysts as an ecological equilibrium between elements recognized by a climax structure (chapter 7) and number-diversity ratio (i.e. increasing entropy under energy constraints). In this case no positive goal is implied and no organized decisions are required (Kuhn 1974). Alternatively, highly organized input-output linkages between groups as special purpose information organizations may lead to increasing differentiation, specialization and negentropy. Market economies are of this form, oriented towards goals of profit maximization, which determine

> how the different kinds of goods shall be allocated among the different individuals. . . . Certain assumptions have to be made about human satisfaction or welfare, including the principle of *diminishing marginal substitutability* between goods . . . the optimum allocation of goods involves the equalization of marginal substitutability [which] . . . is automatically reached by free exchange. . . . [Hence] the best allocation of the different kinds of good among the different kinds of consumers is reached only if the marginal substitutability (*M*) of every good *A* for every other good *B* is not less for every consumer who has some of *A* than it is for any consumer who has some of *B*. (Lerner 1946, p. 14)

The basis of this substitutability is a measure of value based upon *price*, which is based upon a system of *money*. Gains in value (i.e. surplus value) can be made in a number of ways: by exploitation of comparative advantage, by speculation, by arbitrage, and by collusion (Kuhn 1974, pp. 300–95). The conditions for existence of perfect competitive bidding are:

1. Each individual economic unit is so small in proportion to the total market that its behaviour has no influence (e.g. monopolistic) on market price such that there is perfect competition in selling.
2. Market price is determined as a measure of the usefulness of the product to the purchaser and to society by competitive independent bidding between producers and purchasers, such that there is only one market price (Lerner 1946).
3. Each individual purchasing unit is so small in proportion to the total market that its behaviour has no influence on market price, through non-

monopolistic purchasing, such that there is perfect competition in buying.

4 There is perfect spatial mobility of all factors, resources, and final products, all goods are uniform and there is perfect knowledge of price and quality by all buyers and sellers.

Large-scale interaction between numbers of societal units and individuals generates types of transactions which depend upon the nature of sponsors, suppliers and receivers. These large-scale social systems are often special forms of the decision making system discussed in chapter 6 which are organized (i.e. consciously or unconsciously coordinated) to attain goals. Fig. 8.5 summarizes the five major categories of aggregate socio-economic systems defined at the interaction of decision structures and the locus of control (Kuhn 1974, p. 290). *Decision structures* define the rules (e.g. division of labour) by which the system is organized: the rights and responsibilities of each individual or unit (up, down and across); their roles; the line of authority (to grant or withhold rewards); the exercise of authority (delegation and management); and the availability of sanctions (rewards or punishments). The *locus of control* governs what input an individual or unit receives in respect of the benefits he supplies, and is thus an extension of the transaction diagram (fig. 8.4) to incorporate different bargaining positions and multi-individual and multi-organizational settings. One particular example of the formal organization oriented towards facilitating societal interaction is government. This covers a whole society and has monopoly powers with potentially unlimited scope in area and effect upon decision making and transaction. It is based upon a dominant coalition of the people. This can perform one of five functions (Kuhn 1974, pp. 343–6) (see fig. 8.5):

1 *Cooperative*: In this the government acts as the sponsor and the recipients bear costs in order to receive benefits. The Lincoln principle of 'government of the people, by the people, for the people' is a statement of this, whereby there are no strictly private goods since all citizens are sponsors and recipients.
2 *Profit*: Here the government is largely sponsored by subgroup *A* of society who receive the benefits, the costs being born elsewhere. The goal of the sponsors is to maximize the satisfaction of *A*, as in a dictatorship or feudal, class-structured society.
3 *Service*: This involves government when subgroup *A* gives benefits to subgroup *B* in return for promises of general public goods promoted by *B*.
4 *Pressure*: In this subgroup *A* uses government *C* to give outputs to *B* in order to improve *A*, no public goods being produced by government.
5 *Mixed*: Combined effects.

Public goods are, however, a necessary output of most government organizations of society and are assumed necessarily the same for all members

Type of organization	Costs	Benefits	Type of transaction with recipients
1 *Cooperative*: sponsors are recipients of outputs with goal of own welfare, the communist ideal.	Recipient	Recipient	Maximises own welfare
2 *Profit organization*: run in interests of sponsors (which usually does not include the recipients) based on selfish transactions as in the capitalist form.	Recipient	Sponsor	Selfish profit maximiser
3 *Service organization*: output goes to the recipients as for example pure gifts or external economies, the government ideal.	Sponsor	Recipient	Generous
4 *Pressure organization*: organization *C* is given outputs by sponsors *A* to dispose to *B* for use in improving the position of *A* in relation to *B*. The sponsor *A* bears costs and attains most benefits as in housing development, and financial institution monoplies.	Sponsor	Sponsor	Third party strategies
5 *Mixed*: Most real systems	Sponsor and recipient	Sponsor and recipient	Combined

Fig. 8.5 Types of societal organization based on interaction and locus of control.

of the society (e.g. clean air, public roads, etc.). Within any society, minorities not included in the dominant government coalition have three courses of action (Kuhn 1974, p. 335). If the government is satisfactory it can be accepted; if it is unsatisfactory the minority can attempt to gain a dominant position; or if a new coalition of size and power can be found (syndicate) this can influence the government's tenure, stability or operation. Complete government control is only possible if the following conditions are satisfied (Zinam 1969):

1 An omniscient government evolving all economic decisions. These decisions may reflect either the preferences of the population as a whole

(democracy), or the wishes of the government leaders (totalitarianism), or the perceived goal of the population (communism).

2 The incorporation of the economic decisions into a production plan imposed on each production unit.
3 The control of the mobility of all factors of production.
4 Perfect knowledge of the plan by each individual concerned in its implementation.
5 Complete obedience to the plan by all production units.
6 Perfect fulfilment of the plan by all production units. Most important to this form of central control is the structuring of an information and monitoring system capable of disseminating the targets of the plan and determining if they are being achieved. The creation of such perfect information flow has been discussed in chapter 6.

As the aggregation of individual decision making units increases in scale, the locus of control (cost-benefit trade-offs) becomes more complex and indeterminate, and the conflict of negentropy gain (i.e. reduction in uncertainty) and loss of individual freedom of action arises. Buchanan and Tullock (1969) suggest that the optimal level of government should be at that point where the external costs which result from collective action exceed the external benefits to the individual. Fig. 8.6 shows typical graphs of costs and benefits resulting from coalitions of different sizes (N) of groups of individuals. The external cost curve (fig. 8.6A) shows the decline in external costs to the individual (i.e. costs resulting from the actions of others) as the numbers of individuals in the group increases, so that with unanimity there are no external costs since the occurrence of any cost for any individual will lead to non-agreement and no action. The decision costs (fig. 8.6B) increase as the size of the group increases: the familiar phenomenon of increase in the time and effort required by any participating member with group size. The combination of these two cost curves (fig. 8.6C) allows the determination of an optimal point of trade-off between the costs to the individual of participation or non-participation in collective action in a group. These costs are:

> imposed on an individual when his net worth is reduced by the behaviour of another individual or group when the reduction in worth is not specifically recognized by the existing legal structure to be an expropriation of a definite human or property right (Buchanan and Tullock 1969, p. 71).

Hence the existence of these costs is a stimulus to collective action in participation through contracts or constitutions. This approach can be used to define optimal levels of representation (fig. 8.6D) using the ratio of representatives to the number of individuals as a new cost index. When this is applied to defining optimal levels of government (fig. 8.6E) we may find in practice that costs are not smooth functions of group size but have hierarchical steps resulting from moving from different group structure (e.g.

family, neighbourhood, town, community etc.). This cost-based approach differs from the limits on legitimate government action set by Mill (1848) as the residual elements on which agreement can be reached.

The agreement between individuals to achieve common ends through the provision of public goods and services is subject to a number of severe restrictions which arise from the implications of the Arrow (1951) impossibility theorem. Arrow demonstrated that if the two axioms of ordering of the individual's utility (i.e. comparability of alternatives and transitivity; discussed in section 8.1) are subject to five weak and reasonable conditions (See p. 415), then no aggregation of these individuals' utility functions exists that can produce a uniquely ranked aggregate social preference function. The five

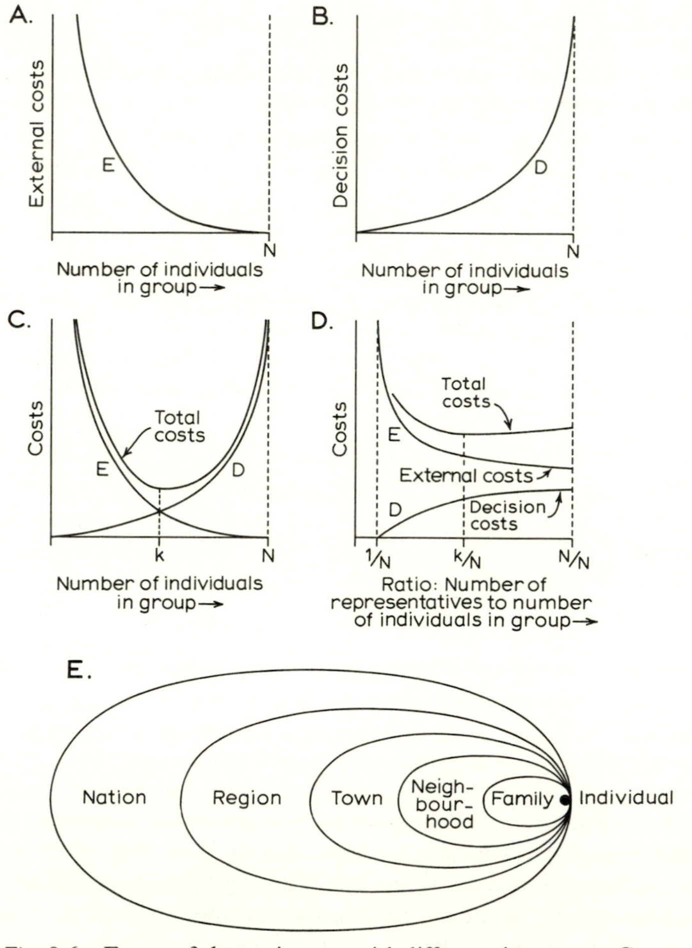

Fig. 8.6 Forms of change in costs with different size groups. Costs are those to the individual resulting from participation or non-participation. (After Buchanan and Tullock 1969.)

conditions imposed by Arrow (1951) are modified by Riker (1961) as:

1 Universal admissibility of individual orderings such that all individuals can choose any set of alternatives in any *order*.
2 Positive association of individual and social values such that if *A* is preferred to *B*, and *B* rises in preference, so does *A* and ordering is retained.
3 Independence from irrelevant alternatives, such that if *A* is preferred to *B*, and *C* preferred to *D* and there is no relation of *A* and *B* to *C* and *D*, then no change results from removing *C* and *D* from consideration of *A* and *B*.
4 Citizen sovereignty, such that social order is not imposed, but results from the sum of individual preference and excludes the Platonic and Marxist ideals of *absolute* social order.
5 Non-dictatorship, such that the individual preference decides the aggregate social ordering.

Under these five conditions no social welfare function can be defined which satisfies the two axioms of utility (order and transitivity). Taken at its face value this means that, firstly, welfare economics has no logical basis, and, secondly, that no *ideal* social order can ever be achieved by democratic collective action but must always be imposed. The response to Arrow's theorem has often been to ignore the implication as an irrelevant special case. For example, Vickrey (1960) notes that the effect of the first and third conditions is to give the cyclical paradox in voting (the so-called Condorcet effect) so that no unique solution exists. If an order restriction is placed on each individual's utility, so that it must conform to a one-dimensional scale, then a 'single peaked' preference function results (Black 1958). Multi-dimensional scaling, factor analysis, and 'unfolding' techniques used in psychology can be employed to construct such a one-dimensional utility scale. It is also argued that 'socialization and agreement on basic normative precepts diminish the probability of multipeaked preferences' (Riker 1961, p. 911). Economists have long sought alternatives in cardinal utility formulations not restricted to ranked alternatives, and in social choice functions. The social choice function has also been proposed as a spatial choice function (Harvey 1973, 1974, Smith 1977). In both cases it represents an ethical or social preference ordering of desired objectives which can then be used by governments as control objectives. The choice function overcomes the inconsistency and lack of tests available for aggregate preference functions (Pattanaik 1971).

Whether the social choice function is an acceptable alternative depends upon our individual philosophical stance. Riker (1961) notes that, as *realists*, we accept that we cannot define the public interest, and hence Arrow's theorem is proof and measure of the extent of our irrationality in attempting to do so. As *idealists*, we would believe that the public interest is not related to individuals' opinions and hence resort to such concepts as the Platonic philosopher ruler, the Marxist dialectical ideal or to other devices. In this case Arrow's theorem merely undermines the democratic process but does not affect our determination of goals. As *rationalists*, we hope that the public interest is discoverable from the summation of preferences, and Arrow's theorem requires that we construct governments of only very limited size according to the collective action principle of Buchanan and Tullock (1969), shown in fig. 8.6. The problems for us as individuals *qua* individuals arise because of our pursuit of social choices as idealists willing to *impose* social values. As we shall see in the next section, this philosophy is often the consequence of treating socio-economic systems as macroscopic systems.

8.3 Systems analysis

The analyst's reaction to complexity in socio-economic systems has been to simplify the conditions of study of any component by removing it from the environment with which it interacts, in a manner similar to the physico-ecological systems described in chapter 7. For large-scale macro-economic and macroscopic social systems a change in the nature of available explanation has usually been accepted which recognizes the indeterminacy of state and indeterminacy of measurement in large-scale systems. As we have already seen, this presents the important additional implication in social systems (which is not present in systems analysis of physico-ecological systems) that there is conflict between individual and aggregate group goals such that the latter can be imposed when analytic (and especially macroscopic statistical) assumptions are employed (See p. 403).

Systems analysis is well expressed in the approach of von Bertalanffy (1950, p. 141):

> if you cannot run after each molecule and describe the state of a gas in a Laplacean formula, take, with Boltzmann, a statistical law describing the average behaviour of many individual molecules. . . . You cannot resolve the individuals within . . . a social unit into cells and finally into physico-chemical processes. Very well, take the individuals as units and eventually you will get a system of laws which is not physics but is of the form of exact physical science – that is, a mathematical hypothetico-deductive system.

Various macroscopic approaches to the analysis of societal systems derive from this approach (Harvey 1969), but few approaches assume precise correspondence of physical and societal process laws. This correspondence can be weakened by assuming that societal processes can be treated *as if* they are subject to given laws (Harvey 1969, p. 428). Alternatively, explanation can be viewed as only possible at the level(s) above – i.e. the search for fundamental psychological and perceptual mechanisms is an example of a possible systems fallacy akin to the genetic fallacy of historicism (Popper

1957). Other approaches rely upon justifying macroscopic explanation in terms of generating simplifications and generalizations of systems structure within given complexity constraints. This may be based upon 'loose' criteria such as parsimony (Tukey 1961, Box and Jenkins 1970) which can be used to create forecasting explanations of systems response but do not attempt to represent physical parameters. Alternatively, generalizations may be determined by searching the matrix of interrelations between the system state variables and by suitable reduction to derive a canonical form (Ashby 1956). Special applications under control conditions of this approach to the dynamic systems described in Chapter 2 are given by Kalman (1960). These generate observability and controllability criteria of a minimal representation and have been extended to space-time dynamic systems by Bennett (1975d). The results of applying these approaches to macroscopic systems have been extremely successful, at least in *simulating* socio-economic behaviour and dynamics.

Pre-eminent in this area has been Keynes's (1936, 1937) model of the economic system (fig. 8.7) from which large macro-economic models have

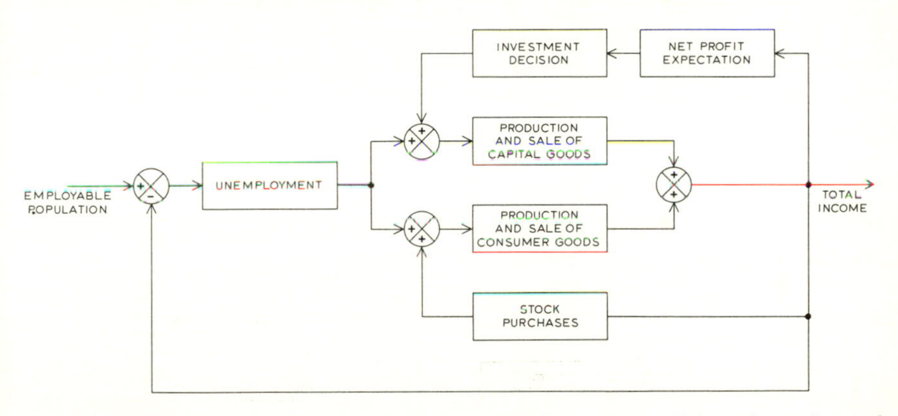

Fig. 8.7 Model of the J. M. Keynes' economic system. (After Saucedo and Schiring 1968, p. 9.)

been derived: for example, the Brookings–SSRC model of the United States economy (fig. 8.8) and the NIESR model of the United Kingdom economy (fig. 8.9) (Duesenberry *et al.* 1965, Surrey 1971), as well as spatial interregional models such as LINK (Ball, 1973) and interregional input-output models (Isard and Cumberland 1961, Mennes *et al.* 1969). The basic aim of these models, which incorporate hundreds or even thousands of equations, is in all cases to reproduce the behaviour and time paths of a set of economic output or endogenous variables from a set of input or exogenous variables using a set of transfer function system operators (i.e. behaviour assumptions). The final aim of these models is thus to produce forecasts which facilitate the formulation of

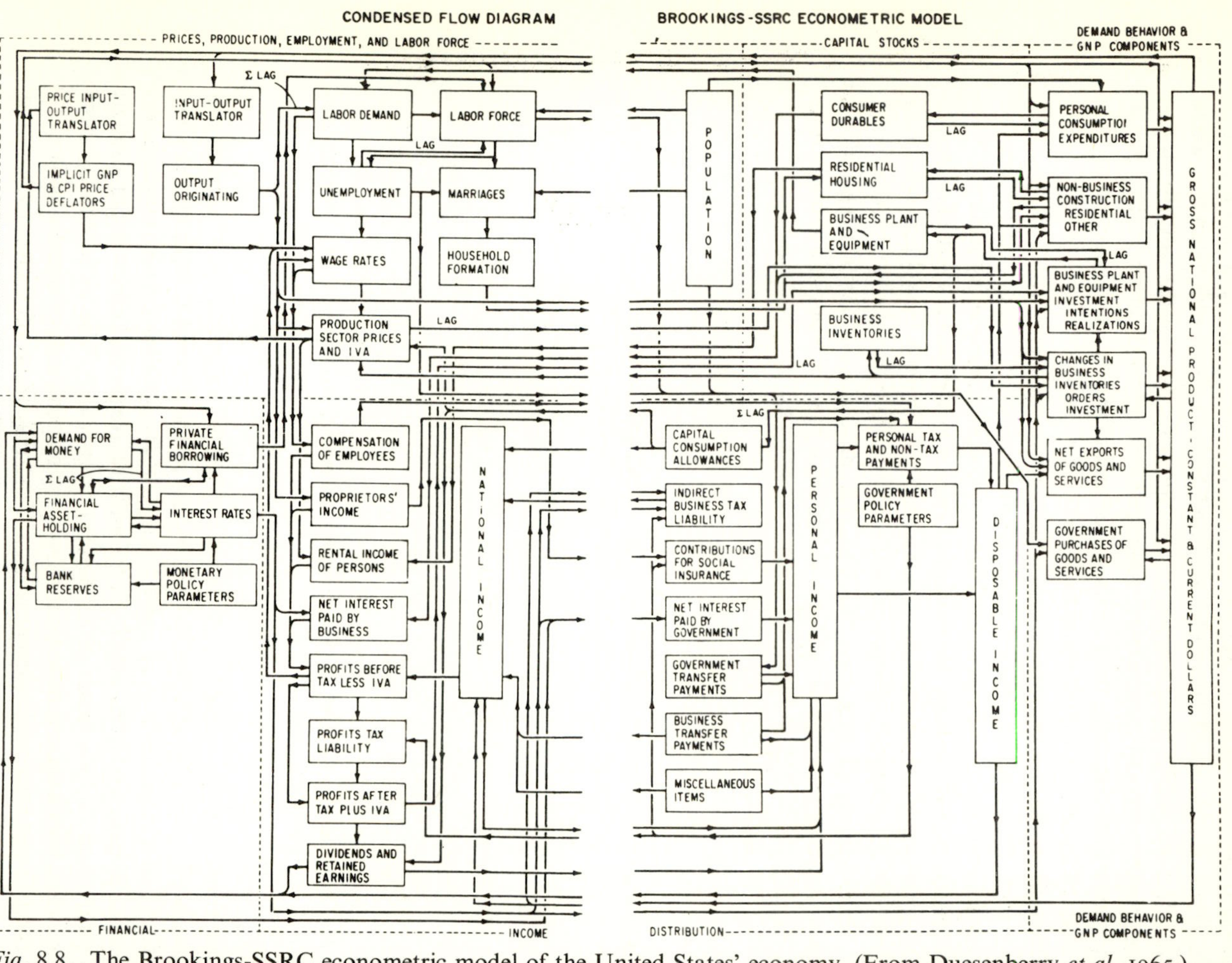

Fig. 8.8 The Brookings-SSRC econometric model of the United States' economy. (From Duesenberry *et al.* 1965.)

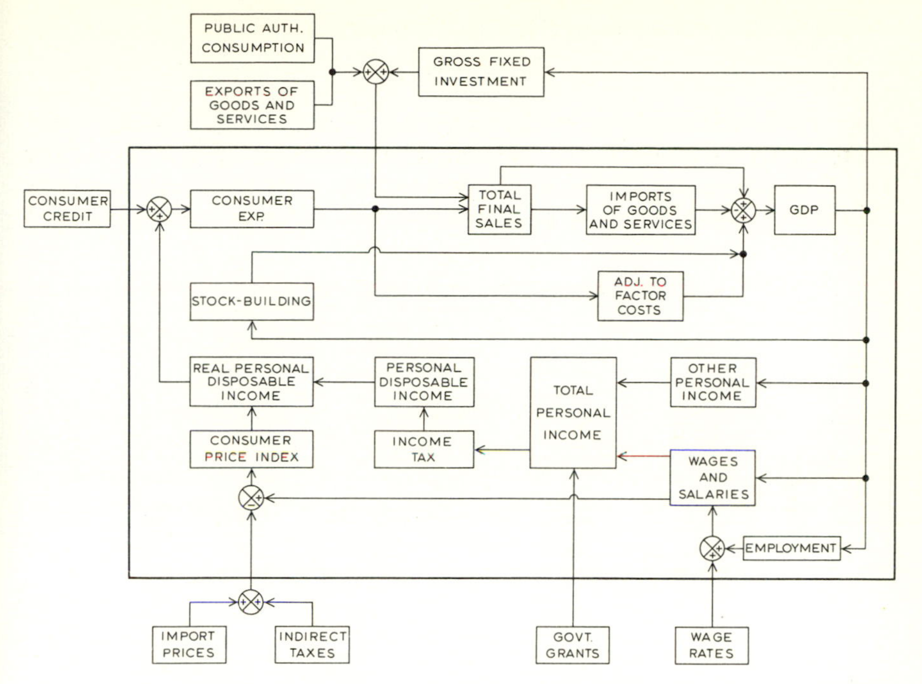

Fig. 8.9 The NIESR econometric model of the United Kingdom economy. GDP = Gross Domestic Product. (Partly after Surrey 1971, Bray 1970.)

policy. The variables normally included in each category are:

1 *Outputs*: real level of production, consumption, investment, rate of interest, price of production elements, labour demand, wage rate, unemployment, demand for money.
2 *Inputs*: limit of money available, value of wage rate, capital state, technology, 'full employment' level, value of household assets, distribution of income.
3 *Transfer functions*:
   (*a*) Consumption function
   (*b*) Investment function
   (*c*) Equity preference function
   (*d*) Labour demand function
   (*e*) Labour supply function
   (*f*) Production function

Each Keynesian model can be simplified to a structure which exhibits the basic hypothesis of the multiplier-accelerator (fig. 8.10). This hypothesis is founded on the presence of two feedback loops: the consumption loop (multiplier) and the investment loop (accelerator). The consumption loop gives the response of total endogenous demand (consumption) in an economy to changes in total output (GNP: real disposable income):

$$C_t = kY_{t-1}. \tag{8.3.1}$$

This is a Robertsonian lag model with c a constant *multiplier* (Kahn 1931) representing the propensity to consume. The investment lag depends on 'the inducement to invest; and the inducement to invest will depend on the relation between the schedules of the marginal efficiency of capital and the complex

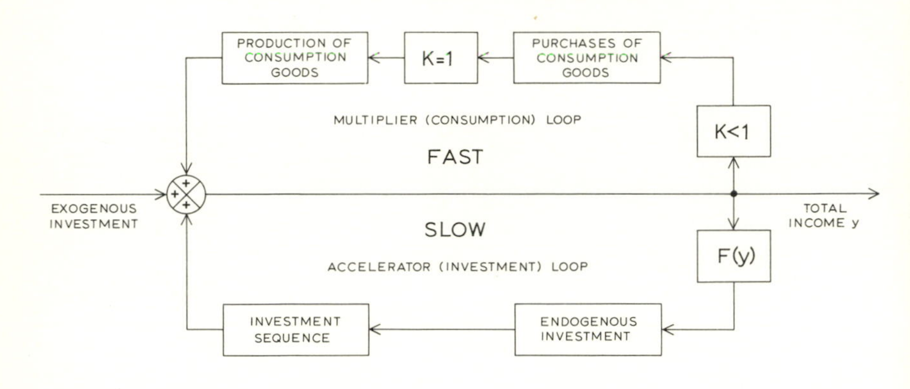

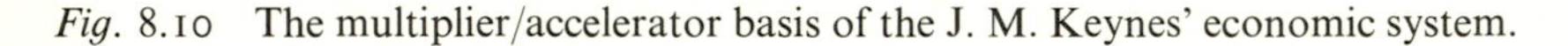

Fig. 8.10 The multiplier/accelerator basis of the J. M. Keynes' economic system.

rates of interest on loans of various maturities and risks' (Keynes 1936; see Young *et al.* 1973, p. 147). Thus:

$$I_t = \frac{\partial K}{\partial t} \approx K_t - K_{t-1}$$

$$= qI_{t-1} + vY_{t-1} = \frac{vz}{1 + qz} Y_t \tag{8.3.2}$$

where z is the unit shift (lag) operator (chapter 2), since investment is a function of output over the previous time period. The constant q is a multiplier on national income termed the *accelerator*. The consumption loop is usually very rapid relative to the investment loop. This marked difference in feedback and response time has important implications for policy control since shifts in overall system inputs are most quickly absorbed by consumption rather than by generating real increases in wealth through investment. Hence control is often concentrated on damping out the fast consumption loop response and redirecting the flows through the slower investment function. An example of this is the monetary policy and fixed limits on government spending discussed in section 8.4.

To these two feedback loops are usually added a system transfer function and government expenditure loop to give a final simplified model which can be represented by eight variables and seven equations, as in fig. 8.11. The system transfer function links total demand to total national output (GNP) subject to a Lundbergian lag (Allen 1968) and production function p:

$$Y_t = pZ_{t-1} \tag{8.3.3}$$

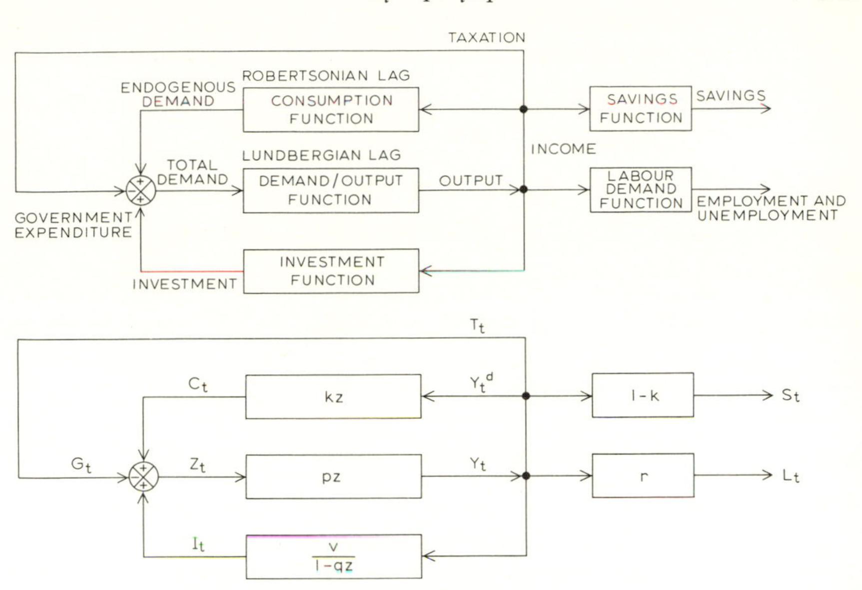

Fig. 8.11 An elaboration of the J. M. Keynes' double-loop economic system.

$$Z_t = C_t + I_t + G_t$$
$$Y_t = f(Z_t) = f(K_t, L_t, t)$$
$$C_t = f(Y_t)$$
$$I_t = f(Y_t) = f(K_t)$$
$$S_t = f(I_t)$$
$$L_t = f(Y_t)$$
$$P_t = f(C_t)$$

Y_t = GNP: the total money value of the output of all industry in the national economy, less stock appreciation, plus income from abroad, at time t.

P_t = Production: the total value of production in consumer goods, at time t.

C_t = Consumption: the total value of purchases of consumer goods, at time t.

I_t = Investment: the total value of investment in industrial production, at time t.

S_t = Savings: the total value of savings, at time t.

L_t = Labour.

K_t = Capital.

Z_t = Aggregate demand.

G_t = Government expenditure.

which, like the Robertsonian lag, is fast. Government expenditure is derived from total national income Y_t (plus investment, savings and external borrowing) and enters into the expression for total demand – i.e.

$$G_t = Y_t - Y_t^d = T_t. \tag{8.3.4}$$

The overall Keynesian economic time path is then derived by reducing the seven equations for each variable. For example, the consumption function time path can be rewritten:

$$C_t' = Y_{t-1} \quad \text{and} \quad Y_t = C_t + I_t + G_t$$

then

$$Y_t = kY_{t-1} + I_t + G_t \qquad Y_t - kY_{t-1} = I_t + G_t$$

and

$$(1 - kz)Y_t = I_t + G_t. \tag{8.3.5}$$

Hence

$$Y_t = \frac{1}{1-kz} I_t + \frac{1}{1-kz} G_t. \tag{8.3.6}$$

The expression $1/(1 - kz)$ is referred to as the *Keynesian multiplier* (k), producing either positive or negative feedback of the system to private and government investment (Tustin 1953). It can be interpreted as a macroscopic homeostatic regulation term. The time path of the Keynesian system subject to an initial disturbance is shown in fig. 8.12 for various values of the multiplier k. In all cases an oscillatory cyclic response is the natural system form, which is explosive for values of $k > 1$. The response with $k = 2$, but with control imposed, is also shown to exhibit how the natural cyclic effects can be damped out (see section 8.4). If k is greater than, or equal to, unity then the consumption loop becomes an explosive feedback mechanism whereby government or private investment generate via consumption an increase in the national output greater than the magnitude of the initial investment. This argument is the key to the economic management of the capitalist economy, by which greater growth can be achieved by increasing investment. The reduced form expression for the accelerator is:

$$Y_t = C_t + \frac{v}{1-q} Y_{t-1}. \tag{8.3.7}$$

If this is combined with the consumption loop, the final reduced form of the multiplier-accelerator is given as:

$$Y_t = kY_{t-1} + \frac{1-k}{1-q} Y_{t-1}.$$

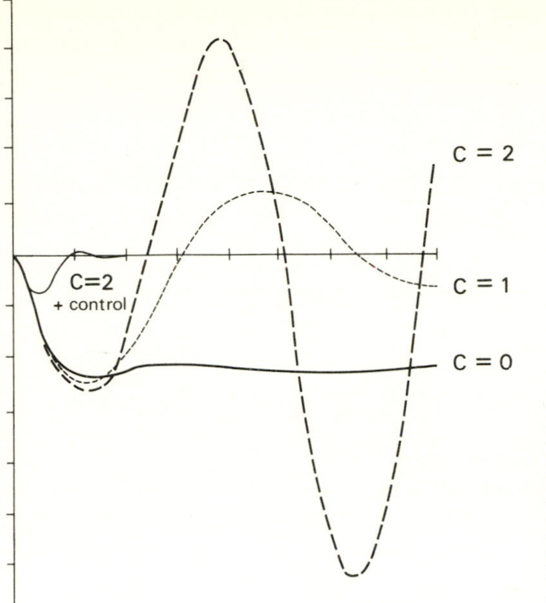

Fig. 8.12
Time path of the Keynes' system following an initial disturbance, demonstrating its self-regulation for various values of propensity to consume (price elasticity c). (C = constant multiplier). (After Phillips 1954.)

This is a first-order difference equation which can be rewritten as:

$$-qY_{t-1} + kqY_{t-2} = 0$$

which has general solution given by using the methods of chapter 2:

$$Y_t = B_1\, e^{\lambda_1 t} + B_2\, e^{\lambda_2 t}$$

where

$$\lambda_1, \lambda_2 = \frac{q}{2} \pm \sqrt{\left(\frac{q}{2}\right)^2 - kq} \qquad (8.3.8)$$

are the roots derived from the solution of the equation $\lambda^2 - q\lambda + kq = 0$. The roots λ_j are real if $(q/2)^2 \geqslant kq$, and if one root is greater than unity there is steady growth. However, if both $\lambda_j < 1$, there is steady decline. The roots λ_j are complex if $(q/2)^2 < kq$ and there is an oscillatory response in that $\lambda_j = (\cos\theta + i\sin\theta)$. This oscillatory response is damped if the roots have real or imaginary parts less than one, but is explosive if either part is equal to or greater than one. The development of the Keynesian multiplier-accelerator in equations (8.3.1) to (8.3.8) is in terms of discrete variables and lag operators z, which is the notation developed in chapter 2. The Keynesian economic system can be described in the alternative continuous (i.e. differential) equation form shown in fig. 8.13 in which the lag operator is replaced by differential rates of change (d/dt).

The micro-economic view has a number of important behavioural

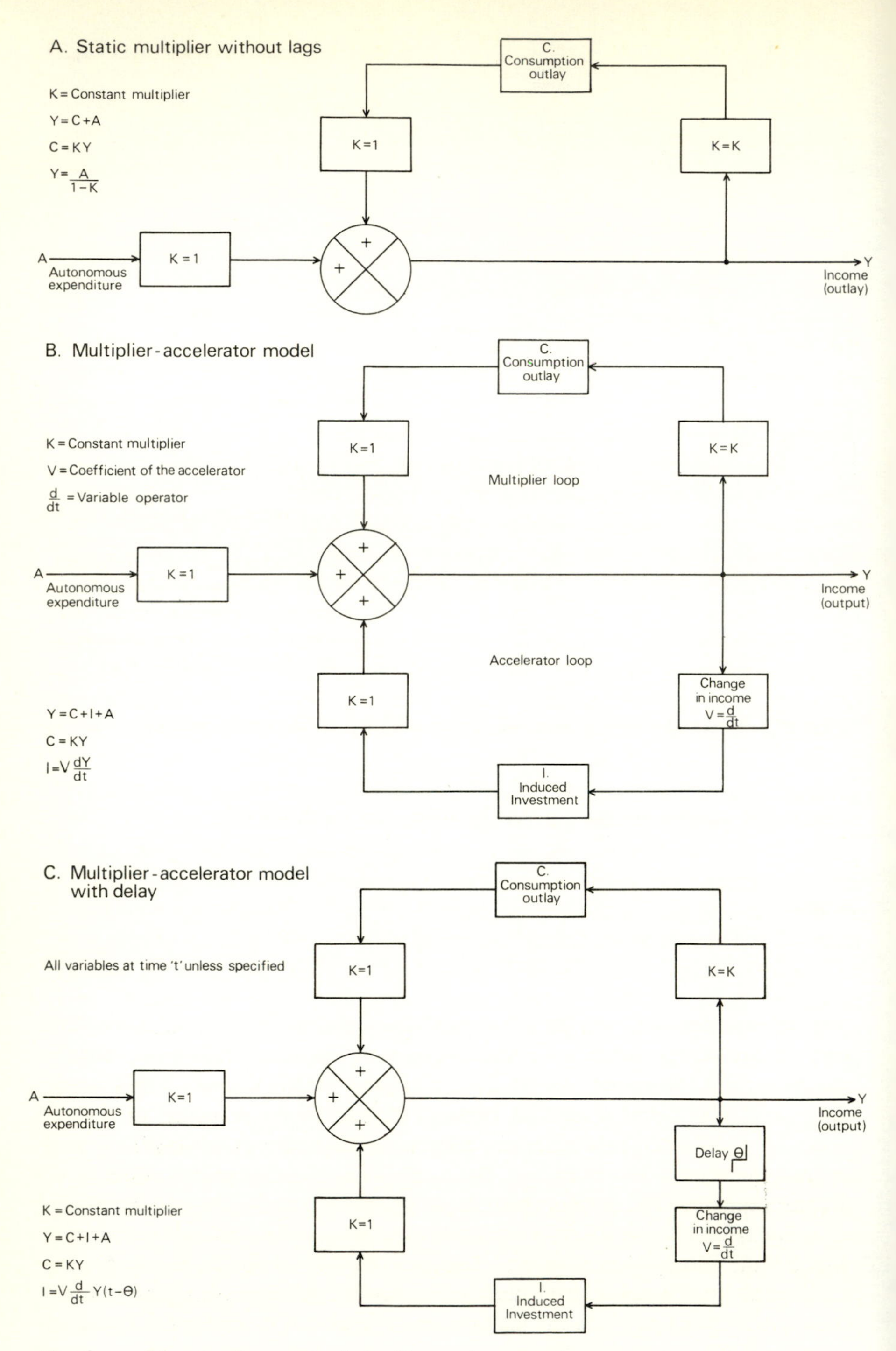

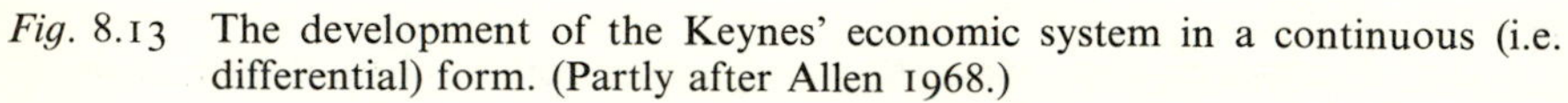
Fig. 8.13 The development of the Keynes' economic system in a continuous (i.e. differential) form. (Partly after Allen 1968.)

implications. Firstly, from equation (8.3.8) the model produces a sum of exponential growth and oscillatory effects. Secondly, government expenditure can stimulate growth, such that in some circumstances short periods of deficit spending might be useful (Keynes 1936). Thirdly, the multiplier takes a time to work equivalent to the lags and multiplier loops. Fourthly, government expenditure is not an autonomous variable and is itself determined by the level of national output. As Young *et al.* (1973) note, the entrepreneur who looks for methods of increasing his wealth must increase his capital investment, but for the overall wealth of the national economy to be increased it may be necessary for the government to initiate investment. The correct investment-expenditure mix can thereby produce 'take-off' to self-sustained growth (Rostow 1960).

Various developments of the Keynesian model are shown in fig. 8.14 (Tustin 1953, Allen 1968). The Samuelson-Hicks (Hicks 1956) model has one-period lags in consumption and investment and is similar to the Keynesian model used above. Goodwin's (1951) model involves the more realistic assumption of a nonlinear investment function, which is elastic to the level of national output. The Kalecki (1954) model is an attempt to refine the structure of the investment lag to incorporate differences in the timing of investment outlays (which will have differential effects in augmenting demand) and the balance of capital stock. Kalecki assumed that the rate of decisions to invest depended positively on the level of output, and negatively on the amount of existing capital equipment and stocks. The Phillips model represents the most general form of Keynesian multiplier-accelerator in which consumption, output and investment functions are subject to complex lags.

The Keynesian hypothesis, particularly its implications for government control action, has been increasingly questioned. However, the greatest weaknesses are not its control implications but, in common with all other macro-economic models, its lack of representation of evolutionary behaviour and its submergence within general equations of the reality of decision making processes, especially their sociopolitical and power bases. We have already noted the difficulties in aggregating individual utility functions to community and social levels. In addition, change in societal systems is a function of more complex processes than can be incorporated in macro-economic models, such as decay, emergence, and learning (Kuhn 1974, pp. 433–44). Decay processes govern the hierarchical breakup of organization levels of societal activity – for example, the decay of social institutions and governance paralleling the decay of animal to vegetable matter. This results in increasing uncertainty and the loss of energy and information (Kuhn 1974). The emergence process is the appearance of new and higher organization levels of societal interaction characterized by information increase, negative entropy and increase of differentiation. Learning is the most fundamental of societal processes and governs the creation of new behaviour modes, new goals, and new methods of achieving goals within existing organizations.

Models of social-cultural emergence and learning can be based upon analogues of the competition-succession process in ecology (Geertz 1963) or on sociopolitical conflict and succession (i.e. the Hegelian thesis and antithesis succession). The cultural balances and checks which limit societal practices within a conservation-balanced community are one of the subjects treated in chapter 9. Within these balanced communities competition and displacement of rival societies may take place through technological advance and diffusion of innovations (Clarke 1972). This will be based on the fitness of each rival community for the particular environment (Harris 1969). Von Mises (1951) characterizes the evolution of cultural norms and regulation (homeostasis) in the face of environmental shifts as the evolution to consensus or *ought*

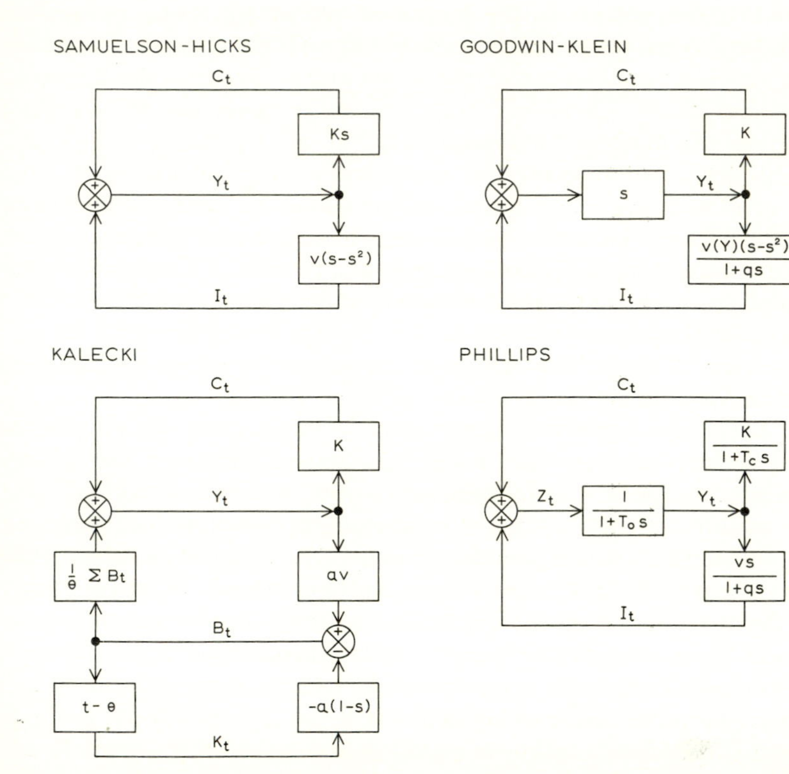

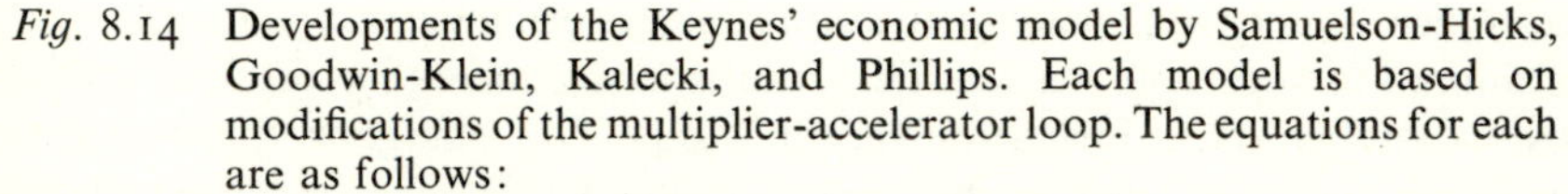

Fig. 8.14 Developments of the Keynes' economic model by Samuelson-Hicks, Goodwin-Klein, Kalecki, and Phillips. Each model is based on modifications of the multiplier-accelerator loop. The equations for each are as follows:

Samuelson-Hicks: Demand $Z_t = C_t + I_t$

$C_t = KY_{t-1}$

$I_t = v(Y_{t-1} - Y_{t-2})$

Output $Y_t = Z_t$

statements. These are not inborn (Piaget 1969) and evolve from communications in groups of shall, shall-not (i.e. good, evil) statements which become abstract in acceptance. Such moral laws evolve from the rank ordering of human conduct, become dogmatized within society and answer the psychological need for certainty, uniform and unassailable valuation (Von Mises 1951). Hence, each set of culture norms and values is society- or group-specific. To this we may adjoin ideas of homeostatic evolution to either ecologically conservationist goals, or to ideal societal internal checks and balances (i.e. dialectic) (Harvey 1974). This relational or dialectical materialist view leads to a conception of the 'totality in which it is neither the part nor the whole, but the relationships within the totality which are regarded as fundamental' (Harvey 1974, p. 265). For Engels the dialectical materialist view and the ecological competition-succession process have no necessary or predictable outcome. This is taken further by von Mises (1951, pp. 322–3):

> the individual orders the influence of his origin, environment, education, and acquired experiences, and arrives at value judgements. The consequences of a part or all of these preconditions within a class of individuals result in a corresponding agreement on valuation. Nowhere can we find criterion for

---

Goodwin-Klein:

Demand
$$Z_t = C_t + I_t$$
$$C_t = KY_t$$
$$I_t = \frac{1}{1 + qs} v(Y)(Y_t - Y_{t-1})$$

Output
$$Y_t = \frac{1}{1 + T_0 s} Z_t$$

Kalecki:

Demand
$$Y_t = C_t + I_t$$
$$C_t = KY_t$$
$$I_t = \frac{1}{\theta} \sum_{t-\theta}^{t} B_t$$
$$B_t = a(vY_t - K_t)$$
$$K_t - K_{t-1} = B_{t-\theta}$$

Output
$$Y_t = Z_t$$

Phillips:

Demand
$$Z_t = C_t + I_t$$
$$C_t = \frac{K}{1 + T_c s} Y_t$$
$$I_t = \frac{vs}{1 + qs} Y_t$$

Output
$$Y_t = \frac{1}{1 + T_0 s}$$

(Partly after Tustin 1953 and Allen 1968.)

> singling out one among many different, mutually contradictory value scales as 'objectively correct'.

Hence, in common with the collective action approach of Buchanan and Tullock (1969), no action occurs without agreement and 'the acceptance or rejection of a normative system by a group occurs through a collective act of resolution, in which every individual acts under the influence of various impacts of society' (Von Mises 1951, p. 370). Ethical codes arise through growth in primitive societies and how these societies organize towards their stability and maintenance (Harris 1969). They are statements of 'ought', 'will' or 'must' which become codified as ethic, value, law or duty; the values evolve and are not made (Von Mises 1951, Hayek 1944).

In contrast to this unpredictable evolution of norms, the Hegelian thesis, as reinterpreted by Marx (1968), evolves through a dialectical clash of ideals which is ontological (see chapter 1). For Marx there is an economic progression from primitive economies, through slavery, feudalism, bourgeois revolution with capitalism, proletarian revolution with socialism, to communism. There is a necessary historical progression inbuilt into the vested power lying within each configuration of society; there is the historicist deterministic desire to make things go in the direction desired or predicted. For Marx the materialist ethic explains social consciousness as the outcome of being, and consciousness is the stimulus to change. Thus Marx (1888, p. 30) states that 'the philosophers have only *interpreted* the world, in various ways; the point, however, is to change it'. Whereas Engels interprets the Hegelian dialectic as leading to no prior definable goals (through the clash of incommensurable values), for Marx systems analysis and systems synthesis come together to yield a unitary end. When reinterpreted by Lenin (see Lenin 1960) the overall behaviour of the economy is constrained to conform to overall planned goals; the descriptions and study of other mechanisms by which the plan might be achieved, or of alternative plans with similar marginal costs, are not in any sense relevant. In particular, there is no interest in attempting to determine the character of socialist enterprises, the comparative efficiency of capitalist enterprises, the nature of entrepreneurial behaviour, or how actual decisions should be made. Marxist economies thus become idealist or ideological, prescriptive rather than descriptive, and even Messianic (Wiles 1962).

Marx's interpretation or dialectical materialism overcomes Arrow's paradox (discussed in section 8.2) by suggesting a necessary social order which must arise because of the inbuilt power and its location within society. The competition-succession approach sees norms established after long periods of cultural evolution along a trajectory which is unconstrained except to retain homeostasis within environmental limitations. On the other hand, the pure idealist approach leads to an evolution towards abstract goals, as did the Socratean concept of good as truth and knowledge which is the basis of

Western conscience, ethic and duty. This may be linked with the religious ethic of Judeo-Christian society or more intuitive concepts of virtuous conduct. An example of this is the Spinoza moderation of power and wealth to preserve one's real nature, to preserve symmetry with nature and satisfy happiness. Each such normative system leads us to a new set of 'ought' and 'must' statements which can then be *imposed* on society. Abstract ethic is the basis of authority. The crux of the abstract idealist approach, like the Marxist, is thus based in the power to control.

Under both an abstract idealist approach or a Marxist dialectic materialist approach we are led to consider the structure of power. Power is determined by control of resources, such that he who has control of resources has the power to accumulate *surplus value* or profit (i.e. the difference between use value and exchange value). The fundamental question of economic response is: Who controls the price rules? Under the Marxist ethic it is inappropriate or not permissible to exploit a power position (e.g. property); surplus value should not be accumulated, and each sector of the economy should have, as far as possible, a precise equilibrium of supply and demand. Fig. 8.15 illustrates the sectors of the capitalist economy divided into five main control areas. Each of these sectors is capable of accumulating surplus value, and hence wealth, by taking a greater income than is justified by outgoings. In each case accumulation will be the result of the relative balances of power and location of ownership (property) in society. Harvey (1973, ch. 5) has discussed the surplus value concepts in the context of urban land use and demonstrated how the roles played by householders, landlords, developers, financial institutions and government institutions distort the traditional competitive free market equilibrium and lead to exploitation and accumulation of surplus value. The *landlord* can accumulate surplus value through profits of charging rent in excess of building, maintenance and running costs and this is a consequence of *ownership monopoly*. The *corporate sector* accumulates surplus value through profit maximization, such that charges for products are in excess of the costs of raw materials and wages, and this results from *employment monopoly*. The *financial institutions* accumulate surplus value by charging interest in excess of administrative cost, thus exploiting a *capital monopoly*. The *household* sector may also accumulate wealth by receiving wages in excess of rent and living requirements. Finally, the state may accumulate surplus value by taking more in tax income than is needed or mandated for the provision of services. The Marxist dénouement is that the financial institutions and corporate sector combine to produce an employment-capital monopoly to exploit labour. More recent capitalist economies evidence the result of labour monopolies through bargaining power to accumulate wealth in the consumption sector (Hayek 1972). Fundamental to the exploitation by any sector is monopoly in the market. Any slight inequality in wealth at initial conditions will lead to wealth accumulation through system feedback and by any power this confers. The Marxist paradigm is a supply-demand equilibrium, a necessary and sufficient

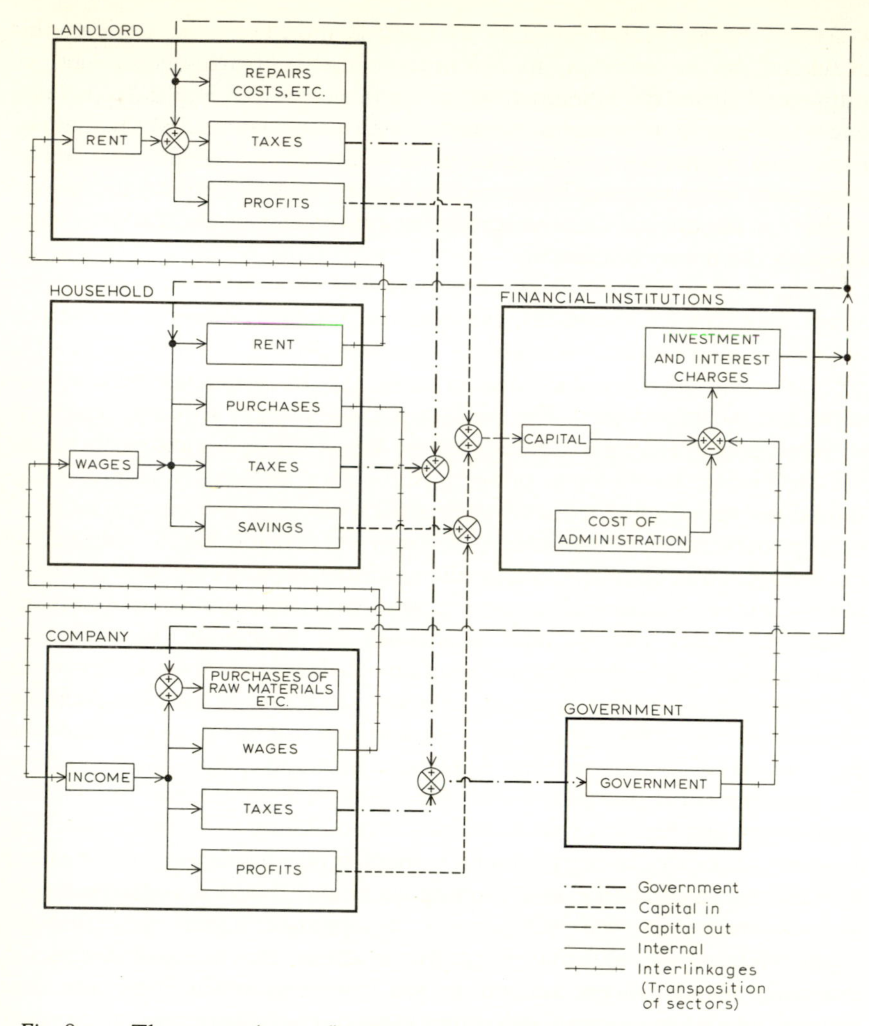

Fig. 8.15 The power (control) sectors of the capitalist economy (arrowed lines show flows).

condition for which is that each societal system transfer function is equal to unity (with each input-output relation expressed in standardized units). In fig. 8.16 the input-output translators (transfer functions) are described by the gain terms K. The aim of Marxist economies has been to achieve an equilibrium in which the output or supply of each section i exactly balances the demand of each sector j. In practice there are considerable problems in achieving the condition $K = 1$. More often $K < 1$, in that one sector generates supply deficiency and via circular feedback determines the overall level of operation of the entire societal system.

In the wider context of the socio-economic system within its physico-ecological environment, supply and demand equilibrium must be based not only upon resolution of competing user demands but upon the environmental 'stock' or supply of resources used in satisfying any set of demands. As we shall see in chapter 9, macro socio-economic systems should contain also a set of complex feedback loops from environmental systems. Increased economic output is associated with increased pollution concentrations and some feedback loop may be necessary to effect control – e.g. tax on pollution

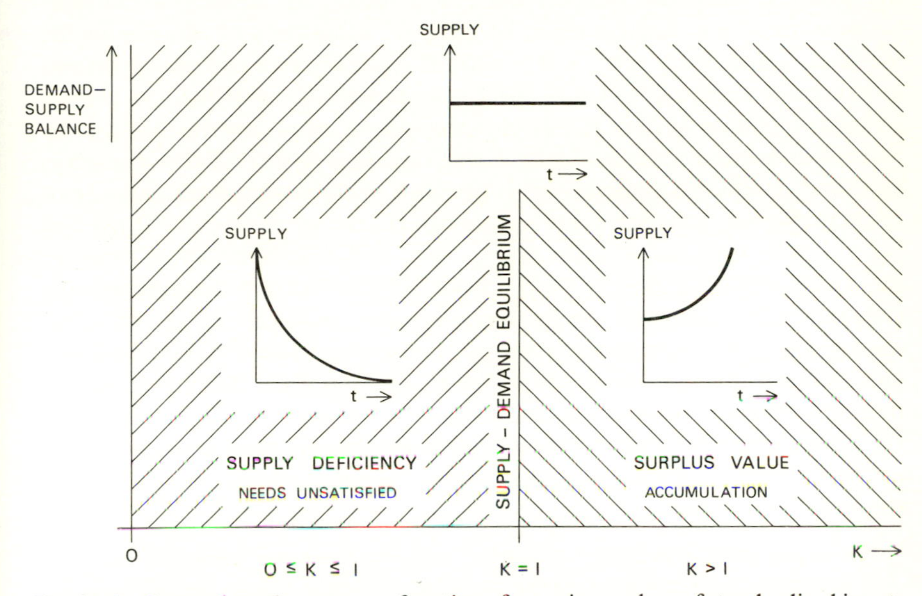

Fig. 8.16 System impulse response functions for various values of standardized input–output translators *K*.

generation. Increased output and growth is also associated with resource use and depletion. Again, feedback is necessary to limit socio-economic demands for resources to either the rate of replenishment in the case of renewable resources, or to the true cost of resources as *capital* (not income) in the case of non-renewable resources. Space, environment and physico-ecological resources are as important in supply-demand equilibrium as purely socio-economic flows. When the rate of socio-economic use of these resources markedly exceeds their replacement rate, then the environmental constraints may become pre-eminent in controlling the supply-demand equation (see chapter 9). Hence, as Harvey (1974, p. 265) notes, resources cannot be understood independently of their relations, especially to sources of demand.

One especially important resource which also affects demand-supply equations is *space*, which is an all-social property that can be consumed and allocated in the same way as any other resource (chapter 4). Every societal decision or behavioural action is spatial. An investment decision is a spatial location decision on *where* production will be situated over the amortization period of that capital. A consumption decision will weigh the costs of production from various sources and and will also be a spatial decision, since the choice will be constrained by the transport costs of obtaining the good, e.g. by the range and threshold of a good (Berry 1962).

The spatial character of macro-economic societal activity derives from the overlapping economic, sociopolitical and administrative networks as cascading systems. Thus the location pattern of economic activity within and between regions is partly a result of the pure economics of location in the Weberian sense (Isard 1956), but is also the result of political boundaries and administrative structure. Under perfect competition, in which all entrepreneurs maximize profits in current production, new investments will be located to minimize site and factor costs, transport costs of factor assembly, and product sales to market. This is the basis of micro-economic theories of the firm (Weber 1909, Isard 1956), of theories of the service sector (Christaller 1933, Lösch 1954, Isard 1956, Berry 1967) and macro-economic location theory (Isard 1975). Isard's (1956) model of the location decision process transfers the traditional profit maximization problem of the firm to encompass substitution between transport inputs each taken two at a time (see equation 6.3.4). The Weberian problem assumes that the factor mix is constant. Hence the transport rates and volumes of inputs are determined and the optimization problem becomes a distance minimization problem which can be solved for any transport input to give the optimal location according to the rule:

> that *at the point of minimum transport cost, the marginal rate of substitution between any two transport inputs, the other(s) held constant, must equal the reciprocal of the ratio of their prices – namely, the corresponding transport rates.* (Isard 1956, p. 224, italics in original)

As Isard (1975) shows, this concept can be extended to apply to service industries, consumer behaviour and most spatial search processes and is equivalent to the general optimization of decision making under economic criteria discussed in chapter 6.

Radical critics of regional science have suggested that Isard's (1956) model and later developments including those by Curry and Olsson have generated a 'regional metaphysics' which is divorced from the real world and has only deductive relevance (Holland 1976, pp. 18–35). This criticism is based mainly on attacks of the gross simplifications of the neoclassical utility maximizing and equilibrium assumptions employed (Kornai 1971, Holland 1976) which ignore spatial monopoly, economies of scale, and the historical, cultural and political influences that govern trade and capital flows as much as does any distance decay function. Holland's analysis sees spatial monopoly as the result of the sectoral concentration of production and not the result of the location of raw materials or the working of comparative advantage. Modern state capitalist intervention and national union wage bargaining have deferred Marx's forecast that capital would displace labour but, Holland argues, spatial disparities are becoming more marked due to the monopolistic meso-economic power of multinational companies. Multinational manipulation of capital flows are particularly important in preventing state manipulation of fiscal instruments and infrastructure investment to correct regional imbalances. Based on evidence of the divergence of regional products and regional incomes, Holland suggests that capital is now being substituted for labour, as predicted by Marx, but at an international level, with labour exploitation in the Third World.

The burden of the radicals' challenge to classical economics and to regional science demonstrates the importance of interfacing political and administrative structures with optimal location decisions based on 'objective' criteria of substitution or discounted cash flow. Fig. 8.17 shows five forms of politico-economic spatial structure that might result from the integration of a three-stage economic process (A = final demand, B = intermediate demand, C = initial demand) in a three-regional economy. Complete spatial specialization results in each stage of the production process being carried out in a different region. In a capitalist economy this would result from each region exploiting to the full its comparative advantage for any particular stage of

A ← B
C

SPATIAL
SPECIALISATION

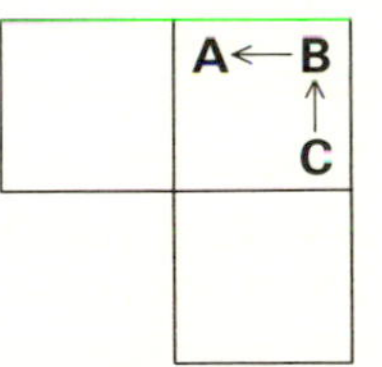

SPATIAL
CONCENTRATION

A←B A←B
C C
A←B
C

SPATIAL
EQUALITY

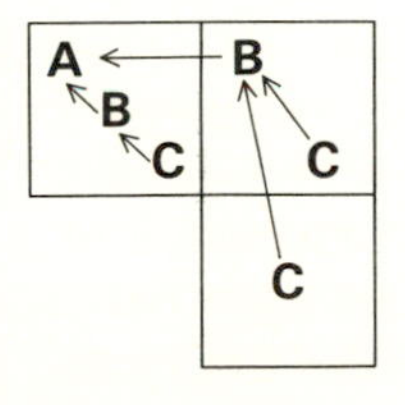

(Equal resources)

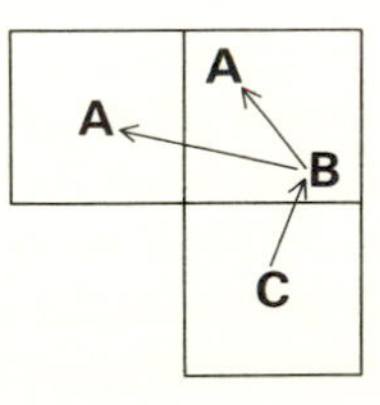

(Unequal resources)

SPATIAL PYRAMIDAL STRUCTURES

Fig. 8.17 Five forms of politico-economic spatial structure resulting from the integration of a three-stage economic process (A = final demand, B = intermediate demand, C = initial demand) in a three-region economy.

production. This form of economic development need not be deleterious provided that the cascade between each region has a standardized transfer function steady state gain of unity (i.e. other values enable the creation of surplus value). Since stage *C* of the production process might involve the processing of raw materials, stage *B* the intermediate production, and stage *A* the marketing, then exploitation by any region will conform to a raw material, labour or marketing monopoly. Complete concentration results in all stages of production being sited in one region. This is the classic precolonial underdevelopment situation in which there is no interaction with the developed world. Complete spatial equality, by contrast, will normally have each stage of the production process in each region (i.e. the white spots model). Various combinations of these forms result in pyramidal structures. With ubiquitous raw material distribution, each region will carry on the stage *C*, but stages *A* and *B* may be concentrated. With concentrations of raw material in one source area, this region will be the only one to carry out stage *C*, but stage *A* and *B* will be possible in other regions. This is the classic colonial underdevelopment situation in which raw material processing is carried out in the Third World, whilst intermediate and final processing is carried out in other regions, with the finished products *A* being sold back to the source region.

In each model a non-unitary value of the standardized transfer function between each production stage will result in the accumulation of surplus value. In the spatial equality model this surplus value will accumulate in one sector in each region so that there will be no gross spatial welfare disparity, as measured by GRP, although there will be social intergroup welfare disparity. With each of the other models there will be spatial disparities of income resulting in capital accumulation in one region, or in a group of regions, at the expense of the others. On a world scale this process has worked to continue the impoverishment of the developing world, a process termed *immiserization* by Bhagwati (1958) (fig. 8.18). This results from output elasticity of supply by which growth in factor stock in the underdeveloped world goes into predetermined lines of investment to promote further output for export to the developed world, but is not available for consumption and other Keynesian leakages into the home underdeveloped economy (c.f. Kahn and Posner 1974). Alternatively, by the exploitation of a raw material monopoly the OPEC countries have been able to secure the accumulation of surplus value. At the extreme, the Third World problem results from the imposition of a 'neocolonialism' in which the developing economies are kept open only to exploitation by capitalist business (Robinson 1970). The profits from native industry go into a middle class or élite in the home country or abroad – for example, by the exploitation of cash crops (i.e. coffee, tea, cocoa, etc.) for export. Aid merely fosters further changes towards the exploitation of the local peasant community with surplus value accruing elsewhere. The only permanent solution seems to be social reorganization and land reform

(Myrdal 1968, Robinson 1970) which will lead arguments about population imbalances to disappear and will shift attention to the social management of food scarcity. At a national level the emergence of increasing conflicts over subnational ambitions is associated with perceived interregional exploitation and imbalance in regional budgets – for example, the Quebec separatist movement in Canada, the Basque movement in Spain, the Scottish, Welsh and English regional devolution proposals in the United Kingdom, the breakup of Pakistan, or tribal rivalry in Africa. In each case the increasing polarization between favoured and unfavoured areas presents an ongoing sociospatial dialectical process of conflict resolution dependent upon where power is vested spatially, and which region is perceived to wield that power and gain from its use.

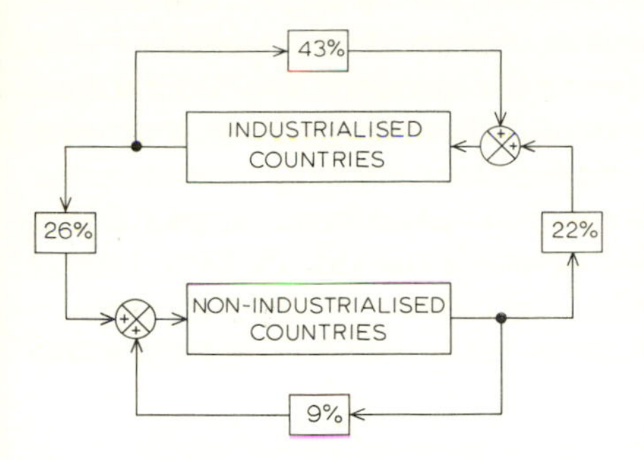

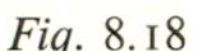

Fig. 8.18
Flows of world trade in approximate percentage value terms in the 1960s, showing immiserization process of cumulative feedback resulting from sectoral concentration in the industrialized world. (After Bhagwati 1958.)

To overcome spatial inequalities, the Marxist approach suggests re-organization of society along spatial lines in accordance with the territorial principle of social justice (Harvey 1973, p. 116):

(1) The distribution of income should be such that (*a*) the needs of the populations within each territory are met, (*b*) resources are so allocated to maximize interterritorial multiplier effects, and (*c*) extra resources are allocated to help overcome special difficulties stemming from the physical and social environment.

(2) The mechanisms (institutional, organizational, political and economic) should be such that the prospects of the least-advantaged territory are as great as they possibly can be.

Hence, Harvey places the traditional Marxist emphasis (to each according to his need) into a spatial context, but recognizes that:

> Deviations in the pattern of territorial investment may be tolerated if they are designed to overcome specific environmental difficulties which would otherwise prevent the evolution of a system which would meet need or contribute to the common good (Harvey 1973, p. 108).

The the Marxist, regional concentration on growth poles is permissible under certain conditions. It is this conflict of common good and Marxist return on investment that is the most difficult to rationalize in real policy decisions. If the main obstacles to international spatial justice are multinational companies, as suggested by Holland (1976), then there will be demands to bring them under greater control. Alternatively we might look to communist countries for guidelines, but developments in these countries show similar irresolvable difficulties in trying to equalize spatial development. For example, Wiles (1962) has recognized three principles which determine the rationalization of location decisions in communist economies. Firstly, there are varying arguments for regional autonomy (autarky) and regional specialization, which are usually conflicting criteria. Secondly, as in capitalist economies, there are attempts to minimize transport costs and maximize economies of scale. Thirdly, there is the total equalization of production principle (or *white spots* criteria) by which the empty areas on the economic map are filled up. This principle has been important in Soviet Russia under Stalinism, and under it the most important determinant of economic location is the political power concentrated at the local level which creates a subordinate autarky (Wiles 1962) (e.g. the Krushchev territorial principle in the USSR). Harvey recognizes that there are considerable difficulties in achieving such a sociospatial optimum. In particular there are a number of unanswered research questions:

1. How do we specify need in a set of territories in accord with socially just principles, and how do we calculate the degree of need fulfilment in an existing system with an existing allocation of resources?
2. How can we identify interregional multipliers and spread effects (a topic which has already some theoretical base)?
3. How do we assess social and physical environment difficulty and when is it socially just to respond to it in some way?
4. How do we regionalize to maximize social justice?
5. What kinds of allocative mechanisms are there to ensure that the prospects of the poorest region are maximized and how do the various existing mechanisms perform in this respect?
6. What kinds of rules should govern the pattern of interterritorial negotiation, the pattern of territorial political power, and so on, so that the prospects of the poorest area are as great as they can be? (Harvey 1973, p. 117)

We look further at spatial and control goals and their achievement in the next section of this chapter, and at the effect of environmental goals in chapter 9.

8.4 Control to objectives

Control in societal systems is concerned with the adjustment of behaviour states and with the allocation of resources. Adjustments to behaviour states arise from the definition by collective action of desirable targets and objectives, or from the imposition of idealistic norms. In small systems agreement between individuals on goals and aggregation of their utility functions can lead to unique social objectives (i.e. the social welfare function). In large groups it is necessary to impose or receive a mandate to impose social choice of defined objectives deriving not from the aggregate of individual utilities, but from a more limited set of 'preferred' social objectives. The allocation of resources is fundamental to the control and decision mechanism, as outlined in chapters 5 and 6. It is a basic assumption of economics that individuals of every society, individually and collectively, desire more goods and services than they produce, and that if more of a given good is available then it will be consumed. Allocation between competing demands for resources in situations of scarcity requires an allocation mechanism based upon an objective function. Again, this objective function may arise through collective action, either organized or disorganized. An example of disorganized collective action for optimal allocation in scarcity is the market economy – the 'hidden hand' of Adam Smith. Alternatively, allocation may be imposed. As with each of the technical control systems discussed in chapters 3 and 4, societal control is a function of the idealistic or collective action targets, goals and objectives to which it is desired to make the system conform, the instruments by which control can be achieved, and the structure of the societal system itself.

8.4.1 CONTROL TARGETS

Control targets are the quantifiable form of policy objectives which can be applied to the optimization of control systems or the achievement of collective action goals. The most common objectives in societal control systems are fivefold:

(1) *Static economic efficiency goal*: This is designed to achieve the proper level of utilization of resources and the optimum allocation of resources between competing users. It demands that: the marginal rates of substitution in consumption should be equal for all users; the marginal rates of transformation and substitution in production must be equal for all firms; and the marginal rate of substitution in consumption must equal the marginal rate of transformation in production (Pickersgill and Pickersgill 1974, ch. 2). Subgoals of static efficiency are full employment and balance of payments equilibrium (Kaldor 1971). Full employment is a special case of maintaining supply-demand equilibrium and has been adopted as a fundamental objective by most Western governments since 1945, as well as by the Marxist system. In the United Kingdom, for example, 'the government accepts as one of their primary aims and responsibilities the maintenance of a high and stable level of employment' (HMSO 1944). The full employment objective has become eroded in the 1970s under labour syndicates (Hayek 1972) and by the recognition that there is an upper asymptote (production capacity) limiting

total output (Paish 1966). The balance of payments target represents an exogenous control on internal allocation and, as noted by Kaldor (1971), need not be simultaneously achievable with the full employment target.

(2) *Growth goal*: This is designed to increase the gross level and quality of output in an economy. Such extensive growth may result from population expansion, expansion into virgin lands, or the increasing volume of investment. In the first two cases, output per man will not increase, whereas increasing investment will raise productivity only up to the marginal utility of capital. Intensive growth has been generally deemed preferable, in which productivity is increased by economies of scale (up to a margin) and by organizational and technological changes (Beckerman 1971). A special feature of the achievement of economic growth is the accumulation of capital. Especially for Marx's theory, the accumulation of capital is fundamental:

> It is capital that destroys the independent peasant and artisan, that creates the propertyless proletariat, that virtually socializes itself by its own concentration in the hands of a few monopolists, that causes ever more damaging swings of the trade cycle until the final crash and revolution – a revolution that consists in a transfer of ownership of capital. And thereafter . . . from a malign tumour capital becomes a beneficent yeast (Wiles 1962, pp. 59–60).

Only by accumulation of capital can technical progress, increases in productivity, and hence economic growth, be achieved; although, as we shall see in chapter 9, the aim of economic growth has dire environmental consequences.

(3) *Social welfare goal*: This is designed to increase the general 'well-being' of society. It is frequently associated with economic growth (Beckerman 1971) in that higher levels of national income provide better opportunities for social well-being, but this will also depend on the distribution of income. Social welfare may not be increased by economic growth, however, if the increased resources go into expenditures such as on defence, or into generating social and environmental costs (e.g. noise, water and air pollution, derelict land and environmental destruction). Much also depends upon *whose* welfare is being improved. The welfare of a country is not solely dependent upon the satisfaction of its material needs. Thus improved social welfare can be achieved by a redistribution of existing economic resources, by political and institutional reform, or by a change in the pattern of production or consumption. It is closely related to the equity goal and has especially important spatial implications (Chisholm and Manners 1971, Manners 1972, J. W. House 1973, Coates and Rawstrom 1971).

(4) *Equity goal*: This is designed to improve the distribution of the final output of the economy (e.g. group, sector and region). Equity can be defined in different ways, however. In a capitalist state income will be equivalent to society's value of the contribution an individual makes to the economy and is

determined by the market mechanism; under socialism income is proportional to labour input; whereas with a communist state the aim is to make 'income' proportional to need. Patterns of ownership and rights to inherit personal property are also important to the equity aim. The achievement of the equity goal is governed by the distributional criteria adopted in the control strategy, viewed in terms of the net balance or proportion of resources going to any particular area (Sen 1970, 1973). Consider a finite stock of resources at any one time. This can be divided between any two sectors, groups or zones *A* and *B* with 100% to group *A*, or 100% to group *B* or any combination in between. If the groups *A* and *B* are placed on two axes, as in fig. 8.19, the 45° line joining the points of 100% utilization of resources by either *A* or *B* defines the possible distributional proportions going to each group. If the existing situation tends more to group *A* than group *B*, then group *A* is obtaining a disproportionate allocation of resources on the basis of equality alone. This is the case in the figure. Four main criteria have been proposed to form the basis of judgements as to the redistributional efficiency of any control strategy. In the first case, any improvement in the present situation is deemed acceptable and is represented by the shaded area to the right of the budget line. This criterion may be combined with certain minimum constraints (or the *padded floor*) which prevents any one group receiving less than a given value of resources. The minimum weekly wage is one example in intergroup income distribution, whilst threshold levels of unemployment have been traditionally used to indicate when significant regional aid is required to equalize welfare across space. The third criterion (i.e. of Pareto optimality) determines that neither group will move towards a worse share of resources. The fourth criterion (i.e. Lorenz or Gini optimality) requires that any change in distribution should be in the direction of greater equality, even to a lower level of resources. Combinations of these four criteria can restrict the Lorenz distribution to being not only more equal but better overall (i.e. more equal but better for both groups). The Marxist ideal of 'from each according to his capacity, to each according to his need' will be achieved by criterion III or one of the combined sets.

(5) *Biological goal*: This is designed to protect nature (flora, fauna, waters, air and earth resources) and to counteract the damaging effect of human intervention. Also important is the management of natural resources such that they increase (O'Riordan 1971) in the case of renewable resources; or so that they are costed in realistic terms of capital (net income) in the case of non-renewable resources (Schumacher 1971). To the biological and other resources is also sometimes added the resource of space which 'must be treated as an all-social property, analogous to, say, solar radiation or air' (Prandecka 1975). The biological goal is frequently ignored in socio-economic studies and in decision making systems oriented to societal control. However, it cannot be ignored where real scarcity (as opposed to scarcity resulting from monopoly) exists. It is an all-pervasive constraint on growth, welfare and productive

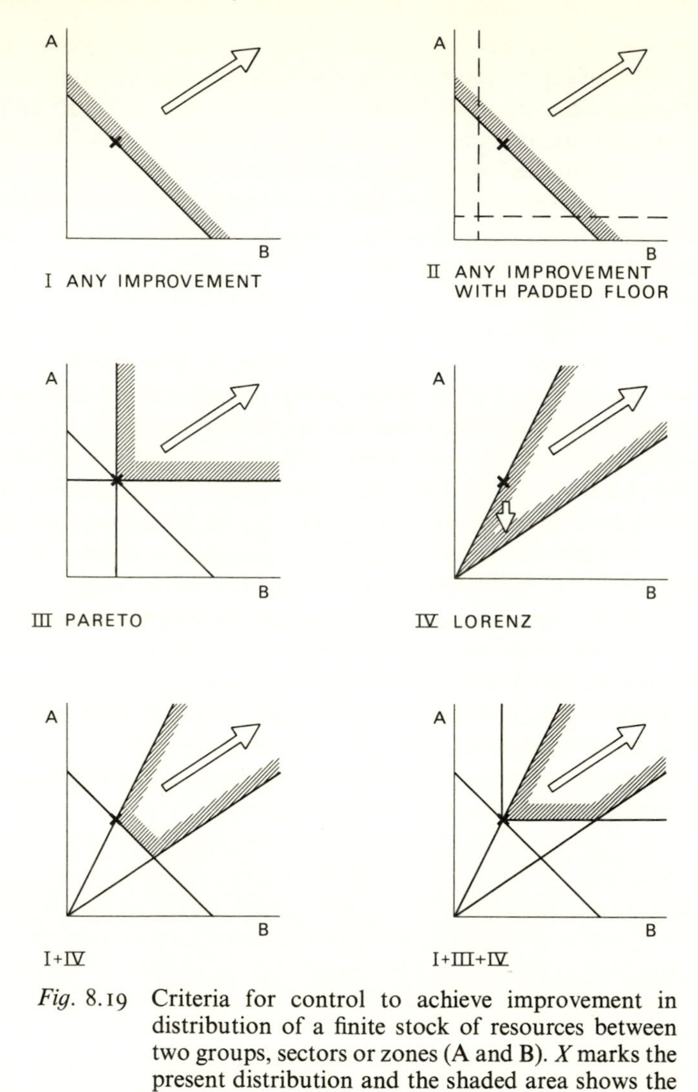

Fig. 8.19 Criteria for control to achieve improvement in distribution of a finite stock of resources between two groups, sectors or zones (A and B). *X* marks the present distribution and the shaded area shows the zone of improvement. (After Smith 1977, Ser 1970.)

efficiency. The effect of this constraint at the interface of socio-economic and physico-ecological systems is treated more fully in chapter 9.

8.4.2 CONTROL INSTRUMENTS

The *control instruments* in socio-economic systems are the policy levers and changes to system state that can be introduced to attain defined aims of collective action or policy goals. At the aggregate and macroscopic level these instruments fall into two classes. Firstly, there is the plan system in which the plan is the target and the administrative bureaucracy is the instrument of policy categorizing all quantities (i.e. inputs, outputs, prices and the identities of sectors suppliers and customers – usually via input-output matrices). Secondly, instruments can be applied to control internal feedback loops (i.e. consumption, investment, government expenditure and external trade).

The control of societal systems by administrative bureaucracies according to a prespecified plan conforms with the monitoring-advisory systems discussed in chapter 6 and is most highly developed in totalitarian states. In the USSR, for example, monitoring is based on the principle of khosraschet – a system of accounting and auditing which checks the inputs and outputs of each sector, industry, individual or household. For Lenin 'accounting and control . . . is the essence of the socialist revolution'; there are enough economic and natural resources 'to satisfy the needs of everyone, provided only labour and its inputs are properly distributed, provided only *businesslike*, practical control over this distribution by the entire people is established' (Lenin 1960, vol. 2, p. 563). For this all that is required are 'the extraordinary simple operations of checking, recording and issuing receipts, which anyone who can read and write . . . can perform' (Lenin 1960, vol. 2, p. 322). Within the overall accounting structure money is manipulated in a passive or semi-active role, and banks become a means not of allocating credit, but of monitoring plan fulfilment by checking each transaction against the plan of transaction laid down for the enterprise. A general planning system has been proposed by Kornai (1967) in the Hungarian context. The central plan acts as a figure parameter and an equation system. This is disaggregated into seven groups of indices (Kornai 1967, pp. 5–11):

1 Synthetic balances: GNP, national income, balance of payments.
2 Product balances: energy resources, materials and finished products.
3 Distribution of investment: quotas on each production regime and sector.
4 Manpower balances: disaggregated by age, sex and qualification.
5 Production cost and profitability: material, wages, taxes and depreciation.
6 Foreign trade plan: import-export needs.
7 Investment actions: significant investment projects.

Each index specifies the balance required and means of achievement.

Control by capitalist manipulation of internal feedback loops involves the use of policy levers which affect consumption, investment, government expenditure and external relations (fig. 8.20) and is more favoured in capitalist states since a higher level of personal freedom is possible. Consumption demands can be manipulated by adjustments to tax rates (and hence disposable income), by credit policy, and by wage controls. Investment is usually manipulated through the market mechanism using interest rates, taxes from company profits and personal unearned income, and by monetary policy. Government expenditure can be used to inject increased demand at times of sharply decreased demand, but becomes progressively less useful as the state assumes greater control of consumption and investment functions (unless a complete command economy is operating). The control of external inputs affects the value of national currency, together with imports and exports through tariff barriers and control of foreign borrowing.

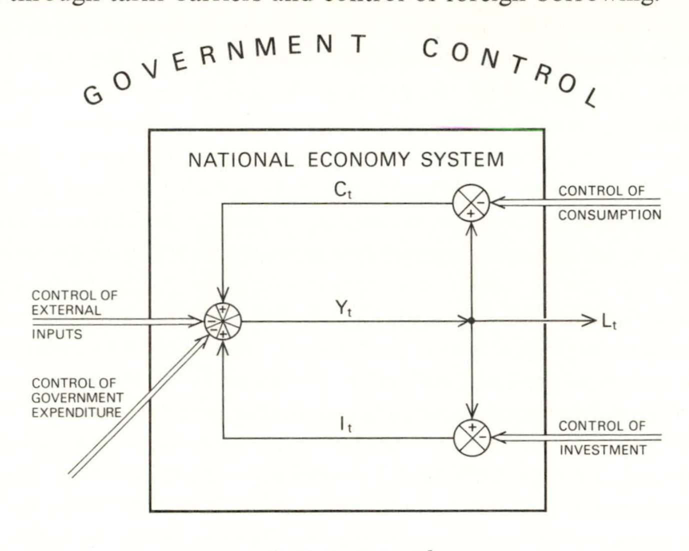

Fig. 8.20 Instruments of government control of a national economy.

The discussion of the form of economic policy instruments is complicated by four factors:

1 Many policy instruments are not available for complete or arbitrary manipulation – i.e. many are only partial control variables.
2 There is little operating experience available with many instruments to know if they are successful in achieving policy targets, and if so, how successful. This makes evaluation difficult.
3 There are frequently not enough policy instruments available to allow the simultaneous attainment of several policy targets. The control system has insufficient degrees of freedom and only partial control can be achieved (Tinbergen 1937).
4 The economic system which it is desired to control is immensely complicated and composed of a large number of feedback loops the structures of which are not completely known. The imposition of control is therefore constrained by theory since erroneous assumptions may be built into the economic model it is desired to control.

In addition to these factors economic policy manipulations are further constrained since even in perfect knowledge, with a sufficient set of instruments available, the resulting policy advice may be politically, socially or ideologically unacceptable (e.g. balance of payments adjustments). The most important consequence of these various features is that the economic control system can neither be looked at partially (either as separate components or as separate policy problems) nor can it be viewed in isolation from political, social and ideological goals, aims and constraints.

8.4.3 CONTROL SYSTEMS

The *control systems* which can be used to manipulate socio-economic systems to achieve desired societal goals can be divided, according to the instruments adopted, into plan control and feedback control. Plan control in economic systems can be classified into seven subcategories (Wiles 1962, ch. 4) depending upon the role of money and the point of control on the demand structure (in initial, intermediate and final demand) (fig. 8.21):

1. *Perfect central allocation without money*: Here no influence is exerted through the market, so that all decisions must be included within the plan. Labour is directed, wages are predetermined and unrelated to effort, and consumers receive predetermined bundles of goods.
2. *Perfect central allocation with money*: Here money is used as a passive accounting device (e.g. 'Rouble control') or as a check on plan fulfilment which does not effect decisions (since prices are meaningless in resource terms) but acts as an auditing or monitoring check.
3. '*Centralized market*' *economy*: Here decisions on intermediate resource allocation are centrally based (e.g. manufacturing production), but consumers, workers and public bodies choose on market principles. Industry is a public monopoly distributing resources on market principles but with central administration. Command systems have central control, but the prices of the intermediate goods will be largely rational and hence control decisions largely rational in a capitalist sense (i.e. the present Polish economy).
4. '*Inverted centralized market*' *economy*: Here the initiative for intermediate resource allocation is centralized, but is not subject to market forces. The central plan determines the allocation of factors of production and persuades or restricts consumer demand; this is mainly achieved by wage differentials, subsidies, and turnover taxes. Money plays a passive role in which prices are centrally determined. Consumer demand is present, but manipulated by the pricing policy (i.e. Soviet and Polish planning).
5. *Capitalist war economy*: This is a weaker form of the inverted centralized market economy in which only the pattern of final production is decided by the planner, the manager being free to choose his intermediate goods

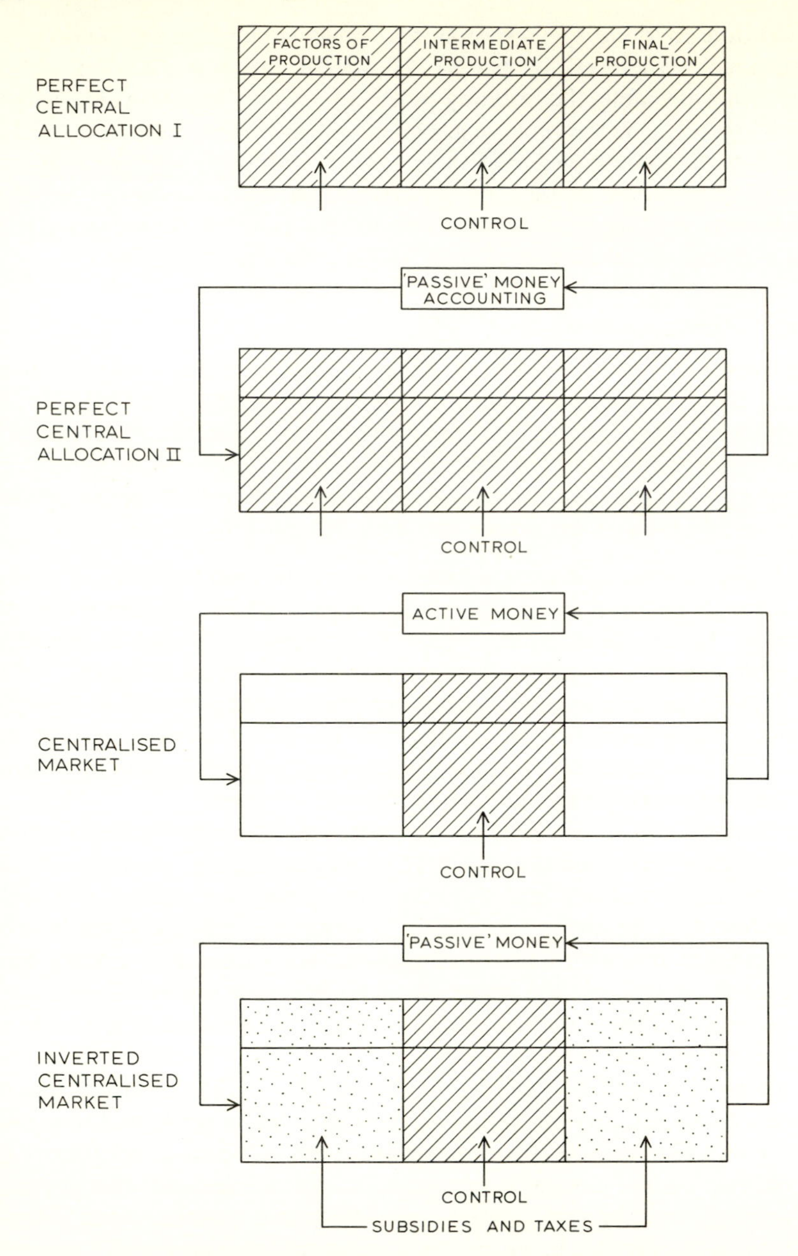

on the basis of the market. The market may be manipulated by the planner by subsidies and taxes, as in the cases of the United Kingdom and the United States of America which offer price support and subsidy schemes for farmers. The price mechanism influences the entrepreneur's outputs, but his inputs are free.

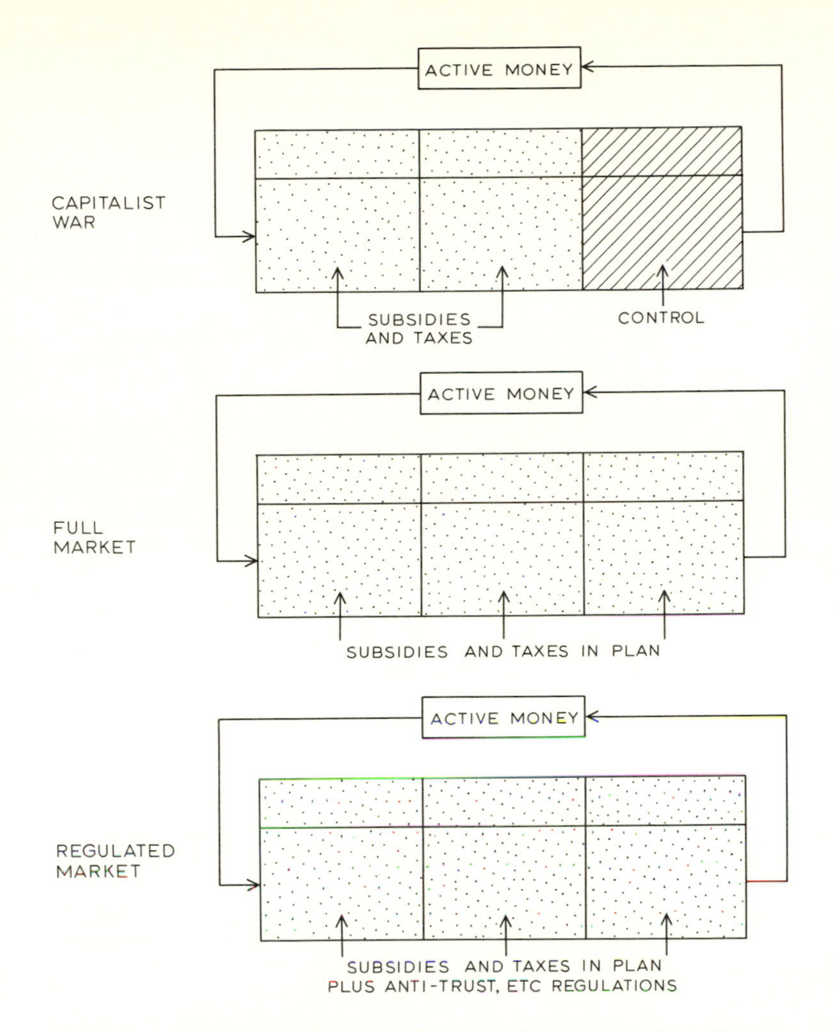

Fig. 8.21 Types of plan control in command economies, depending on the role of money and the point of control of the demand structure (initial, intermediate or final demand). Areas of direct control are shaded and those of partial control dotted. (After Wiles 1962.)

6 *Full market economy*: This is a pure competitive system, as in capitalist economies, but decision making is entirely in public hands. Industry, capital and land are entirely in public ownership, but prices are governed by market forces; although they will be greatly constrained by the desired pattern of resource allocation given in the plan.

7 *Regulated market economy*: This is the same as the full market economy, but with additional indirect controls; e.g. antitrust and rate regulation to prevent antisocial practices (i.e. the Yugoslav form).

For the USSR, the sociopolitical organization which attempts to wield this plan control (fig. 8.22) is based on flows across the administrative structure which serve to create (as far as possible) perfect information and equilibrium of supply and demand within the constraints of planned goals. The administration may be organized according to the client, purpose, process or

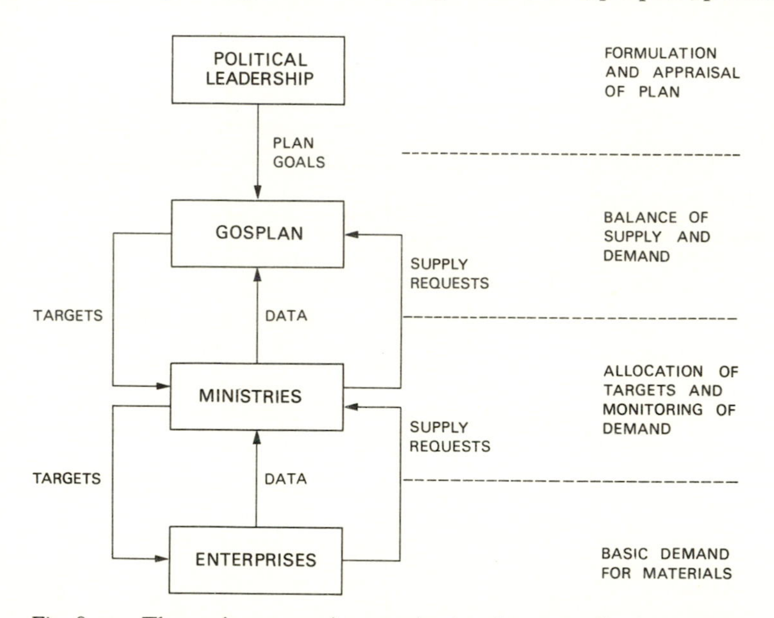

Fig. 8.22 The socio-economic organization for centralized economic control in the USSR. (Partly after Pickersgill and Pickersgill 1974.)

areal principles (see chapter 6). In the USSR two principles have been adopted (fig. 8.23): a ministerial (purpose) system along vertical departmental lines, dominant in the USSR from 1932 to 1957 and from 1965 to the present; and a territorial system (based on geographical boundaries) adopted from 1959 to 1965. The structure of the ministerial principles of organization can be compared with the vertical structure of ministerial decision making in the United Kingdom, and the territorial system with the United States federal structure. In the USSR, however, the administrative structure is not concerned primarily with the allocation of resources, although that is undertaken in Gosplan and within the Council of Ministers (overseen by the Politburo), but with the monitoring and accounting role. However it is organized, the plan control system is based on a method of iterative refinement (Ward 1967) which develops a set of final economic output targets.

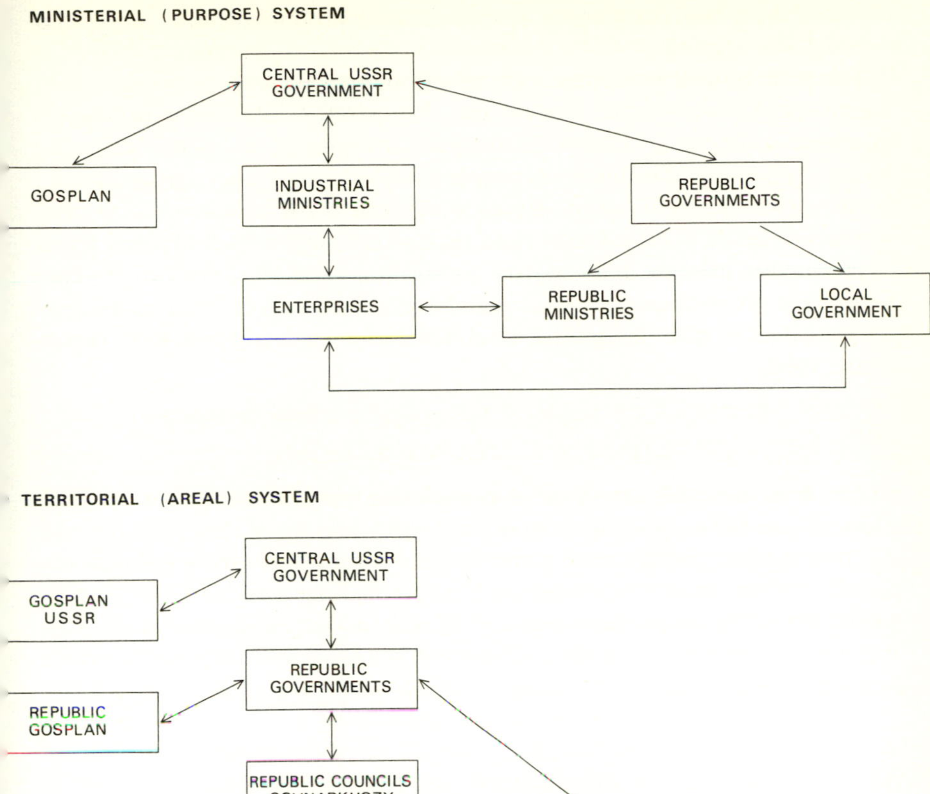

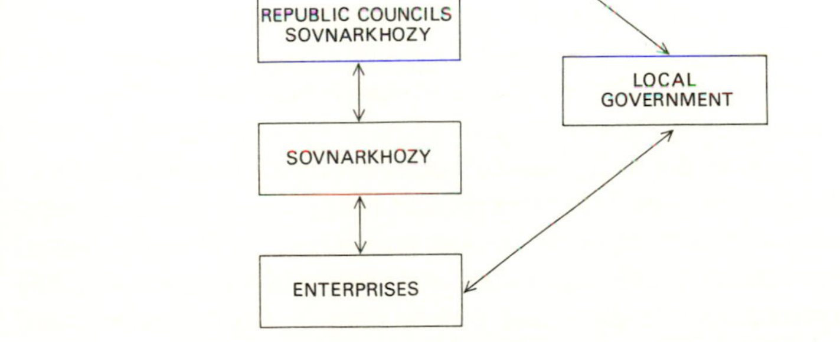

Fig. 8.23 The organization of (A) ministerial and (B) territorial economic institutions in the USSR. (After Pickersgill and Pickersgill 1974.)

As a first stage the central planner defines a set of gross targets for final output which are then transmitted to the production sector. Each productive enterprise then calculates its input factor requirements using a set of technical coefficients which are then transmitted to the planners who add the input requests for any product to their final demand target. The process is repeated with this new set of targets, and further repeated until any modifications are small. Hence final demand targets are transformed into gross output targets which form the final plan targets assigned to each production unit; supply and demand are equated.

The plan control procedure can be automated for each sector and each region using input-output models and the spatial optimization techniques discussed in chapter 4. If temporal dynamics are ignored (as is usually the case in Soviet planning) the allocation problem is a Hitchcock linear programming problem (Mennes *et al.* 1969). A typical example is the allocation of optimal production levels $Y_{ij}(t)$ to sector j and region i at time t, subject to target levels of production increase and capacity constraints when all sectors are mobile. Such linear programming solutions can be combined with interregional input-output tables to give the optimal final demand in each sector in each region (Isard 1960):

$$X_{ij} = a_{ij1}X_{i1} + a_{ij2}X_{i2} + \cdots + a_{ijn}X_{in} + \text{final demands}$$
$$(i = 1, 2, \ldots, N; j = 1, 2, \ldots, m)$$

where: final demands are those of households, government and other final users; X_{ij} are the outputs of the m sectors and N regions of the economy; and a_{ijk} are the input coefficients (i.e. the demand for product i from industry j in region k). Once the set of final demands is specified, the set of equations can be solved to find the output level required of each industry and each region. This may or may not be feasible depending upon the capacity limits of each sector and region and overall constraints of labour, capital and technology.

In the USSR the plan system is oriented around a five-year programme. Begun in 1928 these plans provide a general view of the desired future state of the economy which is then divided into operational one-year plans (a division approximating goals and targets). Output and input levels are set for 200–300 products in the input-output table which is then broken down into about 20,000 subcategories. There is forced saving of income and rigid setting of all intersector flows. In the early stages of economic development, especially in the USSR, this may be useful (Neuberger and Duffy 1976), but after take-off only smaller, marginal changes to the plan are necessary. Thus the central planning system becomes highly redundant and creates blockages, especially as economic expertise also develops and central planning can be abandoned for local-level decision making. Moreover, maintenance of economic growth at the margin must use considerable investment in research and development since imported technology can no longer be used. Again, in later stages of development in the USSR the demand for unskilled labour has fallen (as in the West) and incentives to skilled labour have been required. Outside the USSR, other eastern European economies have encountered problems in central planning at the international level where there is a need to make the economies more effective as earners of foreign exchange, beneficiaries of international division of labour, and exploiters of comparative advantage in trade. The *self-contained* balances of the planned system cannot allow for these international spatial influences of the market.

Feedback control, the alternative to socioeconomic plan control, utilizes full knowledge of the space-time dynamics of societal processes, and, as discussed in chapters 3 and 4, manipulates these by closure of the open-loop system through control feedback. Control feedback policies are more favoured in capitalist economic systems. They do not require the imposition of an overall economic plan since policy targets are achieved by shifting the input into an existing system – the market economy. The response to these instruments is free to find its own level. In a sense, therefore, the economic plan so favoured in communist states is a system of continuous redesign control, which in the Chinese context may be subject to continuous revolutionary redesigns at frequent intervals (e.g. the Cultural Revolution).

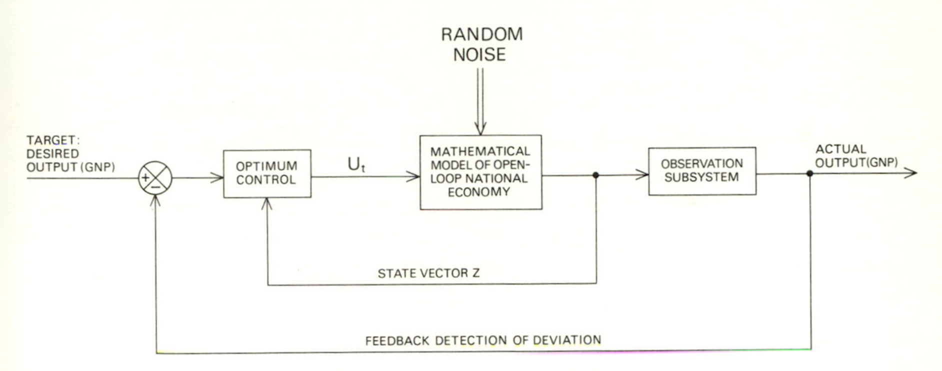

Fig. 8.24 Optimal control of a national economy. (After Buchanan and Noton 1971.)

Optimal feedback can, however, be utilized in fully planned (command) economies but most developments have been under the capitalist value system. The simplest economic control models of this type aim to keep some criterion variable of overall socio-economic performance (usually GNP) as close as possible to a predetermined target value (fig. 8.24). Three main approaches to the formulation of optimal economic policy have been adopted: simulation methods, Monte Carlo methods (Theil 1964, Naylor 1971a, Duesenberry *et al.* 1965), and optimal control theory strategies (Buchanan and Noton 1971, Theil 1964, Livesey 1973, Young *et al.* 1973).

The simulation approach produces predictions forward in time for various values of policy instruments – for example, the simulation of a six-equation macro-economic model used by Naylor (1971a) is shown in fig. 8.25, where the results of four possible fiscal policies aimed at maximizing GNP are compared. Comparing the total, mean and standard deviation of the controlled variable GNP over time, Naylor (1971a) shows that use of increased government spending as a control instrument gives the highest figure of total gross national income. However, this conflicts with the policy of increased government wages (which is optimal in the sense of having the smallest standard deviation). Hence, a minimum deviation control policy will yield a

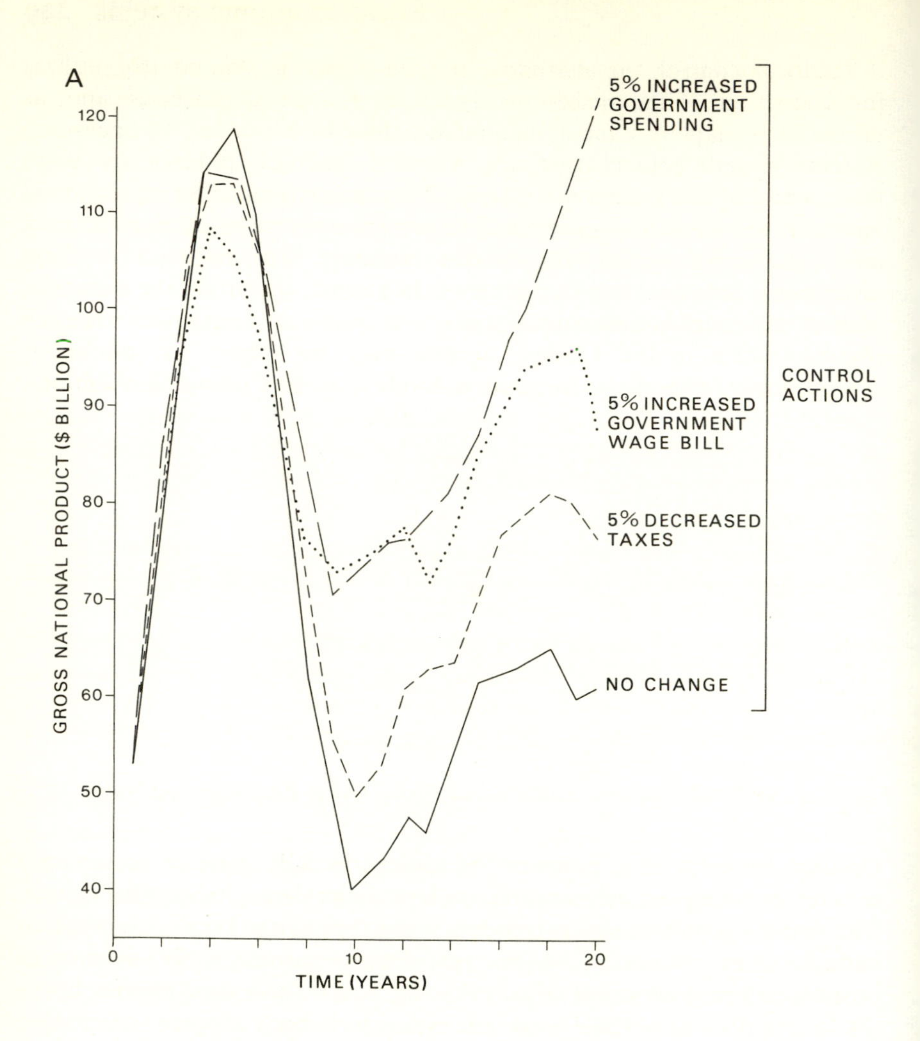

B COSTS OF CONTROL

POLICY	TOTAL	MEAN	ST. DEV.
NO CHANGE	1370·3	68·5	583·8
INCREASED SPENDING	1846·4	92·3	303·1
INCREASED WAGE BILL	1699·9	85·0	174·7
DECREASED TAXES	1539·4	77·0	386·5

Fig. 8.25 (A) simulation and (B) costs of control policies design to maximize gross national product (GNP) by means of a national economic model. (After Naylor 1971a.)

different choice of instruments from a maximum rate of return control policy, in this case. The difficulties of applying this approach are illustrated by studies of the control of spatial dynamics in regional economies. For example Kalman filter methods employed by Bennett (1975c) to the space economy of northwest England exhibit shifts in system equation roots induced by spatial feedback control policies (Regional Employment Premium, labour training subsidies, Industrial Development Certificate restrictions and investment incentives to manufacturing industry) but these effects are confused by shifts in system response due to inbuilt adaptation and learning across the spatial region (fig. 8.16).

A second approach, Monte Carlo evaluation of simulation, is to plot the standard deviation of the GNP over time. This has been adopted by Young *et al.* (1973) and Barrett *et al.* (1973) to evaluate the Livesey (1973) econometric model. An example of the Monte Carlo results under one set of policy criteria corrupted by random noise is shown in fig. 8.27.

The optimal control approach to economic regulation determines which setting of the policy instrument is required to achieve a given target. Again it is in the capitalist context that this method of control has found most application. A number of economic regulators of this form have been discussed by Tustin (1953) and Phillips (1954, 1957). Phillips considered three applications of the PID controller (chapter 3) to socio-economic systems and the combined PID controller and Keynesian macro-economic model is shown in fig. 8.28. Phillips obtained the best controller as a combined PID form ($f_p = 8; f_i = 8; f_d = 1$). The time path of the system output is shown for various PID controllers of the Phillips model in fig. 8.28B. In common with the results of chapter 3, it should be noted that PI control is sluggish and the best control is obtained with inclusion of all three PID elements. In particular, the inclusion of the derivative element removes the tendency of the controllers to initiate cyclical responses. Phillips (1954, 1957) notes that careless choice of proportional and integral control terms may well be the prime amplifier of the 5–7 year economic cycle that characterizes capitalist economies. The construction of true optimal policy models for economic systems derives from Tinbergen (1937), Holt (1962) and Theil (1964) and more recently from Vishwakarma (1974) and Chow (1975). The values of the feedback control terms will not be constants for all time as in the Phillips model. The optimal control law is given (chapter 3) as the inverse of the macro-economic transfer function. Given the economic systems model:

$$Y_t = f(Y_{t-1}, \ldots, U_t)$$

which is an expression, for example, of the Keynesian multiplier – accelerator within $f(\,.\,)$, find the optimal control law

$$U_t = -k(Y_t - Y_t^*)$$

$$k = f^{-1}(Y_{t-1}, \ldots, U_t).$$

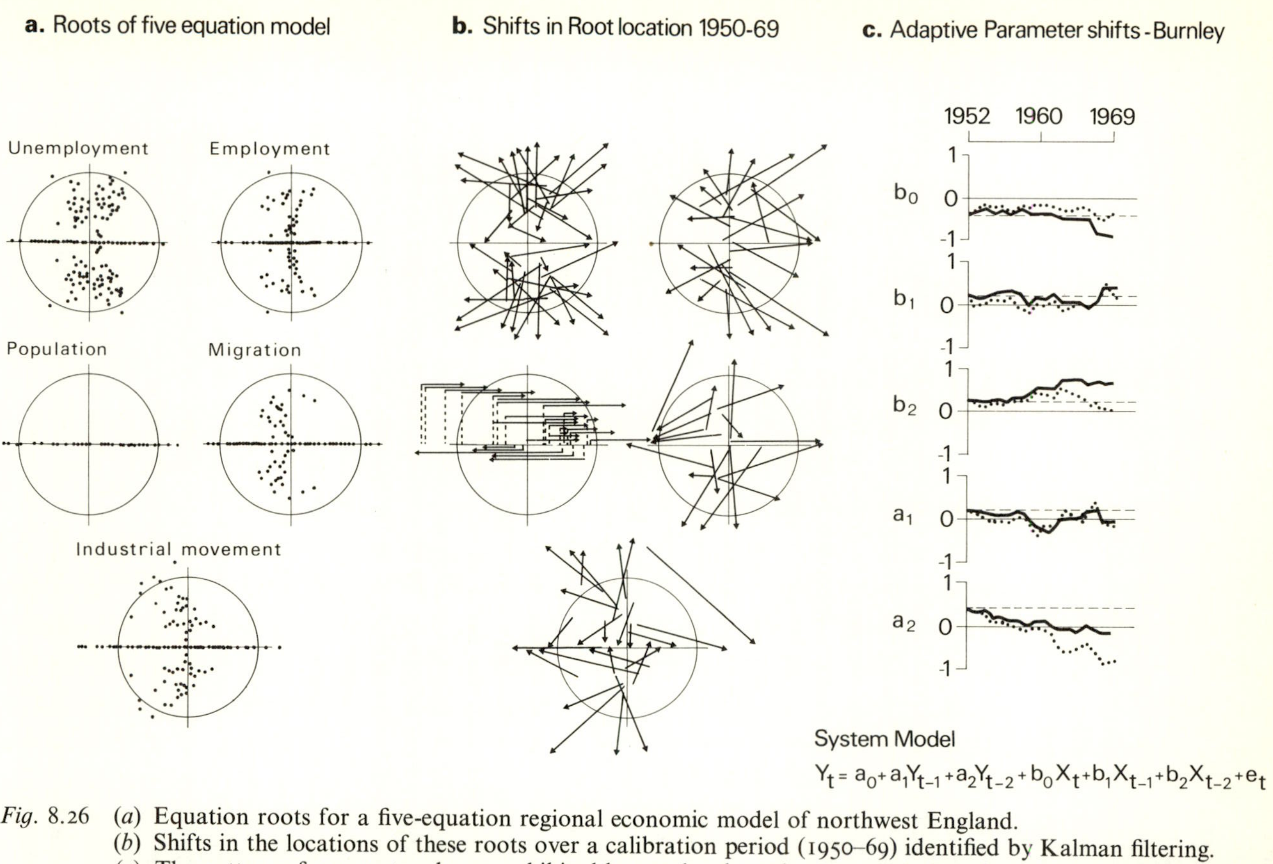

Fig. 8.26 (*a*) Equation roots for a five-equation regional economic model of northwest England.
(*b*) Shifts in the locations of these roots over a calibration period (1950–69) identified by Kalman filtering.
(*c*) The pattern of parameter change exhibited by one local employment area, Burnley, and induced by feedback policies. (After Bennett 1957b, 1957c.)

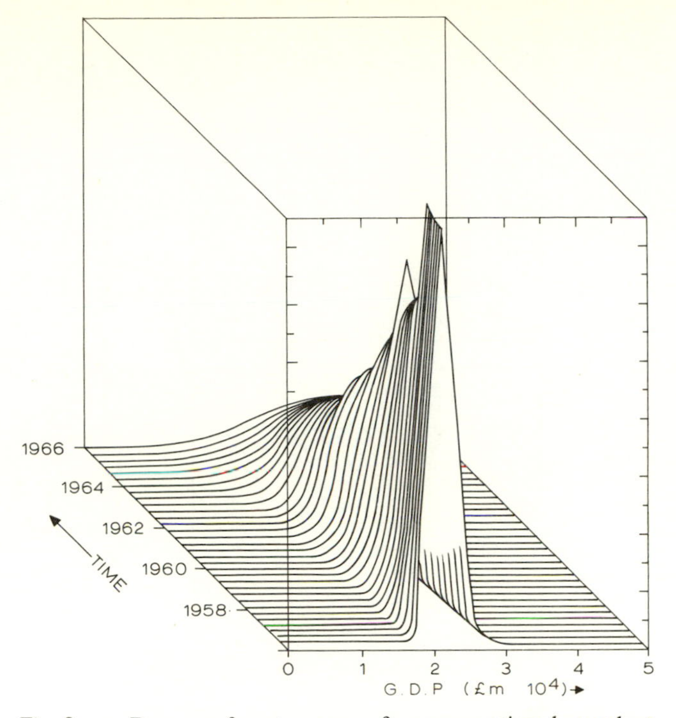

Fig. 8.27 Range of outcomes of gross national product policy target, illustrated by a typical example of a Monte Carlo display of simulations of an economic model with random perturbations. (After Young *et al.* 1973, using the Livesey (1973) model.)

Here k, the controller, is the inverse of the system transfer function and, the control input U_t is governed by the magnitude of the deviation of the output from its desired (target) value Y_t^*.

There have, of course, been many recent criticisms of the whole basis of the Keynesian approach (e.g. Robinson 1964). Despite its many inadequacies, the Keynes model has provided a simple and relatively powerful basis for the development of optimal control models for national economies. Recent applications have employed time varying feedback gains in Kalman filter schemes (e.g. Buchanan and Noton (1971); Chow (1975); Vishwakarma (1974)). In each case the control instruments used derive from tax, money supply and government spending controls. An example of this method (fig. 8.29) derived by Buchanan and Noton (1971) gives the time-varying path of the optimal feedback gains in terms of time-to-go ($T - s$). The gains in the two graphs refer to the weighting of each variable to the feedback controller. These gains are optimal under the quadratic control law (regulator) discussed in chapter 3. In the case of changes in the money supply the main weight is in

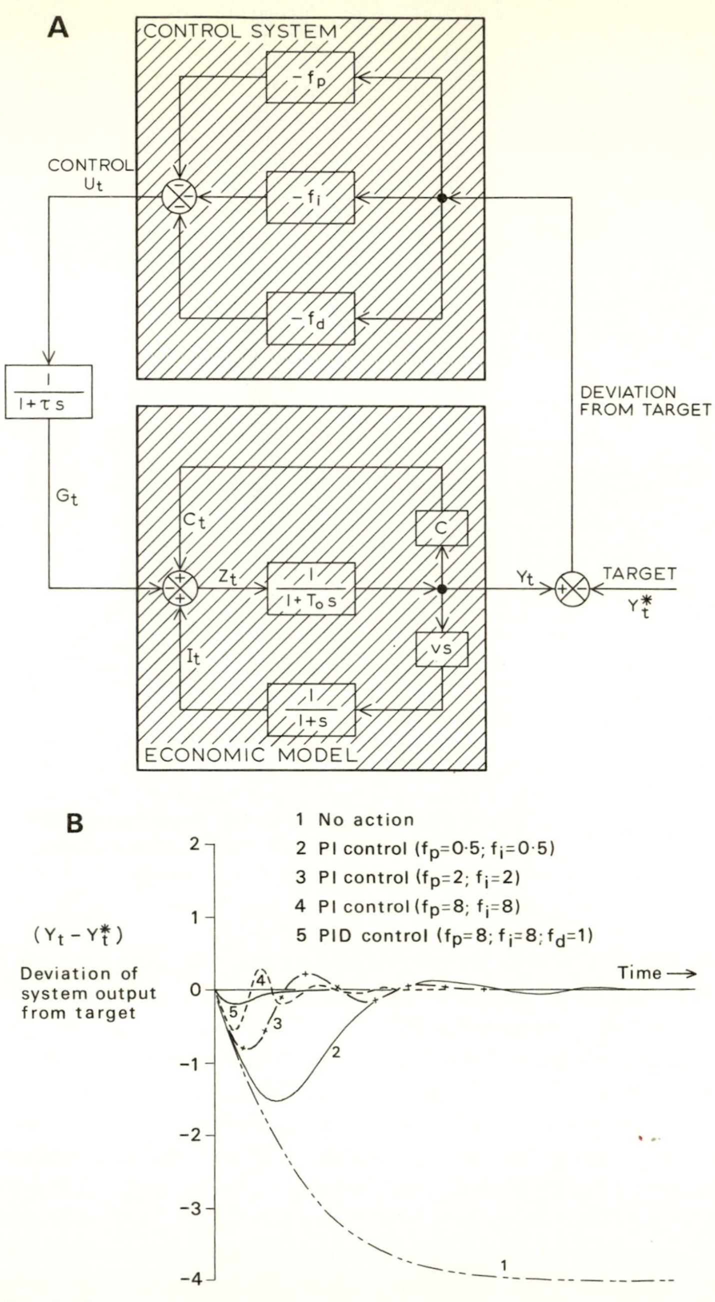

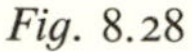

Fig. 8.28

A The Phillips' macro-economic model for combined proportional plus integral plus derivative feedback closed-loop control.

B PI and PID control by the Phillips' controller.

(Partly after Phillips 1954.)

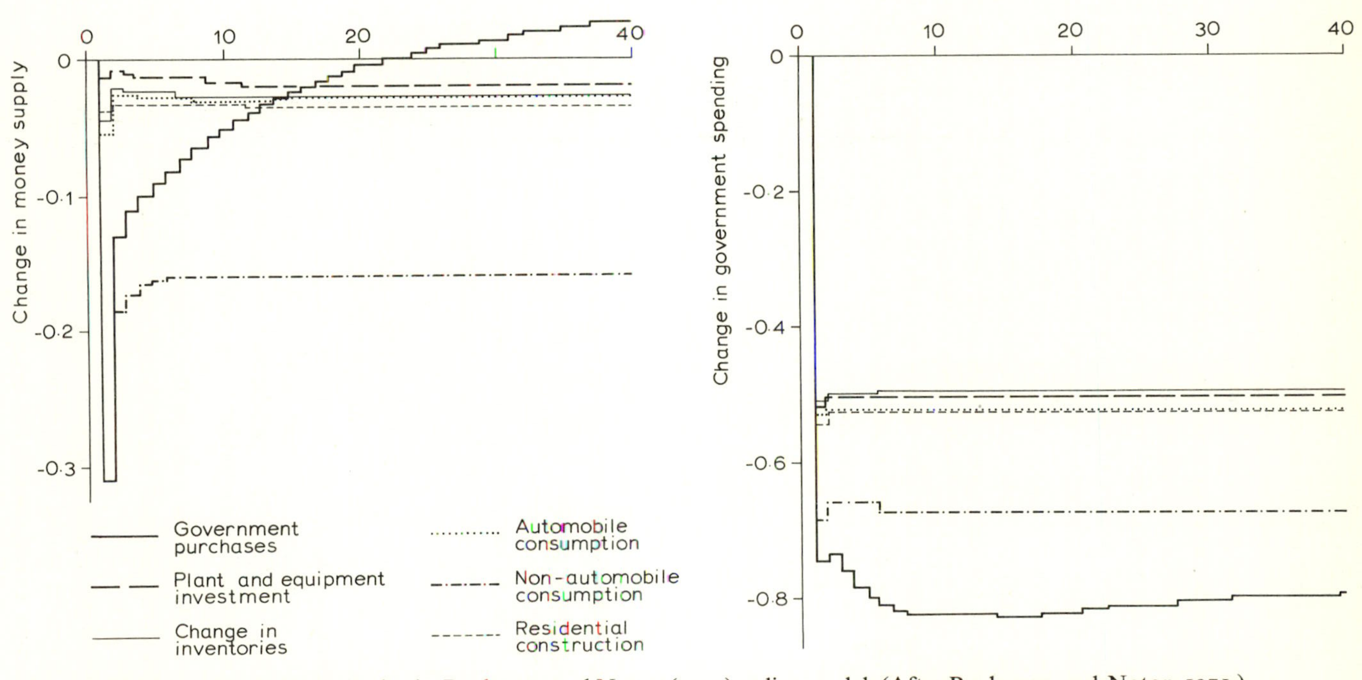

Fig. 8.29 Optional feedback gains in the Buchanan and Noton (1971) policy model. (After Buchanan and Noton 1971.)

the early life of the control policy with little change after the first four quarters. For government spending, however, there is a rise of gains to a plateau after one quarter and maintenance thereafter: a constant subsidy.

The form of these two alternative control solutions, changes in the money supply and changes in government spending, exhibit a number of general points that characterize economic control in practice. It is usual to define economic control policies with certain operational rules in mind; monetary policy, fiscal policy, and balance of payments which defines the overall philosophy and politics of the policy imposed. *Monetary policy* (fig. 8.30) is concerned with influencing the supply of money, the cost of money (interest rate) and the availability of money. The strict monetarist closed-loop control model is based on nineteenth-century economic assumptions, especially of Ricardo, but revised recently by Milton Friedman (1974). It states that control of the economy can be simply achieved through control of the money supply – more specifically, that the rate of growth in the money supply should not exceed the rate of growth of real output. The Radcliffe Report (1959) takes the view that the total demands for money should be regulated not only by the Treasury and Central Banks, but through interest rate and credit policies (i.e. also by liquidity). *Fiscal policy* is affected by control of government expenditure in goods, services and transfer payments (social security, etc.) and through taxation (fig. 8.30). Purchasing power is extracted from the economy through taxation. This can be reapplied to certain sectors of the economy in a discretionary fashion: (i) temporally to even out boom and slump; (ii) socially, through transfer payments; (iii) sectorally, to stimulate demand or investment in some industrial sectors rather than others; and (iv) areally, through investment incentives to industry, labour subsidies, etc., to reduce spatial welfare disparities. The *balance of payments regulator* (fig. 8.31) relies upon four sets of subpolicies: internal deflation, import restrictions, exchange rate adjustments, and discriminatory exchange and trade restrictions (Kaldor 1964, p. 89).

> A simple (necessary and sufficient) condition governs equilibrium in the balance of payments, that the value of total domestic expenditure, including investment, should fall short (or exceed) the 'real' national income by exactly the amount of target surplus (or deficit) on current account (in other words, total domestic expenditure is in line with national income) and that the budget deficit should be set to achieve this. (Godley and Cripps 1974)

Hence the budget is used to determine the balance of payments (which controls demand), and the exchange rate determines the level of economic activity.

The differences and problems in application of these alternative policies are well illustrated by the recent history of economic policy in the United States and Europe. Fiscal control has been uniformly the most popular policy, with shifts in government spending aimed at stimulating growth, stabilizing

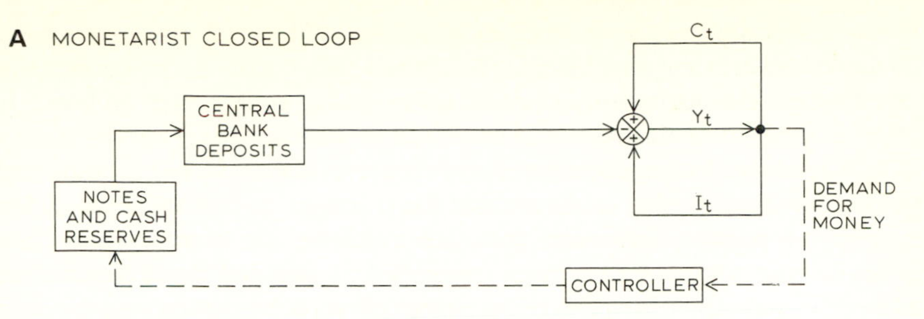

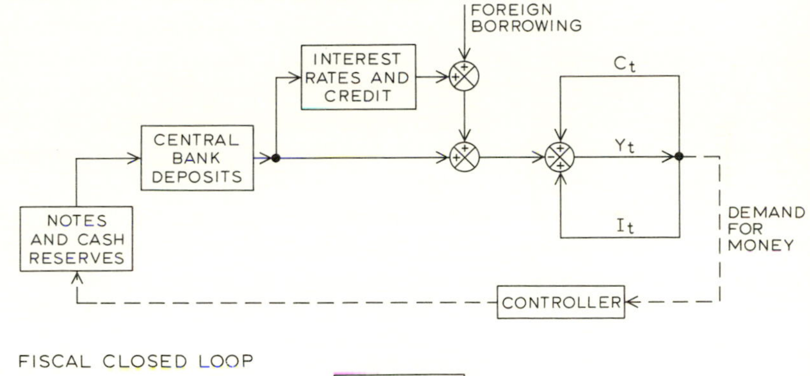

C FISCAL CLOSED LOOP

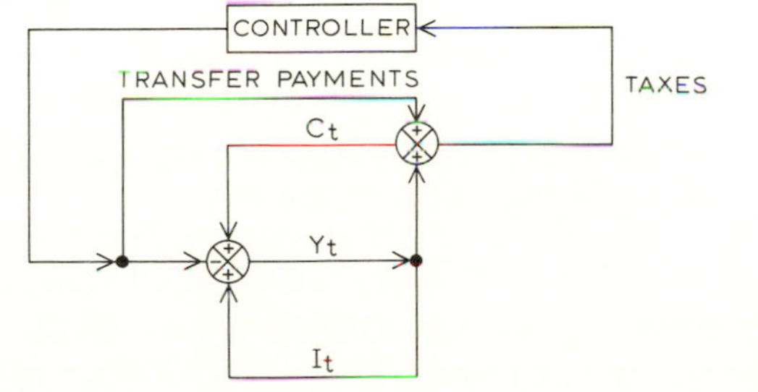

Fig. 8.30 Monetary control (A and B) and fiscal (C) policy closed-loop controllers.

cyclical fluctuations and achieving greater equity. This has usually been achieved by deficit spending, in other words by government spending being in excess of tax revenues and other income. A notable example is the United Kingdom economy within which government spending forms an ever-increasing proportion of the gross national product, but this is true also of most Western economies. The monetarists note that such deficit spending is bound to be inflationary, but the extreme Keynesians justify the policy by accounting for inflation as due to cost pressures external to the economy. Keynes (1936), it will be recalled, advocated deficit spending in times of recession balanced by accumulation of national wealth which remained unspent in boom conditions. It has proved very difficult for any government to reduce government expenditure once it has been initiated, both politically and because the institution or sector taken under government influence is often difficult to reconstruct in private hands subsequently.

A wider difficulty evidenced by recent economic history is that the numbers of economic goals that governments have sought to achieve exceed the number of policy instruments that are available for manipulation. For example, unemployment has been controlled by demand management, the balance of payments by the exchange rate, inflation by incomes policy, and investment (again) by demand management. Demand management has been used to achieve two independent goals and, as we have seen in chapter 3, a

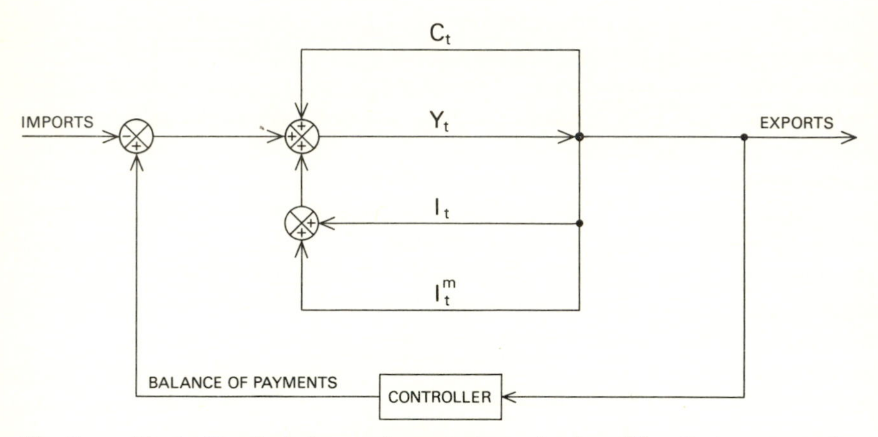

Fig. 8.31 Control by the balance of payments mechanism. I_t^m = the investment in export-based manufacturing industry.

system with more targets than instruments is strictly uncontrollable. In addition, the wrong instruments have arguably been applied to the wrong goals. The monetarist would claim that demand management should be used only to control inflation; unemployment can be controlled only by changing the price for labour in which incomes policy may be helpful; the exchange rate is not a control instrument at all but an expression of the price and productivity of one national labour market relative to another; and achievement of genuine economic growth, increased social welfare and increased government spending are only possible with higher returns to productive investment. In achieving particularly productive investment and realistic price for labour West Germany has developed one of the most successful of the Western economies. In the United States and the United Kingdom monetarist policies have been promoted by more 'right wing' parties – e.g. Conservatives and Republicans who are willing to tolerate higher levels of unemployment in order to release money for investment.

The difficulties in economic policy formulation are connected both with ideology and with the indeterminacy in economic models. As stated at the outset, economic control has arisen from dissatisfaction with the *natural* socio-economic feedback stabilization systems (i.e. 'self-recovery'). Any realistic economic control policy will be a combination of each goal and instrument to suit the particular economic circumstances and ideological goals.

8.5 Conclusion

In this chapter it could be argued that different social models are appropriate at different social and spatial scales, and in particular that mechanism lies at the level below and purpose at the level above (Feibleman 1954). As a necessary consequence, many societal systems models are macroscopic models, as are, in particular, most of the successful (on some criteria) control models. The conflict of mechanism and explanation dates back at least from the controversy between Ostwald and Boltzmann in the 1870s and was exemplified by the differences between Planck's view, that the true goal of explanatory science is to establish the nature of the external world independent of the observer, and that of Mach that science must understand the world not as it is, but as we experience it (Cohen 1968, Toulmin 1953) (see chapter 1). In physico-ecological systems the resolution of the conflict has been achieved through the recognition of the indeterminacy principle and the acceptance of statistical explanation and statistical laws, albeit constrained by the $\sqrt{n}$ rule. In the socio-economic systems the adoption of independent observation and experience has two implications for understanding. Firstly, it analyses only the *status quo* and does nothing to determine the trajectory which society *should* follow (Harvey 1973); further, it determines nothing of a qualitative nature as to whether a pattern of outputs is good or bad, and who gains and who loses by a particular system configuration. Secondly, supposed independence of observation leads to a divorce of the systems analyst from the socio-economic system of which he may be an integral part (see chapter 9). In control situations this leads to an artificial separation of the controller from the control system, such that the product of scientific thought has been purged of its personal characteristics. As we have seen in chapter 1, this accords with Marxist criticisms that there is 'no loss to the scientist personally or to the culture generally to strip human thought of its intimately personal qualities – its ethical vision, its metaphysical resonance, its existential meaning'; as a consequence it is possible to 'break faith with the environment, establish between yourself and it the alienative dichotomy called objectivity, and you will surely gain power. . . . Nothing will inhibit your ability to manipulate and exploit' (Roszak 1972, pp. 151, 159, 168). The link between epistemological objectivity (see chapter 10) and the psychology of alienation creates intense environmental pressures and derives largely from macroscopic views which are treated in chapter 9. The scientific view ignores ontological meaning and dialectic, and is supported by the attenuation of variety within complex administrative-bureaucratic control networks, yielding an approach to control which is not holistic. Control is available only over a subsystem and this at the expense of suppressed freedom, or, as Olsson (1975, p. 500) put it, 'instead of creating a world for becoming, we are consequently creating thingified man'. The important questions of socio-economic control are thus ignored: for whom is control intended, how is justice in resource allocation achieved, how is discipline of acceptance of disagreeable decisions achieved, and who wields the controls? It has been suggested that if we do not ask these questions we allow scientific method to become the 'handmaiden of authoritarian ideology' (Olsson 1975, p. 500, Harvey 1974) – that is, we make the two steps as in equation (8.1.1). In so doing, we explain human action as a macroscopic maximum likelihood (descriptive), and then use our descriptive model to define desired social goals as the expectation (mean) of our statistical distributions (prescriptive). The homeostatic variation which gives rise to the statistical dispersion in the likelihood function is suppressed.

The control of societal systems is thus fundamentally different from control of physical systems, in that the controller is also an individual within the system. Marx observes that the structure of control is the structure of power. Control is aimed at the achievement of given objectives, and, therefore, the nature of objectives and the nature of the control will be dependent upon the nature of power. Power may be exercised by complete individual freedom of choice and the rationalization of competing demands either through collective action or through the mechanism of demand and supply feedback and of competitive bidding through price mechanisms of the market (in *laissez-faire* economies). Alternatively, control may be exercised by decision making bureaucratic systems in which each competing need is weighed and an optimal allocation determined (e.g. through centralized planning in command economies). A mixed market-control system exists in democracies in which central control is allocated in those fields in which true agreement on goals exists.

As we have seen, the definition of goals and objectives in societal control systems is the most important but the most indeterminate area in the allocation of competing demands. Hayek (1944) argues that goals are not made, but evolve or emerge. The organization of socio-economic allocations to achieve ideals is usually defined in terms of achieving objectives of the *common good*, 'social good', or 'general welfare'. However, Hayek (1944, pp. 42–3) recognizes that:

> The welfare and happiness of millions cannot be measured on a single scale of less and more. . . . It cannot be adequately expressed as a single end, but only as a hierarchy of ends, a comprehensive scale of value in which every need of every person is given its place. To direct all our activities

> according to a single plan presupposes that every one of our needs is given its rank in an order of values which must be complete enough to make it possible to decide between all the different causes from which the planner has to choose.

This leads, Hayek (1944, p. 43) suggests, to the need for a complete *ethical* code since only the existence of such a code permits that 'a "social" view about what ought to be done must guide all decisions'. We have already seen that complete orderings of social welfare functions (the general good) cannot be derived by collective action from aggregation of individual utility functions (Arrow's impossibility theorem). Thus we must abandon realist and rationalist positions of control. We are left with the *idealist* position – namely, deductive Aristotelian argument about absolutes and the Kantian *a priori* and dialectic.

One attempt to define an idealist code is the Marxist concept of social justice (Harvey 1973). To Plato justice meant the allocation of one man to one job, the job to which he was most naturally suited, and the freedom of that man to pursue his job without interference from others. This idea of Platonic justice was tempered by the natural corollary of a discipline in which the desires of the less reputable majority were controlled by the desires and the wisdom of the superior minority (i.e. of a philosopher ruler). The freedom of the individual is balanced by discipline towards societal ends:

> First, freedoms are not to be guaranteed by statements of general principle. Secondly, they are residual. Freedom of public assembly, for example, means the liberty to gather wherever one chooses except in so far as others are legally entitled to prevent the assembly from being held or in so far as the holding or conduct of the assembly is a civil wrong or a criminal offence. To define the content of liberty one has merely to subtract from its totality the sum of the legal restraints to which it is subject (de Smith 1971, p. 440).

Marx defined the final 'ideal' in a Platonic form, 'from each according to his capacity, to each according to his need', but ignored the problem of attendant discipline. Whether or not we accept the need for prescriptive idealist approaches to control depends upon the perceived need for control, and the judgement as to whether unfettered freedom produces its own dire consequences. To generalize, there are three main factors which require that freedom must be tempered by a degree of control. Firstly, resources are not limitless, and hence all needs cannot be satisfied. There have been periods, such as the expansion of the American frontier and the growth of the colonial empires, when a legend of inexhaustibility could be supported and pure *laissez-faire* had some realism. However, this cannot be true of the modern world in which population pressure is rapidly reaching the boundaries of available world resource endowments. Much argument surrounds the concept of scarcity. The basic Marxist approach to scarcity is to explain the phenomenon as the result of monopolistic exploitation:

> In sophisticated economies scarcity is socially organized in order to permit the market to function. We say that jobs are scarce when there is plenty of work to do . . . scarcity must be produced and controlled in society because without it price fixing markets cannot function. This takes place through a fairly strict control over access to the means of production and a control over the flow of resources into the productive process. The distribution of output has likewise to be controlled in order for scarcity to be maintained. This is achieved by appropriate arrangements which prevent the elimination of scarcity and preserve the integrity of exchange values in the market place (Harvey 1973, p. 114).

Marx recognized that the price mechanism is also useful in encouraging coordination, facilitating technical innovation and generating overall growth, but his approach is generally ill-adapted to the allocation of resources when *real* scarcity is present. In this new environmental situation it may not be possible to achieve 'a new pattern of organization in which the market is replaced . . . [and] scarcity and deprivation systematically eliminated wherever possible' (Harvey 1973, p. 115).

A second limitation on freedom of action is the high degree of interdependence in the world economy. Hence, if resources are limited then one person's needs satisfied are another's foregone. When interdependence is small, the satisfaction of the desires of one group need not impoverish another; but, generally, the impoverishment process is a function of increasing interdependence and scale. This has been the process of exploitation under colonialism (Nkrumah 1963) and is the present-day process of 'immiserization' by trade described by Bhagwati (1958). Thirdly, the interdependence between individuals, groups and nations has reached such a degree that unregulated environmental catastrophic side effects are increasingly likely (Meadows *et al.* 1971, Dunn 1974) (see chapter 9). Overriding these various factors which encourage the development of controlled societies is the *spatial* dimension. Resources have a particular spatial distribution, such that interdependence must have a spatial expression (see chapters 4 and 6).

Ethical codes to guide control systems will seldom exist by agreement on other than very basic issues. Certainly in large-scale socio-economic systems this represents an absolute limit on democratic decision making for control:

> Planning control creates a situation in which it is necessary for us to agree on a much larger number of topics than we have been used to, and . . . in a planned system we cannot confine collective action to the tasks on which we agree, but are forced to produce agreement on everything in order that any action can be taken at all (Hayek 1944, p. 46).

The societal response is either to abandon or limit attempts at control, or to abandon the democratic ideal of majority agreement on all objectives and actions to achieve it. Thus we have a dialectic between prescriptive action resulting from imposed ideals based upon absolute philosophies (e.g. that of Aristotle) and collective action derived as the sum of individual costs and benefits perceived to derive on the individual from his participation in collective or social organization. A third view, that 'the totality in which it is neither the parts nor the whole, but the relationships within the totality which are regarded are fundamental' (Harvey 1974, p. 265), gives Marx's set of emergent actions in which control goals derive from the conflict of the power vested in monopolistic and class-bounded groups of individuals who attempt to impose their own ethics. The demiurge may arise from *a priori* developments as from Kant, from Hegelian dialectic within the totality, or from the descriptive macroscopic system view, each of which 'places all people on a common noetic level . . . a subversive belief in human equality founded upon the prospect of knowledge available to all on a non-privileged, non-classified basis' (Roszak 1972, pp. 202–3). It is clear that there is no simple solution to the problem of conflict: conflict of the prescriptive and the descriptive; of the subjective individual *qua* individual and the objective individual in society; of the desire for reduction of societal uncertainty of environmental disturbances and for freedom of action; of integration theory (Dahrendorf 1959) based on consensus, and coercion theory based on the dialectic of ongoing social processes involving the imposition of social change. It is appropriate to end with Aristotle's view, which states this conflict, but does not resolve it:

> It is clear, then that constitutions which aim at the common good are *right*, as being in accord with absolute justice; while those which aim only at the good of the ruler are *wrong*. They are all *deviations* from the right standard. They are like the rule of master over slave, where the master's interest is paramount. But the state is an association of free men (Aristotle, *The Politics*, Book III).

Part IV
Systems interfacing

9 Systems interfacing

> Man has been reared by his errors: first he never saw himself other than imperfectly, second he attributed to himself imaginary qualities, third he felt himself in a false order of rank with animal and nature, fourth he continually invented new tables of values and for a time took each of them to be eternal and unconditional, so that now this, now that human drive and state took first place and was, as a consequence of this evaluation, ennobled.
>
> (Friedrich Nietzsche: *The Gay Science.*)

> Man first implanted values into things to maintain himself – he created the meaning of things, a human meaning! Therefore he calls himself 'Man' – that is, the evaluator. . . . Hitherto there have been a thousand goals, for there have been a thousand peoples. Only fetters are still lacking for these thousand necks, the one goal is still lacking. Yet tell me, my brothers: if a goal for humanity is still lacking, is there not still lacking – humanity itself?
>
> (Friedrich Nietzsche: *Thus Spake Zarathustra, Of the Thousand and One Goals.*)

9.1 Physico-ecological and socio-economic systems

It is a commonly held belief among environmental scientists that there exists potentially an area of great difficulty concerned with interfacing 'natural' systems, on the one hand, with 'human' systems on the other. The former are believed to be structured in a manner that causes them to be dominated by stable negative-feedback loops and by homeostatic considerations to do with energy flows, food chains and the like (Chorley 1973). They are believed to be essentially more primitive and less flexible in their operations than socio-economic systems, much more vulnerable to change, and much more imbued with thresholds which, once passed, may not be retraced. On the other hand, socio-economic systems are held to be extremely flexible, largely due to their richness and complexity. They are believed to be potentially sensitive to unstable positive-feedback loops or, at least, to negative feedback loops which encourage explosive growth, but nevertheless to be protected against destruction by the flexibility of their structure and by the feedback mechanisms operated by human control. To some extent the similarity in organization of chapters 7 and 8 has tended to reinforce this notion. However, it has been made clear by the systems structures treated in chapter 2 that an explosive growth is no more the prerogative of positive-feedback systems than homeostasis is of negative feedback ones. Indeed the structure of both types of systems, particularly in terms of the disposition of their gains, delays and storages, provides an overall unification of systems operation which makes any simple division into the stable negative-feedback physico-ecological systems and the unstable positive-feedback socioeconomic ones very difficult to sustain. It will be remembered that a considerable part of chapter 1 was devoted, on a more philosophical level, to considering the essential relationships between 'human' systems and 'natural' systems. From this it was concluded that two broad types of relationships have been held to exist: on the one hand, the idea that man stands apart from the rest of nature on which he operates, and, on the other, the view that man operates as an integral part of nature. This gives rise to the two views of systems interfacing which are presented in this chapter: firstly, the *intervention* view and, secondly, that of *symbiosis*. Some might regard the distinction between intervention and symbiosis interfacing as purely one of scale, in that the former can be sustained on shorter temporal scales, on smaller spatial scales and on smaller scales of the use, utilization or release of mass and energy than can the latter. However, the question is not as simple as this, because, as has been shown in chapters 7 and 8, our structuring of systems models has to do with perception, and the postulated differences between systems structures may be more due to differences in our perception of them than to any intrinsic differences between them. Nevertheless, the concepts of intervention and symbiosis are not without value in viewing the whole problem of the interfacing of socio-economic and physico-ecological systems. The former reduces itself largely to one of the external management of storages within existing systems, or of the forcing of a given system through one or more thresholds to a predictable transformed state. The basic problem of all systems modelling, however, is that it is much easier to model systems exhibiting homeostatic stability rather than those exhibiting some time direction of change in which the system's structure is fundamentally altered.

Hill (1975) has put forward the classic view of equilibrium in relation to ecosystems which can be expressed either in terms of the so-called 'no-oscillation stability', exhibited by the constancy of species or structural relations within an ecosystem, or, secondly, the so-called 'stability resilience' type of stability characterized by the maintenance of, or return to, a stable condition of a system after an external perturbation. Under such a view stability is essentially a function of the complexity of the trophic web, where a large number of interacting pathways provide for the operation of many mechanisms for the adjustment to stress. It is interesting that many workers concerned with socio-economic systems view them as examples of stability-preserving systems of the ecosystem type. (See Bayliss-Smith and Feachem, 1977.)

A further problem, having to do with systems interfacing, has already been introduced in earlier chapters where we have treated the mutual interaction of two loops within a system, each of which is operating at a different rate. Such loops are particularly common in the economic systems described in chapter 8. Koenig *et al.* (1974) have made use of this concept in constructing the essential mass-energy features of an industrial ecosystem which are conceptualized as a system of material transformation, transportation and storage processes, driven by solar and synthetic energy to comprise a material-processing machine (fig. 9.1). They conceive of man's history as partly viewable in terms of his progressive divorce from natural ecosystems, initially by the agricultural revolution and, more recently, by the industrial revolution which was characterized by the tapping of the stored output of ancient

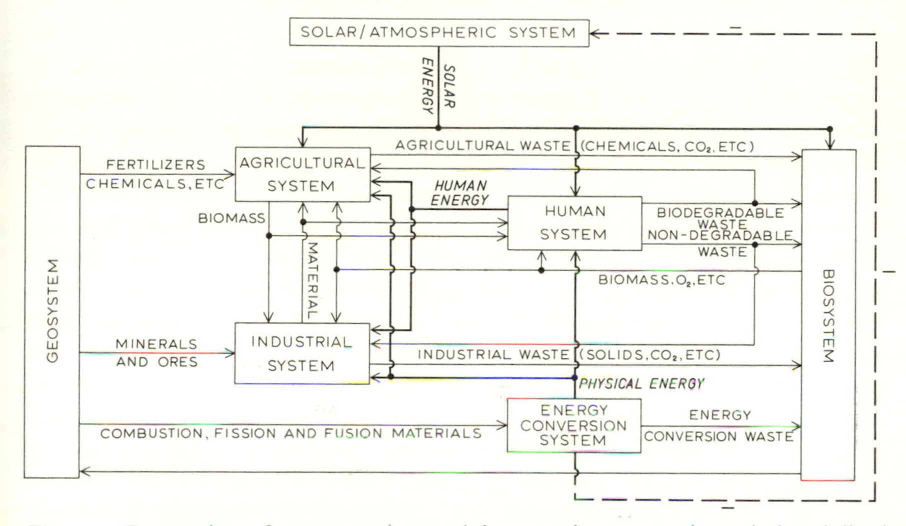

Fig. 9.1 Interaction of energy and material processing sectors in an industrialized ecosystem.
Thick lines: solar, physical and human energy flows.
Thin lines: materials flows.
Dashed lines: feedback loop.
(Partly after Koenig *et al.* 1974.)

ecosystems. They look forward in the future to a possible further revolution whereby much more deeply embedded energy sources associated with nuclear processes are conveniently drawn upon. The different operations of the subsystems of the above industrialized ecosystem are used to pinpoint certain of today's problems and, in particular, the fact that the biosystems have difficulty in adapting sufficiently rapidly to inputs of chemical wastes because their responses are not at the same speed as those of the input changes, and because biosystem transformations are not always reversible. Koenig *et al.* (1974, p. 323) have suggested four possible avenues whereby the interaction of physical and human systems may be made more workable by the improvement of the mass-energy characteristics of industrial ecosystems:

1 By the suitable spatial distribution of human activities so as to achieve maximum utilization of the biosystem for processing bio-degradable wastes.
2 By a more efficient use of earth resources by recycling them.
3 By minimizing the material flow rates by extending the expected life of durable goods required by the human system.
4 By the design of stable ecosystems in terms of trade-offs between optimal states, processing capability and external maintenance costs. This latter might necessitate the allocation of certain ecosystems, such as some lakes or rivers, purely to waste processing, and their management so as to maximize their capacity to carry out this function.

This approach again emphasizes the intervention or utilitarian view of systems interfacing and, in particular, the utilization of storages within the physico-ecological systems, in this case for the processing of waste material. It is clear that there must be some equalization of the rate of operation of the physico-ecological systems on the one hand, and that of the socio-economic ones on the other, such that the equilibrium state of a free economy designed to operate in a finite environment must correspond to the mass-energy equilibrium states that are physically and ecologically feasible (Koenig *et al.* 1974, p. 326). Another way of viewing intervention or exploitation has been presented by Margalef (1968) who sees exploitation as reducing maturity in the natural development of ecosystems, putting a break on succession, by causing the species diversity to drop and by increasing the ratio of primary production to biomass. Thus improved agriculture means lower numbers of species, a simplified soil structure, and a decrease in soil micro-organisms and animals. This has the effect of accentuating exogenous and yearly rhythms, of increasing pests, and, more important, of leading to their extreme fluctuation. Controlling agriculture may be contrasted with the systems view of natural plant succession, whereby the biomass increases along with primary production, but the ratio of primary production to biomass drops, the diversity of species tends to increase, and there is an increase in the proportion of inert and dead matter with a low respiratory rate, fluctuations of all kinds tending to be damped down and rhythms becoming less externally controlled and more endogenous. The increase in diversity associated with natural succession leads to more ecological niches, to larger food chains and more strict specialization, to increased parasitism, to the numbers of organisms becoming more stable and to a greater proportion of the nutrients being stored or retained by the living organisms (Margalef 1968). Ecosystems with more secure storages of mineral elements are thus more protected against destruction by exploitation. The taiga and tundra, with more than 60% of their elements stored in litter, are more buffered against exploitation than tropical rainforests, where more than 60% occurs in the biomass and often less than 1% in the litter. In the same way, salt marshes and floodplains, with more than 90% of mineral elements in the soil layer, are more buffered than the ecosystems of the tundra and taiga (Hill 1975, p. 213). Similarly, ecosystems subject to greater natural fluctuations are more able to resist human exploitation without collapsing or without suffering considerable change. This means, for example, that the taiga is more safely exploitable than the tropical forest, and that the sardine is more exploitable than coral reef fishes without fear of its destruction. Systems of high maturity, such as tropical forest and coral reef ecosystems, are more likely to collapse when excessively exploited than those of lower maturity. Thus ability to be exploited and resistance to exploitation is in the minds of many ecologists connected with evolutionary stages (Margalef 1968, pp. 83–7). In youth great population fluctuations occur which are often more detrimental to the parasites than to the host, and these fluctuations shorten the food chains and favour species with a high rate of potential increase. Life spans are short, there is a high turnover rate, great fecundity and strong competition for food. In maturity the environmental energies tend to be damped and lagged and subordinated to the endogenous rhythms. Under these conditions there is more interdependence between species, more multiplicity of links, a slowing down of dynamics and less brutal and often more subtle forms of competition. The successful species have multiple and delicate relations which, as with tropical rainforest, make the ecosystem structure much more liable to destruction by external influences.

There are many difficulties facing those who have interests both in the natural environment and in the man-made environment together with the problems of their interfacing. One of the currently most intractable is that, having been imbued with the ecosystem model, with the emphasis on balance, equilibrium, cycling and stability, scholars are increasingly faced with the methodological necessity of also accommodating active control involving the impelling of systems on time trajectories through sequences of state, each different, probably non-recoverable and presumably ever more adapted to the evolving needs of man in society. The matter is made more complex by the dynamic nature of these longer-term social needs in relation to population totals and the evolving social goals, and by the increasing difficulty in predicting what will be required of the earth as a home suitable for man's occupancy. In short, scientists are being faced with the basic problem of modelling systems which are stable in the short term under negative feedback mechanisms, yet are capable of long-term changes under the positive feedback evolutionary mechanisms involved in economic and social tendencies.

The mass-energy features of the industrial ecosystem shown in fig. 9.1 include the macrofeatures through which industrialized man relates to physical and biological systems (Koenig *et al.* 1974). The *biosystem* involves the collection of natural ecosystems in the landscape. Subordinate to this, the modern *agricultural system* is similar to the biosystem except that material imports (for example, fertilizers) are required from the *industrialized system*, auxiliary man-made sources of energy are required and recycling has been virtually eliminated. The industrial system is the collection of physical, chemical and technological processes developed by man for the restructuring of materials. The *human system* comprises the natural processing features of the aggregate human population. The *energy conversion system* is the set of all auxiliary physical energy sources which supplement solar energy. Koenig *et al.* (1974) proposed that a comprehensive solution to the environmental problems of industrialized nations requires the development and implementation of scientific, technological and management capability in at least the following four major areas:

1 *Mass-energy rate capacities of environmental components*: This is the capability to assess the material and energy processing capacities of major components of the natural environment as a function of their physical and ecological features and in relationship to projected changes in their quality as measured by their aesthetic, humanitarian, recreational and health characteristics.
2 *Design and management of multispecies communities*: This means the capability to manipulate the ecological communities of restricted portions of the natural environment to increase their mass-energy assimilation capacity and/or other designated characteristics.
3 *Ecosystem design*: This requires the capability to structure the technological and spatial features of the industrial, agricultural and urban sectors so as to retain ecological compatibility with the environment.
4 *Social instruments of control*: This refers to the capability to design and implement comprehensive economic, political and other social instruments of control to manage technological and economic development to be ecologically compatible with the environment and socially desirable. (Koenig *et al.* 1974, p. 329)

An important consideration in the problem of interfacing physico-ecological and socio-economic systems is the energetics of intervention and interaction. This involves the amounts of energy concerned in the operation of the two types of systems and the expenditure of energy in the socio-economic system which is required to release energy in the physico-ecological systems. Fig. 9.2 shows examples of the relationships between the kinetic energy and life span of a number of meteorological systems of varying scales of space and time, together with the equivalent amounts of energy concerned with human activities. It is interesting that the energy which is at present the greatest subject of meteorological control research lies within the energy range 10^{12} to 10^{14} Joules and is of the order of the daily output of the Hoover Dam or produced by burning 7000 tons of coal. We shall see later in a little more detail the manner in which energy can be potentially released from a cumulus cell of this order of magnitude. Wischmeier and Smith (1958) have calculated the kinetic energy versus the intensity of rainstorms and have shown a generalized

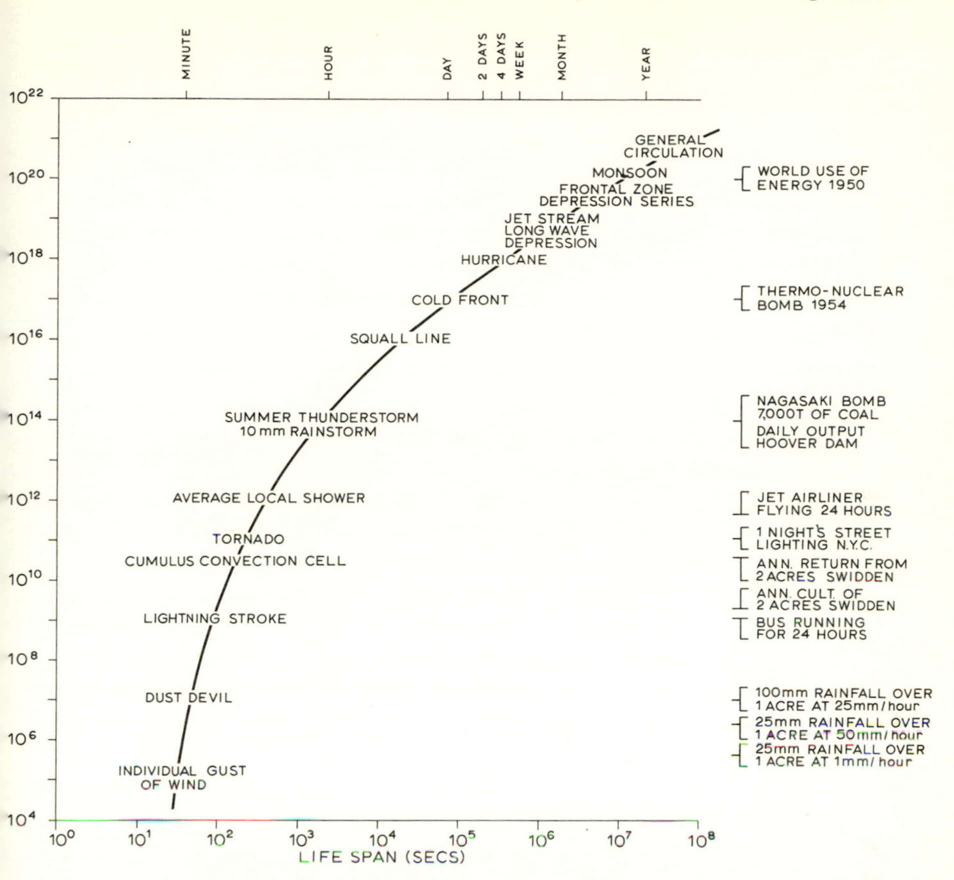

Fig. 9.2 The relationship between life span and kinetic energy of meteorological systems, together with the kinetic energy associated with a number of human activities. (Partly after Hess 1974 and Koppány 1975.)

relationship between soil loss and an index of total rainfall energy combined with its maximum 30-minute intensity (fig. 9.3). More recently Stocking and Elwell (1976) have performed a similar operation for erosion in Rhodesia and have developed a measure of erosivity involving the product of the kinetic energy times the mean annual maximum depth of rainfall lasting for 30 minutes. They have related this to the mean annual rainfall (fig. 9.4). Figures of this kind comparing the energy equivalence of natural and human systems are rather sparse, but we shall have occasion to return to this later when we discuss the energetics of ecosystems involved in agricultural exploitation.

9.2 Strategies of systems interaction

Consideration of the interaction between physico-ecological and socio-economic systems leads us to return to questions involving control, intervention,

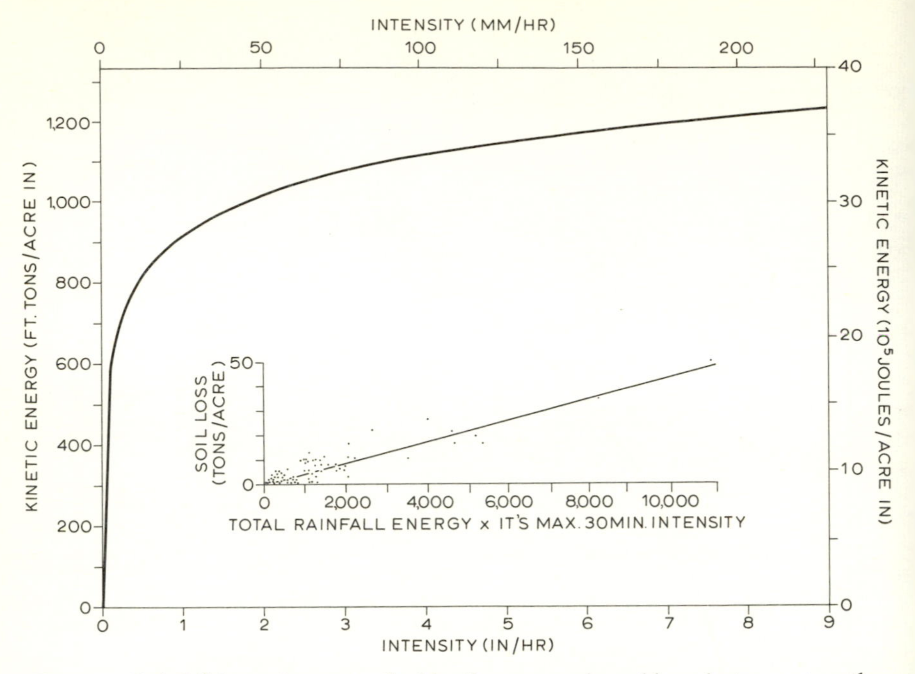

Fig. 9.3 Rainfall intensity versus the kinetic energy released by rainstorms, together with soil loss as a function of total rainfall energy times its maximum minute intensity. (After Wischmeier and Smith 1958.)

monitoring, management and symbiosis. Two of the most popular approaches to such interaction involve ecosystem management, on the one hand, and more advanced socio-economic planning on the other Management, in this sense, is seen to be concerned with the manipulation of the equilibriating operations of the system so as to achieve higher levels of production within given storages, and generally within the existing structure and operation of the system. Ecosystem management is therefore primarily associated with the maximization of existing productivity and the minimization of wastage, by the adoption of a suitable harvesting strategy, pest control, and the scientific cropping of native flora and fauna (Watt 1968, Van Dyne 1969). On the other hand, much advanced socio-economic planning concerns itself with the replacement of one ecosystem by another, or, at the very least, with the impelling of the ecosystem through a trajectory of non-recoverable states. It is significant that when ecological managers refer to the creation of new ecosystems it is commonly within a context, for example, of decrying attempts by early settlers to cultivate semi-arid or humid tropical areas in ways similar to those of temperate cultivators. Clearly, the ecosystem model is of geographical significance in so far as man can be considered to operate in the same manner as other life forms; the model is inappropriate in

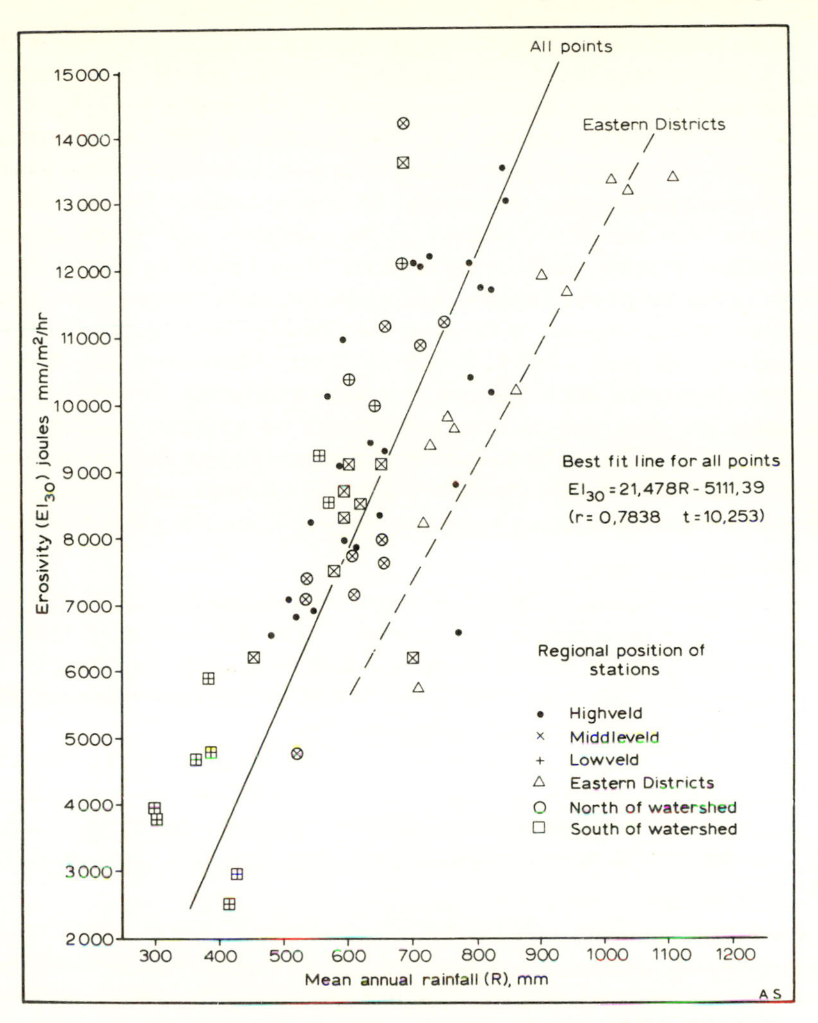

Fig. 9.4 Soil erosivity as a function of mean annual rainfall in Rhodesia. (From Stocking and Elwell 1976.)

so far as man stands apart from the rest of animate nature (see chapter 1). Thus, the general application of the ecosystem concept to the organization of environmental studies raises two fundamental issues: are socio-economic systems so much more complex than the biological parts of ecosystems as to restrict severely the broader applications of this latter model? And, if so, what is the role of man as a controller of the system with which he interacts? (Chorley 1973, pp. 160–1).

In chapter 3 we treated that branch of ergonomics dealing with the problems of human work and control operations (Jones 1967). Certain classes of man-machine systems were identified, all of which fell far short of the kind of systems interaction with which we are concerned in this chapter. Indeed, the most highly sophisticated, computerized systems involving interacting multiple-access man-machine systems are very much simpler in concept than the systems interfacing implied here. Further, the division of interacting systems into two groups, physico-ecological and socio-economic, represents a gross over-simplification. For example, the socio-economic systems treated here make little use of the concept of the voluntary, self-rewarding and collaborative systems which are represented by the family, by certain social groups and by religious affiliation. The whole notion of systems based upon ritual or symbolic actions, of the kind described by Tuan (1973) which we touched on in chapters 1 and 5, should play a very much larger part than we shall give them in the whole question of systems interfacing. However, even if one takes the most limited management view of environmental systems interaction, it is still possible to identify a number of quite distinct strategies whereby this is achieved. In this there is a broad, although not always a clear, division between the exploitation of non-renewable and renewable resources.

1 *Non-renewable resources*:
   (*a*) *Mining*: This can be defined as the use of resources at a rate faster than they can be naturally regenerated within the period of forward planning, or on a scale which will lead to a significant decrease of supply/demand ratio within the period of forward planning. Oil exploitation would be an example of this.
   (*b*) *Tapping*: The use of resources at a rate faster than they can be naturally regenerated but on a scale which will not lead to a significant deterioration of supply/demand ratio within the period of forward planning. The present exploitation of coal resources comes into this group.

2 *Renewable resources*:
   (*a*) *Cropping*: The use of resources at a rate slower than that at which they are naturally regenerated. Primitive gathering in forests would be an example of this.
   (*b*) *Mortgaging*: The use of resources at a rate faster than they can be naturally regenerated, but at a spatial and temporal scale which will allow regeneration within the period of forward planning. Swidden agriculture and the production of rain by the seeding of cumulus clouds would form examples of this class.

3 *Erosion*: The use of resources at rates faster than they can be naturally regenerated and at a scale which will not allow regeneration within the period of forward planning. The use of system storages for the transport and dilution of pollutants leading to permanent damage of the system falls into this class. This would include, therefore, the wholesale dumping of pollutants into stretches of river in which the organized life is disrupted and with it the processes of oxygenation.

4 *Investment*: This involves any action tending to conserve or build up resources. The American government 'Soil Bank' plan – the encouragement of land to lie fallow – would exemplify this, as would the building of dams and schemes which allow for net water storage over a period.
5 *Concentration*: This involves the use of resources at a rate locally which has to be supported by continuous import and concentrated use of resources from other areas. Examples of this would be wet paddy cultivation based upon imports of organically rich water and the whole complex of commercial and plantation monoculture which is based upon imported fertilizers and pesticides.

The human response to resources will differ in the renewable and non-renewable case. For renewable resources the socio-economic aim is to establish a homeostatic equilibrium of rate of use and rate of replacement. Poor management practice (e.g. overstocking and overcropping of land) will lead to degradation of the ecosystem so that eventually no returns to man are possible (fig. 9.5A, curve A). Adoption of immediate reduced use results in curve B of fig. 9.5A. Examples of such curves are a reduced stocking ratio on agricultural land, lower cropping ratio, reduced rate of cutting of forests, reduced rate of fishing and abstraction of water. In some cases, investments in the ecosystem may allow higher levels of return to man. For example, the use of fertilizers in agricultural land, improved crops, land conversion (e.g. reclaimed estuaries), increased mechanization, etc. In each case, improved returns are possible only after a period by which time the rate of amortization of capital is exceeded by rate of return from the investment. Discounted cash flow and the other evaluation techniques discussed in chapter 6 can be used to assess what this period will be. In the exceptional case of new technological developments being available, it may be possible to increase the net returns still further as shown in curve D.

In the case of non-renewable resources it is not possible to achieve a steady state or homeostatic equilibrium of man's rate of use of resources to their rate of replenishment. Conservation policies merely delay the time at which the resource finally runs out (fig. 9.5B). As we shall see, such delays may be very important to allow the development of new technologies which permit better use of resources, and to allow society sufficient time to respond, without catastrophe or collapse, to the pressure of decreased resource availability. It has been noted by Manners (1977) that the effects of social-restructuring will be the main features of the future economics of non-renewable resources resulting mainly from institutional factors: increasing state control and nationalization of resources as in the North Sea and Zambia. This will create its own problems since, although the exploration companies will often obtain the same rate of return on their investments (as in the North Sea), the profits are not recycled for future exploration and development but are used for social ends. This reduces the immediate future known base of resources but

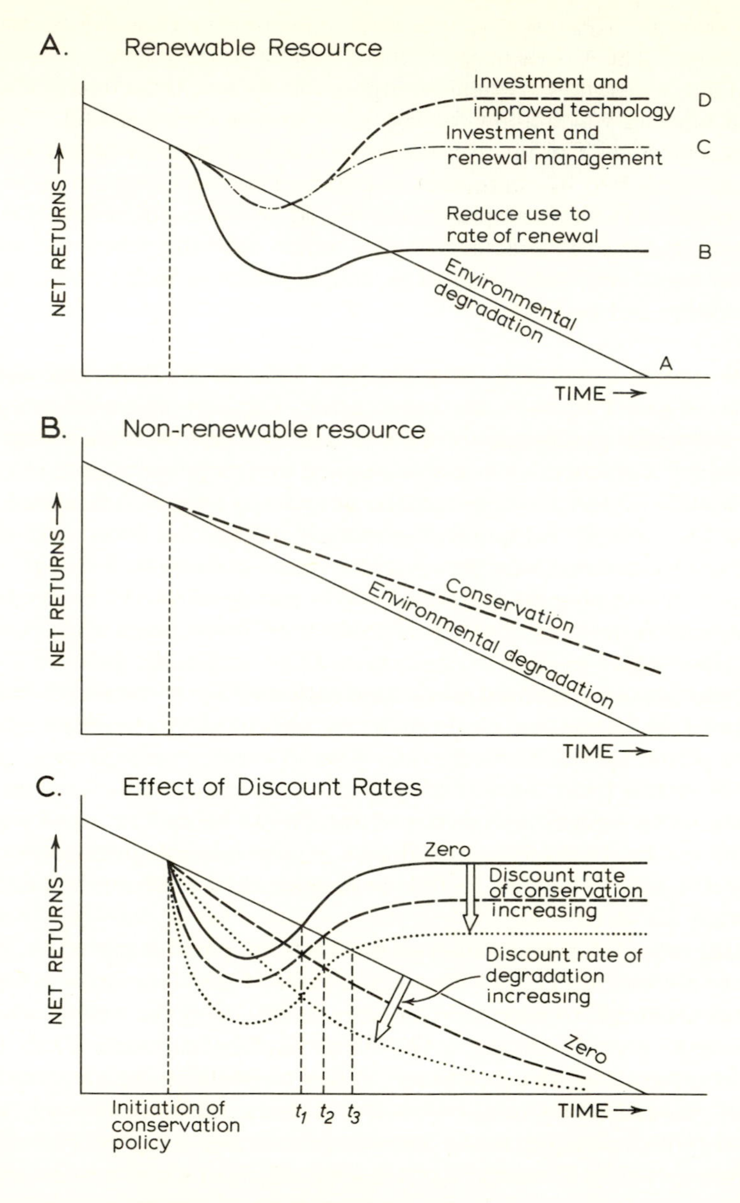

Fig. 9.5 Human responses to resources.
A Renewable resources.
B Non-renewable resources.
C Effect of discount rates.
(Source: Barlowe 1958, partly after Held and Clawson 1965.)

may aid longer-term conservation. This conflict between social goals and those of economic growth is discussed later as a facet of the problems of symbiosis.

In the application of this model to desertification in areas such as Australia and the Sahel, Warren and Maizels (1976) advocate the use of destocking policies to achieve curves such as B in fig. 9.5A and destocking plus investment in fencing and watering points to achieve curve C. More generally, the implementation of investment in conservation policies will be constrained by the rate of discounting of net returns by the decision maker, and this will depend upon his scale of present and future values. When the rate of discounting is equal for conservation or non-conservation policies, then conservation will be adopted provided that the planning period runs well beyond points t_1, t_2 and t_3 in fig. 9.5C at which conservation yields the same net return as depletion at any discount rate. Only for a considerable period after these time points will the increased flow of returns from conservation balance the decreased flows which conservation requires in its early stages. However, it is more usual for discount rates on conservation policies to be higher than depletion since they involve new practices. This has the effect of pushing further into the future the point at which equal returns under either policy are achieved (Barlowe 1958, Held and Clawson 1965, pp. 18–20). Conservation, therefore, tends to impose on cultural systems the necessity for longer planning horizons.

It has been considered expedient in this chapter to examine the interfacing of physico-ecological and socio-economic systems under two main headings. Firstly, there is *environmental intervention* in which socio-economic systems operate apart from physico-ecological systems to cause them to react in a predetermined and advantageous manner. Such intervention systems involve a considerable amount of monitoring and tend to take a rather analytical view of environmental systems, as has been outlined in chapters 7 and 8. This approach is that most nearly equivalent to a view of the operation of a man-machine system as an assemblage of elements that are engaged in the effecting of short- or long-term environmental changes and which are tied together by a common information flow network, the output of the system being a function not only of the characteristics of their elements, but their interactions (Meister and Ribideau 1965, pp. 21 and 23). Fig. 9.6 illustrates the monitoring and compensatory feedback loop adjustments associated with the control of a system element exposed to environmental effects. It is clear that the link between the element and the action is involved first with sensors and then by means of a complicated negative-feedback loop to generate responses which produce the required control of the element. The second, and by far the most complex, type of systems interaction involves the *symbiosis* of physico-ecological and socio-economic systems wherein the latter form an integral part of the whole. This type of system partakes very much more of the systems synthesis and the socio-economic systems form intimate parts of feedforward/feedback cascades containing a complex of interaction, storages and flows.

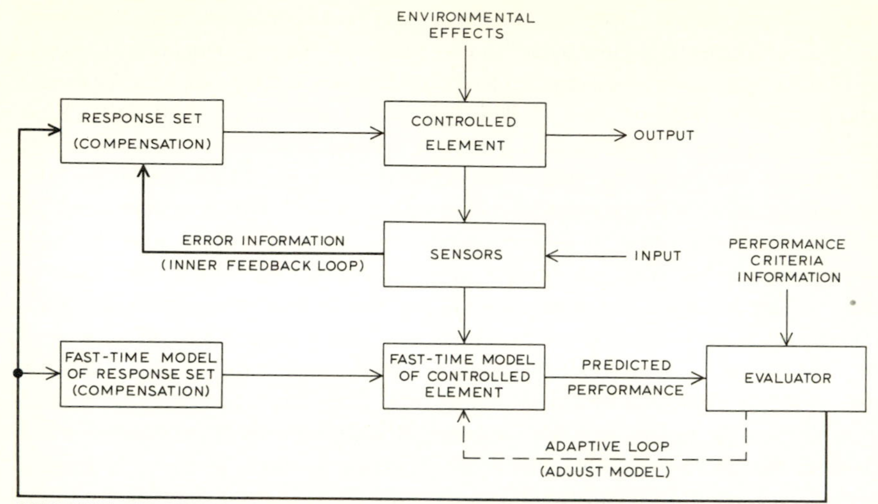

Fig. 9.6 The monitoring and compensatory feedback loop adjustments associated with the control of a system element exposed to environmental effects.

It is clear that in this case the control exercised by the socio-economic system is viewed as being far less free and unfettered than that associated with intervention systems. Figs 9.7 and 9.8 show two different views of pollution systems, the former being concerned with the control of air pollution in a large city. This illustrates the cascade of emissions into the atmosphere producing an air quality which is monitored both at the stack pollution sources within the atmosphere and in terms of public health, each of which produces a negative-feedback loop designed to improve the quality (Savas 1969). Fig. 9.8 depicts a symbiotic system forming the basis for environmental pollution control (Sawaragi *et al.* 1974, p. 198). This consists of a number of intimately associated dynamic systems: (1) the environmental system; (2) the human activity system utilizing the storages of 1; (3) the system model, linked by prediction and observation to 1 and 2; (4) the economic system linked to 1. System 4 plus the human system operates on the results of system 3 to produce an optimal policy which acts by means of legal and technical control associated with direct influence of human beings.

It is clear that whichever model of systems interaction one chooses is partly dependent upon one's view of man–environment relationships (treated in chapter 1) or on the scale of operation of the system. As has been already pointed out, the larger the total environmental system concerned, both in space and time, the greater the tendency there is to model it in a symbiotic, rather than in an intervention, form. Systems intervention is much more popular with engineers and planners than are the large-scale visionary symbiotic types of model – for example, that by Forrester which will be

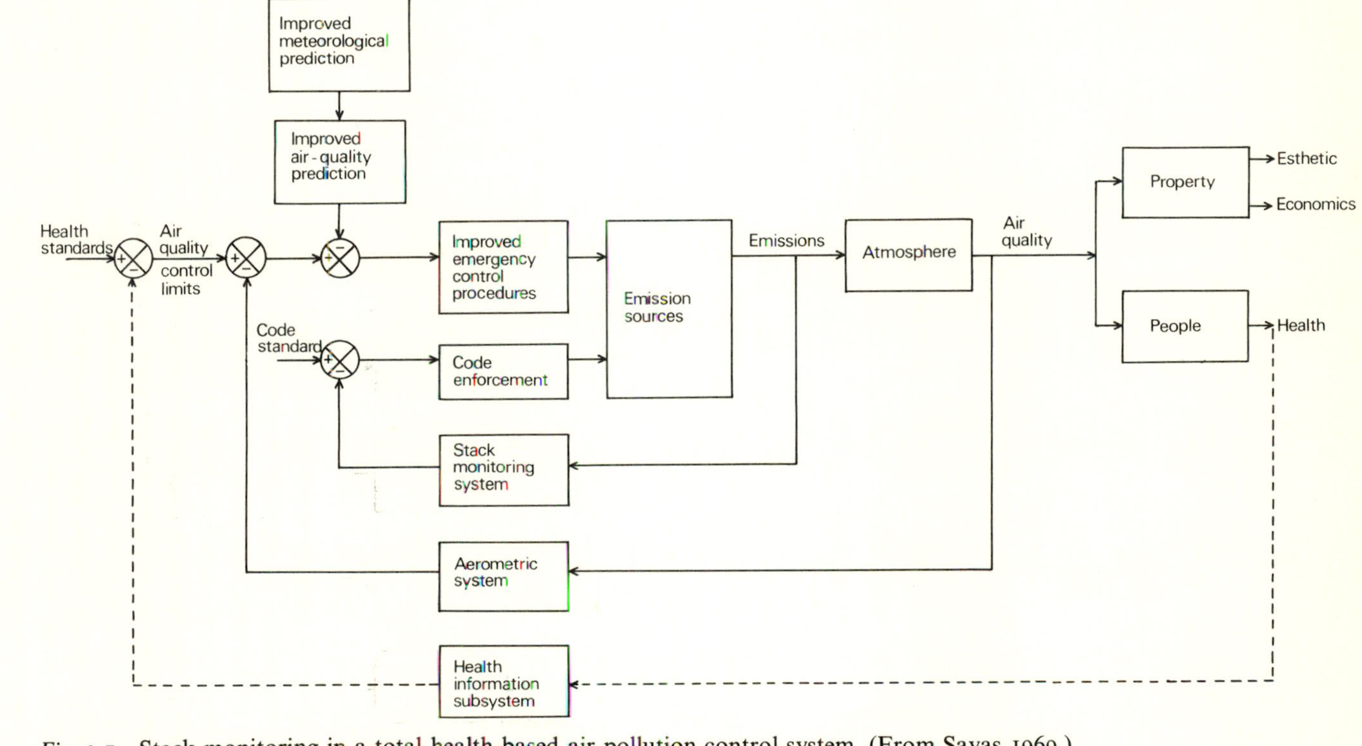

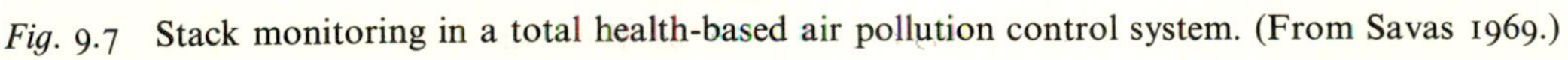

Fig. 9.7 Stack monitoring in a total health-based air pollution control system. (From Savas 1969.)

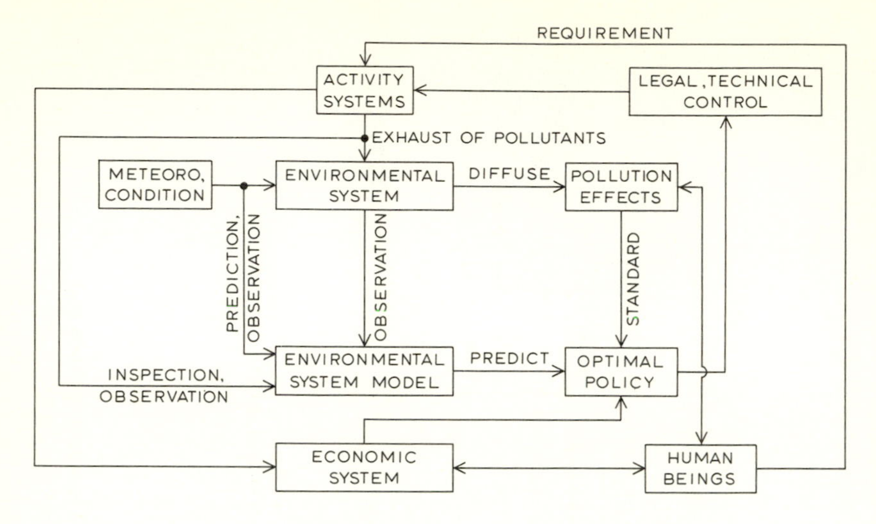

Fig. 9.8 A symbiotic pollution control system. (After Sawaragi *et al.* 1974.)

described later in this chapter. It is also clear that the interfacing of physico-ecological and socio-economic systems depends very largely upon the human perception of the manner in which both these systems operate in detail. It has already been pointed out in chapter 5 that the greatest perceptual problems of the magnitude and frequency of natural events occur when events of relatively high magnitude have a long return period, and fig. 9.9 for example, shows the great range of return periods for daily maximum rainfall in a number of very different climatic regions. One of the most important conclusions which has resulted from studies of the human perception of hazards is that the less probable the hazard, the greater the underestimation of its danger.

The science of animal parasitology provides an interesting analogy with the above view of the possible spectrum of systems interfacing. In the former it is possible to recognize four levels of interaction between the parasite and the host (Baer 1951, Smyth 1972):

1 *Phoresis*: A permanent or transient association between a parasite and the supportive partner in which there is no metabolic dependence, and one wherein the host does not necessarily benefit from the association. In biology this association would be exemplified by one organism providing shelter, support or transport for another.
2 *Commensalism*: This term means literally 'eating at the same table' and involves a loose association between different species which live together to their mutual advantage (although this may be distinctly one sided) without metabolic dependence. The sea anemone living on the shell of the hermit crab is a commensal relationship in that the manner of eating of the latter provides food for the former, and the former helps to

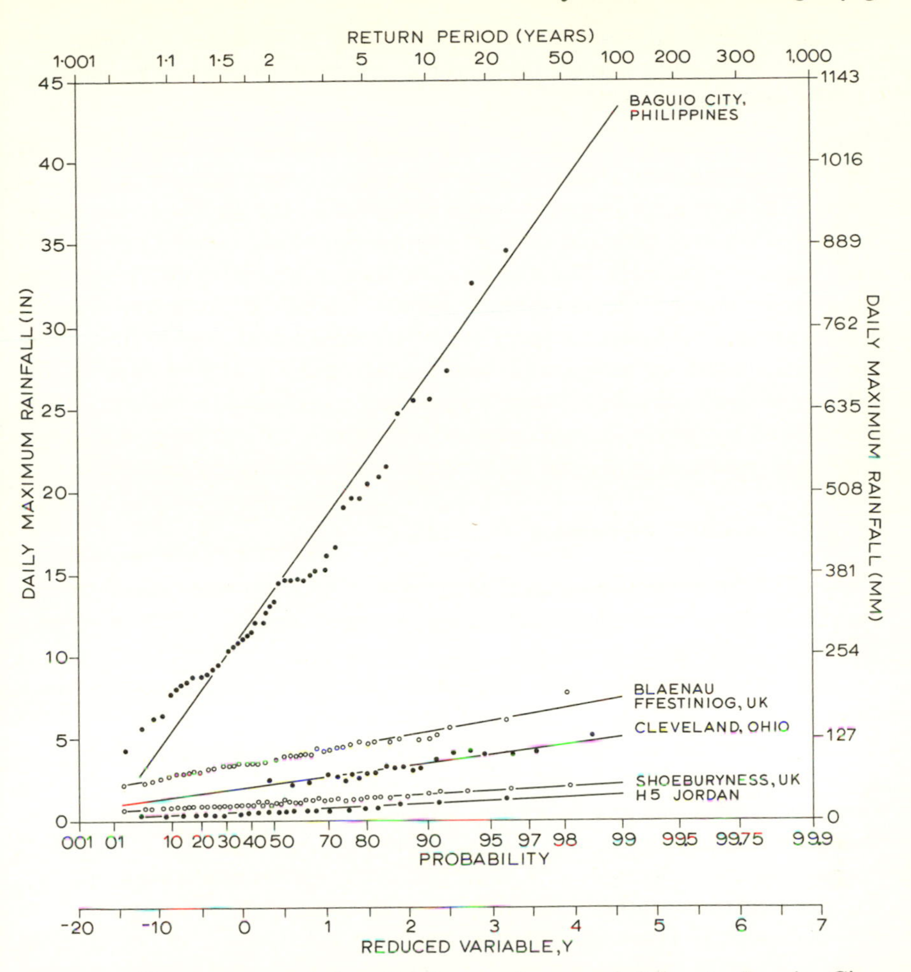

Fig. 9.9 The return periods of daily maximum rainfalls at: Baguio City, Philippines; Blaenau Ffestiniog, UK; Cleveland, USA; Shoeburyness, UK; and H5, Jordan. (Partly from Rodda 1970.)

camouflage the latter. Commensalism is quite distinct from both symbiosis and true parasitism on account of the lack of metabolic dependence.

3 *Symbiosis*: A 'living together' of partners who have lost their physiological independence and both of whom clearly benefit from an association which is necessary to the continuation of their present states. Their mutual dependence therefore maintains their existence, as, for example, when some metabolic byproducts of the parasite are necessary to the host.

4 *True parasitism*: A continuous close association between two species involving a reaction leading to a state of imbalance between them. The

one-sided relationship may develop into a struggle in which the advantage of the parasite (albeit perhaps a short-term one) may be served by the destruction of the host.

Thus the terminology adopted in the present chapter would superficially associate environmental intervention with phoresis and commensalism, and employ the term environmental symbiosis to define the synthetic association between socio-economic and artificial physico-ecological systems, particularly on a large or world scale. The true parasitic state is the analogy of the extreme man – environment degradational spiral feared by conservationists. Intervention would cover a spectrum of environmental relationships from small-scale food gathering and local environmental exploitation to the reliance of socio-economic systems on partially transformed seminatural (Van Der Maarel 1975) physico-ecological systems which, without human management and inputs, would cease to maintain their existing characteristics.

9.3 Environmental intervention

Control of the physico-ecological environment by socio-economic intervention has been the most significant manner whereby these two types of systems have been interfaced in the past. Such interfacing, however, has involved systems which have been small in space and in mass transfers, or ones which have operated over short periods of time. Industrial society has the most dramatic capacity to intervene in physico-ecological systems and in later sections we will describe the kind of environmental symbiosis which is necessary when one considers the interaction of physico-ecological and socio-economic systems on a world scale. However, even in tribal and peasant societies the most elementary technologies, such as stone implements and fire, can have dramatic and extensive effects on the ecosystem if the applications are persevering. Hunting and burning have contributed to mammalian extinction; and pre-Neolithic and Neolithic forest clearance and Bronze Age irrigation in Europe and pre-Columbian North America have initiated environmental degradation and remodelled vegetation, flora and fauna over large areas (Montgomery *et al.* 1973).

Intervention may be considered broadly under three headings:

1 It may be viewed in a purely passive sense in that one is simply prepared for the outputs which may be expected from the system. Such an approach obviously favours the black box model of the system and would be exemplified by the use of the unit hydrograph model in hydrology (chapter 7). Such guarding against expected outputs involves the improvement of prediction by the use of forecasting techniques having to do with interpretation and forward prediction on the basis of previous outputs. It also relates to the mental readjustment necessary to absorb random outputs.

2 A far more important positive approach involves the manipulation of systems storages. Clearly, grey and white box models are specially adapted to this approach. Indeed, many would view environmental control as depending entirely on the ability of man to release energy and mass from systems stores or to transfer them between stores. As we have seen (Koenig *et al.* 1974), the agricultural revolution allowed man greater control over the biomass storages, the industrial revolution allowed him to convert the potential energy of the hydrological cycle and then to release the energy stored in ancient ecosystems, and we are passing now into the use of the stored energy of nuclear fuels and to the direct use of solar energy. This is producing great problems in the depletion of stores for our descendants and in the contamination of stores by waste material.
3 A potentially most important approach, but one which has only been used to a very limited extent up to the present, involves the direct control of inputs into systems. This is very difficult to achieve except over small scales of space and time and requires a considerable input of energy and sophistication of time–space application to make such control effective. As we have seen with problems of cloud seeding and weather manipulation, there are great difficulties relating to the useful scale on which intervention may be necessary to be effective and also that, in tampering with the system input in this manner, we are influencing the system in a most basic way and potentially may produce results which are unpredictable.

9.3.1 PREDICTION OF OUTPUTS

The most ancient manner in which man has attempted to exercise some kind of passive control over environmental systems has been through his own mental adjustment to enable him to cope with a variable and often unpredictable output stream. Religions are very much bound up with a general philosophy toward such outputs and a great deal of human effort and ethical thought has been devoted in order to adapt man and his socio-economic systems to the outputs from physico-ecological ones. Of more recent years studies of environmental and hazard perception, which were touched on in chapter 5, have been concerned with the manner in which socio-economic systems can become adjusted to the outputs of physico-ecological systems in such matters as the occurrence of floods, droughts, environmental catastrophes, and other physical manifestations which control matters to do with life and livelihood. A second type of intervention interfacing involved with the acceptance of output has to do with the whole battery of prediction techniques directed towards the output stream which have been described in the earlier chapters. Such prediction methods form the basis of the widely applied forecasting techniques which have been described and which are designed to permit the implementation of appropriate measures by the socio-economic system to allow for, to combat, or to take advantage of predicted output streams. Such forecasting methods involve magnitude/frequency analysis, such as the statistics of extreme values associated with the work of E. Gumbel (Chorley and Kennedy 1971, ch. 5), together with the whole battery of time-series methods which have been employed earlier in this book, such as autocorrelation, Fourier and Markov methods.

9.3.2 MANIPULATION OF STORAGES

As has been already suggested, by far the most important method of environmental systems interfacing has involved the manipulation of physico-ecological storages by the socio-economic system. This has been concerned with intervention on small and medium scales in which the two systems have remained distinct. There are, however, considerable problems in terms of energetics having to do with transfers of mass and energy between systems storages. In ecology this is expressed in an extreme form by the Darwin-Lotka energy law which states whenever it is necessary to transform and restore a given amount of energy at the fastest possible rate, some 50% of it must inevitably be lost due to the work involved (Odum 1971, p. 31). These rapid transfers of energy, for example, in such operations as the production of precipitation from cumulus cloud seeding or the shifting cultivation of patches of forest, involve large energy losses which are only acceptable if the total area concerned in the potential energy loss is large and if sufficient time is given for local energy amounts to be at least partially restored.

This introduces the whole complex question of the vulnerability of systems when their storages are manipulated. This matter is becoming all the more acute because the amount of energy generated by the socio-economic system is of the same order of magnitude as the energy contained in some physico-ecological systems. The waste heat released by urban dwellers is in some cases more than 20% of the total solar energy flux over the central part of cities. Another example of the scale of human energy can be seen in the extent of the destruction of vascular plant species in the last 100 years. Parts of Poland have lost 10% of their species, the whole of Belgium 5%, Great Britain 8%, and during this period 14% of the species in the Netherlands have become extinct or very rare (Van der Maarel 1975, pp. 263–4). This loss of biotic diversity has been described as the conversion of biotic negentropy. Harte and Levy (1975) have recently examined the stability of ecosystems in terms of the effect on the parameters of their mathematical models, which take the form of sets of coupled nonlinear, first-order differential equations. They recognize three types of ecosystem with regard to their vulnerability to outside intervention.

1 *Open flow without cycling*: An example of such a system would be the energy flow from photosynthesis to top carnivores in which the small fraction of chemical energy removed from the detritus can often be

ignored. When limited in locality, the equations of such ecosystems are neither asymptotically nor structurally stable, and an increase in the number of pathways in the food web (e.g. by introducing more competitors at each tropic level) tends to diminish the domain of stability.

2 *Closed flow with cycling*: These are extremely common systems involving local closed nutrient cycles with little nutrient loss – for example, the carbon flow between different tropic levels. Such a system is not asymptotically stable against perturbations which do not conserve the total amount of nutrient and changes in nutrient amount cause a new steady state to be approached asymptotically.

3 *Open flow with cycling*: These systems involve inputs and outputs with feedback (e.g. the local nitrogen cycle) and are extremely common in nature. If one considers such a system containing six levels – carnivores, herbivores, photosynthesis, inorganic nutrient pool, decomposers, and organic litter – mathematical models of such a system show it to be especially vulnerable to perturbations in the litter, the organic nutrient pool and the decomposers. Thus the sensitivity of the feedback mechanisms, giving stability and protection to the ecosystem, rests primarily on the state of these storages and their protection from degradation. It is generally held that ecosystem diversity provides the most important protection of the system in terms of the vulnerability of its storages to manipulation. Harte and Levy (1975, p. 220) draw attention to the two types of diversity: firstly, the vertical stability involving the number of trophic structural levels; and, secondly, the horizontal stability involving a variety of competitors at each level. They point out that an increase of vertical diversity does not generally increase the domain of asymptotic stability of the system in the normal pyramidal food chain and that an increase in the horizontal diversity leads to a decreasing domain of asymptotic stability. Thus the vulnerability of storages to manipulation must be considered in terms of the comparative rates of storage renewal, instanced by how long pollution will take to clear in a river or in an atmosphere, or how long a forest will take to regenerate after clearing. Further, there is the whole problem of destruction of storages where, for example, ecosystems may be completely degraded, river pollution can destroy the ecological basis of a river to renew its capacity to dissipate pollution and where coastal aquifers when overpumped may be invaded by saline water which leads to their virtual destruction, at least within a long period of forward planning.

Just as certain authors consider that the various categories of parasitism may represent different stages in the domestication of plants and animals, so it is possible to view the origins of agriculture in terms of the manipulation of the storages in natural and increasingly artificial ecosystems. One branch of agricultural origins appears to have developed from tropical cultivation, whereas another may have grown from increasing utilization of arid areas adjacent to permanent water supplies. Tropical forest ecosystems have been described at some length in chapter 7 and have been presented as the most generalized, productive and stable of the major ecosystems. Layering of the vegetation allows very little energy or mineral loss and most energy is used to maintain the complex structure which exists in the vegetation rather than in the soil, such that the effects of external climatic oscillations are reduced, and, as Margalef pointed out, this has created a stable internal climate. Tropical forest ecosystems are of great age and maturity, and their biomass production is very much in equilibrium with energy inputs. Although the total soil thickness may be great, the part involved in the cycling of nutrients is comparatively thin and the *A* horizon and root layer in tropical forests may be only about 1 metre thick. This thin *A* horizon contains at any one time a relatively small proportion of the total ecosystem nutrients, but these are rapidly circulated. Insects are extremely important in the speeding up of decomposition and recycling energy nutrients, and 50 % of the insects feed on dead or decaying matter, although the animal biomass consists of only 0·2 % of the total biomass. The total amount of nutrients is low compared with other vegetational ecosystems, but its rate of exchange is three to four times that of the temperate forest, and the productivity of the tropical forest is two to three times as great (Douglas 1969). The constant microclimate near the forest floor provides a buffer between vegetation and soil, and there is a high nutrient turnover rate within the soil, with the poor soils acting as a small store for rapid cycling. Tropical forest ecosystems are therefore characterized by diverse species (on an average of up to 200 per hectare, as distinct from 20 per hectare in temperate deciduous forests and 10 per hectare in boreal forests); a low density of individual species; a diverse vertical vegetational structure; a uniform microclimate in the lower layers (it has been shown that the clearance of tropical forests causes temperature fluctuations to be initiated down to a depth of 75 cm in the soil); a low animal biomass; an ecosystem which is virtually closed; a closed nutrient system because of the layering; a rapid nutrient cycling; and a high primary productivity (3600–7200 grammes per square metre per year, as distinct from tundra 360, mid-latitude grasslands 180–720, and boreal forest 900). The tropical forest ecosystem is considered to be mature and efficient. Immature systems are identified with high productivity per unit of biomass, but a low total biomass. However, developing maturity is associated with increases in both productivity and in biomass, but with the productivity/biomass ratio decreasing such that it takes progressively less energy to support each unit of biomass. Some authors regard increasing ecosystem maturity as an increase in entropy associated with a more equal distribution of mass and energy throughout the ecosystem. Human intervention, on the other hand, tends to reduce ecosystem maturity, and therefore more energy is needed for its operation (Rappaport 1971, p. 78). Human intervention is thus associated with a decrease in entropy, and there is a much greater concentration of energy and nutrients within the ecosystem due to agricultural intervention.

It has been suggested that various categories of exploitation of tropical rainforest might represent a series of stages of the development of sedentary agriculture. In this scheme the first stage would be that of the small-scale gathering of crops from the existing ecosystem, having relatively no impact on it and simply utilizing energy from various of the biomass storages. The second stage would be that of maintenance involving the scattered mixing of cultivated crops with the natural ecosystem. This maintains the ecosystem structure and operation but it is uneconomical except for very low production requirements. All attempts which involve higher density planting (e.g. rubber and cocoa) mitigate against diversity and cause the ecosystem to degenerate. The third stage in agricultural development may be viewed as that of shifting cultivation (swidden) which is considered to be an intermediate stage between food gathering and sedentary farming. This involves the creation of a miniaturized harvestable tropical forest (Geertz 1963, Harris 1969, Rappaport 1971). Swidden agriculture is carried out by the complete cutting and clearing of a small area of about 1 acre, the burning of the vegetation and the planting in its place of a mixture of varieties of crops from tubers to bananas, up to as many as 100 varieties per field in New Guinea. These simulate something of the structure of the natural ecosystem, maintaining the ground cover and assuring ecological stability; they are harvested for much of the year (perhaps for some eight months), have different life and cropping cycles and keep soil erosion to the minimum. Thus a one to two year cultivation period is followed by up to ten years of fallow to allow regeneration by secondary forest. In the Upper Orinoco forest of Venezuela Harris (1971) has described a ten year swidden cycle in which after one to two years abandonment there is a fairly dense forest canopy of 1·75–2·5 m high with a few bare patches, after two to five years a continuous canopy of 4·5–7·5 m in height, and after ten years the regenerated forest (*rastrojes*) is difficult to distinguish from mature forest at ground level, although easier from the air. The plot size is commonly 1 acre or less for swidden (Harris 1972) and, if there is more rapid cultivation or longer cultivation and shorter fallow periods, there is a more rapid cumulative loss of nutrients by burning, soil erosion, removal of the biomass, etc. Swidden agriculture allows high outputs over short periods of time for low energy inputs (perhaps 25–1), but this can only support a moderate density of population, perhaps up to 50 per square mile with a large amount of available land, although the efficiency of swidden agriculture is very much in question at the moment. In general, however, it gives a good return for relatively little effort, has a high population support relative to the effort input (Rappaport 1971). In systems storage terms swidden involves the rapid release by burning of large amounts of nutrient energy from the biomass. Much of the energy is lost to heat and smoke, the blowing and washing away of ash, and by the introduction of domesticated animals like pigs, but the rest is rapidly transferred to the soil storage which is then exploited by a short period of cropping. Then the system storages are allowed to regenerate during the fallow period. The fertility of the soil, however, may drop 80% between the first and second years of swidden cultivation (fig. 9.10). The regeneration period is increased by drought and the length of the dry season, both of which naturally inhibit vegetation regeneration. The most complete analyses of energy flow in swidden agriculture have been carried out in New Guinea by Rappaport (1971). He describes the mature tropical rainforest as probably 'the most intricate, productive, efficient and stable ecosystem that has ever evolved'. When unmodified it can sustain about one person per square mile but, according to Rappaport, swidden agriculture can support populations comparable with those of industrialized densities, even with a minimum of environmental degradation. In New Guinea only about 10% of any particular area is being used for swidden agriculture at any one time; this is cropped for two years, and supports a population density of some 124 per square mile, with horticulture providing 99% of the human diet. The following table gives the energy input involved in the cultivation of two gardens of 1 acre each (in kilocalories):

Cutting underbrush	56,600
Clearing trees	22,700
Fencing, to keep out pigs	34,200
2 burnings and weedings	19,000
Clearing and working the plots	14,500
Planting and weeding	180,300
Other maintenance	46,000
Harvesting (sweet potatoes, cassava, yams, etc.)	68,300
Cartage	119,800
Total	561,400

The total biomass return is 9,779,500 kilocalories, giving a return of 16–17 to 1 in terms of energy. Of this man consumes 6,151,300 kilocalories and the pigs consume 3,628,200. The pigs are eaten as food, but are inefficient, partly because they are killed prematurely (for religious reasons) and partly because in an ecosystem the energy supply is steadily degraded into heat with each successive step along the food chain such that 80–90% of primary production is available to the herbivore, but the herbivore only provides 10% of this to the carnivore. Despite the inefficiency of the pigs, they do perform a valuable function in providing protein in the balanced human diet as well as in transforming garbage and human faeces into fertilizer.

A fourth level of ecosystem exploitation in the tropics involves the high intensity transformation associated with the sawah system (Geertz 1963) of flooded paddy. This monoculture of irrigated paddy fields maintains fertility by the extensive use of irrigation water plus animal dung. The irrigation water required is far in excess of the paddy requirements for normal transpiration processes which are no greater than for dry land crops, but the excess water is necessary to add nutrients, particularly minerals to the thin soil, in providing nitrogen-fixing algae, in aiding the decomposition of the organic material, and in aerating the soil. Great human effort and large supplies of water are required to support this monoculture, which after a rapid initial decline in

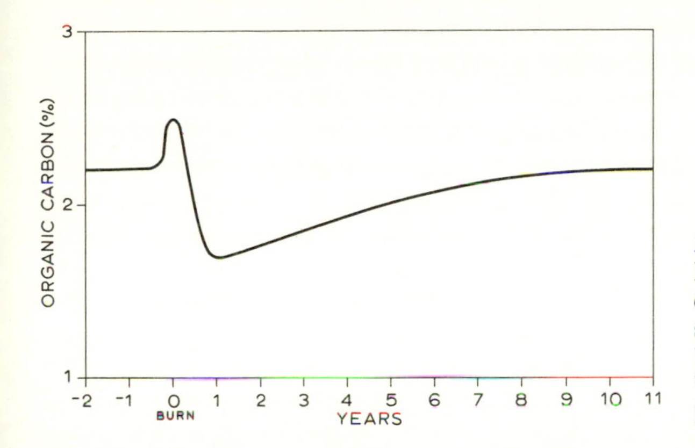

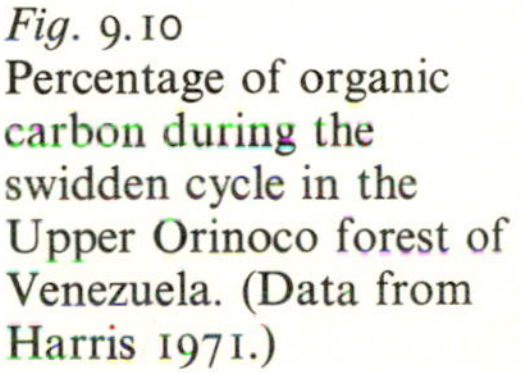

Fig. 9.10
Percentage of organic carbon during the swidden cycle in the Upper Orinoco forest of Venezuela. (Data from Harris 1971.)

productivity achieves a productivity stabilization. Rice as a crop removes few minerals from its soil and a high yield is maintained by a high human input of energy to maintain the ecological stability. The whole agricultural system is dependent upon the very precise control of irrigation water needed to prevent on the one hand, stagnation, and on the other flooding. Human activity is needed all the year round in making good use of the environment without depleting it greatly. The ecological similarity between sawah and swidden agriculture is illustrated by their combined use, as in the chema system of Sri Lanka. As Geertz (1963, pp. 30–1) has put it, 'like swidden, wet-rice cultivation is essentially an ingenious device for the agricultural exploitation of a habitat in which heavy reliance on soil processes is impossible and where other means for converting natural energy into food are therefore necessary.'

The fifth type of ecosystem utilization is the complete transformation of the environment into a specialized artificial ecosystem which does not rely on large amounts of irrigation water, but where the reduction in species-diversity leads to weeds and to the destruction of fertility, necessitating the extensive use of weed killers, pesticides and fertilizers. Thus grain and plantation agriculture requires the input of large amounts of fertilizer together with artificial protection against soil erosion. Sometimes this type of utilization can develop spontaneously by the degradation of the environment by intensive employment of swidden agriculture, leading to the natural development of a savannah grassland. The productivity of transformed ecosystems is very high – that of sugar cane in Hawaii being 600–700 grammes per square metre per year and that of a Minnesota maize field 900 grammes per square metre per year – but this is entirely dependent upon the extensive use of fertilizer and other chemical applications. Thus it is possible to view the development of settled tropical agriculture in terms of a sequence of stages in which more and more reliance is placed upon the inputting of energy into the soil storages, by increasing the rate of turnover of decayed natural organic matter on the one hand, and by the excessive and continual banking of imported fertilizers on the other.

Whilst various forms of cultural evolution have produced societies well adapted to maintaining homeostatic equilibrium within the environment by manipulation of storages, in many cases it has been necessary to *induce* small-scale socio-economic change to favour the maintenance of environmental equilibrium. Policies based on the manipulation of storages have been combined with knowledge of the time evolution of the ecosystem and its response to climatic and other inputs of variable magnitude and frequency as a feedback control system in the forms discussed in earlier chapters. For example, the possible returns from cattle production using forage in an area subject to cyclical drought are shown in fig. 9.11A. Destocking must take place during drought in order to avoid soil erosion and destruction of storage available in the vegetation subsystem. In practice, because of the long lag time in societal response, destocking often will not begin until well after drought conditions begin. Over each drought cycle, this results in successive degradation of the environment, reduction of storages and reduction to net returns, as in fig. 9.11B. In this situation Warren and Maizels (1976) propose a feedback policy based on the forage available (fig. 9.11C). When forage is in surplus, limited restocking is possible; when in deficit, destocking must be undertaken. The difficulty inherent in this policy (as in all feedback schemes) is the decision on the most appropriate base line representing the target level (i.e. reference set point) of 'safe' maximum stocking rate. In a more complex model of desertification in the Sahel, Picardi (1976) raises a number of ethical questions concerning the goals we impose on natural soil and forage systems in order to maintain storage and to achieve a homeostatic equilibrium between man's demands and the environmental constraints. For example: should the Western World desist in giving aid until natural and cultural balances are better understood? What form of cultural response best suits the achievement of maximum net return from non-renewable resources, whilst maintaining the resources into the future? Should we accept lower *per capita*

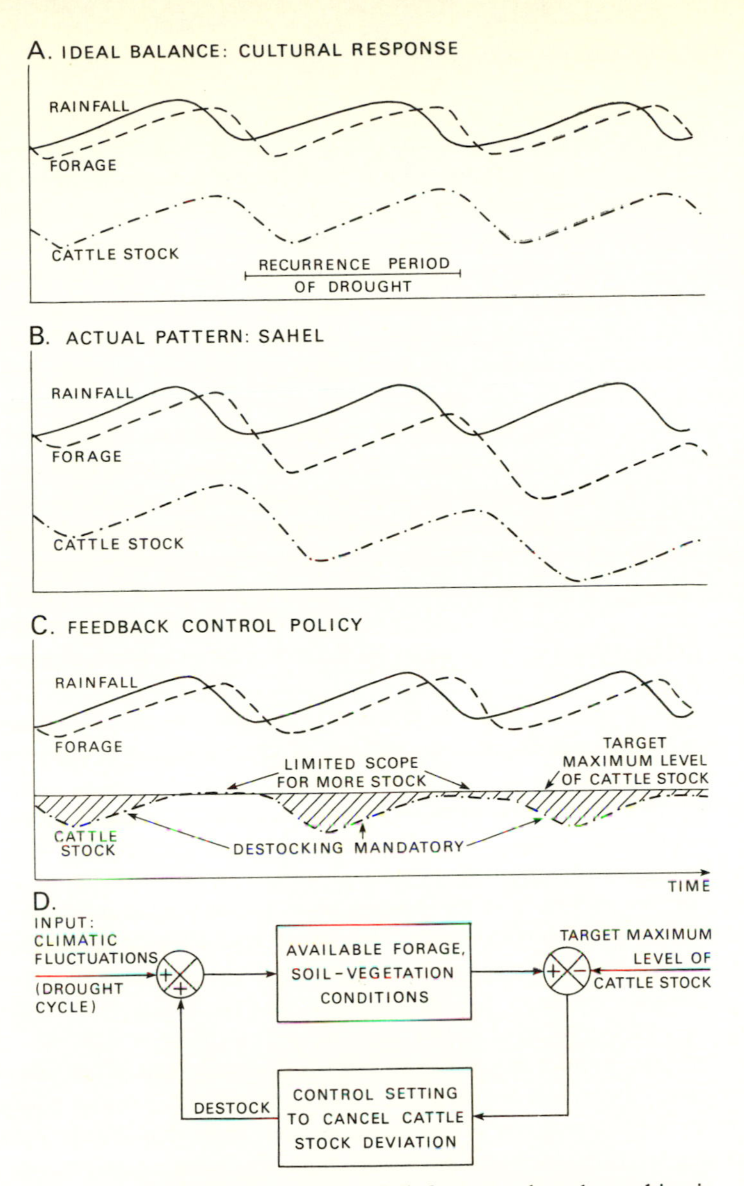

Fig. 9.11 Relations between rainfall, forage and cattle stocking in arid regions.
A An ideal balance of stocking to forage available in areas of cyclical drought.
B The actual response in the Sahel and Australia. Overstocking in drought periods prevents complete recovery of forage between droughts.
C Feedback control policy based on a target level of maximum stocking, making destocking necessary when forage decreases during drought periods.
D The control system shown as a flow diagram.
(Partly after Warren and Maizels 1976.)

income in both the Third World and globally in order to achieve better conservation practices? Which approach should be taken to population regulation? How can the response of feedback control (i.e. stock limitation) be achieved both administratively and politically without destroying the distinctive traditional and cultural values of the pastoralists in marginal areas? Fig. 9.12 shows policy responses which exhibit the effects of various ethical choices. Fig. 9.12A demonstrates simulations of Sahelian population and livestock numbers with and without historic intervention. After 1920 a natural increase in population and livestock began which was associated with elimination of warfare, well digging, and vaccination against human and cattle diseases from 1940; this resulted in marked increases in the number of pastoralists and stock, which decreased the fodder capacity and led to erosion of the soil. The 1969–73 drought merely accelerated the time at which a 'collapse' to much smaller stock and population numbers was inevitable. Fig. 9.12B shows the effect of the feedback stock-limitation policy (as set out in figs 9.11C and 9.11D), together with the effect of this policy combined with cultural change (giving more stock trading and greater material wealth) and with the likely increases in population numbers resulting from improved health and diet.

Through its concern with the control of runoff, much of engineering hydrology also depends on the manipulation of various storages, as in the case of water between various levels of storage in the basin hydrological cycle. Fig. 9.13 shows how a control junction can operate on the error correction between the desired infiltration capacity of a plot or basin and the actual infiltration capacity determined by the state of the soil water storage. This figure should be viewed in conjunction with fig. 7.19 (p. 353). Dams are the most common method of controlling surface runoff but, apart from their obvious effect on the increase of what amounts to 'channel storage', they do have other effects on the disposition of water between storages due to, firstly, their increased evaporation and percolation losses and, secondly, the greater water losses due to increased human usage for irrigation, industry and other purposes associated with the reason for the dam construction. There are also statistical problems in determining the effect of dams and water disposition between storages due to difficulties in comparing the river rating curves before and after the dam construction. Hipel *et al.* (1975) have used intervention analysis (Box and Tiao 1973) to estimate the man-induced intervention on the mean level of the time series of Nile flows during the period 1870–1945, as the result of the completion of the Aswan Dam in 1902. Fig. 9.14 gives the results of this analysis showing a decrease in mean annual discharge of more than 20% after 1902.

Systems storages can also be manipulated in terms of their use as waste dumps and dissipators. The problem of atmospheric pollution has already been referred to in this context and we now return to the use of rivers and estuaries for waste disposal and purification. Wolman (1971) has identified the

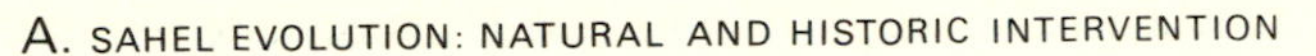

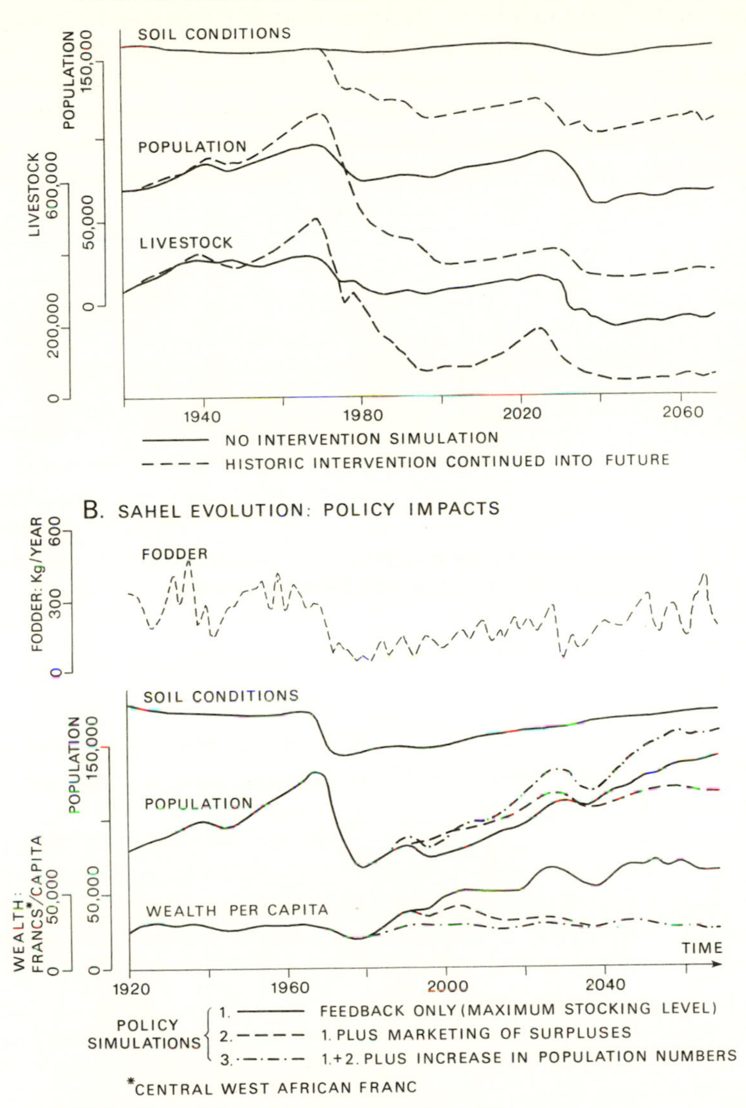

Fig. 9.12 Simulated responses to stock control policies in the Sahel, exhibiting the effects of various ethical choices:

A Without intervention or with historic intervention methods.

B The effects of a feedback stock-limitation policy under three different assumptions.

(Partly after Picardi 1976.)

transport characteristics of water quality as dissolved oxygen, dissolved solids, biochemical oxygen demand (BOD), suspended sediment, pH and temperature. Fig. 9.15 shows a space-time model for BOD along a river profile and there is a clear relation between such processes and those of stream channel diffusion which have been treated in chapter 7. Dissolved oxygen (measured as per cent saturation) varies with seasonal discharge, with position along a river, with reference to pollution sites and with time. Wolman (1971) has given data relating to the pollution of New York harbour, showing a

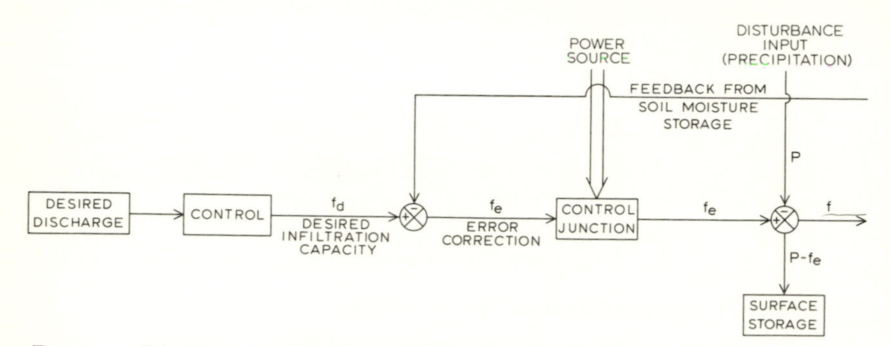

Fig. 9.13 Diagram showing the manipulation of soil water storage through the control of infiltration (see also fig. 7.23, p. 358).

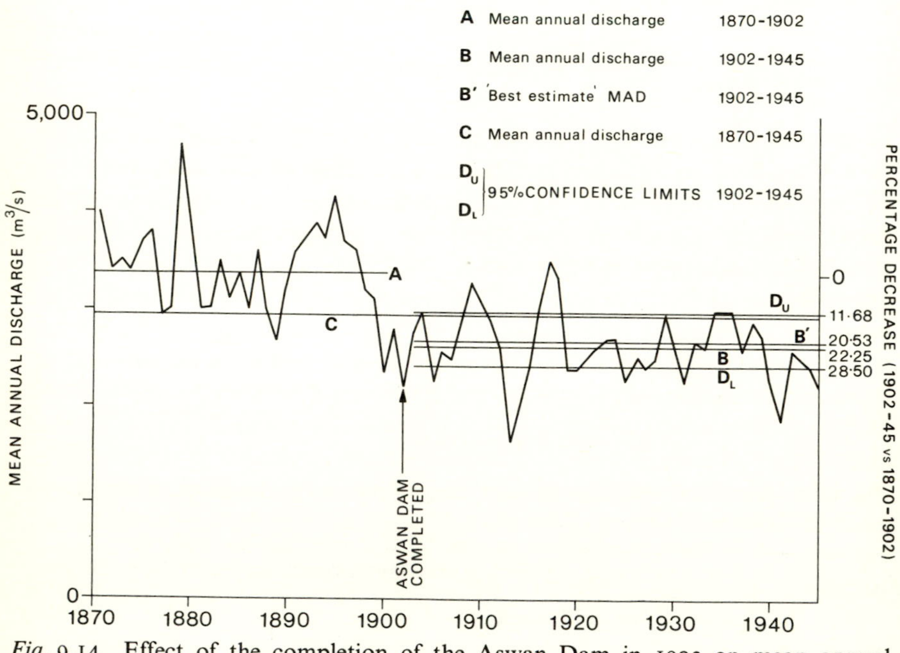

Fig. 9.14 Effect of the completion of the Aswan Dam in 1902 on mean annual discharges. (After Hipel *et al.* 1975.)

decline in dissolved oxygen between 1909 and 1920, since when it has been reasonably stable. On the Mississippi River just below the city of St Paul the construction of sewerage works between 1932 and 1938 beneficially increased the amount of dissolved oxygen, which thereafter was virtually constant. In the Potomac estuary the growth of population, together with the relatively small discharge of the river, has caused the amount of dissolved oxygen to decrease during the last thirty years or so and the zone of oxygen deficiency in the river estuary to increase in length (see chapter 7). Torpey (see Wolman 1971) has suggested a discontinuous inverse relationship between dissolved

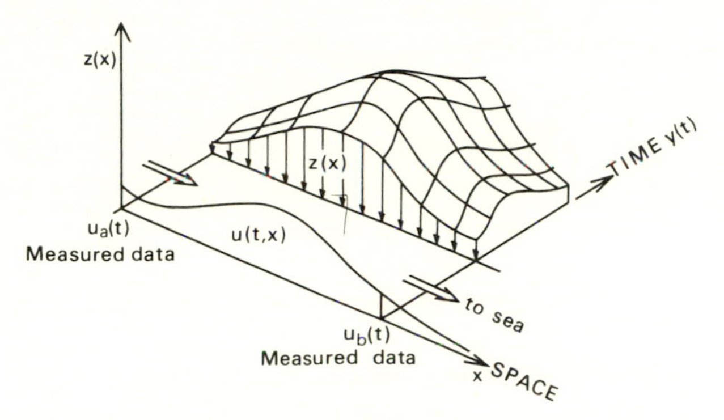

Fig. 9.15 A schematic space-time (*X*–*Y*) model of biochemical oxygen demand (*Z*(*X*)) for a stretch of river ($U_a - U_b$) at varying times (t) (After Sawaragi and Ikeda 1974.)

oxygen and BOD (per acre of surface per day) for estuaries based on data from New York harbour and the Thames (fig. 9.16). Here dissolved oxygen might be primarily controlled by organic activity associated with the nutrients promoting the growth of algae, and the curve shows two catastrophic thresholds separating a plateau within which changes of BOD seem to have little effect on dissolved oxygen. Such curves, if generally substantiated, might have a big influence on the significance of plans for purification, although this is a very complex problem. Clearly, the spatial distribution of pollution along rivers and estuaries is of great importance in connection with the ability of the natural processes of stream channels to renew and dissipate pollution. Graves *et al.* (1972) have dealt with a theoretical estuary divided into three sections (fig. 9.17A) with five pollution discharges and three regional treatment plants. Each of the three sections of the estuary provides information on discharge, volume and length of section, the diffusion coefficient and the re-aeration coefficient. The five dischargers provide information on the discharge of waste water, the biochemical oxygen demand concentration (pounds per milligrammes of effluent), the percentage removal of BOD the cost of removing BOD, together with possible pipeline costs and a set of dissolved oxygen goals.

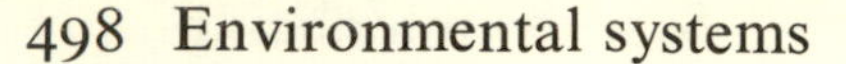

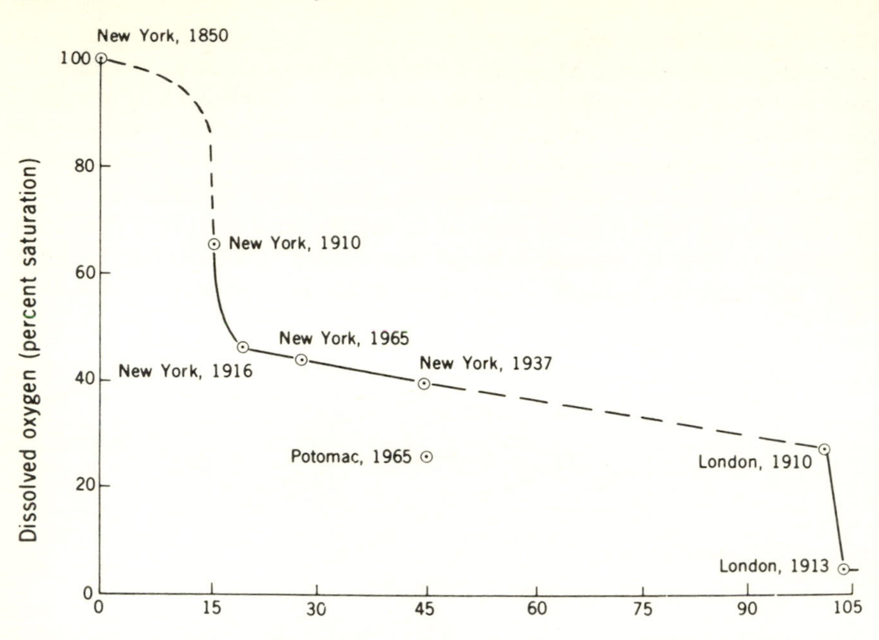

Fig. 9.16 Dissolved oxygen as a function of biochemical oxygen demand for New York Harbour and the Thames estuary, suggesting two important thresholds. (From Wolman 1971.)

The three regional treatment plants are fixed in their location as the result of political and engineering considerations. The problem is first to produce a steady-state dissolved oxygen profile downstream, and from this a profile of desired dissolved oxygen changes (in milligrammes per litre) so as to achieve the required dissolved oxygen changes in each section with a minimum cost by means of nonlinear programming. This gives a problem involving eleven constraints (upper and lower boundary constraints) plus forty-seven variables. These constraints and variables can be grouped into blocks according to principles of: (1) treatment at discharger only; (2) bypassing piping only; (3) treatment at regional plants only; (4) 1 and 2; (5) 1 and 3; (6) 2 and 3; (7) 1, 2 and 3. Priorities can also be introduced which allow a multivariate problem to be progressively solved with only a small subset (e.g. priority class or classes) of variables being active, the rest being zero. The authors use five classes of priority:

1 Waste water treatment restricted to dischargers.
2 A choice between waste water treatment at the discharger or at the regional plant which discharges to the nearest river section.
3 Dischargers can choose between any regional plant, and the regional plants can choose into which river section they discharge treated waste.

4 All bypass pipes open up.
5 Allows for variable treatment effectiveness at the regional plant.

Fig. 9.17 shows four possible solutions to this problem, together with their costings.

Nothing better illustrates the principles of environmental intervention in terms of the manipulation of storages than a consideration of the conjunctive use of water resources. Fig. 9.18 gives a schematic representation of a simple surface-ground water system. Complex and practical examples of this work are provided by the application of linear programming to water control in West Pakistan, where irrigation canal leakage over the past sixty years has raised the water table and caused waterlogging and salinity build-up over

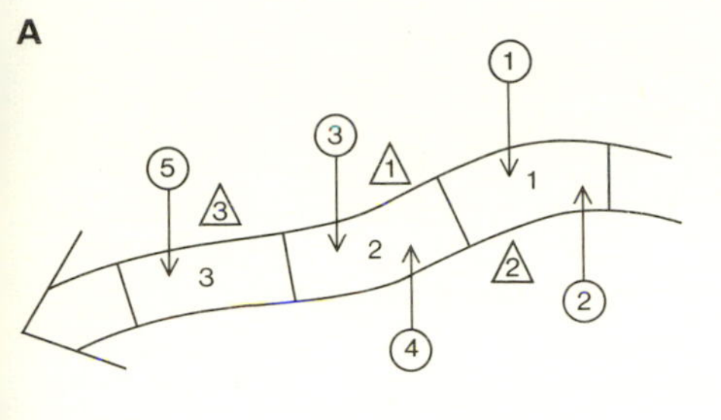

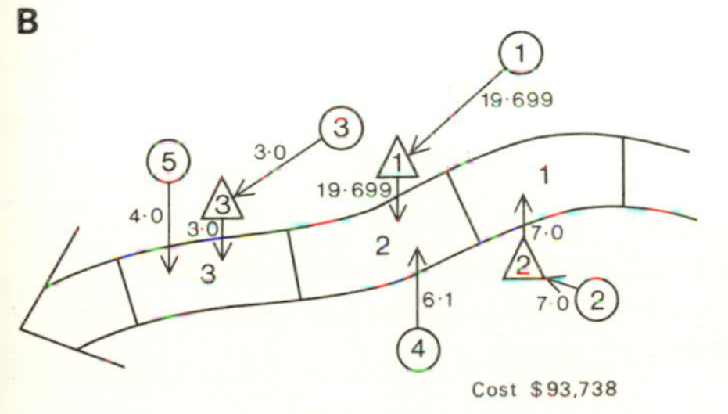

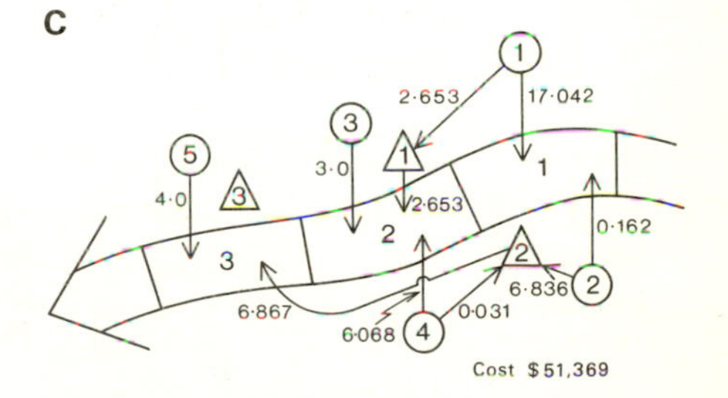

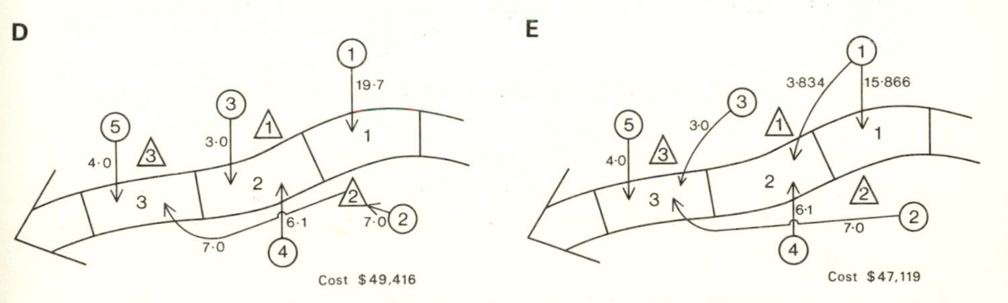

Fig. 9.17 Model of a theoretical estuary divided into three sections, with five fixed pollution dischargers (circles) and three fixed regional treatment plants (triangles). Flows are waste water discharge (mill. galls./day). (After Graves *et al.* 1972.)

more than 5 million acres. The solution to this problem has been to control the water table by sinking tube wells of varying depth to allow the non-saline area to be pumped extensively to recover infiltrating fresh water (750–1000 milligrammes per litre salinity) and to 'mine' the aquifer so as to lower the water table some 100 feet in thirty years. In the area of saline ground water (6000 milligrammes per litre salinity) the infiltrating fresh water was recovered by shallow (skimming) tube wells as well as by a set of deep wells designed to reduce the hydraulic gradient in the direction of the non-saline area and to

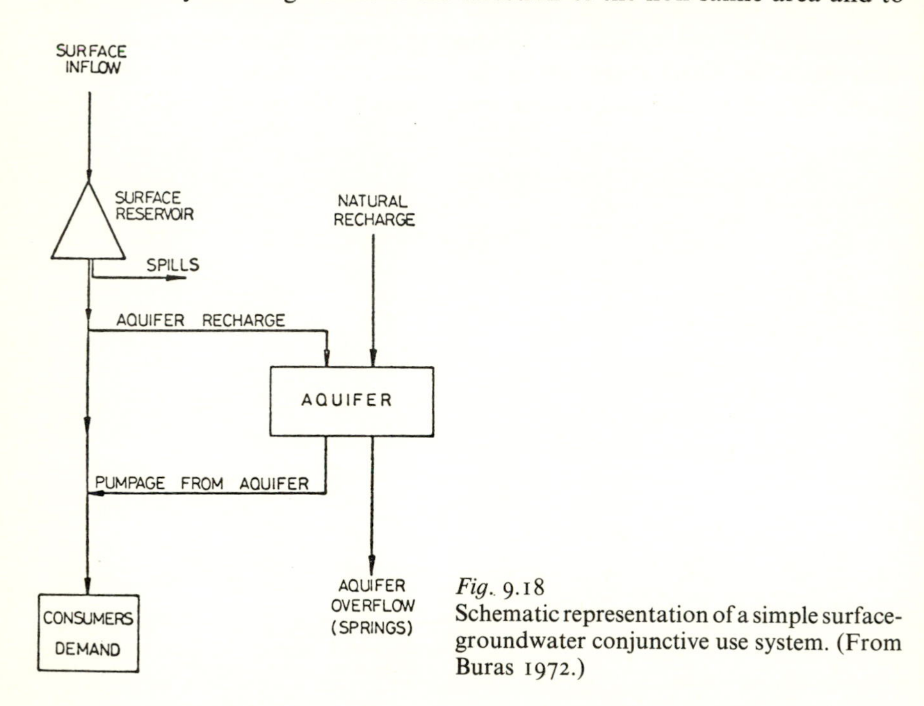

Fig. 9.18
Schematic representation of a simple surface-groundwater conjunctive use system. (From Buras 1972.)

produce saline water which can be diluted with canal water for irrigation. Fig. 9.19 shows the general flow chart of the system in respect of optimum values for:

y = the flow of canal water directed into saline area to dilute deep tube well saline water,

z = saline ground water to be diluted with surface water,

w = rate of mining in saline area, in excess of the 2·9 acre feet per year required to prevent the movement of saline ground water into the non-saline area.

The linear programming problem was set up in terms of a number of constraints: the annual benefits for 1 million acres divided by the annual unit cost of tube well water; the annual unit cost of increasing channel capacity to the saline area divided by the annual unit cost of the tube well water; the

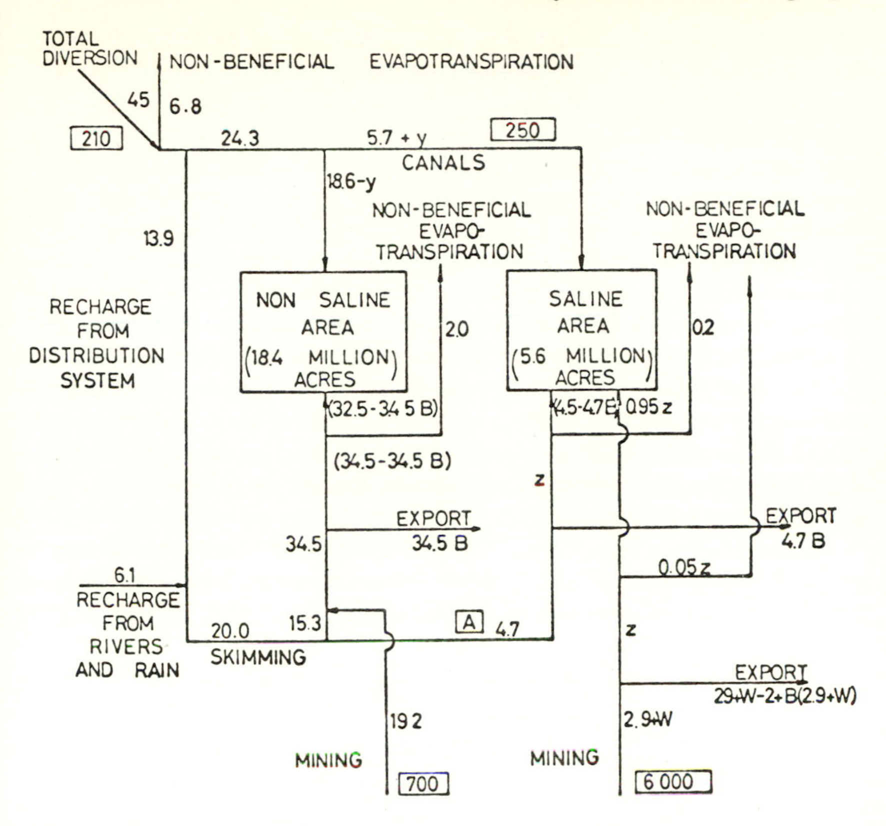

Fig. 9.19 Flow chart for water control in West Pakistan. (From Buras 1972.)

annual unit cost of exporting saline water from the irrigated area divided by the annual unit cost of the tube well water; concentration of recharge water in the saline area; and the upper limit of salinity of applied irrigation water in the saline area (Buras 1972). The problem, as defined, gives seventy-two different possible combinations of these five constraints from which the final design was chosen.

Clearly the constraints involved in the conjuctive use of ground water and surface water depend very much upon the kind of decision making framework adopted, which has been treated in chapter 6. Yu and Haines (1974) have described a multilevel optimization scheme for such conjunctive use by breaking the problem down into subsystems and associating smaller independent subproblems with each subsystem, each of these being optimized before attempts are made to optimize the whole. This needs an iterative procedure utilizing feedback between the regional higher-level solutions and the local lower-level optimizations. In this way each area is autonomous and is served by both pumped ground water and surface water and is divided into polygonal zones based on nodes, the number of which bear a direct relation to the relief of the water table. It is clear that the mobility of ground water makes for interdependence of production conditions between spatially adjacent countries which necessitates high coordination. The alternative to this would be central control and allocation. The regional authority is responsible for all large water control schemes which enhance the public interest (e.g. recharge, water quality, etc.). The revenues are provided by local taxation involving a pumping tax and/or the *ad valorem* tax, and the regional authority is concerned with making an equitable apportionment of pumping rights among local authorities. The regional authority coordinates the optimal management planning for each local area and determines trends of water levels and cross-boundary flows between regions for the entire planning period. This fixes trends of local aquifer boundary conditions at their optimum values. Thus the solution to the manipulation of storages in conjunctive water use proposed by Yu and Haines is a multilevel one (fig. 9.20), requiring the decision techniques of corporate decision making discussed in chapter 6. At level 1 the local authorities optimize independently their own subregional water problem. At the second level the regional authority couples these together to obtain an optimal regional solution, the decision variables being the inter-subregional boundary water levels throughout the planning horizon, the pumping rate and

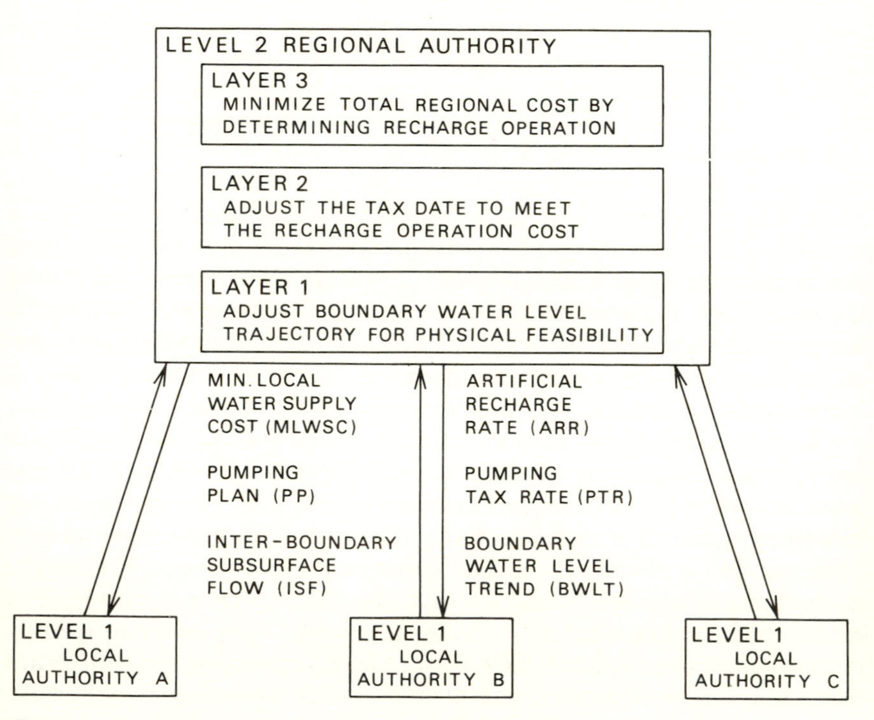

Fig. 9.20 A multilevel organizational scheme for conjuctive water use. (Partly after Yu and Haines 1974.)

the artificial recharge rate. Within level 2 there are three layers of operation by the regional authority. Layer 1 determines BWLT and ISF for all boundary nodes and all years, based on values of PTR and ARR given by layers 2 and 3. Layer 2 determines PTR based on values of ARR given by layer 3 (i.e. tax to provide revenue for the recharge given by layer 3), and layer 3 optimizes the overall regional problem by manipulating the artificial recharge rate.

9.3.3 CONTROL OF INPUTS

So far we have been describing environmental intervention interaction in terms of, firstly, the prediction of output and, secondly, the manipulation of storages. The third type of intervention involves the control over inputs which is an extension of storage manipulation, but a much more difficult operation, particularly so the larger the system. An example of attempts to control the precipitation input into surface hydrological systems is by cloud seeding of cumulus cells (Simpson and Dennis 1974). Cumulus clouds produce more than three-quarters of the total global precipitation and form the energy bases for hurricanes, thunderstorms, tornadoes and squall lines; they are also important in driving the global circulation, and they act as regulators for incoming and outgoing terrestrial radiation. Cumulus cells in the trade winds pick up moisture of which 20% is released by rainfall to drive the trade wind systems and the rest is carried equator-ward to the equatorial trough where some 1500–5000 giant cumulus towers raise and convert the heat energy at a rate of nearly 1 million times the world power consumption. It is therefore clear that the modification of cumulus clouds, even on a comparatively small scale, is potentially very important, particularly as most cumulus clouds evaporate without producing rain which reaches the earth's surface. Much moisture is lost in cumulus cells by evaporation and by particles streaming downwind from the anvil top. Therefore, small thunderstorms in the central United States have a precipitation efficiency of only about 10%, and even the large ones do not exceed 50%. Fig. 9.21 shows a systems diagram of a cumulus cell in which the entrainment effects are included whereby dryer and colder air is drawn into the cloud laterally into the upshear side and detraining occurs on the downshear side by the shedding of heat and moisture, so diluting the cell and cutting down its latent heat release and buoyancy. These effects therefore tend to counteract to some extent the energy input of warm, moist air of the basal inflow, and the introduction of dry ice or other chemical materials into the upper cell or cell top will tend under certain circumstances to increase the amount of precipitation, as distinct from the output of streaming ice. On a global scale such an intervention would be looked upon as the moving of moisture from one storage to another and here we return to the question of scale. The remainder of this chapter deals with the interfacing of physico-ecological and socio-economic systems which is so intimate and on such a large scale as to constitute environmental symbiosis.

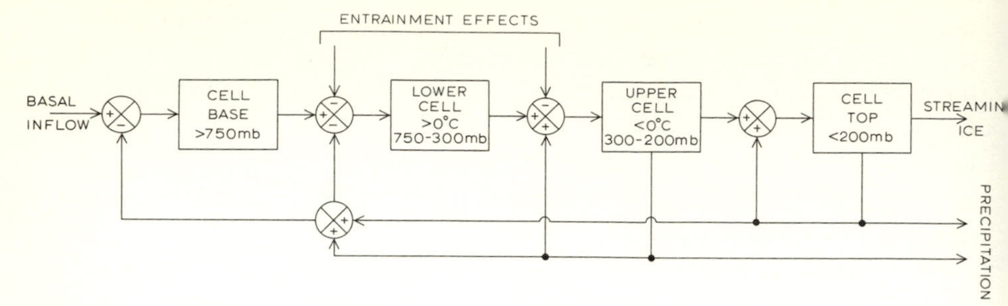

Fig. 9.21 A systems diagram of a cumulus cell.

9.4 Environmental symbiosis

The control of man-environment systems by symbiosis is motivated by the inability of intervention strategies to deal with problems of great time and space extent. The interest in symbiotic control shown by systems analysts has increased rapidly in recent years as it has become evident that many of the most pressing problems of physico-ecological and socio-economic systems interfacing concern large-scale systems. The extent of man's intervention in smaller-scale and lower-order systems and the increasing scale of use of resources (both renewable and non-renewable) require that we can no longer consider any individual control action or system response in isolation or at group or tribal scales. Instead it is necessary to consider the *interdependence* of each man-environment relationship. This has been referred to by some economists as the explosion of externalities. On a philosophical plane, interdependence requires that we must take a holistic view (Mesarovic and Pestel 1975) which Nietzsche sees as both the creation and subjugation of man. This can be viewed on two levels. The first is the consequence of the complexity of the system, such that it is *counter-intuitive* (Meadows *et al.* 1971) due to the multiplicity of space-time feedback loops which characterize large scale man-environment problems. Counter-intuitive behaviour has the effect that no intervention or action can be expected to have only one consequence confined to the subsystem or element at which action is directed. The effect is that complex systems behave as wholes which Lange (1965, p. 1) suggests 'possess attributes distinct from those of their constituent elements and they also have their own modes of action which cannot be derived only from those elements'. The property of wholes derives from the attempt to attain system closure, and hence partly from the non-structuralist view of systems (as discussed in chapter 5). It also derives from the properties of closed loops of negative- and positive-feedback, both natural and induced, which serve to make the properties of system operations difficult to observe (see chapter 3.5).

The result of this 'counter-intuitive' behaviour is that the internal mechanisms of environmental systems appear complex and cannot be reproduced by a simple model which allows individual stimulus-response situations to be followed. Instead a synthetic mathematical structure must often be utilized, as a consequence of which the outputs of the models often become unpredictable. The counter-intuitive effects are produced by the complex delays, memories, reaction and relaxation times which characterize environmental feedback loops and which must be considered as a matrix of simultaneous relationships.

At a second level, interdependence is the consequence of hierarchical organization (Mesarovic and Pestel, 1975). Whatever the origin of the progressive interdependence–symbiosis of the world system. Mesarovic and Pestel suggest that the net result is disorder which results in less understanding and increased counter-intuitive behaviour. This in turn has been taken as a sign that the system is in crisis, namely that various strata cannot cope independently with changes in their environment and we must produce models in which every variable on any level is dependent on 'everything else' (Mesarovic and Pestel 1975, p. 53). Similar conclusions have been drawn for the operation of spatial interdependence in geographical systems (Gould 1970, Tobler 1970), and it may be that concern with interdependence represents an awakening to the difficulties associated with spatial systems and their interaction by systems analysis. Alternatively, interdependence has been represented as a product more of the human response network and its adaptive evolution (Sahlins and Service 1960, Harris 1969). Simon (1969) has characterized this situation as due to the increasing artifice of the man-environment such that we come to live increasingly in a 'man-world' dominated by human needs and human cognition (see chapter 5) in which almost every element of the environment shows evidence of man's artifice, the most significant part of the environment being composed of strings of artifacts called 'symbols' that we receive through our senses as language (Simon 1969, p. 3). In response to this growth of the man-dominated environment, Simon proposes a 'science of the artificial', recognizing that artificial systems differ from natural environmental systems in four characteristics:

1 They are *purposive*, in that they are a synthesis by man designed in order to attain a goal or in order to function (although not necessarily with forethought);
2 They are *interfaces*, in that they imitate the natural but lack its reality at the junction between the inner cognitive environment and the outer real environment;
3 They are *adaptive*, in so far as they are characterized in terms of function and goals so that behaviour can be predicted from goals as homeostatic regulation;
4 They are the result of *design synthesis* (i.e. of imperatives rather than descriptions), concerned with how a rationally-designed system should behave under outer environment conditions.

Despite these differences, the argument here is that man is not apart from nature, but that he adjusts natural environmental behaviour within constraints of natural laws to his own goals and purposes. The interfacing of the man and environment systems necessitates a kind of system shunting through a group organizational set (as shown in fig. 9.22) as an interface of man's

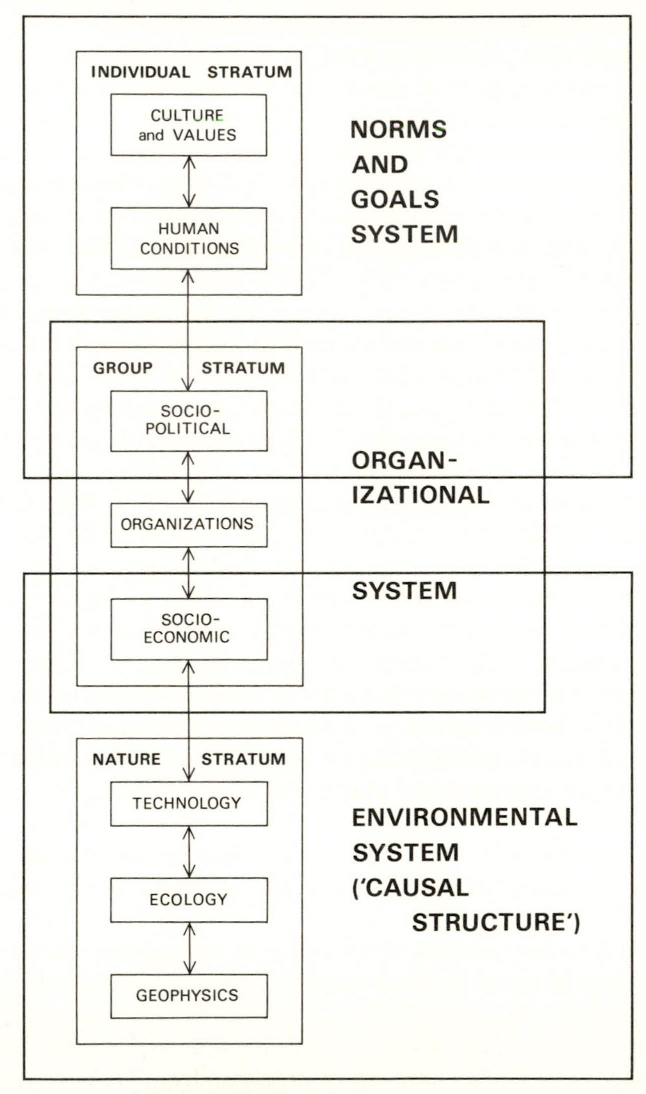

Fig. 9.22 The interfacing of man's cognitive norm and value system with a system of natural environmental processes by means of a group organizational system. (Partly after Mesarovic and Pestel 1975.)

cognitive norm and value system with the physical processes of nature. Hence, symbiosis requires a 'total systems' or 'holistic' approach in which adjustment is extreme in time and spatial extent. Just as in homeostasis, inner systems are insulated from their environment so that invariant stationary relations can be maintained between the inner system and goals independent of variations in the outer environment (Simon 1969, p. 9), so in symbiosis it has been argued that man sets his own goals which the outer environmental system should achieve. Thus the model is reversed and becomes normative in that 'ought' is implied in goal settings (von Mises 1951) although man is still part of the environment, and the environment must still obey natural laws.

Another view is that interdependence is not an end point of social evolution but is inevitably present. There is a Hegelian conflict, which as discussed in chapter 1, leads to social organization in which there appear contradictions which prevent society remaining in a changeless state; these contradictions induce societal changes which lead to readjustments, but these very changes open the way to fresh contradictions which induce further changes and so on as an ongoing socio-environmental evolution oriented towards maintaining man as a species and to improving his life style and social pattern (see Lange 1965, p. 1). The dialectical materialism of this process implies no necessary development sequence but a continuous evolution to maintain and stabilize symbiotic régimes.

Interdependence has also been interpreted as a major challenge to traditional linear mathematics. Using examples such as those of the population dynamics of May and Kolata discussed in chapter 7, Waddington (1977) argues that most mathematical approaches have been more applicable to static equilibria, and hence have influenced interventionist and symbiotic policies to maintain such equilibria. Drawing on adaptive and evolutionary concepts such as those of dialectical materialism, Waddington suggests that man and his mathematics should turn away from creating homeostasis (static equilibria) towards the preservation of system flows, to keep systems altering in the same way as they have altered in the past. This concept of 'evolutionary stability' he terms *homeorhesis*. The concept includes topological concepts drawing from global analysis. Stable growth paths are canalized pathways (or chreods) across generalized surfaces of system development. Using Dickens' words of Mr Micawber, 'annual income £20, annual expenditure £19. 19. 6, result happiness. Annual income £20, annual expenditure £20. 0. 6, result misery', Waddington (1977, chapter 7) demonstrates that the various goals of wealth equality detailed in fig. 8.19 can be extended to encompass pathways of dynamic stability and instability. This approach to man-environment symbiosis helps us to get closer to the structure of interdependence and away from views of environmental complexity which render system behaviour 'counter-intuitive'.

Whichever view of man-environment interdependence one accepts, it is becoming increasingly apparent that in symbiosis man is playing a larger role in setting purposive goals which he wishes natural systems to achieve. But although this presents the possibility of greatly increased human welfare, it also has the added dangers that, through lack of sufficient understanding, man might force the environmental systems away from their stable modes and pathways of behaviour. By presenting new initial conditions and system constraints to the system state space, environmental systems may be pushed across some singularity or threshold in their régime of behaviour and perhaps into a mode of collapse.

It has been argued that the major factor which makes a collapse mode of behaviour possible, and indeed likely, in man-environment symbiosis is the continued growth in the size of the world's population. This creates the following problem for man-environment systems. The present rate of world population growth is exponential, whilst the size of world resources is finite and limited. Exponential growth describes a 'geometrical' progression (e.g. 1, 2, 4, 8, 16, 32, 64, . . .) which is characterized by a doubling time. Since the doubling time of the world population at present is about 35–40 years, world population in 2015 might be expected to be twice the size that it is in 1980. The effect of exponential growth is that early changes have little effect on overall population numbers, but as time progresses, the absolute size of the population increases very rapidly, a phenomenon exhibited in fig. 9.23. Growth in world population is associated with increased demand for food and raw materials to cater for the increased quantitative needs. Additional

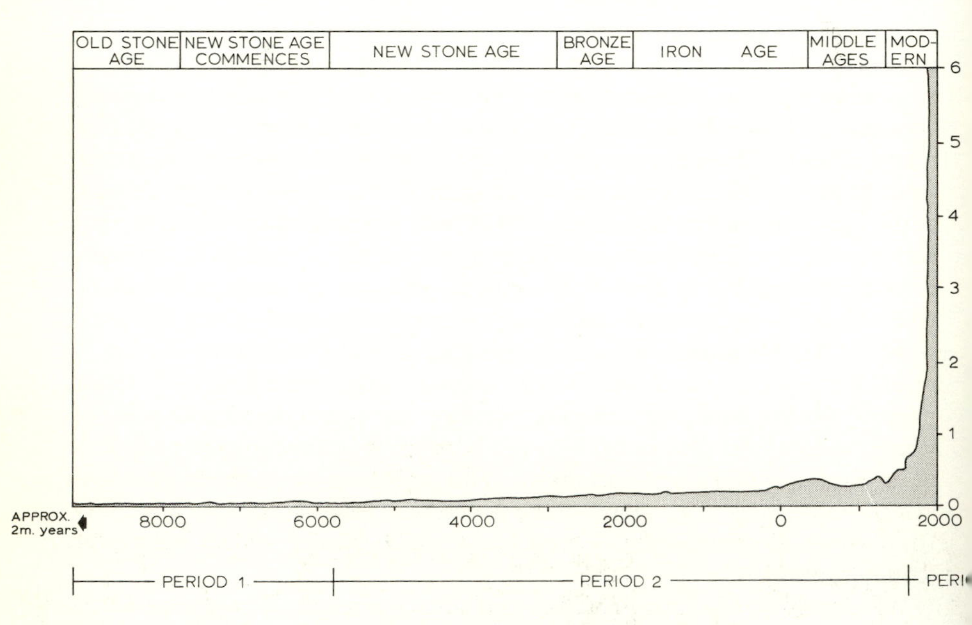

Fig. 9.23 The growth of world population. (After Day 1974.)

qualitative changes in demand for resources are also taking place, however, which produce further quantitative growth in demand for food and resources. These qualitative changes are associated with development in Third World countries and higher levels of standard of living and consumption in the developed world. The general path of qualitative development for the ten major regions of the world economy is shown in fig. 9.24. As the qualitative standard of living (as measured by the gross output, GRP) rises in each region, two paths of quantitative change occur. The first, shown by the upper path in the figure, is characteristic of centrally planned economies with early peaks in energy consumption associated with emphasis on heavy industry. The growth path in fig. 9.24 for the rest of the world (i.e. capitalist and other economies) results in a steady increase in energy usage with rises in GRP to a steady state.

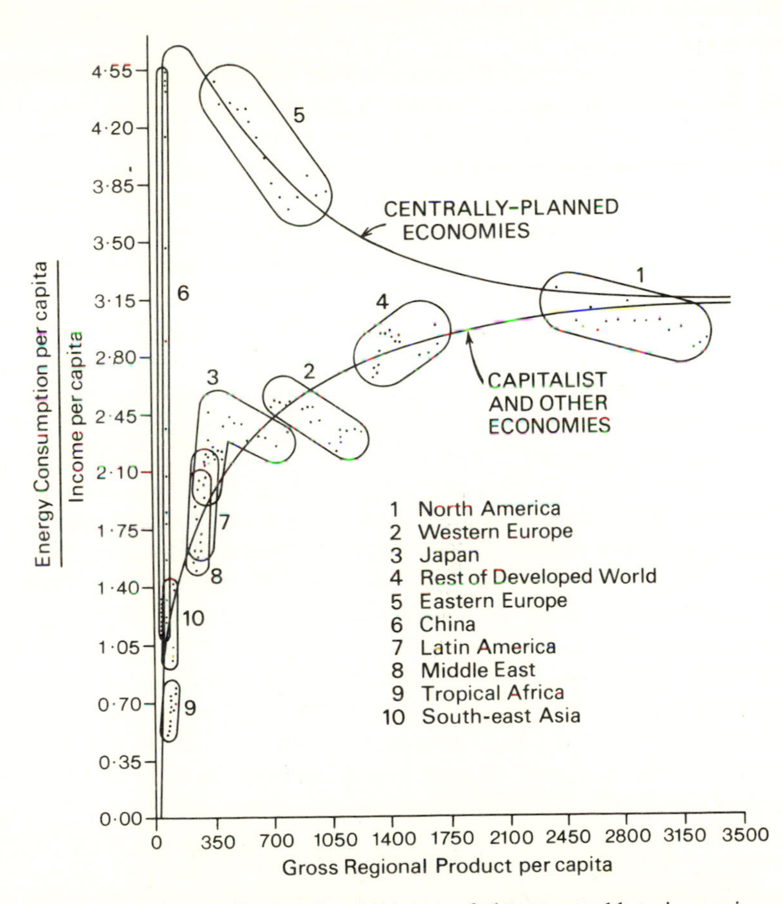

Fig. 9.24 The qualitative development of the ten world major regions, expressed as an energy consumption/income ratio, as a function of gross regional product per capita (the regions are defined in fig. 9.27). (After Mesarovic and Pestel 1975.)

In addition to increasing demand for food and resources, quantitative and qualitative changes in human demands also generate increased interference with the environment, especially through atmospheric, land and water pollution. As we have seen in section 9.2, some resources are non-renewable, at least on the human time scale (for example, coal and oil and minerals). Other resources are renewable, but are being used at a rate faster than they are being replenished. Hence the quantitative level of world resources is finite, either absolutely or because the time lags in the human system are much shorter than those in physico-ecological systems. Without any change in the rate of use of resources *per capita* and without new technology permitting more efficient use of resources, exponential population growth and associated quantitative and qualitative shifts in demand will lead to the exhaustion of world resources and to the collapse of the man-environment symbiosis as we now know it.

The structure of the environmental problem can be simulated by a number of the systems techniques developed in chapters 2, 3 and 4. The major development of large-scale systems models which attempt to describe the nature of man-environment symbiosis in regional and world systems is to be found in work by Forrester (1971), Meadows *et al.* (1971) and Mesarovic and Pestel (1975), and in refinements by Cole *et al.* (1974). Forrester (1971) divides the world system into five major state variables or levels: non-agricultural investment, population, natural resources, pollution, and agricultural investment. There are also a large number of subsidiary level variables. Each level variable is defined as one dependent variable in a vector of system state variables which permit the time evolution of the system to be calculated. The system state matrix is composed of a number of parameters identical in form to the state space structure of equation (2.2.13) (see p. 63), which Forrester terms the *rates*. Some of the rates (i.e. parameters) are nonlinear and are specified in look-up tables (Forrester 1971). They are also allowed to evolve in time as a non-stationary (adaptive) parameter set and are governed by a second set of *rate equations*. A simplified form of Forrester's (1971) model is given in table 9.1, together with notation. This simplified definition of the major causal equations allows a state space representation of the five main level (state) equations, using a reduced form given by Cuypers and Rademaker (1974). The approach of these two authors is to give a linear approximation to Forrester's nonlinear equations using Taylor series expansion so that the final Forrester model becomes identical in structure to equation (2.2.13), i.e.

$$\mathbf{T\dot{Z}} = \mathbf{AZ} + \mathbf{B}$$

in which $\mathbf{Z}$ is the system level (state) variable vector with time derivative $\mathbf{\dot{Z}}$, $\mathbf{A}$ is the state transition matrix, $\mathbf{B}$ is a matrix containing all other inputs (apart from the five level variables) and $\mathbf{T}$ is a vector of time constants. Cuypers and Rademaker (1974) give this equation in full as:

$$\begin{bmatrix} T_1\dot{C}(t) \\ T_2\dot{N}(t) \\ T_3\dot{P}(t) \\ T_4\dot{POL}(t) \\ T_5\dot{A}(t) \end{bmatrix} = \begin{bmatrix} 1 & 0{\cdot}0125 & 0{\cdot}824 & 0 & -14{\cdot}7 \\ -228 & -1 & -49{\cdot}5 & 0 & 1170 \\ 0{\cdot}421 & 0{\cdot}0006 & -1 & -0{\cdot}023 & 2{\cdot}95 \\ 2{\cdot}88 & 0 & -1{\cdot}44 & -1 & 0 \\ -0{\cdot}0185 & 0{\cdot}00008 & 0{\cdot}0926 & 0{\cdot}0009 & -1 \end{bmatrix} \begin{bmatrix} C(t) \\ N(t) \\ P(t) \\ POL(t) \\ A(t) \end{bmatrix} + \begin{bmatrix} 6{\cdot}6 \times 10^9 \\ -10^{12} \\ 0{\cdot}43 \times 10^9 \\ -0{\cdot}95 \times 10^9 \\ 0 \end{bmatrix}$$

in which all state variables are defined in terms of deviations from their 1970 value and **B** is a vector of constants derived from input disturbances prior to 1970. After suitable manipulation, Cuypers and Rademaker show that this system gives a two-level hierarchical ordering (fig. 9.25) in which the capital and natural resource levels at stage 1 define the agricultural, population and pollution levels at stage 2. Using a nonlinear version of this model, Forrester (1971) and Meadows *et al.* (1971) have been able to show that continued exponential growth and finite limits lead to three modes of collapse of the world system which limit the steady progress of world economic and population expansion (fig. 9.26). First, the limits of natural resources are reached which leads to a *resource collapse* (fig. 9.26A). If substitution between various resources is possible and implemented, and if greater stocks of resources become available, this mode of collapse may be delayed sufficiently long for a second mode of collapse to occur. This is *pollution collapse* (fig. 9.26B) in which human and other animate life is choked, drowned or extinguished by the increasing levels of air, land and water pollution derived from continued industrial development, use of agricultural fertilizers and pesticides, and population consumption of space. If this form of collapse is averted – for example, by antipollution legislation derived by pressure from the sociopolitical system in fig. 6.28 – then a third form of *population collapse* (fig. 9.26C) derives from the world population levels finally meeting the limit of available food supply (Forrester 1971). Hence, the model generates very much of a Malthusian view of the future of the world system which is contrary both to

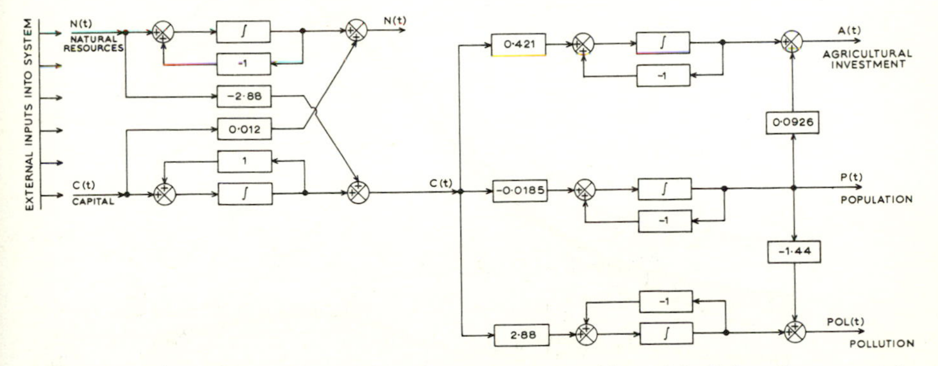

Fig. 9.25 Hierarchical order within the Forrester world model. (After Cuypers and Rademaker 1974.)

A
WORLD MODEL STANDARD RUN

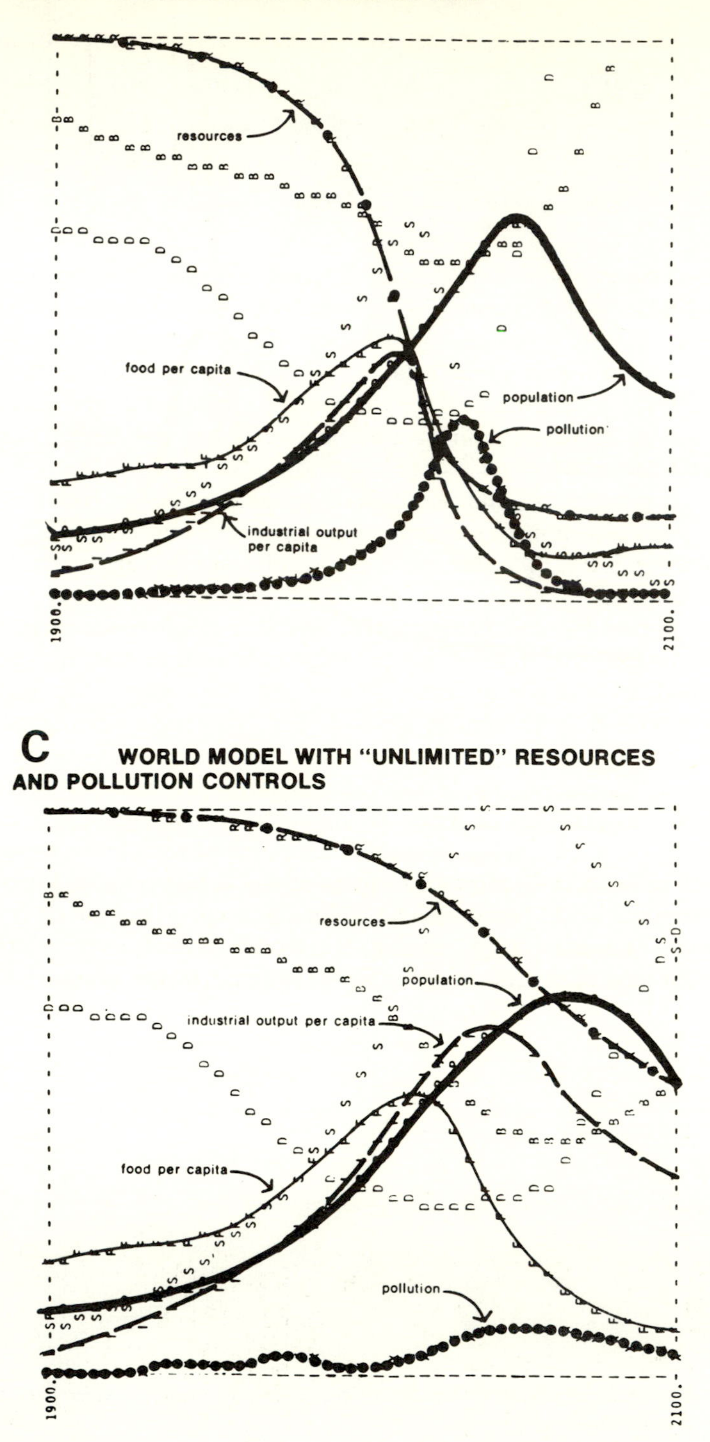
resources
food per capita
population
pollution
industrial output
per capita
1900.
2100.
C
WORLD MODEL WITH "UNLIMITED" RESOURCES
AND POLLUTION CONTROLS
resources
population
industrial output per capita
food per capita
pollution
1900.
2100.

B **WORLD MODEL WITH NATURAL RESOURCE RESERVES DOUBLED**

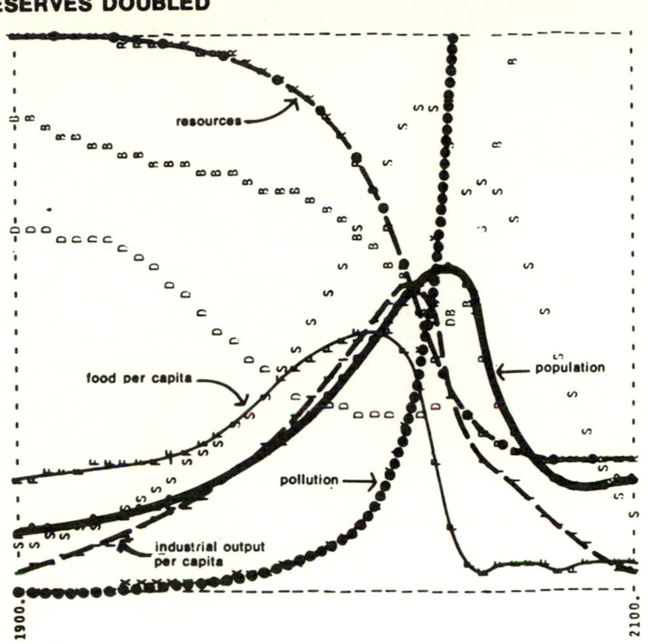

▬▬▬	population (total number of persons)
— — —	industrial output per capita (dollar equivalent per person per year)
———	food per capita (kilogram-grain equivalent per person per year)
●●●●●●●	pollution (multiple of 1970 level)
—●—●—	nonrenewable resources (fraction of 1900 reserves remaining)
B	crude birth rate (births per 1000 persons per year)
D	crude death rate (deaths per 1000 persons per year)
S	services per capita (dollar equivalent per person per year)

Fig. 9.26 Collapse modes in the Forrester-Meadows world system.
A Resources collapse.
B Pollution collapse.
C Population collapse.
(From Meadows *et al.* 1971.)

the classical economic argument, that substitution is always possible and consumption will be regulated by price, and to the strict Marxist view that interprets scarcity to be socially engineered by monopoly (see chapter 8).

Table 9.1

LEVEL EQUATIONS

Capital investment (non-agricultural): $C(t)$

$$\frac{dC(t)}{dt} = g_1 C(t) + g_2 N(t) + g_3 POL(t) - g_4 A(t)$$

Population: $P(t)$

$$\frac{dP(t)}{dt} = [b(t) - d(t)]P(t)$$

Natural resources: $N(t)$

$$\frac{dN(t)}{dt} = h_1\, A(t) - h_2\, C(t) - N(t) - h_3 POL(t)$$

Pollution: $POL(t)$

$$\frac{d\,POL(t)}{dt} = [m_1 C(t) + m_2 A(t) - \delta]m(t)(1 - \text{absorption constant})$$

$$\delta = \text{delay}$$

Agricultural inputs: $A(t)$

$$\frac{dA(t)}{dt} = K_1 N(t) + K_2 P(t) + K_3 POL(t) - A(t) - K_4 C(t)$$

Non-agricultural output: $Q(t)$

$$Q(t) = aN(t)C(t)/P(t) \quad \text{(Harrod-Domar production function)}$$

Food: $F(t)$

$$F(t) = C_1\,\text{land} + C_2 A(t - 1)$$

Quality of life: $q(t)$

$$q(t) = a_1 POL(t) + a_2 \frac{P(t)}{\text{land}} + a_3 \frac{F(t)}{P(t)} + a_4 C(t)$$

RATE EQUATIONS

Birth rate: $b(t)$

$$b(t) = f(\text{population density}, q_t, POL(t), F(t))$$

Death rate: $d(t)$

$$d(t) = f(\text{population density}, q_t, POL(t), F(t))$$

Pollution generation rate: $m(t)$

$$m(t) = f(\text{lifetime of pollutants, ecosystem cycling speed})$$

Later versions of the Forrester and Meadows models by Mesarovic and Pestel (1975) explicitly recognize the importance of the space-time character of the world system and employ space-time simulation, such as was discussed in chapter 4. In the Mesarovic and Pestel model each region evolves at a different speed, resource depletion moves at different rates, and catastrophes, if they occur, are more localized. The ten world-regional divisions adopted by Mesarovic and Pestel are shown in fig. 9.27. A simpler fivefold division of the world has been developed by Connelly and Perlman (1975) in their study of resource depletion. This classification has the advantage, even though it is arbitrary, of concentrating attention on the political elements of importance in interspatial dependence and is based on:

Mesarovic and Pestel regions		*Connelly and Perlman regions*
8, 7	1	Resource-rich developing countries – e.g. Libya and Middle East
1, 2, 3	2	Resource-importing developed countries – e.g. United Kingdom, Western Europe, USA
4	3	Non-communist independents–e.g. Canada, Australia, New Zealand
5, 6	4	Communist countries–e.g. USSR, China, Eastern Europe
9, 10	5	Resource-poor developing countries – e.g. Malawi, India

Although this division includes many countries which fall into two or more classes, it focuses attention on the power of separate groupings. Fig. 9.28 shows the net production and consumption of major non-renewable resources for a range of countries. Although a number of countries are in a position of balanced development, many countries import almost all their raw materials, whereas others export almost all their raw material production. This facet of spatial interdependence is captured in the Mesarovic and Pestel model which also incorporates a two-level regional micro-economic and macro-economic submodel for each spatial region. Combining the conclusions for future evaluation with those of Connelly and Perlman (1975), one finds that the resource-rich developing countries (lower right-hand side of fig. 9.28), especially those in the Middle East with low populations, come into an increasingly strong position to manipulate and control the speed, pattern and location of growth and development, even though their internal *per capita* investment cannot absorb the financial resources available to them. In the resource-importing countries (upper left-hand side of fig. 9.28) high levels of industrial output are achieved at ever-increasing cost, inflation, periodic shortages, and increasingly intolerable levels of pollution and environmental degradation as investment becomes channelled from the resource-rich countries. In the non-communist independents, which are self-sufficient in terms of resources, adaptation to progressive scarcity is fairly smooth since most political processes are 'internalized'. In the communist countries it is thought that industrial development can continue by exchanging investment from the Western world in return for mineral commodities over which there is a large degree of monopolistic control. The resource-poor developing countries with little political bargaining power will be subject to rising costs of both raw materials and manufactured goods which will lead to continuing feedback of real income from resource-poor to resource-rich developing countries.

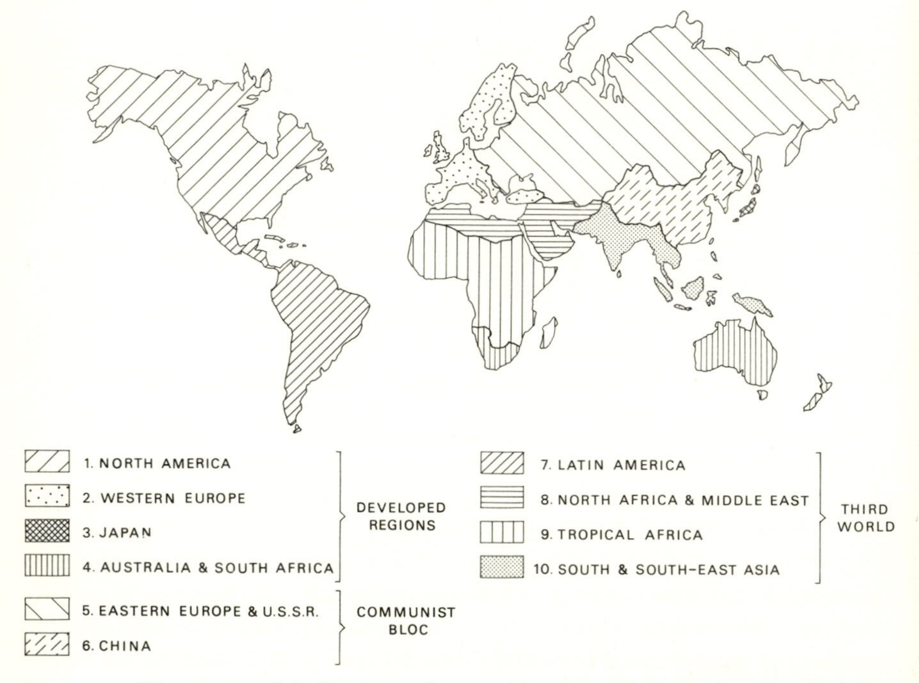

Fig. 9.27 The ten spatial divisions of a model of world dynamics. (Partly after Mesarovic and Pestel 1975.)

In support of the Forrester and Meadows arguments, together with redistribution effects in the spatial model, a considerable degree of evidence can be produced on the rising levels of pollution and resource depletion at world level. Considering water pollution, for example, fig. 9.29 shows the extent of polluted rivers and stretches of open water in 1975. The areas severely affected cover all coastal regions with high concentrations of population and most major waterways. The position is also serious for overall world water resources. Fig. 9.30 itemizes the overall balance of world runoff and use of water for domestic purposes, industry and agriculture for the

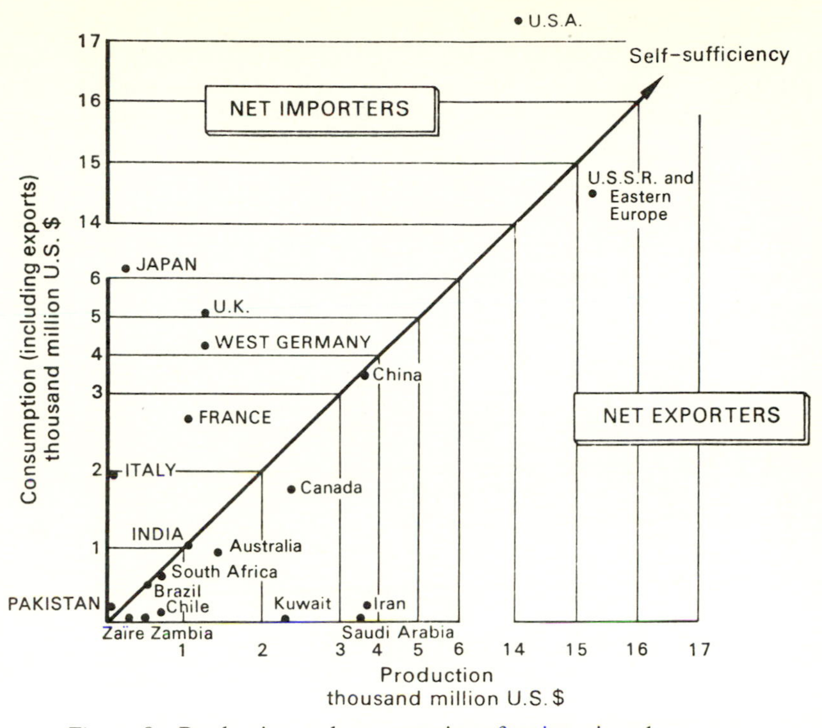

Fig. 9.28 Production and consumption of major mineral resources in 1971. (Petroleum, coal, iron and steel, aluminium and copper.) (After Commodity Research Unit estimates, from Connelly and Perlman 1975.)

present and for the year 2015, when a doubling of 1980 population levels is expected. The resulting increases are in most cases two or more orders of magnitude greater than present demands. The spatial distribution of this potential demand in relation to the supply is shown in fig. 9.31. By the year 2000 most of Asia and the Middle East will be using over 60% of their annual resources, and in some cases this will be as high as 80–90% (Falkenmark and Lindh 1974).

There have been two main reactions to forecasts of collapse of the man-environment symbiosis on a world scale – pessimistic and optimistic. The pessimists take an evolutionary view of man-environment symbiosis. Forrester (1971, p. 12) believes that:

> We are now living in a golden age when, in spite of a widely acknowledged feeling of malaise, the quality of life is, on the average, higher than ever before in history and higher now than the future offers.

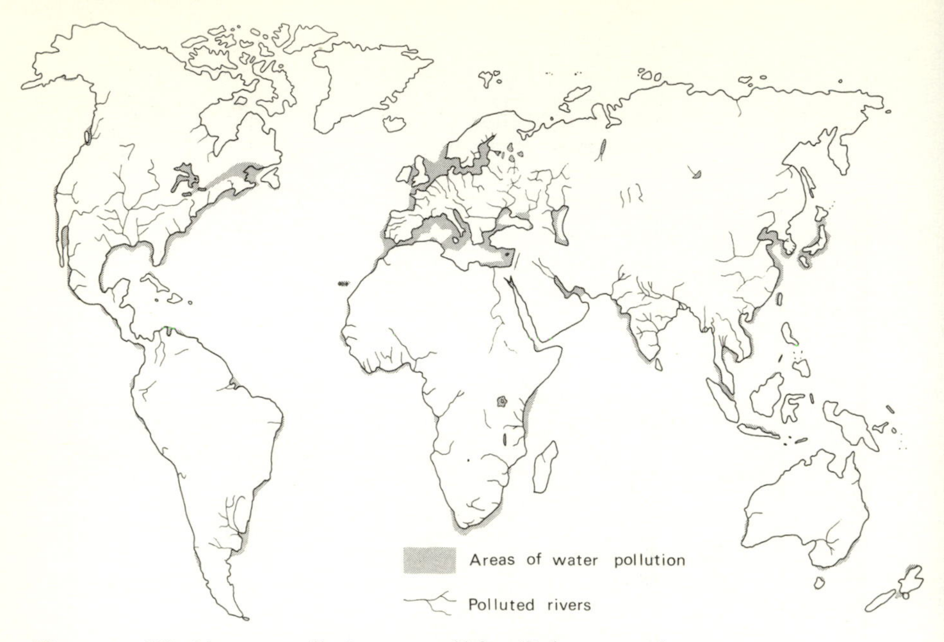

Fig. 9.29 World water pollution 1975. (After Dybern 1974.)

Meadows *et al.* (1971, pp. 23, 145) are even more dogmatic:

> The limits to growth on this planet will be reached sometime within the next one hundred years . . . even the most optimistic estimates of the benefits of technology in the model . . . did not in any case postpone the collapse beyond the year 2100.

As a result, Forrester (1971, p. 120) suggests that the only policy which can be followed to avert collapse is to reduce the usage rate of all natural resources by 75%, to reduce pollution generation by 50%, to reduce gross investment by 40%, to reduce food production by 20% and to reduce birth rates by 30%, all on a world basis from present levels. Only by taking such extreme actions can a global equilibrium (i.e. a true symbiosis) be re-established. Any lesser actions merely and only slightly delay the final day (Sauvy 1975, p. 255):

> Henceforth, man is supposed to wonder, when he breathes, if the gas expelled from his lungs will endanger his environment; when he eats, if it will be possible to restore to the soil the minerals extracted from it; and when he consumes something or other for his momentary pleasure, if he is putting tomorrow's life at risk.

Moreover, because of the massive delay and lead times 'due to the extended dynamics of the world system and the magnitude of current and future change, such actions have to be anticipatory so that adequate remedies can be

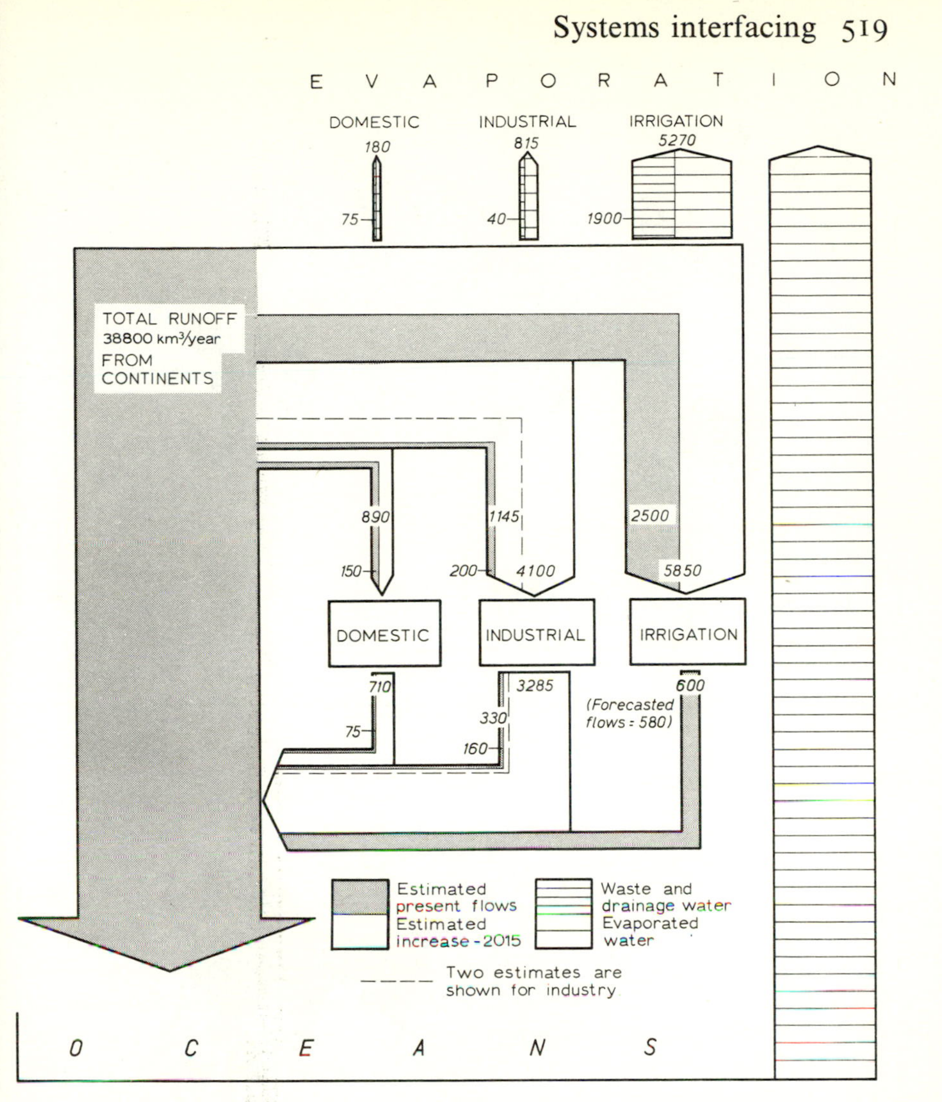

Fig. 9.30 The world hydrological cycle and water-use balance. (After Falkenmark and Lindh 1974.)

operational before the crises evolve to their full scope and force' (Mesarovic and Pestel 1975, p. 31). A second form of pessimism is the belief that the price mechanism link of consumption to scarcity will limit demand for resources, but that the strength of the feedback is too weak and too slow to cut back consumption quickly enough to avoid collapse (Connelly and Perlman 1975, p. 15). An alternative view is that, although consumption cutbacks might be successful, the rate of consumption of resources (especially energy) in extracting low grade ores or in recycling, as resources become limited will act as

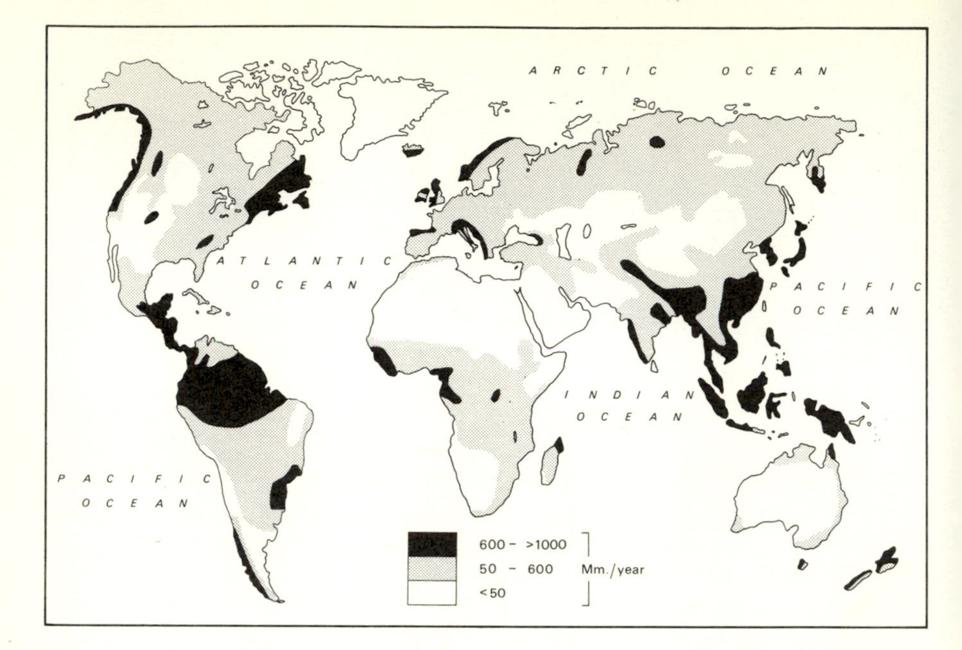

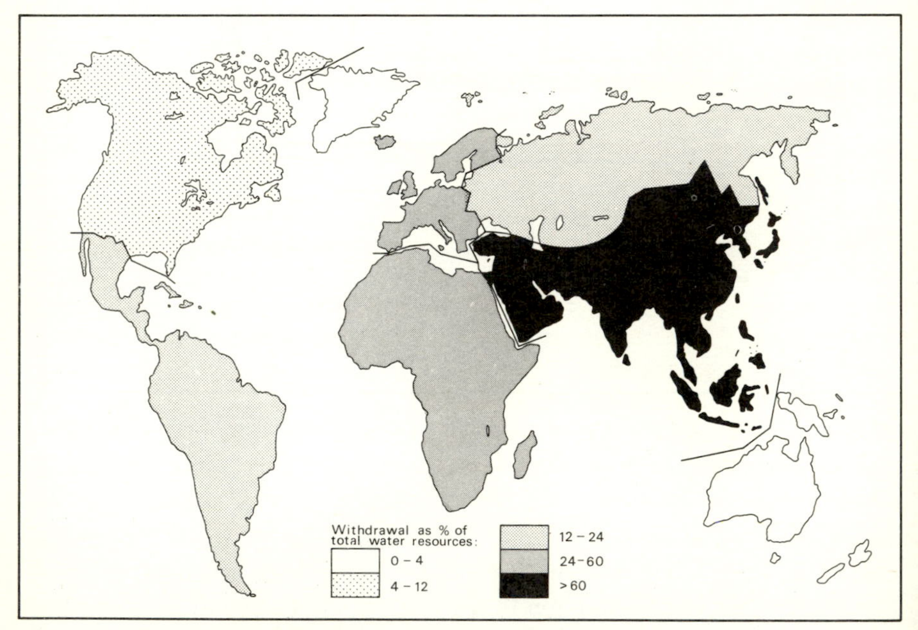

Fig. 9.31 World water resources.
Above Distribution of water resources (river runoff in mm/year).
Below Estimated water withdrawals by the year 2000 (as percentage of available water resources).
(After Falkenmark and Lindh 1974.)

a further feedback to reduce consumption possibilities still further. This is a Ricardian view of rising costs of resource use (Connelly and Perlman 1975). It is also suggested that analysis based on classical economic equilibrium solutions using the price and substitution mechanisms will be largely irrelevant for solving the resulting decision problems when collapse is imminent (Kornai 1971, Adams 1974) since such decisions are essentially political and not economic. Cost-based internalization of externalities as suggested by Buchanan and Tullock (1969) provides the possibility of determining democratic optima but yields only theoretical and not practical solutions. The resulting pressure to place the world system under control yields a more subtle species of pessimism that states that, even if we can avert collapse, this will be at the expense of sufficient loss of freedom that life in the resulting 'global village' will not be worth living (Adams 1972). Whilst sanguine about environmental problems *per se*, Habermas (1973, pp. 2, 44) is pessimistic about the ability of humanity to avoid its own annihilation through conflicts in value systems and resource ownership, perhaps resulting in thermonuclear war. The sociological and critical schools see the environmental crisis as not mainly the result of the effects of nature but as inherent in the incommensurables within humanity itself. In a different context, Foley (1976) notes that the theoretical magnitude or potential of resources is irrelevant if there is unequal spatial distribution (with, for example, 60% of world coal in the USSR, and excessive oil in the Middle East).

Other analysts have suggested that the continued growth of population and industry will upset the climatic balance before resource limits are reached. Pollution is especially important to this argument and it has been calculated that by 2015, when present world population is expected to have doubled, the rise in carbon dioxide would cause a rise in mean global surface temperature of about 0·9 K (Schneider 1974), in the absence of changes in any other factors. Such changes in carbon dioxide levels could trigger positive or negative-feedback and the precise expectation is not known. Similar effects may also result from ozone depletion, increases in sulphur dioxide, dust and other pollutants. The overall world climatic feedback system is shown in fig. 9.32. The feedback of this system structure suggests that when increased CO_2 levels are combined with the presence of other pollutants, when increases in the dust content of the atmosphere occur, when changes in the surface characteristics of the earth by increased agriculture or built-up areas give new albedo values, and when there is a sufficient addition of waste heat to the earth-atmosphere system through increased energy consumption, then it is very likely that significant climatic shifts will occur. It is not known whether these shifts will damp out to give new stable climatic regimes, or whether even slight changes can trigger climatic instability and irreversibility. The pessimists can, however, point to a number of observations of climatic modification (Bryson 1974, Winstanley 1973, Lamb 1966) which tend to support their case.

Against these views, the optimists adopt the belief that man has always

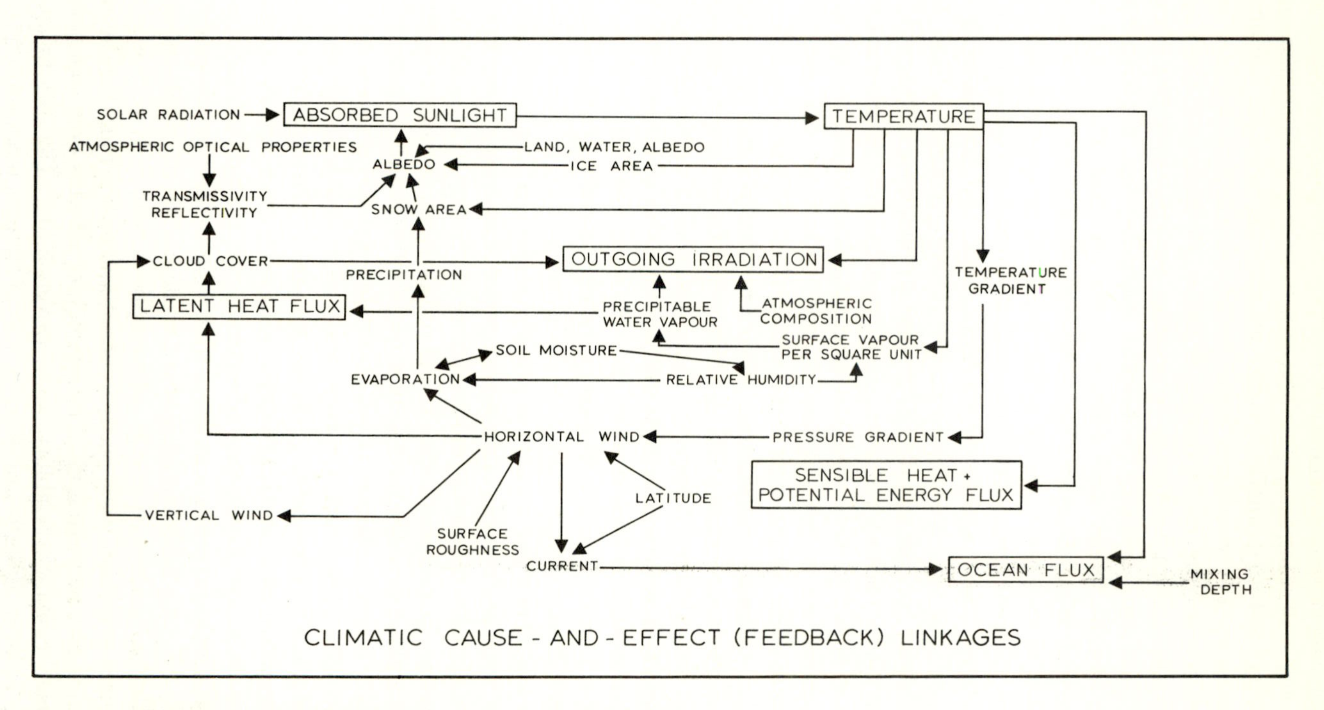

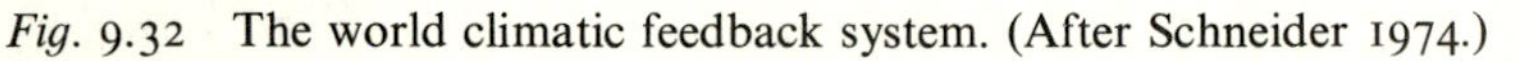

Fig. 9.32 The world climatic feedback system. (After Schneider 1974.)

found a technological solution to resource problems that have been posed in the past, and, indeed, technological development negated Malthus's forecasts. These beliefs rest on criticisms of the model simulation and methodology especially as applied by Forrester and Meadows, but also on the economic principle of the market. Pressures of pollution, food and raw material shortage may well generate socio-economic responses to regulate resource depletion and environmental degradation either by state control or by the price mechanism. Hence, there will always be enough minerals to go round, but at a high price. These economic arguments are adopted by classical economists (Lipsey 1973, Beckerman 1971, 1973) in appraisals of the Forrester-Meadows approach (Connelly and Perlman 1975, Cole *et al.* 1974), and in studies of economic development within an area (Alonso 1971, Richardson 1973). Much of the optimism is grounded in the treatment of the externalities of any human decision. Contrary to the pessimists' view of externalities, the optimists believe that the externalities can be 'internalized'. In this way pollution, radiation, environmental degradation and the reduced resource totals can be given values which allow then to sit alongside other factor costs in a classical equilibrium model or cost-benefit solution. The problem of valuation of such externalities can usually be readily solved, it is argued (e.g. by appropriate tax, subsidy, compensation or definition of legal property rights) (Coddington 1974). The problem with this form of optimism is that the action of the price mechanism results in a large proportion of capital and labour being devoted to obtaining raw materials with resulting lower standard of living, and our lack of ability to get the various financial and legal balances right or to act quickly enough.

Further grounds for optimism are supported by arguments for conservation, recycling of water and obsolete commodities, increasing the lifetime of durable goods, and substituting renewable for non-renewable resources (e.g. solar energy, tides and hydroelectric power for fossil fuels). In addition, it is suggested that problems associated with the disposal of nuclear waste, which takes many thousands of years to decay, can be overcome by setting them in blocks of glass, and then disposing of them into geological formations by drilling holes into stable areas of the crust (Royal Commission on Environmental Pollution 1976). Whilst each of these suggestions is useful, each requires greater capital expenditure and, unless implemented on a large scale, is unlikely to do more than delay potential collapse. To implement these proposals on a large scale would require a reorientation of social systems akin to the pessimists' view.

A further response to the environmental collapse issue which can be espoused by present generations is to ignore the problem, even if it does exist (Beckerman 1973):

> Why on earth should anybody make any sacrifices to keep the human race going any longer than necessary? Extending the duration of the human

> race merely means that a few billions more people will be born in order to die as well. Why is that such a big deal?

This illustrates the problem of actions which are politically difficult today but which are required to safeguard tomorrow, in as much as the perceived time horizons of a single person are normally relatively short and his discounting of the future very rapid. An equally prodigal view (Connelly and Perlman 1975, p. 136) adopts short-term arguments to answer long-term problems. The crucial variable in these arguments is the rate of technological advance which, it is suggested, can have profound repercussions on resource use – i.e. demand creates the technology to provide cheap resources (Barnett and Morse 1963) and resources used are never lost since they can be recycled.

Another, more complex form of optimism is the dialectical materialist approach applying Marxist interpretation to the resource problem. Recognizing that resources can be defined only with respect to a particular technology, culture or stage in development, this view states scarcity to be 'created by social activity and managed by social organization . . . [and is] necessary to the survival of the capitalist mode of production, and it has to be carefully managed, otherwise the self-regulating aspect of the price mechanism will break down' (Harvey 1974, p. 272). Hence, scarcity is a response to the social organization of production and can be eliminated by changes in social organization, changes in technical and cultural appraisals of nature, shifts in accustomed demands, or reduction of our numbers by regulation of population (Robinson 1970, Harvey 1974). In this context population regulation is not the preferable policy response; it is politically and class organized and suggests that overpopulation is inherent in the lower classes (Malthus 1970), who should be regulated. 'The projection of a neo-Malthusian view into the politics of the time appears to invite repression at home and neocolonial policies abroad . . . to preserve the position of the ruling élite' (Harvey 1974, p. 276). Instead we need a shift from the capitalist mode of production, the consumption patterns of which tend to shift costs and environmental degradation onto society as a whole in order to improve its competitive position. We have already seen in chapter 8 how spatial inequality effects are held to exhibit characteristics of 'neocolonialism' (Robinson 1970, Ward 1976) in the Third World. In this context social reorganization means a shift away from a local economy based on cash crops which bring surplus value to a local élite, or to multinationals in the developed world (Myrdal 1968) and away from urbanization. On the basis of this reorganization alone, Robinson (1970) has suggested that enough land will be freed to feed the present and anticipated world population – i.e. the population problem disappears! Hence, Harvey (1974, p. 276) suggests that 'the only kind of method capable of dealing with the complexity of the population-resources relation in an integrated and truly dynamic way is that founded in a properly constituted version of dialectic materialism'. In calling for a reorganization of social values this approach is obviously important, but it contains a number of problems. Firstly, it assumes that the new basis for internal dynamism and adaptation can be constructed rapidly enough. Secondly, it assumes that resource limitations can be removed solely by reorganization of use. This would seem to suggest either that substitution is possible (as with the economic optimist) or that the world must be reduced to a much lower standard of living overall (the pessimistic view).

The dialectic materialist approach may suggest a middle way achievable through reorganization of the world ecosystem and redistribution of its benefits. An alternative middle way recognizes that a problem exists, suggests that the collapse may be delayed for longer than Forrester and Meadows forecast so that new technology might be evolved, but requires that we must take some action now. The action which is suggested is both less radical and less simple than the percentage reductions in growth rates suggested by Forrester (1971). For example, Ward (1976) argues that man has the ability and technology to avoid the worst and establish a better social order. Indeed, this is recognized by Meadows *et al.* (1971, pp. 127–8) in asking 'Is the future of the world system bound to be growth and then collapse into a dismal depleted existence? Only if we make the initial assumption that our present way of doing things will not change.' The cause of collapse is growth, but abandonment of growth is not simple. Sauvy (1975) states that 'zero growth' as a concept is devoid of meaning in a world in which some level of future technical progress is inevitable. Mesarovic and Pestel (1975) suggest that the fault is not in growth *per se*, but in the form of 'undifferentiated' growth that has been followed by the world system in the past. Such growth is quantitative, increasing numbers by replication and is characterized particularly by the accumulation of capital and resources in the developed world. What is required instead is '*organic*' growth which is differentiated so as to allow the parts to differ in structure and function, but with interdependence of the parts such that the growth of any one part depends on the growth or non-growth of others (Mesarovic and Pestel 1975, p. 5). The transition to organic growth would permit the true symbiosis of man and environment to be maintained. In nature, organic growth is maintained by internal regulation according to a master plan which is composed of a set of control targets, goals and objectives. Mesarovic and Pestel (1975, p. 7) see this as 'evolved through the principle of natural selection; it is encoded in the genes and is given from the start . . . so that the development of the organism is specified by it; the plan and the organism are inseparable'. Applications of the control to objective principles discussed in chapters 3 and 6 will be limited, since 'the master plan [in world regulation] has yet to evolve through the options by the people who constitute the world system. To this extent *the options facing humanity contain the genesis of our organic growth.* And it is in this sense that mankind is at a turning point in its history' (Mesarovic and Pestel 1975, p. 9). This requires an internal restructuring of the world system to restore 'normal' conditions of mutual harmony between its components so that each is adjusted to the others. In practice, Mesarovic and Pestel note that this requires two forms of restructuring:

1 *Horizontal restructuring*, in which a change in relations between the spatial elements (nations and regions) is induced. This requires radically reduced standards of living in the developed world with marked redistribution of financial resources to the developing world, but both brought into harmony with a rate of resource use which is equal or less than the rate of resource renewal.
2 *Vertical restructuring*, in which a drastic change in values, goals and norms of man and society is required. This is a prerequisite for horizontal restructuring since lower living standards, reduced levels of consumption and energy are all required.

Such restructuring encompasses both the idealist and dialectical materialist approaches to system change (see chapter 8). We must return to a better balance between man and natural resource renewal and replenishment rates which requires social reorganization.

Neither form of restructuring is likely to occur spontaneously so that, if the environmental problem cannot be ignored, conscious control towards these goals must be assumed in which the control models and decision making structures developed in chapters 3 and 6 are candidates for adoption. For such world-sized systems to be formulated we require a sociopolitical controller which encompasses the man-environment interfacing shown in fig. 9.22. Especially important is the agreement on goals, policies and administrative procedures necessary to accomplish control in the world system:

> To test alternative plans for anticipatory action; to evaluate the realizability of these plans in the face of national, regional and global constraints; to determine whether short-term benefits in one area might not entail great harm in others or even destroy future long-term options . . . credible . . . not only to decision makers and the 'élite', but to the public at large. (Mesarovic and Pestel 1975, p. 156)

Opinion on whether such a control system is possible must be guarded. As we have shown in chapter 8, the agreement on overall goals in society is restricted to an individual or to small groups (Hayek 1944). Moreover, the individual in his development from his origin, environment, training and acquired perceptual experiences may converge in his values with another group (Piaget 1969), but 'nowhere can we find a criterion for singling out one among many different, mutually contradictory value scales as "objectively correct"' (Von Mises 1951, p. 323). In designing objectives in world control systems little agreement will be possible as between individuals and it is likely that goals will be set by political bargaining between nation groups to which no objective solution exists. A form of Buddhist economics advocated by Schumacher

(1973) or any other religious system does offer some indications of desirable directions, as discussed in chapter 5. However, they do not offer any practical resolution to the problems of defining or achieving goals (Von Mises 1951, p. 370) beyond an Arcadian idealism – for example, that a return to a smaller-scale society, or to the earth by virtue of its goodness, are things to be desired in themselves, akin to Meline's law (Sauvy 1975, pp. 32–4). It is to the problems of goal and system construction that we must now turn.

9.5 Problems of environmental symbiosis

In the development of the principles of control systems given in chapter 3 and the interfacing of hard control systems with man as a decision maker developed in chapters 5 and 6, we have seen that control necessitates the definition of five main elements:

1 Control goals, targets or objectives.
2 An information and monitoring system to describe what is occurring.
3 Models of the system under study to generate forecasts.
4 A comparator to assess the significance of past or likely future deviation of the system from target.
5 A mixed-scanning decision response ranging over no action to goal review.

The resulting control system model will be a synthetic device to represent world feedback structure and response. Empirical regularities in world-level relationships of population, resource use and pollution generation are linked by deductive mathematical functions of nonlinear and nonstationary form. On the basis of the web of mathematical relationships we assess, via information and monitoring devices, the most appropriate decision response. The nature of information-monitoring systems and significance assessment has been treated in chapter 6 and syntheses in physico-ecological and socio-economic systems in chapters 7 and 8. We consequently restrict attention in this section to problems of model construction and control response in world-scale symbiotic systems.

The construction of quantitative world symbiotic models is a product mainly of the last ten years, although more qualitative models of man-environment symbiosis have a long history of philosophical development, as discussed in chapter 1. The quantitative models of world symbiosis can be divided into two generations of development (Clark *et al.* 1975). First generation models comprising those of Forrester (1971) and Meadows *et al.* (1971) are comparatively crude models of the world as an aggregate, lumped whole, ignoring internal diversity. Although stimulating, these models have been severely criticized on the basis of their assumptions which are thought to be overpessimistic (especially with reference to reserves of non-renewable resources and to the rate of technological development) and the exclusion of spatial and sociopolitical levels of response. Second generation models are much more diverse in format, hypotheses and approach; Clark *et al.* (1975) recognize four subgroups:

1 *Spatially disaggregated and multilevel models*: The model of Mesarovic and Pestel (1975) allows the internal diversity of the world system to be expressed in terms of interregional and interlevel feedbacks. The overall forecasts are still pessimistic, however, unless 'organic' rather than undifferentiated growth is pursued (see section 9.4). A more limited form has been prepared by the United Kingdom Department of the Environment (Roberts 1973).

2 *Economic-industrial sector models of the Japanese Club of Rome*: The inter-industry flow of goods and services together with initial, intermediate and final demand are explored by the use of world interregional input-output tables in order to determine the optimal future location of industrial production on a world scale (fig. 9.33) (Kaya and Suzuki 1974). The policy responses need now be implemented only at a national level, e.g. by economic stabilization and investment policies to minimize trade balances and demand-supply gaps. Such models have great importance to the planning of production in Third World countries.

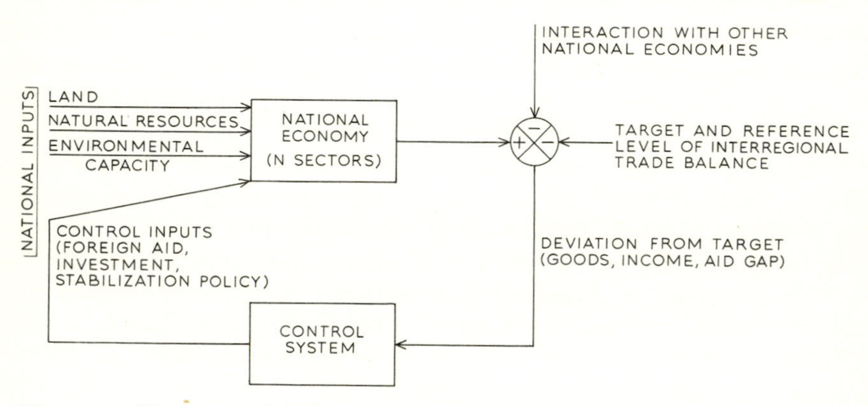

Fig. 9.33 The economic industrial sector model of the Japanese Club of Rome. (Partly after Clark *et al.* 1975.)

3 *Spatial social justice models*: Instead of designing models which merely reproduce present world ecological and economic feedback loops, many workers have attempted to design world symbiotic systems oriented towards the specific goal of equitable distribution of world resources. At an urban level this problem has been discussed by Harvey (1973), but this Marxist approach ignores the role of *real*, rather than monopolistic, scarcity of resources. World-scale social justice models by the Bariloche team (Herrera 1973) use a 'basic human needs function' defined by a

sociopolitical goal formulation system and composed of the amount of material and cultural goods (food, housing and health) a human being needs for his full development without wasteful use of resources (Clark *et al.* 1975, pp. 21–3). The result is a set of alternative scenarios of development by which a 'just' distribution of resources under scarcity is achieved.

4 *Exploratory and ameliorative models*: These are concerned either with testing single or small numbers of loops in the world symbiosis to evaluate intervention strategies, as discussed in section 9.3 (House 1973, Burdekin and Marshall 1972), or with strategies to reduce the severity of collapse modes of behaviour. In these amelioration models a Dutch group has sought to design a strategy in which each individual retains the basic human requirements over the period in which world population is expected to double and before collapse occurs (Linnemann 1973, Clark *et al.* 1975, pp. 23–4). The Batelle Memorial Institute (Thiemann 1973) has adopted a similar '*status quo*' approach by which control policies are used to ameliorate anticipated environmental problems within the constraints of existing system behaviour (Clark *et al.* 1975, p. 24).

Common to each generation of models of world symbiosis are problems connected with any synthetic model: the availability of *suitable data* and information to permit monitoring and forecasting; difficulties of *calibrating* the models with little or no data, together with associated specification and estimation problems (especially over nonlinear and nonstationary effects); and the *sensitivity analysis* of each model to changes in the assumptions, parameters and specifications.

The *data* employed in all world symbiosis models are almost totally inadequate. This has led some commentators to believe that world symbiotic models are not at this stage worth constructing. Nordhaus (1973), for example, criticizes the Forrester-Meadows approach in that not a single variable or relation is drawn from actual data or empirical studies. However, even with severely limited data, valuable simulation results can be obtained which have important policy and control implications for the way man chooses to organize his society. Four main problems characterize the data requirements in world models. Firstly, data are needed for variables which have not been measured before or are difficult to measure; for example, world resources of renewable and non-renewable elements, quality of life, level of technology, etc. Secondly, time series of data are required to permit monitoring of trends when frequently information has been collected for only relatively short periods of the recent past. Thirdly, disaggregated data are required to allow the appraisal of spatially and socially distributed group effects. This particularly allows study of orientation to biological and socially just goals when the system is nonlinear or evolutionary. Fourthly, comparability of disaggregate data measurements is required so that aggregation is possible to obtain regional and world totals. Such aggregations may be performed at great cost since they are only possible when the microrelations are linear (Theil 1954). In the case of the 'demographic transition', for example, Nordhaus (1973) notes that aggregation leads to a serious misspecification bias in Forrester's model. Forrester hypothesizes a positive relation of population growth to increases in *per capita* food consumption, whilst accepted economic thinking suggests a rise followed by a fall in birth rate (and hence population levels) with rising food consumption. Each of these data problems is important in itself, not easily overcome, and limits the inferences that can be drawn from subsequent model calibration and simulations.

Calibration of world symbiotic models involves the scaling of the structure of the proposed model against our observations of the real-world system that it is desired to reproduce. In practice, this should involve the entire sequence of five steps in systems analysis shown in fig. 2.40 (p. 76), from hypothesis generation through system specification, parameter estimation and checking to final forecasting and control. However, in the construction of most world models the first four stages have been collapsed to one set of *a priori* definitions of the structure and magnitude of system transfer functions (with estimation of some parameter values in limited contexts). Most of the difficulties in applying the statistical systems analysis techniques discussed in section 2.3.3 (pp. 76–91) arise from the paucity, and in some cases complete lack, of data. In most world models, however, a purely deductive approach seems to have been preferred. Nordhaus (1973, p. 1183) notes for Forrester's (1971) model that, 'whereas most scientists would require empirical validation of either the assumptions or the prediction of the model before declaring its truth correct, Forrester is apparently content with subjective plausibility'. This creates three main problems for the analysis of symbiotic systems. Firstly, there is no way of checking the validity of the formulations except by purely deductive reasoning or output comparisons which, because of equifinality, are seldom reliable. Secondly, we can obtain no measure of the confidence with which we can view our final model. The statistical identification and estimation techniques discussed in chapters 2 and 4 allow measures of overall model fit (R, F and t), which are useful in testing forecasting accuracy, and parameter standard errors, which aid in testing the inferential meaning of any given equation or of any single term in that equation and also give confidence levels of forecasts (Bennett 1978, chs 4, 8) but these tests cannot be applied when no data are available and when a number of restrictive conditions are not satisfied. A third problem that arises from purely deductive system and parameter specification is that the uncertainty in system estimates is not incorporated into control strategies. Forrester (1971) and Meadows *et al.* (1971) assume certainty equivalence but do not examine the quality of their estimates (Nordhaus 1973). We have seen in chapter 3 that a separation theorem between system estimation and control does not in general exist and that optimal strategies should be designed with self-tuning regulators such as that given by Åstrom and Wittenmark (1971). It must be admitted therefore that severe problems arise in attempting to derive statistical estimates of the multivariate space-time controllers which are necessary for symbiotic systems. Special difficulties arise in particular because of the nested nonlinear loops in world systems. Nevertheless, simulation methods do offer some potential for world system study, since we do not need to know the internal system structure if we can generate the same system output (chapter 2 and the theory of Padé approximants), or, as Simon (1969, p. 17) notes, 'resemblance in behaviour of systems without identity of inner systems is particularly feasible if the aspects in which we are interested arise out of the *organization* of the parts, independently of all but a few properties of the individual components' (i.e. the system can be described macroscopically as discussed in chapter 8). In the future, however, only with refinement of hypotheses by statistical methods and the generation of confidence and uncertainty measures can reliable symbiotic models replace the first and second generation of world systems which are now available, since, as Naylor (1971a, ch. 5) notes: 'simulation models based on purely hypothetical functional relationships and contrived data are void of meaning . . . such a model contributes nothing to the understanding of the system being simulated'. Indeed it has been argued (National Research Council 1973, p. 23) that modification of the social and economic forces of the market would be difficult to achieve even if the underlying natural and economic laws were known, but since chance and uncertainty are the main ingredients of world-scale systems, the only way forward is increased data collection and further analysis and research aimed at reducing the number of unknowns in system structure, increasing the certainty of parameter estimates and further refining assumptions, hypotheses and instruments upon which control can be based. Whilst further refinements are obviously important, it should be remembered that (as noted in chapter 6) information systems can only be defined by the demand which they serve to answer. Particularly in the resource field we should be careful of basing inferences on static inventories of stocks, environmental quality or human satisfaction since not only does our definition of a resource shift with technology, but so also does human need and demand shift with evolution in cultural values. The almost infinite capacity of human adaptation makes no fetish of information systems, and stationary pictures of system throughputs and storages must give way to considerations of wider issues including human epistemology.

The presence of nonlinear feedback and feedforward loops presents severe systems synthesis and design problems. Unless linear transformations exist (e.g. log, Taylor series, etc.), then the application of statistical estimation methods becomes more complex and time consuming, and this result alone would seem to be Forrester's justification for ignoring statistics. Clark *et al.*

(1975, p. 63), on the other hand, would seem to favour the use of linear systems simulation, since:

> the virtue of linear models is that model behaviour is easier to comprehend, interpret and test. The real trade-off being made in deciding on a linear or non-linear form is between modelling 'simplicity' in terms of reducing parameters to a minimum and 'realism' by simulating as closely as possible the observed behaviour.

Linear formulations of world systems can be favoured on the grounds that linearized forms of the Forrester-Meadows model, such as that given in table 9.1, lose little of the character of input-output relations (Cuypers and Rademaker 1974, Rademaker 1974). In addition, Clark *et al.* (1975) note that a 'law of large systems' frequently operates in world systems in which the overall response characteristics of large nonlinear systems frequently behave as if the latter are linear (Young *et al.* 1973). This supposed gain in system simplicity may be of limited value, however, since linear representations lead to a larger number of *singular points* in the system state space (Hirsch and Smale 1974). Both the nonlinear system reduction to linear form, and the space-time system reduction to lumped form, suppress the variety of possible response modes in the resulting system model. The result is that overshoot and collapse modes of behaviour become much more likely. As detailed above, this seems to derive also from the assumptions built into the Forrester-Meadows models, and in addition from the mathematical structure of linear and lumped systems. Referring to fig. 2.35 in the discussion of state space evolution in chapter 2 (pp. 66–9), we see that, in the case of saddles, the path of system evolution will be stable or unstable depending solely upon the system initial state (initial conditions). In the saddle case with system state transition matrix (equation 2.2.13), given for example by

$$\mathbf{A} = \begin{bmatrix} 2 & 0 \\ 0 & -0{\cdot}5 \end{bmatrix},$$

the system is stable for an initial system state of $\mathbf{Z}(0) = [0, K]$ but unstable for an initial system state of $\mathbf{Z}(0) = [K, 0]$, where K is any constant ($-\infty < K < \infty$) and $\mathbf{Z}(0)$ is the state variable vector at time $t = 0$. In more general cases with n state variables and a set of m control variables $\mathbf{U}_t$, the manifolds shown in fig. 4.25 (p. 216) demonstrate similar stable and unstable behaviour depending solely upon the initial conditions. At the point or line dividing the stable from unstable regions (termed saddle and branching points of singularity or universal unfolding; Thom 1975) the system is in a conditionally unstable state. Nonlinear models generate a much reduced set of singularity points than the linearized versions of the same model. The result is that linearized world symbiosis models generate catastrophe and collapse merely because of the way the system is defined, and not because of any inherent structure of the world system itself – i.e. collapse is a result of our synthetic systems view. This is recognized tangentially by Mesarovic and Pestel (1975, p. 53) who state that systems appear conter-intuitive only because we are used to viewing them under 'normal conditions' (i.e. in distinct behaviour regimes rather than at saddle points). 'It is only because strata are more loosely coupled in normal conditions that the world appears comprehensible' (Mesarovic and Pestel 1975, p. 53), but this conclusion should be extended not only to cover systems in crisis but also normal behaviour. Thom's (1975) theory of structural morphogenesis incorporates normality and crisis within the same systems dynamic view (Hirsch and Smale 1974).

Further controversy surrounds the *sensitivity* of world symbiotic models. Forrester (1971) and Meadows *et al.* (1971) state that their type of world model is very insensitive to both the relationships and parameters contained in the system transfer function. However, the space-time development of the model by Mesarovic and Pestel (1975) demonstrates that the incorporation of spatially distributed, specific, and relativistic effects greatly modifies the resulting collapse modes which occur selectively in different places at different times making political and socio-economic response and internal restructuring (horizontal and vertical) more likely. Simulations with Forrester's model using only slightly modified parameters and specifications of relationships also demonstrate that the nature and timing of collapse may be severely modified and may not occur at all (Cole *et al.* 1974, Nordhaus 1973). In historical simulations Cole *et al.* (1974, ch. 9) show that the effect of incorporating technological advance and discovery of natural resources modifies collapse markedly. A 1% *per annum* discovery rate and technology improvement rate, for example, postpones collapse until beyond 2200, whilst slightly more optimistic assumptions of 2% *per annum* rates for both rates postpone collapse indefinitely giving a global dynamic equilibrium. Forrester and Meadows seem to employ one-parameter-at-a-time sensitivity tests, but these will be quite inappropriate in high-order space-time multivariate systems (Clark *et al.* 1975) and are particularly restrictive when few if any of the parameters are known with any degree of certainty. Scolnik (1973) demonstrates that even small levels of variation in initial parameter values of about 5% lead to markedly different output results, and 5% standard error margins are far lower than can be achieved with presently available data in calibration of world symbiosis models.

When we turn our attention to how symbiotic systems can be used for control we find a divergence of views. Kuhn (1974), for example, argues that man's effect on the environment has been to produce a highly constrained uncontrolled system rather than a controlled one. Man has an important dominant conscious effect on the whole system, but this is quite different from controlling the system.

> An ecological system is by definition uncontrolled: it has no goal of its own and maintains no particular equilibrium. If man alters its equilibrium, as

> he does every time he converts a forest to field or town, nature does not respond with negative feedback that automatically offsets man's action. . . . Instead the new equilibrium continues as long as the field or town does. . . . Although man is inside the system, . . . actions by him . . . [are] constraints imposed from outside rather than . . . goals of the whole system taken as a unit. (Kuhn 1974, p. 462)

Again, Clark *et al.* (1975, p. 65) suggest that 'control concepts are often useful for clarifying thinking about objectives [but] . . . the method seems more appropriate to systems with a complexity far below that of global models, such as air traffic control or river authority structure, where the "physical" system described is relatively well defined and unambiguous and the "decisions" to be taken are reasonably straightforward.' In particular, they argue that models based on the systems methods and statistical estimation techniques discussed in chapters 2, 3 and 4 are of little utility for the long time horizons that arise in world symbiosis. 'Although naive extrapolation methods (such as the Box-Jenkins prediction technique (1970)) are atheoretical (in that no explicit theory of the process under observation is employed) and are adequate for short-term forecasting, they may be completely misleading in the longer term and give no insight into the effect of alternative policy options' (Clark *et al.* 1975, p. 59). In symbiosis, therefore, man can follow the intervention type strategies discussed in section 9.3, but must use wider sociopolitical techniques (e.g. dialectical materialism) to obtain answers and cannot rely upon quantitative control models (Harvey 1974).

Other scholars have taken an alternative view based upon how we define the system, in terms of where we choose to draw the system boundaries. At the symbiotic scale these boundaries are of great spatial and temporal extent. The individual, socio-economic group, nation or society as a whole are agents within an environment. The environment depends upon the definition of system closure. At each spatial scale the performance and survival of the system of agents will depend upon their fitness for the environment in which they are embedded. The system will survive if the agent performs a *non-empty* set of *admissible* acts within the environment, but the system will collapse if the agent performs inadmissible acts or has only an *empty* set of acts available. Day and Groves (1975) define the *survival condition* for any act of a set of actions at any time as those actions which form a non-empty set. Symbiotic control action is viable only if the survival condition holds that we have control instruments or actions that can affect the world system and hence avoid collapse. If this condition does not hold, then there are singularities in the action manifold in the control space and the system will 'flip' into uncontrollable modes where the limits of action occur. Natural human behaviour acts by homeostasis, the agents behaving so as to favour the choices of survival where they are available (e.g. risk aversion). At the global scale, and at more local spatial levels, we have the paradox that the chances of triggering the collapse of the action space are growing with man's ability to control. As man becomes a more active decision maker, he is also becoming more passive because of the explosion and saturation of his own information system, his own inability to coordinate actions, and the increasing lag times required in appraisal of complex social and environmental implications of any action. Symbiotic control is an attempt to regulate systems with almost infinite degrees of freedom, which Mesarovic and Pestel (1975) suggest is a function of abandoning lumped for spatial and distributed models (see fig. 9.34):

> the homogeneous world concept resembles a swinging pendulum, which, in order to avoid collision with a limit, has to reduce speed and acceleration immediately. In a similar analogy the diversified view of the world could be represented with a set of balls connected by a spring system; any of the balls might encounter its own limits, but others might still be a distance away from their own constraints. The effect of such a collision will propagate through the system in relation to the strength of interaction developed up to that point – that is, depending on the spring system. (Mesarovic and Pestel 1975, p. 37)

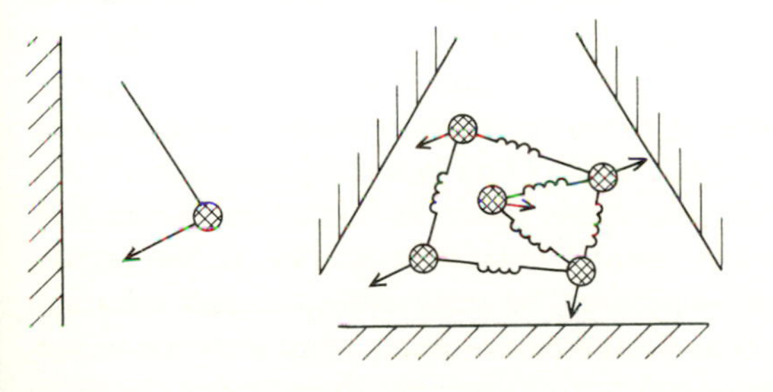

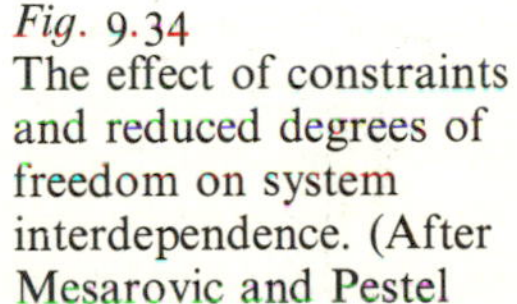

Fig. 9.34
The effect of constraints and reduced degrees of freedom on system interdependence. (After Mesarovic and Pestel 1975.)

The reorientation of man-environment interaction to allow more active control of an expanding decision space with ever-increasing ramifications represents the severest dilemma for environmental symbiosis. In the control of socio-economic systems discussed in chapter 8, it was suggested that, in democratic systems, control could be exercised only in those contexts in which there was agreement on goals through collective action. In large-scale systems such as national economies there would be few areas in which goals can be agreed (Hayek 1944, von Mises 1951). Hence, most attempts to achieve socio-economic control must be without agreement and will not be possible within the collective approach or under complete democracy, but must be imposed, to some extent (Hayek 1944). In environmental symbiosis systems we have immensely large-scale systems in which there is never likely to be agreement on goals. What, therefore, is our response? If we follow the Hayek and *laissez-faire* approach we must not attempt to control, but rely upon what Adam Smith termed the 'hidden hand' of self-regulation. If we rely on collective action we must wait for a collective view to emerge on what should be done (Buchanan and Tullock 1969). However, if we view the impending environmental crisis at least in the terms of the middle way sketched in section 9.4 above, we must accept that some action is necessary, but the necessity of control undermines the liberty and freedom of the individual to choose. Hence, we have two possible approaches to the problem of global control: first, permitting the free play of man-environment interaction; and, second, design synthesis (adapting the man-environment system to more appropriate goals).

The 'natural evolution' approach relies upon the free play of man-environment interactions and the operation of the market as in *laissez-faire* economies by which system fitness is determined by trial and error. The classic example of this is the determination of population characteristics by interaction with the environment through competition and selection of the fittest through the mechanism of the gene pool (i.e. the higher reproductiveness of the fitter organism allows it to contribute more genes to the gene pool) (Simon 1969, Rosen 1975). Also, the high degree of component replication in environmental systems makes them highly reliable (see chapter 3) and adaptive to changes in external environment (Simon 1969). In human populations there is the additional feature of the creation of new rules of adaptation (i.e. information) which may or may not be passed on to the next generation. An *existence condition* for such systems (Day and Groves 1975) is that the population is composed of those agents who have something to do and can do it. This evolution approach is combinatorial, and as such it is inefficient and extravagant (Rosen 1975) since all experiments (and all mistakes) must be made, such that there is much redundancy. It also implies that there are no second chances, which will be an unrealistic model of human behaviour but one that may control man-environment symbiosis. However, any restriction of possible experiments (i.e. of the market) suppresses freedom and attenuates variety. In particular, there is a loss of adaptive ability (i.e. freedom) of the agent whenever he is integrated into a higher-level adaptive (discretionary) system. The bounds on the adaptability of the higher-level (societal) system will in practice be limited (e.g. by principles of *vires* and natural justice; see chapter 6). Hence, the overlay of nested hierarchical control on the man-environment system limits not only the adaptability of the lower-level agents (individuals), but also the adaptability of the higher levels. Each form of the 'hard' nested control systems discussed in chapter 3 has defects in this context (Rosen 1975). Simple feedback systems are particularly vulnerable to incorrect reference point settings and will be slow to respond. Indeed, since they can only respond *after* the event, they will be of little utility in averting socio-environmental collapse. Feedforward control systems permit the anticipation of system and environmental states, but are extremely sensitive to the forecasting accuracy of the anticipatory elements. The accuracy of these will depend upon the accuracy of available environmental models and upon the efficiency of environmental sensor/information systems. The accuracy of current world models and of most man-environment systems is not such that future system and environmental states can be forecast with great accuracy. Moreover, the present network of environmental sensors, though large, is limited by the inability to monitor new control demands as the man-environment system evolves.

Design synthesis is an alternative approach which permits the defects in systems evolution models to be overcome by symbiotic adaptation. One approach to this adaptation is by use of the universal adaptive systems discussed in chapter 3.2. These allow feedback *and* feedforward elements plus the ability to adjust overall goals, set up new sensors and information systems, and modify and extend their modes of coupling with the environment and with other societal systems. This form of symbiotic adaptation attempts to set up mathematical models of the man-environment interface to allow investigation, simulation and learning of the effect of societal networks on environmental feedback structures, and *vice versa*. This requires the development of highly complex monitoring and information systems (as discussed in chapter 6), which act as sensor and effector functions (Kuhn 1974) to enable sufficiently varied societal response and are composed of the five recursive subsets of actions: observing, storing, processing, planning, implementing, and administering (Day and Groves 1975).

A second approach to design synthesis is to induce symbiotic adjustment within society – to accelerate cultural evolution (Sahlins and Service 1960). Dialectical materialism is a special case of this approach requiring a radical adjustment of social values. McInerney (1976) has argued that this adjustment can be made within the framework of economic theory by introducing three new factors: first, that problems be viewed at a societal rather than at an individual level, then focusing attention on the range of available social choice; second, that time considerations must be introduced, especially the goals and ontology of any sociocultural organization; third, that some constraints on social choice, such as resource stocks, are immutably fixed. As a consequence of these factors, the resource and environmental issue can be resolved within the context of an ongoing social choice on use of environmental capital. The fundamental change required is the treatment of resources as capital not income (Schumacher 1973). Thus the choice in use of a resource now, as opposed to the future, reduces in simplified terms to an intertemporal allocation problem (fig. 9.35) between present and future use of the fixed resource stock (McInerney 1976). The resource stock acts as a linear constraint between present and future use (*RR*) and the level of use will be determined by the social preference function (i.e. indifference curve). If we view this figure in conjunction with the effect of various conservation policies (fig. 9.5, p. 478), the desirable course of action should be heavily to discount present costs of conservation in favour of the negative rate of time preference (fig. 9.35C), rather than a positive or zero rate of time preference (fig. 9.35B and 9.35A, respectively). This approach to design synthesis can be criticized on two counts. First, it takes no account of changes in production costs and the price mechanism. Hence, it tells us nothing of how to bring about the desired balance of present and future resource use. A second problem, as recognized by Coddington (1974, p. 464), 'is not that we do not know how, in principle, to achieve a given level of consumer satisfaction with a lower input of resources. It is that we do not have economic institutions which are capable

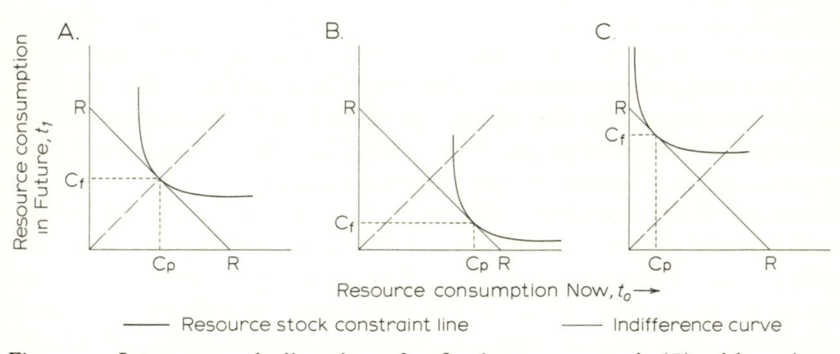

Fig. 9.35 Intertemporal allocation of a fixed resource stock (R) with various social preference functions.
C_f = future consumption.
C_p = present consumption.
(After McInerney 1976.)

of coping with the employment effects of such shifts'. The design synthesis of socio-economic institutions required may thus involve a complete shift from present forms. Robinson (1970) and Ward (1976) have argued that this can be achieved only by shifts from the present capitalist mode of production in the Western world. Certainly conservation alone is not enough since this both reinforces the *status quo* (Robinson 1970) and avoids the basic question of how to maintain aggregate demand and development without shifts in social values.

9.6 Conclusion

The interfacing of socio-economic and physico-ecological systems discussed in this chapter has been treated at various spatial scales and at various levels of intervention. Small-scale intervention in passive form or by manipulation of storages, as discussed in section 9.3, has been shown to be applied fairly readily and commonly to manipulation of the environment for human ends. With potent control of physical systems by manipulation of their inputs as control variables and instruments, we move into another class of control system. Here conscious environmental modification to achieve target and goal values permits potentially large-scale adjustment of the world ecosystem. When these manipulative controls of systems inputs are operated over large spatial and temporal scales the symbiotic modifications to the world system as a whole present possibilities which are as exciting as they are potentially dangerous.

Human intervention in environmental systems is not usually value-laden, but in large-scale man-environment systems the symbiosis of man as part of the environment of the system he wishes to control introduces all the indeterminacies of socio-economic control objectives discussed in chapter 8. In particular, we need to ask what are the political and social implications of control, and for whom and by whom is control intended? Although we may accept in purely socio-economic systems that the degree of control should be limited to those areas in which common agreement on goals exists, in man-environment systems there may be a necessity and urgency to consider implementation of control which is generated by exhaustion of resources, increasing pollution, and the pressure of human needs against the limits of environmental system stability boundaries. This may require that we agree to environmental control as a means of self-preservation even though we may not agree on all the control goals that must be employed. In purely socio-economic systems the limits to intervention suggested by J. S. Mill find few detractors. Mill listed four main objections to the unbounded extension of government agency and control (Mill 1848, bk V, ch. II):

1. There is a limit or discretional boundary of human individual action space defined by that part of the life of the individual that does not affect the interests of others, which government should not interfere with.
2. Continuous extension of government power restricts individual freedom and action spaces. The extension of control, influence and power of government has a natural momentum: 'the public collectively is abundantly ready to impose, not only its generally narrow views of its interests, but its abstract opinions, and even its tastes, as laws binding upon individuals' (Mill 1848, p. 550).
3. The extension of government control results in increasing bureaucratic division of labour and tends to increase inefficiency. 'All the facilities which a government enjoys . . . are not an equivalent for the one great disadvantage of an inferior interest in the result' (Mill 1848, p. 553).
4. The extension of government chokes individual motivation, ability and willingness to take risks. 'A people among whom there is no habit of spontaneous action for a collective interest – who habitually look to their government to command or prompt them in all matters of joint concern – have their faculties only half developed' (Mill 1848, pp. 554–5).

As a result of these perceived limits, Mill suggested that intervention should take place only when there is a strong necessity to justify it. Although we may join the sociologists and critical theorists such as Habermas (1971, p. 1) in being sanguine that the environmental crisis of late capitalism is a purely social phenomenon like a patient who has only temporary deprivation of his powers and recovers from his illness, there seems at least a chance that humanity may die on the operating table: environmental constraints cannot be totally ignored. It may well be that we have now reached a situation in which we must accept a greater level of control, severe limitations on freedom and *laissez-faire*, on private property rights, on the entitlement to pursue individual goals and to partake or not in collective decisions. As we suggest in the next chapter, any technical developments in modelling ability and information bases are overshadowed by this all-embracing need to develop our understanding of the psychological and sociological organization of society and its epistemological bases, especially in terms of the meaning of nature to man, in terms of better understood human goals and indeed a better understanding of the meaning of humanity itself.

10 Conclusion: future problems

> Systems, scientific and philosophic, come and go. Each method of limited understanding is at length exhausted. In its prime each system is a triumphant success: in its decay it is an obstructive nuisance. The transitions to new fruitfulness of understanding are achieved by recurrence to the utmost depths of intuition for the refreshment of imagination. In the end – though there is no end – what is being achieved is width of view, issuing in greater opportunities.
>
> (A. N. Whitehead: *Adventures of Ideas*)

The systems approach has been greeted by some as a universal panacea for all our philosophical and technical problems, by others as a jargon-ridden overstatement of the obvious – it is neither. It is true to say that systems methods have illuminated thought, clarified objectives and cut through the theoretical and technical undergrowth during the third quarter of the twentieth century in a most striking manner. However, the sheer complexity of the systems with which environmental scientists are concerned, together with the in-built bias of the systems which they construct to simulate reality, at present place impediments in the path of the development of systems methods as a general vehicle for understanding, control and forecasting. At the very least, we should view the systems approach as a complex training ground wherein to develop the ringcraft, the deftness of touch, the timing, the incisiveness of action, and the introspective forethought which are increasingly necessary in dealing with our escalating environmental problems. At most, the systems approach provides a powerful vehicle for the statement of environmental situations of ever-growing temporal and spatial magnitude, and for reducing the areas of uncertainty in our increasingly complex decision making control situations.

Despite the success of the application of the systems approach to the study of environmental systems, particularly in a theoretical sense, it is clear that difficulties still lie in the way of the unbridled application of systems techniques to the solution of a wide range of significant environmental problems. The sources of these difficulties can be grouped into four broad categories: technical, demographic, psychological and epistemological.

10.1 Technical problems

The systems approach relies heavily on large amounts of accurate data which are used both to construct and test environmental models and in their manipulation. The more complex the system, the more data are required, both in space and time. Many systems analysts would argue that a significant impediment to their efforts lies in the purely technical matters having to do with such considerations as the lack of basic environmental data, in existing administrative fragmentation resulting in interfacing and compatibility difficulties, in the divorce of the data collector and the systems modeller, in the lack of interdisciplinary integration, in the lack of well-defined social goals, and in the high priority given to value judgments in decision making (De Greene 1970, pp. 551–9 and Foin 1972, pp. 519–21). Also important is the lack of suitable mathematical theory to allow the derivation of optimal decisions that can be applied to spatially distributed systems, to permit the adjoining of system estimation and control solutions (overcoming the breakdown in the separation theorem, as discussed by Tan (1978)), to allow optimization with respect to complex objective functions of non-quadratic and multi-peaked form, and to the incorporation of 'soft system' value data. It can be objected that optimal control models often obscure rather than clarify the scope for social decision, or are not relevant to what is essentially a political debate. As we have seen in chapters 5 and 6, many decisions are constrained more to be 'satisficing' than by 'costs'. Nevertheless, purely technical developments in control, if used judiciously, can only aid in forecasting and simulation, and in laying bare the areas of potential political compromise.

Especially important will be technical developments that permit a more 'holistic' approach to both system description and control in which both exogenous and endogenous elements can be interfaced. This requires techniques for simulating what Shackle terms the diachronic kaleidoscope of social development; for models of Waddington's (1977) homeorhesis of evolutionary stability rather than homeostasis. Thus the analyst will be forced to leave behind developmental models as exemplified in the economic stages of Myrdal and Rostow, in Max Weber's sociological models, in W. M. Davis' cyclic landforms, in H. C. Darby's landscape sequences, and in Keynes' equilibrium models, each stage in the development of which is characterized by an ephemeral equilibrium, stacked as a card house of momentary immobility separated by sudden landslides of readjustment (Shackle 1972, p. 433). It is speculative whether it will prove possible to capture the palimpsest of disjoint evolutions which characterize system evolution by any form of system model. The most promising developments may well emerge from the global analysis approach of Thom, Hirsch and Smale, and Waddington which are capable of treating the whole range of system behaviour and initial conditions and which do not exclude nonlinear, nonstationary, evolutionary, unstable and other mechanisms which have presented difficulties to, or have been excluded from, conventional mathematical analysis. Whatever network of technique results, we will require the replacement of what Latakos (1976) terms formalist and dogmatic mathematics by a more informal heuristics firmly rooted in problems.

Many of these technical difficulties can be circumvented by so-called 'rational' actions involving large-scale integrated data gathering, bigger compartment modelling covering wider spans of time and space in more detail, the introduction of greater varieties of nonlinearities, the development of more meaningful equations, the generation of more probabilistic hierarchical models, the resort to more objective control manipulations, and to more technically-realistic bases for decision making (Clymer 1972, pp. 533–69). In other words, systems analysts may need to adopt a positive view of environmental problems, namely that they are mainly technical ones connected with achieving increased purity and realism of the input streams, a reduction in noise inputs, the enlargement of the system models, the improvement of nonlinear techniques, the increased rejection of spatial and temporal lumping, the achieving of a rational balance between analytical and synthetic systems models, and the like, which have been treated in earlier chapters of this book.

10.2 The demographic dilemma

Many environmentalists take a much more basic and pessimistic view regarding the problems of applying a rational and evolving technically based systems approach to our environmental problems. They point to the present world population growth of three times that necessary for a stable population (see chapter 9) as the key to problems of linkage between the physico-ecological and socio-economic systems. It is postulated that if this population growth continues:

1 The pressure of socio-economic control over physico-ecological systems must increase.
2 Ways will therefore be found to subjugate the latter more rapidly, efficiently and completely to the former.
3 The demands placed by society on the physico-ecological systems will be so heavy and varied that the latter will tend to operate more and more on the margins of efficiency. This will make them more vulnerable to large-scale energy variations, as yet outside the control and prediction of society, so resulting in the Malthusian increase in the effects of natural catastrophes (seismic, volcanic, meteorological, climatic, hydrological and biological) which is already becoming apparent.
4 The number and intensity of conflicts of interest will escalate, making democratic decision making systems more and more unwieldy (see chapter 6). Conflicts of interests, resulting from pressure on resources,

will increase the aggressive competition leading to increased demands for devolution of power towards the most favoured areas, and resulting in the polarization of conflicting power groups between favoured and unfavoured parts of the world. This will introduce a tendency which is in direct opposition to the ever-more widespread planning and control necessary to husband the physico-ecological environment (see chapter 9).

Thus environmental scientists would place population control at the centre of environmental planning strategies and would regard the present demographic explosion as the main impediment to the application of rational control strategies of systems modelling and control. Thus, the population problem resolves itself into the class of difficulties treated in chapters 7, 8 and 9, where the unstable positive-feedback tendencies of socio-economic systems were contrasted with the dominantly stable negative-feedback operation of physico-ecological ones. Other scholars, as we have seen in chapters 8 and 9, have discussed Malthusian and neo-Malthusian arguments for population control, in the context of preservation of *status quo* élites in the Third World and of neocolonial systems in the developed world, and have called for a total reorganization of the capitalist mode of production (Robinson 1970). Such ideas require a reorientation of our psychological and epistemological attitudes.

10.3 Psychological difficulties

Basic as the belief in the demographic impediment is to much environmental thinking, it is possible to isolate a body of even more fundamental fears. These fears relate to what may be identified as a syndrome of group psychological attitudes which place the emphasis upon inevitable and uncontrollable progressions of world events, and which are expressed in a general loss of faith in the kind of logical-positivist approaches of which conventional systems philosophy is an expression. Of course, the demographic doom watchers, drawing on analogies with the behaviour of crowded animal populations, attribute many modern group psychological attitudes to population pressure, but others regard the population and group psychological problems as being distinct, at any rate to some extent. Heilbroner (1974) has clarified the nature of this so-called 'mood of our time' in terms of three dominant facets:

1 *Topical*: This is characterized by such features as a sense of foreboding associated with the increase in world violence, the recognition of the pressure of the rising world population, a decline in the belief in the invulnerability of the West, a belief in the inability of the middle-aged generation to pass on its values to its children, etc.
2 *Attitudinal*: This is dominated by a widespread loss of confidence in the general course of social events, in the efficacy of large social and

environmental engineering and in the desirability of continued economic growth.

3 *Civilizational*: This manifests itself in a loss of faith in the general and continued advance of civilization and in the traditional assumptions that increased material well-being is, in itself, able to satisfy the human spirit.

Associated with this shift in mood has been a tendency to question both the method and implications of science. This questioning has propagated the view that unbridled scientific research and advances lead to a dehumanization of the object of study and of the implications of the results. Wide currency is given to the belief that scientific knowledge is divorced from the subjective penetration of 'really knowing' (Roszak 1972), that science makes the organization of government and society so complex that the ordinary individual has no capacity to participate, and that it makes conventional 'progress' an inevitable consequence of the employment of its methods. For example, scientific advance in economic input-output modelling since Quesnay's (1758) *Tableau Economique* has only served to obscure rather than highlight the fact that nature is the only really productive sector (see Shackle 1972, pp. 250–1). Hence, there is increasing questioning of the need for technological advance, increasing criticism of the technocratic organization of society, and a feeling of alienation of the individual from collectively resolved goals.

In chapters 1 and 5 we have already gone some way towards treating the structure of environmental actions based on faculties other than intellectual or logical thought alone. It is becoming increasingly obvious that in the future the world as 'felt' should also enter into our treatment of environmental systems. This overlaps with considerations of value and legitimacy of actions discussed as epistemological issues in the next paragraphs. But it is worth reiterating at this stage the need to move our demands on systems analysis away from questions purely about things towards questions about flows and about processes. This Waddington (1977) characterizes as a major psychological difficulty dating at least to the difference in view of Democritus that things are fairly stable and definite entities which can be aggregated together, with the view of Heraclitus that things are attributes of processes of change and that mechanisms of change tell us the more interesting information about system behaviour. In relation to the behaviour of man, we can expect to understand the behaviour of environmental systems only by considering the 'wholes' as complexes which can be recognized as more than mere aggregations of physics and chemistry, but are difficult to describe in any meaningful way. This leads the system analyst to the need to accept, in addition to his familiar hard systems and positivist scientific approach, the logic of 'fuzzy' systems with tolerance levels on the potential understanding of behaviour. In mathematical terms the behaviour of such systems which cannot be precisely defined leads us to consider the equations with chaotic solutions discussed by

May (1975) and in earlier chapters. But we also need to consider the cognitive factors discussed by Freud, Jung, Reich and others in some 'holistic' fashion. Thus in attempting to overcome the psychological difficulties to the interfacing of physical and social systems we are, in a very real sense, brought back to the issues raised in chapter 1; of the need to encompass not only abstract or 'true' nature, but also nature as perceived by man on the basis of his experiences.

10.4 Epistemological transitions

The three foregoing groups of impediments to the application of systems approaches to environmental problems have been listed in increasing order of difficulty. They do share, however, to a greater or lesser extent, some ability to be predicted into the future by conventional techniques and therefore to be at least susceptible to premeditated attack by environmental systems scientists. In considering the final group of impediments to the environmental application of systems methods one is forced to question our ability to predict world situations to which systems methods might be called upon to respond. Maruyama (1972, pp. 158–9) has suggested the following typology of such futuristics:

1 *Descriptive*: deals with trends in social change; sociological or historical.

2 *Predictive*: This may be subdivided into:
   (*a*) *Extrapolative*: Inferences regarding future derived from the past rate of change or past rate of acceleration of change.
   (*b*) *Speculative*: Estimates of change resulting from innovations which have no historical precedent.

3 *Pragmatic*: Study of what can and should be done to affect the direction of social change. This can be subdivided into:
   (*a*) *Reactive*: Further subdivided into:
      1 *Defensive*: Mainly concerned with how to preserve the old pattern (for example, family structure, the size of population, clear air over city, etc.).
      2 *Instrumental*: Mainly concerned with how to use new tools for old purposes (for example, how to replace accountants with computers while keeping the accounting system intact).
      3 *Adaptive*: How to modify culture in order to fit it to technological changes – i.e. let culture be dictated by technology.
   (*b*) *Goal-generating*: First ask what the goals of mankind can be, then adapt and develop technology toward these goals. This can be subdivided into:

1 *Uniformistic or homogeneistic*:
   (i) *Constitutional*: Numerical majority imposes its decision uniformly upon minorities (so-called democracy).
   (ii) *Planned Utopia*: A small number of 'planners' design the society 'for' the rest of the population.
2 *Pluralistic or heterogeneistic*: No *a priori* goals; builds on goals generated by people; accommodates diversities and differences; studies intercultural ecology such as cultural symbiosis and cultural parasitism.

We have seen at many stages in the preceding chapters the difficulties in generating goals (descriptive, prescriptive or dialectic), especially those deriving from attempts at constructing social welfare functions. Even when widely agreed goals are available at a high level, they may well be devoid of meaning in terms of approachable operational action.

The most complex group of goal-generating futures involves epistemological transitions-shifts in the foundations, framework and internal structure of our processes of reasoning (Maruyama 1972, 1973). Such transitions are extremely difficult to foresee and to accommodate in terms of existing systems bases, but, as Habermas (1973) notes, are the major causes of overloading of control capacities. Shifts and transitions in goals will affect spatial jurisdictional boundaries, inter-personal legal arrangements and aggregate constitutions, the structure of visual and personal surroundings, and empathy of the state. Clues as to the possible areas of transition may perhaps be presented by existing epistemological discrepancies, such as:

Competition versus Sharing
Technocratic transgression versus Harmony of nature
Material efficiency versus Cultivation of the mind
Hierarchism versus Non-hierarchical mutualism
Concept of leadership versus Interactionism
Majority rule versus Consensus or separatism
Homogenization versus Pluralism

(Maruyama 1972, p. 161)

Of these, the concepts of hierarchism and homogenization are receiving more than their share of attention. Maruyama (1973, p. 436) has questioned the general social applicability of the unidirectional causal thermodynamic model of physics and has suggested that generalizability, universality and homogenization are neither the rules of the universe nor necessarily desirable goals for society. He believes that a more general biological and social rule is heterogenization of processes, such that what survives is not the strongest but the most symbiotic. Pattee (1973, pp. 96, 103) has similarly referred to Koestler's condition for a creative act, the 'juxtaposition of disparate categories', as a justification for the existence and survival of hierarchical systems which may have been generated either as a necessary result of the dynamics of the system or by a 'frozen accident', wherein the organization is generated by accident but remains fixed because any change would be lethal to the system. The Second Report of the Club of Rome (Mesarovic and Pestel 1975) has made the plea for differentiated, organic-like future world growth, rather than undifferentiated exponential replication.

It would be presumptuous to attempt to predict the directions in which epistemological transitions will lead. However, we can identify the present major battlegrounds which highlight possible epistemological futures. In epistemic terms the environmental imperative has been seen variously as crises of capitalism, of positivism, of historicism, and of legitimation.

The crisis of capitalism has been expressed by the call of Joan Robinson (1970) and others for the reorientation of Western social values away from the capitalist mode of production towards more conservation, a better distribution of capital and property, and diffusion of benefits. The future requires the development of a classless individualism, it is argued, which might be found either in Marxist communism, or in Bernstein's Revisionist Marxism of state collectivization, or in late capitalist beneficent institutions (e.g. trusts, financial institutions and international financial syndicates). Thus 'late capitalist' institutions such as the World Bank and the IMF have been severely criticized as maintainers of the neocolonialist *status quo* (Robinson 1964). Hence, Marxists and critical theorists, such as Marcuse, have interpreted late capitalism as unable to maintain the engine of growth. For them the potential for the creation of new demand by innovation of new goods through 'commodity fetishism' is now limited.

The positivist dilemma results from the false isolation in which the analyst finds himself in relation to the social-environmental system of which he is a part, and from the separation of the worker from the results of his labours (i.e. reification). The atomization of man within the framework of macroeconomics and the statistical manipulation of the mass of society by strong central governments (Shackle 1972, p. 362) have, in the opinion of many, produced a 'thingified' man whose abilities (extending even to the depths of his psychic nature) are considered, like resources, to be objectivized properties to be owned and disposed of (Lukács 1971). This is part also of what Berger and Pullberg (1966) see as the end point of an alienation in which man forgets that he lives in a world of his own creation, and in which entities that he has produced, named or classified become things separated from the activities that created them. Much research into resource management has used such concepts, which are then taken as explanations of how people behave in any situation, depicting man as a mere role-player or as an institutional decision maker powerless to change his historical and social circumstances (Keat and Urry 1975, pp. 180–6). Thus the critical social theorists, through the writings of Marcuse, Horkheimer and Habermas, see positivism as unable to move beyond purely technical issues (discussed in section 10.1 above), as noted in

Wittgenstein's famous comment that even when all scientific questions have been answered the problems of life will still remain. The opponents of positivism call for self-reflective understanding to develop both a sense of individual responsibility and of a need for personal action (Habermas 1971, Keat and Urry 1975, pp. 222–7).

The crisis of historicist arguments creates a crucial duality. The positivist critique (as exemplified by Popper 1957) recognizes, through the Hempel and Oppenheim (1948) symmetry thesis, a structural identity between historical laws and those of natural science, wherein individual events in space and time contribute only one element to a generalized map of the kaleidoscope of environmental development. Against this view, nineteenth century radicals and ecologists, in emphasizing the role of social action, have seen decision as inconceivable outside of the social milieu, unique in the time and space in which it is to be applied. Especially influential has been Marx's Theses on Feuerbach (1888) which place ideology in active and conscious interaction with nature in order to transform it (see chapter 1). Environmental action therefore requires historical intuition, empathy and social perspective, or at least what Mesarovic and Pestel (1975) term 'differentiated organic growth'. Hence there has been much recent concern in geography and other social sciences with unique cultural value systems, and with geographies, mind-scapes, and soundscapes of the past. In recognizing the inadequacy of most modern economics based on equilibrium theory, some economists (e.g. Kornai 1971, Dobb 1973, Georgescu-Roegen 1971, Shackle 1972) are now seeking historical insight by biological analogy with Darwinian natural selection and Hegelian synthesis. The implications for environmental epistemology are disturbing since the emergence of new social behaviour is indeterminate, being the result of social mutations which are not knowable in advance. Thus most ecological writings ignore the essential dilemmas of how transition to the Utopian society can be achieved, or the lag time which is required for painful changes in attitudes and the much-needed reeducation of planners and governments. This divorce from reality is variously present in the social energetics of Frederick Soddy, H. G. Wells and Howard Odum (Foley 1976), in the intermediate technology of Schumacher (1971), in Foley's (1976) low energy society based on current institutions and the social capital of cities, and in the suggestions of the 'Blueprint for survival' (*The Ecologist* 1972) involving population control, deurbanization, decentralization of industry, and deindustrialization.

The legitimation crisis concentrates on the central question of values. We have seen in earlier chapters that even if we take a moderate middle way in response to the supposed environmental problems forecast by Forrester and Meadows, then we must change our social and economic goals to a more conservation-oriented system.

Whether these goals derive either from abstract Platonic and Aristotelian idealism or from rationalist communicative socialization of new norms based on democratic collective action, there are increasing difficulties in achieving agreement or in legitimating decisions. If environmental problems are as complex and imperative as is suggested by some commentators, agreed goals are unlikely to be achieved on a large enough scale or with sufficient rapidity. When idealist norms are not available or when there is insufficient time to allow the infinite regress to achieve agreed norms, Habermas (1971, p. 36) suggests that the logical conclusion is to base legitimation of decision on the diffusion of mass loyalty to generalized motives. In this manner the disconnection of motives from agreed norms removes the need to justify individual goals and no legitimation problem exists. Is this, then, the sense in which mankind is at the turning point? Is it the ultimate failure or ultimate triumph of the human species that actions need not be legitimised by participation or democracy; that, as Habermas (1973, pp. 122, 140–3) presages, humanity can be rid of the need for truth? It has often been convenient to invoke totalitarian solutions to overcome perceived inefficiencies in decision making, but we now find ourselves at what is perhaps a fundamental parting of the ways. On the one hand, one may question whether a belief in abstract truth is necessary at all and that, if there is no true world of abstracts, then personal experience and valuations must replace truth (as exemplified by the beliefs of Nietzsche and Sartre). We are left with environmental actions which are determined by the practical considerations of choice of what we desire at any moment (which is the logical extension of Rousseau's social contract) answered in terms devoid of truth but with no possibility of appeal. On the other hand, does the essential ambiguity in the human ethos leave room for a more 'humane' solution? The concept of Platonic right and duty, of eastern familial bonds, of David Hume's limitation of government by law, and the ambiguous role of religion, provide promises of meaning through concepts which are not only of 'value' but perhaps 'true' as well. Modern man has tended to make consolation of environmental impacts mere administrative acts of flood, earthquake, or famine relief, and has often substituted the workings of a 'welfare state' for more naturally evolved cultural checks and balances. Thus we have the spectre of a Thomas Hobbes *Leviathan* of ever-spawning western and communist bureaucracies oriented towards satisfying an ever-expanding set of human demands and aspirations. We may have come to a point of needing a 'cultural shock', or at least a new view of what is essential to the ethic of humanity. But even if, as Polanyi and Teilhard de Chardin suggest, the 'man-world' is now so important that Glacken's (1967) concept of environmental dominance must be rejected, the significance of environmental systems, as conceived in this book, cannot be ignored. Durkheim's concept of the sacred relation between man and his group symbols and the new radical's emphasis on ideological relations are not independent of nature; man, as the proper study of mankind, is not a new totemism, apart from nature and demarked legitimate scientific enquiry (Lévi-Strauss 1962).

Each epistemological crisis discussed above resolves itself into conflicts involving interpretations of utility, value, truth and justice. Oscar Wilde's cynic, described as a man who knew the price of everything and the value of nothing, and Lewis Carroll's 'Humpty Dumpty' might hold that all these words mean as much or as little as one chooses, but many would not. Here, surely, we reach the bounds of our investigation of environmental systems from which we must withdraw to more tractable considerations. Chief among these is a determination of just how *complex* the total environmental system really is, and how *imperative* are decisions to modify it. If environmental systems are as complex and counter-intuitive as has been argued by some commentators, or if the impending environmental collapse is as imminent as has been argued by others, then the environmental imperative requires changes in social values which are needed now. These epistemological changes must be to new sets of values or to new interpretations of old ones. The lag time of mankind's culture behind his technology has been expanding since the Pleistocene, and the argument that we must abandon existing institutions because of their inability to cope with the technological changes necessary for the maintenance of the species is extremely persuasive. However, the degree to which systems are perceived to be too complex, or the environmental imperative too imminent to allow legitimation of actions, is itself an ideological question not devoid of value. Even a modern Joseph would have difficulties in managing the fat years and thin years of the future; stockpiling today for the dearth of tomorrow has profound ideological implications, not least for the Third World who are most likely to suffer today's immiserization. Necessity has often been used by the sophist for repression, and consolation of environmental contingency the apologist for social conscience. The true extent of environmental complexity and imperative will determine the degrees of freedom in answering our epistemological questions, but it is unlikely that there will be one answer to environmental problems. What the form of future emergent societies will be, whether drastic epistemological transitions will occur, and, if so, whether conventional systems approaches or any possible elaborations of them will have relevance to the resulting environmental challenges, is a matter for fascinating speculation.

It has been a criticism of many applications of systems thinking to environmental problems that, while creating a satisfying philosophical milieu for general speculation, they do little towards assisting in the operationalizing of problems in a manner leading to their practical solution (Langton 1972). It is hoped that our attempts to depict a wide range of phenomena as logical sets of transformations which are operational, or might reasonably be expected to become so, have taken the consideration of environmental systems a step in a constructive direction. It would be wrong to end this book, however, without some further explicit reference to the holistic debate raised in chapter 9, and especially to the way in which the behaviour of the whole system controls that of its parts (Phillips 1977).

The complexity of biological and social systems, in particular, stems largely from their very sophisticated capacity for internal reorganization implying 'flips' across thresholds separating systems states which even current non-linear techniques find difficulty in modelling. In social systems such reorganization is particularly striking and it is clear that any developments in environmental systems theory or practice, irrespective of the field of application, must in the future pay increasing attention to this category of problems. As Langton (1972, p. 141) put it: 'Social systems typically, then, progress towards greater organization and greater levels of functional complexity. Such changes can only occur because the maintenance of a functional existence is not generally the *raison d'être* of systems – although this may be true of certain exceptional types of system. . . . They occur because systems function for a *purpose* to which the function itself is subservient'. In other respects, however, the holistic attitude to such systems may be unrealistic. In some sense, parts of systems may be viewed as having come into being and maintained, despite the equilibriating, integrating and entropy-maximizing tendencies of the system as a whole, because they possess some measure of functional autonomy. Thus the opposing tendencies of systems interdependence and functional autonomy produce a tension in the system, and it is the state of this conflict which may be seen as determining much of system structure. A system possessing weak functional autonomy when subjected to drastic change faces either complete destruction and re-creation, or a radical structural reorganization to achieve a new equilibrium state; whereas one with significant functional autonomy can react by surrendering higher levels of integration so that its autonomous parts may regroup on a lower level of complexity (Gouldner 1975, pp. 205–15). Whether such changes originate from without or within the system are matters of crucial importance, and both have been stressed in this book. Gouldner (1975, p. 218) takes the view, however, that the distinction between exogenous and endogenous change is not clear cut and depends to a large extent upon the existing balance between interdependence and functional autonomy displayed by the system. For Marxist sociologists the stresses generated by internal contradictions within social systems represent important endogenous forces, whereas the conventional Parsons' social model is based on the concept of exogenous change. Such concepts may not be restricted to the social sciences, and any developments in systems methods concerned with effecting a realistic interfacing between socio-economic and physico-ecological systems must address questions of the complex interrelations between exogenous and endogenous inputs, mutations of system structure, thresholds and cut-offs of interaction, and hierarchical 'flips'.

One truth which has emerged from this book is that the real-world situations which environmental systems attempt to model are complex. Up to the present efforts to control or influence these systems have been made with reference to aims which were deeply held only by small, articulate and influential groups of society, to produce limited changes in certain aspects of system operation. Thus Marxists have attempted to control production and distribution in order to equalize wealth. Conservationists have attempted to influence government legislation to halt environmental degradation, and Protectionists have tried to control trade in order to preserve local employment. Because environmental systems are complex and form a nested hierarchy up to the global level, the results of such control actions often include those which are broadly detrimental to society as a whole; such as a uniformly-grey police state, the effective exclusion of large sections of the population from the most beautiful parts of the country by ensuring that the facilities are such as to discourage their use, or national recessions. This book has no easy answer for this ultimate control problem except to exhort an increasing realism through systems modelling. Three matters are of particular concern; firstly, the aims of control and intervention, secondly, the results of such control (unplanned as well as planned), and, thirdly, to what degree the aims of control and the intensity and hierarchical level of control must be moderated in order to result in broadly benign effects. On the one hand, it may be that it is merely the inadequacies of current systems conceptions and control techniques which separate a Kafkaesque hell from a Marxist heaven; on the other, it is much more likely that all but the most broadly humane and delicate of controls over environmental systems may be disastrous.

'Nature, with equal mind,
Sees all her sons at play,
Sees man control the wind,
The wind sweep man away.'
(Matthew Arnold: *Empedocles on Etna*)

Appendices

1 Simple matrix algebra

In this appendix we briefly define the four common operations with matrices equivalent to the arithmetic operations of addition, subtraction, multiplication and division. More advanced treatment of matrix methods is given by Graybill (1969) or Gantmacher (1959).

A *matrix* **A** is composed of the $(m \times n)$ elements a_{ij}, where m refers to the number of rows, and n to the number of columns. This is usually written as:

$$\underset{m \times n}{\mathbf{A}} = \begin{bmatrix} a_{11} & a_{12} & \cdots & a_{1n} \\ a_{21} & a_{22} & \cdots & a_{2n} \\ \vdots & \vdots & & \vdots \\ a_{m1} & a_{m2} & \cdots & a_{mn} \end{bmatrix}.$$

The special case of a $(m \times 1)$ matrix is termed a *vector*:

$$\underset{m \times 1}{\mathbf{A}} = \begin{bmatrix} a_1 \\ a_2 \\ \vdots \\ a_m \end{bmatrix}.$$

Vectors are conventionally written as column vectors. A row vector can be defined as the *transpose* of the column vector as:

$$\underset{1 \times m}{\mathbf{A}'} = [a_1 \quad a_2 \quad \ldots \quad a_m].$$

The transpose is given by the interchange of the rows and columns of a matrix. Hence, for the full $(m \times n)$ matrix the $(n \times m)$ transpose matrix is given by:

$$\underset{n \times m}{\mathbf{A}'} = \begin{bmatrix} a_{11} & a_{21} & \cdots & a_{m1} \\ a_{12} & a_{22} & \cdots & a_{m2} \\ \vdots & \vdots & & \vdots \\ a_{1n} & a_{2n} & \cdots & a_{mn} \end{bmatrix}.$$

If $\mathbf{A} = \mathbf{A}'$, then $\mathbf{A}$ is a *symmetric* matrix, and $a_{ij} = a_{ji}$.

Matrix and vector *addition* is carried out by adding the corresponding elements element by element: for the *ij*th element in the two matrices **C** and **B** the addition term is defined by:

$$a_{ij} = b_{ij} + c_{ij}.$$

For example:

$$\begin{bmatrix} 2 & -3 \\ 4 & -2 \end{bmatrix} + \begin{bmatrix} -1 & 0 \\ 4 & -1 \end{bmatrix} = \begin{bmatrix} 1 & -3 \\ 8 & -3 \end{bmatrix}.$$

Subtraction of matrices is defined similarly.

Multiplication of matrices is carried out by taking row by column multiples between the two matrices and summing the resulting terms. Thus,

$$a_{ij} = \sum_{k=1}^{n} b_{ik} c_{kj}.$$

For example, the vector-matrix multiplication gives:

$$\begin{bmatrix} -2 & 4 \\ -3 & 1 \end{bmatrix} \begin{bmatrix} 1 \\ 3 \end{bmatrix} = \begin{bmatrix} (-2 + 12) \\ (-3 + 3) \end{bmatrix} = \begin{bmatrix} 10 \\ 0 \end{bmatrix}.$$

The multiplication of two matrices gives:

$$\begin{bmatrix} -2 & 4 \\ -3 & 1 \end{bmatrix} \begin{bmatrix} 1 & 5 \\ 3 & -1 \end{bmatrix} = \begin{bmatrix} (-2 + 12) & (-10 - 4) \\ (-3 + 3) & (-15 - 1) \end{bmatrix} = \begin{bmatrix} 10 & -14 \\ 0 & -16 \end{bmatrix}.$$

Note that the result in the vector-matrix multiple case is a (2×1) matrix; and in the matrix-matrix case is a (2×2) matrix. Also, it will only be possible to carry out the multiplication when the number of columns of the first matrix equals the number of rows of the second matrix. If this is true, then the two matrices are said to be *conformable* with respect to multiplication (but this does not restrict the number of rows in the first matrix, nor on the number of columns in the second matrix). There is one important difference between matrix and ordinary arithmetic multiplication. In the arithmetic case the multiplication operation will satisfy:

$$a \times b = b \times a$$

and the order of multiplication is not important: the terms a and b are said to be *commutative*, and this holds true for all normal arithmetic multiplication. In the matrix case, however, the commutativity property does not hold (unless the two matrices are diagonal) and thence:

$$\mathbf{AB} \neq \mathbf{BA}.$$

Thus, for the example above,

$$\begin{bmatrix} 1 & 5 \\ 3 & -1 \end{bmatrix}\begin{bmatrix} -2 & 4 \\ -3 & 1 \end{bmatrix} = \begin{bmatrix} -17 & 9 \\ -3 & 11 \end{bmatrix}.$$

The matrix analogue of the arithmetic operation is termed *inversion*. It is more complicated than the previous operations and requires the definition of three new matrix elements: the determinant, cofactor, and adjoint.

The *determinant* of a matrix **A** is written |**A**| or det **A**. It is defined simply for a (2 × 2) matrix as:

$$\det \mathbf{A} = \begin{vmatrix} a_{11} & a_{12} \\ a_{21} & a_{22} \end{vmatrix} = a_{11}a_{22} - a_{12}a_{21}.$$

For a (3 × 3) matrix the determinant is defined by:

$$\det \mathbf{A} = \begin{bmatrix} a_{11} & a_{12} & a_{13} \\ a_{21} & a_{22} & a_{23} \\ a_{31} & a_{32} & a_{33} \end{bmatrix}$$

$$= a_{11}a_{22}a_{33} - a_{11}a_{23}a_{32} - a_{21}a_{12}a_{33} + a_{21}a_{13}a_{32} + a_{31}a_{12}a_{23} - a_{31}a_{22}a_{13}$$

$$= a_{11} \det A_1 - a_{21} \det A_2 + a_{31} \det A_3$$

where the new terms A_i are the submatrices or *minors* formed by deleting the rows and columns contained in the multiplying element. Thus,

$$A_1 = \begin{bmatrix} a_{22} & a_{23} \\ a_{32} & a_{33} \end{bmatrix}.$$

This method of calculating the determinant can be extended to the nth-order determinant by progressively deleting rows and columns until a set of (2 × 2) matrices (minors) is formed. This becomes impractical to carry out by hand for large systems and a computer solution must be sought.

The *cofactors* of a matrix **A** are given by the determinant formed by row and column deletion for each element in the matrix taken in turn. This is then multiplied by $(-1)^{i+j}$. For example, for the (3 × 3) matrix example:

$$[\mathrm{cof}\, a_{ij}] = \begin{bmatrix} \det\begin{pmatrix} a_{22} & a_{23} \\ a_{32} & a_{31} \end{pmatrix} & -\det\begin{pmatrix} a_{21} & a_{23} \\ a_{31} & a_{33} \end{pmatrix} & \det\begin{pmatrix} a_{21} & a_{22} \\ a_{31} & a_{32} \end{pmatrix} \\ -\det\begin{pmatrix} a_{12} & a_{13} \\ a_{32} & a_{33} \end{pmatrix} & \det\begin{pmatrix} a_{11} & a_{13} \\ a_{31} & a_{33} \end{pmatrix} & -\det\begin{pmatrix} a_{11} & a_{12} \\ a_{31} & a_{32} \end{pmatrix} \\ \det\begin{pmatrix} a_{12} & a_{13} \\ a_{22} & a_{23} \end{pmatrix} & -\det\begin{pmatrix} a_{11} & a_{13} \\ a_{21} & a_{23} \end{pmatrix} & \det\begin{pmatrix} a_{11} & a_{12} \\ a_{21} & a_{22} \end{pmatrix} \end{bmatrix}.$$

The *adjoint* of a matrix **A** is given by the transpose of the cofactors:

$$\text{adj}\,\mathbf{A} = [\text{cof}\,a_{ij}]'.$$

Using these three expressions for the determinant, cofactor and adjoint, it is now possible to define the inverse of a matrix by the term $\mathbf{A}^{-1}$ which is given as:

$$\mathbf{A}^{-1} = \frac{\text{adj}\,\mathbf{A}}{\det \mathbf{A}}.$$

This is best understood by reference to an example. For the (2 × 2) matrix:

$$\mathbf{A} = \begin{bmatrix} 2 & 4 \\ 1 & 3 \end{bmatrix}$$

the inverse of **A** can be calculated as follows:

$$\det \mathbf{A} = 2 \times 3 - 4 \times 1 = 2$$

$$\text{cof}\,\mathbf{A} = \begin{bmatrix} 3 & -1 \\ -4 & 2 \end{bmatrix}$$

$$\text{adj}\,\mathbf{A} = \begin{bmatrix} 3 & -4 \\ -1 & 2 \end{bmatrix}$$

$$\mathbf{A}^{-1} = \tfrac{1}{2}\begin{bmatrix} 3 & -4 \\ -1 & 2 \end{bmatrix} = \begin{bmatrix} \frac{3}{2} & -2 \\ -\frac{1}{2} & 1 \end{bmatrix}.$$

For the more complex (3 × 3) example given below, it should be verified by the reader that the following relation is correct:

$$\mathbf{A} = \begin{bmatrix} -2 & 1 & 3 \\ 0 & -1 & 1 \\ 1 & 2 & 0 \end{bmatrix}; \qquad \mathbf{A}^{-1} = \begin{bmatrix} -\frac{1}{4} & \frac{3}{4} & \frac{1}{2} \\ \frac{1}{8} & -\frac{3}{8} & \frac{1}{4} \\ \frac{1}{8} & \frac{5}{8} & \frac{1}{4} \end{bmatrix}.$$

When the matrix to be inverted is not square (equal number of rows and columns), then a method of *pseudo-inversion* must be adopted, as described in Graybill (1969).

2 Derivation of optimum control equations

In the following paragraphs is given the derivation of the optimum control equations for one-stage and N-stage control problems in deterministic and stochastic environments. These results are used in fig. 3.23 (pp. 132–3). More lengthy treatment can be followed in Pontryagin *et al.* (1962).

A2.1 Deterministic control with quadratic costs (W_2)

The quadratic cost equation (3.5.11) can be redefined in terms of the two elements of the costs of control and the costs deviation of the output from the set point:

$$W_2 = M(Y_t - Y_t^*)^2 + N(U_t - U_t^*)^2. \tag{A2.1}$$

The system equation is given by

$$Y_t = A^{-1}(z)B(z)U_t$$

in which all the inputs U_t are considered to be controllable. Substitution of the system equation into the cost equation yields:

$$W_2 = M(FU_t - Y_t^*)^2 + N(U_t - U_t^*)^2$$

where

$$F = A^{-1}(z)B(z).$$

The optimal control law can be derived by differentiating this new definition of the control costs with respect to the policy variable U_t, the optimum being given by that point for which this derivative equals zero – i.e. changes in controls can produce no smaller costs and the gradient of the cost function is zero. Define the optimum control by U_t^0, then:

$$\frac{\partial W_2}{\partial U_t} = 2MF(FU_t - Y_t^*) + 2N(U_t - U_t^*)$$

and

$$0 = (MF^2 + N)U_t^0 - MFY_t^* - NU_t^* + (MF^2U_t^* - MF^2U_t^*)$$
$$= (MF^2 + N)U_t^0 + MF(FU_t^* - Y_t^*) - (MF^2 + N)U_t^*$$

and

$$U_t^0 = U_t^* - \frac{MF}{MF^2 + N}(FU_t^* - Y_t^*). \tag{A2.2}$$

Thus the optimum deterministic controller is a function of the desired set level, and the difference between the set point and the actual level of system output, which is a result identical to Kalman and Koepcke (1958) and Kalman (1960, theorem 4) and is the dual of the forecasting solution. Perfect control yields:

$$Y_t^* = FU_t^*$$

for which $W_2 = 0$, but generally there is an inconsistency between (A2.2) and (A2.1) – i.e. between the desired value of Y_t^* and U_t^*. This inconsistency can be defined as the fall-short in control, or differences between the desired control costs and output deviation costs:

$$D_t = Y_t^* - FU_t^* \tag{A2.3}$$

and so,

$$U_t^0 = U_t^* + \frac{MF}{MF^2 + N} D_t. \tag{A2.4}$$

The term in brackets is the amount by which the control U_t must exceed its desired level U_t^* for the output to equal its desired level Y_t^*. If the control is optimal, then the optimal level of the output is given by:

$$Y_t^0 = Y_t^* + \frac{MF^2}{MF^2 + N} D_t$$

and the cost function becomes, after inserting the optimum values:

$$W_2^0 = \frac{M^2F^2}{MF^2 + N} D_t^2.$$

Since M and N are positive, then the costs will be greater than or equal to zero, and will equal zero only if (A2.3) equals zero. Note that the optimum levels of the output and the control do not depend on the values of the costs M and N, but on the ratio of M to N (of the costs of control to the costs of output deviations). Also if the error criterion is a function of the control targets, then the optimal policy will yield zero costs, but when neither M nor N is equal to zero, then neither goal can be fully achieved, a dilemma noted by economists with reference to exchange rate control.

A2.2 Stochastic control with linear costs (W_1)

Stochastic control problems are the most generally encountered and we discuss here feedforward, feedback, and feedback-feedforward schemes each using the minimum of the one-step ahead errors criterion W_1. In each case the system equation to be controlled is given by

$$A(z)Y_t = B(z)U_t + \frac{C(z)}{D(z)} e_t \,. \tag{A2.5}$$

It is assumed that each polynomial is stable, that the system is controllable, and that $\{n_t\}$ is an unknown independent error sequence.

A2.2.1 STOCHASTIC FEEDFORWARD CONTROL

If the errors are known, then Holt (1962) and Box and Jenkins (1968, 1970) show that their effect can be cancelled by setting:

$$\frac{B(z)}{A(z)} U_t = -\frac{C(z)}{D(z)} e_t$$

which gives the control strategy as:

$$U_t^0 = -\frac{A(z)}{B(z)}\frac{C(z)}{D(z)} e_t. \tag{A2.6}$$

The most common instance in which the errors are known is when there is some delay to their input, so that disturbances are measured before they affect the system. Generally this will not be possible and the error disturbances must be replaced by forecasts of their values in (A2.6). Note that the variance of the output will satisfy:

$$E(Y_t^2) \geqslant E(e_t^2) = \lambda^2$$

and the control will reduce this equation to an equality.

A2.2.2 STOCHASTIC FEEDBACK CONTROL

It is possible to use forecasts of the noise disturbances to implement control (Box and Jenkins 1968, 1970) or to use minimum variance settings given by Åstrom (1965) and Åstrom and Wittenmark (1971). This latter derivation is limited, however, to the case in which $C(z) = 1$, and the derivation is extended here to cover arbitrary forms of the autoregressive noise $C(z)$. Taking the system equation (A2.5), subtract $A(z)e_t$ from each side:

$$AY_t - Ae_t = BU_t + \frac{C}{D} e_t - Ae_t \tag{A2.7}$$

and by interchanging,

$$Y_t = e_t + \frac{B}{A} U_t + \frac{(C/D - A)}{A} e_t \ .$$

The polynomials have been abbreviated for simplicity. Using the system equation (A2.5), lags of e_t can be eliminated to give:

$$Y_t = e_t + \frac{BD}{C} U_t + \left(1 - \frac{DA}{C} \right) Y_t \qquad \text{(A2.8)}$$

which expresses Y_t solely in terms of lags of the input control and the lagged outputs, together with the current value of the errors. If the noise is an uncorrelated sequence, then

$$E(Y_t^2) \geqslant E(e_t^2) = \lambda^2$$

and the equality holds only if the controls are given by:

$$U_t^0 = -\left(\frac{D}{CB} - \frac{A}{B} \right) Y_t = -k(z) Y_t \qquad \text{(A2.9)}$$

which will be realizable if either b_0 or b_1 is non-zero. Moreover, this control is minimum variance since it will reduce the deviation from target to the variance of the unknown noise (equivalent to n_t in (A2.5)). The minimum variance controller can be interpreted by recognizing that the last term in (A2.9) is based on the minimum variance one-step ahead prediction of Y_t based on lagged controls and lagged outputs. This method involves the separation and certainty equivalence principles discussed earlier in chapter 3, and reduces to the same result as the dead beat algorithm when the system noises are not correlated – i.e. $D(z) = C(z) = 1$.

A2.2.3 STOCHASTIC FEEDBACK-FEEDFORWARD CONTROL

The combined controller is given by a simple combination of the previous two algorithms (Holt 1962, Box and Jenkins 1970):

$$U_t = -\frac{A}{B} \left[\frac{C}{D} e_t - \left(\frac{D}{CA} - 1 \right) Y_t \right]. \qquad \text{(A2.10)}$$

A2.3 Stochastic control with quadratic costs

The introduction of the quadratic cost criterion W_2 again requires the certainty equivalence principle for which the derivation is given by Holt (1962), following Theil (1957). The cost criterion now becomes:

$$W_2 = M(e_t + FU_t - Y_t^*)^2 + N(U_t - U_t^*)^2 \qquad \text{(A2.11)}$$

which yields a control law equivalent to (A2.2):

$$U_t^0 = U_t^* + \frac{MF}{MF^2 + N}(Y_t^* - e_t + FU_t^*) \tag{A2.12}$$

which can be transformed to:

$$U_t^0 = U_t^* + \frac{MBA^{-1}}{M(BA^{-1})^2 + N}\left[Y_t^* - \left(e_t + \frac{B}{A}U_t^*\right)\right]. \tag{A2.13}$$

As in the feedback control scheme, the last term inside the square brackets is the minimum mean square one-step ahead predictor; the same result will hold for the N-step ahead predictor, provided that there are no uncertainties in parameter values.

A2.4 Fixed time quadratic cost control

When a planning horizon over n stages into the future is involved, the resulting control algorithms become more complex. The solution is given most readily in state space form:

(1) Given the system equation:

$$\dot{\mathbf{Z}} = \mathbf{AZ} + \mathbf{BU} \tag{A2.14}$$

and cost criterion:

$$\mathbf{J} = \int_{t0}^{t_f} \mathbf{Z'MZ} + \mathbf{U'NU}\, dt. \tag{A2.15}$$

(2) The Hamiltonian is formed by:

$$H(\mathbf{Z}, \mathbf{U}, \nabla J, t) = \nabla\mathbf{J}'(\mathbf{AZ} + \mathbf{BU}) + \mathbf{Z'MZ} + \mathbf{U'NU}. \tag{A2.16}$$

At a minimum of J, $\partial\mathbf{H}/\partial\mathbf{U}$ will equal zero and (A2.16) reduces to:

$$\mathbf{B}'\nabla\mathbf{J} + 2\mathbf{NU}. \tag{A2.17}$$

Hence $U^0(t, \mathbf{Z})$ is given by:

$$\mathbf{U}^0 = -\tfrac{1}{2}\mathbf{N}^{-1}\mathbf{B}'\nabla\mathbf{J}(t, \mathbf{Z}) \tag{A2.18}$$

where $\mathbf{N}$ must be positive definite.

(3) This expression for the optimum control can be substituted into (A2.16) to give the value of H at the optimum point:

$$\begin{aligned} H^0(\mathbf{Z}, \nabla J, t) &= \nabla\mathbf{J}'\mathbf{AZ} - \tfrac{1}{2}\nabla\mathbf{J}'\mathbf{BN}^{-1}\mathbf{B}'\nabla\mathbf{J} + \mathbf{Z'MZ} \\ &\quad + \tfrac{1}{4}\nabla\mathbf{J}'\mathbf{BN}^{-1}\mathbf{NN}^{-1}\mathbf{B}'\nabla\mathbf{J} \\ &= \nabla\mathbf{J}'\mathbf{AZ} - \tfrac{1}{4}\nabla\mathbf{J}'\mathbf{BN}^{-1}\mathbf{B}'\mathbf{B}\nabla\mathbf{J} + \mathbf{Z'MZ}. \end{aligned} \tag{A2.19}$$

(4) Thus the Hamilton-Jacobi equation becomes on substitution into (A2.19):

$$\nabla \mathbf{J}'\mathbf{AZ} - \tfrac{1}{4}\nabla \mathbf{J}'\mathbf{BN}^{-1}\mathbf{B}'\nabla \mathbf{J} + \mathbf{Z}'\mathbf{MZ} + \frac{\partial \mathbf{J}}{\partial t} = 0 \qquad \text{(A2.20)}$$

the solution of which yields the optimum control law.

(5) If we define

$$\mathbf{J}(t, \mathbf{Z}) = \mathbf{Z}'\mathbf{P}(t)\mathbf{Z} = \int_{t0}^{t_f} (\mathbf{Z}'\mathbf{MZ} + \mathbf{U}'\mathbf{NU})\, dt \qquad \text{(A2.21)}$$

then

$$\nabla \mathbf{J} = 2\mathbf{P}(t)\mathbf{Z}; \qquad \frac{\partial \mathbf{J}}{\partial t} = \mathbf{Z}'\frac{\partial \mathbf{P}(t)}{\partial t}\mathbf{Z}. \qquad \text{(A2.22)}$$

(6) Substituting the first of these relations directly into (A2.18) we have the expression for the optimal control law as:

$$U^0 = -\mathbf{N}^{-1}\mathbf{B}'\mathbf{P}(t)\mathbf{Z} = -\mathbf{K}'(t)\mathbf{Z} \qquad \text{(A2.23)}$$

where

$$\mathbf{K}(t) = \mathbf{P}(t) = \mathbf{P}(t)\mathbf{BN}^{-1}. \qquad \text{(A2.24)}$$

This is a standard form of feedback control solution for fixed-time control. Further details and alternative proofs are given by Kalman (1960), Kalman and Koepcke (1958), Schultz and Melsa (1967) and Dreyfus (1965).

(7) To obtain an expression for the covariance of the control strategy, substitute (A2.22) into (A2.20) to give the Hamilton-Jacobi equation in the new form:

$$\mathbf{Z}'\left[2\mathbf{P}(t)\mathbf{A} - \mathbf{P}(t)\mathbf{BN}^{-1}\mathbf{B}'\mathbf{P}(t) + \mathbf{N} + \frac{\partial \mathbf{P}(t)}{\partial t}\right]\mathbf{Z} = 0. \qquad \text{(A2.25)}$$

(8) Expanding $2\mathbf{P}(t)\mathbf{A}$ gives:

$$\text{Sym}\,[2\mathbf{P}(t)\mathbf{A}] = \frac{2\mathbf{P}(t)\mathbf{A} + 2\mathbf{A}'\mathbf{P}(t)}{2} = \mathbf{P}(t)\mathbf{A} + \mathbf{A}'\mathbf{P}(t). \qquad \text{(A2.26)}$$

(9) Expanding into (A2.25) gives:

$$\frac{\partial \mathbf{P}(t)}{\partial t} + \mathbf{M} - \mathbf{P}(t)\mathbf{BN}^{-1}\mathbf{B}'\mathbf{P}(t) + \mathbf{P}(t) + \mathbf{A}'\mathbf{P}(t) = 0 \qquad \text{(A2.27)}$$

which is the *matrix Riccati equation* which expresses the covariance of the state at any time t and can be solved at any time by integrating forward from initial condition $\mathbf{P}(0)$.

3 Notation (chapters 2, 3 and 4)

X	System input.
Y	System output.
$X(t)$	System input as a continuous variable.
$\{X_t\}$	System input as a discrete variable.
S	System transfer function (general form).
T	Time constant.
g	Time delay.
$Z_i(t)$	System state variable.
$\mathbf{Z}$	Vector of state variables $[Z_1, Z_2, \ldots, Z_n]$.
U_t	Control input variable (instrument).
$r(t)$	Reference input (target).
R	Overall control system transfer function.
$\mathbf{J}(U)$	Control objective function.
W, W_i	Performance indicators (costs).
$A(z)$	Transfer function autoregressive elements.
$B(z)$	Transfer function distributed lag elements.
$C(z)$	Transfer function moving-average elements.
$D(z)$	Transfer function autoregressive elements.
a_t	Measurable disturbances.
e_t	Estimation errors.
n_t	Unmeasurable disturbances.
$\mathbf{U}^0(t)$	Optimal control setting.
$\nabla\mathbf{J}(t)$	Variations in $J(t)$.
$\mathbf{H}(t)$	Control Hamiltonian (Pontryagin state function).
$\mathbf{P}(t)$	System covariance matrix.
$\boldsymbol{\theta}$	Parameter vector.
$\mathbf{F}$	System parameter evolution matrix.
$\mathbf{K}_t$	Kalman filter gain matrix.
$g(.)$	Constraint equations (in Langrangian multiplier problem).
λ_i	Lagrangian multipliers.

$\mathbf{M}$	Cost matrix on deviations from target.
$\mathbf{N}$	Cost matrix on control.
k_p	Proportional control gain.
k_I	Integral control gain.
k_D	Derivative control gain.
t_0	Initial time.
t_f	Final time.
p	Order of autoregressive parameter elements.
q	Order of distributed lag parameter elements.
$f(.)$	Function notation.
$E(.)$	Expectation operator.
z	Unit shift operator (Z-transform).
s	Laplace operator.
$\mathscr{Z}\{.\}$	Z-transform.
$\mathscr{L}\{.\}$	Laplace transform.
$R_{yy}(k)$	Autocorrelation function at lag k.
$R_{yx}(k)$	Cross-correlation function of Y on X at lag k.
ϕ_{kk}	Partial autocorrelation function at lag k.
σ_x	Standard deviation of X.
$\hat{Y}_t$, $\hat{\sigma}$	Estimates of $\hat{Y}_t$ or σ.
$S_y(f)$	Spectrum of Y at frequency f.
$G_{yx}(f)$	Gain of Y on X at frequency f.
$\psi_{yx}(f)$	Phase shift of Y from X at frequency f.
$\operatorname{plim}_{n \to \infty}$	Probability in limit as n tends to infinity.
$\lim_{n \to \infty}$	Limit as n tends to infinity.
n	Sample size.
N	Number of regions in a spatial domain of interest.

(*Vector and matrix notation is defined in Appendix* 2.)

References

ADAMS, J. G. U. (1972) Life in a global village. *Environment and Planning*, 4, 381–394.

ADAMS, J. G. U. (1974) Obsolete economics. *The Ecologist*, 4, 280–8.

AGTERBURG, F. P. (1970) Autocorrelation functions in geology. In Merriam, D. F. (ed.) *Geostatistics: A Colloquium*. New York, Plenum.

AIR POLLUTION CONTROL ASSOCIATION (1970) *Recognition of Air Pollution Injury to Vegetation: A Pictorial Atlas.*

AIRY, G. B. (1840) On the regulators of the clockwork for effecting uniform movement of equatorials. *Memoirs of the Royal Astronomical Society*, 11, 249–67.

AKAIKE, H. (1973) *Markovian Representation of Stochastic Processes and its Application to the Analysis of Autoregressive Moving Average Processes*. UMIST Dept of Mathematics (Statistics) Tech. Rep. 41, 28 p.

ALDRED, B. K. (1974) *The Case for a Locational Data Type*. IBM UK Scientific Centre Rep. UKSC-0054. Peterlee.

ALEXANDER, C. (1964) *Notes on the Synthesis of Form*. Cambridge, Mass., Harvard Univ. Press. 216 p.

ALLEN, J. R. L. (1974) Reaction, relaxation and lag in natural sedimentary systems: general principles, examples and lessons. *Earth-Science Reviews*, 10, 263–342.

ALLEN, R. G. B. (1968) *Mathematical Economics*. New York, Macmillan. 812 p.

ALONSO, W. (1971) The economics of urban size. *Papers of the Regional Science Association*, 26, 67–83.

ALTHUSSER, L. (1969) *For Marx* (trans. B. Brewster). London, Allen Lane, The Penguin Press.

AMOROCHO, J. (1963) Measures of the linearity of hydrologic systems. *Journal of Geophysical Research*, 68, 2237–249.

AMOROCHO, J. (1967) The nonlinear prediction problem in the study of the runoff cycle. *Water Resources Research*, 3, 861–89.

AMOROCHO, J. (1969) Deterministic nonlinear hydrologic models. In *Symposium 1969: The Progress of Hydrology, Vol. 1: New Developments in Hydrology*, pp. 420–472. Urbana, Ill.

AMOROCHO, J. (1973) Nonlinear hydrologic analysis. *Advances in Hydroscience*, 9, 203–51.

AMOROCHO, J. and HART, W. E. (1964) A critique of current methods in hydrologic systems investigation. *Transactions of the American Geophysical Union*, 45, 307–21.

AMOROCHO, J. and ORLOB, G. T. (1961) *Nonlinear Analysis of Hydrologic Systems*. Sanitary Engineering Research Laboratory, Univ. California, Water Resources Center, Cont. 40. 147 p.

ANDERSON, T. W. (1958) *An Introduction to Multivariate Statistical Analysis*. New York, Wiley. 374 p.

ANNETT, J. (1969) *Feedback and Human Behaviour*. Harmondsworth, Penguin.

ANTHONY, R. N. (1965) *Planning and Control Systems: A Framework for Analysis*. Cambridge, Mass., Graduate Business School of Administration, Harvard.

ANUCHIN, V. A. (1972) *Theoretical Foundations of Geography*. Moscow (in Russian).

ARBIB, M. A. (1972) *The Metaphorical Brain*. New York, Wiley. 243 p.

ARROW, K. J. (1951) *Social Choice and Individual Values*. New Haven, Conn., Yale Univ. Press. 124 p.

ARROW, K. J. and KURZ, M. (1970) *Public Investment, the Rate of Return and Optimal Fiscal Policy*. Baltimore, Md., Johns Hopkins Press. 218 p.

ASHBY, W. R. (1956) *An Introduction to Cybernetics*. London, Methuen. 295 p.

ÅSTROM, K. J. (1965) *Notes on the Regulation Problem*. Lund, Institute of Technology. 63 p.

ÅSTROM, K. J. and WITTENMARK, B. (1971) Problems of identification and control. *Journal of Mathematical Analysis and Applications*. 34, 90–113.

ÅSTROM, K. J. and WITTENMARK, B. (1972) *On the Control of Constant but Unknown Systems*. Paris, Proceedings of 5th FAC World Congress, Paper 37.5.

ATTINGER, E. O. (1970) *Global System Dynamics*. Basel, S. Karger. (see Attinger, E. O., 130–44).

BAER, J. G. (1951) *Ecology of Animal Parasites*. Urbana, Ill., Univ. Illinois Press. 224 p.

BALL, R. J. (ed.) (1973) *The International Linkage of National Economic Models*. Amsterdam, North Holland. 467 p.

BARLOWE, R. (1958) *Land Resource Economics: The Political Economy of Rural and Urban Land Resource Use*. Englewood Cliffs, N.J., Prentice-Hall. 468 p.

BARNARD, G. A. (1959) Control charts and stochastic processes. *Journal of the Royal Statistical Society*, B, 21, 239–72.

BARNARD, G. A. (1963) New methods of quality control. *Journal of the Royal Statistical Society*, A, 126, 255–8.

BARNETT, H. J. and MORSE, C. (1963) *Scarcity and Growth: The Economics of Natural Resource Availability*. Baltimore, Md., Johns Hopkins Press for Resources for the Future Inc. 288 p.

BARRETT, J. F., COALES, J. F., LEDWICH, M. A., NAUGHTON, J. J. and YOUNG, P. C. (1973) Macro-economic modelling: a critical appraisal. *IEE, London, Conference Papers No. 101, IFAC/IFORS Conference*, 60–80.

BARTHOLOMEW, D. J. (1973) *Stochastic Models for Social Processes*, 2nd ed. London, Wiley. 411 p.

BARTLETT, M. S. (1946) On the theoretical specification and sampling properties of autocorrelated time-series. *Journal of the Royal Statistical Society*, B, 8, 27–41.

BARTLETT, M. S. (1966) *Stochastic Processes*. Cambridge, Univ. Press. 362 p.

BARTLETT, M. S. (1971) Physical nearest neighbour models and non-linear time series. *Journal of Applied Probability*, 8, 222–32.

BASS, J. (1954) Space and time correlations in a turbulent fluid, I. *Univ. California Publications in Statistics*, 5, 55–83.

BASSETT, K. and TINLINE, R. R. (1970) Cross-spectral analysis of time series and geographical research. *Area*, 2, 19–24.

BATTY, M. (1971) An approach to rational design. *Architectural Design*, 41, 436–9, 498–501.

BATTY, M. (1974) A theory of Markovian design machines. *Environment and Planning*, B, 1, pp. 125–46.

BECKERMAN, W. (1971) The desirability of economic growth. In Kaldor, N. (ed.) *Conflicts in Policy Objectives*, pp. 38–61. Oxford, Blackwell.

BECKERMAN, W. (1973) Growth mania revisited. *New Statesman* (October).

BEER, S. (1966) *Decision and Control: The Meaning of Operational Research and Management Cybernetics*. London, Wiley. 556 p.

BEER, S. (1972) *The Brain of the Firm: The Managerial Cybernetics of Organization*. London, Allen Lane. 319 p.

BEER, S. (1974) *Designing Freedom*. London, Wiley. 100 p.

BEKEY, G. A. (1970) The human operator in control systems. In De Greene, K. B. (ed.) *Systems Psychology*, ch. 9, New York, McGraw-Hill. 593 p.

BELLMAN, R. E. (1957) *Dynamic Programming*. Princeton, N.J., Univ. Press. 342 p.

BELLMAN, R. E. (1961) *Adaptive Control Processes: A Guided Tour*. Princeton, N.J., Univ. Press. 255 p.

BENNETT, J. P. (1974) Concepts of mathematical modelling of sediment yield. *Water Resources Research*, 10, 485–92.

BENNETT, R. J. (1974) Process identification for time series modelling in urban and regional planning. *Regional Studies*, 8, 157–74.

BENNETT, R. J. (1975a) Dynamic systems modelling of the North West Region, 1: Spatiotemporal representation and identification. *Environment and Planning*, 7, 525–38.

BENNETT, R. J. (1975b) Dynamic systems modelling of the North West Region, 2: Estimation of the spatiotemporal policy model. *Environment and Planning*, 7, 539–566.

BENNETT, R. J. (1975c) Dynamic systems modelling of the North West Region, 3: Adaptive-parameter policy model. *Environment and Planning*, 7, 617–36.

BENNETT, R. J. (1975d) The representation and identification of spatiotemporal systems: an example of population diffusion in North West England. *Transactions of the Institute of British Geographers*, 66, 73–94.

BENNETT, R. J. (1976) Adaptive adjustment of channel geometry. *Earth Surface Processes*, 1, 131–50.

BENNETT, R. J. (1978) *Spatial Time Series: Analysis, Forecasting and Control*. London, Pion Press.

BENNETT, R. J., CAMPBELL, W. J. and MAUGHAN, R. A. (1976) Changes in atmospheric pollution concentrations. In Brebbia, C. A. (ed.) *Mathematical Models for Environmental Problems*, pp. 221–35. London, Pentach. 537 p.

BENNETT, R. J. and HAINING, R. P. (1976) *Space-Time Models: An Introduction to Concepts*. Department of Geography, Univ. College, London, Occasional Paper 28. 35 p.

BERGER, P. and PULLBERG, S. (1966) Reification and their social critique of consciousness. *New Left Review*, 35, 56–71.

BERRY, B. J. L. (1962) *Market Centers and Retail Distribution*. Englewood Cliffs, N.J., Prentice-Hall. 146 p.

BERRY, B. J. L. (1967) The mathematics of regionalization. *Proceedings, Brno Conference on Economic Regionalization*. Brno, Czech. Academy of Sciences.

BERRY, B. J. L. (1971) Monitoring trends, forecasting change and evaluating goal-achievement in the urban environment: the ghetto expansion versus desegregation issue in Chicago as a case study. In Chisholm, M., Frey, A. and Haggett, P. (eds) *Regional Forecasting*, pp. 93–119. London, Butterworth. 467 p.

BERTALANFFY, L. VON (1950) An outline of general systems theory. *British Journal of Philosophy of Science*, 1.2, 134–65.

BESAG, J. E. (1972) Nearest neighbour systems and the autologistic model for binary data. *Journal of the Royal Statistical Society*, B, 34, 75–83.

BESAG, J. E. (1974) Spatial interaction and the statistical analysis of lattice systems. *Journal of the Royal Statistical Society*, B, 36, 192–235.

BETAK, J. F. (1975) The conceptual basis of a two-dimensional language. *Geographical Analysis*, 7, 1–18.

BHAGWATI, J. (1958) International trade and economic expansion. *American Economic Review*, 48, 941–53.

BLACK, D. (1958) On the rationale of group decision making. *Journal of Political Economy*, 56, 23–34.

BLALOCK, H. M. (1964) *Causal Inferences in Non-Experimental Research*. Chapel Hill, Carolina Univ. Press. 200 p.

BLALOCK, H. M. (1967) *Toward a Theory of Minority Group Relations*. New York, Wiley. 227 p.

BLALOCK, H. M. (1969) *Theory Construction*. Englewood Cliffs, N.J., Prentice-Hall. 180 p.

BLESSER, W. B. (1969) *A Systems Approach to Biomedicine*. New York, McGraw-Hill. 615 p.

BOULDING, K. E. (1955) *Economic Analysis*. London, Harper. 405 p.

BOULDING, K. E. (1956a) General systems theory: the skeleton of a science. *General Systems Yearbook*, 1, 11.

BOULDING, K. E. (1956b) *The Image*. Ann Arbor, Univ. Michigan Press. 175 p.

BOULDING, K. E. (1966) The ethics of rational decision. *Management Science*, 12, B161–9.

BOULDING, K. E. (1968) Business and economic systems. In Milsum, J. H. (ed.) *Positive Feedback*, pp. 101–17. Oxford, Pergamon. 169 p.

BOULDING, K. E. (1973) Forward. In Downs, R. M. and Stea, D. (eds) *Image and Environment*, pp. vii–xi. London, Arnold.

BOX, G. E. P. and JENKINS, G. M. (1968) Discrete models for feedback and feedforward control. In Watts, D. G. (ed.) *The Future of Statistics*, pp. 201–40. New York, Academic Press.

BOX, G. E. P. and JENKINS, G. M. (1970) *Time Series Analysis, Forecasting and Control*. San Francisco, Holden-Day. 553 p.

BOX, G. E. P. and TIAO, G. C. (1973) *Intervention Analysis with Application to Economic and Environmental Problems*. Univ. Wisconsin, Dept of Statistics, Tech. Rep. 335.

BRANDSTETTER, A. and AMOROCHO, J. (1970) *Generalized Analysis of Small Watershed Responses*. Dept Water Science and Engineering, Univ. California, Davis, Rep. 1035. 204 p.

BRAY, J. (1970) *Decision in Government*. London, Tavistock. 320 p.

BRIDGMAN, P. W. (1954) Science and common sense. *Scientific Monthly* (July), 32–39.

BRILLOUIN, L. (1956) *Science and Information Theory*. New York, Academic Press. 320 p.

BROADBENT, A. (1975) The wrong case. *The Planner, Journal of the Royal Town Planning Institute*, 61, 188–9.

BROWN, R. K. (1969) City cybernetics. *Land Economics*, 45, 406–12.

BRUES, A. M. (1974) Models applicable to geographic variation in man. In Dyke, B. and MacCluer, J. W. (eds) *Computer Simulation in Human Population Studies*, pp. 129–41. New York, Academic Press.

BRUNER, J. S. (1970) Constructive cognitions. *Contemporary Psychology*, 15, (2), 81–83.

BRYSON, A. E. and HO, Y.-C. (1969) *Applied Optimal Control*. Waltham, Mass., Blaisdell, 481 p.

BRYSON, R. A. (1974) A perspective on climatic change. *Science*, 184, 753–60.

BUCHANAN, L. F. and NOTON, F. E. (1971) Optimal control applications in economic systems. In Leondes, C. T. (ed.) *Advances in Control Systems*, No. 8, 141–187. New York, Academic Press.

BUCHANAN, J. M. and TULLOCK, G. (1969) *The Calculus of Consent*. Ann Arbor, Univ. Michigan Press. 361 p.

BUCKLEY, W. (1967) *Sociology and Modern Systems Theory*. Englewood Cliffs, N.J., Prentice-Hall. 227 p.

BUNGE, W. (1966) *Theoretical Geography*, 2nd ed. Lund, Lund Studies in Geography. 289 p.

BURAS, N. (1972) *Scientific Allocation of Water Resources*. New York, Elsevier. 208 p.

BURDEKIN, R. and MARSHALL, S. A. (1972) The use of Forrester's systems dynamics approach in urban modelling. *Environment and Planning*, 4, 471–85.

BURKE, R. and HEANEY, J. P. (1975) *Collective Decision Making in Water Resource Planning*. Lexington, Heath. 260 p.

BURT, C. (1953) *The Causes and Treatment of Backwardness*, 2nd ed. London, Univ. Press. 128 p.

BURTON, I., KATES, R. and SNEAD, R. (1969) *The Human Ecology of Coastal Flood Hazard in Megalopolis*. Dept Geography, Univ. Chicago, Res. Pap. No. 115.

BUTKOVSKIY, A. G. (1961) Some approximate methods for solving problems of optimal control of distributed parameter systems. *Automatic and Remote Control*, 22, 1429–38.

BUTKOVSKIY, A. G. (1969) *Distributed Control Systems*. New York, Elsevier. 446 p.

BUTKOVSKIY, A. G. and LERNER, A. YA (1960) The optimal control of systems with distributed parameters. *Automatic and Remote Control*, 21, 472–7.

CALDER, N. (1974) Arithmetic of ice ages. *Nature*, 252, 216–18.

CAMPBELL, D. T. (1961) Methodological suggestions for a comparative psychology of knowledge processes. *General Systems Yearbook*, 6, 15–29.

CAREY, H. C. (1858) *Principles of Social Science*. Philadelphia, Lippincote.

CARNAP, R. (1966) Probability and content measure. In Feyerabend, P. and Maxwell, G. (eds) *Mind, Matter and Methodological Essays in Philosophy*. Minneapolis, Minnesota Univ. Press. 524 p.

CARSLAW, H. S. and JAEGER, J. C. (1941) *Operational Methods in Applied Mathematics*. London, Oxford Univ. Press. 264 p.

CHAPIN, F. S. and HIGHTOWER, H. C. (1965) Household activity patterns and land use. *Journal of the American Institute of Planners*, 31, 222.

CHAPPELL, J. E. (1975) The ecological dimension: Russian and American views. *Annals of the Association of American Geographers*, 65, 144–62.

CHATFIELD, C. (1975) *The Analysis of Time Series: Theory and Practice*. London, Chapman & Hall, 263 p.

CHEN, I. (1954) The environment as a determinant of behaviour. *Journal of Social Psychology*, 39, 115–27.

CHILD, G. I. and SHUGART, H. H. (1972) Frequency response analysis of magnesium cycling in a tropical forest ecosystem. In Patten, B. C. (ed.) *Systems Analysis and Simulation in Ecology*, Vol. 2, pp. 103–35. New York and London, Academic Press. 592 p.

CHISHOLM, M. D. I. and MANNERS, G. (1971) *Spatial Policy Problems of the British Economy*. Cambridge, Univ. Press. 248 p.

CHORAFAS, D. N. (1960) *Stochastic Processes and Reliability Engineering*. New York, Van Nostrand. 438 p.

CHORLEY, R. J. (1973) Geography as human ecology. In Chorley, R. J. (ed.) *Directions in Geography*, pp. 155–70. London, Methuen.

CHORLEY, R. J. (1976) Some thoughts on the development of geography 1950–1975. *Oxford Polytechnic Discussion Papers in Geography*, No. 3, 29–35.

CHORLEY, R. J. and KENNEDY, B. A. (1971) *Physical Geography: A Systems Approach*. London, Prentice-Hall. 370 p.

CHOW, G. C. (1973) Effects of uncertainty on optimal control policies. *International Economic Review*, 14, 632–45.

CHOW, G. C. (1975) *Analysis and Control of Dynamic Economic Systems*. New York, Wiley. 316 p.

CHRISTALLER, W. (1933) *Central Places in Southern Germany* (trans. 1966, C. W. Bastin) Englewood Cliffs, N.J., Prentice-Hall. 230 p.

CHURCHMAN, C. WEST (1971) *The Design of Inquiring Systems: Basic Concepts of Systems and Organizations*. New York and London, Basic Books. 288 p.

CLARK, J., COLE, S., CURNOW, R. and HOPKINS, M. (1975) *Global Simulation Models*. London, Wiley-Interscience. 135 p.

CLARKE, D. L. (ed.) (1972) *Models in Archaeology*. London, Methuen. 1055 p.

CLIFF, A. D., HAGGETT, P. *et al.* (1975) *Elements of Spatial Structure*. Cambridge, Univ. Press. 258 p.

CLIFF, A. D. and ORD, J. K. (1973) *Spatial Autocorrelation*. London, Pion. 178 p.

CLIFF, A. D. and ORD, J. K. (1975) Model building and the analysis of spatial pattern in human geography. *Journal of the Royal Statistical Society*, B, 37, 297–384.

CLYMER, A. B. (1972) Next-generation models in ecology. In Patten, B. C. (ed.) *Systems Analysis and Simulation in Ecology*, pp. 533–69. New York and London, Academic Press.

COATES, B. E. and RAWSTROM, E. M. (1971) *Regional Variations in Britain: Studies in Economic and Social Geography*. London, Batsford. 204 p.

CODDINGTON, A. (1974) The economics of conservation. In Warren, A. and Goldsmith, F. B. (eds.) *Conservation in Practice*, pp. 453–64. London, Wiley. 512 p.

COHEN, R. S. (1968) Ernst Mach: physics, perception and the philosophy of science. *Synthese*, 18, 132–70.

COHEN, S. B. and ROSENTHAL, L. D. (1971) A geographical model for political systems analysis. *Geographical Review*, 61, 5–31.

COLE, H. S. D., FREEMAN, C., JAHODA, M. and PAVITT, K. L. R. (1974) *Thinking About the Future: A Critique of the Limits to Growth.* London, Chatto & Windus, Sussex Univ. Press. 218 p.

COLLINGE, V. K. (1964) The principles of dilution methods for flow measurement. In Collinge, V. K. and Simpson, J. R. (eds) *Dilution Techniques for Flow Measurement*, pp. 9–16. Dept Civil Engineering, Univ. Newcastle-upon-Tyne, Bull. 31. 105 p.

CONNELLY, P. and PERLMAN, R. (1975) *The Politics of Scarcity: Resource Conflicts in International Relations.* London, Oxford Univ. Press and Royal Institute of International Affairs. 162 p.

COOKE, R. U. (1974) The rainfall context of arroyo initiation in southern Arizona. *Zeitschrift für Geomorphologie*, 21, 63–75.

COOKE, R. U. and REEVES, R. W. (1976) *Arroyos and Environmental Change in the American Southwest.* London, Oxford Univ. Press. 213 p.

COWLING, T. M. and STEELEY, G. C. (1973) *Subregional Planning Studies: An Evaluation.* Oxford, Pergamon. 202 p.

CRAWFORD, N. H. and LINSLEY, R. K. (1966) *Digital Simulation in Hydrology: Stanford Watershed Model IV.* Dept Civil Engineering, Stanford Univ., Tech. Rep. No. 39. 210 p.

CRIPPS, E. L. (1970) *An Introduction to the Study of Information for Urban and Regional Planning.* Reading, Urban Systems Research Unit. 46 p.

CUÉNOD, M. and DARLING, A. (1969) *A Discrete-Time Approach for System Analysis.* New York, Academic Press. 221 p.

CURRY, L. (1962) The geography of service centers within towns—the elements of an operational approach. *Proceedings IGU Symposium in Urban Studies, Lund, 1960*, pp. 31–53.

CURRY, L. (1967) Central places in a random spatial economy. *Journal of Regional Science*, 7, 217–38.

CUYPERS, J. G. and RADEMAKER, O. (1974) An analysis of Forrester's world dynamics model. *Automatica*, 10, 195–201.

DACEY, M. F. (1966) A county-seat model for the areal pattern of an urban system. *Geographical Review*, 56, 527–42.

DAETZ, D. and PANTELL, R. H. (eds.) (1974) *Environmental Modeling: Analysis and Management.* Stroudsburg, Pa., Dowden, Hutchenson & Ross.

DAHRENDORF, R. (1959) *Class and Class Conflict in Industrial Society.* London, Routledge & Kegan Paul. 366 p.

DANTZIG, G. B. (1965) *Large Scale System Optimization: A Review.* Operations Research Center, Univ. California, Berkeley, Reprint ORC 65–9.

DAVIES, W. D. T. (1970) *System Identification for Self-Adaptive Control.* London, Wiley-Interscience. 380 p.

DAY, L. H. (1974) The world demographic situation. *Ambio*, 3, 97–106.

DAY, R. H. and GROVES, T. (1975) *Adaptive Economic Models.* New York, Academic Press. 581 p.

DAY, T. J. (1975) Longitudinal dispersion in natural channels. *Water Resources Research*, 11, 909–18.

DEEVEY, E. S. (1949) Biogeography of the Pleistocene. *Bulletin of the Geological Society of America*, 60, 1315–1416.

DE GREENE, K. B. (ed.) (1970) *Systems Psychology.* New York, McGraw-Hill. 593 p. (see chapters 1, 3, 10 and 18 by De Greene).

DELLEUR, J. W. and RAO, R. A. (1971) Linear systems in hydrology: The transform approach, the kernel oscillations and the effect of noise. In Yevjevich, V. (ed.) *Systems Approach to Hydrology*, pp. 116–38. Fort Collons, Colorado, Water Resources Publications.

DE NEUFVILLE, R. and STAFFORD, J. H. (1971) *Systems Analysis for Engineers and Managers*. New York, McGraw-Hill. 353 p.

DE SMITH, S. (1971) *Constitutional and Administrative Law*. Harmondsworth, Penguin. 712 p.

DEUTSCH, K. (1951) Mechanism, teleology and mind. *Philosophy and Phenomenological Research*, 12, 185–222.

DEUTSCH, K. W. (1969) On methodological problems and quantitative research. In Dogan, M. and Rokkan, S. (eds), *Quantitative Ecological Analysis in Social Sciences*, pp. 19–40. Cambridge, Mass., MIT Press.

DEUTSCH, S. (1967) *Models for the Nervous System*. New York, Wiley. 266 p.

DEWEY, J. (1958) *Experience and Nature*. La Salle. 443 p.

DEWEY, J. and BENTLEY, A. F. (1949) *Knowing and the Known*. Boston, Beacon.

DHRYMES, P. J. (1971) *Distributed Lags: Problems in Estimation and Formulation*. Edinburgh, Oliver & Boyd. 414 p.

DICKINSON, H. W. and JENKINS, R. (1927) *James Watt and the Steam Engine*. London, Oxford Univ. Press. 415 p.

DIRECTOR, S. W. and ROHRER, R. A. (1972) *Introduction to Systems Theory*. New York, McGraw-Hill. 441 p.

DI TORO, D. M., O'CONNER, D. J., THOMANN, R. V. and MANCINI, J. L. (1975) Phytoplankton-zooplankton-nutrient interaction model for Western Lake Erie. In Patten, B. C. (ed.) *Systems Analysis and Simulation in Ecology*, Vol. III, pp. 423–74. New York and London, Academic Press, 601 p.

DMOWSKI, R. M. (ed.) (1974) *Systems Analysis and Modelling Approaches in Environmental Systems*. Warsaw, Polish Academy of Sciences. 607 p.

DOBB, M. (1925) *Capitalist Enterprise and Social Progress*. London, Routledge. 409 p.

DOBB, M. (1973) *Theories of Value and Distribution since Adam Smith: Ideology and Economic Theory*. Cambridge, Univ. Press. 295 p.

DODD, S. C. (1950) The interactance hypothesis: a gravity model fitting physical masses and human groups. *American Sociological Review*, 15, 245–56.

DOE (1972) *General Information System for Planning*. Report of Study Team, Department of the Environment. London, HMSO. 119 p.

DOE (1973) *Manual on Point-Referencing Properties and Parcels of Land*. Department of the Environment. London, HMSO.

DOE (1975) *Structure Plans Note 1/75: Principles for Monitoring Development Plans*. Department of the Environment Circular. London, HMSO. 9 p.

DOE (SPNW) (1973) *Strategic Plan for the North-West*. Report of the Joint Planning Team. London, HMSO.

DOE (SPSE) (1973) *Strategic Planning in the South-East: A Final Report of the Monitoring Group*. Department of the Environment. London, HMSO.

DOE (SPSE) (1975) *Population Pressure and Population Change: A Monitoring Report*. Department of the Environment. London, HMSO.

DONOUGHMORE COMMITTEE (1932) *Committee on Ministers' Powers: Report*. London, HMSO.

DOOGE, J. C. I. (1959) A general theory of the unit hydrograph. *Journal of Geophysical Research*, 64, 241–56.

DOOGE, J. C. I. (1968) The hydrologic cycle as a closed system. *Bulletin of the International Association of Scientific Hydrology*, 13, (1), 58–68.

DORFMAN, R. (1969) An economic interpretation of optimal control theory. *American Economic Review*, 59, 817–31.

DOUGLAS, I. (1969) The efficiency of humid tropical denudation systems. *Transactions of the Institute of British Geographers*, 46, 1–16.

DOWNS, R. M. and STEA, D. (eds) (1973) *Image and Environment*. London, Arnold. 439 p. (see Downs and Stea, 1–26).

DREYFUS, S. E. (1965) *Dynamic Programming and the Calculus of Variations*. New York, Academic Press. 248 p.

DRUCKER, P. F. (1969) *The Age of Discontinuity*. London, Pan. 369 p.

DUESENBERRY, J. S., FROMM, G., KLEIN, L. R. and KUH, E. (1965) *The Brookings Quarterly Econometric Model of the United States*. Amsterdam, North-Holland. 776 p.

DUNCAN, O. D. (1975) *Introduction to Structural Equation Models*. New York, Academic Press. 358 p.

DUNN, E. S. (1974) *Social Information Processing and Statistical Systems—Change and Reform*. London, Wiley-Interscience. 246 p.

DURBIN, J. (1960) The fitting of time series models. *Revue Institut Internationale de Statistique*, 28, 233–43.

DYBERN, B. I. (1974) Water pollution: a problem with global dimensions. *Ambio*, 3, 139–45.

EAGLESON, P. S. (1969) Deterministic linear hydrologic systems. In *Symposium, 1969: The Progress of Hydrology, Vol. 1: New Developments in Hydrology*, pp. 400–19. Urbana, Ill.

EASTON, D. (1965a) *A Framework for Political Analysis*. Englewood Cliffs, N.J., Prentice-Hall. 143 p.

EASTON, D. (1965b) *A Systems Analysis of Political Life*. New York, Wiley. 507 p.

EDDISON, T. (1973) *Local Government: Management and Corporate Planning*. Aylesbury, Leonard Hill. 225 p.

EDINGER, J. G. (1973) Vertical distribution of photochemical smog in the Los Angeles Basin. *Environmental Science and Technology*, 1, 247.

ELDERTON, W. P. and JOHNSTON, N. L. (1969) *Systems of Frequency Curves*. Cambridge, Univ. Press. 216 p.

EMERY, F. E. and TRIST, E. L. (1965) The causal texture of organizational environments. *Human Relations*, 18, 21–32.

EMERY, J. C. (1969) *Organizational Planning and Control Systems*. London, Macmillan. 166 p.

ENGELS, F. (1873–83) *Dialectics of Nature*. New York, International Publishers (1940) 383 p.

ENGLISH, P. W. and MAYFIELD, R. C. (eds) (1972) *Man, Space and Environment*. London, Oxford Univ. Press. 623 p.

ETZIONI, A. (1967) Mixed scanning: a 'third' approach to decision making. *Public Administration Review*, 27, 385–92.

EULAU, H. and MARCH, J. G. (eds) (1969) *Political Science*. Englewood Cliffs, N.J., Prentice-Hall.

EYKHOFF, P. (ed.) (1973) *Identification and systems parameter estimation*. Proceedings of 3rd IFAC Symposium. Amsterdam, North-Holland, 2 vols. 1179 p.

EYKHOFF, P. (1974) *System Identification*. New York, Wiley. 555 p.

FALKENMARK, M. and LINDH, G. (1974) How can we cope with the water resources situation by the year 2015? *Ambio*, 3, 114–22.

FALUDI, A. (1973) The 'systems view' and planning theory. *Socio-Economic Planning Theory*, 7, 67–77.

FAYOL, H. (1925) L'administration industrielle et générale-prévoyance, organisation, fondement, coordination contrôle. *Bulletin de la société de l'industrie minerale*. Paris, Dunod.

FEIBLEMAN, J. K. (1954) The theory of integrative levels. *British Journal of the Philosophy of Science*, 5, 59–66.

FOGEL, L. J. (1967) *Human Information Processing*. Englewood Cliffs, N.J., Prentice-Hall. 826 p.

FOIN, T. C. (1972) Systems ecology and the future of human society. In B. C. Patten (ed.) *Systems Analysis and Simulation in Ecology*, pp. 475–531. New York and London, Academic Press.

FOLEY, G. (1976) *The Energy Question*. Harmondsworth, Penguin. 344 p.

FORRESTER, J. W. (1971) *World Dynamics*. Cambridge, Mass. Wright-Allen. 142 p.

FOSTER, E. E. (1949) *Rainfall and Runoff*. New York, Macmillan. 487 p.

FRANKS, O. (1957) *Committee on Administrative Tribunals and Enquiries: Report of the Committee*, Cmnd. 218. London, HMSO.

FREUD, S. (1930) *Civilization and its Discontents* (Vienna). New York, Macmillan (1961). 144 pp.

FRIEDMAN, M. (1974) Inflation, taxation, indexation. In Robbins, Lord *et al.*, *Inflation: Causes, Consequences, Cures*, pp. 71–88. London, Institute of Economic Affairs. 120 p.

FRIEND, J. K. and JESSOP, N. (1969) *Local Government and Strategic Choice: An Operational Research Approach to the Processes of Public Planning*. London, Tavistock. 296 p.

FRIEND, J. K., POWER, J. M. and YEWLETT, C. J. L. (1974) *Public Planning: The Inter-Corporate Dimension*. London, Tavistock. 534 p.

GAGNÉ, R. M. (ed.) (1966) *Psychological Principles in Systems Development*. New York, Holt, Rinehart & Winston. 560 p. (see Gagné, chapter 2).

GAMESON, A. L. H. (ed.) (1973) *Mathematical and Hydraulic Modelling of Estuarine Pollution*. Dept. of the Environment, Water Pollution Research Tech. Paper 13. London, HMSO. 228 p.

GANTMACHER, F. R. (1959) *Applications of the Theory of Matrices*. New York, Wiley-Interscience. 317 p.

GARG, D. P. (1973) A control strategy for spatial trade models incorporating transport costs. *Joint Automatic Control Conference*, Paper 12.3, 321–7.

GARLAND, J. A. (1975) Dry deposition and the atmospheric cycle of sulphur dioxide. In Hey, R. D. and Davies, T. D. (eds) *Science, Technology and Environmental Management*, pp. 145–64. Farnborough, Saxon House. 297 p.

GATES, G. R. and GATES, G. M. (1972) Uncertainty and development risk in *pequeña irrigación* decisions for peasants in Campeche, Mexico. *Economic Geography*, 48, 135–52.

GAZIS, D.-C., MONTROLL, E. W. and RYNIKER, J. E. (1973) Age-specific deterministic model of predator-prey populations: application to Isle Royale. *IBM Journal of Research and Development*, 17, 47–53.

GEARY, R. C. (1954) The contiguity ratio and statistical mapping. *Incorporated Statistician*, 5, 115–45.

GEERTZ, C. (1963) *Agricultural Involution: The Process of Ecological Change in Indonesia.* Berkeley, Univ. California Press. 176 p.

GEORGESCU-ROEGEN, N. (1971) *The Entropy Law and the Economic Process.* Cambridge, Mass., Harvard Univ. Press. 457 pp.

GERASIMOV, I. P. (1969) There is a need for a general plan of the transformation of nature in our country. *Kommunist* (in Russian).

GERASIMOV, I. P. (1976) Problems of natural environmental transformation in Soviet constructive geography. *Progress in Geography*, 9, 73–99.

GHEORGHÏU, A. and THEODORESCU, R. (1956) Observations sur le calcul de la population urbaine et rurale. *Com. Acad. R. P. Romîne*, 6, 521–6 (See *Mathematical Review*, 18 (1957), p. 859.)

GIBSON, J. J. (1966) *The Senses Considered as Perceptual Systems.* Boston, Houghton. 335 p.

GILBERT, E. G. (1963) Controllability and observability in multivariable control systems. *SIAM Journal of Control*, 1, 128–51.

GILBERT, E. G. (1969) The decoupling of multivariable systems by state feedback. *SIAM Journal of Control*, 7, 50–63.

GILLIS, J. D. S., BRAZIER, S., CHAMBERLAIN, K. J., HARRIS, R. J. P. and SCOTT, D. J. (1974) *Monitoring and the Planning Process.* Univ. Birmingham, Institute of Local Government Studies. 104 p.

GLACKEN, C. J. (1967) *Traces on the Rhodian Shore.* Berkeley, Univ. California Press. 763 p.

GLENDINNING, J. W. and BULLOCK, R. E. H. (1973) *Management by Objectives in Local Government.* London, Charles Knight. 255 p.

GLUCKSMANN, M. (1974) *Structuralist Analysis in Contemporary Social Thought.* London, Routledge & Kegan Paul. 197 p.

GODLEY, W. and CRIPPS, F. (1974) *Prospects for Economic Management: 1973–1977.* Cambridge, Department of Applied Economics. 93 p.

GOLDFELD, S. M. and QUANDT, R. E. (1973) The estimation of structural shifts by switching regressions. *Annals of Economic and Social Measurement*, 2, 475–85.

GOLLEDGE, R. G., BRIGGS, R. and DEMKO, D. (1969) The configuration of distances in inter-urban space. *Proceedings of the Association of American Geographers*, 1, 60–5.

GOLLEDGE, R. G., BROWN, L. A. and WILLIAMSON, F. (1972) Behavioral approaches to geography: an overview. *The Australian Geographer*, 12, 59–79.

GOODWIN, B. C. and COHEN, N. H. (1969) A phase shift model for spatial and temporal organization of developing systems. *Journal of Theoretical Biology*, 25, 49–107.

GOODWIN, R. M. (1951) The non-linear accelerator and the persistence of business cycles. *Econometrica*, 19, 1–17.

GOULD, P. (1963) Man against his environment: a game theoretic framework. *Annals of the Association of American Geographers*, 53, 290–7.

GOULD, P. (1965). *On Mental Maps.* Michigan Inter-University Community of Mathematical Geographers, Ann-Arbor, Discussion Paper 9. 53 p.

GOULD, P. (1970) Is *Statistix Inferens* the geographical name for a wild goose? *Economic Geography* (*Supp.*), 46, 439–48.

GOULD, P. (1973) On mental maps. In Downs, R. M. and Stea, D. (eds.) *Image and Environment*. London, Arnold, 182–220.

GOULD, P. (1976) *People in Information Space.* Dept Geography, Univ. Lund, Sweden. 161 p.

GOULDNER, A. W. (1971) *The Coming Crisis of Western Sociology*. London, Heinemann. 528 p.

GOULDNER, A. W. (1975) *For Sociology*. Harmondsworth, Penguin. 465 p.

GRAF, W. L. (1975) The impact of suburbanization on fluvial geomorphology. *Water Resources Research*, 11, 690–2.

GRANGER, C. W. J. (1969) Spatial data and time series analysis. In Scott, A. J. (ed.) *London Papers in Regional Science, Vol. 1: Studies in Regional Science*, pp. 1–24. London, Pion. 216 p.

GRANGER, C. W. J. (1973) Causality, model building and control: some comments. *Conference Publication 101, IEE, London*, pp. 343–55.

GRANGER, C. W. J. and HATANAKA, M. (1964) *The Spectral Analysis of Economic Time Series*. Princeton, N.J., Univ. Press. 300 p.

GRAVES, G. W., HATFIELD, G. B. and WHINSTON, A. B. (1972) Mathematical programming for regional water quality management. *Water Resources Research*, 8, 273–90.

GRAYBILL, F. A. (1969) *Introduction to Matrices with Applications in Statistics*. Belmont, Wadsworth. 273 p.

GREENSLADE, P. J. M. (1968) The distribution of some insects in the Solomon Islands. *Proceedings of the Linnean Society*, 179, 189–96.

GRENE, M. (1963) *A Portrait of Aristotle*. London, Faber & Faber. 271 p.

GRODINS, F. S. (1970) Similarities and differences between control systems in engineering and biology. In Attinger, E. O. (ed.) *Global System Dynamics*, pp. 2–6. Basel, S. Karger.

GROENMAN, S. (1972) Regionalisation as an operational procedure. In Kublinski, A. and Perrella, R. (eds) *Growth Poles and Regional Policies*, pp. 101–12. The Hague, Mouton. 267 p.

GULICK, L. H. and URWICK, L. (1937) *Papers on the Science of Administration*. New York, Institute of Public Administration. 195 p.

GUMBEL, E. J. (1941) Probability-interpretation of the observed return-periods of floods. *Transactions of the American Geophysical Union*, 22, 836–49.

HABERMAS, J. (1971) *Knowledge and Human Interests* (trans. J. J. Shapiro). Boston, Beacon. 356 p.

HABERMAS, J. (1973) *Legitimation Crisis* (trans. T. McCarthy). London, Heinemann. 166p.

HABERMAS, J. (1976) The analytical theory of science and dialectics. In Adorno, T. W., *et al.*, *The Positivist Dispute in German Sociology*, pp. 131–62 (trans. G. Adey and D. Frisby). London, Heinemann. 307 p.

HÄGERSTRAND, T. (1953) *Innovationsforloppet ur Korologisk Synpunkt*. Lund. 334 p.

HÄGERSTRAND, T. (1967) On Monte Carlo simulation of diffusion. In Garrison, W. I. and Marble, D. F. (eds) *Quantitative Geography, Part I: Economic and Cultural Topics*, pp. 1–32. Dept. of Geography, Northwestern Univ., Illinois.

HÄGERSTRAND, T. (1972) Tätortsgrupper Som Regionsamhällen: Tillgången Till Förvärsarbete Och Tjanster Utanför De Större Städerna. In *Regioner Att Leva i*, pp. 141–73. Stockholm, Allmanna Förlaget.

HÄGERSTRAND, T. (1973) The domain of human geography. In Chorley, R. J. (ed.) *Directions in Geography*, pp. 67–87. London, Methuen.

HÄGERSTRAND, T. (1975) Space, time and human conditions. In Karlquist, A. *et al.* (eds) *Dynamic Allocation of Urban Space*, pp. 3–14. Farnborough, Saxon House, D. C. Heath. 383 p.

HAGGETT, P. (1965) *Locational Analysis in Human Geography*. London, Arnold. 339 p.

HAGGETT, P. (1971) Leads and lags in inter-regional systems: a study of the cyclic fluctuations in the southwest economy. In Chisholm, M. and Manners, G. (eds) *Spatial Policy Problems of the British Economy*, pp. 69–95. Cambridge, Univ. Press.

HAGGETT, P. (1972) Contagious processes in a planar graph: an epidemiological application. In McGlashan, N. D. (ed.) *Medical Geography*, pp. 307–24. London, Methuen. 336 p.

HAGGETT, P. and CHORLEY, R. J. (1969) *Network Analysis in Geography*. London, Arnold. 348 p.

HAINING, R. P. (1977) Model specification in stationary random fields. *Geographical Analysis*, 9, 107–29.

HAINING, R. P. (1978) Specification and estimation problems in models of spatial dependence. *Northwestern Studies in Geography*, No. 24. Evanston, Illinois.

HALL, C. S. and NORDBY, V. J. (1973) *A Primer in Jungian Psychology*. New York, Taplinger. 142 p.

HAMMERSLEY, J. M. and CLIFFORD, P. (1971) *Markov Fields on Finite Graphs and Lattices*. Univ. Oxford (mimeo). 26 p.

HANNAN, E. J. (1969) The estimation of mixed autoregressive, moving-average schemes. *Biometrica*, 56, 579–93.

HARRIS, D. R. (1969) Agricultural systems, ecosystems and the origins of agriculture. In Ucko, P. J. and Dimbleby, G. W. (eds) *The Domestication and Exploitation of Plants and Animals*, pp. 3–15. London, Duckworth.

HARRIS, D. R. (1971) The ecology of swidden cultivation in the upper Orinoco rainforest, Venezuela. *Geographical Review*, 61, 475–95.

HARRIS, D. R. (1972) Swidden systems and settlement. In Ucko, P. J., Tringham, R. and Dimbleby, G. W. (eds) *Man, Settlement and Urbanism*, pp. 245–62. London, Duckworth.

HARRIS, R. and SCOTT, D. (1974) The role of monitoring and review in planning. *The Planner, Journal of the Royal Town Planning Institute*, 60, 729–32.

HARRISON, P. J. and STEVENS, C. F. (1971) A Bayesian approach to short-term forecasting. *Operational Research Quarterly*, 22, 341–62.

HART, R. A. and MOORE, G. T. (1973) The development of spatial cognition. In Downs, R. M. and Stea, D. (eds) *Image and Environment*, pp. 246–88. London, Arnold.

HARTE, J. and LEVY, D. (1975) On the vulnerability of ecosystems disturbed by man. In Van Dobben, W. H. and Lowe-McConnell, R. H. (eds) *Unifying Concepts in Ecology*, pp. 208–23. The Hague, Junk B. V.

HARVEY, D. (1969) *Explanation in Geography*. London, Arnold. 521 p.

HARVEY, D. (1973) *Social Justice and the City*. London, Arnold. 336 p.

HARVEY, D. (1974) Population, resources and the ideology of science. *Economic Geography*, 50, 256–77.

HAYEK, F. A. (1944) *The Road to Serfdom*. London, Routledge & Kegan Paul. 184 p.

HAYEK, F. A. (1972) *A Tiger by the Tail: A 40-Years Running Commentary on Keynesianism by Hayek*. London, Institute of Economic Affairs. 124 p.

HEILBRONER, R. L. (1974) *An Inquiry into the Human Prospect*. New York, Norton. 150 p.

HEINE, V. (1955) Models for two dimensional stationary stochastic processes. *Biometrica*, 42, 170–8.

HELD, R. B. and CLAWSON, M. (1965) *Soil Conservation in Perspective*. Baltimore, Md., Johns Hopkins Press. 344 p.

HEMPEL, C. G. (1965) *Aspects of Scientific Explanation*. New York, Free Press. 505 p.

HEMPEL, C. G. (1967) *Philosophy of Natural Science*. Englewood Cliffs, N.J., Prentice-Hall, 116 p.

HEMPEL, C. G. and OPPENHEIM, P. (1948) Studies in the logic of explanation. In Feigl, H. and Brodbeck, M. (eds.) *Readings in the Philosophy of Science*, pp. 319–52. New York, Appleton.

HERBST, P. G. (1961) A theory of simple behaviour systems I and II. *Human Relations*, 14, 71–94, 193–239.

HERMANSEN, T. (1968) Information systems for regional development control. *Papers of the Regional Science Association*, 22, 107–40.

HERRERA, A. (1973) *Latin American World Model: Progress Report* (unpublished report quoted in Clark *et al.* 1975).

HESS, W. N. (ed.) (1974) *Weather and Climate Modification*. New York, Wiley. 842 p.

HEWITT, K. and HARE, F. K. (1973) *Man and Environment: Conceptual Frameworks*. American Geographical Society, Commission on College Geography, Res. Pap. 20. 39 p.

HICKS, J. R. (1956) *A Revision of Demand Theory*. Oxford, Clarendon Press. 198 p.

HILL, A. R. (1975) Ecosystem stability in relation to stress caused by human activity. *Canadian Geographer*, 19, (3), 206–20.

HILL, M. (1968) A goals-achievement matrix for evaluating alternative plans. *Journal of the American Institute of Planning*, 34, 19–29.

HIPEL, K. W., LENNOX, W. L., UNNY, T. E. and MCLEOD, A. I. (1975) Intervention analysis in water resources. *Water Resources Research*, 11, (6), 855–861.

HIRSCH, M. W. and SMALE, S. (1974) *Differential Equations, Dynamical Systems and Linear Algebra*. New York, Academic Press. 358 p.

HMSO (1944) *Employment Policy*. Cmnd 6527. London, HMSO.

HOLLAND, S. (1976) *Capital Versus the Regions*. London, Macmillan. 328 p.

HOLT, C. C. (1962) Linear decision rules for economic stabilization and growth. *Quarterly Journal of Economics*, 76, 20–45.

HOLT, C. C., MODIGLIANI, F., MUTH, J. F. and SIMON, H. A. (1960) *Planning Production, Work Force and Inventories*. Englewood Cliffs, N.J., Prentice-Hall. 419 p.

HORKHEIMER, M. (1972) *Critical Theory: Selected Essays*, trans. M. J. O'Connell and others. New York, Herder and Herder. 290 p.

HOUSE, J. W. (ed.) (1973) *The UK Space: Resources, Environment and Future*. London, Weidenfeld & Nicolson. 371 p.

HOUSE, P. W. (1973) Environmental modelling versus the 'chicken soup' approach. *Simulation*, 20, 181–91.

HUFF, D. L. (1963) A probability analysis of shopping centre trading areas. *Land Economics*, 39, 81.

HULTÉN, E. (1937) *Outline of the History of Arctic and Boreal Biota during the Quaternary Period*. Thule, Stockholm, Akticbolaget.

HUXLEY, A. (1954) *The Doors of Perception*. London, Chatto & Windus. 63 p.

IGU (1975) *Information Systems for State Land Use Planning.* Unpublished Report. International Geographical Union, Commission on Geographical Data Sensing and Processing.

ISACHENKO, A. G. (1973) *Principles of Landscape Science and Physical-Geographic Regionalization* (ed. J. S. Massey, trans. R. J. Zatorski). Melbourne, Univ. Press. 311 p.

ISARD, W. (1956) *Location and Space Economy.* Cambridge, Mass., MIT Press. 350 p.

ISARD, W. (1960) *Methods of Regional Analysis.* Cambridge, Mass., MIT Press. 784 p.

ISARD, W. (1975) *Introduction to Regional Science.* Englewood Cliffs, N.J., Prentice-Hall. 506 p.

ISARD, W. and CUMBERLAND, J. H. (eds) (1961) *Regional Economic Planning: Techniques for Analysis for Less Developed Countries.* Paris, OECD. 450 p.

ITTELSON, W. H. (ed.) (1973) *Environment and Cognition.* New York, Seminar Press. 187 p. (see Ittelson, 1–19).

JACOBS, O. L. R. (1974) *Introduction to Control Theory.* London, Oxford Univ. Press. 365 p.

JACOBY, S. L. S. (1966) A mathematical model for nonlinear hydrologic systems. *Journal of Geophysical Research,* 71, 4811–24.

JAMES, W. (1907) *Pragmatism: A New Name for Some Old Ways of Thinking.* New York, Longmans. 309 p.

JENKINS, G. M. and WATTS, D. G. (1968) *Spectral Analysis and its Applications.* San Francisco, Holden Day. 525 p.

JENKINS, G. M. and YOULE, P. V. (1971) *Systems Engineering: A Unifying Approach in Industry and Society.* London, Watts. 259 p.

JOHNSTON, J. (1972) *Econometric Methods,* 2nd ed. New York, McGraw-Hill. 437 p.

JOHNSTON, R. J. (1971) Mental maps of the city: suburban preference patterns. *Environment and Planning,* 3, 63–72.

JONES, J. C. (1967) The designing of man-machine systems. *Ergonomics,* 10, 101–11.

JORDAN, C. F. and KLINE, J. R. (1972) Mineral cycling: some basic concepts and their application in a tropical rainforest. *Annual Review of Ecology and Systematics,* 3, 33–50.

JUNG, C. G. (1927–31) The structure of the psyche. In *Collected Works, Vol. 8: The Structure and Dynamics of the Psyche* (1960–9). London, Routledge & Kegan Paul.

JUNG, C. G. (1931) Archaic man. In *Collected Works, Vol. 10: Civilization in Transition* (1964–70). London, Routledge & Kegan Paul.

JUNG, C. G. (1936) Wotan. In *Collected Works, Vol. 10: Civilization in Transition* (1964–70). London, Routledge & Kegan Paul.

JUNG, C. G. (1955–6) Mysterium Coniunctionis. In *Collected Works, Vol. 14: Mysterium Coniunctionis* (1963–70). London, Routledge & Kegan Paul.

JURY, E. I. (1964) *Theory and Application of the Z-Transform Method.* New York, Wiley. 330 p.

KAHN, R. F. (1931) The relation of home investment to unemployment. *Economic Journal,* 41, 173–98.

KAHN, R. F. and POSNER, M. (1974) Cambridge economics and the balance of payments. *The Times* (17–18 April).

KALDOR, N. (1964) *Essays on Economic Policy*, Vol. 1. London, Duckworth. 294 p.

KALDOR, N. (ed.) (1971) *Conflicts in Policy Objectives*. Oxford, Blackwell. 189 p.

KALECKI, M. (1954) *Theory of Economic Dynamics*. London, George Allen & Unwin. 178 p.

KALMAN, R. E. (1958) Design of self-optimizing control systems. *Transactions of the American Society of Mechanical Engineers*, 80, 468–78.

KALMAN, R. E. (1960) A new approach to linear filtering and prediction problems. *Transactions of the American Society of Mechanical Engineers, Journal of Basic Engineering*, D, 82, 35–45.

KALMAN, R. E. (1966) On structural properties of linear constant, multivariate systems. *Proceedings IFAC Congress, Prague*, Paper 6A

KALMAN, R. E. and BUCY, R. S. (1961) New results in linear filtering and prediction theory. *Transactions of the American Society of Mechanical Engineers, Journal of Basic Engineering*, D, 83, 95–108.

KALMAN, R. E. and KOEPCKE, R. W. (1958) Optimal synthesis of linear samping control systems using generalized performance indexes. *Transactions of the American Society of Mechanical Engineers*, 80, 1820–6.

KAPLAN, S. (1973) Cognitive maps in perception and thought. In Downs, R. M. and Stea, D. (eds) *Image and Environment*, pp. 63–86. London, Arnold.

KARLQUIST, A., LUNQUIST, L. and SNICKARS, F. (1975) *Dynamic Allocation of Urban Space*. Farnborough, Saxon House, D. C. Heath. 383 p.

KATES, R. W. (1970) Human perceptions of the environment. *International Social Science Journal*, 22, 648–60.

KATES, R. W. and WOHLWILL, J. F. (1966) Man's response to the physical environment: Introduction. *Journal of Social Issues*. 22, 15–20.

KAYA, Y. and SUZUKI, Y. (1974) Global constraints and a new vision for development, II. *Technological Forecasting and Social Change*, 6, 371–88.

KEAT, R. and URRY, J. (1975) *Social Theory as Science*. London, Routledge & Kegan Paul. 278 p.

KELLEY, C. R. (1968) *Manual and Automatic Control*. New York, Wiley. 272 p.

KENDALL, M. G. (1973) *Time-Series*. London, Charles Griffin. 197 p.

KENDALL, M. G. and STUART, A. (1966) *The Advanced Theory of Statistics*, Vol. 3. London, Charles Griffin. 552 p.

KEYFITZ, N. (1968) *The Mathematics of Population*. New York, Addison-Wesley. 450 p.

KEYNES, J. M. (1936) *The General Theory of Employment, Interest and Money*. London, Macmillan. 403 p.

KEYNES, J. M. (1937) The general theory of employment. *Quarterly Journal of Economics*, 51, 209–23.

KING, L. J. (1976) Alternatives to positive economic geography. *Annals of the Association of American Geographers*, 66, 293–308.

KING, L. J., CASETTI, E. and JEFFREY, D. (1969) Economic impulses in a regional system of cities: a study of spatial interactions. *Regional Studies*, 3, 213–18.

KIRKBY, M. J. (1976) Hydrological slope models: the influence of climate. In Derbyshire, E. (ed.) *Geomorphology and Climate*, pp. 247–67. London, Wiley.

KISIEL, C. C. (1969) Time series analysis of hydrologic data. *Advances in Hydroscience*, 5, 1–119.

KOENIG, H. E., COOPER, W. E. and FALVEY, J. M. (1974) Engineering for ecological, sociological and economic compatibility. In Daetz, D. and Pantell, R.

H. (eds) *Environmental Modelling: Analysis and Management*, pp. 330–42. Stroudsberg, Pa., Dowden, Hutchenson & Ross.

KOESTLER, A. (1966) *The Lotus and the Robot*. London, Hutchinson. 296 p.

KOFFKA, K. (1935) *Principles of Gestalt Psychology*. New York, Harcourt-Brace. 720 p.

KOHLBERG, L. (1969) Stage and sequence: the cognitive-developmental approach to socialization theory. In Gashin, D. A. (ed.) *Handbook on Socialization Theory*, pp. 347–480. Chicago, Rand McNally. 1182 p.

KOLATA, G. B. (1975) Cascading Bifurcations: The Mathematics of Chaos. *Science*, 189, 984–5.

KOPPÁNY, GY. (1975) Estimation of the life span of atmospheric motion systems by means of atmospheric energetics. *Meteorological Magazine*, 104, 302–6.

KORMONDY, E. J. (1969) *Concepts of Ecology*. Englewood Cliffs, N.J., Prentice-Hall. 209 p.

KORNAI, J. (1967) *Mathematical Planning of Structural Decisions*. Amsterdam, North-Holland. 526 p.

KORNAI, J. (1971) *Anti-Equilibrium: An Economic Systems Theory and the Tasks of Research*. Amsterdam, North-Holland. 402 p.

KOTLER, P. (1968) Mathematical models of individual buyer behaviour. *Behavioral Science*, 13, 274–87.

KOWAL, N. E. (1971) A rationale for modelling dynamic ecological systems. In Patten, B. C. (ed.) *Systems Analysis and Simulation in Ecology*, Vol. 1, pp. 123–94. New York and London, Academic Press. 607 p.

KOWALIK, P. (1974) Some cybernetic ideas in the theory of control of land ecosystems. In Dmowski, R. M. (ed.) *Systems Analysis and Modelling Approaches in Environmental Systems*, pp. 165–79. Warsaw, Polish Academy of Sciences.

KRIGE, D. G. (1966) Two-dimensional weighted moving average trend surfaces for ore valuation. *Journal of the South African Institute of Mining and Metallurgy*, 7, 13–38.

KUHN, A. (1974) *The Logic of Social Systems*. San Francisco, Jossey-Bass. 534 p.

KUHN, T. (1970) *The Structure of Scientific Revolutions*. Chicago, Univ. Press. 172 p.

LACK, D. (1971) *Ecological Isolation in Birds*. Oxford, Blackwell. 404 p.

LAMB, H. H. (1966) *The Changing Climate: Selected Papers*. London, Methuen. 236 p.

LAMB, H. H. (1972) *Climate: Present, Past and Future*, Vol. 1. London, Methuen. 613 p.

LANGBEIN, W. B. and LEOPOLD, L. B. (1964) Quasi-equilibrium states in channel morphology. *American Journal of Science*, 262, 782–94.

LANGE, O. (1965) *Wholes and Parts: A General Theory of System Behaviour*. Oxford, Pergamon Press. 74 p.

LANGE, O. (1970) *Introduction to Economic Cybernetics*. Oxford, Pergamon Press. 183 p.

LANGEFORS, B. (1973) *Theoretical Analysis of Information Systems*. Philadelphia, Auerback. 489 p.

LANGTON, J. (1972) Potentialities and problems of adopting a systems approach to the study of change in human geography. *Progress in Geography*, 4, pp. 125–79.

LASDON, L. S. (1970) *Optimization Theory for Large Systems*. London, Macmillan. 523 p.

LATAKOS, I. (1976) *Proofs and Refutations: The Logic of Mathematical Discovery* (eds J. Worrall and E. Zahar). Cambridge, University Press. 174 p.

LATHAM, R. E. (1951) Translation of Lucretius 'The Nature of the Universe'. Harmondsworth, Penguin. 256 p.

LAUDON, K. C. (1974) *Computers and Bureaucratic Reform: The Political Functions of Urban Information Systems.* London, Wiley. 325 p.

LAZARUS, R. S. (1966) *Psychological Stress and the Coping Process.* New York, McGraw-Hill. 466 p.

LEACH, E. (1973) Structuralism in social anthropology. In Robey, D. (ed.) *Structuralism: An Introduction*, pp. 37–56. London, Oxford Univ. Press. 153 p.

LEE, E. (1747) *Self Regulating Wind Machines.* British Patents (Old Series) No. 615 (quoted in Mayr 1970).

LEE, H. D. P. (1955) *Plato, The Republic* (translation). Harmondsworth, Penguin. 405 p.

LEE, R. C. K. (1964) *Optimal Estimation, Identification and Control.* Cambridge, Mass., MIT Press. 133 p.

LEIBNIZ, G. W. (1714) Monadology. In *The Monadology and other Philosophical Writings* (ed. and trans. R. Latta, 1898). Oxford.

LENIN, V. I. (1960) *Selected Works*, 3 vols. Moscow, Foreign Languages Publishing House. 935 p, 864 p and 854 p.

LEOPOLD, L. B. and LANGBEIN, W. B. (1963) Association and indeterminacy in geomorphology. In Albritton, C. C. (ed.) *The Fabric of Geology*, pp. 184–92. Stanford, Freeman.

LEOPOLD, L. B. and MADDOCK, T. (1954) *The Flood Control Controversy.* New York, Ronald Press. 278 p.

LERNER, A. P. (1946) *The Economics of Control.* New York, Macmillan. 428 p.

LEROY, P. (1960) Teilhard de Chardin: The Man. In *Le Milieu Divin*, Vol. 4 of the Collected Works of Pierre Teilhard de Chardin. London, Collins.

LÉVI-STRAUSS, C. (1962) *Totemism* (trans. R. Needham). London, Merlin. 116 p.

LEWIN, K. (1936) *Principles of Topological Psychology.* New York, McGraw-Hill. 231 p.

LICHFIELD, N., KETTLE, P. and WHITBREAD, M. (1975) *Evaluation in the Planning Process.* Oxford, Pergamon. 325 p.

LIDICKER, W. E. (1966) Ecological observations on a feral house mouse population declining to extinction. *Ecological Monographs*, 36, 27–50.

LINDBLOM, C. (1968) *The Intelligence of Democracy.* London, Collier-Macmillan. 352 p.

LINNEMANN, H. (1973) *Report to the Club of Rome on the Project 'Problem of Population Doubling'.* Unpublished Report quoted in Clark *et al.* 1975.

LINSLEY, R. K., KOHLER, M. A. and PAULHUS, L. H. (1949) *Applied Hydrology.* New York, McGraw-Hill. 689 p.

LIPSEY, R. G. (1973) *Positive Economics*, 3rd ed. London, Weidenfeld & Nicolson. 740 p.

LITTERER, J. A. (1973) *The Analysis of Organizations*, 2nd ed. New York, Wiley. 757 p.

LIU, C. Y. and GOODIN, W. R. (1976) Shallow-water approximation for the transport of pollutants in an urban basin. In Brebbia, C. A. (ed) *Mathematical Models for Environmental Problems*, pp. 201–19. London, Pentech Press.

LIVESEY, D. A. (1973) Can macro-economic planning problems ever be treated as a quadratic regulator problem? *IEE, London, Conference Papers No. 101, IFAC/IFORS Conference*, pp. 1–14.

LOCKE, J. (1690) *An Essay Concerning Human Understanding*. London, Thomas Basset. 362 p.

LÖSCH, A. (1954) *The Economics of Location*. New York, Wiley. 520 p.

LOTKA, A. J. (1956) *Elements of Mathematical Biology*. New York, Dover. 465 p.

LOUCKS, O. L. (1972) Systems methods in environmental court actions. In Patten, B. C. (ed.) *Systems Analysis and Simulation in Ecology*, Vol. 2, pp. 419–73. New York and London, Academic Press.

LOWENTHAL, D. (1975) Past time, present place: Landscape and memory. *Geographical Review*, 65, 1–36.

LUCE, R. and RAIFFA, H. (1957) *Games and Decisions*. New York, Wiley. 509 p.

LUENBERGER, D. G. (1964) Observing the state of a linear system. *IEEE, Transactions on Military Electronics, MIL-8*, pp. 74–80.

LUKÁCS, G. (1971) *History and Class Consciousness: Studies in Marxist Dialectics*. London, Merlin.

MACARTHUR, R. H. (1972) *Geographical Ecology: Patterns in the Distribution of Species*. London, Harper & Row. 269 p.

MACARTHUR, R. H. and WILSON, E. O. (1967) *The Theory of Island Biogeography*. Princeton, N.J., Univ. Press. 203 p.

MCCLOUGHLIN, J. B. (1973) *Control and Urban Planning*. London, Faber. 287 p.

MCCORMICK, E. J. (1970) *Human Factors Engineering*. New York, McGraw-Hill. 639 p.

MCFARLAND, P. J. (1971) *Feedback Mechanisms in Animal Behaviour*. London, Academic Press. 279 p.

MACFARLANE, A. G. J. (1970) *Dynamic System Models*. London, Harrap. 503 p.

MCINERNY, J. (1976) The simple analytics of natural resource economies. *Journal of Agricultural Economics*, 27, 31–52.

MACKAY, D. M. (1957) Information theory and human information systems. *Impact of Science on Society*, 8.2, 86–101.

MCKINSEY & CO. (1975) *General Review of Local Authority Management Information Systems*. Dept of Environment Report, No. 1. London.

MALONE, D. (1972) Modelling air pollution control as a large scale, complex system. *Socio-Economic Planning Science*, 6, 69–85.

MALTHUS, T. R. (1970) *An Essay on the Principle of Population and a Summary View on the Principle of Population*. Harmondsworth, Penguin.

MANN, H. B. and WALD, A. (1943) On the statistical treatment of linear stochastic difference equations. *Econometrica*, 11, 173–220.

MANNERS, G. (ed.) (1972) *Regional Development in Britain*. London, Wiley. 448 p.

MANNERS, G. (1977) *Three Issues of Minerals Policy*. Ninth Chester Beatty Lecture. London, Royal Society of Arts.

MANNHEIM, M. L. (1966) *Hierarchical Structure: A Model of Design and Planning Process*. Cambridge, Mass., MIT Press.

MARCH, J. G. and SIMON, H. A. (1966) *Organizations*. London, Wiley. 262 p.

MARCH, L. and MARCH, L. (1972) *Urban Space Structures*. Cambridge, Univ. Press. 272 p.

MARGALEF, R. (1968) *Perspectives in Ecological Theory*. Chicago. 111 p.

MARGENAU, H. (1963) Measurement and quantum states, Part 2. *Philosophy of Science*, 30, 138–57.

MARSCHAK, J. (1971) Economics of information systems. In Intriligator, M. D. (ed.) *Frontiers in Quantitative Economics*, pp. 32–108. Amsterdam, North-Holland.

MARSH, G. P. (1864) *Man and Nature: Or Physical Geography as Modified by Human Action*. New York, Scribner. 560 p.

MARTIN, R. L. and OEPPEN, J. E. (1975) The identification of regional forecasting models using space-time correlation functions. *Trans. Inst. Brit. Geog.*, 66, 95–118.

MARUYAMA, M. (1972) Toward human futuristics: trans-epistemological process. *Dialectica*, 26 (3–4), 155–83.

MARUYAMA, M. (1973) A new logical model for futures research. *Futures*, 5, 435–437.

MARX, K. (1888) Theses on *Feuerbach*. In *Selected Works of K. Marx and F. Engels*, pp. 28–30. London, Lawrence & Wishart (1970).

MARX, K. (1967) *Capital: A Critique of Political Economy*, Vol. 3. New York, International Publishers. 356 p.

MARX, K. (1968) *Principles of Political Economy*. New York, Augustus Kelley.

MARX, K. (1971) *The Grundrisse* (*Grundrisse der Kritik der politischen Ökonomie: 1857–58*) (Eng. trans.) London, Macmillan. 156 p.

MASSAM, B. (1975) *Location and Space in Social Administration*. London, Arnold. 192 p.

MATERN, B. (1947) Metoder att uppskatta noggrannheten uid linje-och pro-uytetaxering (summary: Methods of estimating the accuracy of line and sample plot surveys). *Medd. Frän Skogsforkningsinst*, 36 (1).

MATERN, B. (1960) Spatial variation. *Meddelanden Frän Statens Skogsforskningsinstitut*, 49, (5). 144 p.

MATLEY, I. M. (1975) Review of J. Passmore's *Man's Responsibility For Nature*. *Geographical Review*, 65, 533–4.

MATSCHOSS, C. (1908) *Die Entwicklung der Dampfmaschine*, Vol. 1. Berlin. 451 p.

MAXWELL, J. C. (1867–8) On governors. *Proceedings of the Royal Society*, 16, 270–283.

MAY, R. M. (1975) Deterministic Models with Chaotic Dynamics, *Nature*, 256, 165–6.

MAYR, O. (1970) *The Origin of Feedback Control*. Cambridge, Mass., MIT Press. 151 p.

MEAD, R. (1971) Models for interplant competition in irregularly spaced populations. In Patil, G. P., Pielau, E. C. and Waters, W. E. (eds) *Statistical Ecology*, Vol. 2, pp. 13–30. Pennsylvania State Univ. Press.

MEADOWS, D. H., MEADOWS, D. L., RANDERS, J. and BEHRENS, W. W. (1971) *The Limits to Growth*. London, Earth Island, Report of the Club of Rome. 205 p.

MEDITCH, J. S. (1971) Least-squares filtering and smoothing for linear distributed parameter systems. *Automatica*, 7, 315–22.

MEIER, R. (1962) *A Communications Theory of Urban Growth*. Cambridge, Mass., MIT Press. 184 p.

MEISTER, D. and RIBIDEAU, G. F. (1965) *Human Factors Evolution in System Development*. New York, Wiley. 307 p.

MENNES, L. B. M., TINBERGEN, J. and WAARDENBURK, J. G. (1969) *The Element of Space in Development Planning*. Amsterdam, North-Holland. 340 p.

MESAROVIC, M. and PESTEL, E. (1975) *Mankind at the Turning Point: The Second Report to the Club of Rome*. London, Hutchinson. 210 p.

METCALFE, J. L. (1974) Systems models, economic models and the causal texture of organizational environments. *Human Relations*, 27, 639–63.

MILANKOVITCH, M. (1920) *Théorie mathématique des phénoménes theoriques produits par la radiation solair*. Paris, Gauthier-Villars. 339 p.

MILETE, D. S., DRABEK, T. E. and HAAS, J. E. (1975) *Human Systems in Extreme Environments: A Sociological Perspective*. Inst. Behavioral Science, Univ. Colorado, Program on Technology, Environment and Man, Monog. 21. 165 p.

MILL, J. S. (1848) *Principles of Political Economy*. Peoples Edition (1865), in *Collected Works*, 17 vols. London, Routledge & Kegan Paul.

MILLER, G. A. (1966) *Psychology: The Science of Mental Life*. Harmondsworth, Penguin. 415 p.

MILSUM, J. H. (1966) *Biological Control Systems Analysis*. New York, McGraw-Hill. 466 p.

MILSUM, J. H. (ed.) (1968) *Positive Feedback*. Oxford, Pergamon. 169 p.

MONTGOMERY, E., BENNETT, J. W. and SCUDDER, T. (1973) The impact of human activities on the physical and social environment: new directions in anthropological ecology. *Annual Review of Anthropology*, 2, 27–61.

MOORE, W. L. and CLABORN, B. J. (1971) Numerical simulation of watershed hydrology. In Yevjevich, V. (ed.) *Systems Approach to Hydrology*. Fort Collins, Colorado, Water Resources Publications. 464 p.

MORAN, P. A. P. (1948) The interpretation of statistical maps. *Journal of the Royal Statistical Society*, B, 10. 243–51.

MORE, R. J. (1969) The basin hydrological cycle. In Chorley, R. J. (ed.) *Water, Earth and Man*, pp. 65–76. London, Methuen.

MOUSSOURIS, J. (1974) Gibbs and Markov random systems with constraints. *Journal of Statistical Physics*, 10, 11–33.

MURRAY, B. G. (1971) The ecological consequences of interspecific territorial behaviour in birds. *Ecology*, 52, 414–23.

MYRDAL, G. (1968) *Asian Drama: An Inquiry into the Poverty of Nations*. Harmondsworth, Penguin. 388 p.

NASH, J. E. (1957) The form of the instantaneous unit hydrograph. *Publications of the International Association of Scientific Hydrology*, 43 (3), 114–21.

NATIONAL RESEARCH COUNCIL (1973) *Environmental Quality and Social Behaviour: Strategies for Research*. Washington, National Academy of Sciences. 86 p.

NAYLOR, T. H. (ed.) (1971a) *Computer Simulation Experiments with Models of Economic Systems*. New York, Wiley. 502 p.

NAYLOR, T. H. (1971b) Policy simulation experiments with macro-econometric models: the state of the art. In Intriligator, M. D. (ed.) *Frontiers in Quantitative Economics*, pp. 211–27. Amsterdam, North-Holland.

NEDO (1971) *Investment Appraisal*. National Economic Development Office. London, HMSO. 30 p.

NEGEV, M. (1967) *A Sediment Model on a Digital Computer*. Dept Civil Engineering, Stanford Univ. Tech. Rep. 76. 109 p.

NEIDERCORN, J. H. and BECHDOLT, B. V. (1969) An economic derivation of the 'gravity law' of spatial interaction. *Journal of Regional Science*, 9.2, 273–282.

NEISSER, V. (1967) *Cognitive Psychology.* New York, Appleton-Century-Crofts. 351 p.

NEUBERGER, E. and DUFFY, W. J. (1976) *Comparative Economic Systems: A Decision-Making Approach.* Boston, Allyn & Bacon. 378 p.

NIBLETT, G. B. F. (1971) *Digital Information and the Priority Problem.* Paris, Organisation for Economic Cooperation and Development. 58 p.

NKRUMAH, K. (1963) *Africa Must Unite.* London, Mercury-Heineman. 229 p.

NORDHAUS, W. D. (1973) World dynamics: measurement without data. *Economic Journal,* 83, 1156–83.

NOTON, M. *et al.* (1974) A dynamical model of conflict, Parts I and II. In Dmowski. R. M. (ed.) *Systems Analysis and Modelling: Approaches in Environmental Systems,* pp. 95–126. Warsaw, Polish Academy of Sciences. 607 p.

ODUM, E. P. (1963) *Ecology.* London, Holt, Rinehart & Winston. 152 p.

ODUM, H. T. (1957) Trophic structure and productivity of Silver Springs, Florida. *Ecological Monographs,* 27, 55–112.

ODUM, H. T. (1971) *Environment, Power and Society.* New York, Wiley-Interscience. 331 p.

OECD (1973) *Automated Information Management in Public Administration.* Paris, Organisation for Economic Cooperation and Development. 124 p.

OECD (1974) *Information Technology in Local Government.* Paris, Organisation for Economic Cooperation and Development. 168 p.

OLSSON, G. (1975) *Birds in Egg.* Dept. Geography, Univ. Michigan, Ann Arbor, Geographical Pub. 15. 523 p.

ORD, J. K. (1975) Estimation methods for models of spatial interaction. *Journal of the American Statistical Association,* 70, 120–6.

O'RIORDAN, T. (1971) *Perspectives on Resource Management.* London, Pion. 183 p.

PADÉ, H. (1892) Sur la représentation aprochée d'une fonction par des fractions rationelles. *Annales Scièntifiques de l'Ecole Normale Supérieure,* 3rd ser. (supp.), 9, S19–S93.

PAISH, F. W. (1966) *Studies in an Inflationary Economy: The United Kingdom 1945–61.* London, Macmillan. 336 p.

PARKES, J. G. M. (in progress) *Coastal Zone Management in England and Wales: The Case for an Integrated Approach.* Ph.D. thesis, Dept Geography, Univ. College London.

PARSONS, T. (1966) The political aspect of social structure and process. In Easton, D. (ed.) *Varieties of Political Theory,* pp. 71–112. Englewood Cliffs, N.J., Prentice-Hall.

PASSMORE, J. (1974) *Man's Responsibility for Nature: Ecological Problems and Western Traditions.* London, Duckworth. 213 p.

PATTANAIK, P. K. (1971) *Voting and Collective Choice: Some Aspects of the Theory of Group Decision Making.* Cambridge, Univ. Press. 184 p.

PATTEE, H. H. (1973) The physical basis and origin of hierarchical control. In Pattee, H. H. (ed.) *Hierarchy Theory: The Challenge of Complex Systems,* pp. 73–108. New York, Braziller. 156 p.

PATTEN, B. C. (ed.) (1971) *Systems Analysis and Simulation in Ecology,* Vol. 1. New York and London, Academic Press. 607 p. (see Patten 3–121).

PATTEN, B. C. (ed.) (1972) *Systems Analysis and Simulation in Ecology,* Vol. 2. New York and London, Academic Press. 592 p.

PATTEN, B. C. (ed.) (1975) *Systems Analysis and Simulation in Ecology*, Vol. 3. New York and London, Academic Press. 601 p.

PATTEN, B. C., EGLOFF, D. A. and RICHARDSON, T. H. (1975) Total ecosystem model for a cove in Lake Texoma. In Patten (ed.) (1975), pp. 205–421.

PAYNTER, H. M. (1968) Amplification and control technology. In Milsum, J. H. (ed.) *Positive Feedback*, pp. 66–79. Oxford, Pergamon. 169 p.

PESTON, M. (1973) Econometrics and control: some general comments. *Conference Publication 101, IEE, London*, pp. 15–30.

PHILLIPS, A. W. (1954) Stabilisation policy in a closed economy. *Economic Journal*, 64, 290–323.

PHILLIPS, A. W. (1957) Stabilisation policy and the time-form of lagged responses. *Economic Journal*, 67, 265–77.

PHILLIPS, D. C. (1977) *Holistic Thought in Social Science*. London, Macmillan. 149 p.

PIAGET, J. (1969) *The Mechanisms of Perception* (trans. G. N. Seagrim). London, Routledge & Kegan Paul. 384 p.

PIAGET, J. (1971) *Structuralism*. London, Routledge & Kegan Paul. 153 p.

PICARDI, A. C. (1976) Practical and ethical issues of development in traditional societies: insights from a system dynamics study in pastoral West Africa. *Simulation*, 26, pp. 1–9.

PICKERSGILL, G. M. and PICKERSGILL, J. E. (1974) *Contemporary Economic Systems: A Comparative View*. Englewood Cliffs, N.J., Prentice-Hall. 356 p.

PILGRIM, D. H. (1966) Radioactive tracing of storm runoff on a small catchment, I and II: *Journal of Hydrology*, 4, 289–326.

PILGRIM, D. H. (1976) Travel times and nonlinearity of flood runoff from tracer measurements on a small watershed. *Water Resources Research*, 12, 487–96.

PLACKETT, R. L. (1950) On some theorems in least squares. *Biometrica*, 37, 149–57.

POLANYI, M. (1958) *Personal Knowledge: Towards a Post-Critical Philosophy*. London, Routledge & Kegan Paul. 428 p.

POMEROY, L. R. (1970) The strategy of mineral cycling. *Annual Review of Ecology and Systematics*, 1, 171–90.

PONTRYAGIN, L. S., BOLTYANSKI, V. G., GAMKRELIDZE, R. V. and MISHCHENCKO, E. F. (1962) *The Mathematical Theory of Optimal Processes*. New York, Wiley-Interscience. 360 p.

POPPER, K. (1957) *The Poverty of Historicism*. London, Routledge & Kegan Paul. 166 p.

POPPER, K. (1959) *The Logic of Scientific Discovery*. London, Hutchinson. 480 p.

POWERS, W. T. (1974) *Behaviour: The Control of Perception*. London, Wildwood House. 296 p.

PRANDECKA, B. (1975) The four principal problems of spatial policy. In Karlquist, A., Lundquist, L. and Snickars, F. (eds) *Dynamic Allocation of Urban Space*. Farnborough, Saxon-House.

PRED, A. R. (1974) *An Evaluation and Summary of Human Geography Research Projects*. Stockholm, Statens Råd för Samhällsforskning. 46p.

PREST, A. R. and TURVEY, R. (1965) Cost-benefit analysis: a survey. *Economic Journal*, 75, 155–207.

PRESTON, C. J. (1974) *Gibbs States on Countable Sets*. Cambridge, Univ. Press. 128 p.

PROSHANSKY, H. M., ITTELSON, W. H. and RIVLIN, L. G. (eds) (1970) *Environmental Psychology: Man and His Physical Setting*. New York, Holt, Rinehart & Winston. 690 p.

QUENOUILLE, M. H. (1947) A large sample test for the goodness of fit of autoregressive schemes. *Journal of the Royal Statistical Society*, 110, 123–9.

QUENOUILLE, M. H. (1957) *The Analysis of Multiple Time Series*. London, Charles Griffin. 105 p.

QUESNAY, F. (1758) *Tableau Economique avec ses Explications*. Avignon, Quesnay. 149 p.

QUINE, W. V. (1969) *Ontological Relativity and Other Essays*. New York, Columbia Univ. Press. 165 p.

RADCLIFFE, C. J. (1959) *Committee on the Working of the Monetary System*. London, HMSO.

RADEMAKER, O. (1974) The behaviour of pollution in Forrester's world model. In Dmowski, R. M. (ed.) *Systems Analysis and Modelling in Environment Systems*, pp. 67–76. Warsaw, Polish Academy of Science.

RAPPAPORT, R. A. (1971) The flow of energy in an agricultural society. In Flanagan, D. *et al.* (eds) *Energy and Power*, pp. 69–80. San Francisco, Freeman.

RAVENSTEIN, E. G. (1885) The laws of migration. *Journal of the Royal Statistical Society*, 43, 167–235.

RAYNER, J. N. (1971) *An Introduction to Spectral Analysis*. London, Pion Press. 174 p.

REES, P. H. and WILSON, A. G. (1977) *Spatial Population Analysis*. London, Arnold. 324 p.

REICHENBACH, H. (1944) *Philosophical Foundations of Quantum Mechanics*. Cambridge, Univ. Press.

REILLY, W. J. (1929) Methods for the study of retail relationships. *Univ. Texas Bulletin*, No. 2994.

REIRA, B. and JACKSON, M. (1972) *The Design of Monitoring and Advisory System for Sub-regional Planning*. A study by Iscol Ltd. (Notts/Derby M & A Unit, Loughborough). 5 chapters.

RICH, L. G. (1973) *Environmental Systems Engineering*. New York, McGraw-Hill. 448 p.

RICHARDSON, H. W. (1973) *The Economics of Urban Size*. Farnborough, Saxon House. 243 p.

RIKER, W. H. (1961) Voting and the summation of preferences: an interpretative bibliographic review of selected developments during the last decade. *American Political Science Review*, 55, 900–11.

ROBERTS, P. (1973) Models of the future. *Omega*, 1, 5.

ROBINSON, E. and ROBBINS, R. C. (1968) *Sources, Abundance and Fate of Gaseous Atmospheric Pollutants*. Stanford Research Institute, SRI Project PR-6755.

ROBINSON, E. A. (1967) *Multichannel Time-Series Analysis with Digital Computer Programs*. San Francisco, Holden-Day. 298 p.

ROBINSON, J. (1964) *Economic Philosophy*. London, Watts. 147 p.

ROBINSON, J. (1970) *Freedom and Necessity: An Introduction to the Study of Society*. London, Allen and Unwin. 128 p.

RODDA, J. C. (1970) Rainfall excesses in the United Kingdom. *Transactions of the Institute of British Geographers*, 49, 49–60.

ROGERS, A. (1968) *Matrix Analysis of Interregional Population Growth and Distribution*. Berkeley, Univ. California Press. 119 p.

RONNBERG, A., SALOMONSSON, O. and SELANDER, K. (1973) *Allokeringsmetodik distans-och befolkningson andshestömninger och des tillöpningar i praktisk planering.* NIMS Report No. 3. 74 p.

RONNBERG, A., SALOMONSSON, O. and SELANDER, K. (1974) *Geocoding.* National Swedish Building Research Summaries, S2. 2 p.

ROSEN, R. (1975) Biological systems as paradigms for adaptation. In Day, R. H. and Groves, T. (eds) *Adaptive Economic Models*, pp. 39–72. New York, Academic Press.

ROSENBERG, R. L. and COLEMAN, P. J. (1974) 27-day cycle in rainfall at Los Angeles. *Nature*, 250, 481–4.

ROSENBLOOM, R. S. and RUSSELL, J. R. (1971) *New Tools for Urban Management: Studies in Systems and Organizational Analysis.* Boston, Hammond Univ. Press. 298 p.

ROSENBLUETH, A., WIENER, N. and BIGELOW, J. (1943) Behaviour, purpose and teleology. *Philosophy of Science*, 10, 18–24.

ROSTOW, W. W. (1960) *The Stages of Economic Growth.* Cambridge, Univ. Press. 179 p.

ROSZAK, T. (1972) *Where the Wasteland Ends: Politics and Transcendence in Post-Industrial Society.* London, Faber. 492 p.

ROUSSEAU, J.-J. (1750) *Discours sur les Sciences et les Arts.* In Masters, R. D. (ed.) *The First and Second Discourses.* New York, St. Martin's Press.

ROYAL COMMISSION ON THE CONSTITUTION 1969–1973 (1973) Vol. 1: *Report*, Cmnd 5460, Vol. 2: *Memorandum of Dissent by Lord Crowther-Hunt and Prof. A. T. Peacock*, London, HMSO.

ROYAL COMMISSION ON ENVIRONMENTAL POLLUTION (1976) *Nuclear Power and the Environment.* London, HMSO.

RUBIN, M. D. (1968) History of technological feedback. In Milsum, J. H. (ed.) *Positive Feedback*, pp. 9–22. Oxford, Pergamon. 169 p.

RUSHTON, G. (1969) Analysis of spatial behaviour by revealed space preferences. *Annals of the Association of American Geographers*, 59, 391–400.

RUSSELL, B. (1946) *History of Western Philosophy.* London, George Allen & Unwin. 916 p.

RYKIEL, E. J. and KUENZEL, N. T. (1971) Analog computer models of 'The wolves of Isle Royale'. In Patten, B. C. (ed.) *Systems Analysis and Simulation in Ecology*, Vol. 1, pp. 513–41. New York and London, Academic Press. 607 p.

SAGE, A. P. (1968) *Optimum Systems Control.* Englewood Cliffs, N.J., Prentice-Hall. 562 p.

SAGE, A. P. and HUSA, G. W. (1969) Algorithms for sequential adaptive estimation of prior statistics. *Proceedings of the Eighth IEEE Symposium on Adaptive Processes*, Paper 6a, 6 p.

SAHLINS, M. D. and SERVICE, R. (eds) (1960) *Evolution and Culture.* Ann Arbor, Michigan Univ. Press. 131 p.

SAUCEDO, R. and SCHIRING, E. C. (1968) *Introduction to Continuous and Digital Control Systems.* New York, Macmillan. 782 p.

SAUVY, A. (1975) *Zero Growth?* Oxford, Blackwell. 266 p.

SAVAS, E. S. (1969) Feedback controls on urban air pollution. *IEEE Spectrum* (July), 77–81.

SAWARAGI, Y. *et al.* (1974) A large-scale research project in progress on environmental pollution control. In Dmowski, R. M. (ed.) *Systems Analysis and*

Modelling Approaches in Environment Systems, pp. 197–222. Warsaw, Polish Academy of Sciences.

SAWARAGI, Y. and IKEDA, S. (1974) Some problems of methodology of environmental pollution control and application. In Dmowski, R. M. (ed.), *Systems Analysis and Modelling Approaches in Environment Systems*, pp. 223–52. Warsaw, Polish Academy of Sciences.

SAYER, R. A. (1976) A critique of urban modelling: From regional science to urban and regional political economy. *Progress in Planning*, 3, 187–254.

SCHEIDEGGER, A. E. (1975) *Physical Aspects of Natural Catastrophes*. New York, Elsevier. 289 p.

SCHEIDEGGER, A. E. and LANGBEIN, W. B. (1966) *Probability concepts in geomorphology*. USGS Professional Paper 500-C. 14 p.

SCHLESINGER, G. (1959) Two approaches to mathematical and physical systems. *Philosophy of Science*, 263, 240 p.

SCHMIDT, A. (1971) *The Concept of Nature in Marx*. London, New Left Books. 251 p.

SCHNEIDER, S. H. (1974) The population explosion: Can it shake the climate? *Ambio*, 3, 150–5.

SCHOEFFLER, J. D. (1971) On-line multilevel systems. In Wismer, D. A. (ed.) *Optimization Methods for Large Scale Systems, with Applications*, pp. 291–330. New York, McGraw-Hill. 335 p.

SCHRÖDINGER, E. (1967) *What is Life? Mind and Matter*. Cambridge, Univ. Press. 178 p.

SCHULTZ, D. G. and MELSA, J. L. (1967) *State Functions and Linear Control Systems*. New York, McGraw-Hill. 435 p.

SCHUMACHER, E. F. (1971 and 1973) *Small is Beautiful*. London, Earth Island, Blond & Briggs. 255 p.

SCOLNIK, H. (1973) *On a Methodological Criticism of Meadows' World 3 Model*. Unpub. report quoted in Clark *et al*. (1975).

SCOTT, A. J. (1971) *Combinational Programming, Spatial Analysis and Planning*. London, Methuen. 204 p.

SEINFELD, J. H. and KYAN, C. P. (1971) Determination of optimal air pollution control strategies. *Socio-Economic Planning Science*, 5, 173–90.

SELF, P. (1972) *Administrative Theories and Politics: An Inquiry into the Structure and Processes of Modern Government*. London, Allen & Unwin. 308 p.

SELZNICK, P. (1949) *TVA and the Grass Roots: A Study in the Sociology of Formal Organizations*. Berkeley, California University Press. 274 p.

SEN, A. K. (1970) *Collective Choice and Social Welfare*. San Francisco, Holden-Day. 225 p.

SEN, A. K. (1973) *On Economic Inequality*. Oxford, Clarendon Press. 118 p.

SENGUPTA, J. K. and FOX, K. A. (1969) *Economic Analysis and Operations Research: Optimization Techniques in Quantitative Economic Models*. Amsterdam, North-Holland. 478 p.

SHACKLE, G. L. S. (1972) *Epistemics and Economics: A Critique of Economic Doctrines*. Cambridge, Univ. Press. 482 p.

SHANNON, C. (1948) The mathematical theory of communication. *Bell System Technical Journal*, 27, 379–423.

SHERMAN, T. P. (1969) *O and M in Local Government*. Oxford, Pergamon.

SIMON, H. A. (1956) Dynamic programming under uncertainty with a quadratic criterion function. *Econometrica*, 24, 74–81.

SIMON, H. A. (1957) *Models of Man.* New York, Wiley. 287 p.

SIMON, H. A. (1969) *The Sciences of the Artificial.* Cambridge, Mass., MIT Press. 123 p.

SIMPSON, J. and DENNIS, A. S. (1974) Cumulus clouds and their modification. In Hess, W. N. (ed.) *Weather and Climate Modification*, pp. 229–81. New York, Wiley.

SMALE, S. (1967) Differentiable dynamical systems. *Bulletin of the American Mathematical Society*, 73, 747–817.

SMALE, S. (1974) Global analysis and economics. *Journal of Mathematical Economics*, 1, 1–14, 107–18, 119–27, 213–21.

SMITH, D. M. (1973) *The Geography of Social Well-Being in the United States.* New York, McGraw-Hill. 144 p.

SMITH, D. M. (1977) *Human Geography: A Welfare Approach.* London, Arnold. 402 p.

SMYTH, J. D. (1972) *Introduction to Animal Parasitology.* London, English Univ. Press. 470 p.

SOCHAVA, V. B. (1971) Geography and ecology. *Soviet Geography* (May), 277–293.

SONNENFELD, J. (1972) Geography, perception, and the behavioral environment. In English, P. W. and Mayfield, R. C. (eds) *Man, Space and Environment*, pp. 244–51. London, Oxford Univ. Press.

SPIEGEL, M. R. (1965) *Theory and Problems of Laplace Transforms.* New York, McGraw-Hill. 261 p.

SPINOZA, B. DE (1677) *On the Improvement of Understanding of Ethics.* New York, Dover (1955).

STEA, D. and DOWNS, R. M. (1970) From the outside looking in at the inside looking out. *Environment and Behaviour*, 2, 3–12.

STEWART, I. (1975) The seven elementary catastrophes. *New Scientist*, 68 (976), 447–54.

STEWART, J. C. (1947) Empirical mathematical rules concerning the distribution and equilibrium of population. *Geographical Review*, 33, 461–85.

STOCKING, M. A. and ELWELL, H. A. (1976) Rainfall erosivity over Rhodesia. *Transactions of the Institute of British Geographers*, new ser., 1 (2), 231–45.

STODDART, D. R. (1967) Organism and ecosystem as geographical models. In Chorley, R. J. and Haggett, P. (eds) *Models in Geography.* London, Methuen. pp. 511–48.

STRAHLER, A. N. (1965) *Introduction to Physical Geography.* New York, Wiley. 455 p.

STRINGER, J. (1967) Operational research for multi-organizations. *Operational Research Quarterly*, 18, 105–20.

STUDER, R. G. (1970) The dynamics of behavior-contingent physical systems. In Proshansky, H. M. *et al.* (eds) *Environmental Psychology: Man and his Physical Setting*, pp. 56–76. New York, Holt, Rinehart & Winston.

SUNAMURA, T. (1976) Feedback relationships in wave erosion of laboratory rocky coast. *Journal of Geology*, 84, 427–38.

SURREY, M. J. C. (1971) *The Analysis and Forecasting of the British Economy.* Cambridge, Univ. Press, NIESR Occasional Paper XXV. 107 p.

SWAN, J. (1976) *Legislation and Human Settlement*, Brussels, WERC Conference Proceedings. 12 p.

SYMPOSIUM (1969) *The Progress of Hydrology*, Vol. 1: *New Developments in Hydrology.* Urbana, Illinois, Dept of Civil Engineering, Univ. Illinois. 509 p.

SZTOMPKA, P. (1974) *System and Function: Toward a Theory of Society*. New York, Academic Press. 213 p.

TALBOT, S. A. and GESSNER, U. (1973) *Systems Physiology*. New York, Wiley. 511 p.

TAN, K. C. (in preparation, 1978) *Control Applications to Uncertain Spatial Systems*. Ph.D. Thesis, Dept of Geography, University College, London.

TAYLOR, F. M. (1911) *Shop Management*. London, Harper. 207 p.

TERJUNG, W. H. (1976) Climatology for geographers. *Annals of the Association of American Geographers*, 66, 199–222.

THEIL, H. (1954) *Linear Aggregation of Economic Relations*. Amsterdam, North-Holland. 205 p.

THEIL, H. (1957) A note on certainty equivalence in dynamic planning. *Econometrica*, 25, 346–9.

THEIL, H. (1964) *Optimal Decision Rules in Government and Industry*. Amsterdam, North-Holland. 364 p.

THEIL, H. (1966) *Applied Economic Forecasting*. Amsterdam, North-Holland. 474 p.

THIEMANN, H. (1973) *Report on Batelle's Activities to the Club of Rome* (*Tokyo*). Quoted in Clark *et al.* 1975.

THOM, R. (1975) *Structural Stability and Morphogenesis: An Outline of a General Theory of Models* (trans. D. H. Fowler). Reading, Mass., Benjamin. 348 p.

THOM, R. and ZEEMAN, E. C. (1975) *Catastrophe Theory: Its Present State and Future Prospects*. Mathematics Institute, Univ. of Warwick (mimeo). 34 p.

THOMANN, R. V. (1973) Effect of longitudinal dispersion on dynamic water quality response of streams and rivers. *Water Resources Research*, 9, 355–66.

THOMAS, N. (1971) *Computerized Data Banks in Public Administration: Trends and Policies Issues*. Paris, Organisation for Economic Cooperation and Development. 69 p.

THOMLINSON, R. (1972) *Geographical Data Handling*, 2 vols. Ottawa, International Geographical Union, Commission on Geographical Data Sensing and Processing. 1327 p.

THOMPSON, W. R. (1922) Étude mathématique de l'action des parasites entomorphoges. *Comptes Rendus Acad. Sci.*, 174, 1201, 1433.

THORBURN, A. (1975) *Up-to-date Structure Plans: Notes on Monitoring and Reviewing Processes*. Unpub. paper to INLOGOV Seminar, Birmingham. 10 p.

TINBERGEN, J. (1937) *An Econometric Approach to Business Cycle Problems*. Paris, Hermann. 73 p.

TINLINE, R. R. (1971) Linear operators in diffusion research. In Chisholm, M., Frey, A. and Haggett, P. (eds) *Regional Forecasting*, pp. 71–91. London, Butterworth.

TINLINE, R. R. (1972) Lee wave hypothesis for the initial pattern of spread during the 1967–8 foot and mouth epizoatic. In McGlashan, N. D. (ed.) *Medical Geography*, pp. 301–6. London, Methuen. 336 p.

TOBLER, W. R. (1966) *Notes on the Analysis of Geographical Distributions*. Michigan Inter-Univ. Community of Mathematical Geographers, Ann Arbor, Disc. Pap. 8. 15 p.

TOBLER, W. R. (1970) A computer movie simulating urban growth in the Detroit Region. *Economic Geography*, 46, 234–40.

TOMOVIC, R. (1969) On man-machine control. *Automatica*, 5, 401–4.

TOULMIN, S. E. (1953) *The Philosophy of Science: An Introduction.* London, Hutchinson. 176 p.

TRIBUS, M. (1969) *Rational Descriptions, Decisions and Designs.* New York, Pergamon. 478 p.

TRUDGILL, S. T. (1976) Rock weathering and climate: quantitative and experimental aspects. In Derbyshire, E. (ed.) *Geomorphology and Climate*, pp. 59–99. London, Wiley.

TUAN, Y.-F. (1971) Geography, phenomenology, and the study of human nature. *Canadian Geographer*, 15, 181–92.

TUAN, Y.-F. (1973) Environmental psychology: a review. *Geographical Review*, 63, 245–56.

TUAN, Y.-F. (1974) *Topophilia.* Englewood Cliffs, N.J., Prentice-Hall. 260 p.

TUAN, Y.-F. (1976) Geopiety: a theme in man's attachment to nature and to place. In Lowenthal, D. and Bowden, M. J. (eds) *Geographies of the Mind: Essays in Historical Geography in Honour of John Kirkland Wright*, pp. 11–39. New York, Oxford Univ. Press. 263 p.

TUKEY, J. W. (1961) Discussion emphasizing the connection between analysis of variance and spectrum analysis. *Technometrics*, 3, 191–211.

TURNOVSKY, S. J. (1973) Optimal stabilization policies for deterministic and stochastic control. *Review Economic Studies*, 40, 79–95.

TUSTIN, A. (1953) *The Mechanism of Economic Systems.* Cambridge, Mass., Harvard Univ. Press. 191 p.

TYLER, J. S. and TUTEUR, F. B. (1966) The use of a quadratic performance indicator to design multivariable control systems. *Transactions IEEE, Automatic Control*, AC-11, 84–92.

TZAFESTAS, S. G. and NIGHTINGALE, J. M. (1968) Optimal filtering, smoothing and prediction in linear distributed-parameter systems. *Proceedings IEE*, 115, 1207–12.

TZAFESTAS, S. G. and NIGHTINGALE, J. M. (1969) Differential dynamic programming approach to optimal nonlinear distributed-parameter control systems. *Proceedings IEE*, 116, 1079–84.

UEXKÜLL, J. VAN (1926) *Theoretical Biology.* London, Kegan Paul. 362 p.

UREÑA, J. M. DE (1975) *Monitoring in Planning at the Regional Level: Towards a Conceptual Framework.* Unpublished M.Phil. Thesis, Department of Urban Design and Regional Planning, Edinburgh. 240 p.

URWICK, L. (1933) *Management of Tomorrow.* London, Nisbet.

USHER, M. B. (1973) *Biological Management and Conservation.* London, Chapman & Hall. 394 p.

VAN DER MAAREL, E. (1975) Man-made natural ecosystems in environmental management and planning. In Van Dobben, W. H. and Lowe-McConnell, R. H. (eds) *Unifying Concepts in Ecology*, pp. 263–81. The Hague, Junk, B. V

VAN DOBBEN, W. H. and LOWE-MCCONNELL, R. H. (eds) (1975) *Unifying Concepts in Ecology.* The Hague, Junk, B. V. 302 p.

VAN DYNE, G. M. (ed.) (1969) *The Ecosystem Concept in Natural Resource Management.* New York and London, Academic Press. 383 p.

VAN DYNE, G. M. (1974) A systems approach to grasslands. In Daetz, D. and Pantell, R. H. (eds) *Environmental Modelling: Analysis and Management*, pp. 89–101. Stroudsburg, Pa., Dowden, Hutchinson & Ross.

VAN GIERKE, H. E., MEIDEL, W. D. and OESTREICHER, H. L. (eds) (1970) *Principles and Practice of Bionics*. Slough, Technivision Services. 504 p.

VAN GIGCH, J. P. (1974) *Applied General Systems Theory*. New York, Harper & Row. 439 p.

VANYO, J. P. (1971) Law, operations research and the environment. *Journal of Environmental Systems*, 1, 213–36.

VERNEKAR, A. D. (1971) Long-period global variations of incoming solar radiation. *Meteorological Monographs*, 12 (34), 1–20.

VICKREY, W. (1960) Utility, strategy and social decision rules. *Quarterly Journal of Economics*, 74, 507–35.

VISHWAKARMA, K. P. (1974) *Macro-Economic Regulation*. Rotterdam Univ. Press. 316 p.

VITA-FINZI, C. (1973) *Recent Earth History*. London, Macmillan. 138 p.

VON MISES, R. (1951) *Positivism: A Study in Human Understanding*. Cambridge, Mass., Harvard Univ. Press. 404 p.

VON NEUMANN, J. (1952) Probabilistic logic and the synthesis of reliable organisms from unreliable components. In *Collected Works, J. Von Neumann*, Vol. 5, pp. 329–78. Oxford, Pergamon Press.

VON NEUMANN, J. and MORGENSTERN, O. (1944) *Theory of Games and Economic Behavior*. New York, Wiley. 641 p.

WADDINGTON, C. H. (1977) *Tools for Thought*. London, Jonathan Cape. 250 p.

WALMSLEY, D. L. (1973) The simple behaviour system: an appraisal and an elaboration. *Geografiska Annaler*, 55, ser. B., 49–56.

WANG, P. K. C. (1964) Control of distributed parameter systems. In Leandes, C. (ed.) *Advances in Control Systems*, Vol. 1, pp. 75–172. New York, Academic Press.

WARD, B. (1967) *The Socialist Economy: A Study of Organizational Alternatives*. New York, Random House. 272 p.

WARD, B. (1976) *The Home of Man*. Harmondsworth, Penguin. 297 p.

WARREN, A. and MAIZELS, J. K. (1976) *Ecological Change and Desertification: Report to the Desertification Secretariat, UN Environment Programme*. London, Univ. College. 198 p.

WASTLER, T. A. (1973) *Spectral Analysis Applications in Water Pollution Control*. Washington, D.C., Federal Water Pollution Control Administration, Report CWT-3.

WATSON, J. (1913) Psychology as the behaviorist views it. *Psychological Review*, 20, 158–77.

WATT, K. E. F. (1968) *Ecology and Resource Management*. New York, McGraw-Hill. 450 p.

WAY, M. J. (1953) The relationship between certain ant species with particular reference to biological control of the *coreid Thenaptus sp.*; *Bulletin of Entomological Research*, 44 (4), pp. 669–91.

WEBBER, M. J. (1972) *Impacts of Uncertainty on Location*. Cambridge, Mass., MIT Press. 310 p.

WEBER, A. (1909) *Über den Standart der Industrien*. Tübingen. 256 p.

WEBER, M. (1924) *The Theory of Social and Economic Organization* (trans. A. M. Henderson and T. Parsons). New York, Free Press. 404 p.

WEDGWOOD-OPPENHEIM, F. (1970) AIDA and the Strategic Choice Approach. In *Conference Papers IP 25*, pp. 15–29. London, Centre Environmental Studies.

WEDGWOOD-OPPENHEIM, F., HART, D. and COBLEY, B. (1975) An exploratory study in strategic monitoring. *Progress in Planning*, 5, 1–58.

WEISSBROD, R. S. P. (1974) *Spatial Diffusion of Relative Wage Inflation*. Ph.D. Thesis, Northwestern Univ., Evanston, Illinois. 166 p.

WHIPKEY, R. Z. (1965) Subsurface stormflow from forested slopes. *Bulletin of the International Association of Scientific Hydrology*, 10 (2), 74–85.

WHITE, G. F. (1945) *Human Adjustment to Floods*. Dept Geography, Univ. Chicago, Res. Pap. 29.

WHITE, L. (1967) The historical roots of our ecological crisis. *Science*, 155, 3767.

WHITEHEAD, A. N. (1957) *The Aims of Education and Other Essays*. New York, Macmillan. 247 p.

WHITTAKER, R. H. (1953) Consideration of climax theory: the climax as a population and pattern. *Ecological Monographs*, 23, 41–78.

WHITTLE, P. (1952) Tests of fit in time series. *Biometrika*, 39, 309–18.

WHITTLE, P. (1954) On stationary processes in the plane. *Biometrica*, 41, 434–49.

WHITTLE, P. (1963a) Stochastic processes in several dimensions. *Bulletin of the International Statistical Institute*, 34, 974–93.

WHITTLE, P. (1963b) *Prediction and Regulation: By Linear Least Squares Methods*. London, English Univ. Press. 147 p.

WIBERG, D. M. (1967) Feedback control of linear distributed parametric systems. *Transactions ASME, Journal of Basic Engineering*, 89, 379–84.

WIENER, N. (1949) *Extrapolation, Interpolation and Smoothing of Stationary Time Series*. New York, Wiley. 163 p.

WIESLANDER, J. and WITTENMARK, B. (1971) An approach to adaptive control using real time identification. *Automatica*, 7, 211–17.

WILES, P. J. D. (1962) *The Political Economy of Communism*. Oxford, Blackwell. 404 p.

WILLIAMS, H. C. W. L. and WILSON, A. G. (1978) Dynamic models for urban and regional analysis. In Carlstein, T., Parker, P. N. and Thrift, N. J. *Timing Space and Spacing Time in Socio-Economic Systems*. London, Arnold.

WILLIAMS, R. B. (1971) Computer simulation of energy flow in Cedar Bog Lake, Minnesota, based on the classical studies of Lindeman. In Patten, B. C. (ed.) *Systems Analysis and Simulation in Ecology*, Vol. 1, pp. 543–84. New York and London, Academic Press. 607 p.

WILLIAMS, R. B. (1972) Steady-state equations in simple nonlinear food webs. In Patten, B. C. (ed.) *Systems Analysis and Simulation in Ecology*, Vol. 2, pp. 213–40. New York and London, Academic Press. 592 p.

WILSON, A. G. (1970) *Entropy in Urban Regional Modelling*. London, Pion. 166 p.

WILSON, A. G. (1974) *Urban and Regional Models in Geography and Planning*. London, Wiley. 418 p.

WILSON, A. G. and KIRKBY, M. J. (1975) *Mathematics for Geographers and Planners*. London, Oxford University Press. 325 p.

WILSON, E. O. (1971) Competitive and aggressive behaviour. In Essenberg, J. F. and Dillon, W. (eds) *Man and Beast: Comparative Social Behaviour*. Washington, D.C., Smithsonian Institute.

WILSON, E. O. (1975) *Sociobiology: The New Synthesis*. Cambridge, Mass., Belknap, Harvard Univ. Press. 697 p.

WILSON, G. T. (1970) *Modelling Linear Systems for Multivariate Control*. Ph.D. Thesis, Univ. Lancaster. 202 p.

WINSTANLEY, D. (1973) Rainfall patterns and general atmospheric circulations. *Nature*, 245, 190–4.

WISCHMEIER, W. H. and SMITH, D. D. (1958) Rainfall energy and its relation to soil loss. *Transactions of the American Geophysical Union*, 39, 285–91.

WISMER, D. A. (ed.) (1971) *Optimization Methods for Large Scale Systems, With Applications*. New York, McGraw-Hill. 335 p.

WITTENMARK, B. (1969) *On Adaptive Control of Low Order Systems*. Lund Institute of Technology, Control Division, Report 6918. 32 p.

WOLD, H. (1938) *A Study in the Analysis of Stationary Time Series*. Stockholm, Almquist & Wiksell. 196 p.

WOLMAN, M. G. (1971) The nation's rivers. *Science*, 174, 905–18. (Reprinted in Daetz and Pantell 1974, pp. 206–19.)

WOLPERT, J. (1964) The decision process in a spatial context. *Annals of the Association of American Geographers*, 54, 537–58.

WOLPERT, J. (1972) Departures from the usual environment in locational analysis. In English, R. W. and Mayfield, R. C. (eds) *Man, Space and Environment*, pp. 304–315. London, Oxford Univ. Press.

WONHAM, W. M. (1967) Optimal stationary control of a linear system with state-dependent noise. *SIAM Journal of Control*, 5, 486–500.

WOODCOCK, A. E. R. and POSTON, T. (1974) *A Geometrical Study of the Elementary Catastrophes*. Heidelberg, Springer-Verlag. 257 p.

YATSU, E. (1955) On the longitudinal profile of the graded river. *Transactions of the American Geophysical Union*, 36, 655–63.

YEVJEVICH, V. (ed.) (1971) *Systems Approach to Hydrology*. Fort Collins, Colorado, Water Resources Publications. 464 p.

YORE, E. E. (1968) Optimal decoupling control. *Journal of Automatic Control Conference*, 9, 327–36.

YOUNG, P. C. (1973) Macro-economic modelling: a critical appraisal. *IEE, London, Conference Publication* 101, 60–80.

YOUNG, P. C. and BECK, B. (1974) The modelling and control of water quality in river systems. *Automatica*, 10, 455–68.

YOUNG, P. C., NAUGHTON, J., NEETHLING, C. and SHELLSWELL, S. (1973) Macro-economic modelling: a case study. In Eykhoff, P. (ed.) *Identification of Systems*, pp. 145–65. Amsterdam, North-Holland.

YU, W. and HAINES, Y. Y. (1974) Multilevel optimization for conjunctive use of groundwater and surface water. *Water Resources Research*, 10, 625–36.

YUILL, R. S. (1965) *A Simulation Study of Barrier Effects in Spatial Diffusion Problems*. Michigan Inter-Univ. Community of Mathematical Geographers, Ann Arbor, Disc. Pap. 5. 41 p.

ZADEH, L. A. and DESOER, C. A. (1963) *Linear System Theory*. New York, McGraw-Hill, 628 p.

ZEEMAN, E. C. (1973) *Applications of Catastrophe Theory*. Mathematics Institute, Univ. Warwick (mimeo). 30 p.

ZINAM, O. (1969) The economics of command economies. In Prybyla, J. S. (ed.), *Comparative Economic Systems*, pp. 19–46. New York, Appleton-Crofts.

Additional recent references

BAYLISS-SMITH, T. P. and FEACHEM, G. A. (eds.) (1977) *Subsistence and Survival: Rural Ecology in the Pacific*, London, Academic Press. 428 p.

BENNETT, R. J. and TAN, K. C. (1978) Stochastic control of regional economies, In Bartels, C. and Ketelapper, R. H. (eds.), *Exploratory and Explanatory Statistical Analysis of Spatial Data*, Leiden, Nijhoff.

BENNETT, R. J. (1978) Forecasting in urban and regional planning closed loops: The examples of road and air traffic forecasts, *Environment and Planning*, 11, forthcoming.

CAMILLERI, J. A. (1976) *Civilisation in Crisis: Human Prospects in a Changing World*, Cambridge University Press. 303 p.

CASTELLS, M. (1977) *The Urban Question: A Marxist Approach*, London, Arnold. 502 p.

CHAPMAN, G. P. (1977) *Human and Environmental Systems: A Geographer's Appraisal*, London, Academic Press. 422 p.

GREGORY, D. (1978) *Ideology, Science and Human Geography*, London, Hutchinson. 200 p.

HAGGETT, P., CLIFF, A. D. and FREY, A. (1977) *Locational Analysis in Human Geography*, 2nd edn., London, Arnold. 605 p.

O'RIORDAN, T. (1976) *Environmentalism*, London, Pion. 373 p.

TINBERGEN, J., DOLMAN, A. J. and VAN ETTINGER, J. (1976) *Reshaping the International Order: A Report of the Club of Rome*, London, Hutchinson. 325 p.

TRUDGILL, S. T. (1977) *Soil and Vegetation Systems*, Oxford University Press. 192 p.

YOUNG, P. C. and RENNIE, G. (1976) *The Modelling Activities Associated with the Western Port Bay Study*, Report to SCOPE Conference, CRES, Report No. AS/R7, Canberra, Australian National University. 52 p.

Index of persons

(Excluding references)

Adams, J. G. U., 521
Agterburg, F. P., 184
Airy, G. B., 7
Akaike, H., 180
Aldred, B. K., 257
Alexander, C., 301
Allen, J. R. L., 353
Allen, R. G. B., 421, 424, 425, 427
Alonso, W., 523
Althusser, L., 232
Amorocho, J., 37, 41, 320, 322, 323, 324, 325, 326, 327, 328, 329, 331, 361, 362
Anderson, T. W., 182
Annett, J., 249
Anthony, R. N., 265
Anuchin, V. A., 18
Arbib, M. A., 243
Aristotle, 1, 2, 3, 5, 8, 9, 11, 16, 296, 304, 399, 400, 463, 549
Arnold, M., 553
Arrow, K. J., 21, 208, 402, 414, 415
Ashby, W. R., 252, 294, 399, 403, 417
Åstrom, K. J., 89, 135, 137, 145, 146, 147, 148, 531, 560
Attinger, E. O., 306

Bacon, F., 3, 15
Baer, J. G., 482
Ball, R. J., 417
Barlowe, R., 478, 479
Barnett, H. J., 524
Barrett, J. F., 451
Bartholomew, D. J., 387
Bartlett, M. S., 78, 79, 91, 332
Barnard, G. A., 123, 143
Bass, J., 175
Bassett, K., 84
Batty, M., 300, 301, 302
Bayliss-Smith, T. P., 468
Bechdolt, B. V., 402
Beck, B., 334, 373
Beckerman, W., 438, 523
Beer, S., 191, 290, 292, 293, 294, 301
Bekey, G. A., 108
Bellman, R. E., 141
Bennett, J. P., 351
Bennett, R. J., 41, 60, 65, 70, 71, 73, 74, 77, 79, 82, 83, 84, 85, 86, 87, 89, 91, 145, 147, 148, 154, 156, 157, 158, 159, 161, 163, 165, 167, 168, 169, 170, 171, 176, 178, 180, 181, 182, 183, 185, 188, 217, 260, 263, 331, 332, 334, 335, 402, 417, 451, 453, 530
Bentham, J., 21, 296
Bentley, A. F., 15
Berger, P., 548
Berkeley, Bishop, 12
Berry, B. J. L., 260, 263, 403, 432
Bertalanffy, L. Von, 416
Besag, J. E., 184, 187
Black, D., 415
Bhagwati, J., 434, 435, 462
Blalock, H. M., 43, 65, 77, 98, 402, 406
Blesser, W. B., 30, 54
Boulding, K., 224, 237, 243, 267, 280, 402, 410
Boulton, M., 7
Boltzmann, L., 416, 459
Box, G. E. P., 41, 77, 78, 79, 82, 92, 123, 143, 144, 145, 192, 332, 417, 494, 560, 561
Brandstetter, A., 41
Bray, J., 419
Bridgman, P. W., 4
Briggs, R., 237
Brillouin, L., 256
Broadbent, A., 263
Brouwer, D., 381
Brown, L. A., 234
Brown, R. K., 408
Brues, A. M., 389
Bruner, J. S., 247
Bryson, A. E., 129, 145

Bryson, R. A., 331, 332, 379, 521
Buchanan, J. M., 279, 280, 400, 413, 414, 416, 428, 521, 536
Buchanan, L. F., 449, 453, 455
Buckley, W., 286, 407, 408
Bucy, R. S., 129, 263
Budnikova, N. A., 381
Bullock, R. E. H., 251, 283
Bunge, W., 183, 233, 313
Buras, N., 500, 501
Burdekin, R., 529
Burke, R., 261, 265, 267, 268, 269, 278, 279, 284, 285, 293, 315
Burt, C., 20, 21
Burton, I., 237
Butkovskiy, A. G., 154, 189, 190, 192, 196, 197, 201, 202, 218, 219

Calder, N., 382
Calvin, J., 16
Campbell, D. T., 401
Campbell, W. J., 335
Carey, H. C., 403
Carnap, R., 251
Carroll, L. (Dodgson, C. L.), 152, 153, 551
Carslaw, H. S., 48, 59
Chapin, F. S., 402
Chappell, J. E., 15, 19
Chatfield, C., 79
Chebychev, P. L., 92
Chen, I., 245, 246
Child, G. I., 340
Chisholm, M. D. I., 438
Chorafas, D. N., 118, 119
Chorley, R. J., 11, 15, 156, 193, 211, 256, 324, 350, 356, 467, 475, 486
Chow, G. C., 129, 134, 147, 148, 451, 453
Christaller, W., 432
Churchman, C. W., 7, 8, 9, 10
Claborn, B. J., 356
Clark, J., 250, 527, 528, 529, 531, 532, 533, 534
Clarke, D. L., 426
Clawson, M., 478, 479
Clements, F. E., 382
Cliff, A. D., 156, 172, 180, 182, 183, 185, 379, 381, 387
Clifford, P., 187
Clymer, A. B., 543
Coates, B. E., 438
Coddington, A., 523, 538
Cohen, N. H., 398
Cohen, R. S., 459
Cohen, S. B., 296
Cole, H. S. D., 510, 523, 533
Coleman, P. J., 332
Collinge, V. K., 372, 373
Confucius, 19
Connelly, P., 274, 280, 515, 517, 519, 521, 523, 524
Cooke, R. U., 332, 333, 397
Cooper, W. S., 382
Cowling, T. M., 259, 260
Crawford, N. H., 350, 357
Cripps, E. L., 257, 456
Cumberland, J. H., 417
Curry, L., 402, 432
Cuypers, J. G., 510, 511, 532
Cyert, R. M., 251

Dacey, M. F., 171
Dahrendorf, R., 463
Dantzig, G. B., 131, 279
Darby, H. C., 542
Darwin, C., 12, 14, 15, 16, 19, 232, 549
Davies, W. D. T., 110
Davis, W. M., 12, 542
Day, L. H., 508
Day, R. H., 534, 536, 537
Day, T. J., 368
Deevey, E. S., 387
De Greene, K. B., 542
Delleur, J. W., 328
Demko, D., 237
Democritus, 545
de Neufville, R., 271, 277, 278, 280
Dennis, A. S., 503
Descartes, R., 16
de Smith, S., 304, 307, 308, 461
Desoer, C. A., 48, 63, 64
Deutsch, K. W., 38, 237, 238, 242, 404, 407
Dewey, J., 3, 4, 11, 15, 17
Dickens, Charles, 507
Dickinson, H. W., 7
Di Toro, D. M., 382, 384, 385, 386
Dhrymes, P. J., 92
Director, S. W., 61
Dmowski, R. M., 363
Dobb, M., 549
Dodd, S. C., 403
Dooge, J. C. I., 324, 328
Dorfman, R., 96
Douglas, I., 488
Downs, R. M., 224, 225, 226, 234, 235
Drabek, T. E., 304, 305
Dreyfus, S. E., 139, 140, 141, 210, 563
Drucker, P. F., 286
Duesenberry, J. S., 417, 418, 449
Duffy, W. J., 448
Duncan, O. D., 65
Dunn, E. S., 255, 462
Durbin, J., 72, 78
Durkheim, E., 231, 550
Dybern, B. I., 375

Eagleson, P. S., 322
Easton, D., 297, 299
Eddison, T., 250
Edinger, J. G., 376, 378

Einstein, A., 319, 320
Elderton, W. P., 33
Elwell, H. A., 473
Emerson, R. W., 19
Emery, F. E., 120, 245, 246, 247, 256, 258, 271, 272, 286, 287, 288, 290, 291, 292
Emery, J. C., 399
Engels, F., 18, 19, 427, 428
English, P. W., 236
Etzioni, A., 253, 265
Eulau, H., 297
Eykhoff, P., 27, 48, 63, 148

Falkenmark, M., 517, 519, 520
Faludi, A., 296
Fayol, H., 250
Feachem, G. A., 468
Feibleman, J. K., 402, 459
Fogel, L. J., 224, 226, 237, 238
Foin, T. C., 542
Foley, G., 521, 549
Forrester, J. W., 16, 75, 96, 147, 247, 250, 480, 510, 511, 515, 516, 517, 523, 525, 527, 529, 530, 532, 533, 549
Foster, E. E., 325
Fox, K. A., 97, 127, 134
Freud, S., 16, 223, 226, 227, 234, 407, 546
Friedman, M., 456
Friend, J. K., 251, 264, 395

Gagné, R. M., 226, 237, 238, 239, 240, 242, 243
Galbraith, J. K., 15
Galton, F., 21
Gameson, A. L. H., 372, 373
Garg, D. P., 205, 206, 207
Garland, J. A., 379
Gates, G. M., 408
Gates, G. R., 408
Gazis, D.-C., 349
Geary, R. C., 171
Geertz, C., 426, 489, 491
Georgescu-Roegen, N., 233, 549
Gerasimov, I. P., 18
Geosner, U., 28, 29
Gheorghïu, A., 392
Gilbert, E. G., 73, 288, 290
Gillis, J. D. S., 251, 255, 260, 261
Glacken, C. J., 14, 15, 16, 550
Glendinning, J. W., 251, 283
Glucksmann, M., 232
Godley, W., 456
Goldfeld, S. M., 217
Golledge, R. G., 234, 237
Goodin, W. R., 376, 377
Goodwin, B. C., 398
Goodwin, R. M., 425
Gould, P., 156, 236, 237, 273, 408, 505
Gouldner, A. W., 552
Graf, W. L., 350
Granger, C. W. J., 84, 180, 184
Graves, G. W., 497, 499
Graybill, F. A., 554, 557
Greenslade, P. J. M., 390
Grene, M., 2, 5
Grodins, F. S., 94, 95, 100
Groenman, S., 220
Groves, T., 534, 536, 537
Gulick, L. H., 288, 292, 293, 310
Gumbel, E., 33, 363, 486

Haas, J. E., 304, 305
Habermas, J., 232, 407, 521, 539, 547, 548, 549
Hägerstrand, T., 152, 153, 156, 183, 188, 256, 387, 390
Haggett, P., 84, 156, 180, 183, 193, 379, 381
Haines, Y. Y., 501, 502
Haining, R. P., 180, 181, 185, 188
Hall, C. S., 227
Hammersley, J. M., 187
Hannan, E. J., 79, 179
Hare, F. K., 15
Harris, D. R.. 408, 426, 428, 489, 491, 505
Harris, R., 259, 261
Harrison, P. J., 332
Hart, W. E., 37, 322, 324, 326, 327, 361, 362
Harte, J., 486, 487
Harvey, D., 11, 18, 19, 250, 255, 263, 404, 415, 416, 427, 429, 431, 435, 436, 459, 460, 461, 462, 463, 524, 528, 534
Hatanaka, M., 84
Hayek, F. A., 191, 298, 300, 308, 428, 429, 437, 460, 461, 462, 526, 535
Heaney, J. P., 261, 265, 267, 268, 269, 278, 279, 284, 285, 293, 315
Hegel, G., 10, 11, 18, 426, 428, 463, 549
Heilbroner, R. L., 544
Heine, V., 185
Held, R. B., 478, 479
Hempel, G. G., 13, 549
Heraclitus, 545
Herbst, P. G., 245, 246
Hermasen, T., 255
Hermite, C., 92
Herrera, A., 528
Hewitt, K., 15
Hicks, J. R., 425
Hightower, H. C., 402
Hall, A. R., 468, 471
Hill, M., 282
Hipel, K. W., 494, 496
Hirsch, M. W., 68, 69, 214, 532, 533, 542
Ho, Y.-C., 129, 145
Hobbes, T., 296, 550
Holland. S.. 432, 433, 436
Holt, C. C., 129, 131, 134, 451, 560, 561
Horace (Horatius Flaccus, Q.), 15
Horkheimer, M., 232, 548
House, J. W., 438

House, P. W., 529
Hoyt, W. G., 364
Huff, D. L., 402
Hultén, E., 387
Hume, D., 550
Huxley, A., 21
Huxley, J., 239, 241, 242

Ikeda, S., 497
Isachenko, A. G., 18
Isard, W., 417, 432, 448
Ittelson, W. H., 235, 242

Jackson, M., 250, 255, 260
Jacobi, C. G. J., 92
Jacobs, O. L. R., 38, 55
Jacoby, S. L. S., 322, 329, 330, 331
Jaeger, J. C., 48, 59
James, W., 17, 238, 242
Jenkins, G. M., 41, 77, 78, 79, 82, 84, 92, 123, 143, 144, 145, 179, 192, 250, 332, 417, 560, 561
Jenkins, R., 7
Jessop, N., 251, 264, 295
Johnston, J., 90, 91, 179, 182
Johnston, N. L., 33
Johnston, R. J., 237
Jones, J. C., 475
Jordan, C. F., 339
Joseph (Son of Jacob), 551
Jung, C. G., 95, 223, 226, 227, 228, 229, 230, 242, 546
Jury, E. I., 49, 51

Kafka, F., 287, 553
Kahn, R. F., 420, 434
Kaldor, N., 437, 438, 456
Kalecki, M., 425
Kalman, R. E., 26, 73, 96, 122, 129, 145, 148, 149, 263, 332, 404, 417, 559
Kant, E., 10, 11, 18, 20, 463
Kates, R. W., 237
Kaya, Y., 528
Keat, R., 13, 14, 407, 548, 549
Keidel, W. D., 109
Kelley, C. R., 94, 109, 112, 114, 237, 238
Kendall, M. G., 35, 41, 70, 78, 79
Kennedy, B. A., 211, 256, 324, 350, 356, 486
Keyfitz, N., 393
Keynes, J. M., 417, 420, 425, 457, 542
King, L. J., 79, 180
Kirkby, M. J., 140, 350
Kisiel, C. C., 323, 326, 327, 328, 331
Kleine, J. R., 339
Koenig, H. E., 469, 470, 471, 472, 485
Koepcke, R. W., 122, 149, 559, 563
Koestler, A., 241, 547
Koffka, K., 224, 235, 236
Kohlberg, L., 20
Kohler, M. A., 363, 365
Kolakowski, L., 14
Kolata, G. B., 395, 507
Kormondy, E. J., 337, 349
Kornai, J., 432, 441, 521, 549
Kotler, P., 402
Kowal, N. E., 26, 39, 320, 323, 324, 327, 339, 340, 341, 343
Kowalik, P., 339
Krige, D. G., 184
Kuenzel, N. T., 343, 346, 347, 348
Kuhn, A., 111, 235, 236, 239, 399, 401, 407, 409, 410, 411, 412, 425, 533, 534, 537
Kuhn, T., 14
Kurz, M., 208
Kyan, C. P., 376, 377, 378

Lack, D., 392
Lagrange, J. L., 92
Laguerre, E. N., 92
Lamarck, J. B. P. A. de M. de, 232
Lamb, H. H., 332, 379, 387, 521
Langbein, W. B., 324, 364, 404, 407
Lange, O., 115, 116, 118, 119, 120, 122, 504, 507
Langefors, B., 255, 257
Langton, J., 551, 552
La Rochefoucauld, F. duc de, 250
Lasdon, L. S., 131, 279
Latakos, I., 543
Latham, R. E., 3
Laudon, K. C., 284, 285, 302, 312, 313, 314
Lazarus, R. S., 241
Leach, E., 230, 231
Lee, H. D. P., 2
Lee, R. C. K., 142
Legendre, A.-M., 92
Leibniz, G. W., 9, 11, 210
Lenin, V. I., 428, 441
Leopold, L. B., 324, 363, 364, 404
Lerner, A. P., 274, 280, 410
Lerner, A. Ya., 190, 219
Leroy, P., 242
Lévi-Strauss, C., 230, 231, 232, 550
Levy, D., 486, 487
Lewin, K., 235, 236
Lichfield, N., 281, 282, 283
Lidicker, W. E., 206
Lindblom, C., 301
Lindemann, R. L., 349
Lindh, G., 517, 519, 520
Linnemann, H., 529
Linsley, R. K., 350, 357, 363, 365
Lipsey, R. G., 523
Litterer, J. A., 224, 225, 238, 240
Liu, C. Y., 376, 377
Livesey, D. A., 129, 449, 451, 453
Locke, J., 9, 11, 20, 304
Lösch, A., 432

Lotka, A. J., 7, 68, 214, 256, 392, 393, 394, 395, 396, 401
Loucks, O. L., 304
Lowenthal, D., 19, 236
Luce, R., 273, 302
Luenberger, D. G., 145, 149
Lukács, G., 548

MacArthur, R. H., 206, 387, 390, 392, 393
McCloughlin, J. B., 250, 252
McCormick, E. J., 106, 107, 108
McFarland, P. J., 28, 33, 34, 48, 53, 61
MacFarlane, A. G. J., 28, 29, 31, 51, 59
McInerney, J., 537, 538
Mach, E., 459
Mackay, D. M., 404
Maddock, T., 363, 364
Maizels, J. K., 492, 493
Malinowski, B., 231
Malone, D., 376
Malthus, T. R., 21, 523, 524
Mann, H. B., 91
Manners, G., 438, 477
Mannheim, M. L., 266, 267, 274
March, J. G., 285, 297
March, L., 192
March, L., 192
Marcuse, H., 548
Margalef, R., 470, 471, 488
Margenau, H., 404
Marschak, J., 251
Marsh, G. P., 15, 17
Marsh, J. G., 251
Marshall, S. A., 529
Martin, R. L., 173
Maruyama, M., 546, 547
Massam, B., 189, 193, 194
Matern, B., 185
Marx, K., 14, 18, 19, 297, 428, 433, 461, 462, 463, 549 (*See also* Marxism)
Matley, I. M., 15
Matschoss, C., 7
Maughan, R. A., 335
Maxwell, J. C., 7
May, R. M., 395, 507, 546
Mayfield, R. C., 236
Mayr, O., 7
Mead, R., 185
Meadows, D. H., 16, 70, 147, 250, 261, 462, 504, 510, 511, 513, 515, 516, 523, 525, 527, 529, 530, 532, 533
Meditch, J. S., 201
Meier, R., 408
Meister, D., 479
Melsa, J. L., 117, 139, 149, 211, 212, 213, 563
Mennes, L. B. M., 189, 194, 195, 417, 448
Mesarovic, M., 287, 290, 504, 505, 506, 509, 510, 515, 516, 517, 519, 525, 526, 528, 533, 535, 548, 549
Metcalfe, J. L., 287
Micawber, W., 507
Milankovitch, M., 382
Milete, D. S., 304, 305
Mill, J. S., 414, 539
Miller, G. A., 220
Milsum, J. H., 45, 46, 52, 54, 56, 57, 58, 66, 67, 101, 103, 105
Milton, J., 16
Montesquieu, C. L. de S., 304
Montgomery, E., 484
Moore, W. L., 356
Moran, P. A. P., 171
More, R. J., 357
Morgenstern, O., 302
Morris, W., 14
Morse, C., 524
Moussouris, J., 187
Murray, B. G., 389
Myrdal, G., 435, 524, 542

Nash, J. E., 328
Naylor, T. H., 127, 260, 449, 450, 531
Negev, M., 351, 359, 360
Neisser, V., 226
Neuberger, E., 448
Newton, R., 210
Niblett, G. B. F., 258, 307
Niedercorn, J. H., 402
Nietzsche, F., 467, 550
Nightingale, J. M., 192, 201, 202
Nkrumah, K., 462
Nordby, V. J., 227
Nordhaus, W. D., 529, 530, 533
Norton, F. E., 449, 453, 455
Noton, M., 303

Odum, E. P., 387
Odum, H. T., 344, 346, 486, 549
Oeppen, J. E., 173
Oestreicher, H. L., 109
Olsson, G., 403, 432, 460
Oppenheim, P., 549
Ord, J. K., 172, 181, 182, 183, 185
O'Riordan, T., 408, 409, 439
Orlob, G. T., 320, 324, 325, 331
Ostwald, F. W., 459

Padé, H., 92, 93
Paish, F. W., 438
Parkes, M., 295, 297
Parsons, T., 268, 552
Passmore, J., 12, 16, 17
Pattanaik, P. K., 415
Pattee, H. H., 397, 398, 400, 547
Patten, B. C., 324, 326, 337, 340, 341, 344, 345, 346, 361, 362, 363
Paulhus, L. H., 363, 365

Perlman, R., 274, 280, 515, 517, 519, 521, 523, 524
Pestel, E., 287, 290, 504, 505, 506, 509, 510, 515, 516, 517, 519, 525, 526, 528, 533, 535, 548, 549
Peston, M., 96, 129
Phillips, A. W., 125, 127, 423, 451, 454
Phillips, D. C., 551
Piaget, J., 20, 230, 231, 232, 233, 237, 427, 526
Picardi, A. C., 492, 495
Pickersgill, G. M., 437, 446, 447
Pickersgill, J. E., 437, 446, 447
Pilgrim, D. H., 329
Plackett, R. L., 148
Planck, M., 459
Plato, 1, 2, 3, 9, 17, 19, 296, 549
Polanyi, M., 15, 20, 21, 243
Pomeroy, L. R., 339
Pontryagin, L. S., 139, 208, 209, 558
Pope, A., 18, 94
Popper, K., 3, 13, 549
Posner, M., 434
Poston, T., 216
Power, J. M., 295
Powers, W. T., 224, 234, 237, 238, 247
Prandecka, B., 439
Pred, A., 256
Prest, A. R., 282
Preston, C. J., 175, 187
Proshansky, H. M., 243
Pullberg, S., 548

Quandt, R. E., 217
Quenouille, M. H., 72, 79, 179
Quesnay, F., 545

Rademaker, O., 510, 511, 532
Raiffa, H., 273, 302
Rao, R. A., 328
Rappaport, R. A., 489, 490
Ravenstein, E. G., 156, 403
Rawstrom, E. M., 438
Rayner, J. N., 84
Rees, P. H., 393
Reeves, R. W., 332, 333, 397
Reich, W., 546
Reichenbach, H., 404
Reilly, W. J., 403
Reira, B., 250, 255, 260
Ribideau, G. F., 479
Rich, L. G., 338, 368, 369, 371, 372
Richardson, H. W., 523
Riker, W. H., 415
Ritter, K., 12
Robbins, R. C., 379
Roberts, P., 528
Robinson, E., 379
Robinson, E. A., 166
Robinson, J., 434, 435, 453, 524, 538, 544, 548
Rodda, J. C., 483
Rogers, A., 393
Rohrer, R. A., 61
Ronnberg, A., 257
Rosen, R., 101, 110, 111, 536
Rosenberg, R. L., 332
Rosenbloom, R. S., 283
Rosenblueth, A., 407
Rosenthal, L. D., 296
Rostow, W. W., 425, 542
Roszak, T., 315, 459, 463, 545
Rousseau, J.-J., 14, 16, 20, 21, 550
Rushton, G., 402, 403
Russell, B., 3, 4, 11, 223
Russell, J. R., 283
Rykiel, E. J., 343, 346, 347, 348

Sage, A. P., 203, 204
Sahlins, M. D., 505, 537
Santayana, G., 11
Sartre, J. P., 550
Sauvy, A., 518, 525, 527
Savas, E. S., 480, 481
Sawaragi, I., 480, 482, 497
Sayer, R. A., 233
Scheidegger, A. E., 363, 407
Schlesinger, G., 406
Schmidt, A., 19
Schoeffler, J. D., 98, 99, 195, 199, 200
Schrödinger, E., 400, 401
Schultz, D. G., 117, 139, 149, 211, 212, 213, 563
Schneider, S. H., 521, 522
Schumacher, E. F., 19, 439, 526, 537, 549
Scolnik, H., 533
Scott, A. J., 189, 192, 193, 194, 205
Scott, D., 259, 261, 313
Seinfeld, J. H., 376, 377, 378
Self, P., 292, 298
Selznick, P., 312
Sen, A. K., 439
Sengupta, J. K., 97, 127, 134
Service, R., 505, 537
Shackle, G. L. S., 542, 545, 548, 549
Shannon, C., 256
Sharaf, S. G., 381
Sherman, T. P., 283
Shugart, H. H., 340
Simon, H. A., 130, 234, 285, 297, 315, 401, 402, 407, 505, 507, 531, 536
Simpson, J., 503
Skinner, B. F., 21
Smale, S., 68, 69, 214, 254, 332, 403, 532, 533, 542
Smith, A., 437, 536
Smith, D. D., 472, 474
Smith, D. M., 415, 440
Smyth, J. D., 482
Snead, R., 237
Sochava, V. B., 18

Socrates, 428
Soddy, F., 549
Sonnenfeld, J., 235
Spencer, H., 16
Spiegel, M. R., 48, 49, 52, 64
Spinoza, B., 8, 429
Stafford, J. H., 271, 277, 278, 280
Stea, D., 224, 225, 226, 234, 235
Steeley, G. C., 259, 260
Stevens, C. F., 332
Stewart, I., 210, 211, 216, 217
Stewart, J. C., 403
Stocking, M. A., 473, 475
Stoddart, D. R., 18, 336
Strahler, A. N., 363, 364
Stravinski, I., 397
Stringer, J., 287, 294, 295
Stuart, A., 78
Studer, R. G., 243, 245
Sunamura, T., 353
Surrey, M. J. C., 417, 419
Suzuki, Y., 528
Swan, J., 304, 309
Sztompka, P., 14

Talbot, S. A., 28, 29
Tan, K. C., 542
Tansley, A. F., 382
Taylor, F. M., 250
Teilhard de Chardin, 15, 17, 242, 550
Terjung, W. H., 356
Theil, H., 130, 134, 250, 260, 263, 264, 449, 451, 530, 561
Thiemann, H., 529
Theodorescu, R., 392
Thom, R., 7, 66, 210, 211, 214, 215, 332, 363, 532, 533, 542
Thomann, R. V., 368, 370, 371
Thomas, N., 257
Thompson, W. R., 393, 394, 395
Thorburn, A., 252
Tiao, G. C., 494
Tinbergen, J., 96, 97, 98, 250, 281, 442, 451
Tinline, R. R., 84, 173, 185, 378, 380
Tobler, W. R., 156, 173, 505
Tomovic, R., 95, 150
Toulmin, S. E., 459
Tribus, M., 271, 273
Trist, E. L., 245, 246, 247, 286
Trudgill, S. T., 361
Tuan, Y.-F., 4, 5, 11, 19, 232, 236, 245, 476
Tukey, J. W., 89, 92, 404, 417
Tullock, G., 279, 280, 400, 413, 414, 416, 428, 521, 536
Turner, J. M. W., 4
Turnovsky, S. J., 134
Turvey, R., 282
Tustin, A., 422, 425, 427, 451
Tuteur, F. B., 288
Tyler, J. S., 288
Tzafestas, S. G., 192, 201, 202

Uexküll, J. V., 4
Ureña, J. M. de, 287, 294, 296
Urry, J., 13, 14, 407, 548, 549
Urwick, L., 250, 288, 292, 293, 310
Usher, M. B., 338

Van Der Maarel, E., 484, 486
Van Dyne, G. M., 337, 349, 353, 355, 474
Van Gierke, H. E., 109
Van Gigch, J. P., 25, 223, 251
Van Gogh, V., 4
Vanyo, J. P., 306, 308, 309
Van Woerkom, A. J. J., 381
Vernekar, A. D., 382, 383
Vickrey, W., 415
Vidal de la Blache, P., 14
Vishwakarma, K. P., 451, 453
Vita-Finzi, C., 387
Von Mises, R., 20, 426, 427, 428, 507, 526, 527, 535
Von Neumann, J., 117, 118, 119, 302, 303

Waddington, C. H., 507, 542, 545
Wald, A., 91
Walmsley, D. L., 243, 245, 246, 247
Wang, P. K. C., 190
Ward, B., 20, 446, 525, 538
Warren, A., 492, 493
Wastler, T. A., 331, 373, 374
Watson, J. B., 224
Watt, J., 7
Watt, K. E. F., 474
Watts, D. G., 77, 78, 79, 84, 179, 332
Way, M. J., 391
Webber, M. J., 409
Weber, M., 300, 542
Wedgwood-Oppenheim, F., 260, 295
Weissbrod, R. S. P., 177, 179
Wells, H. G., 549
Whipkey, R. Z., 365, 367
White, G. F., 237
White, L., 16
Whitehead, A. N., 4, 25, 541
Whittaker, R. H., 389
Whittle, P., 89, 129, 175, 179, 183, 184, 185, 186
Wiberg, D. M., 201
Wiener, N., 74
Wieslander, J., 148
Wilde, O., 551
Wiles, P. J. D., 428, 436, 443, 445
Williams, H. C. W. L., 123
Williams, R. B., 342, 349, 350, 351, 352
Williamson, F., 234
Wilson, A. G., 123, 140, 188, 273, 321, 388, 389, 393, 403

Wilson, E. O., 20, 21, 206, 320, 321, 322, 387, 388, 389, 390, 393
Wilson, G. T., 181
Winstanley, D., 332, 379, 521
Wischmeier, W. H., 472, 474
Wismer, D. A., 201
Wittenmark, B., 145, 146, 147, 148, 531, 560
Wittgenstein, L. J. J., 3, 14, 549
Wold, H., 72
Wolman, M. G., 494, 496, 497, 498
Wolpert, J., 234, 241
Wonham, W. M., 98, 117, 121
Woodcock, A. E. R., 216

Yatsu, E., 334, 336
Yewlett, E. J. L., 295
Yore, E. E., 288, 289
Youle, P. V., 250, 260
Young, P. C., 334, 373, 420, 425, 449, 451, 453, 532
Yu, W., 501, 502
Yuill, R. S., 156

Zadeh, L. A., 48, 63, 64
Zeeman, E. C., 192, 211
Zinam, O., 412

Subject index

Accelerator, 419, 420
Action, 243, 248; derivative, 125; executive, 308; integral, 124, 125; legislative, 308; on-off, 125; prescriptive, 463; proportional, 124, 125; proportional and integral, 127; proportional, integral and derivative, 127, 143; potential, 245, 246
Actuator, 95, 104, 112
Adjoint of matrix, 557
Administration, 286–96
Administrative, bureaucracy, 441; centre, 192; hierarchy, 284; structure, 432, 433; zoning, 189
Advective, model, 368; term, 370
Aerosol formation, 334
Age structure, 383
Aggregate, demand, 421; levels, 400, 402
Agriculture, 470; origins, 488; plantation, 492; sedentary, 489; swidden, 489, 490
Aid, 434
Aiding, 108
Aircraft, 239, 376
Albedo, 521
Algae, 497
Algorithm, 95, 96; control, 189, 190; learning, 148; recursive least-squares, 148
Alienation, 459
Allelomorph selection, 389, 392
Allocation, optional, 460
Alternative actions, ranking, 402
Alternatives, 415; ranked, 415
American Constitution, 304
Ammonia, 378
Amortization, of capital investment, 281
Amplitude, 53
Analysis of International Decision Areas (AIDA) technique, 295, 298
Anarchy, 398
Anima/animus, 228, 229
Animals, domesticated, 490
Antagonism, 378
Anthropogenesis, 20
Anthropological conditioning, 236
Anthropology, 232; classical, 231; social, 230, 233
Anthropomorphic, 12
Antibiotics, 138
Ants, 387, 391
Apperception, 225, 238
Approximants, 39
Aquifer, 363; coastal, 487; storages, 199
Arbitage, 410
Archetypes, 227, 229
Arrow's impossibility theorem, 280, 414–16, 428, 461
Arroyo cutting, 332, 333, 397
Artifacts, 2, 505
Ashby's law of requisite variety, 252, 291, 296, 399
Assignment problem, 193
Atmospheric, circulation, 201; mixing, 379; processes, 174
Audi-alteram partem, 307
Autarky, 436
Autocorrelation, 33, 78, 89, 334, 486; function, 179, 334; spatial, 381
Autoregression, 42; model, 334, 335
Autoregressive element, 51, 72; loop, 41; process, 79; techniques, 323
Autotrophs, 337, 338, 345, 349

Baboons, 385
Balance of payments, 438, 441, 443; mechanism, 458; regulation, 456
Ballet, 398
Bargaining, 315
Barrier, physical, 190; process, 156, 157
Batelle Memorial Institute, 529
Bayes, forecasting, 263; methods, 408; solution, 271; theorem, 267, 274; utility structure, 256

Beach processes, 350
Behaviour, 3, 4; chaotic, 149; counter-intuitive, 150, 247; function, 211; model, 223, 224; nonlinear, 149
Belief, 237, 243
Bellman's equation, 141, 202
Benefits, 259, 266; public, 280; utility of, 408; welfare, 280
Bernstein's Revisionist Marxism, 548
Bessel function, 175
Bias, 90, 91, 332; perceptual, 75
Billet heating furnace, 196, 197
Biochemical, 217, 361; oxygen demand (BOD), 260, 331, 372, 373, 496, 497, 498
Biogeochemical cycle, 261, 262
Biological, cell division, 211; decay, 379; dispersal, 175; evolution, 232; growth, 398
Biomass, 324, 345, 488; model, 382, 384–6; storage, 485
Bionics, 109
Biosystem, 471
Biotic diversity, 486; biotic subsystem, 378
Birds, 387, 392, 393
Birth rate, 513, 514, 518
Black box, 382, 407; approach, 326, 332, 334, 361; model, 382, 484
Block diagram, 28, 31–3, 103, 341
Blocking, 118
Bode method, 117; plot, 86, 180
Bohemianism, 286
Boolean matrix, binary, 300
Boundary, 195; conditions, 96, 124; control effects, 195; data, 187; political, 203
Box-Jenkins prediction techniques, 534
Breathing space, 305
Bridge, 203
Brookings-SSRC model of U.S. economy, 417, 418
Buddhism, 19
Budget, 456; curve, 277; lines, 278; space, 188
Buffering, 288, 339, 361
Building design, 189, 192
Bumps, 121
Bureaucratic, 287; 'freezing up', 398; structure, 261; models, 312–14
Burning, 484, 489

Cadmium cycle, 261
Calcium cycle, 261
Calculus, of consent, 280; felicific, 21; of variations, 139, 201
Calibration, of models, 76, 529, 530
Canonical analysis, 332; correlation, 402; diagonal structure, 290; factorization, 180, 181; form, 65, 417
Capital, 421, 511; expenditure, 523; investment, 514; state, 419
Capitalism, 428, 540; crisis of, 548; economy of, 422, 430
Carbon, 14 dating, 387; cycle, 261, 338; dioxide, 339, 521; monoxide (CO), 376, 377; organic, 491
Cartesian lattice, 185
Cascade, 31, 65, 118, 121, 177; climatic-ecologic-hydrologic, 256; confined spatial, 363; effects, 372; feedforward-feedback, 479
Cascading, in ecosystem, 342
Cash crops, 524; flow, 282, 433, 477
Catastrophe, 395, 532; butterfly, 215, 216; cusp, 215, 216; elementary, 214, 215; elliptic umbilic, 215; environmental, 485; fold, 215, 216; hyperbolic umbilic, 215, 216; irreversible, 218; manifold, 233; natural, 191, 363, 543; quasi-nonlinear, 218; space-time, 218; swallowtail, 215, 216; theory, 210–219
Catastrophic behaviour, 327; collapse, 118; threshold, 217
Catchment storage, 191
Causal structure, 506
Cell division, 39
Census, 183
Central place, packing, 387; theory, 403
Central tendency, 33
Centralized planning, 460
Certainty, 269, 284; absorption, 256; environment, 270; equivalance principle, 130, 256, 328; equivalence theorem, 147
Chaotic solution, 395
Chema (agricultural system), 491
Chemical gauging, of stream discharge, 373
Chimpanzees, 388
Chipmunks, 387
Chlorophyll production, 386
Chreods, 507
Cicada, 395
Citizen sovereignty, 415
Civil debate, 286
Classification, 341
Client category, system responses, 255
Climatic changes, 262, 322, 379, 383, 397; controls, 199; long-term shifts, 332, 521; regimes, 331, 356; risk, 408
Climax distribution, 389; structure, 410
Closure, 224, 225
Cloud seeding, 485, 503
Club of Rome, 548
Cluster analysis, 402
Coalition, co-operative, 411, 412; mixed, 411, 412; pressure organization, 411, 412; profit organization, 411, 412; service organization, 411, 412
Cofactors of matrix, 556
Cognition, 224, 225, 256, 505
Cognitive, factors, 546; image, 237; issues, 149; norms, 191, 507; perceptual responses, 400; structures, 224
Collapse, 363, 508; reactions to, 517, 518, 525; mode, 532

Collective action, 268, 279, 284, 290, 293, 400, 413, 415, 416, 436, 460, 461, 535, 550; typology of, 269; decision, 280, 400, 540; model, 315
Collective unconscious, 223, 227, 231
Collegial model, 312, 313, 314
Collinearity, 89
Collusion, 285, 302, 303, 410
Combinatorial problem, 193
Commensalism, 482, 484
Common good, 310, 435, 463
Communication, 252, 399
Communism, 428, 436, 515, 516
Community action, 313
Community Land Act, U.K., 307
Commutative terms, 555
Comparator, 102, 103
Compensation, 523
Competition, 285, 411, 426; succession process, in ecology, 426, 428; for biological resources, 389
Competitive bidding, perfect, 410
Complex, 227, 228; plane, 59; system, 255
Compulsory purchase, 307
Condensation, population, 387
Concentration, gradient, 369; of resources, 477
Concensus, 254, 463
Condition, boundary, 48; initial, 48; Jacobi, 210; Kuhn-Tucker, 210; Legendre, 210; Weierstrass, 210
Condorcet effect, 415
Conduction, 174, 190
Confidentiality, 258; of information, 307
Conflict, 286, 290, 308, 409; behaviour, 303; resolution, 301
Conformable matrices, 555
Conjunctive use, groundwater/surface water, 500–3
Conservation, 15, 16, 479, 484, 494, 553; balance, 426
Conservatives, 458
Consistency, 90, 332
Constraints, 131, 140, 279, 308; administrative, 252; legal, 252; political, 252; societal, 252
Consumer behaviour, 402
Consumption, 419, 421
Consumption function, 419
Consumption loop, 419, 420, 422
Contiguity process, 156, 157
Contingent truths, 9
Control, 252, 262, 459, 473; action, 425; active, 471; administrative-bureaucratic, 460; admissible, 136; algorithm, 130, 194; approach, 251; area-integrated, 189; bang-bang, 124; boundary, 189; capitalist war economy, 443, 445; central allocation, 443, 444; centralized economic, 446, 447; 'centralized market' economy, 443, 444; chart, 139, 142, 143, 144; Chinese, 449; closed-loop, 457; composition method, 200; cost-efficiency of, 116; costs, 122, 129; criteria for efficiency, 115–22; criterion, quadratic, 129, 131; dead beat, 127; decomposition, 199; decoupling, 289; design, hierarchical, 195; dynamic, 189–92; econometric, 96; economics of, 266; efficator, 94; engineering, 97, 111; equations, optimum, 558–63; federalist, 189; feedback, 361, 449; fiscal closed loop, 456, 457; fixed time, 138; full market economy, 445; hierarchy of, 197; higher-level, 307; implementation, problems of, 143–50; instruments, 117, 124, 284, 437, 440; integral, 128, 136; interval, 124; intervention, 199; 'inverted centralized market' economy, 443, 444; issues, 254; junction, 106; law, quadratic, 453; legal, 308; levels of, 197; locus of, 411; magnitude, 266; minimum variance, 134, 136; mixed scanning, 142; monetarist closed loop, 456, 457; monetary, 457; nested hierarchical, 247; no action, 253, 265; objectives, 281, 415; on-off, 125, 128; optimal, 123, 127, 263, 451; optimum, 208; parameter, 213; policies, simulation of, 450, 451; precision, 116; plan, 443, 444, 445; proportional, 128, 143; proportional plus derivative, 126; proportional plus integral (PI), 125, 126, 128; proportional plus integral plus derivative (PID), 125, 126; quadratic, 142, 289; range, 116; rationalist, 461; regulated market economy, 445, 446; regulator, 103; reliability of, 116; sensitivity of, 116; by separation of space-time variables, 190; settings, 125; shift, 133; signal, 106; by social instruments, 472; space-time, 195–203; space-time redesign, 190, 203–6; spatial, 189, 192–5; speed of, 115, stability of, 116; stochastic, 134, 144, 561; stop-go, 125; suboptimal, 123; territorial system of, 446; to objectives, 436–459; unlimited, 122
Control problem, fixed time, 139; N single-stage, 141; N-stage, 141; quadratic, 135
Control strategy, 122–50; admissible, 124; dead beat, 137; integral control, 137; minimum variance, 137; N-stage optimal, 131; one-stage optimal, 131; optimal, 124; sub-optimal, 124
Control system, 7, 94–151, 294, 301, 320, 443–459, 462; active closed-loop, 113; active closed-loop feedback, 100, 103–7; active closed-loop man, 106; adaptive closed-loop feedback, 100, 109–10, adaptive feedback, 304, 306; adaptive feedback-feedforward, 111; automated, 153, 208; closed-loop, 125; closed-loop feedback-feedforward, 100, 111; deterministic feedback, 132; environ-

Control system—*continued*
mental, 100, 101, 252; feedback, 492, 493; feedback-feedforward, 254, 289, 308; feedforward, 100, 110–11, 265, 536; frequency responses, 126; lumped, 189; man-machine, 107, 153, 208; man-tool, 106; man-tool powered, 106; mangement, 250; model-reference adaptive, 109; nested, 254, 284, 287, 531, 536; nested feedback, 110–15; open-loop, 99, 100, 101; passive closed-loop feedback, 100, 101, 102, 103; response, 128, 135; restructured, 208; self-adaptive, 109; sensitivity, 120; soft, 315; stabilizing, 320; stochastic feedback, 133; stochastic feedback-feedforward, 133; stochastic feedforward, 132; stochastic quadratic feedback, 133; types of, 98–111; world-based, 191
Control targets, 282, 437–40, 525
Control theory strategies, optimal, 449
Control variables, 275, 532; partial, 442
Control weighting, 124
Controllability, 73, 98, 145, 417; criteria, 73; limitations, 76
Controller, 103, 105, 113; algorithmic, 195; automated, 195; dead beat, 131, 135; feedback, 123, 195; feedback-feedforward, 195, 197; feedforward, 144; fixed time, 136; integral, 135; man-machine, 195; minimum variance, 127, 135; N-stage, 136; on-off, 142, 143; proportional plus integral (PI), 451, 454; proportional plus integral plus derivative (PID), 126, 127, 128, 142, 143, 144, 451, 454; space-time feedback closed-loop, 196; space-time feedback-feedforward, 196, 198; three-term, 125; two-term, 125
Controller-regulation system, regional, 296
Convolution, 36, 92; integral, 328, 331
Coordination, 290; of subsystems, 291, 292
Coral reef, 471; biome, 339
Correlation, 402, 404
Cospectrum, 84
Cost-benefit, criterion, 283; trade-offs, 413
Cost-effectiveness, function, 278; process, 283
Cost-efficiency, 122
Cost-minimization, 276, 290
Costs, 259, 266, 277, 280, 283; and benefits, 408; changes, with groups of different size, 414; curve, 277; disutility, 408; foregone, 408; marginal, 408; opportunity, 408; quadratic, 205; transport, 206, 207
Cournot-Nash equilibrium, 302
Covariance stationary process, 87
Crayfish, 387
Creative act, 547
Credit, 441
Cricket, 398
Crime, 286
Criterion, least squares, 129; minimum error, 129
Critical social theory, 232, 548
Crop improvement, 477
Cropping, 474; of resources, 476
Cross-correlation, 80, 81; effects, 379; function, 179, 180
Cross-coupling, 117
Cross-spectral analysis, 87, 180
Cultivation, shifting (swidden), 489–91
Cultural, artifacts, 232; conditioning, 236; norms, 427; Revolution, 449; specifics, 231; universals, 231
Curve fitting, 326, 332
Cyclic, response, 422; fluctuations, 457

DDT, 138; cycle, 261
Dams, 190, 472, 494, 496
Damping, 363; critical, 60; ratio, 46, 60, 61
Dead time, 340
Death rate, 513, 514
Decay, 425; rate, organic matter, 371
Decentralization, 293, 312, 549
Decision, processes, 402; rules, physico-chemical, 401; situations, typology of, 268; structures, 411; trees, 408
Decision making, 198, 246, 425, 501, 502; economic, 112; economics of, 274–81; hierarchy of, 307; independent, 400; interrelations, 291; environment, 266, 268–74; units, multilevel structure, 387
Decision making system, 250–316; administrative component, 255, 268, 286–96; bureaucratic, 460; components of, 284–310; deterministic, 251; legal component, 255, 268, 304–10; sociopolitical component, 255, 268, 296, 304; spatial structure, 310–13
Decision units, devolved, 310, 311; federalist, 310, 311; separatist, 310, 311
Decisions, 401, 425; social, 401; symbiotic, 293
Decomposers, 337, 338, 345, 349, 353, 487
Decomposition, goal and model, 287
Decoupling, 200, 289; control, 307; of decision areas, 287; state variable, 290
Degrees of freedom, 398, 406, 535; executive, 284
Deformations, family of, 214
Delay, 42, 101, 468; space-time, 155; time, 518
Delta function, 30, 370
Demand, 430; mangement, 458
Democracy, 550; pluralist, 298; populist, 300; representative, 298
Demographic, dilemma, 543, 544; transition, 530
Department of Economic Affairs, U.K., 290
Departmentalization, areal principle, 292; client principle, 292; of decision making

Departmentalization—*continued*
system, 292, 293; process principle, 292; purpose principle, 292
Descriptive model, 269
Desertification, 262, 397
Design, God's, 16; Man's, 16; Nature's, 16; principle, of Von Neumann, 117, 120; synthesis, 537
Destocking, cattle, 492–3; policies, 479
Detector function, 401
Determinant, of matrix, 556
Determinism, 12
Deterministic control with quadratic costs, 558, 559
Developed countries, resource-importing, 515, 516
Developing countries, resource-poor, 515, 516; resource-rich, 515, 516
Developed world, 525
Deviation, 260
Deviations, costs of, 129
Deviative term, 128
Devolution, 310, 311, 435
Dialectical, clash, 428; materialism, 427, 507, 524, 526, 534, 537
Dialectics, 18, 233, 236, 460, 463
Diatom counts, 387
'Differential organic growth', economic, 549
Differentiation, 410, 425; of communities, 392; rural/urban spatial, 392; spatial biological, 388
Differentiator, 30, 39, 42
Diffusion, 174, 256, 369, 370; animal, 201; of ants, 391; biological, 387–96; coefficient, 369, 370; condensation, 387, 390; of disease, 378–82, 395; equations, space-time partial differential, 373; heat, 190; information, 201; of innovations, 426; modelling, 377; models, of Hägerstrand, 387, 390; plant, 201; process, 403; processes, wave, 387; rate, spatial, 377; saturation, 387, 390; simulation of, 389; stream channel, 496, 497; system, 204; waves, 390
Diffusivity, atmospheric, 190; water, 190
Digital signals, 26
Dimensional equivalence (⍝), 368, 369
Direct intervention, 198
Directed graphs, 301
Disaster, 304, 305
Discontinuity, 218, 324; in space and time, 217; of process, 211
Discount rates, 478, 479
Discounting, 524
Discretionary, bodies, 307, 308; powers, 261, 304
Disease, 201; foot and mouth, 378–80
Dispersion, 33
Dissipators, 494
Dissolved oxygen (DO), 372–4
Distance weights, 302
Distributed lag process, 41
Distribution, 268; normal, 72
Disturbance, 98, 99; environmental, 115; operating, 99; parameter, 99; process, 99; stochastic, 74, 134; structural, 99
Disutility cost, 281
Divisibility, assumption of, 280
Division of labour, 383
Domestication, of plants and animals, 487
Donor control, 324, 362
Donoughmore Committee Report, 308
Double feedback loop system, 304, 306
Doubling, time, 508
Downstream concentration, 368
Drainage basin, 39, 262; hydrology, 361
Drought, 485; cyclical, 493; cycle, 492
Dual-feasible method, 199
Dual problem, 140
Duality of estimation and control, 129
Duals, 263
Duty, 428
Dynamic equilibrium, 254, 407; population, 389
Dynamic programming, 139, 141, 142, 192, 201, 279, 308
Dynamics, relativistic, 158; specific, 158
Dysfunctional reactions, 241

Earthquake, 72, 363
Earth's orbit, variations in, 382, 383
Ecological, niche, 387; modelling, 339; stability, 489, 491
Ecology, 14, 17, 486; human, 18; landscape, 18
Economic, equilibrium, 521; growth, 189, 438, 458, 479; trade, 189
Economic-industrial sector models, 528
Economic model, 303; Goodwin-Klein, 425, 426, 427; Kalecki, 425, 426, 427; Keynes, 417, 419, 420, 421, 423, 424, 425, 453; Phillips, 425, 426, 427, 451, 454; regional, 452; Samuelson-Hicks, 425, 426
Economics, 250; capitalist, 509; classical, 433; of location, 432
Economies, Buddhist, 526; centrally planned, 509; command, 444–5, 460; laissez-faire, 460, 461; of scale, 432
Ecosystem, 191, 320, 336–55, 361, 362, 383–7, 474, 475, 485; artificial, 491; Cedar Bog Lake, 349–51; closed flow with cycling, 487; degradation, 477; design, 472; diversity, 487; equilibrium, 468; flows of nutrients and energy in, 189; industrial, 469, 471; island, 206; Isle Royale, 346–9; management, 474; maturity, 488; open flow with cycling, 487; open flow without cycling, 486–7; sensitivity, 340; Silver Springs, 344, 345; stability, 487; structure, 337–53; tropical forest, 488

Ecotome, 262
Eddy forms, 378
Effector, 111; control, 112; function, 401
Efficiency, 115, 332
Ego, 226, 227
Eigenvalues, 51, 117
Elasticity of demand, 39
Electoral representation, 308
Embryonic development, 211
Emergence, 425
Employment, full, 437
Energy, 519; cascades in ecosystems, 337; consumption, 521; consumption/income ratio, 509; conversion, 472; flows, 467; flows, in ecosystem, 344, 345, 346, 349, 350, 351, 353; kinetic, 473, 474; law, Darwin-Lotka, 486; mass cascades, 339; resources, 441; solar, 469; source, 94; synthetic, 469
Entities, operating, 2
Entropy, 5, 230, 269, 273, 400, 410, 488, 489; loss, 256; maximum, 188, 403; nightmare of, 18
Environment, behavioural, 236; causal texture of, 246; geographical, 236; operational, 236; organic, 236; perceptual, 236; placid-clustered, 247; placid-randomized, 246; stochastic, 251
Environmental, degradation, 262, 515, 523, 524; disturbances, 238, 248; exploitation, 470; hazards, 237; information, 227; intervention, 470; reservoirs, 261; symbiosis, 504–27
Enzyme formation, 101
Epicureans, 3
Epidemics, 180; measles, 381, 382
Epistemological, attitudes, 544; issues, 545; objectivity, 459; transitions, 546–53
Equation, difference, 39, 42, 167, 323; differential, 39, 42, 51, 62, 167, 340, 343; first-order difference, 423; first-order differential, 62, 373, 486; integral, 173; linear differential, 323; linear system, 345, 346; log-difference, 39; mass-balance, 370; mass conservation, 382; measurement, 63; nonlinear, 395; nonlinear matrix Riccati, 129; nonlinear system, 345, 346; ordinary differential, 323; p th-order differential, 62; partial differential, 173, 323; second-order differential, 59; state, 63; system state evolution, 123; third-order difference, 395
Equations, 26; with chaotic solutions, 545–6; latitudinal, 381; nonlinear with chaotic solutions, 395
Equilibrium, 187; assumptions, 432; dynamic, 183; free market, 429; global, 212; model, 523; state, critical, 212, 213; static, 183; unstable, 395, 396
Equity, 268; preference function, 419
Ergodic hypothesis, 188
Ergonomics, 106, 475
Erosion, 333, 397; of resources, 476; sheet (overland), 351–2
Error, 72; criterion, 123; measurement, 74, 91, 99, 145; parameter standard, 147; sampling, 395; signal, 102; steady-state, 35, 117; stochastic, 93; terms, 7; type, 2, 79
Estimation, 332; and control problems, 263
Estimators, 332
Estuarine flow, 372
Ethic, developmental, 21; developmental-genetic, 20; intuitionist, 20, 21; Judaeo-Christian, 16, 429; materialist, 428
Ethical, choice, 494, 495; code, 428, 461, 462; questions, 492
Ethics, 19, 308
Ethnicity, 305
Euler's theorem, 275
Eutrophication, 373
Evapotranspiration, 342
Evolution, 14, 397
Evolutionary stability, 507
Executive, 7, 197, 256, 307; action, 198, 199; in government, 304
Explanation, functional, 233; individual, 400; macroscopic, 400
Explanatory and ameliorative models, 529
Explosive growth, 467
Exponential growth, 16, 425, 508
Externalities, 225, 282, 283, 504, 523; internalization of, 521
Extrapolation, 260
Extreme event, 103, 191

Factor analysis, 332, 402, 415
Falb-Wolovich matrix, 290
Feasible method, 200
Feasibility conditions, 124
Feedback, 11, 42, 226, 230, 429, 504, 505; action, 248, 249; autoregressive, 43; autoregressive interregional, 161; climatic, 521, 522; control, 263; control, stochastic, 560; costs of, 144; cumulative, 435; discrete autoregressive, 44; explosive, 422; homeostatic, 363; learning, 248, 249; internal, 60; loop, 41, 304, 441, 442; Lotka-Volterra, 341; mechanisms, 94, 487; negative, 41, 44, 56, 58, 65, 101, 102, 191, 237, 247, 422, 467, 468, 471, 479; on-off, 125; optimal, 449, 455; positive, 41, 44, 55, 57, 58, 65, 343, 422, 467, 468, 471, 479; proportional, 125; self-regulatory, in ecosystems, 339; space-time, 504; stable convergence, 43; stable damped, 43; strategies, 124; transducer, 102, 103; unstable explosive, 43; unstable oscillating, 43
Feedback-feedforward control, stochastic, 561
Feedforward, 41, 42, 230, 337; control, 254; strategies, 124

Feeding chain, 346
Felicific calculus, 296
Fertilizers, 477, 492
Feudalism, 428
Fick's laws of diffusion, 369
Field moisture capacity, 358
Filter, 149
Filtering, 248, 256
Final time, 124
Finite codimension, 214
Fish, 395
Five-year programme, USSR, 448
Fixed time quadratic cost control, 562, 563
Fjord, 373, 375
Flip, 208, 233, 332, 552; catastrophic transition, 211; point, 216; space-time, 211; time-series, 211
Flood, 485; flash, 363; peak, 363; plain, 470; prediction, 35; routing, 363; wave, 364
Flow, graph, 28, 32, 33, 34; meter, 75; mixed, 371, 372; models, 371, 372; plug, 371, 372
Fluid dispersal, 175
Food, 508, 510, 514; chain, 467, 470; production, 518; supply, 511; web, 320, 487
Foot and mouth disease, 185
Forage, 493
Forcing, constant, 347, 348; self-generating, 348; sine, 347
Forecasting, 77, 107, 130, 254, 260, 262–8, 382, 417, 484, 485, 486, 542; model, 260; spatial, 183; system, 264
Forest, Orinoco, 489, 491; clearances, 484
Formulation of policy, 417, 419
Forrester-Meadows world model, 480, 510–15; comments on, 530–4
Fourier, analysis, 322; double, 323; expansion, 190; methods, 486; series, 33; techniques, two-dimensional, 155
Franks Report, 308
Free economy, unfettered, 250
Freedom, 307, 521
Frequency, 33, 53; angular, 53; domain identification, 79; response, 179; response function, 332; response spectra, 85; spectrum, 84
Frontiers, 191
Function, autocorrelation, 77, 79; cosine, 59; cross-correlation, 77–9; partial autocorrelation, 77, 78; sine, 59; social welfare, 461
Functionalist, 263
Futuristics, 546, 547

Gain, 41, 42, 55, 66, 84, 86–8, 101, 430, 468; of controller, 121; feedback, 288, 289; feedforward, 288, 289; loop, 103; optimal feedback, 290; space-time, 155
Galerkin method, 190
Game, against nature, 273, 408; positive or zero-sum, 409; theory, 269, 271, 302
Gamma function, 328
Gas adsorption, 379
Gaussian random sequence, 161
Gene pool, 199, 288, 536
General Information System for Planners (GISP), 257
Genes, 389, 392
Genetic, barrier, 389, 390, 392; information, 238; instincts, 239; mutation, 401
Geographical information systems, coordinate-referenced, 257
Geography, 232
Geomorphology, process-response, 353
Geosophy, 19
Gestalt, 37
Gibbons, 387
Global, analysis, 66, 397, 403, 542; considerations, 323; constraints, 526; maps, 395; minimum, 147; range, 393; response functions, 332
Goal-coordination, algorithm, 200; method, 199
Goals, 21, 103, 105, 111, 246, 247, 252, 254, 308, 408, 410, 467; abstract, 428; biological, 439, 440; conflicting, 95; conservationist, 427; control, 297; desired, 115; equity, 438, 439; exogenously set, 254; externally imposed, 94; group, 416; growth, 438; ideological, 443; interim, 94; intervention, 199; new, 425; not shared, 298; ontological, 264; planned, 428; political, 443; purposive, 508; self-imposed, 94; self-regulatory, 254; setting, 257, 507; shared, 298; social, 260, 443, 471, 479; social welfare, 438; static economic efficiency, 437, 438; subconscious, 95; uniform, 191; wealth equality, 507
Good, common, 460; range of, 403; social, 460; threshold of, 403
Gophers, 387
Gosplan, 446, 447
Government, 280, 411; expenditure, 421, 422, 425, 442, 453, 456; omniscient, 412
Grassland ecosystem, 349, 353, 354, 355
Gravity interaction model, 403
Grey box, approach, 361, 362; model, 485; system, 406
Gross National Product (GNP) (Also Gross Domestic Product, GDP), 421, 441, 450, 453
Ground water, 501–3
Group, attitudes, 238; interaction, 402; pressure, 225, 241
Growth, 456; extensive, 438; intensive, 438; linear, 33; nonlinear, 33; pole, 436
Guilleman-Truxal method of control, 117
Gumbel plot, 33

Hamilton-Jacobi, approach, 192; equation, 139, 140, 145, 202, 563
Hamiltonian, 202, 205, 218; control, 201

Hammersley and Clifford theorem, 187
Harmonic analysis, 322
Harvesting strategy, 474
Hazard, 482; perception, 305, 485
Health, control of, 191
Hearsay, 9
Heat, conduction equation, 197; transfer equations, 174
Heaven, Marxist, 553
Hegelian, conflict, 507; synthesis, 549; thesis, 428
Hell, Kafkaesque, 553
Herbivores, 337, 338, 342, 345, 346, 349, 353
Herding, 383
Heredity, 224
Hermeneutic, 407
Heuristics, 543
'Hidden hand', of Adam Smith, 437, 536
Hierarchical, control, social, 398; order, 511; ordering, biological, 388; organization, 397, 398; policy design, 98; structure, 115; system, 257, 287
Hierarchy, 112; multilevel of responses, 321; nested, 112; process, 156, 157
Historicism, fallacy of, 416
Historicist, crisis of, 549
Holistic view, 504
Homeorhesis, 507, 542
Homeostasis, 102, 254, 467, 468, 477, 492, 507, 534, 542
Homeostatic regulation, 422, 426, 428, 460
Hominid, 229
Homogeneity, 175, 184
Hospitals, 189
Host, 471
Humanist, 19
Humid tropical areas, 474
Hunting, 103, 484
Hydraulic geometry, 324
Hydrodynamics, 13
Hydrograph, basin, 36; unit, 35, 37, 328, 329, 331, 362, 484
Hydrological cycle, 485; basin, 494; world, 519
Hydrological system, 325, 331, 350, 378, 397; basin, 320, 358
Hydrology, 35, 75, 319, 324–31, 356–8, 362–7, 484, 494
Hypothetico-deductive method, 3
Hysteresis, 125

Ice age, 382, 387
Id, 226, 227
Ideal type, functional, 246; goal-directed, 247; ideal seeking, 247; purposive, 247
Idealism, Arcadian, 527
Idealist, 415, 416, 428, 429; design, 254; norms, 550; position, 461
Idealistic, norms, 437; targets, 437
Ideology, 263, 459
Identification, 150, 179, 239, 240
Identifiability criteria, 181
Idiosyncracies, 231
Image, 225, 237, 248; building, 226
Immiserization, 434, 435, 462
Immunity, 109
Impact, degree of, 260; assessment, 237; control, 265
Impulse, 332; response function, 431; responses, space-time, 169; transient, 30
Income, distribution of, 435, 438; national, 441
Indeterminacy, 403, 404; economic, 459; of measurement, 403, 404, 416; of outcomes, 404; of state, 403, 416
Indifference curve, 275–7
Industrial Development Certificate Premiums, 451
Infiltration, 365; capacity, 324, 494; control of, 496
Inflation, 515
Influence decomposition, 197
Information, 248; communication of, 401; display, 256; distributive, 255; filtering, 226, 239; flow, 387; formula, 256; gain, 256; matches, 258; messages, 257; operative, 255; pool, 290; retrieval, 256; soft, 261, 267; storage, 256; system, 254–259, 535; system, spatial, 258; transformation, 256; transmission, 256
Initial, condition, 66, 67, 124; time, 124
Injection, slug, 368
Inputs, 25, 29, 70, 256, 361, 419; agricultural, 514; cascaded, 57, 58; control of, 503, 504; controllable physical, 7; discrete time unit impulse, 55; disturbance, 94, 103; disturbances, identification, 110; emission, 378; exponential, 33; gulp, 371–3; impulse, 52, 61, 371; lagged, 328; meteorological, 377; periodic, 33, 53, 378; ramp, 52; rectangular, 371, 373; reference, 94, 96, 102, 103, 106; regime of, 322; regional, 159, 161; sequence, controlled, 96; sequence, uncontrolled, 96; sinusoidal, 33, 54, 362; stationary, 33; step, 52, 61, 371; stochastic, 33
Input-output, model, 448; model, interregional, 417, 448; relationships, 334
Insecticide, 110
Insects, 394, 395
Instability, 127, 212, 213; conditional, 214
Integral term, 128
Integrator, 29, 30, 42, 45
Interaction, 225, 241
Intercorrelation, 33
Interdependence, 507; between decision making units, 284, 285; market, 284, 285; pooled, 284, 285; reciprocal, 284, 285; sector, 284, 285; sequential, 284, 285; of sub-systems, 292

Interest, rate of, 419
International Monetary Fund (IMF), 548
Interplant competition, 185
Intervention, 468, 472, 473, 488; agricultural, 489; analysis, 494; environmental, 479, 480, 484–503
Intra vires, 307
Intuition, free, 9
Inversion, atmospheric, 377; of matrix, 556
Invertebrates, macroscopic, 375
Investment, 397, 419, 421, 441; agricultural, 510; allocation of, 194; function, 419; gross, 518; lag, 420; loop, 419, 420; non-agricultural, 510; of resources, 477
Irrigation, 408, 491, 499–501
Islands, and migration, 390, 393
Isoquant, 275–7
Isotope dating, 387
Isotropy, 175, 184

Jacobian, 186
Japanese Club of Rome, 528
Jet stream, 332, 382
Joint estimation and control, 148
Jordan normal matrix, 290
Journey, to shop, 273; to work, 273
Judiciary, 304
Justice, 460, 551; social, 436; social, territorial principle of, 435

Kalman, controller, 203; filter, 145, 148, 263, 332, 334, 335, 451, 453; filter estimator, 146; filter parameter estimates, 336
Kernel function, 328, 329
Keynesian, leakages, 434; multiplier-accelerator, 422, 423, 451
Khosraschet monitoring, 441
Knowledge, 2; perfect, 413
Kriging method, 184

L operator method, 176
Labour, 421; demand function, 419; division of, 292; monopolies, 429; supply function, 419; training subsidies, 451
Lag, 42, 55, 353; adjustment dynamics, 184; autoregressive, 88, 159; distributed, 42, 88; economic, 425; polynomial distributed, 159; space-time, 155; temporal, 159; time, 551
Lagrangian multiplier, 139–141, 200, 201, 282
Laguerre, expansion, 331; filter, 330
Laissez-faire approach, 536
Land reform, 434
Landforms, 353, 542
Laplace, operator, 155; transform, 327
Law, 428; of large numbers, 400; of large systems, 361, 532
Laws, moral, 427; of redundancy, 256; statistical, 459; universal, 2
Lead cycle, 261
Lead, process, 42; space-time, 155; time, 518
Leading, 107
Lead-lag, 66; process, 42
Leakage, between regions, 196
Learning, 250, 397, 401, 408, 425, 426, 537; process, 402; system, 267
Least-squares, generalized, 182; regression, 146, 335
Le Châtelier's principle, 254
Lee waves, 379
Legal, constraint, 20; property rights, 523; regulations, interregional, 189; restraints, 397, 461; systems, 304–10
Legendre and Weierstrauss condition, 140
Legislative action, 198, 199
Legislature, 304
Lerner's rule, 274, 279, 400
Liapunov function, 138
Liberty, 461
Libido, 230; progressive, 228, 230; regressive, 228, 230
Limits, to growth, 518; statistical confidence, 70
Lincoln principle of government, 411
Linear, function, 33; mathematics, 507; model, stable, 362; programming, 131, 279, 308, 448, 499, 500, 501; programming problem, Hitchcock, 194, 448; regression, 42; reservoir, 327
Linearity assumption, 331
Litter, organic, 487
Lizards, 387
Location, minimum cost, 192
Location-allocation, problem, 194, 203; shifts, 190
Logical positivism, 3, 4
Love, 211
Luenberger observer, 146, 332
Lumping, 322; spatial, 323
Lundbergian lag, 421

Machiavellian view, 297
Macro-economic, 280; policy, national, 189
Macroscopic scale studies, 403
Magnetism, 190
Magnitude-frequency, of occurrence, 98, 99, 262, 320, 379, 397, 482
Malaria, spread of, 395, 396
Malthusian view, 511, 523, 524, 543
Man, bounded rational, 234; computer, 234; economic, 234; models of, 234; psycho-analytic, 234; satisficing, 234; '-world', 550
Man-environment interface, 262
Man-machine, control, 98; interface, 251, 315
Man-made models, 363
Management, 476; and advisory system, 260; of foreshore, 295, 297; Information Systems (MIS), 260, 261; by objectives, 251, 260, 283; science, 250

Manifold, 393, 395, 396
Marginal, decisions, 280; product, 274, 279; rate of return, 275, 280; rate of substitution, 275, 276, 280, 437
Market, 448; economies, 410, 449; mechanism, of the economy, 439, 441; price, 410; sovereignty of, 300
Markov, assumption, 258, 301; chain, absorbing, 302; chain, disconnected, 302; chain, ergodic, 302; decision machine, 302; decision rule, 141; definition, 187; (lag) operator, spatial, 379, 380; methods, 486; property, 175, 186; property, spatial, 187; random field, 187; techniques, 323, 326; type specification, 99
Marxism, 145, 232, 243, 263, 287, 415, 416, 428, 429, 433, 435–7, 439, 459, 462, 463, 514, 524, 548, 549, 553
Marxist, dialectical materialist approach, 429; economics, 428, 430; sociologists, 552
Mass-balance equation, 370
Mass-energy rate, 472
Material transformation, 469
Mathematics and nature, 319–327
Matrix, 47; algebra, 554–7; convolution integral, 65; determinant, 69; diagonal, 98; exponential function, 63, 65, 166, 168; input-output, 205; polynomials, 154, 158; polynomials, inverse of, 164; rank, 98; Riccati equation, 140, 279, 289, 290, 463; state transition, 63; system, 63; total cost, 138; triangular, 98; transfer function, space-time, 158; vector, 63
Maximization, rule, 123; tolerance, 124
Measles epidemics, 156, 180, 379, 381
Mediation, 225
Meline's law, 527
Memory, 42, 110, 224, 226, 238, 239, 248; folk, 242; long-term, 242; racial, 224; short-term, 242; storage, 247
Mental map, 220, 237
Mescalin, 242
Mercury cycle, 261
Metaphysics, 12
Microclimate, 361, 488
Migration, 189; biological, 387, 390, 393
Mineral resources, 517
Minimal, structures, 402; representation, 417
Minimality, 89, 404
Minimax solution, 277
Minimum, cost criteria, 205; intervention principle, 150
Mining, of resources, 476
Minors (submatrices), 556
Missing data points, 183
Mis-specification, 145
Mixed-scanning, discretion in, 288; process, 95, 253, 265, 266, 312; response, 199
Model, building, 239; control, 116; coordination method, 200; estimation, 187; identification, 187; mis-specification, 75; non-linearity, 75; non-stationarity, 75
Modelling, system, 253; /forecasting systems, 262–8
Money, 441, 445; available, 419; system, 410
Monetarist, 457, 458
Monetary policy, 456–8
Monitoring, 149, 251, 315, 401, 441, 474, 479–481; process, structure of, 259; system, 253, 259–62; unit, 260, 290
Monkeys, 388
Monoculture, 491
Monopoly, 302; capital, 429; employment, 429; spatial, 432
Monopolistic powers, 307
Monte Carlo, methods, 323, 387, 449; modelling, 326; simulation, 451, 453
Mood of our time, 544, 545
Moose, 346–9, 395
Morphogenesis, 217
Mortgaging, of resources, 476
Motorway, 203
Moving-average, model, 336; process, 79; process, first-order, 334; techniques, 323
Multicollinearity, 179
Multidimensional scaling, 415
Multinational corporations, 433, 436, 524
Multi-organizations, 287, 294, 295
Multispecies communities, design of, 472
Multiple control cost minima, 284
Multiplier, 29, 30, 42, 419, 420, 425; autoregressive, 45; back-shift, 31; effects, 435; forward-shift, 31; interregional, 436; process, 41; static without lags, 424
Multiplier-accelerator, 419, 420; economic model, 424
Multitrophic level model, 349
Multivariate maximum likelihood, 182
Music, 397

NIESR model of UK economy, 417, 419
Nannoplankton, 349–352
National economy, 125; optimal control, 449
National income, 422
National Gazetteer, 258
Natural, frequency, 46, 60, 61; hazard, 303, 305; justice, 307, 308, 536; resources, 514; selection, 525, 549
Nearest neighbour, 173
Needs, human, 505; regional, 435
Negative binomial model, 387
Negentropy, 399, 400, 410, 425; biotic, 486
Nemo judex in causa sua, 307
Neocolonialism, 434, 524, 544, 548
Network, 190, 192, 381; linkages, 205; problem, 193; process, 156, 157
New pluralism, 286
Neyman type A model, 387

Nitrogen, cycle, 261, 338, 487; organic, 385
Node, 68; improper, 68
Noise, 74, 99, 239; coloured, 76; corruption, 88; input, 74; levels, 262; level attenuation, 190; measurement, 74; model, 74, 129; output, 75, 181; random, 87; stochastic, 99; transfer function, 75; white, 88
Nomogram, 143
Non-communist countries, 515, 516
Non-structuralist view, 504
Nonlinear impulse response function, 40, 41; nonlinear programming, 498; nonlinear response, 334
Nonlinearity, 323, 331
Non-stationary changes, 321
Noogenesis, 15
Normative models, 269, 315, 507
Notables model, 312, 313, 314
Nottingham/Derbyshire study, 260
Noxiants, 246, 247
Nuclear fuel, 485; waste, 523
Nutrients, 373; cycling, 337, 339; pool, 320, 337, 397, 487
Nyquist, diagram, 88; plot, 84, 86, 87

OPEC, 434
Obedience, complete, 413
Objective, 254, 266; function, 89, 95, 103, 109, 129, 130, 138, 147, 192, 205, 279, 281, 408, 437; function, complex, 542; linear, 131; N-criterion, 209; two-criterion, 209
Observability, 73, 145, 417; criteria, 73; limitations, 76
Observational restrictions, 70
Observer, 149; method, 145
Offset, 127
Oil, 189, 201
Oligopoly, 302
One-stage criterion, 127
Oneness, 230, 238
Ontogenetic maturation, 20, 243
Ontological, 428; meaning, 460; trajectories, 263
Ontology, 285, 537
Ooze, 349–352
Open-loop, 115
Operator, 25, 26, 28, 29; differential, 41; identity, 41; lag, 39, 41; Laplace, 48; linear, 327; space-time lag, 381; summation, 41; Z-, 49, 51
Opinion, 2
Opportunity, cost, 281; function, 409
Optimal, assignment of flows, 193; site selection, 192
Optimality, criterion, 276, 277; Gini, 439; Lorenz, 439; Pareto, 439
Optimization, Bayesian criterion, 269, 273; Laplace criterion, 269, 273; maximum criterion, 269, 273; maximum entropy criterion, 269, 273; multilevel, 501; N-stage, 138; regret criterion, 269, 273
Optimum, allocation, 410; constraint, principle of, 400; control law, 139; levels of government, 413, 414; levels of representation, 413, 414
Ordinal rankings, 282
Ordinary least squares (OLS) estimator, univariate, 181
Organic, carbon, 373; growth, 525
Organization, 285, 401, 411; and methods (OM) approach, 283
Organizational position, 225, 241
Organizing centre, 214
Orthogonal, 179; set, 290; stochastic process, 87
Output, 25, 70, 256, 361, 419; controlled, 103; non-agricultural, 514; regional, 159, 161; prediction of, 485, 486; sinusoidal, 54
Overland flow, 352, 397
Over-sensitivity, 363
Overshoot, 35, 55, 103, 117, 127, 532
Oxidant concentrations, 376, 378
Oxygen, cycle, 261; dissolved, 496, 498
Ozone, 378; depletion, 521

PCB cycle, 261
Padé, approximant, 70, 160, 166, 220, 323, 531; series, 42, 331
Paleobotany, 387
Parallel combination, 31
Parameter, 26, 39, 77, 89, 161, 323, 326; adaptation, 148; autoregressive, 161, 334; behaviour, nonlinear, 217; behaviour, non-stationary, 217; deterministic, 90; estimates, ordinary least square (OLS), 147; estimation, 89, 91, 145, 148, 181, 343, 530; estimation, joint, 147; evolution, 334; evolution model, 335; map, 334; non-stationary 510; standard errors, 89; stochastic, 90, 328; transfer function model, 92; uncertainty, 147
Parasites, 395, 471; invasion by, 394
Parasitism, 487; true, 483, 484
Parasitology, 482–4
Parsimony, 89, 92, 404, 417
Path analysis, 402
Partial correlation function, 179
Payoff matrix, 271, 272
Peace, 211
Pearson curves, 387
Peasant societies, 484
Perceived escape alternatives, 305
Perception, 4, 19, 94, 224, 225, 236, 237, 401, 468; Chicago School of, 237; surface, 237
Perceptual, constancy, 226; readiness, 225, 240; subjectivism, 8; subsystem, 264

Performance, criterion, 130, 138, 208; Evaluation and Policy Review Unit (PEPR), 260; index, 96, 129; indicator, 109, 252, 255, 260, 261
Persona, 228, 229
Personal, unconscious, 227; universe, 242
Personality, 227
Pessimists, 517
Pest control, 470, 474
Pesticides, 138, 477, 492
Phase, 84, 86, 87, 88; lag, 54; lead, 54; shift, 180; transition, 211
Phenomenology, 19, 232
Phoresis, 482, 484
Phosphorous cycle, 261
Photochemical, processes, 334; smog, 377
Photophobic mammals, 110
Photosynthesis, 487
Phylogenetic, 20; heuristics, 243
Physico-chemical processes, 400
Physiological mechanisms, 387
Phytoplankton, 384, 385; biomass, 382, 384
Pithecanthropus, 229
Plankton, 351
Plan, fulfilment, perfect, 413; system, 440
Planned (command) economies, 449
Planning, 227, 273, 282, 283, 476; balance sheet analysis, 283; continuous, 251, 265; environmental, 308; government, 294, 295; process, continuous, 259; structural, 308; time horizon, 129, 136, 479; total, 261
Plantation monoculture, 477
Platonic ideals, 415
Platonic philosopher ruler, 416
Pleistocene, 387
Pluralist model, 312, 313, 314
Plutonium cycle, 261
Poisson model, 387
Polar plot, 60
Poles, 51
Police stations, 189
Policy, instruments, 96, 271, 272, 274, 442, 449, 451; objectives, 252; targets, 266
Political, bargaining, 302, 526; compromise, 542; influences, 432; issues, 149; power, territorial, 436; structure, 261
Pollen analysis, 387
Pollution, 174, 175, 190, 191, 201, 271, 309, 334, 335, 371, 397, 431, 476, 487, 496–9, 510, 511, 514, 515, 518, 521, 539; air, 262, 308, 309, 373, 375–9, 411, 438, 480, 481, 494; closed loop, 376; collapse, 511, 512, 513; control of, 202, 203; control system, 482; emission, 376, 379; estuaries, 372, 373; fjord, 373, 375; generation, 514; legislation against, 511; marine, 373–5; noise, 438; open loop, 376; patterns, 372; river, 365, 368–73; tax on, 431; water, 438, 516, 518
Pollutant, primary, 376; secondary, 377
Polynomial, approximating, 92; characteristic, 45, 51; closed-loop characteristic, 117; detrending, 322; matrix, 159, 167; order, 92; (Padé) series, 331
Pontryagin state function, 201, 211
Population, 471, 514; biological, 388–95; collapse, 494, 495, 511, 512, 513; control, 544; dynamics, 507; growth, 508; human, 472; migration, interregional, 393; problem, 524; regulation, 494, 524
Positivism, 407
Positivist, dilemma, 548; scientific approach, 545
Positivistic, 263
Potassium cycle, 338
Power, 429; groups, 301; series, 343
Pranayama, 241
Precipitation, 322, 327
Predators, lake, 351
Predictions, 148
Preference, 279, 280, 301; conflict, 302; function, 301, 315, 415; function, areal, 300; function, personal, 402, 404; maps, 300; multipeaked, 415
Prescriptive control, 315
Price, 280, 410, 514; elasticity, 423; fixing markets, 462; mechanism, 460, 521, 538; rules, 429
Primordial images, 227
Principal components analysis, 332
Principle, of optimality, 141; of optimum constraint, 398
Prisoners' dilemma, 302
Private property, 540
Probabilities, 269, 281
Probability density, function (pdf), 129; model, 387
'Process adaptive' mechanisms, 407
Process, autoregressive, 169; autoregressive feedback, 170; barrier, 189; spatial, bilateral, 186; contiguity, 175; distributed lag, 170; gain, 169; first-order distributed lead, 170; hierarchy, 189; multilateral, 185; network, 185, 189; purely spatial, 153; quadrilateral, 185; space-time, 176; space-time, barrier, 170, 175; space-time, contiguity, 170; space-time, hierarchy, 170, 175; space-time, network, 170, 175; spatial, multilateral, 186; spatial, unilateral, 186; stationary space-time, 175; stationary spatial, 153; unilateral, 185
Production, 419, 421; capitalist, 524, 544; function, 274, 282, 419
Profit, 429; maximization, 410
Programming, 95, 199, 238; problem, 194
Property, additivity, 177; asymmetry, 177; common, 309; interaction, 177; non-zero diagonal, 177; private, 309; superimposition, 177

Proportional, (gain), 341; term, 128
Prosthetics, 109, 243
Protectionists, 553
Pseudo-inversion of matrix, 557
Psyche, 229
Psychic level, collective unconscious, 228; conscious, 228; personal unconscious, 228
Psychoanalytical model, 223
Psychodynamics, 230
Psychological, actions, deep, 224; behaviourism, 402; difficulties, 544–6; model, 223
Psychology, 13, 415; developmental, 237; environmental, 243; topological, 235
Psychophysical field theory, 224, 235, 236
Public, goods, 411; health, 480
Pumping rate, 502
Pupillary mechanism, 101
Pure delay processes, 127

Quadrat analysis, 75
Quadratic, cost system, 138; form, 209, 211; form, indefinite, 210; form, negative definite, 210; form, positive definite, 210; minimization procedure, 290
Quality of life, 514, 517, 529
Quickening, 108

Race, 305
Racial differentiation, 389
Radar, 239
Rain gauge, 75
Rainfall, 324, 325, 329, 333, 493, 503; excess, 362, 364; frequency, 332; input, 362–4; intensity, 474; maximum, 483; mean annual, 474; -runoff equation, 327
Rainsplash, 351, 352, 397
Rainstorm, 363
Ranked preference, 407
Ranking, 279
Rapid transit system, 308, 309
Rate, of replenishment, 431; of return, 281, 400
Rationalists, 416
Raw materials, 508, 515
Reaction kinetics, 377
Realists, 415
Recharge rate, 503
Recipient control, 324
Recognition, auditory, 237; somatic, 237; visual, 237
Recurrence, period, 363; relation, 141
Recursion, 294
Recursive structure, 164
Recycling, 262, 470
Reference, group, 225, 241; value, 277
Reflex actions, 110, 224
Regional Economic Planning Boards, UK, 290
Regional employment premiums, 451
Regional, resources, concentration of, 434; science, 432, 433; system transfer function matrix, 177
Regression, least-squares, 77; least-squares (OLS), 90, 91; statistical, 343; two-stage least-squares (2 OLS), 91
Regulation, 252; space-time, 189
Regulator, 116, 195, 294, 310; cybernetic, 293, 301; economic, 451; problem, 102; self-tuning, 146, 148, 531; switched, 121
Reification, 407
Relationships, simultaneous, 505
Relativity, 152; effects, 157
Relaxation, 353; time, 153, 397
Reliability, 117, 118, 121; index, 118, 119
Religion, 485, 550
Religious affiliation, 476
Replacement, 262
Republicans, 458
Reputational model, 312–14
Reserve element, 117
Residence time, 372
Residential development, 300
Residuals, 89; tests of, 183
Resource, allocation mechanism 252; collapse, 511–13; management, 111, 548; ownership, 521; renewal, 526
Resources, 431, 504, 515; allocation of, 436, 460; as capital, 283, 537; as income, 283; control of, 429; exhaustion of, 539; nationalization of, 477; non-renewable, 274, 280, 283, 431, 439, 476–8, 510, 523; optimum allocation of, 437; regional, 435; renewable, 261, 262, 431, 439, 476–8, 510, 523; renewable national, 280; scarce, 274
Response, autocorrelation, 82; cross-correlation, 83; forced, 343, 347, 348; free, 343, 347, 348; function, unit impulse, 328, 329; impulse, 35; mixed frequency, 86, non-linear, 39; partial autocorrelation, 82; sinusoidal, 33; spikey, 136; steady-state, 86; step, 35; transient, 86
Restructuring, 526
Revolution, bourgeois, 428; proletarian, 428
Reward system, 225, 241
Ricardian view, 521
Ricardo assumptions, 456
Riccati, equation, 202; matrix differential equation, 192
Rice, 491
Risk, 251, 269; aversion, 534; aversion criteria, 407; aversity, 273; minimum, 408
River, 324; channel erosion, 350, 352; channel morphometry, 331, 334, 336; discharge, 353, 357, 362–5, 368, 372; flow, 174; Mississippi, 497; Nile, 494; Thames, 497, 498
Road, 411; Crossing, 203
Robertsonian lag, 422; model, 420
Role, 225, 241; conflict, 305; playing, 383

Rolling mill, 196
Root-locus method, 117
Roots, 51; complex, 59; equal, 59; mathematical, 423, 452
Ross equations, 395, 396
Rounded chart, 143
Rule of law, 307
Runoff, 324, 325, 329

Saddle, 68, 69; point, 208
Sage-Husa filter, 335
Sahel, 479, 492–5
Salinity, of soil, 499–501
Saltmarsh, 339, 470
Salt solution injection, 371
Salinization, 262
Sampling, 75; theory, N-variate, 182
Sanitation, 262
Satisficer, 234, 302
Saturation, population, 387
Savings, 421
Sawah (agricultural) system, 491
Scale, economies of, 436, 438
Scarcity, 437, 462, 514, 519, 524; of food, 435
Schools, 189, 192
'Science of the artificial', 505
Scientific Committee on Problems of the Environment (SCOPE), UN, 261
Search and selection procedure, 266
Seasonal component, 35
Sector category, system responses, 255
Sediment, 397; suspended, 360, 496; model, 359; movement, 350–353, 357–360; system, 350–3, 359–60, 397; system, space-time cascading, 356; transport, 334, 336, 353
Selectivity, 94, 224, 225
Selector function, 401
Self, 228, 230
Self-contradictions, 9
Self-regulation, of ecosystem, 339
Semi-arid areas, 474
Sensing, 111, 239, 240
Sensitivity, of world models, 533
Sensory classification, 256
Separation, of powers, 307; theorem, 145, 147, 542
Separatist movement, 435
Sequential estimation, 89; methods, 332
Serendipity, invention by, 238
Service centres, 194
Servicing, facility, location of, 192
Services, layout of, 189
Servomechanism, 41, 102, 117, 119, 189, 195, 408
Set point, 102, 264
Settling time, 35
Sewerage works, 497
Shadow, 228, 229
Shift, 42
Ships, 113, 115, 376
Shoal behaviour, 383
Shock waves, 211
Shopping centres, 189
Shortage, 523
Shunting, 226, 238–41, 247, 248
Side effects, catastrophic, 250, 255
Significance level, 77
Simile, of the cave, 2; of the dividing line, 2
Simon's satisficer principle, 315
Simulation, 77, 417, 510, 537, 542; methods, 449; model, 260
Singularity, 214, 216; point, 191
Singular point, 395, 396
Sink, 68; spiral, 68
Sinuosity, of rivers, 334
Skin colour, 389, 392
Slavery, 428
Slope, angle, 72; denudation of, 350; failure, 363
Smog control, 309
Smoke, 334
Smoothing, 322
Social, anthropology, 13; benefit, 310; choice, 537; choice functions, 415; contract, 20, 550; -cultural emergence, 426; engineering, prescriptive, 403; networks, 387; networks, biological, 388; physics, 403; power, 300–2; power networks, 301; preference function, 414, 537, 538; values, 233, 537; welfare, 458; welfare function, 415, 437, 547
Socialism, 406, 428, 439
Socialist legality, 307
Socialization, 415
Societal, objectives, 251; organization, types of, 411, 412; *status quo*, 263
Sociobiological groups, 387
Sociobiology, 20, 21
Sociocultural groups, 255
Sociological forces, 387
Sociology, 13, 17
Sociopolitical model, 303
Sociopolitics, 296–304
Sodium chloride (Na Cl), 368
Soil, 262, 337, 365–7, 471, 488, 491; Bank plan, 477; erosion, 474, 475, 492, 494; fertility, 138; moisture, 320; moisture storage, 324; structure, 470; water, 494, 496
Solar energy, 322, 337, 349, 354, 355; flux, 486
Solar radiation, 262, 382, 383
Solenoidal, 175
Source, 68; spiral, 68
Space, conceptual, 220; dimensionality, 154; perceptual, 220; as resource, 153, 431
Space-time, bubble, 188, 237; dependent outcomes, 257; interactions, 341, 342; model, 263; process, gain (distributed lag), 162, 163; simulation, 515; systems, 152–220;

Space-time—*continued*
transfer function process, general, 162, 163; transformation matrix, 154
Spatial, autocorrelation, 183, 379; averaging, 322; capacity, 256; data, 257; disparities, 434; distribution, unequal, 521; domain, as array, 155; elements, connections between, 154; equilibrium, 233; extrapolation, 183, geometry, 183; indifference curves, 402; interaction, 401; interdependence, 515; justice, 436; lag structure, 175; moving averages, 323; optimization, 448; processes, purely, 182–206; processes, purely, correlation of, 185; separation, 199; social justice models, 528; societal processes, 403; smoothing, 372; stationarity, 258, structure, 403; structure, politico-economic, 433
Spatially disaggregated models, 528
Specialization, 410
Species, counts, 72; diversity, 191, 491
Specification, 343; of state variables, 341
Specificity, 152; effects, 157
Spectral, analysis, 87, 155, 373; factorization, 332; function, 334
Speculation, 410
Squirrels, 387
Stability, asymptotic, 213; condition for, 122; of ecosystem, 339; global, 213, 214; global asymptotic, 212; Liapunov, 211–13; region, 60
Stanford IV basin model, 350–2, 356, 357, 359
Stable, growth, 507; loop, 41
Standard of living, 523
State, function, 212; of nature, 270–2; space, 62, 66, 68; space representation, 393, 394; transition matrix, 510; variable, 326, 340, 341; variable, reconstruction of, 149; variable, of world system, 510
Stationarity, 175, 184; assumption, 184; in space and time, 369
Statistical confidence bands, 145
Statistics of extremes, 486
Status, socio-economic, 305
Steady state, 321, 339, 347, 361; population, 389; stabilization, 361
Stepwise regression, 332
Stimuli, drug, 241; psychological, 241; religious, 242; social, 240; stress, 241
Stimulus-response, 401; model, 224
Stochastic, component, 35, 74; control with linear costs, 560–2; optimization, 129; model, 395; processes, 404, 406; variation, 70
Stock, control subsystems, 288; limitation, 494, 495
Stocking, cattle, 493
Storage, 339; lagged, 320; linear, 363
Storages, 468, 479; man-made changes, 320; manipulation of, 486–503
Storm, 72; runoff, 362
Strain, 245
Strategic, decision, 265; Plan for the South East of England, 260
Strategy, certainty, 271; dead beat, 136; N-stage, 129; risk, 271; uncertainty, 271
Streamflow, 327, 329, 365
Stress, 224, 240, 245
Stress-strain conversion function, 245
Structural decomposition, 195
Structuralism, 20, 230, 232
Structuralist analysis, 233
Structure planning, 260
Structures, Piaget's, 231
Subregional planning control, indicators, 256; long-term forecasts, 256; short-term targets, 256; strategic balances, 256
Subsidies, 206, 456, 523; to railways, 307; regional, 189
Substitutability, marginal, 410
Substitution, 514, 525; analysis, 282; mechanism, 521
Subsurface flow, 363
Subsystem, conscious, 226; linear, 331; unconscious 226
Sulphur, cycle, 261, 378; dioxide (SO_2), 334, 335, 376, 378
Summation operator, 102
Superego, 226
Superimposition, 54; principle of, 323
Superposition, 325; law of, 36, 37
Supply, 430; and demand, 448; -demand equilibrium, 431, 437
Surface water, 501
Surge, 72
Surplus value, 410, 429, 434
Survival condition, 534
Swidden agriculture, 491
Switching, 217
Sylvester's theorem, 209
Symbiosis, 18, 293; environmental, 468, 474, 479, 480, 483, 504–27
Symbols, 505
Synergism, 378
System, agricultural, 471; area-averaged, 156; artificial, 505; automated, 107; automatic, 107; autoregressive moving-average (ARMA), 71, 72, 74, 78, 79, 82, 84, 85, 91; basic, 108; behaviour, 244; behaviour-environment cognitive, 246; biological, 118; bounded, 5; bureaucratic multiple-access, 118; closed loop, 121; cognitive, 110, 223–249, 256, 406, 407; collaborative, 476; compensating, 108; components, active, 284; components, passive, 284; conditionally stable, 51; continuous, 26, 42, 155; control, 94–151; controllability, 115; convolution, 71; cost-limited, 136; counter-intuitive, 70, 504, 551; dead beat, 131; decision making, 110,

Systems—*continued*
122, 250–315, 402; decision making, man-machine, 219; deconvolution, 71; description, 71; deterministic, 131, 149, 213; deviation, 266; discrete, 25, 26, 39, 42, 49, 155; discrete-time first-order, 56; discrete-time second-order, 62; distributed parameter, 154; energy conversion, 472; environmental, 506; equation, solution to, 48; equations, inversion of, 131; estimation, 71, 186, 542; first-order, 45, 51, 55, 91, 125, 128; first-order autoregressive, 167; first-order lead-lag, 67; first-order linear, 363; first-order space-time, 163; forecasting, 71; 'fuzzy', 545; hard, Part I, 25, 98, 223; higher-order, 46, 60; homogeneous and isotropic, 369; human, 467–9, 472; identification, 71, 89, 108, 145; ignored, 181; indeterminate, 70; industrialized, 471; input, 155; inquiring, 8–10; interaction, 157; Jung's psychic, 228; knowns, 71; learning, 110; linear, 35–7, 54, 327; loops, higher-order, 113; low-order transfer function, 83; lumped, 190; lumped parameter, 154; man-environment, 290, 504; man-machine, 95, 104, 142, 150, 220, 479; man-tool, 106; man-tool powered, 105; manual, 106, 107; mathematical representation of, 47; matrix, 68, 69; meteorological, 473; minimum variance, 131; monitoring, 142; monitoring-advisory, 441; morphological, 402; multilevel, 95; multivariate, 46, 47; N-regional, 160; natural, 467, 468; negative feedback, 66; nested closed-loop guidance, 111; nonlinear, 37, 38, 39, 320, 323, 324; non-spatial, 65; non-stationary (time variant), 320, 321; norms and goals, 506; open loop, 121, 290, 449; order, 45; organizational, 506; output, 35, 155; output response, 36; output response trajectory, 69; parasympathetic psychoid, 229; partial differential equation, 190; partially unobservable, 192; perception, 225; physical, 469; point-sampled, 156; political, 299; positive feedback, 66; pure-time-delay, 56; range-limited, 136; rate-aided, 108; redesign of, 143; redundancy, 121; regime conditions, 124; reliability, 119; response to control, 116, 117; sampled data, 25; second-order, 46, 59, 125, 128; semi-automatic, 106, 107; simulation, 71; sluggish, 136; small, invariant, 322; social, 231, 233; soft, Part II, 223; spatial, 363–97, 505; specification, 77, 186; stable, 43, 44, 51, 69, 121; stability-limited, 136; state function, 211; state function, control Hamiltonian, 139; state function, Pontryagin, 139; state matrix, 510, 511; state variables, 417; stochastic, 119, 127, 134, 145, 149, 213, 322; storages, 322; sympathetic psychoid, 229; synthetic, 351; third-order polynomial, 330; trajectory, 66–68; transfer function, 39; uncertain (stochastic), 320, 322, 323; unbounded, 5, 7; uncontrolled, 99, 100, 128; universal adaptive, 111; unknowns, 71; unstable, 44, 51, 69, 121; zero-order, 45

System (space-time), 152–220; contiguity, 201; controllability, 176; estimation, 176; first-order lag, 163–5; first-order lead, 163–5; first-order lead-lag, 163–5; four-region, 171, 172; higher-order, 167; identification, 176; observability, 176, 177; measurement of, 155; interaction, 156; special, 156, 170; stationary, 176; two-region, 157, 158, 165

Systems, analysis, 70, 319, 324, 326–36, 356, 416, 436; analysis, stages, 76; approach, 541; cascading, 120, 402; engineering, 250, 260; fallacy, 416; first-order cascade, 331; hierarchical, 398; high-variety, 291; invariance of, 320; operation, invisible, 150; operation, unobservable, 150; in parallel, 118, 120; philosophy, 25; in series, 118, 120; space-time, cascading, 178; space-time, contiguity, 190; space-time, interaction, 189; space-time, parallel, 178; space-time, partly unobservable, 178; space-time, partly unobservable and uncontrollable, 178; space-time, uncontrollable, 178; synthesis, 319, 326, 327, 336–63, 406–16; synthesis, partial, 362

Taiga, 470, 471
'Take off' to sustained growth, 425
Tapping, of resources, 476
Target, 96, 252, 283; deviation from, 252, 253, 254, 259, 453
Targets, 436; plan, 448
Tariff barriers, 189, 191
Tautologies, 9
Tax, 523; structures, static, 189
Taxation, local, 502
Taxonomy, 238, 239
Taylor series, 331, 510, 531
Technical problems, 542, 543
Technological, advance, 524; change, 438; solution, 523
Technology, new, 477
Teleological bias, 14
Teleology, 3, 11, 14, 16, 17; immanent, 3
Tennessee Valley Authority, land grant colleges, 313
Tension, 409
Territorial principle, of Kruschev, 436
Territoriality, 387, 390; biological, 388–92; interspecific, 387
Temperature gradient, vertical, 378
Theory, 281; -building process, 406, 407; of structural morphogenesis, 533

Theory (basis for), conventionalist, 14; functional, 12; historical, 12; immanent, 12; realist, 13; taxonomic, 12; teleological, 11
Thermodynamics, 13
Thermostat, 94
Thiessen polygon, 322, 387
'Thingified' entity, 407
Third World, 434; development, 509, 524, 544, 551
Threshold, 238, 323, 324, 508
Throughflow, 365–7
Thunderstorm, 363, 503
Tidal mixing, 373
Time, constant, 45; dead, 117; delay, 55, 117; domain specification, 77; lags, 510; as resource, 153; rise, 117; settling, 117
Times series, 98, 373, 529; methods, 486; random walk, 137; regional, 171
Time-space budget, 256
Tolerance level, 260
Topology, 507
Tornado, 363
Total cost function, 278
Totalitarian regime, 254
Totemism, 550
Toxic elements, cycles, 261; accumulation of, 262
Toxic materials, 110
Trace, 69; substances, 262
Tracer, concentration, 373; injection, 372, 373
Tracking, 102, 108; problem, 320
Trajectory, desired, 264; global, 394; state space, 394, 395; system, 394
Transaction diagram, 409, 410
Transfer function, 26, 28, 30, 37, 41, 42, 47, 70, 78, 92, 96, 118, 155, 332, 419, 421, 434; closed-loop, 121; continuous space-time, 173; identification, 77; matrix, 176; photochemical, 377; regulator, 121; second-order, 60; space-time, 173; transport and diffusion, 376
Transform, Laplace, 48, 49, 50, 52, 64; unit shift, 49; Z-, 48
Transport, 189; costs, 432, 436; inquiries, 307; minimum cost, 193; networks, 379; routes, 190; steady state, 384
Transportation, 469; problem, 193
Transaction, 15
Transcendental, 17, 19
Transhipment problem, 193
Transpose, of vectors, 554
Trend, 255, 256, 333; component, 35; surface, 323; surface techniques, 155
Triangularization, 131, 279
Trip distribution, 273
Trophic, energy chain, 156; level feeding rates, 345; (feeding) levels, 337; web, 468
Tropical forest ecosystems, 320, 322, 339
Tropical rainforest, 471, 489, 490
Truth, 551
Tundra, 339, 470, 471
Turbulent flow, 211

Ultra vires, 307
Uncertainty, 129, 179, 251, 263, 264, 269, 284, 287, 300, 301, 304, 322, 400, 408, 425, 530; external, 264; related fields, 264; societal, 463; stochastic, 256; of systems, 255; value judgements, 264
Undamped, 60
'Undifferentiated' growth, 525
Unemployment, 180, 182, 256
'Unfolding' techniques, 415
Unit, function, 30; step, 30
United States, Bureau of Indian Affairs, 292; legal process, 306
Universal unfolding, 214
Urban, land use, 429; hierarchies, 379; and regional planning, 263
Utilitarians, 296
Utility, 21, 269, 404, 415, 551; function, 273, 278, 402, 414, 437; function, objective, 280; function, personal, 279; maximizing, neoclassical, 432; preference table, 270; scale, 415
Utopia, 245, 549

Vague experience, 9
Value, 428, 551; net present, 281; judgements, social, 261; laden decisions, 267; system, 507
Values, 308, 401, 427
Van Der Grinten technique, 109
Variable, continuous, 25; control, 33; dependent, 41; discrete, 25; endogenous, 97; ignored, 75, 91, 99, 129, 395; independent, 41; input, 341, 343; multi-input, 46; multi-output, 46; random, 72; state, 25, 28, 62, 341; system dynamics, 323
Variance, minimum, 90
Variation, stochastic, 76
Variety, 293; amplification, 292, 294; attenuation, 292, 294, 304; control, administrative, 294
Vires, 307, 536
Vector, 46, 90
Volcanic eruption, 363
Volterra equation, 392
Voting behaviour, 188

Wage rate, 419
War, 211; thermoneuclear, 521
Waste, bio-degradable, 470; dumps, 494; heat, 486, 521; load, 371; processing, 470; water treatment, 498
Water, circulation, 201; control, 501; quality, 494; quality control, 334; recycling, 523; reservoirs, 288; resources, 499; resources,

Water—*continued*
world, 520; supply, 262, 397; temperature, of lake, 385; use balance, 519
Waterlogging, 262, 499–501
Watt governor, 7
Waves, in channels, 363
Weather, manipulation, 485; modification, 190
Weathering, 338, 379
Weberian sense, location economics, 432
Weed killers, 492
Weights matrix, 171–3, 185, 379
Welfare, general, 460; economics, 21, 415; state, 550
Welfare test, check list of criteria, 282; cost-benefit analysis, 282; financial investment appraisal, 281; goals-achievement matrix, 282; operational analysis, 283
Weltanschauung, 10, 11
Western society, 286
Wheat yield, 188
White box, approach, 361; model, 485; system, 326, 406
White spots, criteria, 436; model, 434
Wind speed, 376
Wolves, 346–9, 395
World, Bank, 548; economy model, 75; growth, organic-like, 548; model, 527; regional divisions, 515, 516; systems, 510–23; view, 5
Wounds, healing of, 101

Yoga, 241

Z (lag) operator, 155
Zero growth, 525
Zero-one programming problem, 193